The Flora of Canada
Part 2 – Pteridophyta
       Gymnospermae
       Monocotyledoneae

National Museum of Natural Sciences
Publications in Botany, No. 7(2)

Published by the
National Museums of Canada

Staff editor
Bonnie Livingstone

Musée national des Sciences naturelles
Publications de botanique, n° 7(2)

Publié par les
Musées nationaux du Canada

The Flora of Canada

Part 2 – Pteridophyta
Gymnospermae
Monocotyledoneae

H. J. Scoggan

SYSTEMATIC SECTION

KEY TO FAMILIES

1 Plants lacking true flowers or seeds, reproducing by spores (these lacking an embryo) borne in spore-cases (sporangia) on the lower surface or margins of the leaves, in the leaf-axils, or in terminal cone-like spikes (strobiles), or (in *Marsilea*) on short basal peduncles; (Ferns and Fern Allies) . . . . .Division I **PTERIDOPHYTA**

1 Plants with flowers containing stamens or pistils or both, normally reproducing by seeds (each seed containing an embryo); (Seed-Plants; Flowering Plants) . . . . .Division II **SPERMATOPHYTA**

Division I   PTERIDOPHYTA (p. 127)
(Ferns and Fern Allies)

1 Stems conspicuously jointed, simple or branched, mostly hollow; leaves reduced to whorls of scales, these united at base into nodal sheaths; sporangia borne in terminal cone-like spikes (strobiles) . . . . .Equisetaceae (p. 127)

1 Stems neither evidently jointed nor hollow, often a subterranean rhizome (rootstock), bearing green leaves.
   2 Plants aquatic.
      3 Plants free-floating, moss-like, the stems dichotomously branching and producing rootlets on the underside; leaves simple, oblong to broadly ovate or rotund, sessile, closely appressed to spreading-ascending, about 1 mm long; (a single genus, *Azolla*) . . . . .Salviniaceae (p. 175)
      3 Plants rooting in the mud or sand under water.
         4 Stem short and corm-like, crowned by a rosette of linear quill-like leaves (to over 2 dm long) with dilated bases enclosing a single sporangium in a hollow on the inner side; (a single genus, *Isoëtes*) .Isoëtaceae (p. 140)
         4 Stem elongate and creeping, rooting at the nodes; leaves with 4 obdeltoid leaflets (to about 2.5 cm long and broad), solitary at the stem-nodes and on petioles to about 3.5 dm long; spores borne in usually 2 or 3 specialized sporocarps on a short basal or sub-basal peduncle; (a single genus, *Marsilea*) . . . . .Marsileaceae (p. 174)
   2 Plants terrestrial.
      5 Leaves linear to narrowly ovate or oblong.
         6 Stem consisting of a short erect fibrous-rooted subterranean rhizome, bearing a dense tuft of very slender, spiralling and curling, simple (sterile) or terminally pectinate (fertile) leaves (fronds); fertile fronds to about 12 cm long, the terminal blade pectinately dissected into up to 8 pairs of linear ascending segments; sterile blades acuminate to an entire apex, about half as long as the fertile ones; (a single genus, *Schizaea*; s Ont., N.S., Saint-Pierre and Miquelon, and Nfld.) . . . . .Schizaeaceae (p. 148)
         6 Stem elongate, rarely subterranean, closely covered by small scale-like overlapping sessile leaves.
            7 Spores of two sizes, in different sporangia borne in terminal, often 4-sided cones; plants small and often moss-like; (a single genus, *Selaginella*) . . . . .Selaginellaceae (p. 137)
            7 Spores uniformly minute, the sporangia borne either in the upper leaf-axils or in terminal terete cones; plants mostly larger; (a single genus, *Lycopodium*) . . . . .Lycopodiaceae (p. 133)
      5 Leaves (fronds) broader in outline, the fertile ones bearing sporangia beneath or on their margins (or the fronds sometimes more or less specialized and reduced, their segments surrounding or enclosing the sporangia); stem a subterranean rhizome or a short compact crown.
         8 Fertile and sterile fronds essentially alike except sometimes in size (or, in *Onoclea* and *Matteuccia* of the Polypodiaceae, the sporangia covered by the

rolled-up pinnules of the moniliform pinnae); sporangia with an annulus (a partial ring of cells with thin outer walls and thickened inner walls, the cells rupturing osmotically), hence splitting open by a traverse cleft on one side.

    9   Fronds very thin, delicate and translucent, composed (except near the veins) of a single layer of cells, pinnate-pinnatifid, to 5 cm long; stipes to 1.5 cm long, blackish, glabrous except for a small tuft of hair-like scales at the base; indusium 2-valved to base; rhizomes thread-like, extensively creeping and branched; (a single genus, *Mecodium*; SE Alaska and coastal B.C.) . . . . .Hymenophyllaceae (p. 149)

    9   Fronds thin to firm in texture, composed of several layers of cells; stipes often conspicuously scaly and usually longer; indusium various or wanting . . . . .Polypodiaceae (p. 150)

  8 Fertile fronds (or their fertile parts) conspicuously unlike the sterile fronds; sporangia lacking an annulus, opening by a vertical cleft.

    10 Sporangia several cells thick, 2-ranked at the summit of the terminal peduncle, the leaf-like sterile portion lateral or basal; rootstocks practically lacking, the commonly solitary frond arising from above the fleshy roots . . . . .Ophioglossaceae (p. 142)

    10 Sporangia 1 cell thick, densely crowded (but not 2-ranked) on separate modified fronds or on much-modified divisions at the summit or middle of otherwise sterile fronds; fronds crowded on a well-developed rockstock; (a single genus, *Osmunda*; E Canada) . . . . .Osmundaceae (p. 147)

Division II SPERMATOPHYTA
(Seed-Plants; Flowering Plants)

1 Ovules (and seeds) borne on an open carpellary scale or, in Taxaceae, partially surrounded by a fleshy aril; trees or shrubs with scale-like or needle-like, mostly evergreen leaves . . . . .Subdivision I ***GYMNOSPERMAE*** (p. 177)

  2 Fruit red and berry-like; leaves spirally arranged but spreading horizontally in 2 ranks, linear, narrowed to a sharp-pointed apex, tapering to a subsessile base, dark green above, yellowish green beneath; (a single genus, *Taxus*) . . . . .Taxaceae (p. 177)

  2 Fruit a cone (or blue-black and berry-like in *Juniperus*); leaves spirally arranged (but spreading horizontally in 2 ranks in *Abies, Pseudotsuga,* and *Tsuga*, these with leaves whitened beneath) . . . . .Pinaceae (p. 178)

1 Ovules (and seeds) borne in a closed ovary containing 1 or more carpels, the ovary developing into the fruit; herbs or woody plants with commonly deciduous leaves . . . . .Subdivision II ***ANGIOSPERMAE***

  3 Parts of the flower usually in 3's or 6's, never in 5's; embryo with a single cotyledon (seed-leaf); leaves chiefly parallel-veined . . . . .Class I *MONOCOTYLEDONEAE*

  3 Parts of the flower mostly in 4's or 5's or their multiples (rarely in 3's); embryo with a pair of opposite cotyledons; leaves mostly net-veined . . . . .Class II *DICOTYLEDONEAE* (Parts 3 and 4)

Class I *MONOCOTYLEDONEAE* (p. 191)

1 Plants aquatic.

  2 Plants floating or submersed, distinct or forming tangled mats; true leaves wanting, the single frond of each plant bearing the flowers (rare) and suspended roots . . . . .Lemnaceae (p. 456)

  2 Plants rooting in the mud beneath the water surface, with true leaves (these sometimes reduced to scales), the upper leaves sometimes floating on the surface.

    3 Flowers solitary in the axils of sheathing leaves or subtended by a spathe.

      4 Flowers perfect, the very slender thread-like perianth-tube to about 7 cm long, with a pale-yellow spreading ephemeral limb; leaves linear and grass-like, cauline; (*Heteranthera*; s Ont. and s Que.) . . . . .Pontederiaceae (p. 462)

    4   Flowers unisexual.
       5   Pistillate flowers lacking a perianth; staminate flower (a single terminal stamen) enclosed in a sac-like perianth; leaves filiform, opposite on the stem; (a single genus, *Najas*) ....Najadaceae (p. 208)
       5   Pistillate and staminate flowers both with a perianth consisting of 3 small sepals and 0 or 3 petals; stamens 3 – many; leaves opposite or whorled on the stem, or all basal ....Hydrocharitaceae (p. 216)
  3 Flowers in spikes, heads, or clusters.
      6   Leaves all in a basal rosette, linear-subulate, much surpassed by the flowering scape, this terminated by a button-like head of many very small flowers; sepals, petals, and carpels each 2, the last united into a compound ovary; (a single genus, *Eriocaulon*; E Canada) ....Eriocaulaceae (p. 459)
      6   Leaves cauline; carpel 1 or, if more, the carpels separating at maturity.
         7   Flowers in globose heads, unisexual, the upper heads staminate, the lower heads pistillate; perianth of linear or spatulate scales; leaves alternate; (a single genus, *Sparganium*) ....Sparganiaceae (p. 192)
         7   Flowers clustered or in spikes, unisexual or perfect; perianth none; leaves opposite or alternate ....Zosteraceae (p. 195)
1 Plants of dry or marshy terrestrial habitats (if growing in water, the stems at least partly emersed and bearing aerial leaves).
  8 Stems twining or climbing by tendrils; leaves net-veined; flowers unisexual.
      9   Stem climbing by tendrils; fruit a 1–6-seeded blue to black berry; (*Smilax*; Sask. to N.B. and N.S.) ....Liliaceae (p. 484)
      9   Stem climbing by twining, bearing large cordate-ovate, abruptly acuminate, chiefly alternate leaves to about 1 dm long; fruit a 3-winged capsule, the broadly winged seeds to about 2 cm long and 1 cm broad; (a single genus, *Dioscorea*; s Ont.) ....
          ....Dioscoreaceae (p. 515)
  8 Stems not climbing; leaves mostly parallel-veined.
    10 Flowers small, crowded on at least the lower part of a more or less fleshy axis (spadix), unisexual, the staminate ones usually borne above the pistillate.
      11 Spadix usually subtended or surrounded by a rather fleshy petaloid spathe; leaves usually broad (sword-like only in *Acorus*) ....Araceae (p. 453)
      11 Spadix the familiar "cat-tail", naked at base, the spathes being merely deciduous bracts or wanting; leaves linear; (a single genus, *Typha)* ....Typhaceae (p. 191)
    10 Flowers not borne on a fleshy spadix.
      12 Perianth none, the minute unisexual flowers borne in 2 kinds of inflorescences; small scapose annual marsh plant with soft linear-subulate leaves to about 3.5 dm long, these tufted on a very short stem; (a single genus, *Lilaea*; Vancouver Is. and sw Alta.) ....Lilaeaceae (p. 210)
      12 Perianth present; flowers usually perfect.
        13 Perianth consisting of scales, bristles, or chaffy bracts; ovary superior.
          14 Flowers crowded in small flat button-like heads at the summit of a scape, this much surpassing the basal rosette of leaves; (a single genus, *Eriocaulon*; E Canada) ....Eriocaulaceae (p. 459)
          14 Flowers not in button-like heads.
            15 Carpels 3, united into a compound ovary; leaves filiform to linear, flat or terete (if terete, often regularly cross-partitioned or nodulose-septate), often reduced to sheaths ....Juncaceae (p. 463)
          15 Carpels 1 (sometimes 2 in *Sparganium eurycarpum*).
            16 Flowers in axillary or supra-axillary globose heads, the staminate heads in the uppermost leaf-axils; leaves cauline, 2-ranked, elongate and often floating; (a single genus, *Sparganium*) ....
              ....Sparganiaceae (p. 192)
            16 Flowers in spikes, spikelets, or head-like clusters; leaves linear and grass-like, only exceptionally floating.
              17 Stem (culm) commonly hollow (except at the hard nodes) or with readily removable pith, terete or flattened; stem-leaves

2-ranked, their sheaths usually open (split lengthwise on the side opposite the blade); anthers attached near the middle . . . . .Gramineae (p. 218)

    17 Stem usually solid and often 3-angled, with soft nodes; leaves commonly 3-ranked; leaf-sheaths continuous around the stem (or rupturing in age); anthers basally attached . . . . .Cyperaceae (p. 338)

13 Perianth-divisions larger, at least the 3 inner ones petaloid and green or coloured.

  18 Carpels 3 or more, distinct or separating at maturity; ovary superior; leaves all or chiefly basal.

    19 Leaves terete or quill-like; perianth greenish; carpels 3–6, separating from the central axis at maturity; inflorescence spike-like or racemose; (transcontinental species) . . . . .Juncaginaceae (p. 209)

    19 Leaves usually with a broad dilated blade.

      20 Carpels in a single ring or in dense heads, usually forming 1-seeded indehiscent achenes; petals white or roseate; sepals green; leaf-blades linear to ovate (often sagittate at base) . . . . .Alismataceae (p. 211)

      20 Carpels 6, forming usually many-seeded follicles; petals roseate; sepals pink, tinged with green; inflorescence an umbel; leaves grass-like, dilated at base, to about 1 m long and 1 cm broad; (a single genus, *Butomus;* introd.) . . . . .Butomaceae (p. 215)

  18 Carpels mostly 3 (2 in *Maianthemum* of the Liliaceae), united into a compound ovary.

    21 Ovary wholly or partly inferior (appearing below the adnate perianth).

      22 Flowers very irregular, with bilateral symmetry; fertile stamen usually 1 (2 in *Cypripedium*); seeds innumerable and minute . . . . .Orchidaceae (p. 525)

      22 Flowers regular; fertile stamens 3 or 6; seeds larger and fewer.

        23 Stamens 6; anthers facing inward (introrse) . . . . .Amaryllidaceae (p. 516)

        23 Stamens 3.

          24 Anthers facing inward (introrse); flowers dingy yellow, about 1 cm long, in dense woolly corymbs; leaves equitant, in a basal cluster; plant with red juice, slender-stoloniferous, to about 1 m tall; (a single genus, *Lachnanthes;* N.S.) . . . . .Haemodoraceae (p. 514)

          24 Anthers facing outward (extrorse); flowers commonly blue; inflorescence not woolly; juice not red . . . . . . .Iridaceae (p. 518)

    21 Ovary superior, free from the subtending perianth.

      25 Stamens dissimilar, 3 of them fertile and long-exserted, the other 3 sterile and much shorter; flowers irregular.

        26 Sepals green; petals distinct; flowers few, in cymes; leaves alternate on the stem, sessile; plant annual, the finally decumbent stem rooting at the lower nodes; (*Commelina;* introd. in s Ont.) . . . . .Commelinaceae (p. 460)

        26 Sepals and petals violet-blue, united into a 2-lipped funnelform perianth; inflorescence a many-flowered spike; leaves mostly basal, long-petioled; stem erect or nearly so, to about 1 m tall, from a creeping rhizome; (*Pontederia*; Ont. to N.S.) . . . . .Pontederiaceae (p. 462)

      25 Stamens all alike and fertile; flowers regular.

        27 Inflorescence a cone-like head at the summit of a scape, the flowers covered by firm dry overlapping bracts; petals yellow; sepals narrow and chaff-like; leaves basal, linear; (a single genus, *Xyris*; E Canada) . . . . .Xyridaceae (p. 458)

27 Inflorescence otherwise; sepals green or petaloid.
　　28 Sepals green; petals roseate, blue, or purple; flowers in
　　　　umbels; fruit a capsule; leaves alternate, linear; (*Trades-
　　　　cantia*; sw Man. and s Ont.) ....Commelinaceae (p. 460)
　　28 Sepals and petals mostly essentially alike in colour and
　　　　texture (but sepals green in *Calochortus* and *Trillium*),
　　　　commonly greenish white, white, yellow, or orange; inflores-
　　　　cence various; fruit a berry or capsule ....Liliaceae (p. 484)

Class II   *DICOTYLEDONEAE*   (see Part 3)

1　Trees, shrubs, or woody climbers (or epiphytes in Loranthaceae, species of *Arceuthobium*
　　being parasitic on the branches of conifers) ....*GROUP A*
1　Herbs (occasionally semishrubby at base) ....*GROUP B* (p. 104)

GROUP A
(Dicotyledonous trees, shrubs, woody climbers, or epiphytes)

1　Leaves compound (reduced to spiny petioles in *Ulex* of the Leguminosae).
　　2　Leaves 3-foliolate (*Acer negundo* of the Aceraceae may be sought here) or palmately
　　　　5–7-foliolate; ovary superior.
　　　　3　Leaves palmately 5–7-foliolate.
　　　　　　4　Flowers small, in axillary and terminal cymose panicles; fruit a nearly black berry;
　　　　　　　　leaves alternate; vines climbing by tendrils; (*Parthenocissus*) ....Vitaceae (Part 3)
　　　　　　4　Flowers large, in terminal panicles; fruit a leathery globose capsule finally
　　　　　　　　dehiscing by 3 values; leaves opposite; trees; (a single genus, *Aesculus*; introd.) .....
　　　　　　　　...Hippocastanaceae (Part 3)
　　　　3　Leaves 3-foliolate (upper leaves often reduced to a single leaflet in *Cytisus* of the
　　　　　　Leguminosae).
　　　　　　5　Flowers papilionaceous, bright yellow, to 2.5 cm long; stamens 10, united into a
　　　　　　　　column; fruit a several-seeded pod (legume) to 5 cm long; leaves alternate;
　　　　　　　　(*Cytisus*; introd.) ....Leguminosae (Part 3)
　　　　　　5　Flowers regular; petals (when present) and stamens distinct; fruit various.
　　　　　　　　6　Leaves opposite; fruit a much-inflated thin-walled 3-lobed capsule to about 5 cm
　　　　　　　　　　long, dehiscent along the inner side at top; flowers about 1 cm long, in terminal
　　　　　　　　　　panicles; branches green-striped; (a single genus, *Staphylea*; Ont.
　　　　　　　　　　and Que.) ....Staphyleaceae (Part 3)
　　　　　　　　6　Leaves alternate.
　　　　　　　　　　7　Fruit a thin, nearly circular samara to about 2.5 cm long and broad, the broad
　　　　　　　　　　　　wing completely surrounding the 2-locular, 4-seeded body; sepals,
　　　　　　　　　　　　petals, and stamens each 3, 4, or 5; (*Ptelea*; s Ont. and s Que.) ....
　　　　　　　　　　　　...Rutaceae (Part 3)
　　　　　　　　　　7　Fruit not a winged samara; sepals and petals mostly 5.
　　　　　　　　　　　　8　Fruit a 1-seeded drupe with thin dryish exocarp (outer layer); petals
　　　　　　　　　　　　　　small, pale yellow or yellowish-green; stamens 5; plants unarmed; (a
　　　　　　　　　　　　　　single genus, *Rhus*) ....Anacardiaceae (Part 3)
　　　　　　　　　　　　8　Fruit a "raspberry" (a group of 1-seeded drupes on a fleshy receptacle);
　　　　　　　　　　　　　　petals larger, white or red to reddish-purple; stamens
　　　　　　　　　　　　　　numerous; stems and branches armed with stout prickles; (*Rubus*) ....
　　　　　　　　　　　　　　...Rosaceae (Part 3)
　　2　Leaves pinnate.
　　　　9　Fruit a dry winged samara; ovary superior.
　　　　　　10 Leaves alternate, the numerous leaflets subentire; samaras to about 5 cm long and
　　　　　　　　13 mm broad, twisted, the single seed near the centre; petals 5, small, green or
　　　　　　　　yellowish; flowers in ample terminal panicles; (a single genus, *Ailanthus*;
　　　　　　　　introd. in s Ont.) ....Simaroubaceae (Part 3)

      10 Leaves opposite; samaras mostly smaller, not twisted, the seed near one end; petals none.

         11 Fruit a single samara with a symmetrical apical wing; flowers in crowded panicles or racemes on the previous year's twigs; (*Fraxinus*) . . . . .Oleaceae (Part 4)

         11 Fruit a pair of 1-seeded samaras united at base and eventually separating, their apical wings asymmetrical; (*Acer negundo*) . . . . .Aceraceae (Part 3)

   9 Fruit not a winged samara.

      12 Leaves 2–3-pinnate, alternate; flowers greenish.

         13 Leaves 2-pinnate, the ovate or oblong leaflets entire; flowers in terminal racemes or panicles; ovary superior; fruit a pod to about 4 dm long; plant unarmed (*Gymnocladus*) or armed with usually branching thorns (*Gleditsia*) . . . . .Leguminosae (Part 3)

         13 Leaves 2–3-pinnate, the leaflets serrate; flowers small, in numerous umbels in a terminal panicle; ovary inferior; fruit a black berry; stem, branches, and often the leaf-petioles and rachises armed with stout thorns; (*Aralia spinosa*) . . . . .Araliaceae (Part 4)

      12 Leaves 1-pinnate.

         14 Leaves opposite; leaflets finely to coarsely toothed.

            15 Corolla mixed orange and scarlet, 7 or 8 cm long, trumpet-shaped; flowers in corymbs; ovary superior; fruit a dry 2-locular capsule to 2 dm long; plant climbing by aerial rootlets; (*Campsis*; s Ont.) . . . . .Bignoniaceae (Part 4)

            15 Corolla white, yellowish white, or pink, small; flowers in compound corymbs; ovary inferior; fruit a berry-like drupe; (*Sambucus*) . . . . .Caprifoliaceae (Part 4)

         14 Leaves alternate.

            16 Leaves with copious small translucent resin-dots, the plant thus aromatic; stem and branches armed with stout-based thorns; (*Zanthoxylum*; Ont. and sw Que.) . . . . .Rutaceae (Part 3)

            16 Leaves not resin-dotted.

               17 Fruit a thick-husked nut; flowers unisexual, the staminate in aments (catkins), the pistillate solitary or few in a cluster; petals none; tall trees . . . . .Juglandaceae (Part 3)

               17 Fruit otherwise; flowers perfect or unisexual, petals present.

                  18 Fruit a pod (legume); flowers papilionaceous to nearly regular; ovary superior; (*Amorpha; Caragana; Gleditsia; Robinia*) . . . . .Leguminosae (Part 3)

                  18 Fruit otherwise; flowers regular.

                     19 Fruit a group of achenes (these enclosed in a pulpy "hip" in *Rosa*) or small fleshy drupes (the "raspberry" or "blackberry" of *Rubus*); flowers perfect, usually conspicuous, white, yellow, pink, or roseate; ovary superior or inferior; stamens numerous; stem and branches mostly bristly or prickly in *Rosa* and *Rubus* . . . . .Rosaceae (Part 3)

                     19 Fruit a berry or dry drupe; ovary superior.

                        20 Leaflets leathery and evergreen, spine-toothed, usually glossy above; flowers to about 8 mm long, bright yellow; fruit a deep blue, glaucous, usually several-seeded berry; (*Berberis;* s B.C. and sw Alta.) . . . . .Berberidaceae (Part 3)

                        20 Leaflets thinner, not spine-toothed; flowers small, greenish white or yellowish; fruit a 1-seeded drupe; (a single genus, *Rhus*) . . . . .Anacardiaceae (Part 3)

1 Leaves simple.

   21 Leaves all or chiefly opposite (sometimes whorled or apparently so in Cistaceae, Cornaceae, Ericaceae, Loranthaceae, and Rubiaceae; mostly alternate in *Cornus alternifolia* of the Cornaceae).

22 Plants parasitic on conifers and often causing "witches' broom", olivaceous to purplish; stem usually less than 1 dm long, it and the branches jointed; leaves reduced to minute connate scales; ovary inferior; fruit a dry drupe; (a single genus, *Arceuthobium*) . . . . .Loranthaceae (Part 3)

22 Plants not parasitic, rooting in the ground, their leaves green; stem mostly longer, it and the branches not jointed.

   23 Fruit a pair of 1-seeded samaras joined at base and eventually separating, each samara with an asymmetrical apical wing longer than the body; leaves palmately lobed; (a single genus, *Acer*) . . . . .Aceraceae (Part 3)

   23 Fruit an achene, capsule, berry, or drupe.

      24 Flowers very small, in heads subtended by an involucre of appressed-ascending bracts (phyllaries); fruit an achene . . . . .Compositae (Part 4)

      24 Flowers not in heads; fruit a capsule, berry, or drupe.

         25 Leaves more or less distinctly toothed.

            26 Petals united at least at base; fruit a drupe (*Viburnum*) or capsule (*Diervilla; Linnaea*) . . . . .Caprifoliaceae (Part 4)

            26 Petals distinct to base; fruit a capsule.

               27 Leaves closely and finely serrate; petals small, greenish to brownish-purple; style 1; seeds enclosed in a white or red to scarlet aril . . . . .Celastraceae (Part 3)

               27 Leaves remotely low-serrate to subentire (*Philadelphus*) or coarsely serrate (*Hydrangea*); at least the petals of the marginal flowers large and showy, normally white; styles 2–4; seeds naked . . . . .Saxifragaceae (Part 3)

         25 Leaves (and their lobes, when present) entire.

            28 Leaves densely scurfy with rusty scales beneath or with silvery scales on both sides; flowers unisexual; petals none; sepals 4, greenish yellow within; stamens 8; fruit pulpy and berry-like; (*Shepherdia*) . . . . .Elaeagnaceae (Part 4)

            28 Leaves not scurfy; flowers perfect, with petals.

               29 Petals distinct.

                  30 Fruit a small 2-seeded white, red, blue, or blue-black drupe; ovary inferior; petals 4; calyx minutely 4-toothed; stamens 4; (a single genus, *Cornus*) .Cornaceae (Part 4)

                  30 Fruit a capsule; ovary superior; petals usually 3 or 5, yellow or red; sepals 5, often unequal; stamens 5 to many.

                     31 Capsule 3- or 5-locular; styles 3 or 5, united into a long sharp beak, finally becoming separated; petals yellow or orange; leaves dotted with translucent internal glands; (a single genus, *Hypericum*) . . . . .Hypericaceae (Part 3)

                     31 Capsule 1-locular; style single; petals yellow or red; leaves not gland-dotted . . . . .Cistaceae (Part 3)

               29 Petals united (at least at base).

                  32 Leaves evergreen and leathery, from scale-like or linear to narrowly oblong, commonly not over 1 cm long (except in *Kalmia* of the Ericaceae); corolla mostly 5-lobed (usually 4-lobed in *Calluna* of the Ericaceae); ovary superior; fruit a capsule.

                     33 Stamens 5, adnate by their filaments nearly or quite to the sinuses between the 5 corolla-lobes; flowers solitary on scape-like, terminally 3-bracted peduncles; leaves spatulate, crowded and overlapping, commonly about 1 cm long; plant matted and tussock-forming; (arctic, subarctic, and arctic-alpine regions); (a single genus, *Diapensia*) . . . . .Diapensiaceae (Part 4)

                     33 Stamens as many (*Loiseleuria*) or twice as many as the usually 5 corolla-lobes (4 in *Calluna*), free from the

corolla-tube; flowers solitary in the leaf-axils or in terminal umbel-like corymbs; (*Calluna; Cassiope; Kalmia; Loiseleuria*) . . . . .Ericaceae (Part 4)

    32 Leaves deciduous, thinner and softer, mostly much larger.

        34 Corolla somewhat 2-lipped, to about 6 cm broad, white or yellow and more or less spotted or streaked with yellow; orange, or purple, the 5 lobes with undulate-crisped margins; calyx unequally 2-lipped; stamens 5 (2 or sometimes 4 of them fertile, the others sterile and rudimentary); ovary superior; capsule slender, to about 5 dm long; trees or large shrubs; (*Catalpa*; introd. in s Ont.) . . . . .Bignoniaceae (Part 4)

        34 Corolla regular or nearly so (except in several species of *Lonicera*), smaller; stamens all fertile; shrubs.

            35 Flowers densely crowded in globose peduncled heads to about 3 cm thick, white, their styles long-exserted; calyx and corolla each 4-lobed; stamens 4; ovary inferior; leaves occasionally whorled; (*Cephalanthus*; Ont. to N.S.) . . . . .Rubiaceae (Part 4)

            35 Flowers solitary or clustered in the leaf-axils or in terminal panicles or cymes; leaves not whorled.

                36 Calyx and corolla each 4-lobed; stamens 2; flowers in panicles (*Ligustrum* and *Syringa*) or 1–3 in the leaf-axils (*Forsythia*); ovary superior; (introd. species) . . . . .Oleaceae (Part 4)

                36 Calyx and corolla each usually 5-lobed and stamens usually 5 (*Symphoricarpos* sometimes with a 4-lobed corolla and 4 stamens); flowers solitary or in axillary clusters or terminal cymes; ovary inferior . . . . .Caprifoliaceae (Part 4)

21 Leaves all or mostly alternate (subopposite in *Rhamnus cathartica*, with some of the branches usually ending in short thorns).

    37 Plants climbing; leaves palmately veined or lobed; style (or sessile stigma) solitary.

        38 Plants climbing by tendrils; leaves coarsely toothed or lobed, usually cordate; fruit a dark-red to nearly black juicy berry; (*Vitis*) . . . . .Vitaceae (Part 3)

        38 Plants climbing by twining of the stem (or, in *Hedera* of the Araliaceae, by aerial supporting roots).

            39 Fruit a black berry or drupe; flowers small, white or greenish; leaves ovate to rotund, entire to shallowly 3–7-lobed or -angled.

                40 Leaves peltate (the petiole attached a short distance in from the basal margin), deciduous; flowers unisexual, in small loose axillary panicles; fruit a glaucous 1-seeded drupe; stems climbing by twining; (a single genus, *Menispermum*; SE Man. to Que.) . . . . .Menispermaceae (Part 3)

                40 Leaves not at all peltate, coriaceous and evergreen; flowers perfect, in solitary or racemose umbels at the tips of the branches; fruit a usually 5-locular and 5-seeded berry; stems climbing by aerial supporting roots; (*Hedera*; introd. in sw B.C.) . . . . .Araliaceae (Part 4)

            39 Fruit a capsule or orange-red to scarlet berry; leaves neither peltate nor lobed.

                41 Leaves finely serrate, ovate-oblong; flowers small, greenish, in terminal racemose clusters, some of them unisexual; sepals and crenulate petals each 5; ovary superior; fruit an orange or orange-yellow capsule, the seeds covered by a bright-red aril; (the decorative bittersweet, *Celastrus*; SE Sask. to Que. and N.B.) . . . . .Celastraceae (Part 3)

                41 Leaves entire; flowers perfect, on axillary peduncles.

                    42 Corolla funnelform, dull pink-violet to purple, the 5-lobed limb to 1.5 cm broad; calyx 3–5-lobed; ovary superior; fruit an orange-red or scarlet

berry; leaves lanceolate to rhombic-ovate; stem and branches often spiny at the nodes; (*Lycium*; introd.) . . . . .Solanaceae (Part 4)
  42 Corolla none; calyx about 3 cm long, brown-purple, curved like a Dutchman's pipe, abruptly flaring from the contracted mouth into a 3-lobed limb; ovary inferior; fruit a thick-cylindric capsule to about 8 cm long; leaves cordate-ovate to reniform, those on vegetative shoots to 4 dm broad; (*Aristolochia durior*; introd. in s Ont.) . . . .
. . . . .Aristolochiaceae (Part 3)
37 Plants not climbing.
  43 Flowers small, in dense heads subtended by one or more series of involucral bracts (phyllaries); fruit an achene . . . . .Compositae (Part 4)
  43 Flowers not as above.
    44 Leaves (and their lobes, when present) entire or nearly so . . . .*SERIES 1*
    44 Leaves (and their lobes, when present) more or less distinctly toothed . . . .
. . . .*SERIES 2* (p. 103)

SERIES 1
(Non-climbing dicotyledonous trees, shrubs, or semishrubs with entire or merely entire-lobed, simple, alternate leaves)

1 Fruit a fleshy or pulpy 1-seeded drupe (or drupe-like in Elaeagnaceae) or a dryish, usually several-seeded berry (or pome in Rosaceae).
  2 Stem and branches spiny at the nodes; leaves obovate to spatulate; flowers yellow, solitary at the nodes or in small clusters of up to 4; fruit a dryish, red, ellipsoid to globose berry; (*Berberis thunbergii*; introd.) . . . . .Berberidaceae (Part 3)
  2 Stem and branches unarmed; fruit a drupe or drupe-like.
    3 Leaves densely silvery-scurfy on one or both sides; flowers silvery outside, pale yellow within; fruit drupe-like, silvery; (*Elaeagnus*) . . . . .Elaeagnaceae (Part 4)
    3 Leaves, flowers, and fruits not silvery-scurfy.
      4 Leaves less than 1 cm long, crowded and overlapping, scale-like or iinear-subulate to linear-oblong; sepals none or 3 and petaloid; petals none or 3 and greenish or purplish; stamens 3; flowers axillary or in terminal clusters; low bushy heath-like shrubs . . . . .Empetraceae (Part 3)
      4 Leaves longer and mostly relatively broader.
        5 Anthers opening by terminal pores, as many or twice as many as the lobes of the rotate, urn-shaped, or globose corolla (petals distinct only in *Ledum*) . . . . . .
. . . . .Ericaceae (Part 4)
        5 Anthers dehiscing lengthwise.
          6 Leaves sessile or subsessile, or tapering gradually to an obscure margined petiole to about 8 cm long; flowers perfect; petals none; fruit a 1-seeded, green, yellowish, or red drupe; shrubs to about 2 m tall . . . .
. . . . .Thymelaeaceae (Part 4)
          6 Leaves distinctly petioled.
            7 Petals very small and fleshy, or none; flowers mostly unisexual; drupe 1-seeded; leaves elliptic to obovate; shrubs or trees to about 30 m tall; (s Ont.).
              8 Calyx-lobes minute; staminate flowers with 5–12 stamens, borne in peduncled umbels or umbel-like racemes; pistillate flowers (sometimes bearing fertile or sterile anthers on very short filaments) sessile in clusters of up to 8 at the end of a peduncle; drupe dark blue to black; leaves neither lobed nor dotted; (a single genus, *Nyssa*) . . . . .Nyssaceae (Part 4)
              8 Calyx deeply 6-parted; stamens of staminate flowers 9, in 3 rows, the inner ones glandular at base; pistillate flowers with up to 18 staminodia (rudimentary stamens); drupe red or black; leaves dotted with translucent glands (in *Sassafras*, with up to 5 broad deep lobes) . . . . .Lauraceae (Part 3)

7 Petals present and relatively large, greenish white or yellowish; shrubs rarely more than 3 m tall.

9 Flowers borne in axillary bracted racemes, unisexual; sepals and greenish-white petals each 5; stamens 15, borne in 3 distinct series; fruit a 1-seeded blue-black drupe to 1 cm long; (*Osmaronia*; s B.C.) . . . . .Rosaceae (Part 3)

9 Flowers solitary in the leaf-axils or in few-flowered clusters on short spurs or in axillary umbels or cymes; fruit usually with 2 or more seeds, red (often becoming nearly black).

10 Flowers unisexual; petals, calyx-lobes, and stamens each 4–8, or the calyx-lobes wanting; drupe with at least 4 seeds; (E Canada) . . . . .Aquifoliaceae (Part 3)

10 Flowers perfect; sepals and petals each 5.

11 Petals very small (less than 1.5 mm long), greenish; stamens 5; flowers up to 8 in a cluster; drupe becoming nearly black, with 2 or 3 seeds; leaves to about 8 cm long; (*R. frangula*; introd.) . . . . .Rhamnaceae (Part 3)

11 Petals larger, white tinged with pink; stamens about 20; flowers rarely more than 4 in a cluster; fruit a red pome with usually 2 seeds; leaves to 5 cm long; (*Cotoneaster*; introd.) . . . . .Rosaceae (Part 3)

1 Fruit a dry capsule, legume, or nut, or fruit multiple and resembling a dry or fleshy cone; stem and branches unarmed.

12 Fruit multiple, to over 7 cm long; flowers perfect, large, solitary at the ends of the branches; sepals 3; petals 6; stamens numerous; (s Ont.).

13 Petals greenish yellow, tinged with yellow or orange, subequal; fruit resembling a fleshy or dry cone; leaves marked with minute transparent dots, either oblong and pointed (*Magnolia*) or squarish and 4-lobed (*Liriodendron*) . . . . . .Magnoliaceae (Part 3)

13 Petals dull purple, the 3 spreading outer ones longer than the 3 inner nearly erect ones; fruit becoming a yellowish-brown sweet-pulpy edible thick-cylindrical body; leaves obovate, abruptly short-acuminate, gradually narrowed to base; (a single genus, *Asimina*) . . . . .Annonaceae (Part 3)

12 Fruit a capsule, legume, or nut.

14 Fruit the common "acorn", a 1-seeded nut partly enclosed in a cup consisting of more or less consolidated bracts; trees unisexual, the staminate flowers in aments, the pistillate flowers solitary or slightly clustered; (*Quercus*) . . . . .Fagaceae (Part 3)

14 Fruit a capsule or legume.

15 Flowers more or less papilionaceous; fruit a legume; (*Cercis; Genista*) . . . . .Leguminosae (Part 3)

15 Flowers not papilionaceous; fruit a capsule.

16 Seeds tipped with long silky down, numerous in a 2–4-valved capsule; perianth none; plants unisexual, the staminate and pistillate flowers produced in aments (catkins) on separate plants) . . . . .Salicaceae (Part 3)

16 Seeds not tipped with silky down.

17 Anthers mostly dehiscing by a pair of terminal pores (by longitudinal slits in *Cladothamnus, Epigaea,* and *Kalmia*); petals usually united at least at base (free to base in *Cladothamnus* and *Ledum*) . . . . . . .Ericaceae (Part 4)

17 Anthers dehiscing lengthwise.

18 Leaves palmately divided to near base into up to 7 rigid linear-subulate spinulose-tipped segments; corolla to 2.5 cm long, white or lavender-tinged to yellow or almost salmon-colour, the 5 petals united into a long slender tube; stamens 5; (*Leptodactylon*; s B.C.) . . . . .Polemoniaceae (Part 4)

18 Leaves entire; corolla smaller, the 3 red or 5 yellow petals distinct; stamens usually several . . . . .Cistaceae (Part 3)

SERIES 2 (see p. 101)
(Non-climbing dicotyledonous trees or shrubs with finely to coarsely toothed to deeply lobed, simple, alternate leaves, the lobes, when present, themselves toothed)

1 Stem and branches bristly, spiny, or thorny (*Rhamnus cathartica* of the Rhamnaceae, with subopposite leaves, may be sought here).
  2 Flowers small, in numerous head-like umbels on the branches of a terminal raceme or panicle; fruit a red 2-seeded drupe; (*Oplopanax*; Alaska–Yukon–B.C.–Alta. and w Ont.) . . . . .Araliaceae (Part 4)
  2 Flowers solitary in the leaf-axils or in simple racemes or umbels or in umbel-like clusters.
    3 Petals 5, usually showy and much larger than the 5 sepals; stamens usually about 10 or 20; fruit a 1-seeded fleshy drupe (*Prunus*) or a pome with up to 10 seeds; large shrubs or small trees . . . . .Rosaceae (Part 3)
    3 Petals smaller than the often coloured sepals; fruit a berry; relatively low shrubs.
      4 Sepals, petals, and stamens each 6; ovary superior; berries ellipsoid to subglobose, scarlet to red, 1–few-seeded; leaves obovate, finely bristle-toothed, in fascicles; spines commonly branched; (*Berberis vulgaris*; introd.) . . . .
. . . . .Berberidaceae (Part 3)
      4 Sepals, petals, and stamens each 5; ovary inferior; berries globose, many-seeded; leaves commonly rotund to reniform in outline and cordate at base, relatively coarsely toothed and lobed; spines simple; (*Ribes*) . . . .
. . . . .Saxifragaceae (Part 3)
1 Stem and branches unarmed.
  5 Flowers small, in dense heads subtended by one or more series of involucral bracts (phyllaries); fruit an achene . . . . .Compositae (Part 4)
  5 Plants lacking the above combination of characters.
    6 Seeds tipped with long silky down, numerous in a 2–4-valved capsule; perianth none; plants unisexual, the staminate and pistillate flowers produced in aments on separate plants . . . . .Salicaceae (Part 3)
    6 Seeds naked.
      7 Inflorescence cymose, of white or creamy flowers, its long peduncle joined to about the middle of an elongate tongue-shaped membranous bract (an aid to wind-dispersal of the cluster of mature dry 1–2-seeded globose fruits); leaves broadly ovate to rotund-cordate, oblique and often truncate at base; trees to about 40 m tall; (a single genus, *Tilia*) . . . . .Tiliaceae (Part 3)
      7 Inflorescence and other features not all as above.
        8 Flowers (staminate or pistillate or both) in aments or heads; corolla minute or wanting.
          9 Fruit multiple; (s Ont.).
            10 Fruit resembling a blackberry, consisting of small seed-like achenes enclosed (except for the protruding styles) in the swollen juicy mature calyces, these crowded on the axis of a short-cylindric ament; staminate flowers in elongate loose aments; rather small trees with milky juice; (*Morus*) . . . . .Moraceae (Part 3)
            10 Fruit a globose head (to about 3 cm thick) of clavate 1-seeded nutlets; calyx and corolla minute; staminate flowers in spherical heads; tree to about 50 m tall; (a single genus, *Platanus*) . . . . . . .Platanaceae (Part 3)
          9 Fruit a simple nut or nutlet.
            11 Fruit a 1-seeded nut wholly or partly surrounded by the enlarged involucre (forming the familiar "acorn", "chestnut", or "beechnut"); pistillate flowers solitary or in small clusters or spikes; staminate flowers in aments or heads; trees . . . . .Fagaceae (Part 3)
            11 Fruit a winged or wingless nutlet subtended by a more or less woody scale of the pistillate ament; pistillate flowers in dense subglobose to cylindric aments; staminate flowers in longer and looser aments.

>>> 12 Leaves resin-dotted (black-dotted in one species) and often fragrant; fertile flowers 1 to each bract, this unlobed; shrubs . . . . .Myricaceae (Part 3)

>>> 12 Leaves usually not resin-dotted (except in some species of *Alnus*); fertile flowers 2 or 3 to each bract, this usually 3-lobed . . . . .Betulaceae (Part 3)

>> 8 Flowers not in aments or heads.

>>> 13 Leaves asymmetrical, oblique at base, one margin longer than the other.

>>>> 14 Corolla none; fruit an elliptic to rotund broad-winged samara (*Ulmus*) or a 1-seeded dark-red to nearly black drupe (*Celtis*); leaves narrowly to broadly ovate, sharply serrate; trees . . . . .Ulmaceae (Part 3)

>>>> 14 Corolla consisting of 4 yellow strap-shaped petals to 2 cm long; fruit a woody 2-seeded capsule circumscissile near the summit; leaves obovate to rotund, shallowly wavy-toothed; coarse shrub or small tree flowering in autumn; (a single genus, *Hamamelis*; Ont. to N.S.) . . . . .Hamamelidaceae (Part 3)

>>> 13 Leaves essentially symmetrical.

>>>> 15 Anthers 8 or 10, opening by a pair of terminal pores or, if dehiscing lengthwise, often with a pair of single or double horn-like appendages on the back; petals united into an urn-shaped or campanulate 4–5-lobed corolla; (*see* Clethraceae, Part 4) . . . . .Ericaceae (Part 4)

>>>> 15 Anthers dehiscing lengthwise, lacking horn-like appendages; petals separate or united only at base (petals wanting in *Rhamnus alnifolia* of the Rhamnaceae).

>>>>> 16 Stamens 10 or more; fruit a capsule, follicle, pome, berry-like drupe, achene (*Purshia*), or "raspberry" (*Rubus*) . . . . .Rosaceae (Part 3)

>>>>> 16 Stamens usually 4–6; fruit dry and 3-lobed, a berry, or a berry-like or capsule-like drupe.

>>>>>> 17 Stamens alternate with the 4 or 5 calyx-lobes; petals 4 or 5, greenish (none in *Rhamnus alnifolia*); flowers perfect or unisexual; fruit a usually 3-seeded berry-like or capsule-like drupe . . . . .Rhamnaceae (Part 3)

>>>>>> 17 Stamens opposite the 4–6(9) calyx-lobes (alternating with the petals).

>>>>>>> 18 Calyx 5-lobed; styles 2, from nearly distinct to almost completely united; flowers perfect; fruit a many-seeded berry; (*Ribes*) . . . . .Saxifragaceae (Part 3)

>>>>>>> 18 Calyx with 4–9 minute lobes; stigma nearly sessile; flowers unisexual; petals white, yellowish, or greenish; fruit a 4–9-seeded berry-like drupe; (E Canada) . . . . .Aquifoliaceae (Part 3)

GROUP B (see p. 97)
(Dicotyledonous herbs, or plants sometimes with a woody crown or caudex)

> 1 Leaves none or reduced to scales, these usually devoid of chlorophyll.

>> 2 Stems slender and twining, yellowish to reddish-brown; corolla small, short-cylindric to globose, 4–5-lobed; stamens 4 or 5; ovary superior; fruit a usually 4-seeded capsule; plants parasitic, adhering by suckers; (*Cuscuta*) . . . . .Convolvulaceae (Part 4)

>> 2 Stems not twining (sometimes spiralling or even tending to climb in *Bartonia* of the Gentianaceae).

>>> 3 Stems ovoid to globose or composed of flattened joints, spiny and bristly, green, very fleshy; flowers large and showy; stamens numerous; ovary superior; fruit a spiny or smooth berry, (B.C. to w Ont.) . . . . .Cactaceae (Part 4)

>>> 3 Stems relatively slender, unarmed.

>>>> 4 Corolla 2-lipped; flowers solitary or in spikes; stamens 4 (2 long, 2 short); ovary

superior; capsule 1-locular, 2-valved; stems yellowish to brownish or purplish; plants root-parasitic . . . . .Orobanchaceae (Part 4)

    4   Corolla regular (wanting in *Salicornia*, with jointed stems); ovary superior.

        5   Stems jointed, very fleshy, rarely over 3 dm tall; scale-leaves opposite; (*Salicornia*; plants of salt marshes) . . . . .Chenopodiaceae (Part 3)

        5   Stems not jointed; scale-leaves mostly alternate (sometimes opposite in *Bartonia*).

            6   Stems filiform, yellowish green to purplish; corolla campanulate, at most about 4 mm long, the 4 lobes about equalling the tube; stamens 4, dehiscing lengthwise; (*Bartonia*; s Ont. to Nfld. and N.S.) . . . . .Gentianaceae (Part 4)

            6   Stems stouter, waxy-white or pinkish to purplish-brown; petals distinct or united; stamens 8 or 10, their anthers inverted and opening by terminal (morphologically basal) pores; plants saprophytic . . . . .Pyrolaceae (Part 4)

1   Normal leaves present (if rarely scale-like, possessing chlorophyll; leaves reduced to hollow petioles (phyllodia) in *Lilaeopsis* of the Umbelliferae).

    7   Plants distinctly aquatic; leaves linear-elliptic to orbicular, on long weak petioles and evidently adapted to floating (usually raised above the surface of the water in *Nelumbo* of the Nymphaeaceae); or leaves (at least the lower ones) submersed and dissected into narrow segments; (aquatics with leaves neither floating nor conspicuously dissected are included under the contrasting lead); ovary usually superior (inferior in Haloragaceae).

        8   Leaves not dissected, at least the uppermost ones commonly floating, lanceolate to orbicular, entire or with a deep basal sinus extending to the petiole.

            9   Flowers pink to scarlet, small, in dense, ovoid to slenderly cylindric, terminal spikes; petals none, the sepals petaloid; leaves lanceolate to elliptic-oval, their stipules (ocreae) cylindric and sheathing the stem above the swollen joints; (*Polygonum amphibium* and *P. coccineum*) . . . . .Polygonaceae (Part 3)

            9   Flowers white, yellow, or dull purple, often larger; petals present.

                10 Flowers white, 5-merous, to about 1 cm broad, in an umbel arising below the water surface from a petiole, the umbel often with a contiguous whorl of spur-roots; leaves cordate-ovate, to about 7 cm long; (*Nymphoides*; Ont. to Nfld. and N.S.) . . . . .Gentianaceae (Part 4)

                10 Flowers white, yellow, or dull purple, larger, solitary in the leaf-axils; petals 3 to many (if many, passing gradually into the numerous stamens); leaves cordate-ovate to orbicular, often larger . . . . .Nymphaeaceae (Part 3)

        8   Leaves (at least the submersed ones) dissected into mostly filiform or linear segments (*Bidens beckii* of the Compositae will also key out here).

           11 Upper leaves floating, entire or 2-cleft at one end, linear-elliptic, at most about 2 cm long, centrally peltate; submersed leaves capillary-dissected, opposite or whorled; fruit leathery and indehiscent, usually 3-seeded; (*Cabomba*; introd. in s Ont.) . . . . .Nymphaeaceae (Part 3)

           11 None of the leaves at once floating and entire; fruit not as above.

                12 Plants insectivorous, bearing small bladder-traps at the tip of the leaf-segments or on separate branches; corolla usually yellow, 2-lipped, spurred at base, the throat closed by a prominent palate; (*Utricularia*) . . . . .Lentibulariaceae (Part 4)

                12 Plants not insectivorous, lacking bladder-traps.

                    13 Stem attached to stones by fleshy disks; leaves alternate, in 2 vertical ranks, rigid or horny, dilated-sheathing at base; flowers produced within obovoid sessile axillary spathes; (a single genus, *Podostemum*; Ont. to N.S.) . . . . .Podostemaceae (Part 3)

                    13 Stems rooting in the mud or sand; flowers not produced in spathes.

                        14 Leaves dilated-sheathing at base, alternate; sepals usually 3–5, green or yellowish; petals commonly 5, yellow or white with yellowish base;

              stamens commonly numerous; fruit an ovoid to globose head of usually
              beaked achenes; (*Ranunculus*) . . . . .Ranunculaceae (Part 3)
        14 Leaves not dilated-sheathing at base.
            15 Perianth none; stamens 12–16; fruit an achene with 2 or more spines;
               leaves whorled, all submersed; (a single genus, *Ceratophyllum*) . . . . .
               . . . . .Ceratophyllaceae (Part 3)
            15 Perianth (at least the calyx) present; stamens not more than 8.
              16 Calyx present; petals 4 and minute or none; stamens 3, 4, or 8;
                 fruit nut-like and indehiscent or finally separating into 4 mericarps;
                 leaves alternate, opposite, or whorled, the emersed ones usually
                 not dissected . . . . .Haloragaceae (Part 4)
              16 Calyx and corolla both present; petals 4, yellow or white, larger;
                 stamens 6 (4 long, 2 short); fruit an ovoid to linear-cylindric,
                 2-valved, several-seeded capsule; leaves alternate . . . .
                 . . . . .Cruciferae (Part 3)
7   Plants terrestrial (if more or less aquatic, leaves usually neither floating on the surface
  nor conspicuously dissected).
   17 Plants climbing by stem-twining, by bending or clasping of the leaf-petioles, or by
      tendrils.
      18 Leaves compound; ovary superior.
         19 Leaves opposite, 3–9-foliolate, their petioles clasping supporting objects;
           flowers regular; sepals normally 4, petaloid; petals small or none, transitional
           into stamens, these numerous; fruit a cluster of achenes with long plumose
           styles; (*Clematis*) . . . . .Ranunculaceae (Part 3)
         19 Leaves alternate; flowers irregular; stamens not more than 10.
           20 Leaves pinnate-ternately compound, the upper ones greatly reduced,
              their stalks clasping; petals 4, united into a bilaterally symmetrical
              subcordate-ovate spongy corolla, this 2-saccate at base; fruit a capsule;
              (*Adlumia*; sᴇ Man. to N.S.) . . . . .Fumariaceae (Part 3)
           20 Leaves with 3 leaflets (*Amphicarpa, Dolichos, Phaseolus,* and *Stropho-*
              styles, these all climbing by twining) or 1-pinnate; flowers papilionaceous;
              fruit a legume . . . . .Leguminosae (Part 3)
      18 Leaves simple (but sometimes deeply lobed).
         21 Leaves opposite; stem twining; ovary superior.
           22 Leaves deeply 3–7-lobed, serrate, palmately veined, harshly scabrous;
              stems retrorsely spinulose; flowers unisexual, the staminate in loose
              axillary panicles, the pistillate in cone-like spikes or aments; fruit an
              achene; (*Humulus*) . . . . .Cannabinaceae (Part 3)
           22 Leaves entire or sinuately few-toothed; flowers perfect.
             23 Flowers in peduncled umbel-like clusters in the leaf-axils; corolla
               deeply 5-lobed; fruit a slender follicle to 6 cm long; seeds tipped with
               a tuft of hairs; (*Cynanchum*; introd. in s Ont. and s Que.) . . . .
               . . . . .Asclepiadaceae (Part 4)
             23 Flowers in small heads in axillary corymbiform panicles, lilac to
               purplish, the 4-flowered heads subtended by an involucre of 4 linear
               principal bracts (phyllaries), the tubular corollas shallowly 5-lobed;
               fruit an achene crowned by a pappus of numerous capillary bristles;
               (*Mikania*; s Ont.) . . . . .Compositae (Part 4)
        21 Leaves alternate.
          24 Plants climbing by tendrils; flowers unisexual; ovary inferior; fruit a pepo
           (berry-like with more or less rind and lateral placentation); leaves
           cordate-ovate to -rotund in outline, commonly 5-lobed . . . .
           . . . . .Cucurbitaceae (Part 4)
         24 Plants climbing by stem-twining; flowers perfect; ovary superior.
           25 Fruit a red, ovoid or ellipsoid berry; corolla regular, violet or purple,
              rotate, about 1 cm broad; inflorescence a several-flowered cyme;

leaves ovate, entire or deeply 1–few-lobed near base; (*Solanum dulcamara*; introd.) . . . . .Solanaceae (Part 4)
- 25 Fruit a 2–many-seeded capsule.
  - 26 Corolla 2-lipped, blue with yellow palate, to about 1 cm long, with an obtuse-tipped spur 2 or 3 mm long at base; flowers solitary on long slender axillary peduncles; leaves cordate-rotund or reniform, shallowly 5–9-lobed; (*Cymbalaria*; introd.) . . . .
  . . . . .Scrophulariaceae (Part 4)
  - 26 Corolla regular, salverform to campanulate, usually larger, not spurred; flowers 1–few on axillary peduncles; leaves mostly cordate-ovate in outline and entire or with entire lobes . . . .
  . . . . .Convolvulaceae (Part 4)
- 17 Plants not climbing.
  - 27 Flowers in dense heads subtended by one or more series of involucral bracts (phyllaries); fruit an achene, commonly crowned by a pappus of scales, awns, or bristles . . . . .Compositae (Part 4)
  - 27 Flowers and fruits with not all of the above characters.
    - 28 Leaves compound (at least some of them; all or nearly all simple only in *Astragalus spatulatus* of the Leguminosae).
      - 29 Stem-leaves all or chiefly opposite or whorled, or leaves all or chiefly basal . . . .*SERIES 3* (p. 107)
      - 29 Stem-leaves all or chiefly alternate.
        - 30 Leaves 3-foliolate or palmately or digitately 3–7-foliolate . . . .
        . . . .*SERIES 4* (p. 109)
        - 30 Leaves 1–3-pinnate, 2–3-ternate, or pinnately decompound . . . .
        . . . .*SERIES 5* (p. 110)
    - 28 Leaves all simple.
      - 31 Leaves all or mostly alternate (or the involucral ones sometimes opposite or whorled).
        - 32 Leaves entire or essentially so . . . .*SERIES 6* (p. 111)
        - 32 Leaves distinctly toothed to deeply lobed . . . .*SERIES 7* (p. 114)
      - 31 Leaves all or mostly opposite (or some or all whorled); or leaves all basal or nearly so, often in rosettes.
        - 33 Leaves all basal or nearly so (except occasionally for 1 or more involucral leaves), often in rosettes.
          - 34 Leaves entire or essentially so . . . .*SERIES 8* (p. 116)
          - 34 Leaves distinctly toothed to deeply lobed . . . .*SERIES 9* (p. 118)
        - 33 Leaves opposite or whorled.
          - 35 Leaves whorled . . . .*SERIES 10* (p. 119)
          - 35 Leaves, at least some of them, opposite (or the involucral leaves sometimes whorled).
            - 36 Leaves entire or essentially so . . . .*SERIES 11* (p. 120)
            - 36 Leaves distinctly toothed to deeply lobed . . . . .*SERIES 12* (p. 123)

SERIES 3 (see p. 107)
(Terrestrial non-climbing dicotyledonous herbs with compound leaves, these either all basal or sub-basal, or all or mostly opposite or whorled)

1 Leaves all or chiefly basal or sub-basal.
- 2 Leaflets finely dissected into mostly linear to narrowly oblong ultimate segments, the leaf rather fern-like in appearance.
  - 3 Fruit consisting of 2 seed-like mericarps separating at maturity; flowers small, in umbels; calyx-teeth, petals, and stamens each 5 (or the calyx-teeth obsolete); ovary inferior . . . . .Umbelliferae (Part 4)
  - 3 Fruit a capsule; flowers relatively large; sepals 2 (rarely 3), early deciduous; petals usually 4.

      4   Flowers solitary or occasionally in racemes, regular, none of the 4 (sometimes 6) petals spurred; stamens numerous; juice milky or coloured . . . .
. . . . .Papaveraceae (Part 3)

      4   Flowers in racemes or panicles, irregular, the 2 outer petals spurred or saccate at base; stamens in 2 sets of 3 each; juice watery; (*Dicentra*) . . . .
. . . . .Fumariaceae (Part 3)

  2 Leaflets entire to toothed or divided into mostly oblong to ovate ultimate segments or secondary leaflets.

     5 Leaflets 2 or 3.

        6 Leaflets 2, obliquely semi-ovate, normally entire or shallowly sinuate-dentate, to 1.5 dm long; flower solitary, white, 2 or 3 cm broad; sepals usually 4, deciduous; petals usually 8; ovary superior; fruit capsular, the upper part opening like a lid; (*Jeffersonia*; s Ont.) .Berberidaceae (Part 3)

        6 Leaflets 3 (or leaves often 2-ternate or pinnate-ternate in *Vancouveria* of the Berberidaceae).

          7 Leaflets entire or at most broadly emarginate at summit; flowers white or pinkish, 5-merous; fruit a capsule.

            8 Leaflets obcordate; peduncles 1-flowered, filiform, from slender scaly rhizomes; ovary superior; (a single genus, *Oxalis*) . . . . . .Oxalidaceae (Part 3)

            8 Leaflets oblanceolate to oblong or obovate, thickish; peduncles stout, terminated by a raceme and arising from thick rhizomes; (*Menyanthes*; transcontinental) . . . . .Gentianaceae (Part 4)

          7 Leaflets sinuate-toothed to deeply lobed.

            9 Plant pubescent; flowers in cymes, white or yellow; petals and sepals (or calyx-lobes) each 5; fruit a cluster of achenes; (*Duchesnea; Fragaria; Sibbaldia; Waldsteinia*) . . . . .Rosaceae (Part 3)

            9 Plants glabrous.

              10 Flowers solitary (or 2 or 3 in *Coptis*); fruit a cluster of several-seeded follicles . . . . .Ranunculaceae (Part 3)

              10 Flowers in terminal spikes or panicles; (*Achlys* and *Vancouveria*; s B.C.) . . . . .Berberidaceae (Part 3)

     5 Leaflets more than 3.

       11 Flowers papilionaceous; fruit a legume; leaves 1-pinnate, the leaflets entire; (*Oxytropis* and certain species of *Astragalus*) . . . . .Leguminosae (Part 3)

       11 Flowers regular; leaflets toothed or lobed.

         12 Flowers to 2.5 cm long, the petaloid sepals blue or purplish, each with a slender basal spur to over 1 cm long; fruit a cluster of follicles; (*Aquilegia jonesii*; sw Alta.) . . . . .Ranunculaceae (Part 3)

         12 Flowers very small, spurless.

           13 Inflorescence consisting of 2 or more umbels in a terminal corymb; sepals, petals, and stamens each 5; ovary inferior; fruit a purplish-black drupe; leaf solitary, ternate-pinnate; (*Aralia nudicaulis*; transcontinental) . . . .
. . . . .Araliaceae (Part 4)

           13 Inflorescence a terminal raceme; sepals 4 or 5, petaloid; petals none; stamens numerous; ovary superior; fruit a cluster of prominently ribbed achenes; leaves several, 2-pinnate; (*Thalictrum alpinum*) . . . .
. . . . .Ranunculaceae (Part 3)

 1 Leaves opposite on the stem or in a whorl of 3 (basal leaves also often present, the 2 or 3 stem-leaves then often more or less involucrate).

   14 Leaves pinnate, the stem bearing several opposite pairs; petals distinct, they and the sepals each usually 5.

     15 Leaflets (up to 16) entire, oblong, at most about 1.5 cm long; flowers pale yellow, solitary on slender axillary peduncles; fruit beakless, the 5 mature 3–5-seeded carpels each with 2 stout divergent spines and a row of tubercles; plant hirsute; (a single genus, *Tribulus*; introd. in s Ont.) . . . . .Zygophyllaceae (Part 3)

     15 Leaflets toothed or incised; flowers pink to purple, paired or in umbels; fruit 1-seeded, beaked by the slender twisted or coiled style, this separating the 5

carpels from the base of the central column at maturity; (*Erodium*; introd.) . . . .
. . . . .Geraniaceae (Part 3)
14 Leaves palmate.
  16 Stem bearing several pairs of opposite leaves.
    17 Fruits 1-seeded, beaked by the slender style, this strongly upcurving and
      separating the 5 carpels from the base of the central column at maturity; petals
      5, distinct; stamens 10, often united at base; (*Geranium*) . . . . .Geraniaceae (Part 3)
    17 Fruit a several-seeded capsule dehiscing along 4 sutures; corolla white or
      blue, gamopetalous, with a short tube and a somewhat 2-lipped spreading limb;
      stamens 4, distinct; (*Tonella*; sw ?B.C.) . . . . .Scrophulariaceae (Part 4)
  16 Stem bearing an involucre-like single pair of opposite leaves or a whorl of 3 below
    the inflorescence.
    18 Flowers small, white or greenish, rather numerous in a solitary terminal
      long-peduncled simple umbel; petals and stamens each 5, the 5 calyx-lobes
      minute or rudimentary; ovary inferior; fruit a 2–3-seeded yellow or bright-red
      berry; basal leaves wanting; (*Panax*; E Canada) . . . . .Araliaceae (Part 4)
    18 Plants without the above combination of characters.
      19 Flowers small, gamopetalous, yellowish or greenish, in head-like clusters,
        the rotate corolla 4–6-cleft, each sinus bearing a pair of distinct or slightly
        united stamens; ovary inferior; stem-leaves a single pair of 3-parted or
        merely 3-cleft obovate leaves, the 1–3 long-petioled basal leaves
        1–3-ternate; (a single genus, *Adoxa*; B.C. to Man.) . . . . .Adoxaceae (Part 4)
      19 Flowers larger, polypetalous, solitary or in few-flowered cymes, racemes,
        or 2–3-rayed umbels; basal leaves long-petioled, the involucral leaves
        a single pair or a whorl of 3.
        20 Flowers in racemes or corymbs; sepals and petals each 4; stamens 6
          (4 long, 2 short); fruit a 2-locular capsule with an internal septum;
          (*Dentaria*; Ont. to N.S.) . . . . .Cruciferae (Part 3)
        20 Flowers solitary on long peduncles or few in an umbel; sepals 4 to many,
          petaloid; petals none; stamens numerous; fruit a head of achenes;
          (*Anemone; Anemonella*) . . . . .Ranunculaceae (Part 3)

SERIES 4 (see p. 107)
(Terrestrial or subaquatic non-climbing dicotyledonous herbs with 3-foliolate or palmately or
digitately compound alternate stem-leaves)

1 Flowers papilionaceous (except in *Cassia*); stamens mostly 10, monadelphous or
  diadelphous (distinct only in *Baptisia, Cassia,* and *Thermopsis*); ovary superior; fruit a
  continuous or jointed legume (pod) . . . . .Leguminosae (Part 3)
1 Flowers not papilionaceous.
  2 Petals none; flowers unisexual, in panicles (staminate panicles loose, pistillate ones
    dense); stamens 5; ovary superior; fruit a glandular achene; leaves 3–7-digitate (the
    lower ones opposite), the linear-lanceolate leaflets coarsely toothed, pubescent, to
    about 1.5 dm long; (*Cannabis*; introd.) . . . . .Cannabinaceae (Part 3)
  2 Petals usually present; flowers perfect.
    3 Flowers in simple or compound umbels, small; ovary inferior; fruit consisting of 2
      seed-like mericarps separating at maturity . . . . .Umbelliferae (Part 4)
    3 Flowers not in umbels (except sometimes in Oxalidaceae); fruit various.
      4 Fruit consisting of achenes or follicles; ovary superior; stamens 5 to many.
        5 Sepals and petals each usually 5; flowers regular . . . . .Rosaceae (Part 3)
        5 Sepals 3–5; petals 2–10 or wanting or converted into staminodia; flowers
          regular (*Ranunculus; Trollius*) or irregular (*Aconitum; Delphinium*) . . . .
          . . . . .Ranunculaceae (Part 3)
      4 Fruit a capsule or, in Malvaceae, a ring of carpels separating at maturity.
        6 Sepals and petals each 4; ovary superior.
          7 Capsule (a silique) with an internal thin septum separating its 2 locules;

petals spreading; stamens 6 (4 long, 2 short); leaflets 3, coarsely toothed
or incised; (*Dentaria*; Ont. to N.S.) .... Cruciferae (Part 3)

    7   Capsule without an internal septum, 1-locular; petals erect; stamens
4 to many, uniform; leaflets 3, 5, or 7, entire or serrulate ....
.... Capparidaceae (Part 3)

  6 Sepals and petals each 5.

    8   Flowers yellow; stamens 10 or 15, the filaments united at base into a
short tube; ovary superior; fruit a capsule; leaves 3-foliolate, the obcordate
leaflets entire; (a single genus, *Oxalis*) .... Oxalidaceae (Part 3)

    8   Flowers not yellow.

      9   Stamens numerous, united into a column; ovary superior; fruit a ring
of carpels separating at maturity; leaflets usually 5 .... Malvaceae (Part 3)

      9   Stamens 10, distinct; ovary partly or wholly inferior; fruit a capsule;
leaflets 3 (but the leaf sometimes appearing 5-foliolate by incision of
the lateral leaflets); (*Lithophragma; Tiarella*) ...... Saxifragaceae (Part 3)

SERIES 5 (*see* p. 107)
(Terrestrial or subaquatic non-climbing dicotyledonous herbs with compound (1–3-pinnate,
ternate-pinnate, or pinnately decompound) alternate leaves)

1   Fruit a capsule (or 1-seeded and indehiscent in *Fumaria* of the Fumariaceae); ovary
superior.

  2 Leaves pinnately decompound, delicate; flowers irregular, zygomorphic (the corolla
laterally compressed and with 1 or 2 of the outer pair of petals spurred or saccate at
base), in racemes or panicles; stamens in 2 sets of 3 each .... Fumariaceae (Part 3)

  2 Leaves usually 1-pinnate (1–2-pinnate in Rutaceae); flowers regular.

    3 Petals distinct.

      4 Leaves strongly gland-dotted, 1–2-pinnate; petals 4 or 5, greenish yellow;
stamens 8 or 10; capsule 4–5-lobed, lacking an internal septum; (*Ruta*; introd.
in ?Alta. and Ont.) .... Rutaceae (Part 3)

      4 Leaves not gland-dotted, 1-pinnate; petals 4, spreading, white or yellow;
stamens 6 (4 long, 2 short); capsule 2-locular by an internal thin septum ....
.... Cruciferae (Part 3)

    3 Petals united, the 5-lobed corolla white to pink-purple or blue; stamens 5.

      5 Calyx-lobes distinct nearly to base, more or less hirsute or bristly-ciliate;
flowers in scorpioid cymes (uncoiling from the tip); style shallowly to very
deeply 2-cleft .... Hydrophyllaceae (Part 4)

      5 Calyx-lobes at most about equalling the calyx-tube, glabrous or nearly so; flowers
in a dense terminal head or corymb or in a compact to loose cyme or cymose
panicle; style 3-lobed; (*Gilia; Navarretia; Polemonium*) ..... Polemoniaceae (Part 4)

1   Fruit not a capsule; flowers regular.

  6 Fruit drupe-like or a berry (or berry-like).

    7 Flowers in umbels (the umbels solitary in the upper leaf-axils or numerous in
terminal racemose panicles), 5-merous, white or greenish; ovary inferior; leaves
2-pinnate or ternate-pinnate, the leaflets serrate; (*Aralia*) .... Araliaceae (Part 4)

    7 Flowers in cymes or panicles; ovary superior.

      8 Flowers 5-merous, in cymes, the petals united at the base of the rotate corolla;
stems leafy, the leaflets entire to coarsely toothed or rather deeply lobed;
(*Lycopersicon; Solanum*) .... Solanaceae (Part 4)

      8 Flowers 6-merous, in a terminal panicle; stem-leaf solitary and sessile above
the middle of the stem, 3-ternate, the leaflets 2–5-lobed above the middle;
(*Caulophyllum*; s Man. to N.S.) .... Berberidaceae (Part 3)

  6 Fruit neither drupe-like nor berry-like.

    9 Flowers in simple or compound umbels, small; ovary inferior; fruit consisting of 2
seed-like mericarps separating at maturity .... Umbelliferae (Part 4)

    9 Flowers not in umbels; ovary usually superior (often inferior in Rosaceae); fruit
not as above.

10 Fruit deeply cleft and separating into up to five 1-seeded nutlet-like segments; flowers solitary on axillary peduncles; low to erect decumbent-based annuals . . . . .Limnanthaceae (Part 3)

10 Fruit an achene or follicle (a berry in *Actaea*); stamens 4 to many.

    11 Sepals 3–5, often petaloid; petals often none (or 4–10, small and deciduous, in *Actaea*; 5 and long-spurred in *Aquilegia*; 2–10 and with a nectariferous spot or pit at base in *Ranunculus*) .Ranunculaceae (Part 3)

    11 Sepals commonly 5 (rarely 4), usually green and herbaceous; petals commonly 5 (4 in *Potentilla anglica* and *P. erecta*; 4–7 in *Filipendula*; none in *Sanguisorba*) . . . . .Rosaceae (Part 3)

SERIES 6 (see p. 107)

(Terrestrial or subaquatic (if aquatic, leaves not dissected except sometimes the basal ones) non-climbing dicotyledonous herbs with simple alternate entire leaves)

1 Corolla usually distinctly irregular (*Echium* and *Lycopsis* of the Boraginaceae may be sought here; corolla nearly regular in certain genera or species of the Scrophulariaceae); fruit a capsule.

  2 Petals 3, the lowest one keel- or boat-shaped and with a fringe-like crest; sepals 5, the 2 inner ones much the largest and coloured like the petals; ovary superior; flowers rather small, in usually crowded spikes or spike-like racemes; (a single genus, *Polygala*) . . . . .Polygalaceae (Part 3)

  2 Petals or corolla-lobes 4 or 5.

    3 Corolla nearly regular at summit (but with the lowest petal larger and gibbous or saccate at base), greenish white, 5-merous; anthers united into a sheath around the pistil; ovary superior; flowers solitary or few in the leaf-axils, the strongly recurved peduncles jointed above the middle; (*Hybanthus*; s Ont.) . . .Violaceae (Part 3)

    3 Corolla with unequal petals and often more or less 2-lipped.

      4 Sepals and yellow petals each 4, the petals irregularly incised, one pair larger than the other; ovary superior; flowers in a terminal spike-like raceme; (*Reseda luteola*; introd. in N.S.) . . . . .Resedaceae (Part 3)

      4 Sepals and corolla-lobes each 5, the flowers usually more or less 2-lipped (upper lip 2-lobed, lower lip 3-lobed); fruit a capsule.

        5 Stamens 5, the anthers united into a tube around the style, 2 of them shorter than the others and bearded at summit; ovary inferior; flowers usually blue (sometimes purple or white; scarlet in *Lobelia cardinalis*); juice milky . . . . .Lobeliaceae (Part 4)

        5 Stamens 2 or 4 (2 long, 2 short), distinct, a fifth sterile one often present (except in *Verbascum*); ovary superior; juice not milky . . . . .Scrophulariaceae (Part 4)

1 Corolla (when present) regular or nearly so.

  6 Corolla wanting.

    7 Calyx wanting.

      8 Flowers borne in a more or less cup-shaped cyathium resembling a calyx or corolla, its margins bearing 4 or 5 nectariferous glands with or without coloured appendages; fruit a 3-lobed and 3-seeded capsule; plants with milky juice; (*Euphorbia*) . . . . .Euphorbiaceae (Part 3)

      8 Flowers lacking the above characters, in 1 or 2 peduncled spikes to about 1.5 dm long, these terminal but often surpassed by axillary branches; leaves cordate-ovate, to about 1.5 dm long; stem jointed; juice not milky; (a single genus, *Saururus*; s Ont. and sw Que.) . . . . .Saururaceae (Part 3)

    7 Calyx present.

      9 Calyx about 3 cm long, yellow, very oblique at summit by an ovate lobe on one side; stamens 6, the anthers sessile on the short 6-angled style; leaves broadly cordate-ovate to reniform, to about 1 dm long and broad; (*Aristolochia clematitis*; introd. in s Ont.) . . . . .Aristolochiaceae (Part 3)

      9 Calyx much smaller, regular; anthers not sessile.

      10 Fruit a juicy dark-purple berry derived from a ring of up to 15 carpels; calyx
          greenish white or pinkish; flowers in racemes; leaves oblong-lanceolate to
          ovate, to about 3 dm long; (a single genus, *Phytolacca*; s Ont. and sw Que.) . . . . .
          . . . . .Phytolaccaceae (Part 3)
      10 Fruit usually dry, an achene, utricle, or 1-seeded nut (a juicy scarlet drupe
          in *Geocaulon* of the Santalaceae).
          11 Calyx 5-lobed, fused to the ovary (ovary and 1-seeded nut or drupe wholly
              inferior); stamens 5; flowers commonly 3–5 in small cymes, these solitary
              in the upper leaf-axils (*Geocaulon*) or in a terminal panicle or corymb
              (*Comandra*) . . . . .Santalaceae (Part 3)
          11 Calyx free from the ovary (ovary and 1-seeded achene or utricle superior).
              12 Fruit an achene often enclosed in the mature calyx.
                  13 Stems with swollen joints subtended by sheathing stipules
                      (ocreae); styles or stigmas 2 or 3; flowers solitary in the leaf-axils
                      or in spikes, racemes, or panicles . . . . .Polygonaceae (Part 3)
                  13 Stems with neither swollen joints nor sheathing stipules; stigmas
                      solitary, sessile; flowers in cymose axillary clusters; (*Parietaria*;
                      B.C. to Que.) . . . . .Urticaceae (Part 3)
              12 Fruit a bladdery 1-seeded utricle often enclosed in the calyx; stipules
              none.
                  14 Flowers subtended by scarious bracts, greenish or reddish; calyx
                      scarious . . . . .Amaranthaceae (Part 3)
                  14 Flowers not subtended by bracts, greenish; calyx herbaceous or
                      fleshy . . . . .Chenopodiaceae (Part 3)
 6 Corolla present.
    15 Inflorescence an umbel; flowers 5-merous; stamens 5.
      16 Umbel simple; corolla with a crown (corona) composed of 5 fleshy hooded
          bodies seated on the tube of stamens; ovary superior; fruit a pair of large
          follicles; seeds with a tuft of long silky hairs at summit; juice often milky . . . .
          . . . . .Asclepiadaceae (Part 4)
      16 Umbel compound; corolla lacking a corona; stamens distinct; ovary inferior;
          fruit a pair of dry seed-like mericarps separating at maturity; plant glaucous;
          (*Bupleurum*) . . . . .Umbelliferae (Part 4)
    15 Inflorescence not an umbel; flowers often solitary.
      17 Fruit a berry or consisting of achenes, utricles, nutlets, or a ring of carpels; ovary
          superior.
          18 Fruit a ring of up to about 15 pubescent carpels; flowers yellow, to 2.5 cm
              broad, the 5 petals distinct or nearly so; stamens numerous, their filaments
              united into a column around the style; (*Abutilon*; introd.) . . . . .Malvaceae (Part 3)
          18 Fruit not a ring of carpels; other characters not all as above.
              19 Fruit a head of mostly beaked achenes; petals yellow, distinct; stamens
                  mostly at least 10; (*Ranunculus*) . . . . .Ranunculaceae (Part 3)
              19 Fruit otherwise; stamens 5, alternating with the petals or corolla-lobes.
                  20 Fruit a many-seeded berry; flowers often showy . . . . .Solanaceae (Part 4)
                  20 Fruit a utricle or cluster of nutlets; flowers mostly smaller and numerous.
                    21 Fruit a somewhat 3-sided indehiscent purplish-black utricle
                      about 1 mm long; flowers very small, numerous and crowded in
                      axillary and terminal paniculate-cymose clusters; plants glabrous
                      and glaucous; (*Corrigiola*; introd. in sw B.C.) . . . .
                      . . . . .Caryophyllaceae (Part 3)
                  21 Fruit consisting of usually 4 (occasionally fewer) nutlets at the
                    base of the style; flowers in scorpioid (tip-coiled) racemiform or
                    spike-like cymes; plants commonly rough-hairy . . . .
                    . . . . .Boraginaceae (Part 4)

17 Fruit a usually dehiscent capsule (in Crassulaceae, a group of follicles).
    22 Petals united into a mostly 5-lobed corolla (usually 4-lobed in *Centunculus* of the Primulaceae); stamens 5 (sometimes 4 in *Centunculus*), inserted on the corolla.
        23 Stamens opposite the corolla-lobes; flowers small, in terminal racemes and the white corolla 5-lobed (*Samolus*) or nearly sessile in the leaf-axils and the pink corolla usually 4-lobed (*Centunculus*); ovary superior . . . . .Primulaceae (Part 4)
        23 Stamens alternating with the corolla-lobes.
            24 Ovary (and capsule) inferior, fused with the calyx-tube; flowers commonly blue, sometimes nearly white . . . . . . .Campanulaceae (Part 4)
            24 Ovary (and capsule) superior, free from the calyx.
                25 Leaves linear to oblong-lanceolate, sessile; flowers mostly in terminal head-like leafy-bracted cymes (solitary or in small part paired in the leaf-axils in *C. tenella*), whitish to lilac-purple, with a long slender tube; (*Collomia*) . . . . .Polemoniaceae (Part 4)
                25 Leaves mostly broader and petioled; flowers solitary in the leaf-axils or in racemose or paniculate clusters . . .Solanaceae (Part 4)
    22 Petals distinct (*Croton* of the Euphorbiaceae will key out here).
        26 Fruit a cluster of at least 4 follicles; ovary superior; calyx-lobes and petals each at least 4; stamens 8 to many; leaves fleshy; (*Sedum; Sempervivum*) . . . . .Crassulaceae (Part 3)
        26 Fruit a capsule.
            27 Petals and sepals (or calyx-lobes) each 4; flowers solitary or in racemes.
                28 Ovary (and capsule) superior, free from the calyx, the capsule usually 2-locular by an internal thin septum; stamens usually 6 (4 long, 2 short) or rarely only 4 or 2 . . . . .Cruciferae (Part 3)
                28 Ovary (and capsule) completely inferior and fused with the calyx-tube, the capsule usually 4-locular (2-locular but 4-valved in *Gayophytum*); stamens usually 8 (4 in *Ludwigia*) . . . . .Onagraceae (Part 4)
            27 Petals and sepals (or calyx-lobes) usually 2, 3, or 5 (but as many as 7 or sometimes none in Portulacaceae); ovary (and capsule) usually superior and free from the calyx (or partly inferior in Saxifragaceae or in *Portulaca* of the Portulacaceae).
                29 Sepals 2, usually unequal in size; petals mostly 5 (but as few as 2 and as many as 7); leaves more or less fleshy or succulent; (*Calandrinia; Montia; Portulaca; Talinum*) . . . . .Portulacaceae (Part 3)
                29 Sepals 5.
                    30 Sepals unequal, the outer pair bract-like and much smaller than the others (or sometimes wanting); petals 3 and reddish (*Lechea*) or 5 and yellow; stamens few to many . . . . .Cistaceae (Part 3)
                    30 Sepals equal; petals 5; stamens 5 or 10.
                        31 Stamens 5; capsule 5-locular, finally splitting into 5 united carpels; petals white, yellow, or blue; stems leafy, lacking basal rosettes of leaves . . . . .Linaceae (Part 3)
                        31 Stamens 10 (but 5 of them reduced to apically fringed staminodia in *Parnassia*); capsule 1-locular (*Parnassia*) or 2-locular (*Saxifraga*; the capsule usually splitting into nearly distinct follicles at maturity); petals white; principal leaves mostly in a basal rosette; (*Parnassia; Saxifraga*) . . . . .Saxifragaceae (Part 3)

SERIES 7 *(see* p. 107)
(Terrestrial or subaquatic non-climbing dicotyledonous herbs with simple, alternate, toothed to deeply lobed leaves; basal leaves also sometimes present)

1 Corolla usually none (sometimes represented by staminodia in Ranunculaceae).
  2 Fruit dry and 1-seeded, an achene or bladdery utricle.
    3 Fruit a utricle; stamens 1 to many; leaves not subtended by stipules . . . .
    . . . . .Chenopodiaceae (Part 3)
    3 Fruit an achene or cluster of achenes.
      4 Leaves coarsely pinnate-toothed, ovate, subtended by 2-cleft stipules; flowers unisexual, the staminate ones with 5 sepals and 5 stamens; plant beset with stinging bristles; *(Laportea;* Sask. to N.S.) . . . . .Urticaceae (Part 3)
      4 Leaves shallowly to very deeply palmately lobed.
        5 Leaves exstipulate, round-reniform in outline, to about 4 dm broad, very deeply lobed (the lobes themselves variously toothed or incised), those of the stem commonly only 1 or 2; sepals distinct, soon falling; pistils several, the several achenes each beaked with a hooked style; stems erect, to over 1 m tall, from widely spreading rhizomes; *(Trautvetteria;* B.C.) . . . .
        . . . . .Ranunculaceae (Part 3)
        5 Leaves subtended by broad stipules, fan-shaped to round-reniform in outline, shallowly to deeply lobed; sepals united into a tube with persistent lobes; pistils usually solitary, the achene beakless; stems lower; *(Alchemilla)* . . . .
        . . . . .Rosaceae (Part 3)
  2 Fruit a capsule, a group of follicles, or a head of small berries.
    6 Fruit a capsule.
      7 Sepals 2, cream-coloured; capsule 2-locular, 4–6-seeded, splitting to base; stamens many; stigma nearly sessile, 2-lobed; flowers in elongated terminal panicles; leaf-blades round-cordate, deeply lobed (the lobes sinuate or dentate), white-glaucous beneath, to 3 dm long; plant to 2.5 m tall, with saffron-coloured juice; *(Macleaya;* introd. in sw Que.) . . . . .Papaveraceae (Part 3)
      7 Sepals (calyx-lobes) 3, 4, or 5; capsule 1- or 3-locular; stamens few; styles 2 or 3; plants lower.
        8 Capsule 1-locular, 2-lobed, many-seeded; calyx 4-parted; stamens usually 4; styles 2; leaves reniform; juice watery; plants of moist habitats; *(Chrysosplenium)* . . . . .Saxifragaceae (Part 3)
        8 Capsule 3-locular, 3-lobed, 3-seeded; calyx 3–5-parted (or wanting in *Euphorbia*); stamens 1 *(Euphorbia;* juice milky) or 8–18 *(Acalypha;* juice watery); styles 3; plants of mostly dry habitats . . . . .Euphorbiaceae (Part 3)
    6 Fruit a group of follicles or (in *Hydrastis* of the Ranunculaceae) a raspberry-like head of dark-red 1–2-seeded small berries.
      9 Flowers rather small and numerous, yellowish green, in 2–4-branched cymes; sepals 5; stamens 10; fruit a flattened circular group of strongly beaked follicles; leaves finely serrate; *(Penthorum;* SE Man. to N.B.) . . . . .Crassulaceae (Part 3)
      9 Flowers relatively large and mostly solitary; sepals variable in number; stamens numerous; fruits various . . . . .Ranunculaceae (Part 3)
1 Corolla present.
  10 Calyx or corolla or both usually more or less irregular (nearly regular in *Digitalis, Verbascum, Veronica,* and *Veronicastrum* of the Scrophulariaceae).
    11 Calyx or corolla with 1 of its segments spurred or saccate at base; ovary superior.
      12 Sepals petaloid, surpassing the petals.
        13 Sepals 5, the upper one prolonged at base into a long spur enclosing the long spurs of 2 of the 4 petals; stamens many; fruit consisting of up to 5 many-seeded follicles; leaves deeply palmately cut; *(Delphinium)* . . . .
        . . . . .Ranunculaceae (Part 3)
        13 Sepals apparently 4, the largest one forming a spurred sac (except in *I. ecalcarata*); petals 2, 2-lobed (each a pair united), spurless; stamens 5;

fruit a 5-locular capsule bursting elastically; leaves finely to coarsely
  toothed; (a single genus, *Impatiens*) ....Balsaminaceae (Part 3)
 12 Sepals or calyx-lobes 5, green and herbaceous, not spurred; 1 of the petals
  spurred; fruit a capsule.
   14 Petals distinct; sepals auricled at base; stamens 5, their anthers connivent
    about the united styles; capsule opening by 3 valves; cleistogamous flowers
    often present; (*Viola*) ....Violaceae (Part 3)
   14 Petals more or less united; sepals not auricled; stamens 4 (2 long, 2 short),
    separate; capsule usually opening by terminal pores or chinks; cleistogamous
    flowers none; (*Chaenorrhinum; Cymbalaria; Kickxia; Linaria*) ....
    ....Scrophulariaceae (Part 4)
 11 Flowers spurless.
  15 Sepals 5, petaloid, the upper one (the helmet) strongly concave on the back;
   petals included, the 2 upper ones small spur-shaped bodies on long claws, the
   others much reduced or wanting; stamens numerous; ovary superior; fruit
   consisting of 3–5 several-seeded follicles; (*Aconitum*) ......Ranunculaceae (Part 3)
  15 Sepals green or whitish, not petaloid; fruit a capsule (or consisting of more or less
   separated follicles in Saxifragaceae).
   16 Petals distinct.
    17 Petals commonly 6 (in 3 pairs of different sizes), irregularly incised,
     sordid- to greenish-white or yellowish; calyx-lobes subequal; stamens
     commonly at least 10; capsules with 3 or more carpels; leaves deeply
     pinnatifid; glabrous annuals or biennials; (a single genus, *Reseda*;
     introd.) ....Resedaceae (Part 3)
    17 Petals 4, equal, linear-subulate, chocolate-colour; calyx-lobes unequal, 2
     lobes smaller than the other 3; stamens 3, opposite the larger calyx-lobes;
     capsules 2-locular; leaves shallowly lobed and irregularly once or twice
     crenate-serrate; hirsute and somewhat glandular perennial; (*Tolmiea*;
     s Alaska–B.C.) ....Saxifragaceae (Part 3)
   16 Petals united into a more or less 2-lipped corolla.
    18 Ovary (and capsule) superior, free from the calyx-tube; stamens 2 or 4
     (2 long; 2 short), distinct, a fifth one often present, this sterile except in
     Verbascum; anthers bearded or beardless; juice not milky ....
     ....Scrophulariaceae (Part 4)
    18 Ovary (and capsule) inferior, fused with the calyx-tube; stamens 5, 2 of
     them bearded; juice milky ....Lobeliaceae (Part 4)
10 Calyx and corolla both regular or nearly so (corolla slightly oblique at summit in
 Hyoscyamus of the Solanaceae).
 19 Petals united.
  20 Stamens numerous, exceeding the number of corolla-lobes, their filaments
   united into a column attached at base to the short claws of the petals; ovary
   superior; fruit usually a ring of carpels around a central axis (in *Hibiscus*, a
   5-locular capsule); flowers large and showy ....Malvaceae (Part 3)
  20 Stamens 5, free (but converged into a tube in *Solanum*), alternating with the 5
   corolla-lobes; fruit a capsule or (in several genera of Solanaceae) a berry.
   21 Ovary (and capsule) inferior, fused with the calyx-tube; corolla generally
    blue and showy; juice milky ....Campanulaceae (Part 4)
   21 Ovary (and capsule) superior, free from the calyx; juice not milky.
    22 Corolla funnelform, white, dull yellowish, lavender, pink, or violet, to about
     5 cm broad; style 1; flowers solitary in the leaf-axils or few in spike-like
     racemes or in panicles ....Solanaceae (Part 4)
    22 Corolla cylindric to campanulate, white, blue, or purple, rather small; style
     2-cleft; flowers solitary or 1-sided in terminal scorpioid (tip-coiled) cymes ...
     ....Hydrophyllaceae (Part 4)
 19 Petals distinct or united only at base.
  23 Ovary (and capsule) inferior, fused at least basally with the calyx-tube.

24 Fruit consisting of 2 seed-like mericarps separating at maturity; flowers in
a compound umbel and leaves palmately lobed (*Sanicula crassicaulis*) or
flowers in a head-like cluster subtended by spinose bracts, the leaves
pinnately toothed or lobed (*Eryngium*) . . . . .Umbelliferae (Part 4)
24 Fruit and flowers otherwise.
  25 Petals 4, white, yellow, pink, or purplish; stamens 8; style 1, the stigma
2–4-lobed or capitate; capsule 4-locular, wholly fused with the calyx-tube
(the calyx usually projecting above it as a free hypanthium) . . . .
. . . . .Onagraceae (Part 4)
  25 Petals 5, mostly white or yellow (sometimes purplish or dotted with red
or yellow in Saxifragaceae).
    26 Ovary (and 1-locular capsule) completely inferior, fused with the
calyx-tube (the calyx projecting above it as a short, often flared,
5(4)-lobed hypanthium); stamens numerous; petals cream-colour to
yellow; leaves brittle and readily detached, adherent by the scabrous
barbed pubescence; (a single genus, *Mentzelia*; w Canada) . . . .
. . . . .Loasaceae (Part 4)
    26 Ovary (and capsule) at most fused with the calyx to near the middle,
the capsule often separating at maturity into 2 (rarely 3) nearly distinct
follicles; stamens usually 5 or 10 (3 in *Tolmiea*) . . . .Saxifragaceae (Part 3)
23 Ovary (and capsule) superior, free from the calyx.
  27 Fruit usually a ring of often pubescent carpels around a central axis (fruit a
5-locular capsule in *Hibiscus*); stamens numerous, their filaments united
into a column . . . . .Malvaceae (Part 3)
  27 Fruit and stamens otherwise.
    28 Stamens numerous, usually more than 10.
      29 Sepals mostly 2, sometimes 3; petals usually 4, showy; fruit a capsule;
juice usually orange-yellow (watery in *Eschscholtzia*) . . . .
. . . . .Papaveraceae (Part 3)
      29 Sepals 3, 4, or 5 (sometimes more in Ranunculaceae); fruit not a
capsule; juice watery.
        30 Fruit a cluster of fleshy small drupes in a "raspberry"-like head
(these red-tinged to amber-colour, finally yellowish, in *Rubus
chamaemorus*, red in *R. odoratus* and *R. parviflorus*) or 5 or more
dryish seed-like drupes (*Dalibarda*); petals 5 (white to rose-purple
in *Rubus*; yellow in *Dalibarda*) . . . . .Rosaceae (Part 3)
        30 Fruit consisting of achenes; petals 5 or 6 (sometimes fewer or
more), white, yellow, orange, or red; leaves usually deeply lobed
or dissected; (*Adonis; Ranunculus*) . . . . .Ranunculaceae (Part 3)
    28 Stamens not more than 10.
      31 Fruit a head of achenes; petals yellow, with a nectariferous spot or
pit at base; stamens mostly numerous; (*Ranunculus*) . . . .
. . . . .Ranunculaceae (Part 3)
      31 Fruit a capsule (sometimes indehiscent) or a cluster of follicles.
        32 Stamens 6 (4 long, 2 short) or sometimes 4 or only 2; sepals and
petals each 4; fruit a capsule usually 2-locular by a thin internal
septum . . . . .Cruciferae (Part 3)
        32 Stamens 8 or 10; sepals and petals each 4 or 5; fruit a cluster of
4 or 5 follicles; (*Sedum*) . . . . .Crassulaceae (Part 3)

SERIES 8 (see p. 107)
(Terrestrial or subaquatic non-climbing dicotyledonous herbs with simple entire leaves, these all
basal or nearly so; *Astragalus spatulatus* of the Leguminosae may be sought here)

1 Inflorescence a simple or twice-forking umbel.
  2 Leaves reduced to hollow linear-oblanceolate petioles (phyllodia) with transverse
partitions visible externally; flowers white, very small, few in simple umbels, the sepals

obsolete; ovary inferior; fruit consisting of 2 dry seed-like mericarps separating at maturity; stems creeping, rooting at the nodes; (*Lilaeopsis*; tidal shores and salt-marshes of Vancouver Is. and w N.S.) . . . . .Umbelliferae (Part 4)

  2 Leaves with normal dilated blades; ovary superior; fruit otherwise; stems not creeping.

      3 Inflorescence a simple umbel of white to lilac, rose-pink, or bluish-purple 5-merous flowers; stamens 5, inserted on the corolla opposite its lobes; fruit a capsule; (*Androsace; Dodecatheon; Douglasia; Primula*) . . . . .Primulaceae (Part 4)

      3 Inflorescence a twice-forking umbel of small yellow flowers; calyx commonly 6-lobed; petals none; stamens 9; fruit an achene; (*Eriogonum*; w Canada) . . . . .Polygonaceae (Part 3)

1 Inflorescence usually not a distinct umbel (but often scorpioid-umbellate in *Spraguea* of the Portulacaceae).

  4 Scapes 1-flowered.

      5 Mature inflorescence spike-like, consisting of very numerous achenes on a slender receptacle to about 6 cm long; sepals and petals each 5, the petals spurred at base; ovary superior; leaves linear-spatulate; (*Myosurus*; B.C. to s Ont.) . . . . .Ranunculaceae (Part 3)

      5 Mature inflorescence not spike-like; sepals not spurred.

         6 Corolla of 5 separate white or greenish-veined petals, each petal subtending a deeply 3–9-cleft staminodium; fertile stamens 5; ovary partially inferior; fruit a 4-valved capsule; (*Parnassia*) . . . . .Saxifragaceae (Part 3)

         6 Corolla gamopetalous, small; stamens 4; ovary superior.

            7 Flowers perfect, the calyx and corolla 5-lobed; fruit a many-seeded capsule; stems creeping, rooting at the nodes; (*Limosella*; transcontinental) . . . . .Scrophulariaceae (Part 4)

            7 Flowers unisexual, the pistillate ones with minutely 4-lobed corollas; fruit an achene enclosed in the persistent calyx; plant stemless, the leaves and scapes arising from filiform rhizomes and stolons; (*Littorella*; Ont. to Nfld. and N.S.) . . . . .Plantaginaceae (Part 4)

  4 Scapes mostly few- to many-flowered.

      8 Plants insectivorous, the linear to orbicular leaf-blades clothed with reddish gland-bearing hairs or tentacles that exude a glistening glutinous insect-trapping and -digesting clear secretion; flowers several, in a 1-sided nodding terminal raceme-like cyme; ovary superior; (a single genus, *Drosera*) . . . . .Droseraceae (Part 3)

      8 Plants not insectivorous.

         9 Petals distinct or united only at the base.

           10 Sepals and petals each 4; stamens 6 (4 long, 2 short); ovary superior; fruit a capsule usually 2-locular by a thin internal septum . . . . .Cruciferae (Part 3)

           10 Plants without the above combination of characters.

              11 Sepals 2; petals 2–9; stamens variable in number; style 1; ovary superior; capsule 1-locular . . . . .Portulacaceae (Part 3)

              11 Sepals (or calyx-lobes) and petals each 5; stamens 10.

                 12 Anthers dehiscing lengthwise; styles 2; ovary partly inferior; fruit a 2-locular capsule finally splitting into 2 nearly distinct follicles; petals narrow, small, often deciduous; (*Saxifraga*) . . . . . .Saxifragaceae (Part 3)

                 12 Anthers inverted and opening by a pair of terminal (morphologically basal) pores; style 1; ovary superior; fruit a 5-locular capsule; petals broad, persistent . . . . .Pyrolaceae (Part 4)

         9 Petals united (corolla gamopetalous).

           13 Corolla 2-lipped, white to pale blue, usually about 1.5 cm long; ovary inferior; racemes few-flowered; leaves linear, fleshy, hollow, submersed, to 9 cm long; (*Lobelia dortmanna*; transcontinental) .Lobeliaceae (Part 4)

           13 Corolla regular; ovary superior; plants terrestrial.

              14 Flowers 4-merous, in compact to interrupted spikes; stamens usually 4 (sometimes 5 or 6; 2 in *P. virginica*); fruit a 2–many-seeded capsule circumscissile near the middle; (*Plantago*) . . . . .Plantaginaceae (Part 4)

14 Flowers 5-merous, in heads (*Armeria*) or panicles (*Limonium*); stamens 5; fruit 1-seeded, indehiscent or circumscissile . . . . .Plumbaginaceae (Part 4)

SERIES 9 (*see* p. 107)
(Terrestrial or subaquatic non-climbing dicotyledonous herbs with simple, shallowly toothed to deeply lobed leaves, these all or chiefly basal)

1 Inflorescence a simple umbel; flowers 5-merous, regular; stamens 5.
  2 Fruit a 5-valved many-seeded capsule; ovary superior; flowers white to lilac or blue-purple, their petals united; leaves tapering to base; stems ascending or erect; (*Androsace; Primula*) . . . . .Primulaceae (Part 4)
  2 Fruit consisting of 2 seed-like dry mericarps separating at maturity; ovary inferior; flowers small, whitish, their petals distinct; leaves peltate or cordate-rotund; stems prostrate and rooting at the nodes; (*Hydrocotyle*) . . . . .Umbelliferae (Part 4)
1 Inflorescence not an umbel; petals not united (except in Gentianaceae; petals wanting in *Hepatica* of the Ranunculaceae).
  3 Perianth irregular (either the corolla or the calyx); stamens 5.
    4 Corolla irregular, white, yellow, blue, or violet, the lower of the 5 petals spurred at base, much longer than the 5 auricled sepals; ovary superior; flowers solitary, nodding; cleistogamous flowers often produced; plants relatively low; (*Viola*) . . . .
    . . . . .Violaceae (Part 3)
    4 Calyx-tube greenish, strongly oblique at summit, gibbous at base; petals spatulate, about equalling the calyx-lobes; ovary partially inferior; flowers in narrow panicles; scapes to over 6 dm tall; (*Heuchera*) . . . . .Saxifragaceae (Part 3)
  3 Perianth regular or nearly so, spurless.
    5 Flowers few to many in spikes, racemes, cymes, or panicles.
      6 Sepals and petals each 4; stamens 6 (4 long, 2 short); ovary superior; fruit a capsule usually 2-locular by a thin internal septum; flowers in racemes . . . .
      . . . . .Cruciferae (Part 3)
      6 Sepals (or calyx-lobes) and petals each usually 5; stamens 5 or 10, equal; ovary partially inferior.
        7 Petals united into a tube to 4 mm long, white, the corolla-lobes spreading, to 6 mm long, their midnerves and often their margins with erect toothed flanges running lengthwise; capsule 1-locular; flowers in a loose cyme; (*Fauria*; s Alaska and w B.C.) . . . . .Gentianaceae (Part 4)
        7 Petals distinct to base; capsule often separating into 2 nearly distinct beaked follicles; flowers in racemes or panicles (sometimes solitary in *Saxifraga*) . . . . .Saxifragaceae (Part 3)
    5 Flowers solitary on long peduncles; ovary superior.
      8 Plant insectivorous, the hollow horn-shaped insect-trapping leaves winged along the inward-facing suture, the outer margin at summit expanded into a broad hood; petals usually dark purple-red, arching over the style; (a single genus, *Sarracenia*; essentially transcontinental) . . . . .Sarraceniaceae (Part 3)
      8 Plants not insectivorous, the leaves not hollow.
        9 Leaves deeply lobed (*Ranunculus* of the Ranunculaceae will often key out here).
          10 Sepals 5–12, white to pink, lilac, or bluish; petals none; fruit a cluster of pubescent achenes; leaves deeply 3-lobed, the lobes entire; (*Hepatica*; Man. to N.S.) . . . . .Ranunculaceae (Part 3)
          10 Sepals 2, soon falling; petals normally 8 (up to about 16), soon falling; fruit a capsule; leaves irregularly 3–9-lobed, undulate to coarsely crenate; rhizome with abundant red-orange juice; (*Sanguinaria*; SE Man. to N.S.) . . . . .Papaveraceae (Part 3)
        9 Leaves merely crenate (*Caltha* and *Ranunculus* of the Ranunculaceae will often key out here).
          11 Sepals and white petals each 5; stamens 10, their deflexed anthers

opening by a pair of terminal (morphologically basal) pores; fruit a capsule; (*Moneses*; transcontinental) . . . . .Pyrolaceae (Part 4)
   11 Sepals and yellow or white petals each 5–10; stamens numerous, dehiscing longitudinally; fruit a cluster of achenes tipped by the plumose styles (*Dryas*) or 5–10 seed-like drupes (*Dalibarda*) . . . . . . .Rosaceae (Part 3)

SERIES 10 (see p. 107)
(Terrestrial or subaquatic non-climbing dicotyledonous herbs with simple stem-leaves, these wholly or partly in whorls)

1 Leaves distinctly toothed to deeply lobed.
  2 Flowers strongly irregular, bilaterally symmetrical, the reddish, bluish, or purplish petaloid sepals saccate and prolonged into a bent spur at base; inflorescence irregularly umbelliform, the peduncles elongate; valves of capsule opening elastically; (*Impatiens glandulifera*; introd.) . . . . .Balsaminaceae (Part 3)
  2 Flowers regular or nearly so, solitary or in racemes or spikes.
    3 Stamens 2, dehiscing longitudinally; corolla 4–5-lobed, white, pink, or blue; capsule 2-locular; flowers in terminal spikes or spike-like racemes; plants relatively tall; (*Veronica; Veronicastrum*) . . . . .Scrophulariaceae (Part 4)
    3 Stamens 10, deflexed and opening by terminal (morphologically basal) pores; petals 5, distinct, white or pink; capsule 5-locular; flowers solitary (*Moneses*) or umbellate or corymbose (*Chimaphila*); plants commonly less than 2 dm tall . . . .
. . . . .Pyrolaceae (Part 4)
1 Leaves entire or essentially so.
  4 Stem square in cross-section (except in *Asperula glauca*); flowers commonly 4-merous (often 3-merous in *Galium*), the petals united at least at base; stamens commonly 4, sometimes 3; ovary inferior; fruit twinned, consisting of 2 dry 1-seeded carpels (or only 1 by abortion in *Asperula*); (*Asperula; Galium; Sherardia*) . . . . .Rubiaceae (Part 4)
  4 Stem terete or nearly so (sometimes squarish in Lythraceae); fruit otherwise.
    5 Corolla with a central crown (corona) of 5 hooded bodies seated on the tube of 5 stamens; petals 5, reflexed and concealing the 5 small reflexed sepals; ovary superior; fruit a pair of large many-seeded follicles, the seeds bearing a tuft of long silky hairs; inflorescence an umbel; plants with milky juice; (*Asclepias quadrifolia; A. verticillata*) . . . . .Asclepiadaceae (Part 4)
    5 Plants lacking the above combination of characters.
      6 Petals none.
        7 Stem strict, unbranched, densely leafy; leaves sessile, to about 3.5 cm long (or submersed ones to about 6 cm long); flowers sessile in whorls in the upper leaf-axils; ovary inferior; fruit nut-like and indehiscent, 1-seeded; (a single genus, *Hippuris*) . . . . .Hippuridaceae (Part 4)
        7 Stems prostrate and mat-forming; leaves narrowly to broadly oblanceolate, to about 3 cm long, long-tapering to a subpetiolar base; flowers 2–5 from each node, on pedicels to 1.5 cm long; ovary superior; fruit a 3-locular capsule; (a single genus, *Mollugo*; introd. from Ont. to N.S.) . . . .
. . . . .Aizoaceae (Part 3)
      6 Petals present; ovary superior.
        8 Fruit a cluster of 4 or 5 many-seeded follicles; sepals and petals each 4 or 5; stamens 8 or 10; flowers in cymes; leaves fleshy; (*Sedum*) . . . .
. . . . .Crassulaceae (Part 3)
        8 Fruit a capsule.
          9 Calyx or corolla (or both) irregular; sepals 5; petals 3; (E Canada).
           10 Petals white, greenish, or pinkish, the lowest one keel-shaped or boat-shaped, with a fringed crest; the 2 inner sepals much larger than the outer 3; style elongate; flowers short-pedicelled in slender taper-pointed racemes; (*Polygala verticillata*; S Man. to SW Que.) . . . . .
. . . . .Polygalaceae (Part 3)

        10 Petals dark red, minute; the 3 inner sepals broader than the outer 2;
stigmas 3, sessile, plumose; flowers in large leafy panicles; (*Lechea*;
Man. to N.S.) ....Cistaceae (Part 3)

  9 Flowers regular.

      11 Leaves filiform, to about 5 cm long, up to 16 in a whorl; flowers rather
small, in loose dichotomously branching cymes, 5-merous, the petals
about equalling the sepals; plant often viscid; (*Spergula*; introd.) ....
....Caryophyllaceae (Part 3)

      11 Leaves broader; flowers larger, 5–7-merous, on axillary peduncles
or in terminal leafy spikes, racemes, or panicles.

         12 Calyx-teeth alternating with appendages in the sinuses; petals
usually 5 or 6, purplish; flowers whorled in the upper leaf-axils;
(*Decodon; Lythrum*) ....Lythraceae (Part 4)

         12 Calyx-teeth lacking appendages in the sinuses; petals 5, yellow
(*Lysimachia*) or white and commonly 7 (*Trientalis*); flowers long-
peduncled ....Primulaceae (Part 4)

SERIES 11 (see p. 107)
(Terrestrial or subaquatic (if aquatic, leaves not dissected) non-climbing dicotyledonous herbs with
simple entire stem-leaves, these wholly or chiefly opposite or subopposite, rosette-leaves also
often present)

1   Calyx and corolla (or both) more or less irregular (or corolla nearly regular in *Isanthus* of the
Labiatae and *Veronica* of the Scrophulariaceae).

  2 Stem square in cross-section; stamens 4 (2 long, 2 short; a fifth sterile filament present in
Scrophularia of the Scrophulariaceae); ovary superior.

    3 Fruit a 2-locular many-seeded capsule; (*Mimulus; Scrophularia*) ....
....Scrophulariaceae (Part 4)

    3 Fruit consisting of 4 small seed-like nutlets in the bottom of the persistent calyx ....
....Labiatae (Part 4)

  2 Stem terete or angled (but not distinctly square in cross-section).

    4 Fruit a drupe or drupe-like; flowers 5-merous.

      5 Flowers 1–4 in the leaf-axils; corolla pubescent, gibbous on one side at base;
fertile stamens 5; ovary inferior; fruit a dry greenish-orange to orange-red drupe
with 3 bony ribbed 1-seeded nutlets; perennials with erect stems; (*Triosteum*;
Ont. to N.S.) ....Caprifoliaceae (Part 4)

      5 Flowers in a terminal raceme to about 5 cm long, dull whitish or yellowish and
purple-spotted; fertile stamens 4; ovary superior; fruit several-seeded, densely
glandular-pubescent, the outer fleshy part falling away as 2 valves, the inner
woody part with 2 long incurved terminal horns; clammy-pubescent annual with
spreading stems; (a single genus, *Proboscidea*; introd. in s Ont.) ....
....Martyniaceae (Part 4)

    4 Fruit a capsule.

      6 Flowers 6-merous, solitary or in racemes; petals red-purple, 2 about 8 mm long,
the other 4 smaller; calyx 12-ribbed, gibbous or spurred at base, 16-toothed;
stamens usually 11 or 12, unequal; ovary partially inferior; densely viscid-hairy
annual; (*Cuphea*; introd. in s Ont.) ....Lythraceae (Part 4)

      6 Flowers 5-merous; stamens 2 or 4 (2 long, 2 short; a fifth one sometimes present
in Scrophulariaceae but usually sterile); ovary superior.

        7 Stamens 2; flowers white to pale violet, opposite in dense long-peduncled
axillary spikes to about 3 cm long, the spikes often overtopped by the
developing leafy tip; plants of mud or shallow water; (a single genus,
Justicia; s Ont. and sw Que.) ....Acanthaceae (Part 4)

        7 Stamens usually 4 (only 2 in *Veronica* and *Veronicastrum*, with nearly regular
corolla, and in *Gratiola* and *Lindernia*, with flowers solitary in the leaf-axils);
plants usually of drier habitats ....Scrophulariaceae (Part 4)

1   Calyx and corolla regular or nearly so (or the calyx merely unequally cleft and the corolla sometimes wanting.).
    8   Corolla gamopetalous, with a central crown (corona) composed of 5 fleshy hooded bodies seated on the tube of stamens (*Asclepias*) or a 5-lobed ring or disk (*Cynanchum*); stamens 5, their filaments united into a tube enclosing the pistil; ovary superior; fruit a pair of large follicles, the seeds with a tuft of long silky hairs; flowers 5-merous, in umbels; juice milky (except in *Asclepias tuberosa*) . . . . .Asclepiadaceae (Part 4)
    8   Plants lacking the above combination of characters, the flowers never with a corona.
        9   Fruit a pair of slender many-seeded follicles (the seeds of *Apocynum* with a tuft of long silky hairs); ovary superior; flowers 5-merous, solitary (*Vinca*) or in cymes (*Apocynum*); stamens 5; juice milky . . . . .Apocynaceae (Part 4)
        9   Fruit otherwise; juice not milky (except in *Euphorbia* of the Euphorbiaceae).
            10 Corolla none (calyx petaloid in Nyctaginaceae, wanting in Callitrichaceae; in *Euphorbia* of the Euphorbiaceae, flowers borne in a more or less cup-shaped cyathium resembling a calyx or corolla); (*Ludwigia* of the Onagraceae, with petals sometimes wanting, and *Glaux* of the Primulaceae, with petals wanting, are treated under the contrasting lead 10).
                11 Plants aquatic and completely submersed except for often floating leaves; leaves at most about 1.5 cm long, the submersed ones linear, the floating ones (when present) spatulate to obovate; flowers unisexual, minute, sessile or subsessile, 1–3 in the leaf-axils; ovary superior; (a single genus, *Callitriche*) . . . . .Callitrichaceae (Part 3)
                11 Plants not aquatic.
                    12 Flowers borne in a somewhat cup-shaped cyathium resembling a calyx or corolla, its margins bearing 4 or 5 nectariferous glands with or without coloured appendages; stamens generally 10; styles 3; ovary inferior; fruit a 3-lobed 3-seeded capsule; plants with milky juice; (*Euphorbia*) . . . .
                        . . . . .Euphorbiaceae (Part 3)
                    12 Flowers not as above; stamens various; styles or stigmas 1 or 2; ovary superior; fruit an indehiscent achene or utricle; stems with swollen joints; juice not milky.
                        13 Flowers small (the calyx green to brownish-yellow but scarcely petaloid), sessile or pedicelled in the upper leaf-axils or in diffuse leafy-bracted cymes; sepals distinct or united only at base; styles or capitate stigmas 2; leaves linear-subulate to narrowly oval, scarcely fleshy; (*Paronychia* and *Scleranthus*; other atypically apetalous genera may key out here) .Caryophyllaceae (Part 3)
                        13 Flowers rather showy, greenish white to yellow, orange, pink, or purple, sessile in terminal heads or short-pedicelled in cymes or panicles; sepals united into a funnelform to broadly campanulate basal tube; style solitary; leaves mostly elliptic to broadly ovate, rather thick and fleshy . . . . .Nyctaginaceae (Part 3)
            10 Corolla usually present at least in youth (except in *Ludwigia* of the Onagraceae and *Glaux* of the Primulaceae).
                14 Flowers in dense heads subtended by an involucre of bracts (as in the Compositae, but the usually 4 stamens distinct and the corollas usually 4-lobed); ovary inferior; receptacle chaffy or more or less hairy; (introd.) . . . .
                    . . . . .Dipsacaceae (Part 4)
                14 Flowers otherwise (if in somewhat composite-like heads, stamens commonly 3 and distinct).
                    15 Corolla of separate petals (petals sometimes connate at base in *Montia* of the Portulacaceae); fruit usually a capsule or follicle (a 1-seeded utricle in *Paronychia* and *Scleranthus* of the Caryophyllaceae).
                        16 Ovary inferior or apparently so, the mature capsule adnate to the calyx-tube (Onagraceae) or surrounded by it (Lythraceae), the stamens inserted in the throat or at the summit of the calyx-tube; style 1.

17 Ovary (and capsule) completely inferior, fused with the calyx-tube (the calyx usually projecting above the capsule as a free hypanthium) . . . . .Onagraceae (Part 4)

17 Ovary (and capsule) surrounded by the calyx-tube but free from it . . . . .Lythraceae (Part 4)

16 Ovary usually superior (partially inferior in *Saxifraga* of the Saxifragaceae), the mature capsule or follicle neither adnate to, nor surrounded by, the calyx.

18 Stamens usually of the same number as the petals or corolla-lobes and opposite them (if more than 6 in Papaveraceae, some alternating with the petals); capsule 1-locular.

19 Petals yellow or orange, sometimes purple-dotted; calyx 5–6-parted; stamens inserted on the corolla-tube . . . . .Primulaceae (Part 4)

19 Petals white to pink; sepals 2 or 3, distinct; stamens adhering to the petal-claws or borne on the receptacle.

20 Flowers solitary on slender erect axillary and terminal peduncles to 4 cm long; petals white, at most 4 mm long, commonly 6 (sometimes 4 or 5), the stamens up to twice as many; capsule linear; slender glabrous and glaucous annual to about 1.5 dm tall; (*Meconella*; sw B.C.) . . . . .Papaveraceae (Part 3)

20 Flowers usually several in racemes; petals whitish or pink, often with deeper stripes, commonly longer and usually 5 in number; capsule ovoid or obovoid; perennials with rhizomes, thick taproots, or corms; (*Claytonia; Montia*) . . . . .Portulacaceae (Part 3)

18 Stamens not of the same number as the petals or, if of the same number, alternating with them (stamens sometimes more than 10 in Hypericaceae).

21 Stamens inserted on the calyx; leaves often fleshy.

22 Fruit a 2-beaked 2-locular capsule; sepals and lilac to purple petals 5; stamens 10; leaves 4-ranked, usually less than 5 mm long; (*Saxifraga oppositifolia* and *S. nathorstii*) . . . . .Saxifragaceae (Part 3)

22 Fruit consisting of 3–5 follicles; sepals and petals 3–5; stamens 3 or 4 (in the aquatic or subaquatic *Tillaea*) or 8 or 10 (in the terrestrial *Sedum*); leaves usually longer . . . . .Crassulaceae (Part 3)

21 Stamens inserted on the receptacle (Caryophyllaceae), on the base of the perianth (Elatinaceae; Hypericaceae), or on a hypogynous disk at the base of the calyx-tube (Linaceae); leaves fleshy only in some species of Caryophyllaceae.

23 Stems swollen at the joints; stamens 2–10, inserted on the receptacle; styles 2, 3, or 5; fruit usually a capsule (a 1-seeded bladdery utricle in *Paronychia* and *Scleranthus*) . . . . .Caryophyllaceae (Part 3)

23 Stems not swollen at the joints; fruit a capsule.

24 Leaves dotted with translucent glands; stamens 5 to many, often united at base into 3–5 clusters; sepals and petals each 4 or 5, the petals usually yellow or orange; (a single genus, *Hypericum*) . . . . .Hypericaceae (Part 3)

24 Leaves not translucent-dotted; stamens 2, 3, or 5, free.

25 Sepals, petals, stamens, styles or sessile capitate stigmas, and capsule-locules each 2 or 3; flowers small, whitish or pinkish, solitary and sessile in the

leaf-axils; plants of muddy shores and shallow waters; (a single genus, *Elatine*) . . .Elatinaceae (Part 3)

25 Sepals, etc., each 4 or 5 (4 in *Millegrana*, with 8-seeded capsules; 5 in *Linum*, with 1-seeded capsules); flowers white or yellow, in corymbs or cymes; plants of dry habitats . . . . .Linaceae (Part 3)

15 Corolla with petals more or less united; stamens inserted on the corolla-tube (in *Glaux*, with no corolla, stamens inserted on base of calyx).

26 Ovary (and mature fruit) partly or wholly inferior, partly or wholly fused with the calyx-tube.

27 Stamens and styles each 1–3; corolla usually 5-lobed, whitish, pinkish, or bluish; fruit indehiscent, dry and 1-seeded; flowers in clustered or panicled cymes . . . . .Valerianaceae (Part 4)

27 Stamens 3–6 and of the same number as the corolla-lobes; style 1; fruit a capsule with its upper half projecting from the calyx-tube (*Houstonia*, with solitary or cymose flowers) or a twinned 4-seeded berry-like red drupe (*Mitchella*, with paired terminal flowers) . . . . .Rubiaceae (Part 4)

26 Ovary (and mature fruit) superior, entirely free from the calyx; stamens 4, 5, or 6 and of the same number as the corolla-lobes.

28 Stamens 5 or 6, opposite the corolla-lobes; flowers solitary or in racemes . . . . .Primulaceae (Part 4)

28 Stamens usually 4 or 5, alternate with the corolla-lobes (*Sabatia* of the Gentianaceae with up to 12 stamens and corolla-lobes).

29 Fruit a berry enclosed in a globose fruiting calyx to 2 cm thick; corolla rotate, white or violet-tinged, with a yellow eye, to 5 cm broad; stem viscid-villous; (*Chamaesaracha*; Sask. to E Que.) . . . . .Solanaceae (Part 4)

29 Fruit a capsule; flowers smaller; stem not viscid.

30 Flowers in dense ellipsoid to subglobose spikes; stamens 4; capsule circumscissile near middle; leaves linear; (*Plantago psyllium*; introd.) . . . . .Plantaginaceae (Part 4)

30 Flowers not in spikes; capsule not circumscissile.

31 Ovary and 2-valved many-seeded capsule 1-locular; stamens and corolla-lobes each 4 or 5 (up to 12 in *Sabatia*); flowers solitary or variously grouped . . . . .Gentianaceae (Part 4)

31 Ovary and usually 3-valved and 3-seeded capsule 3-locular; stamens and corolla-lobes each 5; flowers in cymes; (*Microsteris; Phlox*) . . . . .Polemoniaceae (Part 4)

SERIES 12 (see p. 107)
(Terrestrial non-climbing dicotyledonous herbs with simple shallowly toothed to deeply lobed stem-leaves, these wholly or chiefly opposite)

1 Flowers in dense heads subtended by an involucre of bracts (as in the Compositae, but the usually 4 stamens distinct and the corollas usually 4-lobed); receptacle chaffy or more or less hairy; ovary inferior; (introd.) . . . . .Dipsacaceae (Part 4)

1 Flowers not in heads (if somewhat so in Valerianaceae, stamens commonly 3 and distinct).

2 Corolla more or less irregular and 2-lipped (or nearly regular in *Elsholtzia, Lycopus,* and *Mentha* of the Labiatae, these with square stems, and in *Buchnera, Collinsia, Mimulus, Verbascum, Veronica,* and *Veronicastrum* of the Scrophulariaceae).

3 Fruit a 2-locular and usually many-seeded capsule; ovary superior; stamens usually 2 or 4 (2 long, 2 short; sometimes a fifth, this sterile except in *Verbascum*) . . . . .Scrophulariaceae (Part 4)

3 Fruit indehiscent (an achene in Phrymaceae, achene-like in Valerianaceae, or 2 or 4 small dry nutlets in Labiatae and Verbenaceae).

      4   Ovary (and fruit) inferior, fused with the calyx-tube, a plumose pappus
          sometimes present on the achene-like fruit; stamens 3 (rarely 2); corolla
          5-lobed; flowers in panicled or capitate cymes . . . . .Valerianaceae (Part 4)
      4   Ovary (and fruit) superior, free from the calyx; stamens 2 or 4 (2 long, 2 short).
          5   Ovary 4-lobed around the style; stamens commonly 4 or sometimes only 2;
              flowers in terminal spikes or racemes or in axillary whorls; plants with
              square stems and aromatic foliage . . . . .Labiatae (Part 4)
          5   Ovary not lobed, the style apical; stamens 4; flowers in spikes; stems terete
              or squarish; foliage not aromatic.
              6   Ovary 1-locular, forming an achene at the bottom of the tightly closed
                   calyx-tube, this 2-lipped, its long upper teeth becoming hooked at tip and
                   indurated; mature roseate or purplish flowers strongly deflexed in the
                   spike; leaves ovate; (a single genus, *Phryma*; s Man. to N.B.) . . . .
                   . . . . .Phrymaceae (Part 4)
              6   Ovary 2–4-locular, the fruit consisting of 2 or 4 small dry nutlets; flowers
                   white to pale blue or purple, not deflexed; calyx regular or with one of its
                   teeth shorter than the others . . . . .Verbenaceae (Part 4)
  2   Corolla regular or wanting.
      7   Corolla none; flowers small, greenish, unisexual.
          8   Calyx none; flowers borne in a somewhat cup-shaped cyathium resembling a
              calyx or corolla, this bearing on its margins 4 or 5 nectariferous glands with or
              without coloured appendages; ovary inferior; fruit a 3-lobed 3-seeded capsule;
              plants with milky juice; (*Euphorbia*) . . . . .Euphorbiaceae (Part 3)
          8   Calyx present; flowers not as above; juice not milky.
             9   Fruit a capsule; ovary wholly or partly inferior.
               10   Flowers in interrupted axillary spikes; calyx 3-parted; stamens 8–20;
                   capsule hispid, globose, 2-locular, 2-seeded; leaves lanceolate to
                   narrowly ovate, rather closely crenate-serrate, to about 6 cm long;
                   (*Mercurialis*; introd. from Ont. to N.S.) . . . . .Euphorbiaceae (Part 3)
               10   Flowers usually solitary in the leaf-axils; calyx 4-parted; stamens not
                   more than 8; capsule smooth, obcordate, 1-locular, many-seeded;
                   leaves ovate to rotund, obscurely and irregularly toothed, the blades
                   at most about 2 cm long; (*Chrysosplenium*) . . . . .Saxifragaceae (Part 3)
             9   Fruit an achene; ovary superior.
               11   Flowers relatively large, solitary or in few-flowered umbels; sepals few to
                   many, petaloid; stamens numerous; (*Anemone*) . . . . .Ranunculaceae (Part 3)
               11   Flowers small, in axillary clusters or spikes, the latter often disposed in
                   panicles; calyx 3–4-parted; stamens 3 or 4, often in separate staminate
                   flowers; (*Boehmeria; Pilea; Urtica*) . . . . .Urticaceae (Part 3)
  7   Corolla present.
      12   Stem square in cross-section and wing-angled, setose at the nodes, otherwise
          glabrous or sparingly glandular-hirsute; flowers showy, crimson, in terminal
          cymes; calyx-lobes, petals, and capsule-locules each 4; style 1; stamens 8; ovary
          inferior; (a single genus, *Rhexia*; s Ont. and N.S.) . . . . .Melastomataceae (Part 4)
      12   Stem neither markedly square nor wing-angled.
          13   Flowering stems with a single pair of opposite lobed leaves below the
              inflorescence; flowers white.
              14   Stem-leaves large, peltate on long petioles, very deeply 3–7-parted;
                   flowers large, solitary on long terminal peduncles; sepals 6, early
                   deciduous; petals 6 or 9, to about 4 cm long, the stamens twice as
                   many; ovary superior; fruit a yellow fleshy berry to about 5 cm long;
                   (*Podophyllum*; s Ont. to N.S.) . . . . .Berberidaceae (Part 3)
              14   Stem-leaves smaller, sessile, rather shallowly 3–5-lobed; flowers small,
                   short-pedicelled in a terminal raceme; calyx-lobes and petals each 5;
                   stamens 10; ovary inferior; fruit a 1-locular 2-beaked few-seeded
                   capsule; (*Mitella diphylla*; Ont. and Que.) . . . . .Saxifragaceae (Part 3)

13 Flowering stems with more than a single pair of leaves.
    15 Fruit an achene fused with the calyx-tube, sometimes with a plumose pappus; ovary inferior; stamens 3 (rarely 2); corolla 5-lobed; flowers in capitate clusters disposed in cymes ....Valerianaceae (Part 4)
    15 Fruit not an achene.
        16 Sepals (or calyx-lobes) and petals each 2 or 4; inflorescence a raceme.
            17 Sepals, petals, and stamens each 2; ovary (and capsule) completely inferior, fused with the calyx-tube (the calyx usually projecting above it as a free hypanthium); fruit a 4-valved many-seeded linear capsule (*Epilobium*) or small, pear-shaped, indehiscent, covered with hooked bristles, and with only 1 or 2 seeds (*Circaea*) ....Onagraceae (Part 4)
            17 Sepals and petals each 4; stamens usually 6 (4 long, 2 short; rarely only 4 or 2); ovary (and capsule) superior, free from the calyx; fruit a 2-valved capsule usually 2-locular by a thin internal septum ....Cruciferae (Part 3)
        16 Sepals (or calyx-lobes) and petals (or corolla-lobes) each 5; stamens 5 (Polemoniaceae) or 10 (only 5 of them anther-bearing in *Erodium* of the Geraniaceae); ovary superior.
            18 Fruit consisting of 5 beaked and finally distinct 1-seeded carpels suspended from the top of the elongated central axis by their tightly twisted (*Erodium*) or outwardly upcurving or coiling (*Geranium*) styles; flowers pink to purple; leaves deeply dissected ....
            ....Geraniaceae (Part 3)
            18 Fruit a several-seeded capsule.
                19 Leaves palmately divided nearly to base into up to 9 linear segments to 2 cm long (or the uppermost leaves entire); corolla salverform to campanulate, with a white to deep pink or purplish, spreading or rotate, 5-lobed limb; stamens 5, inserted in the corolla-throat, their anthers dehiscing longitudinally; inflorescence capitate or cymose; glabrous to scabrous-puberulent taprooted annuals to 3 dm tall; (*Linanthus*; B.C. to sw Sask.) ....Polemoniaceae (Part 4)
                19 Leaves merely shallowly toothed; petals distinct, white to pink or roseate; stamens 10, their anthers inverted and opening by terminal (morphologically basal) pores; flowers solitary or corymbose-umbellate; low glabrous perennials with slender rhizomes; (*Chimaphila; Moneses*) ....Pyrolaceae (Part 4)

Division I PTERIDOPHYTA (Ferns and Fern Allies) (see p. 93)
(Ref.: Broun 1938; Cody 1956; T.M.C. Taylor 1963; Marie-Victorin 1923)

EQUISETACEAE (Horsetail Family)

EQUISETUM L. Horsetail. Prêle

Rush-like plants with extensively creeping, jointed, branched, often blackish rhizomes bearing whorls of felted roots at the nodes. Aerial stems linear-cylindric, grooved, usually roughened with rows of siliceous dots on the ridges, jointed and separating easily at the solid nodes (the internodes usually hollow at centre except in *E. scirpoides*), the nodes subtended by a sheath of rudimentary, scale-like, appressed leaves usually lacking chlorophyll. Fruiting structure a terminal cone-like spike consisting of closely packed whorls of shield-shaped stalked scales, each bearing on the under surface several sporangia (spore-cases) that open on the inner side. Spores uniform. [Species all transcontinental except *E. telmateia* of coastal B.C.].

(Ref.: Marie-Victorin 1927*a*)
1   Cones tipped with a short rigid point, subsessile or short-stalked from the uppermost
    sheath; stems stiff and mostly evergreen, the sterile and fertile ones similar and without
    regularly whorled branches; (transcontinental).
    2   Stems slender, tufted, low and flexuous, ascending or prostrate, solid throughout or
        with up to 4 longitudinal side cavities (vallecular canals), one under each ridge;
        sheaths 3-toothed; cones to about 5 mm long . . . .*E. scirpoides*
    2   Stems stouter and taller, ascending to erect, to about 3 dm tall, with a central longi-
        tudinal cavity (the centrum) at least 1/4 the total diameter and several smaller side
        cavities; sheaths with 3 or more teeth.
        3   Stems annual, soft and easily crushed, to over 1 m tall, their convex ridges bearing
            a single row of silica-dots; cones to about 2 cm long, with rounded summits; sheaths
            enlarged upward, pale green with a narrow blackish band at summit (only the lower
            sheaths sometimes with a black band at base), their blackish teeth jointed at base
            and promptly deciduous . . . .*E. laevigatum*
        3   Stems typically evergreen; cones tipped with a firm point.
            4   Teeth of sheaths persistent (not articulated at base), whitish or white-margined;
                sheaths darkened at summit; stems commonly less than 3 dm tall, their 2-angled
                ridges each bearing a pair of marginal rows of silica-dots; cones rarely over
                1 cm long . . . .*E. variegatum*
            4   Teeth of sheaths usually soon deciduous (in ours); sheaths black-banded at
                summit and typically also at base; stems to over 2 m tall; cones to about 2 cm
                long . . . .*E. hyemale*
1   Cones rounded at summit (*E. laevigatum* may be sought here), long-stalked; stems
    relatively soft and pliant, annual, at least the sterile ones usually with regularly whorled
    branches.
    5   Stems very coarse, the whitish succulent fertile stems to about 6 dm tall and 2.5 cm
        thick, the sterile stems becoming 3 m tall and 2 cm thick; branches 4–6-angled, their
        sheaths with 2-keeled teeth; cones to nearly 1 dm long and 2.5 cm thick (coastal B.C.) . . . . .
        . . . .*E. telmateia*
    5   Stems relatively slender and low; cones smaller; (transcontinental).
        6   Stem soft and papery, very strongly flattened under pressure, the central cavity at
            least 4/5 the total diameter; side cavities (vallecular canals) absent or invisible to
            the naked eye; sheaths tight, with up to 20 dark-brown, distinct teeth . . . .*E. fluviatile*
        6   Stem firmer, the central cavity smaller, the side cavities visible to the naked eye.
            7   First internode of the branches mostly shorter than or merely equalling the
                subtended stem-sheath, the branches unbranched.
                8   Branches 4–6-angled, few and irregular to many and whorled; fertile and
                    sterile stems alike, lacking tubercules or spicules on the 5–10 ridges . . . .
                    . . . .*E. palustre*

8   Branches mostly 3-angled and regularly whorled; stems beset with siliceous tubercles or short ridge-crests on the 10–18 ridges, somewhat dimorphic, the fertile ones with longer sheaths and teeth than the sterile ones, at first simple and pale, later producing whorls of simple green branches and often themselves becoming green ....*E. pratense*

7   First internode of the branches much longer than the subtended stem-sheath (at least in the upper part of the plant), the branches mostly regularly whorled and themselves often branched.

9   Fertile stems permanently whitish or brownish, unbranched, soon withering; sterile stems green, their sheaths scarcely inflated, their 3–4-angled branches sometimes themselves branched ....*E. arvense*

9   Fertile stems commonly with longer sheaths and teeth than the sterile ones, at first simple and pale, later producing whorls of green, mostly branched, 4–5-angled branches and often themselves becoming green; sterile stems green and compoundly branched from the first, their sheaths conspicuously inflated, the papery reddish-brown teeth cohering in 3 or 4 compound lobes .....
....*E. sylvaticum*

E. arvense L.   Common or Field-Horsetail.   Queue de renard
/AST/X/GEA/ (Grh) Open or lightly wooded habitats (a common pioneer and often the dominant species in dry, eroded, or denuded places), the aggregate species throughout Alaska–Canada–Greenland (N to northernmost Ellesmere Is. and northernmost Greenland), s throughout most of the U.S.A.; Iceland; Eurasia; N Africa. MAPS and synonymy: *see* below.

The following key outlines the distinguishing characters of the numerous subspecific taxa of *E. arvense* reported from Alaska–Canada. These may have little taxonomic significance, it being known that 3-angled branches (as distinct from branchlets), characteristic of var. *boreale*, and 4-angled branches, characteristic of var. *arvense*, may occur on the same plant. R. L. Hauke (Am. Fern J. 55(3): 134. 1965) favours the recognition of the various phases as fluctuations ("fl.") rather than varieties or forms, believing them to be ecological responses to the environment that, under normal conditions of soil and temperature, would all have the usual growth-form of var. *arvense*. They are keyed out below, however, because the same argument undoubtedly applies to many other species for which subspecific taxa are generally recognized.

Also keyed out below are *E. fluviatile* and *E. litorale,* the latter evidently a hybrid between *E. arvense* and *E. fluviatile.* Fassett (1940) notes that *E. palustre* may also be one of the parents in some nothomorphs.

1   Stems dimorphic, the fruiting ones succulent, whitish to pinkish-brown or tawny, appearing before the green and branching sterile ones and soon wilting ....*E. arvense*

2   Branches 3-angled, simple or essentially so, their sheaths 3-toothed; [*E. boreale* Bong., the type from Sitka, Alaska; transcontinental, the common form northwards] ....
....var. *boreale* (Bong.) Ledeb.

3   Stem erect or nearly so, with relatively remote nodes and symmetrical whorls of branches.

4   Branches loosely spreading, to about 3 dm long; [incl. f. *atratidens* Lepage and f. *pseudonemorosum* Boivin] ....f. *boreale*

4   Branches appressed-ascending, rarely over 5 cm long; [type from Longueuil, near Montreal, Que.] ....f. *pseudovarium* Vict.

3   Stem depressed, with crowded nodes and ascending branches; [incl. f. *caespitosum* Rousseau; type from the Mingan Is., E Que.] ....f. *pseudoalpestre* Vict.

2   Branches 4(5)-angled, their sheaths usually 4-toothed ....var. *arvense*

5   Branches forking.

6   Sterile stem depressed or prostrate, its branches ascending ....
....f. *diffusum* (Eat.) Clute

6   Sterile stem erect or nearly so, its branches in symmetrical whorls.

7   Stem less than 4 dm tall, the short internodes mostly hidden by the strongly ascending branches ....f. *ramulosum* (Rupr.) Klinge

7   Stem to about 6 dm tall, the internodes commonly over 3 cm long, the branches spreading ....f. *pseudosilvaticum* (Milde) Luerss.

5 Branches simple.
    8 Cones commonly less than 1.5 cm long, terminating slender green branches of the otherwise sterile stem; [f. *?arcticum* (Rupr.) Braun; f. *decumbens* (Mey) Klinge; f. *?irriguum* Milde; var. *serotinum* Mey.; *E. ?calderi Boivin*] . . . .
. . . . .f. *campestre* (Schultz) Klinge
    8 Cones to over 3 cm long, solitary at the summit of the mostly unbranched flesh-coloured stem, the separate green branching sterile stems appearing later.
        9 Plants very tall and lax, with prolonged internodes and spreading branches to about 3 dm long .f. *nemorosum* (A. Br.) Klinge
        9 Plants relatively low and compact, the branches relatively short.
            10 Sterile stem depressed, its branches ascending . . . .
. . . . .f. *alpestre* (Wahlenb.) Luerss.
            10 Sterile stem erect or nearly so, its branches in symmetrical whorls.
                11 Branches often unequal, rarely over 5 cm long, strongly appressed-ascending . . . . .f. *varium* (Milde) Klinge
                11 Branches mostly subequal, to over 1 dm long, spreading; [transcontinental, the common form southwards; MAPS (aggregate species): Porsild 1957: map 6, p. 161; Raup 1947: pl. 13; Meusel, Jaeger, and Weinert 1965: 7; Hultén 1968*b*:38, and 1962: map 98, p. 107] . . . .
. . . . .f. *arvense*
1 Fertile and sterile stems similar, simple or branching, green, both present simultaneously.
    12 Central cavity 4/5 the diameter of the main stem; vallecular canals small or wanting; sheaths tight, their dark-brown teeth distinct; spores fertile . . . .*E. fluviatile*
    12 Central cavity at most 2/3 the diameter of the main stem; vallecular canals well developed; sheaths becoming loose, their narrowly white-margined teeth often united in 2's or 3's; spores abortive . . . . ×*E. litorale* Kuhl.
        13 Branches relatively numerous, whorled.
            14 Stem erect, to about 9 dm tall, its internodes clearly visible; [transcontinental, as also both putative parents] . . . . .f. *litorale*
            14 Stem depressed or slightly ascending, commonly less than 3 dm tall, its internodes partly hidden by the strongly ascending branches . . . .
. . . . .f. *arvensiforme* (Eat.) Vict.
        13 Branches few and scattered or wanting.
            15 Branches few, irregularly scattered . . . . .f. *humile* (Milde) Vict.
            15 Branches none, the stem very slender . . . . .f. *gracile* (Milde) Vict.

E. fluviatile L.   Water-Horsetail.   Pipes
/aST/X/EA/   (Grh (Hel))   Shores, shallow water, and wet places from N Alaska–Yukon to Great Bear L., L. Athabasca (Alta. and Sask.), northernmost Man., N James Bay, and N Ungava–Labrador to s Wash., ?Oreg. and Va.; Iceland; Eurasia. MAPS and synonymy: *see* below.
1 Stems bearing numerous whorls of branches.
    2 Cone solitary at the summit of the main stem; [incl. f. *natans* (Vict.) Broun; *E. limosum* f. *verticillatum* Döll; transcontinental; MAPS: Hultén 1968*b*: 36, and 1962: map 96, p. 105; Raup 1947: pl. 13] . . . . .f. *fluviatile*
    2 Cones terminating many stiffly ascending branches . . . . . . .f. *polystachyum* (Brückn.) Broun
1 Stem simple or with a few scattered slender branches.
    3 Stem to 7 or 8 mm wide when flattened, its sheaths usually closely appressed and with linear-lanceolate teeth mostly over 2 mm long; [var. *limosum* (L.) Gilbert; *E. limosum* L. and its f. *linnaeanum* Döll] . . . . .f. *linnaeanum* (Döll) Broun
    3 Stem to about 3 mm thick when flattened, its sheaths usually looser and with relatively broad teeth mostly less than 2 mm long . . . . .f. *minus* (A. Br.) Broun

E. hyemale L.   Scouring-rush.   Prêle des tourneurs
/sT/X/EA/   (Grh)   Dry to moist, open to lightly wooded habitats, the aggregate species from cent. Alaska and N B.C. to Great Slave L., L. Athabasca, s Ont., s Que., Nfld., N.B., and N.S. (the report from P.E.I. by John Macoun 1890, requires confirmation), s to s Calif., Mexico, and N Fla.; Iceland; Eurasia. MAPS and synonymy: *see* below.

Following is a key outlining the major diagnostic features of *E. hyemale* and its subspecific taxa reported from Alaska–Canada, together with those of *E. laevigatum* and *E. ferrissii,* this a purported hybrid between the other two. The subspecific taxa may actually have little taxonomic significance, R. L. Hauke (Am. Fern J. 52: 60. 1962) noting that "variation in any one of these characters is so gradual that lines cannot be drawn to separate definitely one condition from the other . . . . The differences are the end result of this gradual variation." According to C. V. Morton (Am. Fern J. 43(4): 171. 1953), the type of *E. hyemale* var. *affine* is a specimen whose sheaths bear persistent teeth and "is obviously a small specimen of the southern variety, and not distinguishable from var. *robustum.*" He proposes the name var. *pseudohyemale* to designate the northern plant with deciduous teeth to which the name var. *affine* has, in his opinion, been so commonly misapplied. Var. *robustum* he includes as a synonym of the earlier-published var. *elatum.* Concerning *E. laevigatum,* R. L. Hauke (Am. Fern J. 48(2): 68–72. 1958, and 50(2): 185–93. 1960) has produced evidence that many collections identified as *E. hyemale* var. *intermedium* have abortive spores (× *E. ferrissii*), whereas collections that fit the description of *E. laevigatum* have fertile spores. *E. laevigatum* is thus separable from *E. hyemale* but merges with it through hybridization.

1 Sheaths distinctly flaring toward the black-banded summit, green or greenish, longer than broad, without (or the lower sheaths sometimes with) a black band at base; stems relatively smooth and easily compressed.

    2 Spores fertile . . . .*E. laevigatum*

    2 Spores abortive; [*E. hyemale* vars. × *E. laevigatum; E. hyemale* var. *intermedium* Eat.; *E. inter.* (Eat.) Rydb.; probably the range of the putative parents; reported from Lambton Co., s Ont., by Gaiser and Moore 1966] . . . .× *E. ferrissii* Clute

1 Sheaths cylindric, tight, drab or greyish-brown, about as broad as or broader than long, black-banded at base and summit; stems rough and firm; [*Hippochaete* Farw.; *E. hiemale,* the original spelling] . . . .*E. hyemale*

    3 Stem-ridges bearing 2 lateral rows of silica-dots.

        4 Teeth of sheaths promptly deciduous; [Eurasia only, if var. *calif.* is actually distinct; MAP (aggregate species): Hultén 1962: map 174, p. 185] . . . . .var. *hyemale*

        4 Teeth of sheaths more or less persistent; [Aleutian Is.–Alaska (*see* Hultén 1941: map 36, p. 117), Yukon Territory (Porsild 1951a), w Dist. Mackenzie, and B.C.; MAPS: Hultén 1962: map 174 ("*E. laev.* var. *calif.*"), p. 185, and 1968*b*:34] . . . . .var. *californicum* Milde

    3 Stem-ridges bearing only 1 row of silica dots.

        5 Stem to about 3 m tall and 1.5 cm thick, simple; teeth of sheaths tardily deciduous; [var. *robustum* (A. Br.) Eat.; *E. rob.* A. Br.; B.C. and Alta.; s Ont.: Essex, Lambton, Welland, and York counties] . . . . .var. *elatum* (Engelm.) Morton

        5 Stem at most about 1 m tall and 1 cm thick; teeth of sheaths mostly promptly deciduous . . . . .var. *pseudohyemale* (Farw.) Morton

            6 Stem short-branching above, some or all of the branches terminated by a small cone . . . . .f. *polystachyum* (Prager) Scoggan

            6 Stem simple, terminated by a solitary cone.

                7 Stem less than 5 dm tall and 5 mm thick . . . . .f. *pumilum* (Eat.) Scoggan

                7 Stem to about 1 m tall and 5 mm thick; [var. *affine* of most Canadian reports, not *E. affine* Engelm.; *E. prealtum* Raf., *E. ramosissimum sensu* John Macoun 1890, not Desf., transcontinental; MAPS: Hultén 1962: map 174 (aggregate species), p. 185; J. H. Schaffner, Am. Fern J. 29(2): map 2 (*E. "praealtum"*), p. 46. 1939] . . . . .f. *pseudohyemale*

E. laevigatum A. Br. Smooth Scouring-rush

/T/X/ (Grh) Very similar to *E. hyemale* and often confused with it (*see* discussion under that species), the area thus uncertain but apparently extending from s B.C. (*see* B.C. map by T.M.C. Taylor 1963:153) to s Que., s to Baja Calif. and Tex. [*E. hyemale* var. *intermedium* of auth., not Eat.; *E. funstonii* Eat.; *E. kansanum* of auth., perhaps not Schaffner]. MAP: combine the maps by J. H. Schaffner, Am. Fern J. 29(2): map 1 (*E. laev.*), map 3 (*E. kans.*) and map 4B (*E. funst.*), p. 46. 1939.

A hybrid with *E. variegatum* (× *E. nelsonii* (Eat.) Schaffner; *E. varieg.* var. *nels.* Eat.) is known from Ont. and Que.

E. palustre L.   Marsh-Horsetail.   Prêle des marais
/aST/X/EA/   (Grh)   Marshes and wet meadows, woods, and shores from N Alaska and Great Bear L. to NE Alta., Sask. (N to Nipawin, ca. 53°20'N), Man. (N to Churchill), Ont. (N to W James Bay at 53°35'N), Que. (N to ca. 58°N), SE Labrador (N to the Hamilton R. basin), Nfld., N.B., and N.S., s to Wash., Idaho, Nebr., Ohio, and N.J., Iceland; Eurasia. MAPS and synonymy: *see* below.
1   Branches irregularly scattered or wanting.
    2   Primary stem elongate, to about 5 mm thick . . . . .f. *simplex* Milde
    2   Primary stem very short, the numerous secondary stems erect, only 1 or 2 mm thick . . . .
       . . . . .f. *filiforme* Lacks.
1   Branches regularly whorled or tufted.
    3   Nodal branches (at least the lower ones) with short branchlets . . . .
       . . . . .f. *ramulosum* (Milde) Klinge
    3   Nodal branches all simple.
        4   Stem prostrate, its mostly paired branches strongly ascending; [type from Longueuil, near Montreal, Que.] . . . . .f. *fluitans* Vict.
        4   Stem erect or strongly ascending, its branches whorled.
           5   Cones borne from the tips of many ascending branches . . . .
            . . . . .f. *polystachion* (Weigel) Duval-Jouve
           5   Cones borne only at tip of stem.
              6   Branches widely spreading to horizontally divergent; [incl. f. *luxurians* Vict. and var. *ramosissimum* Peck] . . . . .f. *arcuatum* Milde
              6   At least the middle and upper leaves strongly ascending.
                 7   Branches to over 3 dm long, very strongly ascending . . . .f. *verticillatum* Milde
                 7   Branches commonly less than 6 cm long.
                    8   Teeth of primary sheaths blackish throughout; [var. *nig.* St. John, the type from the Côte-Nord, E Que.] . . . . .f. *nigridens* (St. John) Vict.
                    8   Teeth of primary sheaths white-margined; [var. *americanum* Vict., transcontinental but not known from P.E.I., reports from there being probably based upon *E. fluviatile*, according to D. S. Erskine 1960; MAPS (aggregate species): Hultén 1962: map 89, p. 99, and 1968*b*: 37; Raup 1947: pl. 13] . . . . .f. *palustre*

E. pratense Ehrh.   Meadow-Horsetail.   Prêle des prés
/aST/X/EA/   (Grh)   Moist woods, thickets, and meadows from N Alaska to S -cent. Yukon, cent. Dist. Mackenzie (Fort Simpson), cent. Sask. (Meadow L.), cent. Man. (Gillam), N Ont. (Caribou Rapids on a branch of the Severn R. at ca. 55°10'N), Que. (N to ca. 58°N), S Labrador, Nfld., N.B., and N.S. (not known from P.E.I.), s to S B.C., Mont., Iowa, and N.J.; Iceland; Eurasia. [Incl. f. *nanum* Milde; *E. umbrosum* Mey., not Lapeyrouse]. MAPS: Hultén 1968*b*:38, and 1962: map 83, p. 93.

E. scirpoides Michx.   Dwarf Scouring-rush.   Prêle faux-scirpe
/aST/X/GEA/   (Grh)   Damp woods, thickets, and springy banks from N Alaska and Banks Is. to S Baffin Is., Nfld., N.B., P.E.I., and N.S., s to Wash., S.Dak., and N.Y.; w Greenland N to ca. 71°N; Eurasia. [Incl. var. *minor* Lawson]. MAPS: Hultén 1968*b*:36, and 1962: map 30, p. 37; Porsild 1957: map 7, p. 161; *Atlas of Canada* 1957: map 3, sheet 38; Raup 1947: pl. 13.

E. sylvaticum L.   Wood-Horsetail.   Prêle des bois
/aST/X/GEA/   (Grh)   Woods and thickets, the aggregate species from N Alaska and the Mackenzie R. Delta to Great Slave L., s Dist. Keewatin, northernmost Ungava–Labrador, Nfld., N.B., P.E.I., and N.S., s to Idaho, Mont., S.Dak., and Va.; w Greenland N to ca. 69°N; Eurasia. MAPS and synonymy: *see* below.
1   Branches scabrous at least along the lower internodes; [incl. var. *squarrosum* Eat.; *E. umbrosum* Lapeyr., not Mey.; *E. silvaticum* L., orthographic variant; the Eurasian phase, transcontinental in Canada but much less common than the following ones; MAPS

(aggregate species): Hultén 1962: map 86, p. 95, and 1968b:37; Meusel, Jaeger, and Weinert 1965:7; Raup 1947: pl. 13] .... var. *sylvaticum*
1 Branches smooth or nearly so .... var. *pauciramosum* Milde
  2 Branches copiously forking; [transcontinental; the common form southwards] ....
    .... f. *multiramosum* Fern.
  2 Branches simple or sparingly branched.
    3 Cone exserted from the upper sheath only after the complete vegetative development of the plant; [type from the Roggan R., Ungava, at ca. 54°N, 78°W] ....
    .... f. *opsistachyum* Lepage
    3 Cone exserted at an early stage; [transcontinental; the common form northwards] ......
    .... f. *pauciramosum*

E. telmateia Ehrh.   Giant or Ivory Horsetail
/t/D/E/ (Grh) Woods, thickets, and borders of streams from w B.C. (N to N Queen Charlotte Is.; see B.C. map by T.M.C. Taylor 1963:153) to s Calif., with an isolated area on the Keweenaw Pen., L. Superior, Mich.; Europe. [Incl. vars. *frondescens* A. Br. and *serotinum* Milde; our material is otherwise referable to var. *braunii* Milde, differing from typical *E. telmateia* (*E. maximum* Lam.) of Europe in several minor characters]. MAPS: Hultén 1958: map 258, p. 277; Meusel, Jaeger, and Weinert 1965:7; T.M.C. Taylor 1970: 29 (N. American area).

E. variegatum Schleich.   Variegated Horsetail.   Prêle panachée
/AST/X/GEA/ (Grh) Damp sands, shores, and bogs, the aggregate species from N Alaska to northernmost Ellesmere Is. and northernmost Labrador, s to Calif., Colo., Ill., and N.J.; circumgreenlandic; Iceland; Eurasia. MAPS and synonymy: see below.
1 Teeth of sheaths incurved, either uniformly black or with narrow white margins; [Aleutian Is.–Alaska (type from Yukatat Bay) and B.C.; MAPS: on the below-noted 1962 map by Hultén, and Hultén 1968b:35] .... var. *alaskanum* Eat.
1 Teeth of sheaths straight, with broad white margins.
  2 Stem to 4.5 dm tall and 4.5 mm thick, with up to 15 ridges; sheaths tight, their teeth retaining their bristle-tips; cone commonly about 1 cm long; [*E. hyemale* var. *jes.* (Eat.) Vict.; *E. ?trachyodon* A. Br., this perhaps a hybrid between *E. hyemale* and *E. variegatum*; s Ont. to N.B. and N.S.; MAP: on the below-noted 1962 map by Hultén] ....
    .... var. *jesupii* Eat.
  2 Stem to about 2.5 dm tall and 3 mm thick, with rarely more than 10 ridges; sheaths loose, their teeth soon losing their bristle-tips; cone less than 1 cm long; [incl. var. *anceps* Milde, the slender extreme closely resembling *E. scirpoides*; transcontinental; MAPS (aggregate species); Hultén 1968b:35, and 1962: map 45, p. 53; Porsild 1957: map 8, p. 161; Raup 1947: pl. 14] .... var. *variegatum*

LYCOPODIACEAE (Club-moss Family)

LYCOPODIUM L.   Club-moss.   Lycopode

Low evergreen plants, often trailing but with erect or ascending fertile branches. Leaves scale-like or linear to oblong, small, 1-nerved, entire or minutely toothed, usually crowded and overlapping. Spores uniform, the sporangia borne either in the axils of ordinary leaves or in terminal, sessile or stalked, terete, cone-like spikes. [Species all essentially transcontinental, only *L. alpinum* and *L. tristachyum* with extensive gaps in the range].

(Ref.: Marie-Victorin 1925)
1 Sporangia borne in the axils of normal green leaves.
  2 Leaves essentially uniform, commonly appressed or ascending, entire or minutely toothed, plump and hollow at base, to 8 mm long; plant with a short slender rooting base to about 8 cm long . . . .*L. selago*
  2 Leaves in alternating zones of longer and shorter ones (the latter subtending sporangia), widely spreading or reflexed, flat-based, to 1.5 cm long; plant with a prostrate rooting marcescently leafy base to 4 dm long . . . .*L. lucidulum*
1 Sporangia borne in terminal cone-like spikes.
  3 Sporophylls (bracts of the spike) green and leaf-like; sporangia subglobose; sterile branches creeping or ascending only at the tips . . . .*L. inundatum*
  3 Sporophylls firm, yellowish and scale-like; sporangia reniform; sterile branches erect or strongly ascending from the creeping primary stems.
    4 Leaves linear to lanceolate, commonly 6 or 7 mm long.
      5 Leaves tipped with a soft hair-like bristle; stems creeping and forking, the ascending branches dichotomously branching; cones peduncled . . . .*L. clavatum*
      5 Leaves not bristle-tipped; cones sessile.
        6 Ascending branches freely forking and bushy; free portion of leaves commonly only 4 or 5 mm long; prostrate stems deeply subterranean . . . .*L. obscurum*
        6 Ascending branches simple or few-forked; free portion of leaves longer; prostrate stems trailing or near the surface of the ground . . . .*L. annotinum*
    4 Leaves small and scale-like, in 4 or 5 rows, adnate for more than half their length, their free tips not over about 3 mm long; ascending branches bushy or fan-like.
      7 Sterile branchlets not strongly flattened or concave beneath; leaves uniform; cones mostly solitary . . . .*L. sabinaefolium*
      7 Sterile branchlets flattened or concave beneath; leaves 4-ranked, those of the under surface smaller than or unlike the lateral ones; cones commonly at least 2.
        8 Cones sessile, rarely over 2 cm long; leaves of sterile branchlets trimorphic, those of the under side trowel-shaped but not reduced. . . . .*L. alpinum*
        8 Cones or groups of cones usually peduncled, to about 5 cm long; leaves of sterile branchlets dimorphic, those of the under side subulate, much reduced.
          9 Creeping stems usually deeply subterranean; sterile branches less than 2 mm broad; leaves of under surface of branchlets about equalling the marginal ones; cones usually less than 3 cm long. . . . .*L. tristachyum*
          9 Creeping stems on or near the surface of the ground; sterile branches to 4 mm broad; leaves of under surface of branchlets much smaller than the lateral ones; cones to about 5 cm long . . . .*L. complanatum*

L. alpinum L.   Alpine Club-moss
/aST/X/GEA/   (Ch)   Cool woods and subalpine meadows from the Aleutian Is., N-cent. Alaska, and Great Bear L. through the mts. of B.C. and w Alta. to Wash. and Mont.; Keweenaw Co., N Mich.; SE Hudson Bay and E James Bay; northernmost Labrador to the mts. of the Gaspé Pen., E Que. (an early report from Nfld. requires confirmation or may refer to *L. sabinaefolium*); s half of Greenland; Iceland; Eurasia. [*Diphasium* Rothm.; *Lepidotis* Beauv.; incl. f. *umbrosum* Porsild]. MAPS: Hultén 1968*b*:30, and 1958: map 215, p. 235; Meusel, Jaeger, and Weinert 1965:8; Raup 1947: pl. 14.

L. annotinum L.   Stiff Club-moss
/aST/X/GEA/   (Ch)   Woods and clearings, the aggregate species from N Alaska–Yukon–Dist.
Mackenzie to s Dist. Keewatin, s-cent. Baffin Is., and northernmost Ungava–Labrador, s to Oreg.,
Colo., Minn., and Va.; w and E Greenland N to ca. 74°N; Iceland; Eurasia. MAPS: *see* below.
1   Leaves mostly spreading or reflexed, to over 1 cm long.
    2   Leaves lanceolate to linear-oblong, distinctly serrate .var. *annotinum*
        3   Cone solitary, [*Lepidotis* Beauv.; transcontinental; MAPS: Hultén 1962: map 62,
            p. 71, and 1968b:27; Raup 1947: pl. 14] . . . . .f. *annotinum*
        3   A second cone added during the second season, separated from the first one by
            a tuft of leaves; [type from Wood Harbour Is., SE James Bay] . . . . . . .f. *proliferum* Lepage
    2   Leaves narrower, entire or nearly so . . . . .var. *acrifolium* Fern.
1   Leaves strongly ascending to tightly appressed, mostly not over 6 mm long; [transcontinen-
    tal; the common forms northwards].
    4   Leaves lanceolate to lance-oblong, flat, obscurely serrate . . . . .var. *alpestre* Hartm.
    4   Leaves narrower, rounded on the back, entire; [*L. pungens* La Pylaie; *L. dubium*
        Zoega; MAPS: Hultén 1962: map 62, p. 71, and 1968b:27; Porsild 1957: map 9, p. 162] . . . . .
    . . . . .var. *pungens* (La Pylaie) Desv.

L. clavatum L.   Common or Running Club-moss.   Courants verts
/aST/X/GEA/   (Ch)   Dry woods, thickets, and clearings, the aggregate species from the
Aleutian Is. and N Alaska to cent. Yukon, sw Dist. Mackenzie, L. Athabasca (Alta. and Sask.), Man.
(N to the Cochrane R. at 58°13′N), Ont. (N to w James Bay at 52°N), Que. (N to Ungava Bay),
Labrador (N to the Hamilton R. basin), Nfld., N.B., P.E.I., and N.S., s to NW Calif., Idaho, Minn., and
N.C.; s Greenland; Iceland; Eurasia. MAPS and synonymy: *see* below.
1   Terminal bristle of leaf deciduous during the first season; [Alaska–Yukon and B.C. (Queen
    Charlotte Is.; near Cassiar, ca. 59°15′N)] . . . . .var. *integerrimum* Spring
1   Terminal bristle of leaf persistent for usually 4 or 5 seasons; (transcontinental).
    2   Peduncles mostly bearing a solitary spike; [MAPS: Hultén 1962: map 61, p. 71, and
        1968b:29; Raup 1947: pl. 14] . . . . .f. *monostachyon* (Desv.) Clute
    2   Peduncles mostly bearing at least 2 cones; [*Lepidotis* Beauv.; incl. f. *brevipedunculatum*
        Louis-Marie (peduncles at most 3 cm long) and f. *decapitatum* Louis-Marie (peduncles
        very long, 2–3-branched, lacking spikes), var. *megastachyon* Fern. & Bissell and its
        f. *furcatum* (Luerss.) Vict., and vars. *laurentianum* and *subremotum* Vict.; *L.* ?*integri-*
        folium Goldie; MAPS: Hultén 1962: map 61, p. 71, and 1968b:28] . . . . .f. *clavatum*

L. complanatum L.   Ground-Cedar
/aST/X/GEA/   (Ch)   Dry woods, thickets, and clearings, the aggregate species from N-cent.
Alaska to SE Yukon, Great Bear L., s Dist. Keewatin, Que. (N to Ungava Bay), cent. Labrador, Nfld.,
N.B., P.E.I., and N.S., s to Wash., Idaho, and S.C.; w Greenland N to ca. 70°N; Eurasia. MAPS and
synonymy:*see* below.
1   Horizontal stems mostly subterranean; erect stems loosely and remotely forking, the
    indeterminate branches with annual constrictions; peduncles commonly bearing 1 or 2
    cones; [*Diphasium* Rothm.; *Lepidotis* Beauv.; incl. var. *canadense* Vict.; *L. tristachyum*
    var. *boreale* Vict.; transcontinental; MAPS: Hultén 1968b:29, and 1962: map 109, p. 118;
    Raup 1947: pl. 14 (aggregate species); Meusel, Jaeger, and Weinert 1965:8 (incl. *L.*
    tristachyum)] . . . . .var. *complanatum*
1   Horizontal stems mostly on the surface of the ground; erect stems more regularly branched,
    the branches determinate, lacking annual constrictions . . . . .var. *flabelliforme* Fern.
    2   Peduncles bearing a solitary cone; [N L. Superior watershed] . . . .f. *wibbei* (Haberer) House
    2   Peduncles bearing up to 9 cones; [*L. flab.* (Fern.) Blanch,; incl. vars. *elongatum* Vict.
        and *gartonis* Boivin, *L. habereri* House, and *L. tristachyum* var. *laurentianum* Vict.;
        Ont. to Nfld. and N.S.; MAPS: Hultén 1962: map 109, p. 119; Marie-Victorin 1925:
        fig. 8, p. 63] . . . . .f. *flabelliforme*

L. inundatum L.   Bog-Club-moss
/t/X/EeA/   (Ch)   Damp shores, swamps, or bogs, the aggregate species from SE Alaska through
B.C. (*see* B.C. map by T.M.C. Taylor 1963:155) to NW Calif. and Idaho; Ont. (N to the N shore of L.

Superior) to Que. (N to the Marten R. at ca. 51°N and the Côte-Nord), SE Labrador (Goose Bay), Nfld., N.B., P.E.I., and N.S., S to Tex. and Fla.; Eurasia. MAPS and synonymy: *see* below.
1 Cones to 4 cm long and 12 mm thick, their bracts loosely ascending or spreading; leaves to 0.7 mm broad, mostly entire; leading creeping terminal shoot elongating to 1(1.5) dm beyond the 1 or 2(3) fertile erect stems (branches), these to 1 dm long and 1 mm thick; [*Lepidotis* Börner; *Plananthus* Beauv.; S Alaska Panhandle (*see* Hultén 1941: map 47, p. 119) and B.C. (*see* T.M.C. Taylor 1963: map, p. 155); Sask. (Boivin 1966b); Ont. (N to the N shore of L. Superior), Que. (N to the Marten R. at ca. 51°N and the Côte-Nord), SE Labrador (Goose Bay), Nfld., N.B., P.E.I., and N.S.; MAPS: Hultén 1958: map 198, p. 217, and 1968b:26(aggregate species); Meusel, Jaeger, and Weinert 1965:8] . . . . .var. *inundatum*
1 Cones to 11 cm long and 8 mm thick, their bracts appressed to ascending; leaves to 1.2 mm broad, mostly ciliate-denticulate; leading creeping terminal shoot elongating to 4 dm beyond the last of the 2–10 fertile erect stems, these to about 3 dm long and 3 mm thick; [N.B. and N.S.; MAP: on the above-noted map by Hultén] . . . . .var. *bigelovii* Tuck.

L. lucidulum Michx.   Shining Club-moss
/T/EE/A/ (Ch) Cool woods from SE Man. (Bissett) to Ont. (N to Kapuskasing), Que. (N to Anticosti Is. and the Gaspé Pen.; reports from Jigger Is., S Labrador, by Delabarre (1902) and from Labrador at ca. 53°N by Hustich and Pettersson (1943) may refer to *L. selago*), Nfld., N.B., P.E.I., and N.S., S to Mo., Tenn., and S.C.; the Himalayas; China; Japan. [*L. porophilum* Lloyd & Underw.; *Huperzia selago* ssp. *lucidula* (Michx.) Löve & Löve].

L. obscurum L.   Ground-Pine.   Petits-pins
/sT/X/eA/ (Ch) Damp woods and clearings, the aggregate species from the westernmost Aleutian Is. (Attu Is.) and cent. Alaska to L. Athabasca (Alta. and Sask.), Man. (N to Reindeer L. at 57°49'N), Ont. (N to Big Trout L. at ca. 54°N), Que. (N to ca. 57°N), Nfld., N.B., P.E.I., and N.S., S to Wash., Idaho, S.Dak., and N.C.; E Asia. MAP and synonymy: *see* below.
1 Leaves linear-lanceolate, about 1 mm broad, the lower (and often upper) series appressed, the lateral series spreading; cones to about 15 in number, to 6.5 cm long and 6 mm thick; branchlets spreading or recurving at tip; [transcontinental] .var. *obscurum*
1 Leaves linear-attenuate, mostly less than 1 mm broad, all incurved-ascending or uniformly slightly spreading; cones to over 25 in number, to 8.5 cm long and 4.5 mm thick; branchlets more ascending or erect . . . . .var. *dendroideum* (Michx.) Eat.
  2 Cone solitary, appearing to terminate the principal axis; [Que., the type from St-Jérôme, Terrebonne Co.; also known from Québec Co.] . . . . .f. *monostachyon* Vict.
  2 Cones several to many.
    3 Cones sessile or nearly so.
      4 Cones normal; [*L. dendroideum* Michx.; transcontinental; MAP: Hultén 1968b: 28]
         . . . . .f. *dendroideum*
      4 Cones terminated by a small cluster of normal leaves; [Que. (St-Jérôme; Ste-Agathe) and N.S. (type from Belleville, Yarmouth Co.)] . . . . .f. *proliferum* Vict.
    3 Cones more or less exserted on peduncles 2–5 cm long.
      5 Cones normal; [Que.: N to the type locality, Kondiaronk, about 35 mi S of L. St. John, and the Côte-Nord] . . . . .f. *exsertum* Vict.
      5 Cones, or some of them, 2-forked (sometimes to base); [type from Kondiaronk, Que.] . . . . .f. *exsertumfurcatum* Vict.

L. sabinaefolium Willd.   Ground-Fir
/ST/X/eA/ (Ch) Woods, thickets, and clearings (ranges of Canadian taxa outlined below), S to Oreg., Mont., N Mich., Pa., and N New Eng.; E Asia. MAPS and synonymy: *see* below.
1 Leaves of the erect branchlets mostly 4-ranked, their free tips shorter than the decurrent base; cones to about 4.5 cm long, on peduncles to about 8 cm long; [*Diphasium* D. Löve; incl. vars. *patens* and *superfertile* Vict.; Ont. (N shore of L. Superior), Que. (L. Mistassini; L. St. John; Rimouski and Charlevoix counties; Côte-Nord; Gaspé Pen.), Nfld., N.B., P.E.I., and N.S.] . . . . .var. *sabinaefolium*
1 Leaves of erect branchlets in 4 or 5 ranks, their free tips usually longer than the decurrent base; cones less than 3.5 cm long, sessile or on peduncles to about 3 cm long;

[*L. (Diphasium) sitchense* Rupr., the type from Sitka, Alaska; incl. f. *decipiens* Lepage; coastal Aleutian Is.–Alaska–B.C. and southernmost interior B.C.; Alta. N to ca. 54°N; isolated at L. Athabasca, Alta. and Sask.; Ont. (N shore of L. Superior; isolated at Pickle Lake, ca. 51°30′N), Que. (N to the George R. at ca. 58°30′N), Nfld., N.B., and N.S.; MAPS: Hultén 1968*b*:30; Porsild 1966: map 3, p. 67] . . . . .var. *sitchense* (Rupr.) Fern.

L. selago L.   Mountain-Club-moss
/AST/X/GEA/   (Ch)   Cold woods, mossy rocks, and barrens, the aggregate species from the Aleutian Is. and N coast of Alaska to Ellesmere Is. at ca. 80°N and northernmost Labrador, s to L. Athabasca (Alta. and Sask.), cent. Man., James Bay, E Que. (Côte-Nord and Gaspé Pen.), Nfld., N.B., and N.S. (not known from P.E.I.), and in the mts. to Oreg., Mont., Mexico, and N.C.; an isolated area in Ont. (L. Superior), Minn., Wisc., and Mich.; S. America; circumgreenlandic; Iceland; Eurasia; New Zealand and Tasmania. MAPS and synonymy: *see* below.
1   Leaves spreading or even reflexed, to about 8 mm long; stems loosely ascending; [incl. var. ?*miyoshianum* Makino; *Plananthus patens* Beauv.; *L. ?chinense* Chr.; *L. lucidulum* f. *occidentale* Clute; Alaska–B.C.; Ont. to E Que. and Nfld.; MAPS: Hultén 1968*b*:26 (ssp. *chinense*) and 1962: map 46, p. 53] . . . . .var. *patens* (Beauv.) Desv.
1   Leaves ascending or appressed; stems densely tufted; (transcontinental).
  2   Leaves mostly narrowly ovate-lanceolate, appressed, to about 5 mm long; [type from St-Pierre and Miquelon–Nfld.; MAPS: Hultén 1968*b*:25, and 1962: map 46, p. 53] . . . .
      . . . . .var. *appressum* Desv.
  2   Leaves lance-attenuate, ascending, to about 8 mm long; [*Huperzia* Bernh.; *Plananthus* Beauv.; MAPS (the last three of the aggregate species): Hultén 1968*b*:25, and 1962: map 46, p. 53; Porsild 1957: map 10, p. 162; Raup 1947: pl. 14; Meusel, Jaeger, and Weinert 1965:8] . . . . .var. *selago*

L. tristachyum Pursh   Ground-Cedar
/sT/EE/E/   (Ch)   Dry woods and clearings from Ont. (N to the N shore of L. Superior) to E Que. (Côte-Nord; Gaspé Pen.), s Labrador (Goose Bay), Nfld., N.B., P.E.I., and N.S., s to Minn., Tenn., and N.C.; Europe. MAP: Hultén 1958: map 39, p. 59.
    Hultén's map indicates stations around L. Athabasca in Alta. and Sask., but these are probably referable to *L. complanatum*. His dot for s James Bay is based upon *L. complanatum* (relevant collection from Charlton Is. in CAN), as, also, his Great Bear L. stations for *L. tristachyum* var. *boreale*.

SELAGINELLACEAE (Spikemoss Family)

SELAGINELLA Beauv.   Spikemoss.   Selaginelle

Similar to *Lycopodium*, but plants dwarf and more moss-like, spikes sessile and usually distinctly 4-angled, and spores of two kinds. Macrosporangia commonly borne in the lower part of the spike, each with usually 3 or 4 globose-angular macrospores; microsporangia minute, commonly borne in the upper part of the spike, each filled with powdery orange-red microspores.

(Ref.: Tryon 1955)
1   Leaves soft, lanceolate to narrowly ovate, neither they nor the sporophylls (spike-leaves) bristle-tipped; stem and branches very delicate.
  2   Leaves spirally arranged, all alike, lance-acuminate, sparsely ciliate, to 4 mm long and 1 mm broad; spike subcylindric, to over 3 cm long, its sporophylls commonly in about 10 ranks; (transcontinental) . . . .*S. selaginoides*
  2   Leaves 4-ranked, pellucid-membranaceous (plants forming pale or whitish-green mats), those of the two lateral ranks the largest, broadly ovate or oval-oblong, abruptly acutish to broadly rounded at summit, spreading, 2 or 3 mm long, those of the dorsal and ventral ranks smaller and stipule-like, appressed (or the dorsal ones spreading) and pointed; spikes strongly 4-angled, rarely over 1.5 cm long.
    3   Larger (lateral) leaves ciliate at base, oval-oblong, very blunt to broadly rounded at summit; (?B.C.) . . . .[*S. douglasii*]
    3   Larger (lateral) leaves merely minutely serrulate (not at all ciliate), rather narrowly ovate to oblong, mostly less rounded or even acutish at summit; (Ont. and sw Que.) . . . .*S. apoda*
1   Leaves firm, subulate, bristle-tipped, spirally arranged, ciliate or eciliate; stem and branches mostly firm; spikes strongly 4-angled, their bristle-tipped sporophylls 4-ranked.
  4   Leaves distinct from the stem in colour, their bases abruptly adnate (sometimes slightly decurrent on the leader stems); leafy stems rarely over 2 dm long, mostly radially symmetrical, the upper and under leaves on the same portion of the stem subequal in length and degree of spreading; megaspores pale orange; (B.C. and sw Alta.) . . . .*S. wallacei*
  4   Leaves with strongly decurrent bases (this character best observed between the 1st and 3rd branches back from the apex of the stem), thus less markedly distinct from the stem in colour.
    5   Plants usually epiphytic, their long slender free stems (to several dm long) pendent in festoons from mossy trunks or branches of trees (particularly *Acer macrophyllum*) and rooting only at the attached basal portion; branches strongly curled when dry (often forming ringlets); megaspores pale yellow; (?B.C.) . . . .[*S. oregana*]
    5   Plants terrestrial in rocky or sandy habitats, the short stems commonly rooting throughout their length, the branches not or only slightly curled when dry.
      6   Apex of the upper leaves scarcely narrowed, usually truncate in profile; branches tending to bear scattered roots; (Aleutian Is. to NW Dist. Mackenzie and N B.C.) . . . .*S. sibirica*
      6   Apex of the upper leaves distinctly narrowed; plant rooting only at base.
        7   Upper and under leaves on same portion of stem equal or subequal; bristles at tips of sporophylls scarcely broadened and flattened at base; stems forming open spreading mats with intricate and commonly intertwining branches; megaspores bright orange; (NE Alta. to N.S.) . . . .*S. rupestris*
        7   Upper leaves distinctly shorter than the under ones on the same portion of stem; base of sporophyll-bristles often strongly broadened and flattened; stems forming compact mats with discrete branches; megaspores pale to bright orange; (B.C. to sw Man.) . . . .*S. densa*

S. apoda (L.) Fern.
/T/EE/   (Ch)   Damp woods, meadows, and wet rocks from Wisc. to Ont. (N to the Ottawa dist.) and sw Que. (N to the Montreal dist.), s to Tex. and Fla.; S. America. [*Lycopodium* L.; *S. apus*

Spring). MAP: the greater part of the Canadian distribution is shown in the Ont. map by J. H. Soper (Am. Fern J. 53(3): fig. 8, p. 114. 1963].

S. densa Rydb.
/sT/WW/ (Ch) Dry rocks, sandy prairies, and exposed hillsides, the aggregate species from southernmost Alaska and B.C. to sw Man., s to Calif., N.Mex., and Tex. MAPS and synonymy: *see* below.
1 Sporophylls (cone-leaves) ciliate to tip; [B.C. (N to ca. 55°N; *see* B.C. map of the aggregate species by T.M.C. Taylor 1963:164; occurring only east of the coastal mountains according to Taylor), Alta. (N to Fort Saskatchewan), Sask. (N to ca. 55°N), and sw Man. (Boivin 1966*b*); MAPS: Tryon 1955: map 4, p. 69; R. M. Tryon, Brittonia 23: figs. 3 and 4 (aggregate species) and 5 (var. *densa*), p. 93, and fig. 6 (aggregate species), p. 95. 1971] . . . . . .var. *densa*
1 Sporophylls ciliate only below the middle; [incl. *S. scopulorum* and *S. standleyi* Maxon; *S. columbiana* Eat.; southernmost Alaska, B.C. (Vancouver Is.; Ootsa L. at ca. 53°45'N; Windermere), and sw Alta. (Waterton Lakes); MAPS: as indicated for the above-noted maps by Tryon; Hultén 1968*b*:32] . . . . .var. *scopulorum* (Maxon) Tryon

[*S. douglasii* (H. & G.) Spring]
[The inclusion of B.C. in the range assigned to this species of the w U.S.A. (Wash.–Oreg.–Idaho) by various authors (Abrams 1923; Jepson 1951; Broun 1938) is believed by Jones (Am. Fern J. 54:84. 1964) to have resulted from the erroneous addition of the word "British" to Hooker's "Columbia, Douglas", indicating the vague source of a collection in the Royal Botanic Gardens, Kew, England.]

[*S. oregana* Eat.]
[This species of the w U.S.A. (N Wash. to N Calif.; *see* Tryon 1955: map 40, p. 63) is reported from w B.C. by L. M. Underwood (Bull. Torrey Bot. Club 25: 132. 1898; Nootka Sound, Vancouver Is., and Observatory Inlet, ca. 55°N) in synonymy under the name *Selaginella struthioloides* (Presl) Underw. However, Tryon believes that a confusion of labels must be involved in the Observatory Inlet report. Such is also probably the case for the Nootka Sound collection, said by Underwood to represent the type material of *Lycopodium struthioloides*, now, however, recognized as a true *Lycopodium* restricted to the Old World.]

S. rupestris (L.) Spring
/aST/X/G/ (Ch) Dry rocks, gravels, and sands from Alta. (N to Fort Fitzgerald, 59°55'N) to Sask. (N to L. Athabasca), Man. (N to Norway House, off the NE end of L. Winnipeg), Ont. (N to Sandy L. at ca. 53°N), Que. (N to the Côte-Nord), N.B., and N.S. (not known from P.E.I. or Nfld.), s to S.Dak., N Tex., and N Ga.; sw Greenland at ca. 61°N. [*Lycopodium* L.]. MAPS: Tryon 1955: map 42, p. 63; T. W. Böcher, Medd. Gronl. 147(3): fig. 2, p. 9. 1948; R. T. Clausen, Am. Fern J. 36(3): fig. 1, p. 68. 1946; R. M. Tryon, Jr., Brittonia 23: fig. 7, p. 95, and figs. 8 and 9, p. 96. 1971.

S. selaginoides (L.) Link
/aST/X/GEA/ (Ch) Damp shores, rocks, and mossy banks from the Aleutian Is. and N Alaska to cent. Yukon, Great Bear L., N Sask., Man. (known only from Churchill), Ont. (N to near the mouth of the Black Duck R., Hudson Bay, at ca. 56°45'N), Que. (N to the Larch R. at ca. 57°N), Labrador (N to ca. 56°30'N), Nfld., N.B., and N.S. (not known from P.E.I.), s in the West to the mts. of Nev., Wyo., and ?Colo. and in the East to Minn. and Maine; w and E Greenland N to ca. 65°N; Iceland; Eurasia. [*Lycopodium* L.; *S. spinosa* Beauv.]. MAPS: Hultén 1968*b*:31, and 1958: map 222, p. 241; Porsild 1966: map 4, p. 67; Raup 1947: pl. 14; Meusel, Jaeger, and Weinert 1965:9; the northernmost stations in E Canada are indicated in a map by Lepage 1966: map 2, p. 212.

S. sibirica (Milde) Hieron.
/aST/W/A/ (Ch) Dry rocks, gravels, and sands of Alaska–Yukon and the Mackenzie R. Delta region, s to the E Aleutian Is. (Unalaska) and mts. of N B.C. (s to Mt. Selwyn at ca. 56°N); Japan; Sakhalin; N Korea; E Siberia. [*S. rupestris* f. *sib.* Milde]. MAPS: Hultén 1968*b*:31; Porsild 1966: map 5, p. 67; Tryon 1955: map 47, p. 73.

S. wallacei Hieron.
/t/W/ (Ch) Cliffs and dry grassy or rocky slopes of s B.C. (N to ca. 50°15′N; *see* B.C. map by T.M.C. Taylor 1963:164) and sw Alta. (Waterton Lakes), s to N Calif. and w Mont. [*S. montanensis* Hieron.]. MAP: Tryon 1955: map 24, p. 42.

ISOËTACEAE (Quillwort Family)

ISOËTES L. Quillwort. Isoète

Small aquatic or amphibious plants with a crown of elongate-subulate quill-like leaves from a short corm-like base. Spores of two kinds, borne in a solitary sporangium in a pit in the dilated leaf-base, the megasporangia and microsporangia usually borne in alternating cycles of leaf-bases.

(Ref.: Soper and Rao 1958; Pfeiffer 1922)
1 Surface of megaspore irregularly jagged-crested or reticulate; (eastern species).
  2 Megaspores regularly and coarsely honeycomb-reticulate on the rounded lower half, the three faces of the upper pyramidal half with thin, more or less parallel-walled reticulation; leaves commonly 2 or 3 dm long, lacking bast-bundles; (Ont. to Nfld. and N.S.) . . . .
    . . . .*I. macrospora*
  2 Megaspores conspicuously jagged-crested, the crests with sharp, distinct or slightly confluent peaks.
    3 Plants completely submersed; leaves lacking stomata, less than 2 dm long, to about 40 in number; microspores smooth; (?B.C.; southernmost Greenland) . . . . . . *I. lacustris*
    3 Plants amphibious; stomata present; microspores minutely roughened to decidedly tuberculate or spinulose; (s Ont. to N.S.) . . . .*I. riparia*
1 Surface of megaspore spiny, tuberculate, or papillate.
  4 Megaspores spiny; (transcontinental) . . . .*I. echinospora*
  4 Megaspores tuberculate or papillate (the tubercles sometimes confluent); (western species).
    5 Corm slightly 3-lobed; leaves often more than 30, with numerous stomata; megaspores densely covered by small, usually glistening, distinct papillae; velum (the membranous extension on the inner face of the leaf) completely covering the sporangium; leaves with 3 peripheral strands of bast-bundles; plant amphibious, growing in damp terrestrial places or in shallow water; (Vancouver Is.) . . . .*I. nuttallii*
    5 Corm 2-lobed; megaspores covered with tubercles frequently confluent into wrinkles; velum usually covering not more than 1/3 of the sporangium; leaves usually not more than 30.
      6 Amphibious, growing in damp terrestrial places or in shallow water; stomata numerous; leaves with usually 4 peripheral strands of bast-bundles in addition to the central fibro-vascular bundles; (?B.C.) . . . .[*I. howellii*]
      6 Aquatic, growing submersed in water; stomata not numerous; peripheral strands obsolete or nearly so; (s B.C.–sw Alta.) . . . .*I. bolanderi*

I. bolanderi Engelm.
/T/W/ (HH) Shallow water or wet shores from s B.C. (N to Kamloops and Sicamous) and sw Alta. (Waterton Lakes; Breitung 1957b) to Calif., Ariz., and Colo. MAP: Taylor 1970:37 ("Pacific Northwest" area).

I. echinospora Dur.
/aST/X/GEA/ (HH) Shallow water or wet shores, the aggregate species from the Aleutian Is. and s Alaska to s Yukon, Great Bear L., L. Athabasca, s Dist. Keewatin, Que. (N to s Ungava Bay), Labrador (N to 53°27′N), Nfld., N.B., and N.S., s to Calif., Colo., Ind., Pa., and N.J.; w Greenland N to ca. 68°N and southernmost E Greenland; Iceland; Eurasia. MAPS and synonymy: see below.
1 Stomata absent; sporangia unspotted.
  2 Velum (the membranous tissue more or less covering the sporangium) very narrow, covering less than 1/3 of the sporangium; [Eurasia; MAP: Hultén 1958: map 235, p. 255] . . . . .var. *echinospora*
  2 Velum covering 2/3–3/4 of the sporangium; [*I. asiatica* Makino; tentatively reported from Prince William Sound, Alaska, by Bernard Boivin, Am. Fern J. 51(2):85. 1961, but see Hultén 1968b:33, under *I. truncata*] . . . . .var. *asiatica* Makino
1 Stomata present; sporangia spotted.
  3 Velum covering 1/4–1/2 of the sporangium; megaspores covered with blunt spines;

stomata numerous; [*I. maritima* Underw. and its var. *flettii* (Eat.) Reed; *I. braunii* var.
maritima (Underw.) Pfeiffer; *I. macounii* Eat.; *I. ?truncata* (Eat.) Clute; Aleutian
Is.–Alaska–Vancouver Is. (type from Alberni); MAP: Hultén 1958: map 235, p. 255, and
1968*b*:33] . . . . .var. *maritima* (Underw.) Eat.
3   Velum covering 1/2–3/4 of the sporangium; megaspores covered with sharp or blunt
spines; stomata relatively few; [vars. *braunii* (Dur.) Engelm. and *robusta* Engelm.; ssp.
muricata var. *savilei* Boivin; *I. braunii* Dur., not Unger, which is a fossil species;
I. setacea var. *mur.* (Dur.) Holub; *I. muricata* Dur. and its vars. *braunii* (Dur.) Reed
and *?hesperia* Reed; transcontinental; MAPS: Hultén 1958: map 235, p. 255, and
1968*b*: 32; Meusel 1943: fig. 28a] . . . . .var. *muricata* (Dur.) Engelm.

[Isoëtes howellii Engelm.]
[The inclusion of B.C. in the range of this species of the w U.S.A. (Wash. and Mont. to Calif.) by
Rydberg (1922) requires clarification; the MAP by Taylor (1970:41) indicates no Canadian stations.]

I. lacustris L.
/aST/W/GEA/   (HH)   Shallow water or wet shores, the typical form in s Greenland, Iceland, and
Eurasia, var. *paup.* from B.C. to Calif. and Colo. MAPS and synonymy: *see* below.
1   Crests of megaspores somewhat confluent but little anastomosing; [southernmost
Greenland; fossil remains in Europe; reports from Sask. and Que. by John Macoun 1890,
presumably refer to phases of *I. echinospora*; MAPS: Hultén 1958: map 247, p. 267;
Meusel, Jaeger, and Weinert 1965:9; Meusel 1943:26] . . . . .var. *lacustris*
1   Crests of megaspores forming an irregular network on the basal face; [*I. paup.* (Engelm.)
Eat.; *I. occidentalis* Hend.; B.C. N to ca. 54°30′N; MAPS (*I. occid.*): on the above-noted
map by Hultén; Taylor 1970:44 ("Pacific Northwest" area) . . . . .var. *paupercula* Engelm.

I. macrospora Dur.
/T/EE/   (HH)   Shallow water or wet shores from N Minn. and Ont. (N to near Chalk River,
Renfrew Co.) to Que. (N to the Côte-Nord, Gaspé Pen., and George R. at ca. 55°N), ?Labrador (the
report from Makkovik, ca. 55°N, by Hustich and Pettersson 1943, requires confirmation), Nfld.,
N.B., and N.S. (not known from P.E.I.), s to Wisc., s Ont., and N.J. [Incl. *I. tuckermanii* A. Br.].
MAPS: Hultén 1958: map 247 (*I. lacustris* var. *mac.*), p. 267; Meusel, Jaeger, and Weinert 1965: 9;
Soper and Rao 1958: map 2, p. 100.

Forma *hieroglyphica* (Eat.) Pfeiff. (ridges of the megaspores more rounded and prominent than
in those of the typical form) is known from Que. (Marie-Victorin 1925) and N.S. (North Sydney,
Cape Breton Is., where taken by John Macoun in 1883; CAN).

I. nuttallii A. Br.
/t/W/   (HH)   Shallow water or wet shores from sw B.C. (reported from near Nanaimo, Vancouver
Is., by John Macoun 1890, this taken up by Henry 1915, and Pfeiffer 1922) to N Baja Calif. and
Idaho. MAP: Taylor 1970:43 (N area).

I. riparia Engelm.
/T/EE/   (HH)   Shallow water or wet shores and tidal pools from s Ont. (N to Georgian Bay, L.
Huron) to Que. (N to the Montreal dist.), ?Nfld., N.B., ?P.E.I., and N.S. [Incl. var. *canadensis*
Engelm., var. *robbinsii* (Eat.) Proctor, and *I. dodgei* Eat.]. The report of *I. dodgei* from B.C. by
Henry (1915) is probably erroneous. MAP: Soper and Rao 1958: map 3, p. 100.

OPHIOGLOSSACEAE (Adder's-tongue Family)

(Ref.: Clausen 1938)

Frond consisting of a terminal fertile spike or panicle and a lower sterile blade, the common stipe from an erect rhizome with fleshy roots, its base containing the bud for the next year's frond (*Botrychium*) or the bud situated at one side of the base (*Ophioglossum*). Sporangia several cells thick, lacking an annulus.

1 Sterile blade deeply lobed to 1–2-pinnate, sessile or stalked, its veins free; sporangia separate, short-stalked, free from the leaf-tissue, in panicles (rarely in simple spikes) . . . . *Botrychium*
1 Sterile blade entire, oblong-elliptic to ovate, sessile or short-stalked, to about 1 dm long, net-veined; sporangia fused with the leaf-tissue and coherent in a simple 2-rowed spike; (s B.C.; Ont. to Nfld. and N.S.) . . . . *Ophioglossum*

BOTRYCHIUM Sw.   Moonwort, Grape-Fern.   Botryche

(Ref.: Clausen 1938)

1 Sterile blade ternately compound, sessile above the middle of the plant, deltoid in outline, relatively large; bud hairy, partly exposed by the sheathing base of the stipe (common stalk), this open on one side; plant to about 7 dm tall; (transcontinental) . . . .*B. virginianum*
1 Sterile blades various, not entirely as above; bud hairy or smooth, completely enclosed by the sheathing base of the stalk; plants commonly lower.
  2 Sterile blade ternately decompound, fleshy or leathery and often evergreen, slightly pilose at least in bud, mostly relatively long-stalked from near the base of the plant, its margins whitish under magnification; buds commonly hairy, their sterile and fertile blades completely reflexed; spores pitted.
    3 Ultimate segments of sterile blade lanceolate to lance-ovate, acute or acutish, commonly 4 or 5 times as long as broad; plant not very fleshy, usually fruiting in autumn, the sterile blade then usually bronze or purple; (Ont. to N.S.) . . . . .*B. dissectum*
    3 Ultimate segments of sterile blade ovate to obovate or rotund, obtuse or rounded at tip, not more than 3 times as long as broad; sterile blades remaining green over winter; (transcontinental) . . . .*B. multifidum*
  2 Sterile blade usually smaller, commonly less (pinnately) divided (sometimes simple in *B. simplex*), glabrous even in bud, sessile or short-stalked, parting from the common stalk at various heights, its margins not whitish; buds glabrous; spores tuberculate.
    4 Sterile blade deltoid, 1-pinnate-pinnatifid, to about 4.5 cm long and 4 cm broad, sessile or very short-stalked above the middle of the plant; fertile blade paniculate, both it and the sterile blade completely reflexed in the expanding bud; (transcontinental) . . . .*B. lanceolatum*
    4 Sterile blade relatively narrower, oblong or ovate in outline, sessile or stalked, the fertile one erect or nearly so in the expanding bud, the upper part of the sterile one either slightly inclined over it or bent down and more or less embracing or covering it.
      5 Sterile blade borne at or below middle of plant (except in *B. simplex* var. *tenebrosum*), its divisions commonly entire or only shallowly toothed, (transcontinental).
        6 Sterile blade sessile or nearly so, to about 9 cm long, with up to 7 or more pairs of overlapping to slightly remote fan-shaped segments; fertile blade paniculate, to about 1.5 dm long . . . .*B. lunaria*
        6 Sterile blade commonly distinctly stalked, usually less than 4 cm long, simple or with at most about 3 pairs of approximate, obliquely ovate segments; fertile blade simple or but slightly compound, usually less than 5 cm long . . . .*B. simplex*
      5 Sterile blade borne above the middle or near the summit of the plant.
        7 Sterile blade ovate to broadly ovate-oblong, to 4 cm broad, usually sessile, pinnate, the primary lobes palmately lobed or crenate; fertile blade simple or

paniculate, to 5 cm long, its stalk to 2 cm long; (Alaska–Yukon–B.C.–Alta.) . . . . . .
. . . .*B. boreale*
7   Sterile blade commonly oblong, to 7 cm broad, sessile or on a stalk to
      2.5 cm long, usually pinnately (sometimes ternately) divided; fertile blade
      usually paniculate, to 8 cm long, its stalk to 5 cm long; (transcontinental) . . . .
      . . . .*B. matricariaefolium*

B. boreale (Fries) Milde

/aST/D/GEeA/ (Grh) Grassy tundra and exposed slopes, the aggregate species from the
Aleutian Is., s Alaska, sE Yukon, and N B.C., s in the mts. through B.C. (s to Nelson) and sw Alta.
(Brazeau L., 52°25′N) to N Oreg. and N Nev.; s Greenland; Scandinavia; E Siberia. MAPS and
synonymy: *see* below.
1   Segments of the sterile blade pinnately or palmately lobed, acute at apex; [*B. lunaria* var.
      bor. Fries; Unalaska Is., Alaska (Clausen 1938); N Europe; E Asia; MAPS: Hultén 1968*b*: 40
      (aggregate species); Clausen 1938: fig. 13, p. 73; Taylor 1970: 71 ("Pacific Northwest"
      area) . . . . .var. *boreale*
1   Segments of sterile blade pinnately lobed, obtuse at apex; [*B. crassinervium* var. *obtus.*
      Rupr., the type from Unalaska, Alaska; s Alaska (the Alaska map for *B. boreale* given by
      Hultén 1941, applies here), sE Yukon (Porsild 1951*a*), B.C., and w Alta. (Waterton Lakes;
      Brazeau L. and Maligne L., both ca. 52°30′N); s Greenland; Siberia; MAP: on the
      above-noted map by Clausen (the occurrence in s Greenland should be indicated)] . . . .
      . . . . .var. *obtusilobum* (Rupr.) Broun

B. dissectum Spreng.

/T/EE/ (Grh) Pastures, sterile fields and meadows, open thickets, and woods from Minn. and
Ont. (N to near Thunder Bay) to Que. (N to St-Raymond, near Quebec City), N.B., and N.S. (not
known from P.E.I. or Nfld.), s to E Tex. and Fla. MAPS and synonymy: *see* below.
1   Segments of the sterile blade mostly broadly ovate to roundish, remaining green over
      winter; [*B. ternatum (dissectum; multifidum*) var. *oneid.* Gilbert; *B. oneid.* (Gilbert) House;
      s Ont., s Que., and N.B.; MAP: Clausen 1938: fig. 3, p. 25] . . . . .f. *oneidense* (Gilbert) Clute
1   Segments of the sterile blade narrowly lanceolate to lance-oblong, becoming bronze or
      purple.
      2   Division of the blade deeply and finely lacerate or divided; [*B. ternatum* var. *dissectum*
            (Spreng.) Eat.; Ont. to ?N.B. and N.S.; MAP: Clausen 1938: fig. 8, p. 55] . . . . . . .f. *dissectum*
      2   Divisions of the blade not deeply lacerate.
            3   Basal segments of the basal pinnules of the lowest pinnae not greatly surpassing
                  the others; [*B. obl.* Muhl.; Ont. to N.B. and ?N.S.; MAP: Clausen 1938: fig. 8, p. 55] . . . . . .
                  . . . . .f. *obliquum* (Muhl.) Fern.
            3   Such segments up to about half the length of the pinnules; [*B. obliquum* var. *el.*
                  Gilbert & Haberer; sw Que. (Marie-Victorin 1923; Hatley)] . . . .
                  . . . . .f. *elongatum* (Gilbert & Haberer) Weatherby

B. lanceolatum (Gmel.) Ångstr.

/aST/X/GEA/ (Grh) · Dry to swampy meadows, slopes, and woods, or sandy open places, the
aggregate species from the Aleutian Is., s Alaska, and cent. Yukon through B.C. and sw Alta. to
Wash., Utah, and Colo.; Great Lakes region; Great Whale R., E Hudson Bay, ca. 55°20′N; Que.
(Gaspé Pen.; Côte-Nord; Montreal dist.) to Nfld., N.B., P.E.I., and N.S., s to Ohio, Pa., and N.J.; w
and E Greenland N to ca. 69°N; Iceland; Eurasia. MAPS and synonymy: *see* below.
1   Sterile blade relatively thick and fleshy, the lobes of the rather crowded pinnae mostly
      rounded at apex and up to 5 mm broad toward base; [*Osmunda* Gmel.; *B. matricariaefolium*
      var. *lan.* (Gmel.) Watt; Alaska–B.C.–sw Alta.; Ont. to Nfld. and N.S.; MAPS (the last three of
      the aggregate species): Hultén 1958: map 237, p. 257; Clausen 1938: fig. 17, p. 83; Hultén
      1968*b*:41; Porsild 1966: map 2, p. 67; Meusel, Jaeger, and Weinert 1965:10] . . . .
      . . . . .var. *lanceolatum*
1   Sterile blade thinner, the lobes of the less crowded pinnae mostly acutish at apex and not
      over 2.5 mm broad toward base; [*B. ang.* (Pease & Moore) Fern.; Ont. (Peninsula, N shore

of L. Superior) to w Nfld. and N.S.; MAPS: on the above-noted maps by Hultén and Clausen] ....
....var. *angustisegmentatum* Pease & Moore

B. lunaria (L.) Sw.   Moonwort
/aST/X/GEA/   (Grh)   Open fields and meadows, ledges, or gravelly slopes and shores from the Aleutian Is. and N Alaska to SE Yukon (Porsild 1951a), Great Bear L., N Man. (Churchill), northernmost Ont., Que. (N to s Ungava Bay, L. Mistassini, and the Côte-Nord), Labrador (N to ca. 59°N), Nfld., and N.S. (not known from N.B. or P.E.I.), s to s Calif., Colo., Minn., and Maine; s half of w Greenland; Iceland; Eurasia. MAPS and synonymy: *see* below.
1   Sterile blade often inserted below the middle of the plant, firm and leathery, much longer than broad, the toothed segments usually not markedly fan-shaped; spores to 40 microns in diameter; [*B. minganense* Vict., the type from the Mingan Is., E Que.; Yukon–B.C.–Alta.; NE Man. (Churchill); Que. (N to the Côte-Nord); s Labrador; MAP: Marie-Victorin and Rolland-Germain 1969: fig. 19, p. 405] ....f.*minganense* (Vict.) Clute
1   Sterile blade usually inserted toward middle of plant, commonly broadly oblong in outline; spores mostly less than 35 microns in diameter.
  2   Segments of sterile blade deeply toothed to somewhat incised; [E Que.: L. Mistassini, Côte-Nord, Anticosti Is., and Gaspé Pen.] ....f.*subincisum* (Roeper) Milde
  2   Segments of sterile blade entire or only shallowly toothed or notched.
    3   Blade membranous, the fan-shaped or oblong divisions rather remote; [var. *onondagense* (Underw.) House; *B. onon.* Underw.; eastern part of the range] ....
....f.*gracile* (Schur) Aschers. & Graebn.
    3   Blade firm and leathery, the fan-shaped divisions often overlapping; [*Osmunda* L.; *Botrypus* Richards; transcontinental; MAPS: Hultén 1968b: 40; Clausen 1938: figs. 10 and 11, p. 63; Meusel, Jaeger, and Weinert 1965: 10; N.C. Fassett, Ann. Mo. Bot. Gard. 28(3): map 33, p. 365. 1941] ....f.*lunaria*

B. matricariaefolium A. Br.
/T/X/EA/   (Grh)   Dry to moist woods, thickets, and old fields, the aggregate species from s B.C. to Alta. (Moss 1959), cent. Sask., Man. (N to Norway House, off the NE end of L. Winnipeg), Que. (N to SE Hudson Bay at ca. 55°15′N; reported from Anticosti Is. by Saint-Cyr 1887), Nfld., N.B., P.E.I., and N.S., s to Idaho, S.Dak., Ohio, and Va.; s S. America; Eurasia. MAPS and synonymy: *see* below.
1   Sterile blade sessile or short-stalked, membranous or fleshy, usually oblong in outline, its major divisions pinnately divided or crenately lobed; [*B. neglectum* Wood; *B. ramosum* of auth., not Roth; transcontinental (the report from Unalaska, Alaska, by John Macoun 1890, probably refers to some other species); MAPS: Clausen 1938: fig. 16, p. 83; Meusel, Jaeger, and Weinert 1965:10] ....var.*matricariaefolium*
1   Sterile blade sessile or nearly so, leathery, oblong-ovate in outline, the rhombic or oblong, obtuse or acutish divisions entire or coarsely and bluntly lobed, often overlapping; [Sask.: reported from Beechy, near Swift Current, by A. J. Breitung, Am. Midl. Nat. 61(2):510. 1959, who transfers his 1957 report from Amisk Lake to the typical form; also known from the Cypress Hills; MAPS: on the above-noted maps by Clausen and Hultén] ....
....var.*hesperium* (Maxon & Clausen) Broun

B. multifidum (Gmel.) Rupr.   Leathery Grape-Fern
/sT/X/EA/   (Grh)   Dry to moist, grassy or sandy fields or open thickets and woods, the aggregate species from the E Aleutian Is. and B.C. to sw Dist. Mackenzie, N Alta. (L. Athabasca), Sask. (N to McKague, 52°37′N), Man. (N to The Pas), Ont. (N to sw James Bay at 52°11′N), Que. (N to SE Hudson Bay at ca. 55°15′N), s Labrador (N to the Hamilton R. basin), Nfld., N.B., P.E.I., and N.S., s to Calif., Wyo., Ill., Mich., and N.J.; Eurasia. MAPS and synonymy: *see* below.
    Forma *dentatum* Tryon (segments of the sterile frond-blade distinctly toothed) is reported from s Ont. by Landon (1960; Norfolk Co.).
1   Plant to over 4 dm tall, the blade to about 1.5 dm long and 2 dm broad, its ultimate divisions usually rather remote and not overlapping; [ssp. *silaifolium* (Presl) Clausen (*B. sil.* Presl); *B. ternatum* var. *int.* Eat.; *B. occidentale* Underw.; transcontinental; MAP: Clausen 1938: fig. 1, p. 25] ....var.*intermedium* (Eat.) Farw.

1 Plant to about 2.5 dm tall, the blade less than 1 dm long and broad, its ultimate divisions usually rather crowded and sometimes overlapping.

    2 Divisions of the sterile blade mostly obtuse or rounded at tip; plant mostly glabrous; [*B. matricariae* (Schrank) Spreng.; *B. ternatum* var. *rutaefolium* (Br.) Eat.; transcontinental] . . . . var. *multifidum*

    2 Divisions of the sterile blade acutish; plant sparingly hairy; [*B. rob.* (Rupr.) Underw.; *B. rutaefolium* var. *rob.* Rupr.; Alaska; MAP: Hultén 1968*b*:41] . . . . . .var. *robustum* (Rupr.) Chr.

B. simplex E. Hitchc.

/aST/X/GEeA/ (Grh) Meadows, pastures, and shores: s B.C. to Sask., s to Calif., N.Mex., and Colo.; Ont. (N to the N shore of L. Superior) to Que. (N to the Gaspé Pen.), Nfld., N.B., and N.S. (not known from P.E.I.), s to s Calif., N.Mex., Ind., Pa., and Md.; sw Greenland; Iceland; Europe; E Asia. MAPS and synonymy: *see* below.

1 Sterile blade borne near the summit of the stem; [*B. tenebrosum* Eat.; ?Alaska (Boivin 1966*b*); Alta. and Sask.; Ont. to N.B.; sw Greenland; MAP: Clausen 1938: fig. 12, p. 73] . . . . .var. *tenebrosum* (Eat.) Clausen

1 Sterile blade borne from near the base to slightly above the middle of the stem . . . .var. *simplex*

    2 Sterile blade with up to 6 pairs of remote, thinnish, lateral lobes, at least the lower of these slender-based or short-stalked; plant to 2 dm tall; [s Ont.; Soper 1949] . . . . .f. *laxifolium* (Clausen) Fern.

    2 Sterile blade with at most 3 pairs of close, firm, broad-based lateral lobes; plant usually less than 1.5 dm tall; [s B.C. (Mt. Benson, Vancouver Is.; Selkirk Mts. at ca. 51°30'N); Alta. (Waterton Lakes); Sask. (Boivin 1966*b*); Ont. to Nfld., N.B., and N.S.; MAPS: on the above-noted map by Clausen; Hultén 1958: map 193, p. 213] . . . . .f. *simplex*

B. virginianum (L.) Sw.   Rattlesnake-Fern

/sT/X/EA/ (Grh) Dry to moist woodlands, thickets, and clearings, the aggregate species from the E Aleutian Is. and sw Alaska (*see* Hultén 1941: map 34, p. 116) to SE Yukon, sw Dist. Mackenzie, Man. (N to Gillam, about 165 mi s of Churchill), Ont. (N to the Severn R. at ca. 55°45'N), Que. (N to James Bay at 52°37'N), southernmost Labrador (near Forteau), Nfld., N.B., P.E.I., and N.S., s to Wash., ?Oreg., Colo. (irregularly to Mexico), S.Dak., Minn., Mich., and Fla.; Eurasia. MAPS and synonymy: *see* below.

1 Sterile blade lax and membranous, the pinnae divided to midrib, the ultimate divisions toothed or lobed, not overlapping; sporangia at most 1 mm thick, the valves recurving widely in dehiscence . . . . .var. *virginianum*

    2 Sterile frond with none of its pinnae bearing sporangia; [*Osmunda* L.; *O. (Botrychium; Botrypus) virginica* L.; *B. gracile* Pursh; incl. var. *intermedium* Butters; transcontinental; MAPS: Clausen 1938: fig. 18 (aggregate species), p. 95; Hultén 1962: map 179 (ssp. *virg.* and the aggregate species), p. 191] . . . . .f. *virginianum*

    2 Sterile frond with some of its pinnae bearing sporangia; [type from near Braeside, Renfrew Co., Ont.] . . . . .f. *anomalum* Cody

1 Sterile blade compact and leathery, the pinnae less deeply divided, the ultimate divisions less toothed or lobed, often crowded and overlapping; sporangia to 1.5 mm thick, the valves less recurving . . . . .var. *europaeum* Angstr.

    3 Sterile frond with none of its pinnae bearing sporangia; [incl. var. *laurentianum* Butters; transcontinental, the common form northwards; MAPS: on the above-noted map by Hultén, and Hultén 1968*b*: 42] . . . . .f. *europaeum*

    3 Sterile frond with some of its pinnae bearing sporangia; [known from the type locality, McKague, Sask., and from Old Factory, s James Bay, Que., at 52°37'N] . . . . .f. *heterodoxum* Cody

OPHIOGLOSSUM L.   Adder's-tongue.   Herbe sans couture

O. vulgatum L.

/sT/X/EA/ (Grh) Damp meadows, sterile pastures, shores, or wet thickets and woods: E Aleutian Is. (Unalaska, the type locality of *O. alaskanum* Britt.); s B.C. (Monashee Pass, E of Vernon) to Mont., Nebr., Minn., Ont. (N to the mouth of the Rainy R. and Cochrane, ca. 49°N), s

Que., N.B., P.E.I., and N.S., s to Ariz., Tex., Mexico, Miss., N.C., and ?Fla.; Iceland; Eurasia. [Incl. vars. *alaskanum* (Britt.) Chr. and *pseudopodum* (Blake) Farw.]. MAPS: Clausen 1938: fig. 23, p. 115; Hultén 1968b:39, and 1962: map 91, p. 101; Meusel, Jaeger, and Weinert 1965:10.

OSMUNDACEAE (Flowering Fern Family)

OSMUNDA L.   Flowering Fern.   Osmonde

Coarse ferns with compound scaleless long-stalked fronds spirally arranged along stout creeping scaleless rhizomes. Blades wholly dimorphic or divided into fertile and sterile parts, the fertile lacking green tissue. Sporangia naked, reticulated, becoming 2-valved by a longitudinal slit but the annulus none or reduced to a cluster of cells near the summit.

1   Fronds wholly dimorphic, the vernal cinnamon-brown densely woolly fertile ones borne
     within a crown of green 1-pinnate-pinnatifid sterile blades, soon withering; pinnae of
     sterile fronds to about 1.5 dm long, bearing a tuft of brownish hairs at base, deeply
     pinnatifid into laxly ciliate segments; (Ont. to Nfld. and N.S.) . . . .*O. cinnamomea*
1   Fronds fertile at summit or middle, sterile elsewhere, glabrous or soon glabrate.
     2   Fronds 1-pinnate-pinnatifid, the inner ones fertile on the median pinnae (these
          becoming blackish); sterile pinnae to about 1.5 dm long, deeply pinnatifid into entire
          segments; (SE Man. to s Labrador, Nfld., and N.S.) . . . .*O. claytoniana*
     2   Fronds 2-pinnate, the pinnae divided into distinct pinnules, the upper fertile pinnae
          forming a terminal greenish (finally brownish) panicle; sterile pinnules frequently
          lobed at the oblique base, otherwise subentire or minutely serrate; (Ont. to Nfld.
          and N.S.) . . . .*O. regalis*

O. cinnamomea L.   Cinnamon-Fern.   Osmonde cannelle
/T/EE/eA/   (Grh)   Damp woods, thickets, and swampy ground from Ont. (N to the Ottawa dist.) to Que. (N to the Côte-Nord), Nfld., N.B., P.E.I., and N.S., s to N.Mex., Tex., Fla., and S. America; (varieties in E Asia). MAP and synonymy: *see* below.

1   Fertile fronds partly leafy; [range of f. *cinnamomea*] . . . . .f. *frondosa* (T. & G.) Britt.
1   Fertile fronds not at all leafy.
     2   Basal lobes of the lower half of the pinnae greatly prolonged and acutely toothed;
          [s Ont.; Soper 1949] . . . . .f. *auriculata* (Hopkins) Kittredge
     2   Basal lobes of pinnae not greatly prolonged.
          3   Lobes of sterile pinnae themselves lobed or toothed; [s Ont.; Soper 1949].
               4   Pinnae-lobes obtuse . . . . .f. *bipinnatifida* Clute
               4   Pinnae-lobes acute . . . . .f. *incisa* (Hunt.) Gilbert
          3   Lobes of sterile pinnae unlobed.
               5   Pinnae-lobes relatively narrow, acutish; [Nfld. and E Que.] . . . . .f. *angustata* Clute
               5   Pinnae-lobes broadly oblong, obtuse; [*O. alata* Goldie; Ont. to Nfld. and N.S.;
                    MAP: Hultén 1962: map 157, p. 167] . . . . .f. *cinnamomea*

O. claytoniana L.   Interrupted Fern
/T/EE/E(fossil)A/   (Grh)   Damp woods and thickets from SE Man. (N to Victoria Beach, about 50 mi NE of Winnipeg) to Ont. (N to Sioux Lookout, about 175 mi NW of Thunder Bay), Que. (N to E James Bay at 53°23′N, L. Mistassini, and the Côte-Nord), s Labrador (N to the Hamilton R. basin), Nfld., N.B., P.E.I., and N.S., s to N Ark. and Ga.; a variety in Asia. [*O. interrupta* Michx.]. MAP: Hultén 1962: map 158, p. 167.

O. regalis L.   Royal Fern
/T/EE/EA/   (Grh)   Low woods and damp or wet ground from Ont. (N to near Thunder Bay) to Que. (N to L. Mistassini, Anticosti Is., and the Gaspé Pen.), Nfld., N.B., P.E.I., and N.S., s to Mexico, Tex., and Fla.; Eurasia. [Incl. var. *spectabilis* (Willd.) Gray (*O. spectabilis* Willd., the N. American plant lacking the black scales on the rachises of the panicle-branches said to characterize the Eurasian plant) and its dwarf extreme, f. *nana* Fern.]. MAPS: Hultén 1958: map 244, p. 263; Polunin 1960: fig. 64, p. 203; Meusel, Jaeger, and Weinert 1965:11.

Forma *anomala* (Farw.) Harris (some branches of the fertile panicle leafy) is known from Ont. (Russell Co.; Cody 1956) and Nfld. (Rouleau 1956).

SCHIZAEACEAE (Curly-grass Family)

SCHIZAEA Sm.   Curly-grass

Densely tufted grass-like fern with short linear spiralling sterile fronds crowded at the base of a short erect fibrous-rooted rhizome. Fertile fronds much longer (to about 12 cm), terminated by a 1-sided fruiting portion of up to 8 obliquely ascending pinnae to 4 mm long. Sporangia in a double row along the single vein of the pinnae, with a subapical transverse annulus and opening by a vertical slit.

S. pusilla Pursh
/T/EE/   (Hr)   Hummocky bogs, low mossy open woods, and ledgy shores: s Ont. (apparently now extinct at Sauble Beach, Bruce Co.); N.S. (to the N.S. map by Roland 1947: map 14, p. 145, add a dot for Guysborough Co.); Nfld.; St-Pierre and Miquelon; Long Island, N.Y.; Pine Barrens of N.J. [*S. filifolia* La Pylaie]. MAPS: *Atlas of Canada* 1957: map 14, sheet 38; Fernald 1933: map 7, p. 86; Roland 1941:109.

HYMENOPHYLLACEAE (Filmy Fern Family)

MECODIUM Presl

Ferns with delicate translucent pinnate-pinnatifid fronds to 5 cm long composed (except near the veins) of a single layer of cells. Veins free, the sori marginal at their ends. Indusia 2-valved to base. Stipes to 1.5 cm long, blackish, glabrous except for a small tuft of hair-like scales at base. Rhizomes thread-like, extensively creeping and branching.

M. wrightii (Bosch) Copeland
/T/W/eA/ (Grh) This species, the first representative of the Filmy Fern Family ever found in w N. America (*Trichomanes boschianum* Sturm occurs in the SE U.S.A.), was first discovered in B.C. in 1957 by Persson (shady vertical cliffs of Graham Is., Queen Charlotte Is.; *see* Hermann Persson, Bryologist 61(4): 359–61. 1958). It was later reported by T.M.C. Taylor (Am. Fern J. 57(1): 1–6. 1967; *see* his Alaska–B.C. MAP, pl. 1, p. 2) from the s Alaska Panhandle (Biorka Is. at 56°54'N; gametophytes only) and from additional wet rock and crevice habitats of Queen Charlotte Is. (Graham, Chaatl, and Moresby islands) and the B.C. mainland coast (12 and 20 miles E of Prince Rupert at ca. 54°15'N; gametophytes only). Male gametophytes growing on bark and decaying wood of Sitka spruce were recently reported from w Vancouver Is. by L. D. Cordes and V. J. Krajina (Am. Fern J. 58(4): 181. 1968). The species is otherwise known only from E Asia (Sakhalin, Japan, and s Korea). [*Hymenophyllum* Bosch]. *See* Kunio Iwatsuki (Am. Fern J. 51(3): 141–44. 1961). MAPS: Hultén 1968*b*: 43; T.M.C. Taylor 1970: 33 (N. American area).

POLYPODIACEAE (Fern Family)

(Ref.: Roland 1941; Cody 1956; T.M.C. Taylor 1963; Marie-Victorin 1923)

Leafy plants from erect or creeping rhizomes. Fronds (leaf-like blades) stalked, uncoiling from the tip (circinate in vernation), simple to decompound, bearing on their backs or margins dots, lines, or variously shaped clusters (sori) of spore-cases (sporangia). Sori either naked or partly or completely covered or surrounded by a membrane (the indusium) or by the reflexed margin (false indusium) of the frond or its segments. Sporangia at maturity opening elastically on the inner side by a transverse slit produced by the drying-out tension of a vertical incomplete ring of specialized cells (the annulus) on the outer side.

1   Fronds entire or merely undulate, deeply cordate at base, evergreen; sori elongate, their indusia attached by one side to the lateral veins.
  2   Blades narrowly triangular-lanceolate, to about 3 cm broad at base, tapering to prolonged slender rooting tips; stipe naked except at the scaly base; inner sori straight, the outer tending to converge in V-shaped pairs along the netted veins; outer veinlets free; (Ont. and s Que.) . . . .*Camptosorus*
  2   Blades tongue-shaped, to about 3.5 cm long and 4 cm broad above the middle, rather abruptly narrowed to apex; stipe densely scaly; sori to 2 cm long, straight; veins free; (s Ont.) . . . .*Phyllitis*
1   Fronds pinnatifid or pinnate (sometimes ternate).
  3   Fertile and sterile fronds dissimilar in length or appearance, or both.
    4   Sori enclosed in subglobose bead-like segments of the much modified and contracted, narrowly lanceolate or oblanceolate, fertile fronds, these shorter than the sterile fronds and becoming blackish in age.
      5   Fronds scattered along the creeping rootstock, the sterile ones net-veined, to 4 dm long or more, broadly triangular in outline, deeply 1-pinnatifid (the rachis broadly winged except at base), the lower segments moderately to deeply lobed; fertile fronds 2-pinnate, their ultimate subglobose segments distinct; (Man. to s Labrador, Nfld., and N.S.) . . . .*Onoclea*
      5   Fronds borne in a crown, the free-veined pinnate-pinnatifid sterile ones to 2 m long or more, oblong-lanceolate, abruptly tapering to apex, more gradually tapering to base, forming a circle around the 1-pinnate fertile ones; fertile pinnae moniliform by confluence of the ultimate segments; (transcontinental) . . . .
. . . .*Matteuccia*
    4   Sori not enclosed in bead-like segments; fertile fronds longer than the sterile ones.
      6   Indusia marginal and soon confluent as a marginal band more or less covering the sori.
        7   Pinnae sessile (their broad bases mostly confluent except toward the base of the fertile fronds), entire, the fertile ones mostly narrowly linear, the sterile ones mostly narrowly lanceolate; fronds linear to linear-oblanceolate, tapering at both ends, deeply 1-pinnatisect, the sterile ones numerous, evergreen, short-stipitate, in a circular crown, to about 1 m tall, their strongly scaly stipes to 3 dm long; indusium true, often finally reflexed; (Alaska–B.C.) . . . .*Blechnum*
        7   Pinnae either short-stalked or their bases at least rounded and not confluent; fertile fronds at most about 3 dm tall, their blades lanceolate to deltoid in outline; indusium false (formed by the reflexed margin of the frond).
          8   Pinnae (and pinnules, if present) of frond jointed at base; fronds 1-pinnate, or often 2-pinnate near base; fertile fronds lanceolate to narrowly ovate; stipes purplish brown, mostly shorter than the blades; rhizome short, erect . . . .*Pellaea*
          8   Pinnae and pinnules not jointed at base; stipes mostly longer than the 2–3-pinnate, ovate to deltoid-ovate blade.
            9   Reflexed indusial margins yellowish or greenish, herbaceous or

barely scarious; pinnules of sterile fronds shallowly dentate or crenate . . . . *Cryptogramma*

    9 Reflexed indusial margins whitish and scarious; pinnules of sterile fronds sharply serrate or incised-serrate; fertile fronds 3-pinnate, their pinnules linear-lanceolate to narrowly oblong; stipes densely tufted from a short, ascending, rigid, reddish brown to purplish rhizome; (*C. siliquosa*) . . . . *Cheilanthes*

6 Indusia attached to the lateral veins and therefore neither marginal nor false.

    10 Sterile blades net-veined (only the short outer veinlets free), deltoid-ovate in outline, to about 2 dm long, very deeply 1-pinnatifid, with up to 10 pairs of lanceolate, minutely serrate segments; fertile blades longer and narrower, with almost separate linear divisions; sori linear to linear-oblong; stipes usually longer than the blades; rhizomes long-creeping; (*W. areolata*; N.S.) . . . . *Woodwardia*

    10 Sterile blades free-veined, narrower in outline; pinnae serrate, the fertile not much narrower than the sterile; stipes usually shorter than the blades; rhizomes short-creeping.

        11 Indusium narrowly elliptic, fragile, attached by its lower edge; fronds 1-pinnate, linear-oblanceolate, tapering at both ends but more gradually toward base; smaller pinnae asymmetrically deltoid or ovate about their midribs; larger pinnae lanceolate to narrowly oblong, auricled on the upper edge near base; sterile blades spreading, commonly less than 1 dm long; fertile blades stiffly erect, to 4.5 dm long and 4.5 cm broad; stipes dark brown and shining; (*A. platyneuron*; s Ont. and s Que.) . . . . *Asplenium*

        11 Indusium reniform, rather firm, centrally attached; fronds 1-pinnate-pinnatifid, lanceolate to narrowly oblong, moderately narrowed toward base, gradually acuminate to apex; fertile blades to 7.5 dm long and 3 dm broad, their pinnae often twisted on the rachis in a subhorizontal plane; stipes with pale-brown scales to 2 cm long; (*D. cristata*; transcontinental) . . . . *Dryopteris*

3 Fertile and sterile fronds nearly identical in both appearance and length (but upper fertile pinnae distinctly reduced in *Polystichum acrostichoides*; and fertile pinnae with revolute margins in *Thelypteris palustris*).

    12 Blade of frond deeply 1-pinnatifid (the lanceolate to narrowly oblong or deltoid-oblong, mostly entire segments confluent at base), truncate at base; sori large, round, lacking an indusium . . . . *Polypodium*

    12 Blade 1–3-pinnate or 1–3-pinnate-pinnatifid, at least the primary segments distinct.

        13 Fronds partly net-veined, the oblong areolae forming a usually single-layered row along each side of the midveins of the pinnae and their entire or minutely serrate lobes, the long outer veinlets free; blades 1-pinnate-pinnatifid, scarcely reduced at base; indusia oblong, finally reflexed, one on each areole . . . . *Woodwardia*

        13 Fronds entirely free-veined.

            14 Stipe forking at summit into 2 or 3 primary rachises (*Gymnocarpium dryopteris* may be sought here); indusium (false) formed by the somewhat modified revolute margins of the ultimate segments; rhizomes long-creeping.

                15 Sori oblong, whitish, distinct, to about 5 mm long; stipe purplish black, 2-forked at summit, the primary subrectangular branches recurved and each bearing up to 9 pinnae (with up to 25 pairs of pinnules) on the upper side only, the whole blade thus typically broadly fan-shaped; (*A. pedatum*; Alaska–B.C.–Alta.; Ont. to Nfld. and N.S.) . . . . *Adiantum*

                15 Sori mostly continuous as a brownish marginal band; stipe brownish,

2–3-forked at summit into pinnate-pinnatifid pinnules on both sides of
the rachises; (transcontinental) . . . .*Pteridium*

14 Stipe simple (or 3-forked in *Gymnocarpium dryopteris*, with small round
naked sori and delicate stipe), continuing into a single main rachis.

16 Indusium none or soon shrivelling; sori dorsal upon the veins (rarely
marginal); blade of frond triangular-ovate to broadly deltoid or deltoid-
pentagonal, broadest toward base, nearly as broad as or broader than
long; stipe usually longer than the blade.

17 Sori following the course of the branched veinlets throughout,
usually confluent; fronds clustered on a stoutish, short-creeping or
ascending rhizome, the blades to about 1.5 dm long and broad,
deltoid-pentagonal in outline, covered beneath with a white to
deep-yellow powder, the pinnae all sessile; stipe stout, dark brown,
glossy, about twice as long as the blade, naked except for
a few small rigid brownish scales at base; (sw B.C.) . . . .*Pityrogramma*

17 Sori borne singly on the veinlets; rhizome slender, long-creeping.

18 Fronds with the lowest pair of opposite pinnae and the remaining
terminal segment long-stalked; rachis not winged, only the
reduced upper pinnae or segments confluent; (transcontinental
species) . . . .*Gymnocarpium*

18 Fronds with all of the principal segments sessile, all of the pinnae
except often the lowest pair confluent by a narrow wing along
the rachis . . . .*Thelypteris*

16 Indusium present (or rudimentary in *Athyrium distentifolium*).

19 Sori marginal.

20 Indusium shield-shaped and centrally attached, covering the
top of the sorus; blades firm, glabrous, 2-pinnate, the pinnules
entire, rounded at apex (or those of the lower pinnae often
somewhat pinnatifid); fronds to about 1 m long; rhizome
covered with long scales, stout, ascending; (*D. marginalis*;
Ont. to N.S.) . . . .*Dryopteris*

20 Indusium cup-like or formed of the revolute margin of the
pinnules.

21 Indusium cup-like, whitish, opening on the outer side at top;
blades soft, minutely glandular and pilose, sweet-scented when
dry, the acutish or obtuse pinnules incised; rhizome
lacking scales, slender and long-creeping; (Ont. to Nfld.
and N.S.) . . . .*Dennstaedtia*

21 Indusium formed of the revolute thinnish margins of the
pinnules; blades 2–3-pinnate; rhizomes bearing scales.

22 Sori clearly separate, borne on the under side of sharply
reflexed membranous lobes; fronds lax, to 5 dm long, their
pinnules fan-shaped or rhomboid; rhizomes elongate and
creeping; (*A. capillus-veneris*; introd. in s B.C.) . . . . . .*Adiantum*

22 Sori continuous, not borne on the back of reflexed lobes;
fronds firm or rigid, to about 3 dm long, their pinnules
narrowly lanceolate to broadly oblong or roundish;
rhizome multicipital, with numerous short tufted
branches; (mts. of B.C.–Alta.; *C. siliquosa* also in s Ont.
and E Que.) . . . .*Cheilanthes*

19 Sori borne on the back of the frond; rhizomes with chaffy scales.

23 Indusium borne below and surrounding the sorus, deeply lacerate
into ascending hair-like (sometimes scale-like) lobes; fronds
1–2-pinnate-pinnatifid, mostly less than 2 dm long, densely tufted
on a short ascending rhizome, in some species deciduous by a
jointed stipe . . . .*Woodsia*

23 Indusium attached laterally or centrally, not obviously lacerate; stipe never jointed.

    24 Indusia linear or elongate, with a linear attachment to the upper side of a veinlet along one edge, opening inward toward the midvein along the other edge.

        25 Indusium nearly straight and not crossing the vein; fronds mostly firm or evergreen, their blades mostly 1–2-pinnate; rhizome mostly short-creeping, its firm scales composed of short broad thick-walled cells . . . .*Asplenium*

        25 Indusium straight to hooked or horseshoe-shaped and crossing the vein; fronds deciduous; rhizome suberect to long-creeping, its thin scarious scales composed of elongate narrow cells . . . .*Athyrium*

    24 Indusia either orbicular to reniform and centrally attached or hood-like and laterally attached.

        26 Indusium hood-like, soon withering, laterally attached, its open end at right angles to the veinlet and facing toward a tooth or sinus; rhizome long-creeping . . . .*Cystopteris*

        26 Indusium orbicular or reniform, centrally attached and covering the sorus, opening along its whole circumference.

            27 Indusia orbicular, peltate, lacking a sinus; fronds 1–2-pinnate, spinulose-toothed (except in *P. aleuticum*), coriaceous and evergreen, tufted on a stout rhizome; stipe and frond-rachis usually densely chaffy . . . .*Polystichum*

            27 Indusia reniform, attached slightly off centre at the tip of the single deep sinus; fronds 2-pinnate to 2-pinnate-pinnatifid, their teeth at most bristle-tipped.

                28 Rhizomes slender, long-creeping (relatively stout and short-creeping in *T. limbosperma*), dark brown to black, their scales commonly cilitate; veins reaching the often-ciliate margins of the pinnules; fronds annual, their upper surfaces with acicular unicellular hairs on the veins . . . .*Thelypteris*

                28 Rhizomes stout, suberect or short-creeping, their scales not ciliate; veins ending short of the margin in elongate hydathodes; fronds annual or evergreen, lacking acicular hairs on the midvein above . . . .*Dryopteris*

ADIANTUM L. Maidenhair. Adiante

1 Stipe not forked, continuous with the rachis of the frond-blade; blade narrowly ovate, longer than broad (to about 1.5 dm broad), the ultimate segments fan-shaped or rhomboid; rhizome-scales less than 0.5 mm broad; [s B.C., where probably introd.] . . . .*A. capillus-veneris*

1 Stipe equally 2-forked (rarely 3-forked) at summit, the secondary branches borne along the upper side of the 2 curved rachises, the blade thus typically reniform in outline, to about 4 dm broad; ultimate segments subrectangular; rhizome-scales 1 or 2 mm broad; [Alaska, B.C., and sw Alta.; Ont. to N.S. and Nfld.] . . . .*A. pedatum*

A. capillus-veneris L. Venus'-hair-Fern

This species, a native of the s U.S.A. and subtropical parts of both hemispheres, is known as a common escape from cult. in greenhouses as far N as the N U.S.A. According to Eastham (1947), it has been known since about 1915 as abundant around salt-springs at Fairmont Hot Springs, on the w slope of the Rocky Mts. of the Columbia Valley, SE B.C., and T.M.C. Taylor (1963) also reports it from Saltspring Is., off SE Vancouver Is. Further studies are necessary before it can be accepted as native in these localities, the nearest stations being some hundreds of miles south in Baja Calif. and Utah. MAP: Hultén 1962: map 139, p. 149. The "Pacific Northwest" map by T.M.C. Taylor (1970:93) indicates the above Fairmont Hot Springs station.

A. pedatum L.   Maidenhair Fern.   Capillaire
/sT/D/eA/   (Grh)   Rich woods and rock-crevices (ranges of Canadian taxa outlined below), s to Calif., Utah, Okla., La., Miss., and Ga. MAP and synonymy: *see* below.
1   Pinnae strongly recurved, their green pinnules with rounded teeth; sori linear or short-oblong, usually more than twice as long as broad; stipes to about 5 dm long, from a horizontal rhizome; [incl. var. *rangiferinum* Burgess; Ont. (N to Algonquin Park, w to Pembroke) to Que. (N to Ste-Angèle, Matane Co.; *see* Que. map by Doyon and Lavoie 1966: fig. 22, p. 820), N.B., and N.S. (apparently extinct in P.E.I.)] . . . . .var. *pedatum*
1   Pinnae ascending, their bluish-green pinnules relatively deeply cleft, the teeth more acute; sori markedly crescent-shaped, rarely over twice as long as broad; stipes very stiff, glaucous, commonly not over 2 dm long, from a suberect rhizome; [Aleutian Is.–s Alaska (*see* Hultén 1941: map 26, p. 115; type material from Unalaska and Kodiak) to B.C. (N to Prince Rupert; *see* B.C. map by T.M.C. Taylor 1963:157) and the mts. of sw Alta. (Waterton Lakes); w Ont. (Fernald *in* Gray 1950); Que. (Megantic, Stanstead, and Gaspé counties); Nfld.; NE Asia; MAP: Hultén 1968*b*:42] . . . . .var. *aleuticum* Rupr.

ASPLENIUM L.   Spleenwort.   Doradille

1   Fronds dimorphic, 1-pinnate, narrowly oblanceolate, the minutely serrate pinnae auricled on the upper margin near base, the lower ones broadly deltoid, the upper narrowly oblong; sterile fronds spreading, to about 1 dm long, with up to 22 pairs of pinnae at most about 1 cm long; fertile fronds stiffly erect, to about 6 dm long, with up to 50 pairs of pinnae to about 6 cm long; stipe and rachis chestnut-purple, shining; (s Ont. and s Que.) . . . . . .
. . . .*A. platyneuron*
1   Sterile and fertile fronds essentially alike; pinnae at most about 12 mm long, broadly oblong to round-ovate or rhombic, shallowly few-toothed or lobed.
  2   Blades 2-pinnate or 2-pinnate-pinnatifid, ovate or rhombic in outline, the 4–8 primary segments alternate, long-stalked, each with up to 7 short-stalked rhombic pinnules; rachis and stipe green throughout; (s Ont.) . . . .*A. ruta-muraria*
  2   Blades 1-pinnate, linear, with up to 20 or more pairs of sessile or subsessile pinnae; (essentially transcontinental).
    3   Rachis and stipe purplish brown and shining; fronds to about 2.5 dm long; pinnae oval to round-oblong . . . .*A. trichomanes*
    3   Rachis green; stipe green above, brown at base; fronds to about 1.5 dm long; pinnae round-ovate or rhombic-ovate . . . .*A. viride*

A. platyneuron (L.) Oakes   Ebony-Spleenwort
/T/EE/   (Hr.)   Open woods, rocky banks, or crevices of ledges from Kans. to s Ont. (N to Georgian Bay, L. Huron; a collection in OAC from Carleton Co. has been placed here but the species is not listed for the Ottawa dist. by Cody 1956, or Gillett 1958) and sw Que. (N to Montreal), s to E Tex. and Fla. [*Acrosticum* L.; *Aspl. ebeneum* Ait.]

A. ruta-muraria L.   Wall-Rue
/T/EE/EA/   (Hr)   Calcareous cliffs and ledges from N Mich. to s Ont. (Manitoulin Is. and Bruce Pen., L. Huron; *see* s Ont. map by J. H. Soper, Am. Fern J. 45(3):100. 1955) and Vt., s to Ark., Tenn., and Ala.; Eurasia. [Incl. var. *cryptolepsis* (Fern.) Wherry (*A. crypt.* Fern.), differing from the typical Eurasian form in a few minor and intergrading characters]. MAP (*A. crypt.*): Fernald 1935: map 5, p. 209.

A. trichomanes L.   Maidenhair-Spleenwort
/T/X/EA/   (Hr)   Shaded, often calcareous, rock-crevices from SE Alaska, B.C., and Alta. (N to the Peace River dist.; reports from Man. require confirmation) to Ont. (N to the N shore of L. Superior),

Que. (N to Anticosti Is.), Nfld. (20 mi N of Corner Brook; W. J. Cody, Am. Fern J. 58(4):179. 1968), N.B., and N.S. (not known from P.E.I.), s to Oreg., Ariz., Colo., N.Dak., Wisc., Tex., Ala., and Ga.; Iceland; Eurasia. [*A. melanocaulon* Willd.]. MAPS: Hultén 1968*b*:46, and 1962: map 130, p. 139; Meusel, Jaeger, and Weinert 1965:13.

A. viride Huds.   Green Spleenwort
/aST/X/GEA/   (Hr)   Shaded calcareous rocks: cent. Alaska–Yukon–sw Dist. Mackenzie through B.C. and Alta. (reports from Sask. require confirmation) to Wash., Nev., Utah, and Colo.; Black Hills of S.Dak.; Mexico; Ont. (N to the SE end of L. Superior) to Que. (James Bay and Hudson Bay N to ca. 55°20′N; L. Mistassini; Côte-Nord; Reed Mt., ca. 52°N, 68°W; Rimouski Co.; Anticosti Is.; Gaspé Pen.), Nfld., N.B. (Gloucester, Restigouche, and St. John counties; not known from P.E.I.), and N.S., s to N.Y. and Vt. (reported from Penn.); w Greenland N to ca. 62°N, E Greenland N to ca. 68°N; Iceland; Eurasia. [Possible basis of the report of the European *A. marinum* L. from N.B. by Hooker 1840]. MAPS: Hultén 1968*b*: 47, and 1962: map 92, p. 101; Meusel, Jaeger, and Weinert 1965:13.

ATHYRIUM Roth

1   Fronds 1-pinnate, elongate-lanceolate, the sterile ones with linear acute pinnae, the
      fertile ones with lance-linear pinnae; indusia linear, in 2 rows oblique to the midrib of the
      pinna; (Ont. and s Que.) . . . .*A. pycnocarpon*
1   Fronds pinnate-pinnatifid or 2–3-pinnate.
      2   Fronds pinnate-pinnatifid, elliptic-lanceolate, tapering to apex, the basal pinnae only
            slightly reduced, the pinnae all deeply pinnatifid, their lobes entire; each lobe of the
            fertile pinnae bearing 2 rows of whitish or silvery linear-oblong indusia oblique to the
            midvein; (Ont. to N.S.) . . . .*A. thelypteroides*
      2   Fronds 2–3-pinnate; basal pinnae much reduced.
            3   Indusia obsolete or rudimentary; sori roundish; fronds elliptic-lanceolate to
                  lance-ovate, commonly 1/3 as broad as long or broader; (s Alaska–B.C.–w Alta.; E
                  Que. and Nfld.) . . . .*A. distentifolium*
            3   Indusia mostly crescent- or horseshoe-shaped; fronds narrowly to broadly
                  laceolate; (transcontinental) . . . .*A. filix-femina*

A. distentifolium Tausch
/aST/D/GEA/   (Hr)   Wet rocky slopes or subalpine meadows and open woods. Ranges, MAPS, and synonymy: *see* below.
1   Fronds 2-pinnate to somewhat 3-pinnatifid, their ultimate segments oblong, mostly
      broad-based and approximate; sori median or submedian, the larger ones to 1.4 mm
      broad; [*Aspidium (Athyrium; Phegopteris) alpestre* Hoppe, not *Ath. alp.* Clairv.; NW Nfld.
      (M. L. Fernald, Rhodora 30(351): 48. 1928); sw Greenland and E Greenland N to ca. 66°N
      (Hultén 1958); Iceland; Eurasia: MAPS: Hultén 1958: map 223 (*A. alp.*), p. 243 (also noting
      a 1955 map by Saxer), and 1968*b*: 47] . . . . .var. *distentifolium*
1   Fronds more copiously dissected, their ultimate segments linear to linear-lanceolate,
      distant; sori submarginal, mostly not over 0.8 mm broad; [*A. alpestre* var. *americanum*
      Butters (*A. amer.* (Butters) Maxon), the lectotype from Rogers Pass, B.C.; incl. *A. alp.*
      var. *gaspense* Fern.; s Alaska through B.C. and sw Alta. to Calif., Nev., and Colo.; E Que.
      (Shickshock Mts. of the Gaspé Pen., Labrador (N to ca. 56°N), and ?Nfld.; E Asia; MAPS:
      on the above-noted maps by Hultén] . . . . .var. *americanum* (Butters) Boivin

A. filix-femina (L.) Roth   Lady-Fern.   Fougère femelle
/aST/X/EA/   (Hr)   Damp thickets, meadows, swamps, and subalpine slopes (ranges of Canadian taxa outlined below), s to Calif., Idaho, S.Dak., Mo., and Va. MAPS and synonymy: *see* below.
1   Rhizome usually erect or ascending, the young growth surrounded by the older fronds;
      stipe commonly less than 1/3 the length of the frond, its chaffy scales to 3 mm broad; sori
      less than 1 mm long, their indusia usually long-ciliate; [*Polypodium* L.; *Nephrodium*
      Michx.; *Aspidium* Sw.; *Asplenium* Bernh.; Eurasia only; MAP: Hultén 1962: map 168A

(this and his map 168B also indicating the area of the aggregate species)] . . . . .var. *filix-femina*
1   Rhizome horizontal or nearly so, the young growth appearing at the end in advance of the bases of the older fronds; stipe often half as long as the frond, its scales at most 1.5 mm broad; sori mostly over 1 mm long, their indusia usually toothed or with a few short cilia.
    2   Fronds broadest near middle, mostly about 3 times as long as the stipe; basal chaffy scales of stipe often pale, at least 1 cm long; indusia mostly less than 1 mm long, often as broad as long, long-ciliate; [*A. cyclosorum* Rupr.; incl. the immature phase, f. *hillii* (Gilbert) Butters and the extreme sun-form with narrow revolute-margined pinnules, f. *strictum* (Gilbert) Butters; Aleutian Is., Alaska (type from Sitka), the Yukon, and B.C.; E Que. (Shickshock Mts., Gaspé Pen.) and NW Nfld.; MAPS: Hultén 1962: map 168A, p. 179, and 1968*b*:48 (ssp. *cyclosorum*); Fernald 1933: map 13, p. 129] . . . .
    . . . . .var. *sitchense* Rupr.
    2   Fronds broadest below middle, mostly about twice as long as the stipe; basal scales often blackish, rarely over 6 mm long; indusia mostly about 1 mm long and about half as broad, usually short-ciliate; [SE Man. to S James Bay, Labrador (N to the Hamilton R. basin), Nfld., N.B., P.E.I., and N.S.; MAPS (*see* f. *michauxii*)] . . . .
    . . . . .var. *michauxii* (Spreng.) Farw.
        3   Fronds dimorphic, the fertile ones more coriaceous than the sterile and somewhat contracted or plaited lengthwise; sori confluent at maturity.
            4   Pinnae of fertile fronds not over 12 cm long, their pinnules simple, at most 12 mm long; sori straightish, not crossing the adjacent vein; pinnules of sterile frond only slightly toothed or lobed, obtuse; [ssp. *angustum* (Willd.) Hult. (*Aspidium angustum* Willd.); *Ath. ang.* (Willd.) Presl and its var. *boreale* Jennings; *Asplenium michauxii* Spreng.; MAPS (aggregate species): Hultén 1962: map 168A, p. 179; Meusel, Jaeger, and Weinert 1965:15; Meusel 1943: fig. 30c] . . . . .f. *michauxii*
            4   Pinnae of fertile fronds to about 2 dm long, their pinnatifid pinnules to about 1.5 cm long; sori often strongly curved; pinnules of sterile fronds strongly toothed or pinnatifid, acutish . . . . .f. *elatius* (Link) Clute
        3   Fronds neither strongly dimorphic, coriaceous, nor plaited; sori mostly not confluent at maturity.
            5   Pinnules oblique to the pinna-rachis, their bases prominently decurrent but not connected by a wing, their teeth acute . . . . .f. *elegans* (Gilbert) Clute
            5   Pinnules widely to subhorizontally divergent from the pinna-rachis.
                6   Pinnules varying irregularly in size and irregularly toothed or lobed, joined by a broad rachis-wing, their broad lobes overlapping; [*A. angustum* f. *con.* Butters, the type from Tabletop Mt., Gaspé Pen., E Que.] . . . .
        . . . . .f. *confertum* (Butters) Fern.
                6   Pinnules regularly diminishing in size toward tip of pinna.
                    7   Pinnules lanceolate, subacute, regularly and coarsely toothed or pinnatifid, their segments toothed, their rachises scarcely winged; [incl. *A. angustum* var. *glanduliferum* Jennings] . . . .f. *rubellum* (Gilbert) Farw.
                    7   Pinnules oblong, obtuse, only obscurely toothed, their rachises strongly winged; [*A. angustum* var. *laur.* Butters, the type of Tabletop Mt., Gaspé Pen., E Que.] . . . . .f. *laurentianum* (Butters) Fern.

A. pycnocarpon (Spreng.) Tidestr.   Glade-Fern
/T/EE/   (Hr)   Rich (mostly calcareous) woods and ravines from Minn. to S Ont. (N to the Ottawa dist.) and SW Que. (N to the Montreal dist.), S to E Kans., La., Ala., and Ga. [*Asplenium* Spreng.; *Diplazium* Broun; *Asplenium (Athyrium) angustifolium* Michx.]. MAP: the restricted Canadian area is indicated in a map by Raymond 1950*b*: fig. 33, p. 85.

A. thelypteroides (Michx.) Desv.   Silvery Spleenwort
/T/EE/   (Hr)   Rich woods and shaded slopes from Minn. to Ont. (N to Current R., near Thunder Bay), E Que. (N to Mt. Nicolabert, Gaspé Pen.), N.B., P.E.I., and N.S., S to E Mo., La., and Ga. [*Asplenium* Michx.; *Diplazium* Presl].

Forma *acrostichoides* (Sw.) Gilbert (*Asplenium (Athyrium; Diplazium; Nephrodium) acrost.* Sw.; ultimate segments of the fronds coarsely toothed and rather gradually tapering to the blunt or acutish tip, rather than minutely toothed and with sides nearly parallel, the tip rounded) occurs throughout the area.

BLECHNUM L.

B. spicant (L.) Sm.   Deer-Fern
/sT/W/EA/   (Hr)   Swamps, moist woods, cliffs, and springy banks from the Aleutian Is. (Atka Is.) and s-cent. Alaska (*see* map by Hultén 1941: map 23, p. 114) through coastal B.C. (*see* B.C. map by T.M.C. Taylor 1963:158; also known from Revelstoke) to Calif.; Iceland; Eurasia; N Africa. [*Osmunda* L.; *Lomaria* Desv. and its var. *elongata* Hook.; *Struthiopteris* Weiss; *Osmunda (Blechnum; Lomaria) borealis* Salisb.]. MAPS: Hultén 1968b:58, and 1962: map 143, p. 153; Meusel, Jaeger, and Weinert 1965:12; Fernald 1929:1488; T.M.C. Taylor 1970:113 ("Pacific Northwest" area).
Forma *bipinnatum* Clute (the frond-segments themselves pinnatifid) is reported from the type locality, Vancouver Is., B.C., by W. N. Clute (Fern Bull. 15: 19. 1907).

CAMPTOSORUS Link

C. rhizophyllus (L.) Link   Walking Fern.   Fougère ambulante
/T/EE/   (Hr)   Shaded rocks and boulders (often calcareous) from Minn. to s Ont. (N to Manitoulin Is., N L. Huron, and the Ottawa dist.; *see* s Ont. maps by J. H. Soper, Bull. Fed. Ont. Nat. 98:16. 1962, and Am. Fern J. 53(3): fig. 7, p. 112. 1963), s Que. (N to the Montreal dist.; tentatively reported as far N as Sorel by Marie-Victorin 1923), and w-cent. Maine, s to Okla., Ark., and Ga. [*Asplenium* L.]. MAPS: the restricted Canadian distribution is shown in maps by Rouleau 1945: fig. 1, p. 22, and Raymond 1950b: fig. 32, p. 85.
Forma *auriculata* Hoffm. (basal auricles of at least some of the fronds divergent, acute to long-tapering and sometimes, like the frond-tips, themselves rooting, is reported from sw Que. by Marcel Raymond (Contrib. Inst. Bot. Univ. Montréal 48:72. 1943; St-Armand, Missisquoi Co).

CHEILANTHES Sw.

1   Frond and stipe glabrous, the frond broadly ovate to deltoid-oblong or subpentagonal, to about 8 cm long and 5 cm broad; indusia narrowly linear, joined in a continuous line to the tightly reflexed margins of the frond, the sori confluent; fertile and sterile fronds somewhat dissimilar, the fertile ones long-stalked, to 3 dm long, the sterile ones smaller and more dissected, soon darkening or shrivelling; (mts. of B.C.; s Ont. and s and E Que.) . . . .*C. siliquosa*
1   Frond densely hairy beneath with long brownish or rusty hairs, narrower in outline, to about 13 cm long; stipe at first pilose, soon glabrate; indusia formed of the scarcely modified recurved or reflexed tips or margins of the ultimate frond-segments, these roundish or oval and somewhat bead-like; fertile and sterile fronds similar; (western species).
　　2   Fronds thinly villous above the lax flexuous white hairs, lacking scales, linear-oblong to ovate or deltoid-ovate; indusia slightly interrupted, concealed by the dense pubescence of the under surface; (mts. of B.C.–Alta.) . . . .*C. feeii*
　　2   Fronds with a few minute stellate scales, otherwise glabrous above, linear to oblong-lanceolate; indusia continuous, not concealed; (mts. of B.C.) . . . .*C. gracillima*

C. feeii Moore   Slender Lip-Fern
/T/WW/   (Hr)   Dry crevices of calcareous cliffs and ledges from s B.C. (*see* B.C. map by T.M.C. Taylor 1963:158) and sw Alta. (N to Banff and Morley) to s Calif., N Mexico and Tex. [*C. lanuginosa* Nutt.; *C. vestita sensu* Hooker 1840, not *Adiantum vestitum* Spreng., basionym]. MAP: T.M.C. Taylor 1970:115 ("Pacific Northwest" area).

C. gracillima Eat.   Lace-Fern
/t/W/   (Hr)   Exposed crevices of cliffs and ledges from s B.C. (N to Spences Bridge in the Fraser

Valley; Henry 1915; *see* B.C. map by T.M.C. Taylor 1963:158) and sw Alta. (a 1970 collection from Waterton Lakes, detd. T.M.C. Taylor) to s Calif. and Mont. MAP: T.M.C. Taylor 1970:117 ("Pacific Northwest" area).

C. siliquosa Maxon   Indian's Dream
/T/D/   (Hr)   Exposed crevices of cliffs and ledges: s B.C. (N to Shuswap L. E of Kamloops; *see* B.C. map by T.M.C. Taylor 1963:158; a 1901 report by Clute from Alaska is considered by Hultén 1941 to probably refer to *Cryptogramma crispa* var. *acrostichoides*) to s Calif., N Utah, and NW Wyo.; s Ont. (Durham, Bruce Pen., L. Huron, where taken in 1883 by Ami and probably now extinct); serpentine formations of Que. (Black Lake, Megantic Co.; Mt. Albert, Gaspé Pen.). [*Onychium (Aspidotis; Cheil.; Cryptogramma; Pellaea) densum* Brack.]. MAPS: Wynne-Edwards 1937: map 2, p. 24; Fernald 1925: map 8, p. 251.

CRYPTOGRAMMA   R. Br.   Rock-Brake

1   Rhizomes slender but short and erect, their scales brown with dark centres, the fronds clustered, thick and opaque; sterile fronds 2-pinnate or 2-pinnate-pinnatifid, their cuneate-elliptic segments 4 or 5 mm long; fertile fronds 2–4-pinnate, their linear segments to about 12 mm long; stipes pale brown, firm; (Alaska–B.C. to Ont.) . . . .*C. crispa*
1   Rhizomes slender and long-creeping, their pale-brown ovate-lanceolate scales not over 2 mm long, the fronds scattered and remote, thin, translucent; sterile fronds 2-pinnate, their fan-shaped segments to about 12 mm long; fertile fronds 2–3-pinnate, their narrowly lanceolate to narrowly oblong segments to about 2 cm long; stipes greenish or brownish, fragile; (transcontinental) . . . .*C. stelleri*

C. crispa (L.) R. Br.   Mountain-Parsley
/aST/(X)/EA/   (Hr)   Barren rocky soil, cliffs, ledges, and talus slopes, the aggregate species from the Aleutian Is., cent. Alaska, and SE Yukon through B.C. and Alta. to Baja Calif. and N.Mex., eastwards locally to Great Bear L., N Alta., Sask. (N to L. Athabasca), and Ont. (N to Sandy L. at ca. 53°N), s to cent. Sask., SE Man. (Lake of the Woods), L. Superior (Mich. and Ont.), and s Ont. (Manitoulin Is., N L. Huron); S. America; Iceland; Eurasia. MAPS and synonymy: *see* below.
   The inclusion of Dist. Keewatin in the range given by Fernald *in* Gray (1950) is probably based upon collections taken before 1912, at which date the boundaries of Manitoba were extended to embrace about half of the area of the former Dist. Keewatin. The reports of *C. acrostichoides* from Labrador and Ungava by Marie-Victorin (1923) are undoubtedly erroneous.
1   Sterile fronds herbaceous or subherbaceous, their pinnules with translucent, sharp-toothed, cuneate-based ultimate segments, their veins not enlarged; pinnules of fertile fronds mostly not over 1 cm long; basal stipe-scales mostly uniformly brown; [*Osmunda* L.; Eurasia only; MAPS: Hultén 1958: map 225, p. 245; Meusel, Jaeger, and Weinert 1965:12; Fernald 1935: map 12 (aggregate species), p. 245] .ssp. *crispa*
1   Sterile fronds coriaceous, their pinnules with opaque, crenate or incised, narrowly elliptic to oblong ultimate segments, their vein-tips (in dried material) obviously pitted; pinnules of fertile fronds to 2 cm long; basal stipe-scales with chestnut-brown centres and paler margins . . . . .ssp. *acrostichoides* (R. Br.) Hult.
   2   Sterile fronds narrowly to broadly ovate, with elliptic to oblong or obovate ultimate segments; [*C. acrostichoides* R. Br., the type from w Canada at ca. 56°N; Aleutian Is.–Alaska–B.C. to Great Bear L., L. Athabasca, Man., and Ont. (shores of L. Superior; Manitoulin Is., L. Huron); MAPS: on the above-noted maps; Hultén 1968*b*: 44] . . . . .var. *acrostichoides*
   2   Sterile fronds broadly deltoid, with relatively small obovate ultimate segments; [*Allosorus sit.* Rupr., the type from Sitka, Alaska; also known from the Aleutian Is., SE Yukon (Porsild 1951*a*), sw Dist. Mackenzie (Brintnell L. at 62°05′N), and B.C. (s to Wells (Gray Provincial Park); MAPS: on the above-noted maps by Hultén and Meusel, Jaeger, and Weinert; Hultén 1968*b*:44] .var. *sitchensis* (Rupr.) Chr.

C. stelleri (Gmel.) Prantl   Slender Cliff-Brake
/aST/X/A/   (Hr)   Crevices of cool and shaded calcareous cliffs or springy slopes from N-cent.

Alaska, w-cent. Yukon, and NW Dist. Mackenzie (Porsild and Cody 1968) to s Alta., N-cent. Sask. (Porter L.; not known from Man.), Ont. (N to the Mattagami R. at ca. 50°N), Que. (N to Hudson Bay at ca. 56°N, L. Mistassini, and the Côte-Nord, this last the probable basis of reports from s Labrador), Nfld., N.B., and N.S. (reported from P.E.I., where probably now extinct), s to Wash., Utah, Colo., N Iowa, N Ill., Mich., W.Va., and N.J.; Asia. [*Pteris* Gmel.; *Pellaea* Watt; *Allosorus* Rupr.; *Pteris* (C.; Pellaea) *gracilis* Michx.]. MAP: Hultén 1968b:45.

CYSTOPTERIS Bernh.   Bladder-fern.   Cystoptéride

1   Fronds broadly triangular-ovate, scattered along slender cord-like rhizomes, each consisting of three 2-pinnate-pinnatifid primary divisions (the lower 2 of these slightly narrower and shorter than the upper); stipe at least twice as long as the blade; (Alaska–Dist. Mackenzie–B.C.–Alta.; Ont. to Labrador and Nfld.) . . . . *C. montana*
1   Fronds lanceolate to lance-oblong, 2-pinnate or 2-pinnate-pinnatifid, tufted from stoutish rhizomes; stipe shorter than the blade.
  2   Fronds broadest at base, often bearing bulblets beneath, somewhat dimorphic, the fertile ones narrower than the sterile and prolonged-attenuate to tip; veins mostly ending at the sinuses between the teeth; (SE ?Man. to Nfld. and N.S.) . . . . *C. bulbifera*
  2   Fronds broadest above the base, the sterile and fertile ones similar; veins mostly extending to the teeth; (transcontinental). . . . . *C. fragilis*

C. bulbifera (L.) Bernh.   Bulblet-Fern
/T/EE/   (Hr)   Moist woods and rocks (chiefly calcareous) from Ont. (N to Kapuskasing, 49°24′N; concerning reports from Man., *see* Scoggan 1957) to Que. (N to Anticosti Is. and the Gaspé Pen.; reported from the Côte-Nord by Saint-Cyr 1887), Nfld., N.B., and N.S. (not known from P.E.I.), s to Ariz., Ark., and Ga. [*Polypodium* L.; *Aspidium* Sw.; *Filix* Underw.; *Nephrodium* Michx.].

Forma *horizontalis* (Lawson) Gilbert (the shorter, normally sterile fronds becoming fertile) is known from Niagara Falls and Collin's Bay, s Ont., the type region.

C. fragilis (L.) Bernh.   Fragile Fern
/AST/X/GEA/   (Hr)   Moist slopes, ledges, rich open woods, and wooded talus slopes, the aggregate species from the Aleutian Is. and N Alaska to Banks Is., Boothia Pen., northernmost Ellesmere Is., northernmost Ungava–Labrador, Nfld., N.B., and N.S. (reported from P.E.I., where probably now extinct; *see* D. S. Erskine 1960), s to Calif., Tex., Mo., and Va.; S. America; circumgreenlandic; Iceland; Eurasia. MAPS and synonymy: *see* below:
1   Indusia about 0.5 mm long, round-ovate, nearly entire or at most shallowly lobed; lower segments of larger pinnae narrowly ovate to oblong, cuneate-based; [Ont. (Sibley Pen., near Thunder Bay; Carleton, Lanark, and Russell counties), s Que., N.B., and N.S. (type from near Pictou)] . . . . var. *mackayii* Lawson
1   Indusia to 1 mm long, attenuate to slender tips when young, soon deeply cleft at apex; lower segments of larger pinnae ovate-oblong to deltoid.
  2   Indusia minutely glandular on the back; frond to over 1 dm broad; [incl. × *C. laurentiana* (Weath.) Blasdell (*see* note below); Ont. (N to Moosonee, sw James Bay), E Que. (type from Bic, Rimouski Co.; Gaspé Pen.; Côte-Nord; Magdalen Is.), Nfld., and N.S. (Cape Breton Is.)] . . . . var. *laurentiana* Weath.
  2   Indusia glabrous; frond to 8 cm broad.
    3   Frond and rachis minutely stipitate-glandular; [sw Alta.; Boivin 1967b] . . . .
     . . . . var. *huteri* (Hausman) Luerss.
    3   Frond and rachis nonglandular . . . . var. *fragilis*
      4   Surface of spores merely rugose-warty-reticulate; [*C. dickieana* Sim; transcontinental; MAPS: Hultén 1962: map 56, p. 63, and 1968b:49; Löve and Freedman 1956: fig. 6, p. 163] . . . . f. *dickieana* (Sim) Boivin
      4   Surface of spores spinulose.
        5   Basal pinnules of the larger pinnae of the frond relatively narrow and often cuneate to a subpetiolar base.
          6   Pinnae tapering to unforked tips; [Ont., Que., and Nfld.; Fernald *in* Gray 1950] . . . . f. *angustata* (Hoffm.) Clute

6   Pinnae or some of them forking at tip; [Whycocomagh, Inverness Co., N.S.; Roland 1947] ....f. *cristata* (Lowe) Weath.

5   Basal pinnules of the larger pinnae ovate, ovate-oblong, or deltoid to suborbicular, sessile at their broad or curving bases; [*Polypodium* L.; *Aspidium* Sw.; *Athyrium* Spreng.; *Filix* Gilib.; *Nephrodium (Aspidium) tenue* Michx.; transcontinental; MAPS (aggregate species): Hultén 1962: map 55, p. 63, and 1968*b*:49; Porsild 1957: map 4, p. 161; Raup 1947: pl. 13] ....f. *fragilis*

Concerning var. *laurentiana,* N. H. Wagner, Jr., and D. J. Hagenah (Am. Fern J. 46(4):139. 1956) note the interesting fact that *C. fragilis* may occasionally bear bulblets on the lower surface of the frond as in *C. bulbifera* and that some of the material referred by Weatherby to var. *laurentiana* has since proven to have bulblets (as in collection No. 9,333 of Marie-Victorin and Rolland-Germain from Magdalen Is., E Que.). The bulblets seem to occur, as far as yet known, only on specimens growing on vertical walls of rock cliffs (limestone or sandstone) or on artificial walls such as those of canals or iron furnaces. On the basis of their appearing to be a hybrid between *C. bulbifera* and *C. fragilis,* R. F. Blasdell (Mem. Torrey Bot. Club 21(4):51. 1963) has named such plants × *C. laurentiana* (Weath.) Blasdell.

In addition to the taxa keyed out above, var. *woodsioides* Christ. is reported from E Que. by Marie-Victorin and Rolland-Germain (1969; Mingan Is. of the Côte-Nord). From its description by Gustav Hegi (*Illustrierte Flora von Mittel-Europa.* 2nd. rev. ed. Hanser, Munich. Vol. 1, p. 10. 1935), it appears to be a reduced form (at most about 1 dm tall) with the general aspect of a *Woodsia.*

C. montana (Lam.) Bernh.   Mountain Bladder-Fern
/aST/D/GEA/   (Grh)   Wooded talus slopes, ledges, and rocky banks: N Alaska to cent. Yukon and sw Dist. Mackenzie, s through the mts. of B.C. and sw Alta. (Waterton Lakes; also known from Lesser Slave L.) to N Mont.; isolated in the mts. of Colo.; Ont. (shore of L. Superior near Thunder Bay and Schreiber and on the Slate Is.); Que. (N to the Koksoak R. at ca. 57°45′N; L. Mistassini; Côte-Nord; Shickshock Mts., Gaspé Pen.); s Labrador at ca. 53°N; NW Nfld.; southernmost Greenland; Eurasia. [*Polypodium* Lam.; *Filix* Underw.]. MAPS: Hultén 1968*b*:50, and 1958: map 226, p. 245; Porsild 1966: map 1, p. 67; Meusel, Jaeger, and Weinert 1965:15.

DENNSTAEDTIA Bernh.

D. punctilobula (Michx.) Moore   Hay-scented Fern
/T/EE/   (Hr)   Dry open woods, rocky slopes, and sterile meadows and pastures from Ont. (N to the Ottawa dist.) to Que. (N to Montmagny Co.), N.B., and N.S. (not known from P.E.I.), s to Ark. and Ga. [*Nephrodium* Michx.; *Aspidium* Sw.; *Dicksonia* Gray; *Polypodium* Poir.].

DRYOPTERIS Adans.   Shield-Fern, Wood-Fern

1   Fronds rarely over 3 dm long and 5 cm broad, coriaceous or submembranaceous, evergreen, glandular and spicy-aromatic, the pinnae to about 2.5 cm long and 8 mm broad; indusia glandular-margined, often overlapping, their sinuses often obscure; basal scales lustrous, brown or reddish, often glandular-puberulent, to 1.5 cm long; rhizome covered with numerous persistent old fronds; (transcontinental) ....*D. fragrans*

1   Fronds (and pinnae) commonly larger, rarely glandular; indusia not overlapping, their sinuses distinct; basal scales often larger.

    2   Sori marginal; fronds 2-pinnate, blue- or grey-green, firm to coriaceous, evergreen, their ultimate segments subentire or crenate (or some of them pinnatifid); (Ont. to N.S.) ....*D. marginalis*

    2   Sori not marginal; fronds mostly softer, the margins of their ultimate segments mostly distinctly toothed.

        3   Teeth of pinnules sharp, spine-like or bristle-tipped, salient.

            4   Blade of frond triangular (broadest near the base), 2-pinnate-pinnatifid to 3-pinnate or even 3-pinnate-pinnatifid; indusia glabrous or sparingly glandular

(if glandular, the frond with the extreme type of dissection); (transcontinental) . . . .
. . . .*D. austriaca*
    4   Blade of frond lance-deltoid to oblong or lance-ovate (broadest near the
        middle), 2-pinnate to 2-pinnate-pinnatifid; indusia glandular; (sw B.C.) . . . . .*D. arguta*
  3 Teeth of pinnules merely acute, not bristle-tipped; fronds 2-pinnate.
    5   Fronds more or less strongly dimorphic, the fertile ones much taller than the
        sterile ones, their pinnae often twisted on the rachis; (transcontinental) . . .*D. cristata*
    5   Fronds scarcely dimorphic, plane.
      6   Fronds ovate to ovate-oblong, commonly over half as broad as long; lowest
          pinnae with up to 31 pairs of segments, these mostly at least 2 cm long;
          basal scales firm, with a glossy dark-brown centre and pale margins; (Ont.
          to N.S.) . . . .*D. goldiana*
      6   Fronds lanceolate to lance-oblong, mostly less than half as broad as long;
          lowest pinnae with at most about 20 pairs of segments, these mostly less
          than 1 cm long; basal scales papery, uniformly pale brown; (transcon-
          tinental) . . . .*D. filix-mas*

D. arguta (Kaulf.) Watt   Coastal Shield-Fern
/t/W/ (Hr (Grh)) Rocky ledges and woods along the coast from sw B.C. (*see* below) to Calif.,
locally inland to Ariz. [*Aspidium argutum* Kaulf.; *A. rigidum* var. *arg.* (Kaulf.) Eat.]. MAPS: T.M.C.
Taylor 1963:159, and 1970:129 (reporting the species from Denman southwards in B.C.).

    This species was considered by early N. American authors to be identical with the Eurasian
Aspidium rigidum Sw. (*D. rigida* (Sw.) Gray), currently known as *D. villarii* (Bell.) Woynar, but the
two reports of this from Mt. Finlayson, Vancouver Is., by John Macoun (1890) are referred by
Boivin (1967a) to *D. filix-mas* and *D. austriaca*, respectively. On the other hand, T.M.C. Taylor
(1963) reports *D. arguta* from Hornby Is. and Nanaimo, Vancouver Is., and he also (1970) reports it
from Denman Is. southwards in B.C. C.-J. Widén and D. M. Britton (Can. J. Bot. 49(9):1590. 1971)
recognize the species in ten specimens taken by Bonnell in 1970 on Hornby Is. (their citation in
Table 6 of "Ontario" as the source of a collection should probably read "Oregon"). According to
Hultén (1941), the report by Trelease of *Aspidium rigidum* var. *argutum* from Alaska is
undoubtedly erroneous.

D. austriaca (Jacq.) Woynar   Spinulose Shield-Fern
/aST/X/GEA/ (Hr (Grh)) Damp woods, thickets, and swamps, the aggregate species from
N-cent. Alaska, SE Yukon, and w Dist. Mackenzie to cent. Man., James Bay, Ungava–Labrador (N
to ca. 58°N), Nfld., N.B., P.E.I., and N.S., s to Calif., Wyo., Iowa, Tenn., and N.C.; w Greenland N to
69°14′N, E Greenland N to 61°32′N; Iceland; Eurasia. MAPS and synonymy: *see* below.
1 Basal pinnules of the lowermost pinnae to 1.5(2) cm apart along the rachis; first lower
    pinnule to 1 dm long, usually surpassing the adjacent one; fronds generally deciduous,
    they, the rachises, and indusia nonglandular; [*D. (Aspidium) spinulosa* var. *dilatata*
    (Hoffm.) Watt; *D. (Polypodium; Thelypteris) dil.* (Hoffm.) Gray; incl. *D. spin.* var.
    americana (Fisch.) Fern. and *D. campyloptera* (Kunze) Clarkson; transcontinental; MAPS:
    Hultén 1958: map 156, p. 175, and 1968b: 55 (*D. dil.* and its ssp. *amer.*); Meusel, Jaeger,
    and Weinert 1965:18] . . . . .var. *austriaca*
1 Basal pinnules of the lowermost pinnae subopposite (rarely over 4 mm apart); first lower
    pinnule to 6 cm long, rarely surpassing the adjacent ones; fronds generally evergreen.
    2   Fronds (and indusia) nonglandular; pinnae obliquely ascending and asymmetrical in
       outline, the basal pinnule of the lower side of the lowermost pinnae longer than the
       adjacent ones and about twice as long as the opposing pinnule; [*Polypodium (D.;*
       Aspidium; Thelypteris) spinulosum Muell.; *D. carthusiana* Gray; transcontinental;
       MAPS (*D. spin*): Hultén 1958: map 155, p. 175; Meusel, Jaeger, and Weinert 1965:18] . . . . . .
       . . . . .var. *spinulosa* (Muell.) Fiori
    2   Fronds glandular, especially on the rachises and indusia; pinnae more regular in
       outline, the basal pinnule shorter than to rarely surpassing the adjacent ones; [Ont. to
       Nfld. and N.S.].
      3   Rhizome short-creeping; pinnae obliquely ascending, gradually tapering to apex;
         mature indusia to nearly 1.5 mm broad; [*D. spin.* var. *fr.* (Gilbert) Trudell; MAP:

Hultén 1958: map 155 (*D. spin.* var. *fr.*), p. 175] . . . . .var. *fructuosa* (Gilbert) Morton
3 Rhizome suberect; pinnae mostly horizontally spreading, usually with prolonged tips; mature indusia less than 1 mm broad; [*Aspidium intermedium* Muhl.; *D. spin.* var. *int.* (Muhl.) Underw.; MAP: Hultén 1958: map 155 (*D. spin.* var. *int.*), p. 175. For a varying treatment of this and other members of the complex, *see* C.-J. Widén and D. M. Britton (Can. J. Bot. 49(9):247–58. 1971).] . . . .var. *intermedia* (Muhl.) Morton

D. cristata (L.) Gray   Crested Wood-Fern
/T/X/EwA/   (Hr)   Wet woods, thickets, and swampy ground, the aggregate species from s B.C. (Kitchener; Clanwilliam) to northernmost Alta., Sask. (N to Meadow L. at ca. 54°N), Man. (N to Porcupine Mt.), Ont. (N to Nipik L. at ca. 53°N, 92°W), Que. (N to Anticosti Is. and the Gaspé Pen.), Nfld., N.B., P.E.I., and N.S., s to Idaho, Mo., La., Tenn., and N.C.; Europe; w Asia. MAPS and synonymy: *see* below.
1 Pinnae to about 8 cm long and 2.5 cm broad, strongly reduced toward the base of the strongly dimorphic frond; sori about midway between margin and midvein of segments; [*Polypodium* L.; *Aspidium* Sw.; *Thelypteris* Nieuwl.; *Nephrodium* Michx.; SE B.C. (Kitchener; Clanwilliam) to Alta. (N to ca. 59°50′N according to Hultén's map), Sask. (N to Meadow Lake, 54°08′N), Man. (N to Porcupine Mt.), Ont. (N to Nikip L., ca. 53°N, 92°W), Que. (N to the Gaspé Pen., Anticosti Is., and Magdalen Is.), Nfld., N.B., P.E.I., and N.S.; MAPS: Hultén 1958: map 40, p. 59; Meusel, Jaeger, and Weinert 1965:17]. *D. boottii* (Tuck.) Underw. (*Aspidium boottii* Tuck.), intermediate between *D. austriaca* var. *intermedia* and *D. cristata* and probably a somewhat fertile hybrid of this parentage (fronds intermediate in outline; indusia glandular), has been found in Ont., Que., N.B., P.E.I., and N.S. and reported, no doubt erroneously, from Alaska. *D. uliginosa* (A. Br.) Druce, a hybrid with *D.* (*austriaca* var.) *spinulosa,* is reported from s Ont. by C.-J. Widén and D. M. Britton, Can. J. Bot. 49(9):1144. 1971 . . . . .var. *cristata*
1 Pinnae to nearly 1.5 dm long, mostly over 2.5 cm broad, the lower ones only slightly reduced; sori slightly closer to the margin; fronds not strongly dimorphic; [*Aspidium cr.* var. *clint.* Eat.; *D. clint.* (Eat.) Dowell; Ont. (N to near North Bay) and sw Que. (N to near Montreal); MAP: on the above-noted map by Hultén]. This taxon is perhaps of hybrid origin between *D. cristata* and *D. goldiana.* Apparent hybrids between it and *D. goldiana* and *D. austriaca* vars. *intermedia* and *spinulosa* are reported from s Ont. by Soper (1949). A putative hybrid with *D. marginalis* (× *D. burgessii* Boivin; type from Georgeville, sw Que.) is reported from Ont. and Que. by Boivin (1966b) . . . . .var. *clintoniana* (Eat.) Underw.

D. filix-mas (L.) Schott   Male Fern
/aST/X/GEA/   (Hr)   Rich woods, rocky slopes (often calcareous), and upland pastures from B.C. (N to the Nass R. at ca. 56°N) to sw Alta. (Waterton Lakes; not known from Sask. or Man.), s Ont. (Michipicotin Is., L. Superior; Manitoulin Is., N L. Huron; Bruce and Grey counties), Que. (N to Anticosti Is.), Nfld., N.B., and N.S. (not known from P.E.I.), s to s Calif., Ariz., Mexico, Okla., S.Dak., N Mich., and Maine; SE Greenland at ca. 60°30′N; Iceland; Eurasia; N Africa [*Polypodium* L.; *Aspidium* Sw.; *Thelypteris* Nieuwl.]. MAPS: Hultén 1962: map 110, p. 119; Meusel, Jaeger, and Weinert 1965:17.
Forma *incisa* (Moore) Hayek (pinnules relatively coarsely toothed) is reported from Nfld. and N.S. by Roland (1947). Putative hybrids with *D. austriaca* var. *spinulosa* have been collected near Owen Sound, Grey Co., s Ont.

D. fragrans (L.) Schott   Fragrant Cliff-Fern
/AST/X/GEA/   (Hr)   Dry ledges and cliffs (often calcareous), the aggregate species from N Alaska, s Yukon, and N Dist. Mackenzie to cent. Ellesmere Is. and northernmost Labrador, s to N B.C. at ca. 57°N, L. Athabasca (Alta. and Sask.), N Man. (Reindeer L. at 57°19′N), Ont. (N shore of L. Superior), Minn., Wisc., Mich., and N New Eng.; w Greenland N to ca. 80°N; an isolated station in E Greenland at ca. 70°N; an isolated station in N Finland; Asia. MAPS and synonymy: *see* below.
1 Pinnae overlapping and often inrolled; fronds coriaceous, mostly less than 2 dm long, heavily chaffy beneath; [incl. var. *aquilonaris* (Maxon) Gilbert; *Polypodium* L.; *Aspidium* Sw.; *Thelypteris* Nieuwl.; *Nephrodium* Rich.; transcontinental; MAPS (aggregate species): Hultén 1968b:56, and 1962: map 27, p. 35; Porsild 1957: map 5, p. 161; Raymond 1950b:

fig. 13, p. 23; Raup 1947: pl. 13]. Var. *aquilonaris* (Maxon) Gilbert is reported from near
Nome, Alaska, by Hultén 1941, who describes it as differing from the typical form in
the lower pinnae not being gradually reduced and in possessing more dissected pinnules . . . . .
. . . . .var. *fragrans*
1   Pinnae mostly flat and not overlapping; fronds relatively soft, to about 3 dm long, only
sparingly chaffy beneath; [*Thelypteris fr.* var. *hookeriana* Fern.; s Yukon (Porsild 1951a);
Ont. (N to the N shore of L. Superior), Que. (N to s James Bay), Nfld., N.B., and N.S. (early
reports from P.E.I. require confirmation); a comparison of Raup's map with the others
cited will illustrate the more southern range of this taxon] . . . . .var. *remotiuscula* Komarov

D. goldiana (Hook.) Gray   Goldie's Fern

/T/EE/   (Hr)   Rich woods and calcareous ledges from Minn. to Ont. (N to the Ottawa dist.), Que.
(N to Lac Trois-Saumons, l'Islet Co.; *see* Que. map by Doyon and Lavoie 1966: fig. 19, p. 819;
reported N to Cacouna, Temiscouata Co., by Penhallow 1891), N.B., and N.S. (not known from
P.E.I.), s to Iowa, Tenn., and N.C. [*Aspidium goldianum* Hook., the type from near Montreal, Que.;
Thelypteris Nieuwl.]. MAP: Meusel, Jaeger, and Weinert 1965:17.

D. marginalis (L.) Gray   Marginal Shield-Fern

/T/EE/   (Hr)   Woods, clearings, and rocky slopes from Ont. (N to L. Nipigon; *see* Ont. map by D.
M. Britton and J. H. Soper, Can. J. Bot. 44(1): fig. 11, p. 65. 1966), Que. (N to Bic, Rimouski Co.),
N.B., and N.S. (not known from P.E.I.), s to Okla., Ark., Ala., and Ga. [*Polypodium* L.; *Aspidium*
Sw.; *Nephrodium* Michx.; *Thelypteris* Nieuwl.].

Reports from B.C. (Eastham 1947), Sask. (Rydberg 1922), and SE Man. (John Macoun 1890)
probably refer to other species. The typical form has deeply pinnatifid to 1-pinnate (or
pinnate-pinnatifid) pinnae. Other forms, with the pinnae more extensively dissected, are: f. *elegans*
(Rob.) Gray (most of the pinnae 2-pinnatifid; Que. (Hatley, Stanstead Co.; Bic Mt., near St-Fabien,
Rimouski Co.) and N.B. (St. Andrews)) and f. *tripinnatifida* (Clute) Weath. (pinnae still more
dissected, the ultimate, very narrow segments themselves deeply toothed; reported from s Ont. by
C. A. Weatherby, Am. Fern J. 31(2): 61. 1941, and from N.S. by Roland 1947).

GYMNOCARPIUM Newm.

1   Fronds glabrous, the two lower broadly triangular main divisions much more than half as
long as the terminal one; pinnules acutish, the first one on the lower side of the basal
pinna about 1/3 as long as the pinna-rachis or longer . . . .*G. dryopteris*
1   Fronds glandular-puberulent, the two lower narrowly triangular main divisions not much
more than half as long as the terminal one; pinnules obtuse, the first one on the lower side
of the basal pinnae rarely more than 1/4 as long as the pinna-rachis . . . .*G. robertianum*

G. dryopteris (L.) Newm.   Oak-Fern

/aST/X/GEA/   (Grh)   Cool mossy or rocky woods from the Aleutian Is. and cent. Alaska to SE
Yukon, s Dist. Mackenzie, SE Hudson Bay, N Labrador (ca. 57°40'N; also indicated on Raup's map
at ca. 60°N), Nfld., N.B., P.E.I., and N.S., s to Oreg., Ariz., Iowa, and Va.; w Greenland N to ca.
69°30'N, E Greenland N to ca. 66°20'N; Iceland; Eurasia. [*Polypodium* L.; *Carpogymnia* Löve &
Löve; *Nephrodium* Michx.; *Phegopteris* Fée; *Thelypteris* Slosson; *Dryopteris disjuncta* (Rupr.)
Morton; *D. linnaeana* Chr.] MAPS (*Dry. linn.*): Hultén 1968b:56, and 1962: map 108, p. 117; Meusel,
Jaeger, and Weinert 1965:16; Raup 1947: pl. 13.

Possible hybrids between *G. dryopteris* and *G. robertianum* are reported from Alaska, Nfld.,
and Greenland by E. E. Root (Am. Fern J. 51(1): 21. 1961). W. H. Wagner, Jr. (Rhodora 68(774):
132–37. 1966; *see* his map, fig. 3, p. 124) also reports such apparent hybrids from L. Superior,
Ont., and believes that they actually represent an apomictic species, to which he gives the name *G.
heterosporum* Wagner.

G. robertianum (Hoffm.) Newm.   Northern Oak-Fern

/ST/X/EA/   (Grh)   Damp to dry calcareous ledges, cliffs, and talus from cent. Alaska–Yukon to
sw Dist. Mackenzie, s Dist. Keewatin, Ont. (N to Lac Seul and the Missinaibi R. at ca. 50°N), Que.
(N to the Kaniapiscau R. at 58°18'N, L. Mistassini, and the Côte-Nord), Nfld., and N.B. (not known

from N.S.; an early report from P.E.I. not accepted by D. S. Erskine 1960), s to B.C., Idaho, Iowa, and Pa.; Eurasia. [*Polypodium* Hoffm.; *Dryopteris* Chr.; *Phegopteris* A. Br.; *Thelypteris* Slosson; *G. dryopteris* var. *pumilum* (DC.) Boivin; *D. linnaeana* f. *glandulosa* Tryon; *Polypodium* (*Phegopteris*) *calcareum* Sm.]. MAPS (*Dry. rob.*); Hultén 1968*b*:57, and 1962: map 116, p. 125; Meusel, Jaeger, and Weinert 1965:16.

MATTEUCCIA Todaro

M. struthiopteris (L.) Todaro   Ostrich-Fern.   Ptérétide
/sT/X/EA/   (Grh)   Low open ground, alluvial thickets, and rich woods from cent. Alaska to Great Slave L., Sask. (N to ca. 54°N), Man. (N to Hill L. N of L. Winnipeg), Ont. (N to the Fawn R. at ca. 54°30′N), Que. (N to the Côte-Nord), Nfld., N.B., P.E.I., and N.S., s to s B.C., S.Dak., Mo., and Va.; Eurasia. [*Osmunda* (*M.; Onoclea*) *struthiopteris* L.; *Struthiopteris filicastrum* All.; *S. germanica* Willd.]. MAPS: Hultén 1968*b*:52, and 1962: map 115, p. 125.

The typical form of Eurasia has the chaff-scales of the rhizome and stipe-bases with a conspicuous blackish central band. The N. American plant has uniformly pale-brown scales, and may be distinguished as var. *pensylvanica* (Willd.) Morton (*Struthiopteris* (*M.; Pteretis*) *pen.* Willd.; *S. germanica* var. *pen.* (Willd.) Lowe; *Pteretis struth.* var. *pen.* (Willd.) Farw.; *Onoclea* (*M.; Pteretis; Struth.*) *nodulosa* Michx.). Var. *pubescens* (Terry) Clute (frond-rachis minutely canescent-tomentose rather than glabrous and shining) is known from Matane Co., Gaspé Pen., E Que.

ONOCLEA L.   Sensitive Fern.   Onoclée

O. sensibilis L.
/sT/EE/   (Grh)   Low open ground, alluvial thickets, and low woods from s Man. (N to Sasaginnigak L. at ca. 52°N; reports from Sask. require confirmation) to Ont. (N to L. Nipigon and the Albany R. at 51°32′N), Que. (N to E James Bay at 52°32′N and the Côte-Nord), s Labrador (N to the Hamilton R. basin), Nfld., N.B., P.E.I., and N.S., s to Tex. and Fla.

Forma *obtusilobata* (Schk.) Gilbert (*O. obtus.* Schk.; most or many of the pinnules of the fertile frond foliaceous and sterile) occurs throughout the range.

PELLAEA Link   Cliff-Brake

1   Stipe and rachises of fronds dark brown or purplish to blackish, long-hairy, the fronds
    distinctly dimorphic, the fertile ones to about 5 dm long, much longer than the sterile
    ones; stalks of pinnae diverging at a broad angle from the rachis of the frond; pinnae
    with up to 15 pinnules, all except the upper pair of these distinctly stalked; (B.C. to Que.) . . . . . .
    . . . .*P. atropurpurea*
1   Stipe and rachis of fronds golden- or reddish-brown to dark brown, glabrous or sparsely
    hairy, the fertile and sterile fronds subequal, to about 3.5 dm long; stalks of pinnae
    strongly ascending; pinnae with up to 7 sessile or short-stalked pinnules; (B.C. and sw
    Dist. Mackenzie to Que.) . . . .*P. glabella*

P. atropurpurea (L.) Link   Purple Ciff-Brake
/sT/X/   (Hr)   Calcareous ledges and crevices from SE B.C. (Columbia Valley) and sw Alta. (Banff) to Sask. (known only from L. Athabasca; not known from Man.), s Ont. (Manitoulin Is. and Bruce, Leeds, Lincoln, and Welland counties; *see* s Ont. map by J. H. Soper, Am. Fern J. 53(2): fig. 3, p. 73. 1963), and s Que. (Pontiac and Montmorency counties), s to Ariz., S.Dak., Mich., and Fla. [*Pteris* L.; *Allosurus* Kunze; *Notholaena* Keyserl.]. MAP: S. J. Rigby and D. M. Britton, Can. Field-Nat. 84(2): fig. 1, p. 140. 1970.

P. glabella Mett.   Smooth Cliff-Brake
/sT/X/   (Hr)   Chiefly calcareous ledges and crevices (ranges of Canadian taxa outlined below), s to Wash., Utah, N N.Mex., Okla., Tenn., and Va. MAPS and synonymy: *see* below.
1   Fronds to about 3.5 dm long, their pinnae with up to 7 pinnules; stipe and rachises of
    frond reddish brown to dark brown; [var. *?bushii* Mack.; Ont. (N to Lake of the Woods and

the NW shore of L. Superior) and SE Que. (Richmond Co.); MAPS: S. J. Rigby and D. M. Britton, Can. Field-Nat. 84(2): fig. 2, p. 141. 1970; A. F. Tryon and D. M. Britton, Evolution 12(2): fig. 3, p. 142. 1958] . . . . .var. *glabella*

1 Fronds to about 2 dm long.

    2 Stipe and rachises of frond reddish brown to dark brown; pinnae with up to 5 pinnules; [*P. suksdorfiana* Butters; B.C. (N to ca. 54°30′N) and sw Alta. (N to Jasper); MAPS: on the above-noted maps by Rigby and Britton and by Tryon and Britton] . . . . .var. *simplex* Butters

    2 Stipe and rachises of frond golden-brown to dark brown; pinnae with only 1 pinnule (this may be 2–3-lobed); [*P. occidentalis* (Nels.) Rydb.; *P. pumila* Rydb.; sw Dist. Mackenzie (N to ca. 64°N), sw Alta. (Banff), Sask. (N to L. Athabasca), and Man. (N to ca. 55°N); MAPS (var. *occid.*): on the above-noted maps by Rigby and Britton and by Tryon and Britton] . . . . .var. *nana* (Rich.) Cody

PHYLLITIS Hill

P. scolopendrium (L.) Newm.   Hart's-tongue
/T/EE/EA/ (Hr) Crevices, sink holes, and cool shaded slopes of calcareous formations along or near the Niagara escarpment of s Ont. (Welland, Simcoe, Peel, Halton, Grey, Dufferin, and Bruce counties; also reported from Manitoulin Is., N L. Huron; *see* Ont. map by J. H. Soper, Am. Fern J. 44(4): fig. 1, p. 131. 1954); cent. N.Y. (largely exterminated); Marion Co., Tenn.; var. *lindenii* (Hook.) Fern. in s Mexico; Europe; Japan; N Africa. [*Asplenium* L.; *Scolopendrium vulgare* Sm.; *S. officinarum* Sw.]. MAPS: Hultén 1962: map 147, p. 157; Fernald 1935: map 2, p. 200; A. Löve 1954: fig. 1, p. 215; Meusel, Jaeger, and Weinert 1965: 12; T.M.C. Taylor 1970:160 ("Pacific Northwest" area).

The typical form of Eurasia is a relatively large plant, the tips of the veins (hydathodes) being elliptic and reaching nearly to the margin. The smaller N. American plant has the tips of the veins more elongate and ending somewhat farther in from the margin, and may be distinguished as var. *americana* Fern. (*P. fernaldiana* Löve). The plant has been widely cultivated and reports from Canada other than from s Ont. (as from Vancouver Is. and N.B.) are probably based upon such material.

PITYROGRAMMA Link

P. triangularis (Kaulf.) Maxon   Gold-back Fern
/t/W/ (Hr) Ledges, talus slopes, and hillside thickets from sw B.C. Vancouver Is. and adjacent islands and mainland; CAN; V; *see* B.C. map by T.M.C. Taylor 1963:161; a report from Nome, Alaska, by Underwood not accepted by Hultén 1941) to Baja Calif. and Ariz. [*Gymnogramma* Kaulf.; *Gymnogramme* Hook. & Grev.; *Gymnopteris* Underw.; *Ceropteris* Underw.; *Gymnogramme oregana* Nutt.].

POLYPODIUM L. Polypody. Polypode

1 Segments of the deltoid-ovate frond rigidly coriaceous, narrowly oblong, obtuse or rounded at apex, the margins cartilaginous, the midrib at first scaly beneath, some the the veins anastomosing; sori to 5 mm broad, crowded against the midrib; rhizome not licorice-flavoured, to 1 cm thick; (B.C.) . . . .*P. scouleri*

1 Segments of the lanceolate to narrowly oblong or deltoid-oblong frond herbaceous to membranous, the margins not cartilaginous, the midrib not scaly, the veins all free; sori smaller, separated from the midvein.

    2 Rhizome firm, licorice-flavoured, to 1 cm thick, its pale-cinnamon to castaneous scales uniformly coloured (or darker toward base), peltately attached slightly above base, to 1 cm long; stipes to 3 dm long and 3 mm thick; fronds to over 5 dm long; pinnae opposite, subopposite, or alternate, the lowest commonly shorter than the middle ones, these to 2 cm broad, their midribs commonly curving at base; sori mostly midway between the margin and midvein; (B.C. and Alta.) . . . .*P. vulgare*

2 Rhizome rather soft and spongy, not licorice-flavoured, to 7 mm thick, its scales darkened on the back, cordate at base, less than 5 mm long; stipes to 2 dm long and less than 2 mm thick; fronds to about 2.5 dm long; pinnae alternate (or the lowest subopposite), the lowest usually about as long as or slightly longer than the middle ones, these less than 1 cm broad, their midribs and those of the upper pinnae straight; sori nearly marginal; (transcontinental) ....*P. virginianum*

P. scouleri Hook. & Grev.   Leathery Polypody
/t/W/   (Grh)   Cliffs, talus slopes, and mossy tree trunks from sw B.C. (Vancouver Is. and adjacent islands and ?mainland N to s Queen Charlotte Is.; *see* B.C. map by T.M.C. Taylor 1963:161) to Baja Calif.; Guadaloupe Is. (Broun 1938). MAP: T.M.C. Taylor 1970:169 ("Pacific Northwest" area).

P. virginianum L.   Rock-Polypody.   Tripe-de-roche
/sT/X/eA/   (Grh)   Ledges, boulders, and rocky woods from N B.C. (collections in CAN from the Beaton R. at ca. 57°N and Mt. Selwyn, ca. 56°N) and Great Bear L. to L. Athabasca (Alta. and Sask.), Man. (N to Churchill), Ont. (N to ca. 53°N), Que. (N to SE Hudson Bay at 56°10′N, L. Mistassini, and the Côte-Nord), Nfld., N.B., P.E.I., and N.S., s to Ark. and Ga.; E Asia. [*P. vulgare* vars. *virg.* (L.) Eat. and *americanum* Hook.]. MAPS: Hultén 1962: map 167A, p. 177 (*P. vulg.* ssp. *virg.*); Meusel, Jaeger, and Weinert 1965:18. The following forms occur in Canada:
1 Fronds lance-oblong, the basal segments not conspicuously longer than the median ones.
  2 Frond firm to coriaceous, its segments obtuse to round-tipped; [transcontinental] ....
  ....f. *virginianum*
  2 Frond relatively thin, its principal segments acutish to attenuate; [Stanstead Co., sw Que.; Henry Mousley, Am. Fern J. 15(3):89. 1925] ....f. *acuminatum* (Gilbert) Fern.
1 Fronds deltoid-lanceolate or broader, the basal segments often conspicuously longer than the median ones.
  3 Frond deltoid-lanceolate, the upper half narrow and merely undulate-lobed, the lower segments blunt and unlobed; [sw Que.] ....f. *elongatum* (Jewell) Fern.
  3 Frond relatively broader, at least the lower segments pinnatifid or auricled.
    4 Fronds deltoid, gradually tapering to tip, the lower segments hastate or auricled and often pinnatifid; [s Ont. and sw Que.] ....f. *deltoideum* (Gilbert) Fern.
    4 Fronds broadly oblong, abruptly short-tipped, their broad segments deeply cleft into many long and often toothed lobes; [sw Que. and Charlotte Co., N.B.] ....
    ....f. *cambricoides* F. W. Gray

P. vulgare L.   Licorice-Fern
/aST/W/GEA/   (Grh)   Mossy rocks, talus slopes, and tree trunks (ranges of Canadian taxa outlined below), s to Baja Calif., N.Mex., and the Black Hills of S.Dak.; southernmost Greenland; Iceland; Eurasia; N Africa. MAPS and synonymy: *see* below.
1 Fronds broad in outline, ovate or ovate-oblong, to 2.5 dm long and 2 dm broad, firm, many or all of their pinnae deeply and irregularly pinnatifid or lacerate; [*P. cambricum* L.; sw B.C.: T.J.W. Burgess, Proc. Trans. R. Soc. Can. 4 (Sect. iv): 10. 1887] ....
  ....var. *cambricum* (L.) Willd.
1 Fronds narrower, lanceolate to narrowly oblong, their pinnae less deeply lobed.
  2 Frond usually thin in texture, to 5.5 dm long and 2 dm broad; pinnae acute or attenuate, narrowly lanceolate, finely serrate or serrate-dentate; [*P. glycyrrhiza* Eat.; *P. falcatum* Kellogg, not L. f.; Aleutian Is., cent. Alaska–Yukon (*see* Hultén 1941: map 28, p. 115), and B.C. (the common coastal form according to Henry 1915); MAPS: Hultén 1962: map 167B, p. 177, and 1968b:58; T.M.C. Taylor 1970:163 (*P. gly.*; "Pacific Northwest" area)]................................var. *occidentale* Hook.
  2 Frond firm, at most about 2.5 dm long and 1 dm broad.
    3 Fronds narrowly oblong, to about 2 dm long and 5 cm broad, their oblong-elliptical to narrowly obovate pinnae round-tipped, obscurely to deeply crenate, to about 2 cm long; [*P. hesperium* Maxon; Alaska–Yukon (Hultén 1968a), B.C., and sw Alta.; MAPS: Hultén 1962: map 167B, p. 177, and 1968b:57] ....var. *columbianum* Gilbert

3 Fronds lanceolate, to about 2.5 dm long and 1 dm broad, their oblong pinnae subacute to obtuse, crenate or minutely serrulate, to about 4.5 cm long; [var. *commune* Milde; var. *rotundatum* Milde, not Britt., which is var. *columbianum*; incl. *P. montense* Lang; Aleutian Is.–Alaska–B.C.; MAPS: Hultén 1962: map 167B, p. 177 (reporting on p. 176 its occurrence in s Greenland but not indicating this on his map); F. A. Lang, Madroño 20(2): fig. 1 (combine the areas of the collective *P. glycyrrhiza-hesperium-montense*), p. 58. 1969] . . . . .var. *vulgare*

POLYSTICHUM Roth Shield-Fern. Polystic

1 Blade of frond 2-pinnate or at least pinnately lobed or divided at base; pinnules scarcely auricled at base.
  2 Fronds at most about 3 dm long and 5 cm broad, tapering about equally to base and apex, the upper part simply pinnate, the middle and lower parts pinnate-pinnatifid or 2-pinnate; pinnules overlapping; (mts. of B.C. and E Que.) . . . .*P. mohrioides*
  2 Fronds larger, 2-pinnate nearly to tip, the pinnules not overlapping; (Alaska–B.C.; Ont. to Nfld. and N.S.) . . . .*P. braunii*
1 Blade of frond simply pinnate throughout, the pinnae variously serrate to regularly incised (but never pinnatifid to pinnately lobed), inequilateral, strongly auricled on the upper side at base.
  3 Teeth of pinnae not sharp-pointed (neither spinose nor aristate); fronds narrowly oblanceolate, the blades to about 1 dm long, the whole plant at most about 1.5 dm tall, with the general appearance of *Woodsia alpina*; (Aleutian Is.) . . . .*P. aleuticum*
  3 Teeth of pinnae sharp-tipped; fronds to over 6 dm long.
    4 Upper fertile pinnae abruptly reduced; fronds narrowly lanceolate, acuminate, to about 7.5 dm long, their linear-oblong pinnae to about 8 cm long, the lower ones only slightly reduced; stipe and rachis (except upper part) rather densely chaffy, the stipe usually at least 1/4 the length of the blade; (Ont. to N.S.) . . . .*P. acrostichoides*
    4 Upper fertile pinnae only gradually reduced; fronds tapering about equally from base to apex.
      5 Stipe very short; pinnae mostly oblong-lanceolate (the basal ones deltoid), salient-toothed; fronds to about 6 dm long; (Alaska–B.C.–w Alta.; Ont. to Nfld. and N.S.) . . . .*P. lonchitis*
      5 Stipe much longer; pinnae linear-attenuate (the basal ones at least twice as long as broad), their teeth incurved; fronds to over 1 m long; (Alaska–B.C.) . . . .
. . . .*P. munitum*

P. acrostichoides (Michx.) Schott Christmas Fern
/T/EE/ (Hr) Wooded rocky slopes and rock-crevices (ranges of Canadian taxa outlined below), s to Mexico, E Tex., and N Fla.
1 Tip of frond more or less forking; [Grey Co., s Ont.] . . . . .f. *cristatum* Clute
1 Tip of frond not forking.
  2 Pinnae of the relatively pale frond coarsely serrate to incised; [s Ont., sw Que., and N.S.] . . . . .f. *incisum* (Gray) Gilbert
  2 Pinnae merely minutely serrate with appressed bristle-tipped teeth; [*Nephrodium* Michx.; *Aspidium* Sw.; Ont. (N to near Ottawa), Que. (N to Cap-Rouge, near Quebec City), N.B., P.E.I., and N.S.]. × *P. hagenahii* Cody (× *P. marginale sensu* Boivin 1966b, not *P. lonchitis* f. *marginale* McColl; see W. J. Cody, Am. Fern J. 58(1): 30–31. 1968), a hybrid between *P. acrostichoides* and *P. lonchitis,* is known from the type locality, Cape Croker, L. Huron, s Ont. . . . . .f. *acrostichoides*

P. aleuticum C. Chr.
/s/W/ (Hr) Known only from Bering Is., Bering Strait, and the type locality, Atka Is., Aleutian Is. MAPS: Hultén 1968b:53; 1960: map 6 (at end); and 1941: map 14, p. 113.
    According to Carl Christensen (Am. Fern J. 28(4):111. 1938), this species is totally different from all other American ones of the genus, being very closely related to several forms from western China, perhaps closest to *P. lachenense* (Hook.) Bedd.

P. braunii (Spenner) Fée   Braun's Holly-Fern
/sT/(X)/EA/   (Hr.)   Rocky woods, shaded talus, and ravines (ranges of Canadian taxa outlined below), s in the West to N Wash.–Mont. and in the East to N Mich. and Pa.; Eurasia. MAPS and synonymy: see below.
1   Stipe-chaff relatively firm and short-pointed; teeth of pinnules relatively short; fronds relatively slender; [*Aspidium* Spenner; *P. aculeatum* var. *br.* (Spenner) Döll; incl. *P. alaskense* Maxon; s Alaska (see Hultén 1941: maps 16a and 16b (var. *alaskense*), p. 113) and B.C. (Stikine R. at ca. 55°N; Salmon R. near Ymir; see B.C. map by T.M.C. Taylor 1963: 161); MAP: Hultén 1962: map 180 (*P. braunii* and its var. *alaskense*; area of the aggregate species also indicated), p. 191] . . . . .var. *braunii*
1   Stipe-chaff thinner and longer-pointed; pinnule-teeth longer; fronds mostly stouter.
   2   Fronds commonly bearing proliferous buds toward summit; basal pinnule on the upper side of the pinnae usually distinctly longer than the adjacent one on the same side; [*P. andersonii* Hopkins, the type from Strathcona Park, Vancouver Is.; Aleutian Is. (Attu Is.), SE Alaska (see Hultén 1941: map 15, p. 113), and B.C. (Mt. Waddington; Alice Arm at ca. 55°N; Smithers, ca. 54°N; a station at ca. 51°30'N; see B.C. map by T.M.C. Taylor 1963:161); MAPS: Hultén 1968*b*:55; T.M.C. Taylor 1970:173 (*P. and.*; "Pacific Northwest" area)] . . . . .var. *andersonii* (Hopkins) Hultén
   2   Fronds lacking proliferous buds; basal pinnule not markedly longer than the adjacent one on the same side; [*Aspidium aculeatum* of E Canadian reports, not Sw.; *P. acul.* var. *br.* subvar. *pur.* (Fern.) Farw.; Ont. (N to the N shore of L. Superior), Que. (N to Anticosti Is.), Nfld., N.B., and N.S.; MAP: Hultén 1962: map 180, p. 191] . . .var. *purshii* Fern.

P. lonchitis (L.) Roth   Holly-Fern
/aST/(X)/GEA/   (Hr)   Cool shaded talus slopes and wooded hillsides: E Aleutian Is. and w-cent. Alaska through B.C. and sw Alta. (Waterton Lakes) to Calif., Utah, and Colo.; N Mich. to Ont. (Bruce, Grey, Simcoe, and Welland counties; Batchewana, E shore of L. Superior; see s Ont. map by J. H. Soper, Am. Fern J. 44(4): fig. 2, p. 144. 1954); E Que. (Rimouski Co. and Gaspé Pen.), Nfld., and N.S. (not known from N.B. or P.E.I.); w and E Greenland N to ca. 70°N; Iceland; Eurasia. [*Polypodium* L.; *Aspidium* Sw.]. MAPS: Hultén 1968*b*:53, and 1958: map 219, p. 239; Meusel, Jaeger, and Weinert 1965:18; Fernald 1935: map 3, p. 207.

P. mohrioides (Bory) Presl
/T/D/   (Hr)   Shaded talus slopes and cool crevices of cliffs: s B.C. (Bridge River mts.; Coquihalla; Tulameen; see B.C. map by T.M.C. Taylor 1963:162 (*P. scop.*) to Mont., s to Calif., Idaho, and Utah; E Que. (gulches and serpentine tableland of Mt. Albert, Gaspé Pen., between ca. 600 and 1000 m; see Scoggan 1950:21); the typical form and varieties s along the Andes of S. America to subantarctic regions. [Incl. var. *scopulinum* (Eat.) Fern. (*P. scop.* (Eat.) Maxon; *P. kruckebergii* Wagner); *P. lemmonii sensu* Henry 1915, probably not Underw.]. MAPS (var. *scop.* or *P. scop.*): Fernald 1918*b*: map 17, pl. 13, 1929: map 10, p. 1490, and 1925: map 4, p. 251. For a discussion of the distinguishing characters of the scarcely separable Canadian representative, var. *scopulinum, see* M. L. Fernald, Rhodora 26(305):89–93. 1924.

P. munitum (Kaulf.) Presl   Western Sword-Fern
/sT/W/   (Hr)   Moist woods and shaded talus slopes from SE Alaska (see Hultén 1941: map 18, p. 114) through w B.C. (see B.C. map by T.M.C. Taylor 1963:162) to Baja Calif. and Mont. [*Aspidium* Kaulf.; incl. vars. *imbricans* and *incisoserratum* (Eat.) Maxon]. MAPS: Hultén 1968*b*:54; T.M.C. Taylor 1970:183 ("Pacific Northwest" area).

PTERIDIUM Gled.   Bracken

P. aquilinum (L.) Kuhn   Brake, Bracken.   Fougère d'aigle or Grande fougère
/sT/X/EA/   (Grh)   Open woods, thickets, burns, clearings, and abandoned fields, the aggregate species from SE Alaska and B.C. to sw Alta., s Man. (N to The Narrows of L. Winnipeg; not known from Sask.), Ont. (N to L. Nipigon), Que. (N to SE James Bay, L. Mistassini, and the Côte-Nord (an 1895 collection in CAN by Low, labelled "s Labrador", was probably taken here)), Nfld., N.B.,

P.E.I., and N.S., s to Calif., Tex., S.Dak., Mo., Tenn., and N.C. (isolated in Miss.); Eurasia. MAPS
and synonymy: *see* below.
1   Indusia glabrous (rarely the fertile ones slightly pubescent or ciliate); fronds usually
     ternate, their pinnules more or less oblique to the rachis of the primary segments;
     ultimate segments essentially glabrous except for the moderately pubescent margins and
     the usually pubescent midrib beneath; [*Pteris lat.* Desv.; var. *champlainensis* Boivin in
     part; SE Alaska–B.C.–sw Alta.; s Man. to Nfld. (type locality) and N.S.; MAPS: Hultén 1962:
     map 131, p. 141; R. M. Tryon, Rhodora 43(506): map 8, p. 42. 1941] . . . .
     . . . . .var. *latiusculum* (Desv.) Underw.
1   Indusia ciliate and sometimes also pubescent on the outer surface; fronds not ternate
     (usually 2–4-pinnate-pinnatifid), their pinnules usually nearly or quite at right angles to the
     rachis; ultimate segments usually densely pubescent beneath.
     2   Upper surface of the ultimate segments usually glabrous or nearly so (even along the
          margins); [*Pteris aquilina* L.; Eurasia; MAPS: Hultén 1962: map 131, p. 141; Meusel,
          Jaeger, and Weinert 1965:12 (aggregate species); Tryon, loc. cit., map 1, p. 14]
          . . . . .var. *aquilinum*
     2   Upper surface of the ultimate segments moderately pubescent along the margins and
          also often pubescent over the back; [var. *champlainense* Boivin in part; var.
          lanuginosum (Bong.) Fern., not *Pteris languinosa* Bory; SE Alaska–B.C.–sw Alta.;
          Ont. (NW shore of L. Superior; Bruce Pen., L. Huron); Que. (Megantic Co.; near L.
          Chibougamau, about 100 mi NW of L. St. John); MAPS: Hultén 1962: map 131, p. 141,
          and 1968*b*:43; Tryon, loc. cit., map 3, p. 24] .var. *pubescens* Underw.

THELYPTERIS Schmidel   Thélyptéride

1   Fronds broadly deltoid, their longest pinnae at or near the base; rachis winged; indusia
     none; sporangia bearing spine-like processes.
     2   United upper and lower basal segments of the opposite pinnae forming a
          fiddle-shaped wing, the confluent wings extending down to the lowest pair of pinnae
          (these contracted at base, somewhat asymmetrical, rarely deflexed); frond commonly
          as broad as or broader than long, minutely glandular-puberulent but rarely hairy
          beneath, pilose above, with or without slender whitish chaffy scales; (s Ont. and
          s Que.) . . . .*T. hexagonoptera*
     2   United basal segments of pinnae not forming a fiddle-shaped wing, the confluent
          wings not extending down to the somewhat remote lowest pair of opposite pinnae
          (these not much contracted at base, nearly symmetrical, projected downward and
          forward); frond commonly about 2/3 as broad as long, more or less hairy on both
          surfaces, usually with lanceolate to ovate brownish scales beneath; (transcon-
          tinental) . . . .*T. phegopteris*
1   Fronds lanceolate or elliptic-lanceolate, the lower pinnae shorter than to barely equalling
     the middle ones; rachis not winged; indusia present, at least before maturity; sporangia
     not spiny.
     3   Fronds truncate at base, the lowest pair of pinnae rarely less than 1/3 as long as the
          middle ones; sori finally confluent.
          4   Basal pair of pinnae mostly at least 1/2 as long as the middle ones; fronds
               somewhat dimorphic, the ultimate segments of the fertile ones narrower than
               those of the sterile and commonly revolute-margined; lateral veins of sterile
               pinnae mostly forking; indusia glabrous or with a few long cilia; (E Man. to Nfld.
               and N.S.) . . . .*T. palustris*
          4   Basal pair of pinnae commonly 1/3 as long as the middle ones; fronds essentially
               uniform; lateral veins of both sterile and fertile pinnae simple; indusia minutely
               glandular-ciliate; (s Ont., s Que., and N.S.) . . . .*T. simulata*
     3   Fronds tapering about equally at both ends, the lowest pair of pinnae many times
          shorter than the middle ones; sori rarely confluent; indusia minutely glandular-ciliate.
          5   Rhizome relatively stout and short-creeping; ultimate segments of frond 4 or 5 mm
               broad; veins often forked; (Alaska–B.C.) . . . .*T. limbosperma*

5 Rhizome slender and long-creeping; ultimate segments of frond about 2 mm broad; veins mostly simple.
    6 Species of Vancouver Is., B.C. .... *T. nevadensis*
    6 Species of Ont. to Nfld. and N.S. .... *T. noveboracensis*

T. hexagonoptera (Michx.) Weath.   Broad Beech-Fern
/T/EE/ (Grh) Wooded slopes and open rocky thickets from E Kans. to Minn., s Ont. (N to the N shore of L. Ontario), sw Que. (Monteregian Hills: Mt. St. Bruno, about 14 mi E of Montreal, and Mt. Johnson, about 22 mi SE of Montreal), and New Eng., s to E Tex. and N Fla. [*Polypodium* Michx.; *Dryopteris* Chr.; *Phegopteris* Fée]. MAP: Hultén 1962: map 107 (*Dry. hex.*), p. 117.
According to Marcel Raymond and James Kucyniak (Am. Fern J. 37(4): 97–99. 1947), the two Monteregian stations are the first authentic records of the plant in Que., other reports (N to Quebec City), as far as could be traced, proving referable to *T. phegopteris,* to which the report from St-Pierre and Miquelon by Louis Arsène (Rhodora 29(346): 205. 1927) also probably refers.

T. limbosperma (All.) Fuchs   Mountain-Fern
/sT/W/EA/ (Hr) Rocky slopes and banks of streams from the Aleutian Is. and SE Alaska through coastal B.C. (*see* B.C. map by T.M.C. Taylor 1963:162) to Wash.; Eurasia. [*Polypodium* All.; *Dryopteris* (*Aspidium; Polypodium; Polystichum*) *oreopteris* (Ehrh.) Maxon and its var. *hesperia* (Slosson) Broun]. MAPS: Hultén 1968*b*:45, and 1962: map 144, p. 153 (*Dry. oreo.*); Meusel, Jaeger, and Weinert 1965:17; T.M.C. Taylor 1970:193 (N. American area).

T. nevadensis (Eat.) Clute   Sierra Wood-Fern
/t/W/ (Grh) Moist meadows and wooded slopes from sw B.C. (known only from Sooke, s Vancouver Is.; *see* B.C. map by T.M.C. Taylor 1963:162) to Calif. and Nev. [*Aspidium* Eat.; *Dryopteris* Underw.; *D. oregana* Chr.]. MAP: T.M.C. Taylor 1970:195 ("Pacific Northwest" area).

T. noveboracensis (L.) Nieuwl.   New York Fern
/T/EE/ (Grh) Moist woods, thickets, and swamps from Minn. to Ont. (N to the E shore of L. Superior and the Ottawa dist.), Que. (reported N to Quebec City by John Macoun 1890), Nfld. (Fernald *in* Gray 1950), N.B., P.E.I., and N.S., s to Ark., Miss., and Ga. [*Polypodium* L.; *Aspidium* Sw.; *Dryopteris* Gray].

T. palustris (Salisb.) Schott   Marsh-Fern
/T/EE/EA/ (Grh) Low woods and swampy ground from SE Man. (N to Riverton, about 75 mi N of Winnipeg) to Ont. (N to Big Trout L. at ca. 53°45′N and sw James Bay), Que. (N to the Gaspé Pen. and Magdalen Is.), Nfld., N.B., P.E.I., and N.S., s to ?Tex., Okla., Tenn., and Ga.; Eurasia. [*Polypodium* Salisb.; *Acrostichum (Aspidium; Dryopteris) thelypteris* L.]. MAP: Hultén 1962: map 170, p. 181.
The plant of N. America and NE Asia has been separated as var. *pubescens* (Lawson) Fern. (*Lastrea thelypteris* var. *pub.* Lawson, the type from Odessa, Addington Co., Ont.; indusia glabrous or long-ciliate (rarely glandular-ciliate) rather than margined with coarse gland-tipped teeth, veins all or nearly all simple, rather than segments of the median fertile pinnae with about half of their veins forking). Var. *pubescens* f. *suaveolens* (Clute) Prince, the fresh plant glandular-aromatic, is known from N.S. (Cape Breton Is.).

T. phegopteris (L.) Slosson   Long Beech-Fern
/aST/X/GEA/ (Grh) Cool rocky banks and woodlands from the Aleutian Is. and cent. Alaska to SE Yukon, NW Dist. Mackenzie (Porsild and Cody 1968), N Sask. (L. Athabasca), Man. (N to Tod L. at ca. 57°N), Ont. (N to Sioux Lookout, about 175 mi NW of Thunder Bay), Que. (N to Chimo, s Ungava Bay), Labrador (N to Bowdoin Harbour, 60°24′N), Nfld., N.B., P.E.I., and N.S., s to Oreg., Iowa, Ind., Tenn., and N.C.; w and E Greenland N to ca. 66°N; Iceland; Eurasia. [*Polypodium* L.; *Dryopteris* Chr.; *Phegopteris polypodioides* Fée; *Polypodium (Pheg.) connectile* Michx.]. MAPS (*Dry. pheg.*): Hultén 1968*b*:46, and 1962: map 107, p. 116; Meusel, Jaeger, and Weinert 1965:16.

T. simulata (Davenp.) Nieuwl.   Massachusetts Fern

/T/EE/   (Grh)   Shaded margins of sphagnum bogs and moist woods: Ont. (reported by Dore and Gillett 1955, from near Cornwall, Stormont Co.), sw Que. (Megantic and Missisquoi counties), and N.S. (known from 6 of the southwestern counties; reported from P.E.I. by Broun 1938, and Roland 1947), s to Wisc., Pa., and Va. [*Aspidium* Davenp.; *Dryopteris* Davenp.]. MAPS (dots should be added to indicate the Ont. and Que. localities): T. G. Hartley, Rhodora 67(772): 401. 1965; H. K. Svenson, Am. Fern J. 38(4): 196. 1948.

WOODSIA R. Br.   Woodsia

1   Stipes jointed near base, the persistent old bases mostly 2 or 3 cm long; fronds not glandular; rhizome-scales uniformly brown; segments of indusia filamentous; (transcontinental).
  2   Stipe densely brown-chaffy near base; fronds lanceolate, firm, pilose and usually chaffy beneath; segments of indusia conspicuously overtopping the sporangia . . . .*W. ilvensis*
  2   Stipe nearly or quite naked; fronds glabrous or promptly glabrate.
    3   Segments of indusium conspicuously overtopping the sporangia; stipes firm; fronds linear to linear-oblanceolate . . . .*W. alpina*
    3   Segments of indusium only slightly overtopping the sporangia; stipes green or straw-coloured, delicate; fronds linear . . . .*W. glabella*
1   Stipes not jointed, remotely chaffy, the old bases of irregular length; fronds mostly minutely glandular; rhizome-scales often with a dark central stripe; indusia more or less concealed beneath the sporangia.
  4   Indusia cup-like, splitting into about 6 broad, toothed segments; fronds lanceolate, to about 4 dm long, minutely glandular or granular beneath; (sw Que.) . . . .*W. obtusa*
  4   Indusia splitting nearly to base into narrow segments; fronds usually less than 3 dm long; (B.C. to NW Sask.; Ont. and E Que.).
    5   Frond and stipe minutely hispid with white hairs mixed with the glandular pubescence; indusial segments linear- to lance-attenuate . . . .*W. scopulina*
    5   Frond and stipe not hispid; indusial segments filiform to linear . . . .*W. oregana*

W. alpina (Bolton) S. F. Gray   Northern Woodsia

/AST/X/GEA/   (Hr)   Crevices and talus of slaty or calcareous rocks from N Alaska–Yukon–Dist. Mackenzie to cent. Dist. Keewatin, cent. Ellesmere Is., and cent. Baffin Is., s to w-cent. B.C., Great Slave L., Man. (s to Norway House, off the NE end of L. Winnipeg; John Macoun 1890, as *W. hyperborea*), N Minn., Ont. (N Great Lakes region), Que. (s to Rimouski Co. and the Gaspé Pen.), cent. Labrador, Nfld., s N.B. (Fernald *in* Gray 1950; not known from P.E.I.), N.S., and Vt.; w Greenland N to ca. 79°N, E Greenland N to ca. 70°N; Iceland; Eurasia. *Acrostichum* Bolton; *W. hyperborea* (Liljebl.) R. Br.; incl. var. *bellii* Lawson (*W. bellii* (Lawson) A. E. Porsild)]. MAPS: Hultén 1968*b*:51, and 1958: map 210, p. 229; Porsild 1957: map 2, p. 161; Meusel, Jaeger, and Weinert 1965:16.

A hybrid with *W. ilvensis* (× *W. gracilis* (Lawson) Butters; *W. ilvensis* var. *gr.* Lawson) is known from the type station along the Dartmouth R., Gaspé Pen., E Que., and is reported from Rivière-du-Loup and Saguenay, Que., by R. M. Tryon, Jr. (Am. Fern J. 38(4): 165. 1948). It is also reported from w Ont. by Boivin (1966*b*), who also tentatively reports a hybrid between *W. alpina* and *W. glabella* from the same region.

W. glabella R. Br.   Smooth Woodsia

/AST/X/GEA/   (Hr)   Calcareous cliffs and ledges from N Alaska to Banks Is., Somerset Is., northernmost Ellesmere Is., northernmost Ungava–Labrador, Nfld., N.B., and N.S. (not known from P.E.I.), s to s B.C., E-cent. Sask., cent. Man., N Minn., Ont. (N shore of L. Superior), N.Y., and N New Eng.; circumgreenlandic; Iceland; Eurasia. The type is from NW Canada. MAPS: Hultén 1968*b*: 52, and 1962: map 38, p. 45; Porsild 1957: map 3, p. 161.

A hybrid with *W. ilvensis* (later named × *W. tryonis* Boivin; type from Sleeping Giant, Thunder Cape) is reported from Ont. by R. M. Tryon, Jr. (Am. Fern J. 38(4): 167. 1948; Sleeping Giant, Thunder Bay).

W. ilvensis (L.) R. Br.   Rusty or Fragrant Woodsia
/aST/X/GEA/   (Hr)   Dry rocks, cliffs, and talus from N Alaska to SE Yukon, Great Bear L., the coast of Dist. Mackenzie (Coronation Gulf), cent. Dist. Keewatin, Hudson Strait (Nottingham Is.), cent. Baffin Is., northernmost Ungava–Labrador, Nfld., N.B., and N.S. (not known from P.E.I.), s to s B.C., N Alta., cent. Sask., s Man., N Iowa, N Ill., Mich., and N.C.; w and E Greenland N to ca. 75°N; Iceland; Eurasia. [*Acrostichum* L.; *Nephrodium* (*Aspidium*) *rufidulum* Michx.]. MAPS: Hultén 1968b: 51, and 1962: map 49, p. 57; Porsild 1957: map 1, p. 161; Raup 1947: pl. 13; Meusel, Jaeger and Weinert 1965:15.

W. obtusa (Spreng.) Torr.   Blunt-lobed or Large Woodsia
/T/EE/   (Hr)   Rocky woods, ledges, and dry slopes from Minn. to Nebr., Ohio, sw Que. (known only from limestone ledges at Frelighsburg and St-Armand, Missisquoi Co., near the N end of L. Champlain; CAN; MT; reports from N.S. and B.C. by John Macoun 1890, undoubtedly refer to other species), and cent. Maine, s to E Tex., Ala., and Ga. [*Polypodium* Spreng.].

W. oregana Eat.   Oregon Woodsia
/T/D/   (Hr)   Calcareous or siliceous ledges and cliffs (ranges of Canadian taxa outlined below), s in the West to Baja Calif., N.Mex., and w Okla., with an inland area in the U.S.A. from N Iowa to N Wisc. MAP and synonymy: see below.
1   Stipe and frond-blade sparingly pubescent and glandular, also sparingly scaly; [*W. abbeae* Butters; *W. confusa* Taylor; Sask. (type from Beaver L., near the Manitoba boundary at ca. 55°N) and Ont. (N shore of L. Superior)] . . . . .var. *squammosa* Boivin
1   Stipe and frond-blade glabrous or glandular (but neither pubescent nor scaly) . . . .var. *oregana*
   2   Stipe and frond-blade copiously glandular throughout; [f. *glandulosa* Taylor; *W. cath.* Rob.; range of f. *oregana*]. A hybrid with *W. scopulina* (× *W. maxonii* Tryon) is known from the type station, Sleeping Giant, near Thunder Bay, Ont. . . . .
. . . . .f. *cathcartiana* (Rob.) Boivin
   2   Stipe and frond-blade glabrous or slightly glandular only toward the insertion of the pinnules; [B.C. (see B.C. map by T.M.C. Taylor 1963:163; a dot should be added for an isolated station at Liard Hot Springs, ca. 59°10′N), Alta. (N to L. Mamawi, ca. 58°35′N), and Sask. (L. Athabasca); Ont. (Manitoulin Is., L. Huron; Blackwater R. SE of L. Nipigon); E Que. (Cap-Enragé, near Bic, Rimouski Co.); MAP: Fernald 1925: map 17 (incomplete northwards), p. 255] . . . . .f. *oregana*

W. scopulina Eat.   Rocky Mountain Woodsia
/sT/D/   (Hr)   Rock-crevices, ledges, and talus from s Alaska (see Hultén 1941: map 4, p. 111) through B.C. and sw Alta. (N to Jasper) to s Calif., N.Mex., and S.Dak.; N shore of L. Athabasca, Sask.; Ont. (NW shore of L. Superior; Algonquin Park, Renfrew Co.), locally to N Minn., N Wisc., and N Mich.; E Que. (Christie, Tourelle, and Marten R., Gaspé Co.). [*W. obtusa* var. *lyallii* Hook.]. MAPS: Hultén 1968b:50; Fernald 1925: map 9 (incomplete northwards), p. 251.

WOODWARDIA Sm.   Chain-Fern

1   Fronds dimorphic, very deeply 1-pinnatifid (with a winged rachis) into rarely more than 10 pairs of segments; sterile blades net-veined (only the outermost veinlets free), deltoid-ovate, to about 2 dm long, their minutely serrulate segments lanceolate or narrowly oblong; fertile blades taller and narrower, their linear divisions almost distinct; (N.S.) . . . .*W. areolata*
1   Fronds uniform, longer, pinnate-pinnatifid (primary rachis wingless), the commonly 15 or more sessile pinnae deeply pinnatifid; net-venation confined to 1 or 2 rows of areolae along each side of the midveins of the pinnae and their segments.
   2   Frond to 3 m long, the stipe short, the linear-oblong to oblong-ovate blade to 5 dm broad, its veins resinous-dotted; middle pinnae to 3 dm long, linear-oblong to narrowly ovate, long-acuminate; ultimate segments long-acuminate, spinulose-serrate, commonly 3 or 4 cm long; sori 3 or 4 mm long; (s B.C.) . . . .*W. fimbriata*
   2   Frond to about 1.5 m tall, the stipe relatively long, the oblong-lanceolate blade to 8 dm long and 3 dm broad, its veins not resinous-dotted (but the surface more or less

stipitate-glandular); pinnae commonly not over 1 dm long and 2 cm broad, acute or short-acuminate; ultimate segments acutish or rounded, subentire or minutely serrulate, about 1 cm long; sori about 1.5 mm long; (Ont. to N.S.) . . . .*W. virginica*

W. areolata (L.) Moore   Netted Chain-Fern

/T/EE/   (Grh)   Acid peat, swampy ground, and boggy woods: sw Mich.; sw N.S. (Yarmouth, Shelburne, and Queens counties; *see* N.S. map by Roland 1947:134); SE N.H., s on or near the Coastal Plain of Fla. and Tex., inland to Minn. and Tenn. [*Acrostichum* L.; *Lorinseria* Presl; *W. onocleoides* Willd.]. MAPS: *Atlas of Canada* 1957: map 14, sheet 38; M. L. Fernald, Rhodora 33 (386): map 30, p. 55. 1931; W. J. Cody, Am. Fern J. 53(1): pl. 1, p. 18. 1963.

W. fimbriata Sm.   Great Chain-Fern

/t/W/   (Grh)   Moist canyons, stream banks, and springy places from sw B.C. (Lasqueti Is.; Texada Is.; reported from Saanich Arm, Vancouver Is.; *see* B.C. map by T.M.C. Taylor 1963:164) to Calif. and Ariz. [*W. chamissoi* Brack.; *W. spinulosa sensu* Henry 1915, not Mart. & Gal.; *W. radicans* var. *americana* Hook. in part]. MAPS: W. J. Cody, Am. Fern J. 53(1): pl. 3, p. 26. 1963; T.M.C. Taylor 1970: 209 ("Pacific Northwest" area).

W. virginica (L.) Sm.   Virginian Chain-Fern

/T/EE/   (Grh)   Acid peat, swamps, and moist thickets from Ill. and s Mich. to s Ont. (N to L. Nipissing and the Ottawa dist.), s Que. (N to L. St. Peter, about 45 mi NE of Montreal; the report from near Gaspé Basin, E Que., by John Macoun and T.J.W. Burgess (Proc. Trans. R. Soc. Can. 2 (Sect. 4): 190. 1885) probably refers to some other species), N.B., P.E.I., and N.S., s to Tex. and Fla.; Bermuda. [*Blechnum* L.; *Anchistea* Presl]. MAP: W. J. Cody, Am. Fern J. 53(1): pl. 2, p. 22. 1963.

MARSILEACEAE (Marsilea Family)

MARSILEA L. Pepperwort

Aquatic plants with elongate stems creeping and rooting at the nodes. Leaves 4-foliolate, solitary at the nodes, on petioles to about 3.5 dm long. Leaflets obdeltoid, to about 2.5 cm long and broad, sessile. Spores borne in specialized, thick-walled, basally 2-toothed spore-cases (sporocarps), these at maturity splitting into two valves and extruding the elastic elongating central receptacle, this bearing 1-spored megasporangia at the summit and many-spored microsporangia on the sides.

1   Pedicel or peduncle elongate, adnate to the base of the petiole; sporocarps to 5 mm long, becoming glabrate, commonly 2 or 3 on a peduncle, the basal teeth low and blunt, the lowest one the largest; leaflets to about 2.5 cm long; (introd. in s Ont.) . . . .*M. quadrifolia*
1   Pedicel of the solitary sporocarp short, arising directly from the rhizome and apparently axillary; sporocarp to 7 mm long, coarsely and persistently strigose, the basal teeth sharp and conspicuous, the upper one the longest; leaflets rarely as much as 1.5 cm long, usually appressed-hairy; (B.C. to Sask.) . . . .*M. vestita*

M. quadrifolia L. Pepperwort
Eurasian; natzd. in the E U.S.A. and reported by Bert Miller (Am. Fern J. 46(2):90. 1956) from Nanticoke Creek, s of Nanticoke, Haldimand Co., s Ont. where already established for many years and forming a dense stand over about half an acre of water surface. It is also reported from s Ont. by Landon (1960; Woodhouse Twp., Norfolk Co.).

M. vestita Hook. and Grev. Clover-Fern
/T/X/ (HH) Shallow ponds, pools, and wet shores from s B.C. (Kamloops) to s Alta. (Redcliff; Spur Creek; Hand Hills; Cypress Hills), s Sask. (Glen Kerr; Juniata; Trossachs; Wiseton; Yeomans; the report from Man. by Burman 1910, requires confirmation), and w Minn., s to Calif., N Mexico, Tex., and Fla. [*M. mucronata* A. Br.]. MAP: T.M.C. Taylor 1970:65 ("Pacific Northwest" area).

SALVINIACEAE (Salvinia Family)

AZOLLA Lam.   Water-fern

Small, moss-like, free-floating aquatics with dichotomously branching leafy stems producing rootlets on the under side. Leaves simple, oblong to broadly ovate or roundish, sessile, spreading-ascending, to 1 mm long. Spores borne in pairs of specialized thin-walled indusia or spore-cases (sporocarps) on a common peduncle beneath the stem, one of each pair usually a small narrowly acorn-shaped or subfusiform megasporocarp containing a single basal megaspore, the other usually a larger globose microsporocarp containing many microsporangia, the microspores agglutinated into several globose masses armed with numerous barbed bristle-like appendages (glochidia).

1   Glochidia with scattered internal septa or cross-walls; leaves crowded and overlapping; plant dichotomously branched, to about 1.5 cm in diameter; (?B.C.) . . . .[*A. mexicana*]
1   Glochidia lacking cross-walls (or rarely with 1 or 2 septa near the apex).
   2   Plant elongate, to 6 cm long, the oblong to ovate, papillose leaves closely appressed and overlapping, about 1 mm long; microsporangia up to about 100 in each indusium; (introd. in ?Alaska and B.C.) . . . .*A. filiculoides*
   2   Plant dichotomously branched, to about 1 cm in diameter, the nearly orbicular, smooth leaves not closely overlapping, about 0.5 mm long; microsporangia commonly less than 40 in each indusium; (s Ont.) . . . .*A. caroliniana*

A. caroliniana Willd.   Mosquito-Fern
/T/EE/   (HH)   Quiet waters, the main area from N.C. to La. and Fla., locally northwards to Wisc., Ind., Ohio, s Ont. (beach at Hamilton, Wentworth Co., where taken by Logie in 1862; CAN; Soper 1949, also notes a 1934 collection on the N.Y. side of the lake; reported from L. Ontario by Pursh 1814; reports from B.C. by John Macoun 1890, are based upon *A. filiculoides,* relevant collections in CAN; the report from Alaska by Broun 1938, requires clarification), N.Y., and Mass.; W.I.; Mexico to Patagonia.

A. filiculoides Lam.
Native in the w U.S.A. and S. America; spread from cult. or introd. elsewhere, as in Shuswap L., B.C., near Salmon Arm and Sicamous; reported from Alaska by H. K. Svenson (Am. Fern J. 34(3): 79. 1944) but not listed by Hultén (1941; 1968*b*).

[A. mexicana Presl]
[Reports of this species from Sicamous, B.C. (Svenson, loc. cit.; Eastham 1947) probably refer to *A. filiculoides*, from which it appears scarcely separable.]

Division II SPERMATOPHYTA (Seed-plants)

NOTE: the numbers in square brackets following the generic-name headings are those of K. W. von Dalla Torre and Hermann Harms *(Register zu Genera Siphonogamarum.* Graz, Akademishe Druck-u. Verlagsantalt. 1958). In the few cases of genera not listed by them *(Podagrostis; Schizachne; Sphenopholis; Torreyochloa; Geocaulon; Idahoa),* an upper-case "A" has been added to the number of the apparently most closely related genus, followed by the name of that genus. In using these numbers, it must be borne in mind that the present treatment of some genera actually comprises a combination of species often assigned to more than one genus. In the Polemoniaceae, for example, *Collomia* and *Microsteris* are both numbered 7015 and *Gilia, Leptodactylon, Linanthus,* and *Navarretia* are all numbered 7016.

Subdivision I GYMNOSPERMAE (Gymnosperms) (see p. 94)

TAXACEAE (Yew Family)

TAXUS L.   [18]   Yew.   If

Leaves linear, flat, rigid, pointed, 1 or 2 cm long and 1 or 2 mm broad, yellowish green beneath, spirally arranged but spreading in two ranks. Staminate flowers solitary in the axils. Pistillate flowers in pairs, each subtended by a pair of scales. Fruit globular, red and berry-like, consisting of a seed surrounded by the fleshy aril except at the open summit.

1 Tree usually 5 or 10 m tall (up to 20 m), with a trunk to 1 m thick; (B.C. and sw Alta.) . . . .
. . . .*T. brevifolia*
1 Straggling, diffusely branched shrub rarely over 1 m tall; (SE Man. to Nfld. and N.S.) . . . .
. . . .*T. canadensis*

T. brevifolia Nutt.   Pacific or Western Yew
/T/W/ (Mc (Ms)) Low elevations near the coast up to about 4,000 ft inland from the southernmost Alaska Panhandle *(see* Hultén 1941: map 54, p. 120) through coastal B.C. (and in the Selkirks and Rocky Mts. of s B.C.) and sw Alta. (Waterton Lakes) to Calif. and w Mont. MAPS: Hosie 1969:110; Hultén 1968*b*:59; Canada Department of Northern Affairs and Natural Resources 1956: 2; Meusel, Jaeger, and Weinert 1965:19; Preston 1961:118, and 1947:92; Munns 1938: map 60, p. 64; Little 1971: map 86-N.

T. canadensis Marsh.   American Yew, Ground Hemlock.   Buis de sapin
/T/EE/ (N(Ch)) Rich woods and thickets from SE Man. (N to the shore of L. Winnipeg at 51°40′N; concerning a report from near York Factory, Hudson Bay, *see* Scoggan 1957) to Ont. (N to Sandy L., ca. 53°N, 94°W), Que. (N to the Nottaway R. at 50°57′N, L. Mistassini, and the Côte-Nord), Nfld. (reports from Labrador, as in the Que.–Labrador map by Marie-Victorin 1927*b*: fig. 38, p. 137, require confirmation), N.B., P.E.I., and N.S., s to Iowa and Va. [*T. baccata* of Canadian reports, not L.; *T. bacc.* vars. *can.* (Marsh.) Gray and *minor* Michx.]. MAPS: *Atlas of Canada* 1957: map 13, sheet 38; Meusel 1943: fig. 33B; Meusel, Jaeger, and Weinert 1965:19; Little 1971: map 86.1-N.

PINACEAE (Pine Family)

Trees (or *Juniperus* often a shrub) with resinous juice, mostly evergreen (except *Larix*). Leaves scale-like, awl-shaped, or needle-shaped, entire, solitary or clustered. Flowers unisexual, borne in scaly aments, of which the pistillate ones become woody-scaled cones (or berry-like cones in *Juniperus*).

(Selected references for maps: Tolmachev 1952; *Atlas of Canada* 1957; Canada Department of Northern Affairs and National Resources 1956; Fowells 1965; Halliday and Brown 1943; Hosie 1969; Hough 1947; Hustich 1953; Munns 1938; Preston 1947 and 1961; Little 1971).

1   Leaves scale-like or awl-shaped, at most about 1 cm long, opposite or whorled.
    2   Fruit berry-like, dark blue or blue-black; branchlets scarcely flattened . . . .*Juniperus*
    2   Fruit a dry cone (commonly about 1 cm long) with 4 to 6 pairs of thin brown scales; leaves scale-like, rarely over 4 mm long; ultimate branchlets very soft and flat.
        3   Leaves sharp-pointed, prickly to the touch; cones subglobose, falling during the winter, their scales peltate, not overlapping; tree to about 40 m tall, with a trunk to about 2 m thick, the branchlets arranged in flat drooping sprays; (coastal B.C.)
            . . . .*Chamaecyparis*
        3   Leaves blunt; cones and their oval or oblong scales overlapping, the cones persisting during the winter . . . .*Thuja*
1   Leaves linear or needle-like, mostly over 1 cm long, alternate or in clusters; fruit a dry cone.
    4   Leaves in clusters of 2 or more.
        5   Leaves at most 5 in a cluster, evergreen, the clusters sheathed at base; cones at least 3 cm long . . . .*Pinus*
        5   Leaves many in a cluster, deciduous, at most 2.5 cm long, the cluster lacking a sheath; cones 1 or 2 cm long . . . .*Larix*
    4   Leaves solitary at the nodes, spirally arranged.
        6   Leaves mostly 4-angled and easily rolled between the fingers (except in *P. sitchensis*), spreading in all directions, their persistent woody bases roughening the branches; cones drooping, their scales persistent . . . .*Picea*
        6   Leaves flat, appearing 2-ranked, whitened along two lines beneath.
            7   Cones erect, their scales deciduous; bracts (subtending seeds) normally shorter than and concealed by the scales; leaves sessile, leaving the branches smooth . . . .*Abies*
            7   Cones pendulous, their scales persistent.
                8   Leaves blunt or round-pointed, distinctly petioled; bracts much shorter than the cone-scales . . . .*Tsuga*
                8   Leaves sharp-pointed, only slightly narrowed at base; bracts much surpassing the cone-scales, 3-forked; (s half of B.C. and w Alta.) . . .*Pseudotsuga*

ABIES Mill. [29] Fir. Sapin or Sapin baumier

1   Cones yellow-green to green-purple, to about 18 cm long, their scales much broader than long; leaves to about 5 cm long, dark green and shining above, with two broad whitish stomatal bands beneath, distinctly 2-ranked, the branches forming flat sprays; (s B.C.) . . . .*A. grandis*
1   Cones darker purple, their scales about as long as or slightly longer than broad; leaves rarely over 3 or 4 cm long, the lateral and upper ones commonly ascending.
    2   Bark unbroken except on very old trunks; leaves dark green and shining above, with two broad whitish stomatal bands beneath; cones to about 1.5 dm long; (chiefly coastal mts. of B.C.) . . . .*A. amabilis*
    2   Bark on old trunks usually shallowly furrowed or somewhat roughened with reddish plates or scales; cones to about 1 dm long.
        3   Leaves dark green and shining above, with two broad whitish stomatal bands beneath; (Alta. to Labrador, Nfld., and N.S.) . . . .*A. balsamea*

178

3 Leaves blue-green and glaucous, with whitish stomatal lines on both surfaces;
   (mts. of B.C. and sw Alta.) ....*A. lasiocarpa*

A. amabilis (Dougl.) Forbes   Amabilis or Pacific Silver Fir
/T/W/   (Ms (Mg))   Lowland to subalpine elevations from the southernmost Alaska Panhandle (see Hultén 1941: map 62, p. 122) through coastal B.C. along the w slope of the Coast Range to N Calif. [*Pinus* Dougl.; *Picea* Dougl.; *Pinus grandis* Don, not *A. grandis* Lindl.]. MAPS: Hultén 1968*b*: 63; Fowells 1965:31; Preston 1961:82; Hosie 1969:92; Canada Department of Northern Affairs and Natural Resources 1956:68; Benson 1962: fig. 5–27, p. 201; Halliday and Brown 1943: fig. 6, p. 365; Munns 1938: map 45, p. 49; Little 1971: map 1-N.

A. balsamea (L.) Mill.   Balsam-Fir.   Sapin or Sapin baumier
/sT/(X)/   (Ms)   Woods at low to fairly high elevations, the aggregate species from N-cent. Alta.–Sask.–Man. to Ont. (N to Big Trout L. at ca. 53°45′N, 90°W), Ungava–Labrador (N to ca. 58°N), Nfld., N.B., P.E.I., and N.S., s to NE Iowa, Ohio, and Va. MAPS and synonymy: see below.
1 Bracts subtending seeds shorter than the cone-scales ....var. *balsamea*
   2 Tree; leaves to about 2.5 cm long; [*Pinus* L.; *Picea* Loud.; *A. (Pinus) balsamifera* Michx. in part; B.C. (Boivin 1966*b*) to Nfld. and N.S.; MAPS: in all of the map-reference publications listed under the Pinaceae family description except that by Preston 1947] ....f. *balsamea*
   2 Low or prostrate shrub with shorter and relatively broader leaves; [Anticosti Is., E Que.; J. Rousseau 1950] ....f. *hudsoniana* (Bosc) Fern. & Weath.
1 Bracts conspicuously exserted beyond the cone-scales ....var. *phanerolepis* Fern.
   3 Tree; [Ont. to Labrador (N to ca. 55°45′N), Nfld., Que. (type from Percé, Gaspé Co.), and N.S.] ....f. *phanerolepis*
   3 Prostrate or trailing shrub; [type from Mt-Blanc, Matane Co., E Que.] .....f. *aurayana* Boivin

A. grandis (Dougl.) Lindl.   Grand or Lowland Fir
/T/W/   (Ms (Mg))   Stream bottoms, valleys, and mountain slopes from s B.C. (N to ca. 51°N, w to the Kootenay Valley) to N Calif. and Idaho [*Pinus grandis* Dougl., not Don]. MAPS: Fowells 1965:19; Benson 1962: fig. 5-27, p. 201; Preston 1961:80, and 1947:64; Canada Department of Northern Affairs and National Resources 1956:70; Halliday and Brown 1943: fig. 6, p. 365; Munns 1938: map 43, p. 47; Little 1971: map 6-W; Hosie 1969:94.

A. lasiocarpa (Hook.) Nutt.   Subalpine, Alpine, or Rocky Mountain Fir
/sT/W/   (Ms)   Lowland to subalpine elevations from the Alaska Panhandle (see Hultén 1941: map 63, p. 122), the Yukon (N to ca. 64°N), and sw Dist. Mackenzie (N to Brintnell L. at ca. 62°N) through B.C. and the s half of w Alta. to Oreg., N Nev., Ariz., and N.Mex. [*Pinus* Hook.; *Picea* Murray; *A. subalpina* Engelm.; *A. balsamea* var. *fallax* (Engelm.) Boivin]. MAPS: Hultén 1968*b*:64; Fowells 1965:37; Benson 1962: fig. 5-27, p. 201; Preston 1961:76, and 1947:58; *Atlas of Canada* 1957: sheet 41; Hosie 1969:90; Canada Department of Northern Affairs and National Resources 1956:66; Raup 1947: pl. 14; Munns 1938: map 41, p. 45; Whitford and Craig 1918: map facing p. 62; Halliday and Brown 1943: fig. 6, p. 365; Little 1971: map 7-N; W. J. Cody, Nat. can. (Que.) 98(2): fig. 12, p. 149. 1971.

CHAMAECYPARIS Spach   [44]   False Cypress

C. nootkatensis (Lamb.) Spach   Nootka, Alaska, or Yellow Cypress or Cedar
/sT/W/   (Ms)   Low to medium elevations from s Alaska (see Hultén 1941: map 65, p. 122) through the islands and coast of B.C. w of the Coast Range (Fowells 1965, also notes, and indicates on his below-noted map, isolated stands in SE B.C. about 450 mi inland) to N Calif. [*Cupressus* Lamb.; *Thuja excelsa* Bong.]. MAPS: Hosie 1969:102; Hultén 1968*b*:65; Fowells 1965:146; Preston 1961:104; *Atlas of Canada* 1957: sheet 41; Canada Department of Northern Affairs and National Resources 1956:78; Munns 1938: map 56, p. 30; Little 1971: map 12-N.

JUNIPERUS L.   [45]   Juniper.   Genévrier

1   Leaves linear or linear-subulate, appressed-ascending to spreading, mostly in whorls of
3, the berry-like fruits sessile in their axils; (transcontinental) ....*J. communis*
1   Leaves of adult branches scale-like, crowded and strongly appressed, mostly opposite;
berry-like fruit terminal on the branchlets.
2   Prostrate shrub with trailing branches; fruits on short recurved pedicels; (transcontinental) ....*J. horizontalis*
2   Erect trees to over 10 m tall; fruits on straightish erect pedicels.
3   Fruit maturing the first year; leaves in whorls of 3; (Ont. and sw Que.) ......*J. virginiana*
3   Fruit maturing the second year.
4   Leaves in 3's, minutely toothed, conspicuously glandular on the back;
heartwood brown; (?B.C.) ....[*J. occidentalis*]
4   Leaves usually in 2's, entire, not glandular; heartwood red; (s B.C. and sw
Alta.) ....*J. scopulorum*

J. communis L.   Common Juniper.   Genévrier commun
/aST/X/GEA/   (N (Ch))   Dry rocky soil and sterile pastures and fields, the aggregate species
from N Alaska–Yukon–Dist. Mackenzie to s Dist. Keewatin, northernmost Ont., Que. (N to s Ungava
Bay), Labrador (N to ca. 59°N), Nfld., N.B., P.E.I., and N.S., s to Calif., N.Mex., Nebr., N Ill., N Ind., N
Ohio, and Ga.; w and E Greenland N to ca. 70°N; Iceland; Eurasia. MAPS and synonymy: *see* below.
1   Main trunk erect or nearly so; leaves to about 2 cm long, subulate; [var. *erecta* Pursh;
?N.S.; reported from Man. by Fernald *in* Gray 1950; MAPS: Hultén 1962: map 66, p. 75,
and 1968b: 65 (aggregate species); Meusel, Jaeger, and Weinert 1965: 22; Little 1971:
map 22-N] ....var. *communis*
1   Main trunk prostrate or the lower branches decumbent.
2   Leaves subulate, to over 1.5 cm long; fruits to 1 cm thick; [transcontinental; MAP: on
the above-noted 1962 map by Hultén] ....var. *depressa* Pursh
2   Leaves less than 1 cm long but up to 2 mm broad.
3   Fruits to 13 mm thick; seeds commonly 6 or 7 mm long; [Que. (type from
Magdalen Is.) to Nfld. and N.S.; MAP: on the above-noted 1962 map by Hultén] ....
....var. *megistocarpa* Fern. & St. John
3   Fruits less than 1 cm thick; seeds at most 6 mm long; [var. *alpina* L.; *J. nana* Willd.;
J. sibirica Burgsd.; var. *saxatilis* (var. *hemisphaerica* (J. & C. Presl) Parl.) of N.
American reports, not Pallas (*see* D. R. Hunt and H. J. Welch, Taxon 17(5): 545.
1968); transcontinental; MAPS: on the above-noted maps by Hultén; Hustich 1953:
fig. 6 (N limits), p. 157; Raup 1947: pl. 15] ....var. *montana* Ait.

J. horizontalis Moench   Creeping Savin.   Savinier
/sT/X/   (Ch)   Rocky, sandy, or boggy places from s Alaska, s-cent. Yukon, and NW Dist.
Mackenzie to Great Bear L., L. Athabasca, northernmost Ont., Que. (N to James Bay and the
Côte-Nord), Nfld., N.B., P.E.I., and N.S., s to s B.C., Wyo., Colo., Iowa, Ill., and N.Y. [*Sabina*
Rydb.; *J. prostrata* Pers.; *J. sabina sensu* Michaux 1803, not L., and its vars. *humilis* Hook. and
procumbens Pursh; *Cupressus ?thyoides senus* Hooker 1838, not L.]. MAPS: Hultén 1968b:66;
Little 1971: map 22.1-N.
Forma *alpina* (Loud.) Rehd., differing from the typical form in its more upright or ascending
habit and its juvenile acicular foliage, is reported from E Que. by Marie-Victorin (1927a; Côte-Nord,
Gaspé Pen., and Magdalen Is.), from Nfld. by Rouleau (1956), and from N.S. by Jacques
Rousseau, Nat. can. (Que.) 65: 301. 1938; Guysborough). A presumed hybrid between
J. horizontalis and *J. scopulorum* (× *J. fassettii* Boivin; *J. scopulorum* var. *patens* Fassett) is
reported from s B.C. by Boivin (1966b) and from Alta. by N. C. Fassett (Bull. Torrey Bot. Club
72(1): 46. 1945; Banff).

[*J. occidentalis* Hook.]
[Reports of this species of the w U.S.A. (Wash. and w Idaho to s Calif.) from B.C. by John Macoun
1886, and Rydberg 1922, presumably refer to either *J. horizontalis* or *J. scopulorum*. MAPS (no

Canadian stations indicated): F. C. Vasek, Brittonia 18(4): fig. 5, p. 367. 1966; Fowells 1965:223; Preston 1961:114, and 1947:76; Munns 1938: map 58, p. 62; Little 1971: map 26-W.

J. scopulorum Sarg.   Rocky Mountain Juniper
/T/WW/   (Ms)   Drier foothills and lower mts. of B.C. (N to ca. 53°N; isolated stations N to ca. 56°N) and sw Alta., s to Ariz., N.Mex., and w ?Tex. [*Sabina* Rydb.; *J. excelsa* Pursh, not Willd.]. MAPS: Fowells 1965:217; Preston 1961:110, and 1947:78; Canada Department of Northern Affairs and National Resources 1956:84; Little 1971: map 30-W: Hosie 1969:108.

J. virginiana L.   Red Cedar or Savin.   Cèdre rouge
/T/EE/   (Ms)   Dry open woods or rocky slopes and barrens (often calcareous) from Ont. (N to near Parry Sound and Ottawa) and sw Que. (N to Hull) to N.H. and Maine, s to Tex. and Fla. [*Sabina* Antoine]. MAPS: Hosie 1969:106; Fowells 1965:212; Preston 1961:108; *Atlas of Canada* 1957: map 12, sheet 38; Canada Department of Northern Affairs and National Resources 1956:82; Hough 1947:45; Munns 1938: map 59, p. 63; Little 1971: maps 31-W and 31-E.

The Canadian plant may be distinguished as var. *crebra* Fern. & Grisc., more columnar than the typical form, the branches rarely drooping, the bases of the seeds shallowly (rather than conspicuously) pitted. The typical form is reported from N.B. by C. S. Sargent (*A catalogue of the forest trees of North America,* Government Printing Office, Washington, D.C. 1880) and is indicated for that province and for N.S. on the above-noted maps by Preston, Hough, and Munns. As pointed out by M. L. Fernald (Rhodora 52(624): 274. 1950), however, neither it nor var. *crebra* occur in those provinces.

LARIX Mill.   [24]   Larch.   Mélèze

1   Cones 1 or 2 cm long, ovoid, their scales commonly about 20, longer than the bracts; leaves 3-angled, commonly less than 2.5 cm long; (transcontinental in moist or wet lowland habitats) . . . .*L. laricina*
1   Cones usually over 2.5 cm long, their bracts with awn-tips much surpassing the numerous scales; leaves usually over 2.5 cm long; (mts. of s B.C. and sw Alta.).
    2   Leaves 3-angled, rarely more than 30 in a cluster; young twigs pubescent, becoming glabrate; cone-scales usually villous below the middle; seeds about 6 mm long . . . .
    . . . .*L. occidentalis*
    2   Leaves 4-angled, commonly more than 30 in a cluster; twigs white-villous for about two seasons; cone-scales fringed and covered with matted hairs on the lower surface; seeds about 3 mm long . . . .*L. lyallii*

L. laricina (Du Roi) Koch   American Larch, Tamarack.   Épinette rouge
/ST/X/   (Ms)   Chiefly in bogs and black spruce muskeg from N-cent. Alaska, N Yukon, and the Mackenzie R. Delta to Great Bear L., s Dist. Keewatin, Que. (N to Ungava Bay), northernmost Labrador, Nfld., N.B., P.E.I., and N.S., s to E-cent. B.C., s Man., Minn., Ill., and N N.J. [*Pinus* DuRoi; *L. alaskensis* Wight; *L. (Abies) americana* Michx.; *L. (Pinus) microcarpa* Lamb.; *L. (Pinus) pendula* Lamb.]. MAPS: Fowells 1965:227; Preston 1961:44, and 1947:30; Dansereau 1957: map 2B, p. 33; *Atlas of Canada* 1957: sheet 41; Canada Department of Northern Affairs and National Resources 1956:28; Hough 1947:21; Meusel 1943: fig. 3a; Munns 1938: map 27, p. 31; Little 1971: map 32-N; Hosie 1969:56.

Forma *depressa* Rousseau (dwarf, stunted, depressed, more or less gnarled and often forming tangled mats; type from Charlevoix Co., Que.) is known from Que. (N to the Korok R. at 58°35′N, the Côte-Nord, and Anticosti Is.), Nfld., N.S. (Guysborough Co.), and Maine (Mt. Kahtadin).

L. lyallii Parl.   Alpine, Mountain, or Lyall's Larch or Tamarack
/T/W/   (Ms)   Usually on open grassy slopes and forming the upper timberline at altitudes to about 7,000 ft from the mts. of sw and SE B.C. (N to ca. 50°N) and sw Alta. (N to Banff) to Wash., Idaho, and Mont. MAPS: Hosie 1969:60; Canada Department of Northern Affairs and National Resources 1956:32; Preston 1947:34; Little 1971: map 33-W.

L. occidentalis Nutt.   Western Larch or Tamarack
/T/W/   (Mg)   Usually in well-drained soils at mid-altitudes (2,000 to 4,000 ft) from s B.C. (N to ca. 50°N, E of the Coast and Lillooet ranges) and sw Alta. (Moss 1959: "locally near Crow's-nest; a single tree known at Kananaskis") to Oreg., Idaho, and Mont. MAPS: Fowells 1965:235; Preston 1961:46, and 1947:32; *Atlas of Canada* 1957: sheet 41; Canada Department of Northern Affairs and National Resources 1956:30; Munns 1938: map 28, p. 32; J. R. Anderson 1925: facing p. 32; Whitford and Craig 1918: facing p. 56; Little 1971: map 34-W; Hosie 1969:58.

PICEA Dietr.   [26]   Spruce.   Épinette

1   Twigs and buds glabrous or nearly so.
    2   Leaves flattened in cross-section, green above, whitened along two lines beneath; cones to about 1 dm long, their scales toothed or erose above the middle; branchlets frequently pendulous; (coastal B.C.) . . . .*P. sitchensis*
    2   Leaves 4-angled and rhombic in section.
        3   Cones to about 5 cm long, their scales entire or nearly so; branchlets rarely pendulous; (transcontinental) . . . .*P. glauca*
        3   Cones to over 1 dm long, their scales irregularly toothed at apex; branchlets usually pendulous; (introd.) . . . .[*P. abies*]
1   Twigs and buds more or less puberulent; leaves 4-angled and rhombic in section.
    4   Cones to about 8 cm long, their scales more or less rhombic; leaves bluish green, lacking resin-ducts; (s half of B.C. and sw Alta.) . . . .*P. engelmannii*
    4   Cones at most about 4.5 cm long, their scales rounded at summit; leaves with two resin-ducts visible in cross-section.
        5   Leaves bluish green, more or less glaucous, commonly less than 1.5 cm long; cones usually persistent for many years, dark purple, becoming greyish brown, to about 3.5 cm long, their scales erose-margined; tree columnar or slenderly pyramidal in outline; (transcontinental) . . . .*P. mariana*
        5   Leaves green, not glaucous, acute or with firm sharp tips, commonly 2 or 3 cm long; cones promptly deciduous, green or purplish-green, becoming clear brown or reddish-brown, to 4.5 cm long, their scales entire or minutely denticulate; tree pyramidal in outline; (Ont. to N.S.) . . . .*P. rubens*

[P. abies (L.) Karst.]   Norway Spruce
[European; occasionally planted as an ornamental or windbreak and reported from s Ont. by Soper (1949) and F. H. Montgomery (Can. Field-Nat. 62(2): 94. 1948), from Nfld. by Rouleau (1956), and from N.S. by Roland (1947); also known from Vaudreuil, near Montreal, Que. Fernald *in* Gray (1950) notes that it "spreads slightly from planted trees northw.", but there are as yet no definite indications that it has been found wild in Canada. (*Pinus* L.).]

P. engelmannii (Parry) Engelm.   Engelmann Spruce
/T/W/   (Mg)   Common throughout the interior mts. of B.C. E of the Coast Range (N to ca. 55°N), w to the E slope of the Rocky Mountains of w Alta., s to N Calif., Ariz., and N.Mex. [*Abies* Parry; *P. glauca* var. *eng.* (Parry) Boivin]. MAPS: Hosie 1969:66; Fowells 1965:300; Preston 1961:56, and 1947:42; *Atlas of Canada* 1957: sheet 41; Canada Department of Northern Affairs and National Resources 1956:44; Halliday and Brown 1943: fig. 2, p. 358; Munns 1938: map 32, p. 36; J. R. Anderson 1925: facing p. 32; Whitford and Craig 1918: maps facing p. 58 and 62; Little 1971: map 37-N.

A hybrid with *P. glauca* is reported from near Wells Gray Provincial Park, E B.C., by L. Hamet-Ahti (Ann. Bot. Fenn. 2(2): 145. 1965).

P. glauca (Moench) Voss   White Spruce.   Épinette blanche
/ST/X/   (Ms)   The dominant species of the drier, usually upland, areas of the transcontinental boreal-forest region shown on the maps by Halliday (1937) and Rowe (1959; map dated 1957), the aggregate species ranging from N Alaska–Yukon–Dist. Mackenzie to s Dist. Keewatin, Que. (N to Ungava Bay), Labrador (N to 58°13'N), Nfld., N.B., P.E.I., and N.S., s to B.C., Wyo., S.Dak., Minn.,

Wisc., Mich., N N.Y., and Maine (isolated stations s to Mont., Wyo., and S.Dak.). MAPS and synonymy: see below.

A hybrid with *P. sitchensis* (× *P. lutzii* Little) is reported from Alaska by E. L. Little, Jr. (J. Forestry 51: 745–47. 1953) but Hultén (1968b: 62) considers the alternate parent to be *P. mariana*.

1 Bark smooth or smoothish, light-grey, with visible resin-blisters (resembling bark of *Abies*); [Alaska–Yukon–Dist. Mackenzie (type from Brintnell L., sw Dist. Mackenzie) and N B.C. (s to the Beatton R. at 57°05′N; CAN)] . . . . var. *porsildii* Raup
1 Bark scaly, darker, the resin-blisters not obvious.
   2 Cones broadly ovate, mostly less than 3.5 cm long, often nearly as broad as long, their stiff rigid scales relatively dark-coloured; [*P. albertiana* Brown, the type from Bankhead, Alta.; Alaska–B.C.–w Alta. to s Dist. Keewatin; Cypress Hills of sw Sask.] . . . . var. *albertiana* (Brown) Sarg.
   2 Cones narrowly ovate, to about 5 cm long, their thin flexible scales commonly longer than broad and relatively light-coloured . . . . var. *glauca*
      3 Trunk depressed and more or less prostrate, at most 1 or 2 m tall but up to 2 dm thick, the apex commonly divergent laterally, the very numerous branches crowded; [*P. canadensis* f. *parva* Vict., the type from E Que.; the typical "krummholz" phase, forming an impenetrable tangle, as on the Shickshock Mts. of E Que. (see photograph of krummholz on Mt. Albert, E Que., in Scoggan 1950: pl. IV-A, p. 375) and probably throughout the range of the species in subarctic and subalpine localities] . . . . f. *parva* (Vict.) Fern. & Weath.
      3 Trunk erect, to over 40 m tall; [*Pinus gl.* Moench; *Pinus (Abies; Picea) alba* Ait.; *Abies (Picea; Pinus) canadensis* Mill., not Michx.; transcontinental but evidently largely replaced in the West by the above varieties; MAPS: in all of the map-reference publications listed following the Pinaceae family description; additional maps: Cain 1944: fig. 22C, p. 171; Dansereau 1957: map 1A, p. 33; N.C. Fassett, Ann. Mo. Bot. Gard. 28(3): map 30, p. 365. 1941; Little 1971: map 39-N; Hultén 1968b:61] . . . . f. *glauca*

P. mariana (Mill.) BSP.   Black Spruce.   Épinette noire
/ST/X/   (Ms)   Essentially the same Canadian range as *P. glauca* throughout the transcontinental boreal-forest region (s to B.C.–Alta.–Sask.–Man., Minn., and Pa.; compare the N limits of these species in the map by Hustich 1953: fig. 2, p. 152). [*Abies* Mill.; *Pinus* Du Roi; *Pinus (A.; Picea) nigra* Ait.; *A. denticulata* Michx.; incl. f. *squamea* (Prov.) Vict.]. MAPS: in all of the map-reference publications listed following the Pinaceae family description; additional maps: A. E. Porsild and Howard Crum, Natl. Mus. Can. Bull. 171: fig. 2, p. 146. 1961; Hultén 1968b:62; Little 1971: map 38-N; W. J. Cody, Nat. can. (Que.) 98(2): fig. 10, p. 149. 1971.

The typical form has erect trunks to over 20 m tall and leaves to over 1 cm long. The following forms (leaves averaging less than 6 mm long) are doubtless merely an expression of severe climatic factors: f. *empetroides* Vict. & Rousseau (plant trunkless, the branches trailing; type from Mt. Sterling of the Shickshock Mts., Gaspé Pen., E Que., where a component of the "krummholz"; see *P. glauca* f. *parva*); f. *semiprostrata* (Peck) Blake (a trunk present but prostrate or depressed; like f. *empetroides*, probably occurring throughout the range in subarctic and subalpine habitats).

P. rubens Sarg.   Red Spruce.   Épinette rouge
/T/EE/   (Ms)   Woods, chiefly on well-drained loams in moist valleys: an isolated area in and about the Ottawa dist., Ont. (Gillett 1958; see the below-noted map of the Canada Department of Northern Affairs and National Resources); the main area from easternmost Ont. through sw Que. (N to Laval, about 10 mi N of Quebec City) to N.B., P.E.I., and N.S. (reports from St-Pierre and Miquelon and Nfld. require confirmation), s to Tenn. and N.C. [*Pinus mariana rubra* Du Roi; *Picea (Abies; Pinus) rubra* (Du Roi) Link, not Dietr.; *Picea nigra* var. *rubra* (Du Roi) Engelm.]. MAPS: Fowells 1965:306; *Atlas of Canada* 1957: sheet 41; Canada Department of Northern Affairs and National Resources 1956:42; Dansereau 1957: map 5B, p. 35; Preston 1961:54; Hough 1947:27; Munns 1938: map 30, p. 34; Little 1971: map 41-N; Hosie 1969:70.

Because of characters more or less intermediate between those of *P. glauca* and *P. mariana*, this taxon is considered by some authors to be a hybrid of this parentage. In this connection, see S. A. Cain (Asa Gray Bull. (n.s.) 2(3): 303–08. 1953).

Pinaceae

P. sitchensis (Bong.) Carr.   Sitka or Tideland Spruce
/sT/W/   (Mg)   This, the largest of the spruces, grows in a narrow strip over 1,800 mi long on the Pacific coast from s Alaska and the southernmost Yukon (*see* Hultén 1941: map 59, p. 121; type from Sitka) through the islands and coast of B.C. (seldom at elevations exceeding 1,000 ft but recorded up to 2,500 ft; to 3,000 ft in s Alaska) to NW Calif. [*Pinus* Bong.] MAPS: Hosie 1969:68; Hultén 1968b:61; R. Daubenmire, Can. J. Bot. 46(6): fig. 1, p. 790. 1968; Fowells 1965:312; *Atlas of Canada* 1957: sheet 41 and map 10, sheet 38; Canada Department of Northern Affairs and National Resources 1956:46; Preston 1961:58; Halliday and Brown 1943: fig. 2, p. 358; Munns 1938: map 34, p. 38; J. R. Anderson 1925: facing p. 32; Whitford and Craig 1918: facing p. 62; Little 1971: map 42-N; W. J. Cody, Nat. can. (Que.) 98(2): fig. 14, p. 149. 1971.

PINUS L.   [22]   Pine.   Pin

(Ref.: Critchfield and Little 1966; Mirov 1967)
1   Leaves 5 in a cluster, their basal sheaths deciduous; cone-scales not spine-tipped; (*Soft Pines*).
  2   Cone-scales greatly thickened; cones short-stalked; seeds much longer than the short wings; leaves to about 8 cm long; (mts. of B.C. and sw Alta.).
    3   Cones subglobose, rarely over 7.5 cm long, remaining closed at maturity . . .*P. albicaulis*
    3   Cones subcylindrical, to about 2.5 dm long, opening at maturity . . . .*P. flexilis*
  2   Cone-scales thin and flexible; cones subcylindrical, long-stalked, opening at maturity; seeds much shorter than the wings.
    4   Leaves usually less than 1 dm long; cones to 3.5 dm long (averaging 2 dm); (s B.C. and sw Alta.) . . . .*P. monticola*
    4   Leaves often over 1 dm long; cones rarely over 2 dm long; (SE Man. to Nfld. and N.S.) . . . .*P. strobus*
1   Leaves 2 or 3 in a cluster, their basal sheaths persistent; cone-scales greatly thickened; (*Hard Pines*).
  5   Leaves 3 in a cluster (sometimes only 2 in *P. ponderosa*); cones opening at maturity.
    6   Cones at most about 9 cm long, their scales tipped with a short stiff prickle; leaves less than 1 dm long; (s Ont. and s Que.) . . . .*P. rigida*
    6   Cones to about 1.5 dm long, their scales usually tipped with a sharp prickle; leaves to over 2.5 dm long; (dry belt of s B.C.) . . . .*P. ponderosa*
  5   Leaves 2 in a cluster; cone-scales unarmed or tipped with a small reflexed spine.
    7   Leaves to over 1.5 dm long; cones ovoid-conic, reddish brown, sessile, to about 8 cm long; (SE Man. to Nfld. and N.S.) . . . .*P. resinosa*
    7   Leaves usually less than 7 cm long, commonly twisted; cones tawny or greyish.
      8   Cones dehiscent at maturity, to about 6 cm long, reflexed on a short peduncle, their scales not spinulose-tipped; leaves to about 7 cm long; (introd.) . . .*P. sylvestris*
      8   Cones indehiscent for many years, to 4 or 5 cm long, their scales often spinulose-tipped; (native).
        9   Cones spreading or somewhat reflexed, their scales with evident terminal spines; tree to over 30 m tall; (B.C. to sw Sask.) . . . .*P. contorta*
        9   Cones erect, more or less incurved, their scales unarmed or with a minute terminal spine when young; tree commonly less than 20 m tall; (essentially transcontinental) . . . .*P. banksiana*

P. albicaulis Engelm.   White-bark Pine
/T/W/   (Ms)   This tree grows sparingly at altitudes of 3,000 to 7,000 ft from B.C. (Vancouver Is.; Coast, Selkirk, and Rocky Mts. N to ca. 55°N) and w Alta. (N to the Jasper dist.) to cent. Calif., Nev., and w Wyo. [*Apinus* Rydb.] MAPS: Mirov 1967: fig. 3–3, p. 138; Critchfield and Little 1966: map 5, p. 36; Preston 1961:10, and 1947:8; Canada Department of Northern Affairs and National Resources 1956:10; Munns 1938: map 5, p. 9; Little 1971: map 43-N; Hosie 1969:42.

P. banksiana Lamb.   Jack- or Scrub-Pine.   Pin gris or Cyprès
/ST/X/   (Ms)   Barren, sandy, or rocky soil from w-cent. Dist. Mackenzie to L. Athabasca, Man. (N to the Nelson R. at ca. 56°30'N), Ont. (N to James Bay), Que. (N to E Hudson Bay at ca. 56°N, L.

Mistassini, and the Côte-Nord), N.B., P.E.I., and N.S., s to cent. Alta.–Sask., s Man., Minn., N Ill., N N.Y., and N New Eng. [*P. divaricata* (Ait.) Dum., the correct name, according to G. W. Argus (Can. J. Bot. 49: 575. 1971; *P. hudsonia* Poir.; *P. hudsonica* Parl.; *P. rupestris* Muchx. f.]. MAPS: Hultén 1968*b*:60; Mirov 1967: fig. 3–14, p. 169; Critchfield and Little 1966: map 56, p. 90; Fowells 1965:338; Hosie 1969:50; Preston 1961:18, and 1947:26; *Atlas of Canada* 1957: sheet 41; Canada Department of Northern Affairs and National Resources 1956:16; Hustich 1953: fig. 4 (N limits), p. 154; Hough 1947:13; Halliday and Brown 1943: fig. 4, p. 362; Munns 1938: map 24, p. 28; Little 1971: map 46-N.

Low bushy individuals have been named f. *procumbens* Rousseau, the type from Charlevoix Co., Que. A hybrid with *P. contorta* var. *latifolia* (× *P. murraybanksiana* Righter & Stockwell; *P. divaricata* var. × *musci* Boivin) is reported from Alta. by Bernard Boivin (Nat. can. (Que.) 93(3): 272. 1966).

P. contorta Dougl.   Shore Pine
/ST/WW/   (Ms)   Low to fairly high elevations (var. *latifolia* usually between 2,000 to 6,000 ft) from s-cent. Yukon (N to ca. 64°N) and the Alaska Panhandle, sw Dist. Mackenzie, and w Alta. (an isolated area in the Cypress Hills of SE Alta. and sw Sask.) to Baja Calif., Utah, Colo., and S.Dak. [Var. *hendersonii* Lemmon; *P. inops sensu* Bongard 1833, not Ait.]. MAPS (some including var. *latifolia*): Hosie 1969:52; Hultén 1968*b*:59; Mirov 1967: fig. 3–14, p. 169; Critchfield and Little 1966: map 56, p. 90; Fowells 1965:373; Preston 1961:34, and 1947:24; *Atlas of Canada* 1957: sheet 41; Canada Department of Northern Affairs and National Resources 1956:22 and 24; Halliday and Brown 1943: fig. 4, p. 362; Munns 1938: map 15, p. 19; Little 1971: map 50-N.

The inland phase is commonly distinguished as var. *latifolia* Engelm. (*P. divaricata* var. *latifolia* (Engelm.) Boivin; *P. murrayana* Balf.), known as the lodgepole pine by reason of its relatively slender trunk (to about 100 ft tall) with very thin, grey to orange-brown, slightly scaly bark. Its leaves and cones also average somewhat longer than those of the typical coastal form, the shore pine, a smaller tree (to about 50 ft tall) of more scrubby growth, the branches twisted and generally much forked.

P. flexilis James   Limber or Rocky Mountain Pine
/T/WW/   (Ms)   Rocky Mountains of SE B.C. and sw Alta. (N to near Jasper at ca. 52°30′N) at altitudes of 5,000 to 6,000 ft, s to s Calif., N Ariz., N N.Mex., and Nebr. [*Apinus* Rydb.]. MAPS: Mirov 1967: fig. 3–4, p. 140; Critchfield and Little 1966:39; Preston 1961:10, and 1947:10; Canada Department of Northern Affairs and National Resources 1956:12; Munns 1938: map 4, p. 8; Little 1971: map 56-N; Hosie 1969:40.

P. monticola Dougl.   Western White Pine
/T/W/   (Mg)   Low to medium elevations (seldom exceeding 2,500 ft along the coast or 3,500 ft in the interior) from B.C. (N to ca. 53°N in the Coast, Selkirk, and Rocky Mts. region; largely absent in the intervening Dry Interior) and sw Alta. (Waterton Lakes) to s-cent. Calif., w Nev., cent. Idaho, and w Mont. [*Strobus* Rydb.; *P. strobus* var. *mont.* (Dougl.) Nutt.]. MAPS: Hosie 1969:38; Mirov 1967: fig. 3–6, p. 145; Critchfield and Little 1966: map 6, p. 37; Fowells 1965:478; Preston 1961:8, and 1947:4; *Atlas of Canada* 1957: sheet 41; Canada Department of Northern Affairs and National Resources 1956:8; Munns 1938: map 2, p. 6; Little 1971: map 62-W.

P. ponderosa Dougl.   Ponderosa or Western Yellow Pine
/T/WW/   (Mg)   Well-drained valleys and slopes of the drier parts of the southern interior of B.C. (N to Vavenby, on the North Thompson R. at ca. 51°N, w to near the Alta. boundary), s to Baja Calif., Mexico, N.Mex., Colo., and Tex. [Incl. var. *scopulorum* Engelm.]. MAPS: Hosie 1969:44; Mirov 1967: fig. 3–12, p. 165; Critchfield and Little 1966: map 47, p. 80; Fowells 1965:418; Preston 1961:30, and 1947:18; *Atlas of Canada* 1957: sheet 41; Canada Department of Northern Affairs and Natural Resources 1956:20; Cain 1944: fig. 29, p. 232; Munns 1938: map 19, p. 23; J. R. Anderson 1925: facing p. 32; Whitford and Craig 1918: facing p. 65; Little 1971: map 64-W.

The report from Alta. by Moss (1959; "found in the form of small trees close to the B.C. border") requires confirmation. It may be based upon an 1881 Dawson collection in CAN labelled "Missouri R.", this followed by "Alta." in another person's handwriting. In this connection, *see* Boivin (1967*b*:154).

P. resinosa Ait.   Red Pine.   Pin rouge
/T/EE/   (Mg)   Dry woods from SE Man. (N to Black Is., L. Winnipeg, ca. 51°30'N) to Ont. (N to ca.
52°N; *see* W. R. Haddow, J. Arnold Arbor. Harv. Univ. 29: 217–26 and appended maps. 1948),
Que. (N to L. St. John and the s Gaspé Pen.), Nfld., N.B., P.E.I., and N.S., s to Minn., Wisc., Mich.,
Pa., Conn., and N Mass. [*P. rubra* Michx. f., not Mill.; *P. ?inops sensu* Cochran 1829, not Ait.;
P. ?mitis sensu Reeks 1871, not Michx.]. MAPS: Mirov 1967: fig. 3–17, p. 178; Critchfield and Little
1966; map 26, p. 58; Fowells 1965:432; Preston 1961:16; *Atlas of Canada* 1957: sheet 41; Canada
Department of Northern Affairs and National Resources 1956:14; Hough 1947:11; Munns 1938:
map 12, p. 16; Nichols 1935: fig. 5(I), p. 408; M. L. Fernald, Rhodora 13(151): map 8, facing p. 139.
1911; Little 1971: map 69-N; Hosie 1969:48.

P. rigida Mill.   Pitch-Pine
/T/EE/   (Ms)   Sandy and barren soil from Ky., Ohio, Pa., and N.Y. to s Ont. (NE shore of L.
Ontario along the Thousand Islands and adjacent mainland) and sw Que. (according to Ernest
Rouleau, Rhodora 57(682): 299. 1955, the discovery of a natural stand at Cairnside, Chateauguay
Co., just s of Montreal, covering an area of about one square mile, removes doubt as to the
occurrence of the species in the province; the report from the St. John R. valley of w N.B. by John
Macoun 1886 requires confirmation; not known from P.E.I. or N.S.), s to N Ga., Va., Md., and Del.
MAPS: Hosie 1969:46; Mirov 1967: fig. 3–25, p. 193; Critchfield and Little 1966: map 44, p. 77;
Fowells 1965:404; Dansereau 1957: fig. 4A, p. 34; Canada Department of Northern Affairs and
Natural Resources 1956:18; Preston 1961:20; Hough 1947:9; Munns 1938: map 19, p. 23; Little
1971: map 71-E.

P. strobus L.   White Pine.   Pin blanc
/T/EE/   (Mg)   Chiefly on well-drained sandy soil from extreme SE Man. to Ont. (N to ca. 52°N),
Que. (N to the Côte-Nord and Anticosti Is.), Nfld., N.B., P.E.I., and N.S., s to NE Iowa, Ky., N Ga.,
and NW S.C. [*P. alba canadensis* Prov.] MAPS: Mirov 1967: fig. 3–16, p. 175; Critchfield and Little
1966: map 6, p. 37; Fowells 1965:329; Dansereau 1957: map 3B, p. 34; *Atlas of Canada* 1957:
sheet 41; Canada Department of Northern Affairs and Natural Resources 1956:6; Hosie 1969:36;
Preston 1961:8; Hustich 1953: fig. 4 (N limits), p. 154; Hough 1947:3; Munns 1938: map 1, p. 5;
Nichols 1935: fig. 5E, p. 408; Little 1971: map 73-N.
     Forma *prostrata* (Mast.) Fern. & Weath. (plant of exposed habitats, depressed, the branches
trailing) is reported from the serpentine mts. of w Nfld. by M. L. Fernald and C. A. Weatherby
(Rhodora 34 (404): 168. 1932).

P. sylvestris L.   Scotch Pine
European; much cult. and locally natzd. in E N. America. W. T. Macoun (Ont. Nat. Sci. Bull. 3: 11.
1907) reports this tree as reproducing naturally from seed at Ottawa, Ont., and Baldwin (1958)
notes its occurrence in Ont. as far N as Iroquois Falls, 48°46'N, where, "It is significant that
reproduction by seeding into the surrounding native young jack-pine forest has been successful." It
has also been reported under cultivation in Que., Nfld., and N.S.

PSEUDOTSUGA Carr.   [27]

P. menziesii (Mirb.) Franco   Douglas Fir
/T/WW/   (Mg)   Low to fairly high elevations (in B.C. to about 4,000 ft, chiefly on south-facing
slopes; southwards to about 6,000 ft, chiefly on north-facing slopes), attaining its best development
(to over 200 ft tall; recorded at 385 ft in Wash.) in moist well-drained soils with abundant
precipitation, from B.C. (N to ca. 55°N near Babine L. and Stuart L. in the interior) and the slopes
and foothills of the Rocky Mts. of sw Alta. (N to the Athabasca R. at ca. 53°N) to Calif., Mexico, and
w Tex. [*Abies* Mirb.; *A.(Abietia; Picea; Pinus; Pseudotsuga; Tsuga) douglasii* Lindl.; *A.
(Pseudotsuga) mucronata* Raf.; *Pinus (Abies; Pseudotsuga) taxifolia* Lamb., not Salisb.;
Pseudotsuga vancouverensis Flous]. MAPS: Fowells 1965:547; Preston 1961:62, and 1947:53;
Atlas of Canada 1957: sheet 41; Canada Department of Northern Affairs and Natural Resources
1956:58 and 60; Halliday and Brown 1943: fig. 8, p. 367; J. R. Anderson 1925: facing p. 32;
Whitford and Craig 1918: facing p. 56; Little 1971: map 80-N; Hosie 1969:84.

The depressed, more or less straggling phase has been named f. *alexidis* Boivin (type from Waterton Lakes, sw Alta.). Two races are commonly recognized, the typical coast form with bright-green leaves, the trunk commonly attaining a height of 200 ft and a diameter of 6 ft, occurring in B.C. on Vancouver Is. and the adjacent mainland; and var. *glauca* (Beissn.) Franco of the interior, rarely over 130 ft tall and with a bluish cast, the leaves bright blue-green. Their respective areas are shown in the above-noted maps by Fowells and the Canada Department of Northern Affairs and Natural Resources.

THUJA L.   [42]   Arbor Vitae.   Thuier

1   Leaves not glaucous beneath; cones commonly about 1 cm long, their scales pointless; tree to about 20 m tall; (Man. to N.S.) . . . .*T. occidentalis*
1   Leaves usually glaucous beneath; cones to over 1.5 cm long, their scales frequently spine-pointed; tree to about 70 m tall; (coast and interior wet belts of B.C. and wet belt of w Alta.) . . . .*T. plicata*

T. occidentalis L.   White Cedar.   Cèdre or Balai
/T/EE/   (Ms)   Swamps, wet woods, and cool rocky banks (chiefly calcareous), the main area from SE Man. (N to L. Winnipeg; an isolated stand on Cedar L., N of L. Winnipegosis at ca. 53°30′N; reports from Sask. require confirmation) to Ont. (N to ca. 51°30′N; an isolated stand near Winisk, s Hudson Bay, ca. 55°20′N), Que. (N to s James Bay and Anticosti Is.; cult. in Nfld.), N.B., P.E.I., and N.S., s to Minn., Tenn., and N.C. MAPS: Hosie 1969:98; Fowells 1965:679; Preston 1961:96; *Atlas of Canada* 1957: sheet 41; Canada Department of Northern Affairs and Natural Resources 1956:74; Hough 1947:41; Cain 1944: fig. 22(B), p. 171; Munns 1938: map 53, p. 57; M. L. Fernald, Rhodora 13 (151): map 12, facing p. 142. 1911. The northern limits are indicated in a map by Hustich (1953: fig. 6, p. 157).

Plants with the foliage confined to near the tips of the branchlets have been named f. *gaspensis* Vict. & Rousseau (type from along the Little Cascapedia R., Gaspé Pen., E Que.). Plants with decumbent gnarled trunks forming carpets to about 1 m tall have been named f. *prostrata* Vict. & Rousseau (type from Mont-St-Pierre, Gaspé Pen., E Que.); Little 1971: map 89-N.

T. plicata Don   Western Red Cedar
/T/W/   (Mg)   Low to rather high elevations, preferably in deep moist porous soils, from the s Alaska Panhandle (*see* Hultén 1941: map 64, p. 122) through coastal B.C. (type from Nootka Sound, Vancouver Is.; upper limit in the Coast Range about 2,400 ft) to N Calif., and from interior B.C. at ca. 54°30′N and sw Alta. (N to the Jasper dist.; upper limit in the B.C. Selkirks about 4,500 ft) to Calif. and Mont. [*T. occidentalis* var. *pl.* (Don) Loud; *T. gigantea* Nutt.; *T. menziesii* Dougl.]. MAPS: Hultén 1968*b*:64; Fowells 1965:686; Preston 1961:96, and 1947:66; *Atlas of Canada* 1957: sheet 41, and map 10, sheet 38; Canada Department of Northern Affairs and Natural Resources 1956:76; Munns 1938: map 54, p. 58; J. R. Anderson 1925: facing p. 32; Whitford and Craig 1918: facing p. 56 and 58; Little 1971: map 90-N; Hosie 1969:100.

TSUGA (Endl.) Carr.   [27]   Hemlock.   Pruche

1   Leaves densely radiating on all sides of the twig and more or less curved, nearly semicircular in cross-section, bluish green on both sides, to 2.5 cm long; cones to 7 or 8 cm long; tree to about 50 m tall; (coastal and interior wet belts of B.C.) . . . .*T. mertensiana*
1   Leaves always appearing 2-ranked, flattened in cross-section, marked with two white bands below, usually less than 2 cm long.
    2   Cones to 2.5 cm long, light brown; leaves of markedly differing lengths; bark scaly or relatively shallowly grooved; tree to over 70 m tall, the trunk to 2 or 3 m thick; (coastal and interior wet belts of B.C.) . . . .*T. heterophylla*
    2   Cones at most about 2 cm long, at first pale green, turning red-brown at maturity; leaves of essentially equal length; bark furrowed longitudinally; tree to about 30 m tall; (Ont. to N.S.) . . . .*T. canadensis*

T. canadensis (L.) Carr.   Eastern Hemlock
/T/EE/   (Ms)   Chiefly in hilly or rocky woods from NE Minn. to Ont. (N to the NE end of L. Superior near Michipicoten, ca. 48°N), Que. (N to Kamouraska Co.), N.B., P.E.I., and N.S., s to Ala., Ga., and N.C. [*Pinus* L.; *Abies canadensis* Michx., not Mill.; *A. (T.) americana* Mill.]. MAPS: Fowells 1965:703; *Atlas of Canada* 1957: sheet 41; Canada Department of Northern Affairs and Natural Resources 1956:50; Preston 1961:66; Hough 1947:31; Munns 1938: map 35, p. 39; Nichols 1935: fig. 5(D), p. 408; Little 1971: map 91-N; Hosie 1969:76.

Depressed, bushy plants forming carpets to about 1 m tall have been named f. *parvula* Vict. & Rousseau (type from Montmagny, Montmagny Co., Que.).

T. heterophylla (Raf.) Sarg.   Western or Pacific Hemlock
/sT/W/   (Mg)   Low to fairly high elevations (in the Selkirks of B.C. to about 5,000 ft), preferably on deep porous soils with abundant precipitation, from s Alaska and sw ?Yukon (*see* Hultén 1941: maps 60a and 60b, p. 121) through coastal B.C. to Calif. and from the Selkirks and Rocky Mts. of B.C. at ca. 55°N (close to the Alta. boundary but evidently not yet known from that province) to N Calif. and NW Mont. [*Abies* Raf.; *Pinus canadensis sensu* Bongard 1833, not L.]. MAPS: Hultén 1968b:62; Fowells 1965:717; Preston 1961:68, and 1947:48; *Atlas of Canada* 1957: sheet 41, and map 10, sheet 38; Canada Department of Northern Affairs and Natural Resources 1956:52; Halliday and Brown 1943: fig. 7, p. 366; Munns 1938: map 36, p. 40; J. R. Anderson 1925: facing p. 32; Whitford and Craig 1918: facing p. 62 and 63; Little 1971: map 92-N; Hosie 1969:78.

T. mertensiana (Bong.) Sarg.   Mountain or Black Hemlock
/sT/W/   (Ms)   Essentially the same range and ecology as *T. heterophylla* (but ascending to about 6,000 ft in coastal B.C.), from s Alaska and southernmost ?Yukon (*see* Hultén 1941: map 61, p. 121) through coastal B.C. to cent. Calif. and from the Selkirk Mts. of SE B.C. at ca. 52°N (elbow of the Big Bend, N of Revelstoke) to Idaho and w Mont. [*Pinus (Abies; Hesperopeuce) mert.* Bong., the type from Sitka, Alaska; *Abies (T.; Picea; Pinus) hookeriana* Murr.; *Abies (T.; Hesperopeuce; Pinus) pattoniana* Murr.]. MAPS: Hultén 1968b:63; Fowells 1965:712; Preston 1961:70, and 1947:50; Canada Department of Northern Affairs and Natural Resources 1956:54; Munns 1938: map 37, p. 41; Little 1971: map 93-N; Hosie 1969:80.

[TAXODIACEAE] (Redwood Family)

[SEQUOIA Endl.]   [32]   Redwood

The redwoods are endemic to the humid Pacific coast of the U.S.A. but are noted here because of their great interest and the fact that one species, *S. sempervirens,* is grown in B.C., as in Stanley Park, Vancouver. They differ from the Pinaceae (2 ovules borne on each cone-scale) in their cone-scales each bearing 5 or more ovules (later, seeds). The two species may be distinguished as follows:

1   Leaves short-lanceolate or awl-shaped, ascending all around the stem and largely adhering to it, only their tips free; cones commonly over 5 cm long, ripening the second year; (Giant Sequoia, Big Tree, or Sierra Redwood; forming a narrow belt about 260 mi long on the w slopes of the Sierra Nevada Range in cent. Calif.; this species is probably the longest-lived organism, Fowells (1965) noting ring-counts on felled trees showing ages up to 3,200 yrs. and an unsubstantiated record of over 4,000 yrs. He also notes that "Trees with an average basal diameter of 20 feet and a height of 275 feet are common in the southern groves where the best stands are found".) . . . .[*S. gigantea* (Lindl.) Dcne.]
1   Leaves mostly linear, sharp-pointed, 2-ranked and forming flat sprays (adult trees commonly with leaves at the top resembling those of *S. gigantea*); cones less than 3 cm long, ripening the first year; (Coast or California Redwood; extreme sw Oreg. to cent. Calif.; Fowells (1965) notes that ring-counts have shown ages of nearly 2,200 yrs. and that a maximum height of 368.7 ft has been recorded) . . . .[*S. sempervirens* (Don) Endl.]

Subdivision II *ANGIOSPERMAE* (Angiosperms)

Class I *MONOCOTYLEDONEAE* (see p. 94)

TYPHACEAE (Cat-tail Family)

TYPHA L. [49] Cat-tail. Massette or Quenouille

Aquatic or marsh herbs with long linear sheathing sub-basal leaves and narrowly cylindrical tough stems to about 3 m tall. Inflorescence terminal, consisting of a narrow staminate spike above and contiguous with or separated from a lower broader pistillate spike. Perianth none. Fruit a linear or narrowly fusiform achene subtended by copious white hairs.

1 Pistillate part of spike commonly less than 1.5 cm thick, cinnamon-brown without black markings, it and the upper staminate part usually separated by a gap of at least 2 cm; denuded spike-axis bearing stout blunt compound papillate pedicels to 0.7 mm long; seed usually slightly above the middle of the achene; pollen-grains single; leaves slightly rounded on the back, less than 1 cm broad; stem slender; (SE Man. to N.S.] . . . . .*T. angustifolia*
1 Pistillate part of spike up to 2.5 cm thick, dark brown with black markings, it and the upper staminate part normally contiguous; denuded spike-axis bearing slender pedicels 1 or 2 mm long; seed near the middle of the achene; pollen-grains in 4's; leaves flat, to 2.5 cm broad; stem stout; (transcontinental) . . . .*T. latifolia*

T. angustifolia L. Narrow-leaved Cat-tail
/T/(X)/EA/ (Hel) Swampy ground and shallow water from SE Man. (Löve and Bernard 1959) to Ont. (N to the Ottawa dist.), SW Que. (N to near Sorel), N.B., P.E.I., and N.S., s to Calif., ?Colo., ?Utah, Nebr., Mo., and S.C.; Eurasia. MAPS: Hultén 1962: map 132, p. 141; S. G. Smith, Am. Midl. Nat. 78(2): fig. 1, p. 260. 1967.

The tentative report from B.C. by Boivin (1967a) requires confirmation. Reports from the U.S.A. from Calif. to Utah may possibly refer to *T. domingensis* Pers. Robust individuals with perhaps some genetic infiltration from *T. latifolia* have been named var. *elongata* (Dudl.) Wieg. Plants with more definite indications of an *angustifolia-latifolia* parentage have been named *T. glauca* Godr. Such apparent hybrids have been reported from SE Man. (Löve and Bernard 1959), s Ont. (Fernald *in* Gray 1950), and Que. (near L. St. John; Père Louis-Marie, Inst. Agric. d'Oka. Rev. 34: 11. 1960).

T. latifolia L. Common Cat-tail
/ST/X/EA/ (Hel) Swampy ground and shallow water from cent. Alaska (N to ca. 67°N), s-cent. Yukon, and Great Bear L. to Alta. (L. Athabasca), N Sask., Man. (N to Nueltin L. at 57°16′N), Ont. (N to Sandy L. and James Bay at ca. 53°N), Que. (N to Anticosti Is. and the Gaspé Pen.), Nfld., N.B., P.E.I., and N.S., s to s Calif., N.Mex., Mexico, Tex., La., and Fla.; Eurasia; N Africa. MAPS: *see* below.
1 Staminate and pistillate parts of spike separated as in *T. angustifolia;* [s Man.: Rufford, about 20 mi N of Brandon; Whitewater L., near Boissevaine; near Otterburne, about 30 mi s of Winnipeg] . . . . .f. *ambigua* (Sonder) Kronf.
1 Staminate and pistillate parts of spike contiguous.
　　2 Inflorescence 3-parted; [type from La Trappe, Deux-Montagnes Co., SW Que.] . . . .
　　. . . . . .f. *divisa* Louis-Marie
　　2 Inflorescence unparted.
　　　　3 Pistillate part of spike only 5 or 6 cm long (but still to 2.5 cm thick); [Pontiac Co., SW Que.] . . . . .f. *media* (Coss. & Germ.) Louis-Marie
　　　　3 Pistillate part of spike to 2.5 dm long; [transcontinental; MAPS: Raymond 1950b: fig. 23, p. 56; S. G. Smith, Am. Midl. Nat. 78(2): fig. 1, p. 260. 1967; Neil Hotchkiss and H. L. Dosier, Am. Midl. Nat. 41(1): 251. 1949; Hultén 1968b:66; W. J. Cody, Nat. can. (Que.) 98(2): fig. 1, p. 146. 1971 . . . . .f. *latifolia*

SPARGANIACEAE (Bur-reed Family)

SPARGANIUM L. [54] Bur-reed. Rubanier

Aquatic or marsh plants with alternate sessile linear 2-ranked sheathing leaves. Staminate and pistillate flowers separate in dense globose heads, these sessile or short-peduncled in the leaf-axils or above them, the staminate heads near the ends of the branches. Perianth consisting of 3–6 sepals. Ovary superior. Fruit a usually distinctly beaked achene.

(Ref.: P. A. Rydberg, North Am. Flora 17: 5–10. 1909)
1   Stigmas 2; achenes sessile, to 8 mm thick at the abruptly rounded to slightly retuse
    beaked summit; leaves flat, to about 12 mm broad; (transcontinental) . . . .*S. eurycarpum*
1   Stigma 1; achenes more or less stipitate, rarely over 3 mm thick, gradually rounded or
    tapering to summit.
    2   Achene-beak wanting or not over 1.5 mm long; staminate head commonly solitary.
        3   Achenes beakless; staminate head and upper pistillate head contiguous; one or
            more of the 1–3 pistillate heads borne above the leaf-axils; leaves thickish,
            commonly less than 5 mm broad; (transcontinental) . . . .*S. hyperboreum*
        3   Achenes with beaks up to 1.5 mm long.
            4   Pistillate heads all axillary, the terminal staminate head remote; leaves thin, to
                8 mm broad; (transcontinental) . . . .*S. minimum*
            4   Some of the pistillate heads borne above the left-axils, the 1 or 2 staminate
                heads not remote from the upper pistillate one; leaves thickish, to about 2 cm
                broad; (s Yukon; ᴇ Que.) . . . .*S. glomeratum*
    2   Achene-beak to about 6 mm long; staminate heads commonly at least 2.
        5   Sepals borne chiefly along middle of stipe, rarely reaching middle of achene; beak
            and outer layer of achene firm; anthers less than 0.8 mm long; leaves thin, flat,
            ribbon-like, to about 1 cm broad; (essentially transcontinental) . . . .*S. fluctuans*
        5   Sepals borne chiefly at summit of stipe, commonly surpassing the middle of the
            achene; outer layer of achene thin, its beak fragile; anthers at least 0.8 mm long.
            6   Heads or branches of inflorescence all borne directly in the leaf-axils.
                7   Leaves stiffish, strongly ascending, keeled, to 1.5 cm broad; fruiting heads
                    to 3.5 cm thick; bracts strongly ascending; achenes lustrous, pale brown,
                    the body to 7 mm long, the beak to 6 mm long; branches of inflorescence
                    with up to 8 staminate heads and usually no pistillate ones; (Que.) . . . .
                    . . . .*S. androcladum*
                7   Leaves thin and soft, flat, loosely ascending or sometimes floating, to 2 cm
                    broad; fruiting heads at most 2.5 cm thick; bracts spreading or
                    spreading-ascending; achenes dull, sordid brown, the body to 5 mm long,
                    the beak to 4.5 mm long; branches of inflorescence (when present) with up
                    to 6 staminate heads and 3 pistillate heads; (essentially transcontinental) . . . .
                    . . . .*S. americanum*
            6   Heads (at least one of them) borne above the leaf-axils; leaves flat or only
                slightly keeled; (transcontinental).
                8   Beak of achene about equalling the body; tips of sepals appressed; leaves
                    erect or strongly ascending, to 12 mm broad . . . .*S. chlorocarpum*
                8   Beak of achene much shorter than the body; tips of sepals loosely
                    ascending or spreading; leaves very elongate, mostly submersed or
                    floating.
                    9   Leaves at most 5 mm broad, rounded on the back; fruiting heads
                        commonly less than 2 cm thick . . . .*S. angustifolium*
                    9   Leaves to 12 mm broad, flat and ribbon-like; fruiting heads to 2.5 cm
                        thick . . . .*S. multipedunculatum*

S. americanum Nutt.
/T/X/   (Hel (HH))   Muddy or peaty shores and shallow water from B.C. to Sask. (Boivin 1967a; not known from Man.), Ont. (ɴ to the ɴ shore of L. Superior), Que. (ɴ to the Gaspé Pen.), Nfld.

(Fernald *in* Gray 1950), N.B., P.E.I., and N.S., s to Tex. and Fla. [*S. simplex* var. *nuttallii* Engelm.]. MAP: E. O. Beal, Brittonia 12(3): map 2, p. 177. 1960.

S. androcladum (Engelm.) Morong
/T/EE/ (Hel (HH)) Muddy or peaty shores and shallow water from Minn. to Que. (Quebec Co.; Fernald *in* Gray 1950), s to Okla., Mo., Ill., Tenn., and Va. [*S. lucidum* Fern. & Eames]. MAP: E. O. Beal, Brittonia 12(3): map 3, p. 177. 1960.

This species is scarcely separable from *S. americanum,* to which reports from B.C., Ont., Nfld., N.B., and N.S. (as well as the above one from Que.) may eventually prove referable.

S. angustifolium Michx.
/aST/X/GEeA/ (HH) Shallow to deep water or wet shores from the Aleι "ian Is. and N-cent. Alaska to NW Dist. Mackenzie, s Dist. Keewatin, northernmost Ont., Que. (N to s Hudson Bay and the Côte-Nord; type from L. Mistassini), Labrador (N to ca. 55°N), Nfld., N.B., ?P.E.I. (D. S. Erskine 1960), and N.S., s to Calif., Colo., Minn., and N.J.; w Greenland N to ca. 69°N, E Greenland N to 65°37′N; Iceland; Europe; Kamchatka and N Japan. [*S. affine* Schnitzl.; *S. ?natans sensu* Hooker 1838, not L.]. MAPS: Hultén 1968*b*:67, and 1958: map 195, p. 215; Meusel, Jaeger, and Weinert 1965:23.

S. chlorocarpum Rydb.
/ST/(X)/ (Hel (HH)) Muddy or peaty soil and shallow water from sw Alta. and Sask. (Boivin 1967*a*; not known from Man.) to N Minn., Ont. (N to the L. Nipigon region and Winisk, s Hudson Bay, ca. 55°20′N), Que. (N to the Larch R. at ca. 57°35′N), Nfld., N.B., P.E.I., and N.S., s to N ?Calif., ?Wyo., S.Dak., Iowa, W.Va., and NW N.C. [*S. diversifolium* of Canadian reports in large part, not Graebn.]. MAP: E. O. Beal, Brittonia 12(3): map 1, p. 177. 1960.

Forma *acaule* (Beeby) Voss (*S. simplex acaule* Beeby, the type from Lake Verde, P.E.I.; *S. acaule* (Beeby) Rydb.; staminate heads at most 5, rather than up to 9, the pistillate heads at most 3, rather than sometimes 4) is common throughout the range.

S. eurycarpum Engelm.
/sT/X/ (Hel) Shallow water from B.C. (N to Vanderhoof, ca. 53°30′N) to w-cent. Dist. Mackenzie (Porsild 1943), Alta. (N to Wood Buffalo National Park), Sask. (N to McKague, 52°37′N), Man. (N to Hill L., N of L. Winnipeg), Ont. (N to Albany, sw James Bay, ca. 52°10′N), Que. (N to Rupert House, SE James Bay, ca. 51°30′N, and the Gaspé Pen.), Nfld. (Rouleau 1956), N.B., P.E.I., and N.S., s to Baja Calif., Kans., Mo., Ind., Ohio, and ?Fla. MAP: Hultén 1968*b*:67.

The tentative report of *S. greenei* Morong (*S. eury.* var. *gr.* (Morong) Graebn.) from B.C. by Boivin (1967*a*) may be referable here.

S. fluctuans (Morong) Robins.
/sT/X/ (HH) Cold lakes and ponds from ?B.C. (tentatively reported from Garibaldi and Creston Flats by Eastham 1947) to L. Athabasca (Alta. and Sask.; Moss 1959, and Breitung 1957*a*; not known from Man.), Ont. (N to Sachigo L. at ca. 54°N, 92°W), Que. (N to Fort George, E James Bay, ca. 53°50′N, L. Mistassini, and the Gaspé Pen.), Nfld., N.B., P.E.I., and N.S., s to Minn., Pa., and New Eng. [*S. androcladum* var. *fl.* Morong].

S. glomeratum Laest.
/ST/(X)/EA/ (HH) Shallow pools: s-cent. ?Yukon (Dawson; A. E. Porsild, Can. Field-Nat. 56(7): 112. 1942; *see* Hultén 1968*a*); L. Superior, N Minn.; E Que. (Natashquan, Saguenay Co.; Lewis 1931; probable basis of the report from Labrador by Boivin 1967*a*, who also reports it as introd. in B.C. and Alta.); Eurasia. MAP: Hultén 1962: map 70, p. 79.

S. hyperboreum Laest.
/aST/X/GEA/ (HH) Ponds and lakes from N Alaska–Yukon–Dist. Mackenzie to cent. Dist. Keewatin, northernmost Ungava–Labrador, and Nfld., s to N B.C. (Hultén 1962), Alta. (Boivin 1967*a*; not known from Sask.), SE Man., s James Bay, Que. (s to Anticosti Is. and the Gaspé Pen.), and N.S. (Guysborough, Inverness, and Cape Breton counties; not known from N.B. or P.E.I.); w

Greenland N to 69°25′N, E Greenland N to ca. 64°N; Iceland; Eurasia. [Incl. var. *americanum* Beeby]. MAPS: Hultén 1968*b*:69, and 1962: map 32, p. 39; Porsild 1966: map 6, p. 67.

S. minimum (Hartm.) Fries
/ST/X/EA/   (HH)   Shallow pools and streams from cent. Alaska–Yukon–Dist. Mackenzie to L. Athabasca (Alta. and Sask.), s Dist. Keewatin, Que. (N to Sims L. at 54°05′N and the Côte-Nord), Labrador (N to the Hamilton R. basin), Nfld., N.B., P.E.I., and N.S., s to Oreg., Utah, Colo., Minn., Pa., and N N.J.; Iceland; Eurasia. [*S. natans* var. *min.* Hartm.]. MAPS: Hultén 1968*b*:68, and 1962: map 93, p. 103; Meusel, Jaeger, and Weinert 1965:23.

S. multipedunculatum (Morong.) Rydb.
/sT/X/   (HH)   Lakes, ponds, and pools from SE Alaska to Great Bear L., L. Athabasca (Alta. and Sask.), Man. (N to Churchill; Schofield 1959), Ont. (N to the N shore of L. Superior), James Bay (Charlton Is., ca. 52°N), Que. (N to the Côte-Nord), southernmost Labrador, Nfld., N.B., P.E.I., and N.S., s to Calif., Colo., s Ont., and N New Eng. [*S. simplex* Huds. var. *mult.* Morong; *S. emersum* var. *mult.* (Morong) Reveal (*see* Taxon 19(5):796–97. 1970); *S. simplex* of Canadian reports in large part, not Huds.; *S. ramosum sensu* Hooker 1838, in part, not Huds.]. MAPS: Hultén 1968*b*:68, and 1962: map 153, p. 163.

ZOSTERACEAE (Pondweed Family)

Submersed aquatics with jointed stems and mostly filiform to linear, entire leaves (*Potamogeton* often with broader floating leaves; leaves of *P. crispus* and *P. robbinsii* minutely serrulate), the leaves sheathing or with sheathing stipules. Flowers small, in clusters or spikes, perfect or unisexual, lacking a perianth. Stamens 1, 2, or 4. Style or sessile stigma solitary. Fruit an achene or follicle. (Incl. Potamogetonaceae, Ruppiaceae, and Zannichelliaceae).

1 Flowers axillary, subtended by a cup-shaped involucre; stamen 1; fruit slightly curved; leaves opposite, filiform, to about 1 dm long; (coastal waters of B.C. and E Canada) . . . .*Zannichellia*
1 Flowers in spikes or heads; leaves alternate (or the uppermost sometimes subopposite).
  2 Leaves ribbon-like, to over 1 m long; spikes hidden in a long leaf-like spathe; stamen 1.
    3 Fruit deeply cordate-sagittate at the base; plants unisexual (dioecious); leaves to about 2 m long and 4 mm broad; (tidal waters of B.C.) . . . .*Phyllospadix*
    3 Fruit narrowly ovoid, rounded at base; plants bisexual but the flowers unisexual (monoecious); leaves to about 1.5 m long and 8 mm broad; (coastal waters of B.C. and E Canada) . . . .*Zostera*
  2 Leaves not ribbon-like, relatively short; flowers perfect; mature spike exserted; stigma 1, sessile.
    4 Flowers commonly 2 on a short spadix within a leaf-sheath; stamens 2; fruits ovoid, pointed, each on a slender stalk (podogyne) which elongates after flowering; leaves capillary, often in fan-shaped groups, to about 2 dm long; plants of saline or alkaline ponds . . . .*Ruppia*
    4 Flowers (and fruits) sessile in usually many-flowered terminal spikes; stamens 4, the anther-connectives each bearing a sepal-like outgrowth; leaves filiform to very broad; plants of freshwater ponds and streams . . . .*Potamogeton*

PHYLLOSPADIX Hook. [56] False Eelgrass, Surf-grass

1 Spadix (spike of flowers enclosed in the spathe) usually solitary, basal; leaves to about 4 mm broad, with 3 primary nerves . . . .*P. scouleri*
1 Spadices several, cauline; leaves at most 2 mm broad, with 1–3 primary nerves (sometimes nerveless) . . . .*P. torreyi*

P. scouleri Hook.
/T/W/ (HH) Coastal waters of the Pacific Ocean from SE Alaska (*see* Hultén 1941: map 73, p. 123) and B.C. to s Calif. MAP: Hultén 1968*b*:70.

P. torreyi Wats.
/t/W/ (HH) Coastal waters of the Pacific Ocean from sw B.C. (Vancouver Is. and adjacent islands) to Baja Calif.
    According to Hultén (1941), several authors have reported intermediate forms between this species and the broader-leaved *P. scouleri*.

POTAMOGETON L. [58] Pondweed. Potamot or Herbes à Brochets

(Ref.: Fernald 1932; Ogden 1943)
1 Leaves with thin cartilaginous margins, usually minutely serrulate toward apex, uniform, stiffly 2-ranked, linear to lanceoolate, auricled at base, to about 1 dm long; fruits (rarely produced) 4 or 5 mm long, with a narrow acute dorsal keel and 2 rounded lateral ones, the nearly central beak about 1 mm long, (s B.C.; Ont. to N.S.) . . . .*P. robbinsii*
1 Leaves without cartilaginous margins.
  2 Leaves sharply serrate, undulate-margined, uniform, broadly linear to oblong, sessile, to 8 cm long; fruit 5 or 6 mm long, with 3 rounded dorsal keels and an elongate conical beak about 2 mm long; (introd. in Alta., Ont., Que., and N.S.) . . . .*P. crispus*

2   Leaves entire or merely undulate; beak of fruit much shorter than the body.

     3   Submersed leaves slenderly linear, semiterete, firm, phyllodial (with no differentiation between blade and petiole), to about 2 dm long; floating leaves ovate to oblong-ovate, shining, at most about 12 cm long, attached to the long petioles by a brownish, often curved, joint-like section; fruits to 4 or 5 mm long, rounded or slightly keeled on back, with a short broad marginal beak; (trans-continental) . . . .*P. natans*

     3   Submersed leaves with definite soft blades.

        4   Leaves uniform, with no differentiation between submersed and floating ones.

           5   Leaves at least 4 mm broad; (transcontinental).

              6   Stem wing-flattened; leaves linear, not over 5 mm broad, to 2 dm long, with 1 or 3 main nerves and up to about 35 fine nerves; fruit 4 or 5 mm long, with a prominent, more or less undulate-dentate dorsal keel and a marginal beak to 1 mm long . . . .*P. zosterifolius*

              6   Stem terete, not wing-flattened; leaves mostly broader and shorter.

                 7   Leaves tapering to sessile bases, to 2 cm broad; fruit with a hard smooth tawny-olive outer layer, about 3 mm long, with a narrow sharp dorsal keel and 2 rounded lateral ones, the short beak lateral . . . .*P. alpinus*

                 7   Leaves all rounded or cordate at the more or less clasping base; fruit with a softer, often shallowly pitted, grey-green to dark-green outer layer.

                    8   Stipules conspicuous, rigid, to 1 dm long, usually persistent; leaves lance-oblong to ovate-oblong, to about 4 dm long and 4 cm broad, more or less hood-shaped at tip; fruits 4 or 5 mm long, with a narrow acute dorsal keel and 2 obscure or rounded lateral ones, the short thick beak marginal; rhizome with rust-coloured spots . . . .*P. praelongus*

                    8   Stipules delicate and soon disappearing or, if coarse, soon shredding into white fibres; fruits 3 or 4 mm long, with 3 obscure dorsal keels, the beak marginal; rhizomes not noticeably spotted . . . .*P. perfoliatus*

           5   Leaves not over 4 mm broad.

              9   Stipules adnate to the base of the leaf, only their summits free; leaves all setaceous or linear-filiform, septate their whole length; fruits with 1 or 3 obscure rounded dorsal keels; (transcontinental).

                10 Leaves of the branches sharply acute, not over 1 mm broad, their sheaths close; fruits commonly 3 or 4 mm long, with a very short marginal beak . . . .*P. pectinatus*

                10 Leaves shallowly notched, blunt or merely short-pointed; fruits at most 3.5 mm long, with a nearly central wart-like beak.

                   11 Sheaths mostly loose and slightly inflated, much thicker than the stem; leaves 1 or 2 mm broad, blunt; spike with up to 12 nearly equidistant whorls, to 8 cm long . . . .*P. vaginatus*

                   11 Sheaths close, only slightly thicker than the stem; spike with commonly not more than 5 whorls, the lower of these more or less remote . . . .*P. filiformis*

           9   Stipules mostly free to base; leaves to 4 mm broad, not regularly septate.

              12 Stem wing-flattened, to 3 mm broad; leaves to 2 dm long and 4 mm broad; fruit 4 or 5 mm long, with a prominent, more or less undulate-dentate dorsal keel and a marginal beak to 1 mm long; (transcontinental) . . . .*P. zosterifolius*

              12 Stem not wing-flattened or, if flattened, slender.

                 13 Stem and leaves flaccid, the latter not over 0.5 mm broad; stipules essentially nerveless; peduncle terminal, usually solitary, to about 2.5 dm long; spike thick-cylindric or sub-

globose, to 12 mm long; fruits 2 or 3 mm long, with a prominent dorsal keel and 2 obscure lateral ones, the short beak lateral; rootstock extensively creeping; (Ont. to s Labrador, Nfld., and N.S.) . . . .*P. confervoides*

    13 Stem firm; leaves usually firmer, mostly wider; stipules nerved; peduncles usually in the upper forks, at most 9 cm long . . . . *GROUP A*

4 Leaves dimorphic, the floating ones mostly long-petioled and wider than the sessile or short-petioled submersed ones.

    14 Submersed leaves less than 4 mm broad, sessile.

        15 Submersed leaves at most 2 mm broad.

            16 Floating leaves to over 5 cm long and 3 cm broad, with up to 25 nerves; submersed leaves to 1 mm broad; stipules to over 5 cm long, delicately fibrous; fruiting spike to 3.5 cm long; fruits to about 3.5 mm long, keelless or with an obscure dorsal keel, the short beak marginal; (Ont. to s Labrador, Nfld., and N.S.) . . . .*P. oakesianus*

            16 Floating leaves at most about 3 cm long and 12 mm broad, with rarely more than 15 nerves; fruits at most about 2.5 mm long.

                17 Stipules adnate to the base of the leaf, the resulting connate leaf-sheaths much longer than the free stipular tip; submersed leaves to 2 mm broad, 3-nerved, obtuse; floating leaves to 12 mm broad, with up to 15 nerves; spikes dimorphic or trimorphic, those in the axils of the linear submersed leaves 1–6-flowered, subglobose, rarely exserted beyond the sheathing leaf-base, those in the axils of floating leaves to about 1.5 cm long, on peduncles to 3 cm long; beak of fruit obsolete; (Ont. to Nfld. and N.S.) . . . .*P. spirillus*

                17 Stipules free; submersed leaves at most 1 mm broad, 1-nerved, acute to almost bristle-tipped; floating leaves at most about 8 mm broad, with up to 9 nerves, present on fruiting plants (sterile plants with only submersed bristle-tipped leaves to 0.5 mm broad); fruits with a distinct dorsal keel and a slender marginal recurved beak; (Ont., Que., and N.B.) . . . .*P. vaseyi*

        15 Submersed leaves to 4 mm broad; (transcontinental).

            18 Stem compressed; submersed leaves ribbon-like, with nearly parallel sides, more or less strongly 2-ranked; fruits about 3 or 4 mm long, strongly flattened laterally, distinctly 3-keeled when dry, with a central wart-like beak . . . .*P. epihydrus*

            18 Stem terete; submersed leaves not 2-ranked; fruits 2 or 3 mm long, with rounded sides, more or less keeled, the short beak nearly central . . . .*P. gramineus*

    14 Submersed leaves at least 4 mm broad.

        19 Stem compressed; submersed leaves ribbon-like, with nearly parallel sides, more or less strongly 2-ranked; fruits strongly flattened laterally, 3-keeled when dry, with a central wart-like beak; (transcontinental) . . . .*P. epihydrus*

        19 Stem round in cross-section; submersed leaves not 2-ranked, mostly broader, their sides not parallel; fruits with rounded sides . . . .*GROUP B*

GROUP A

1 Margins of stipules united at least 2/3 their length from base; (essentially transcontinental).

    2 Rootstock filiform, freely branching; fruit to 2.5 mm long, with a thin, undulate or dentate dorsal keel and a short marginal beak; spikes subcapitate, at most about 5 mm long, on clavate peduncles; leaves glandless and somewhat tapering at base . . . .*P. foliosus*

    2   Rootstock not elongate; fruits rounded dorsally; spikes to 1.5 cm long, interrupted; leaves scarcely tapering at base, often with a pair of basal glands.

        3   Stipules not strongly fibrous, olivaceous; leaves 3-nerved; peduncles filiform to tip; fruits to 2.8 mm long, with a low broad rounded dorsal keel and a rather short marginal beak . . . .*P. pusillus*

        3   Stipules strongly fibrous, becoming whitish; peduncles broadened at tip; fruits 2 or 3 mm long, with a low acute dorsal keel and a small marginal beak.

            4   Leaves thin, 5–7-nerved, obtuse to rounded at the minutely cuspidate tip; peduncles at most about 5 cm long . . . .*P. friesii*

            4   Leaves firm, usually 3-nerved; peduncles to 9 cm long . . . .*P. strictifolius*

1   Margins of stipules often inrolled but not united.

    5   Leaves 5–17-nerved, to 2 mm broad; stipules firm, to 3 cm long.

        6   Leaves 5–9-nerved, subrigid, attenuate to bristle-tips; stipules coarsely fibrous; fruit unknown; (s Ont. and NW Nfld.) . . . .*P. longiligulatus*

        6   Leaves 9–17-nerved, flaccid, rather abruptly mucronate; stipules subrigid; (Alaska–Yukon–Dist. Mackenzie; cent. Ont.–Que.) . . . .*P. subsibiricus*

    5   Leaves 1–3-nerved; stipules delicate or, if coarsely fibrous, less than 2 cm long.

        7   Leaves tapering to long bristle-tips, 1-nerved (or obscurely 3-nerved), to 0.4 mm broad; stipules linear-attenuate, subherbaceous; peduncles filiform, to 3.5 cm long; spike elongate, remotely flowered; (s Que.) . . . .*P. gemmiparus*

        7   Leaves obtuse to acute but not bristle-tipped; stipules obtuse; (transcontinental).

            8   Leaves to 4 mm broad, obtuse, the lateral nerves joining the midrib at tip; fruits 3 or 4 mm long, with a very low sharp dorsal keel; fruiting spikes to 1.5 cm long . . . .*P. obtusifolius*

            8   Leaves less than 2.5 mm broad, the lateral nerves joining the midrib well below the tip or evanescent; fruits less than 3 mm long, rounded on back; fruiting spikes less than 1 cm long . . . .*P. berchtoldii*

GROUP B

1   Floating leaves (when developed) delicate, translucent, tapering to the petiole; submersed leaves sessile, to 2 cm broad; fruit with a hard smooth tawny-olive outer layer, about 3 mm long, with a narrow sharp dorsal keel and 2 rounded lateral ones, the short beak lateral; (transcontinental) . . . .*P. alpinus*

1   Floating leaves firm and opaque, cuneate or cordate at base; outer layer of fruit soft and porous.

    2   Fruit beakless or nearly so and nearly or quite keelless, at most 2.5 mm long; fruiting spike about 6 mm thick; submersed leaves (often wanting) to 1.5 cm broad; floating leaves to 4 cm broad; (Nfld. and N.S.) . . . .*P. polygonifolius*

    2   Fruit definitely beaked, to 5 mm long; fruiting spike to 1.5 cm thick.

        3   Stem and petioles usually conspicuously black-dotted; submersed leaves to 3.5 cm broad, 9–21-nerved, sessile or short-petioled; floating leaves to over 8 cm broad and with up to 35 nerves, broadly rounded or cordate at base; fruiting spikes at most about 3.5 cm long; fruits 3 or 4 mm long, with an often prominent dorsal keel and a prominent central beak; (s Ont. and N.S.) . . . .*P. pulcher*

        3   Stem and petioles not conspicuously black-dotted; (essentially transcontinental).

            4   Submersed leaves arcuate, to 7.5 cm broad, with up to 37 nerves, tapering to petioles up to 6 cm long; floating leaves to 5 cm broad, with up to 51 nerves, rounded or tapering at base; spikes to 8 cm long; fruits to 5 mm long, with 3 low rounded dorsal keels and a prominent conical central beak . . . .*P. amplifolius*

            4   Submersed leaves not arcuate.

               5   Submersed leaves tapering to petioles up to 13 cm long, the blade to 3.5 cm broad, with up to 15 nerves; floating leaves to 4.5 cm broad, with up to 31 nerves, acutish or rounded at base; fruiting spikes to 7 cm long; fruits about 4 mm long, with a sharp but narrow, often tuberculate dorsal keel, the 2 obscure lateral keels also often tuberculate, the short beak central . . . .*P. nodosus*

    5   Submersed leaves sessile or on petioles not over 4 cm long.
        6   Stipules to 8 cm long; submersed leaves sessile or tapering to petioles
           up to 4 cm long, the blade to 4.5 cm broad, with up to 19 nerves; floating
           leaves (often wanting) to over 6 cm broad; fruiting spikes to 7 cm long;
           fruits to 3.5 mm long, with a prominent acute dorsal keel and a deltoid
           central beak about 0.5 mm long . . . .*P. illinoensis*
        6   Stipules at most about 3 cm long; submersed leaves sessile, to about
           1.5 cm broad and with up to 11 nerves; floating leaves commonly less
           than 3 cm broad; fruiting spike less than 3 cm long; fruits less than 3 mm
           long, more or less keeled and with a short central beak . . . .*P. gramineus*

P. alpinus Balbis

/aST/X/GEA/ (HH) Shallow pools and slow streams (the ranges of the N. American phases given below; for discussions of the distinguishing characters of the typical Eurasian form, *see* M. L. Fernald, Rhodora 32(376): 76–83. 1930, and Ogden 1943), s to Calif., Colo., Minn., Pa., and New Eng.; Greenland; Iceland; Eurasia. MAPS and synonymy: *see* below.

1   Submersed leaves oblong to oblong-ovate, at most about 1 dm long, rounded and often
    slightly hooded at tip; [*P. microstachys* var. *sub.* Fern.; *P. tenuifolius* var. *sub.* Fern.; *P.*
    rufescens of Canadian reports in part, not Schrad.; s B.C. (Shuswap L., E of Kamloops);
    Sask. (L. Athabasca); Ont. (N to the Shamattawa R. at 54°24'N); Charlton Is., s James
    Bay; E Que. (Côte-Nord; Anticosti Is.; Gaspé Pen.; type from Magdalen Is.), Nfld., N.B.,
    and N.S.; MAP: Ogden 1943: map 3, p. 91] . . . . .var. *subellipticus* (Fern.) Ogden
1   Submersed leaves oblong-linear to linear-lanceolate, to 2.5 dm long, tapering to blunt or
    acutish tips; [*P. tenuifolius* Raf., the type from along the Mistassini R., Que.;
    P. microstachys Wolfg.; *P. rufescens* of Canadian reports in part, not Schrad.; *P. lucens*
    sensu Michaux 1803, not L.; transcontinental: Aleutian Is.–N Alaska–NW Dist.
    Mackenzie-sw Dist. Keewatin and B.C. to Man. (N to Churchill), Ont. (N to s Hudson Bay),
    Que. (N to near s Ungava Bay), Labrador (N to 58°54'N), Nfld., N.B., P.E.I., and N.S.; w
    Greenland N to ca. 67°N; E Asia; MAPS: Ogden 1943: map 2, p. 91; Hultén 1962: map 160,
    p. 169, and 1968*b*:72] . . . . .var. *tenuifolius* (Raf.) Ogden

P. amplifolius Tuckerm.

/T/(X)/ (HH) Lakes and streams from s-cent. B.C. (Prince George; not known from Alta.) to cent. Sask. (Waskesiu Lake, ca. 54°N; Breitung 1957*a*), Man. (Boivin 1967*a*, validating earlier reports by Burman 1909, and Lowe 1943), Ont. (N to the N shore of L. Superior), Que. (N to the Gaspé Pen.), Nfld., N.B., and N.S. (not known from P.E.I.), s to Calif., S.Dak., Okla., and Ga. MAP: Ogden 1943: map 5, p. 127.

A presumed hybrid with *P. illinoensis* (× *P. scoliophyllus* Hagstr.) has been reported from Cedar L., Ont., and Buckingham, Que., and plants with apparent genetic infiltration with varieties of *P. alpinus* and with *P. gramineus* have been taken in N.S.

P. berchtoldii Fieber

/aST/X/GEA/ (HH) Quiet waters, the aggregate species from cent. Alaska–Yukon and B.C.–?Alta. to Sask. (N to L. Athabaska), s Man., Ont. (N to w James Bay at ca. 53°N), Que. (N to L. St. John and the Côte-Nord), Nfld., N.B., and N.S. (not known from P.E.I.), s to Calif., Nebr., Okla., Minn., Ind., Ky., and Va.; s half of Greenland; Eurasia. MAPS and synonymy: *see* below.

1   Leaves mostly rounded or obtuse at tip.
    2   Leaves mostly green, their midribs bordered on each side by 1 row of lacunae.
        3   Principal leaves to 7 cm long; [*P. groenlandicus* Hagstr.; *P. pusillus* var.
           mucronatus (Fieb.) Graebn., not *P. mucronatus* Schrad.; transcontinental; MAPS:
           Hultén 1968*b*:76, and 1962: map 117 (as *P. pusillus*), p. 127; Fernald 1932: map
           21, p. 88] . . . . .var. *berchtoldii*
        3   Principal leaves less than 3 cm long; [NE Alta., w James Bay; Gaspé Co., E Que.;
           Cape Breton Is.; MAP: Fernald 1932: map 22, p. 90] . . . . .var. *polyphyllus* (Morong) Fern.
    2   Leaves tawny, their midribs bordered on each side by 2–4-rows of lacunae; [*P.*
        pusillus var. *colpophilus* Fern., the type from near the mouth of the Dartmouth R.,
        Gaspé Co., E Que.; MAP: Fernald 1932: map 23, p. 91] . . . . .var. *colpophilus* Fern.

1 Leaves subacute to sharp-pointed.
    4 Midrib of principal leaves bordered on each side by 3–5 rows of lacunae; [*P. lacunatus* Hagstr.; s Vancouver Is.; s Man.; s Ont.; N.S.; MAP: Fernald 1932: map 20, p. 86] . . . . .var. *lacunatus* (Hagstr.) Fern.
    4 Midrib of principal leaves bordered on each side by 1 or 2 rows of lacunae.
        5 Principal leaves to 1.5 mm broad, mostly with 2 rows of lacunae on each side of the midrib; [*P. pusillus* var. *typicus sensu* Fernald 1932:81; transcontinental; MAP; Fernald 1932: map 18 (*P. pusillus* var. *typicus*), p. 82] . . . . .var. *acuminatus* Fieber
        5 Principal leaves at most 1 mm broad, with 1 (or no) row of lacunae on each side of the midrib; [*P. pusillus* vars. *capitatus* Benn. and *tenuissimus* Mert. & Koch; transcontinental; MAP: Fernald 1932: map 19, p. 84] . . . . .var. *tenuissimus* (Mert. & Koch) Fern.

P. confervoides Rchb.
/T/EE/ (HH) Sandy or peaty ponds and pools from Wisc. to Mich., Ont. (Boivin 1967*a*), Que. (Senneterre, E. Abitibi Co.; Laurentide Park, Charlevoix Co.), SE Labrador (Goose Bay, 53°18′N), Nfld., N.B., and N.S. (not known from P.E.I.), s to Pa., N.Y., and N.J. MAPS: Fernald 1932: map 1, p. 33, and Rhodora 33(386): map 33, p. 57. 1931.

P. crispus L.
Eurasian; locally aggressive in ponds and streams (usually brackish or calcareous) throughout the U.S.A. and known in Canada from Alta. (Calgary), Ont. (N to Sault Ste. Marie and Ottawa; *see* Ont. map by Montgomery 1956: fig. 1, p. 92), Que. (N to Montreal), and N.S. (Windsor, Hants Co.; J. S. Erskine, Acadian Nat. 1: 147. 1944; reported from P.E.I. by McSwain and Bain 1891, but not listed by D. S. Erskine 1960). MAPS: Hultén 1962: map 128, p. 137; Ogden 1943: map 1, p. 91; Meusel, Jaeger, and Weinert 1965: 24.

P. epihydrus Raf.
/ST/X/E/ (HH) Lakes and streams, the aggregate species from the southernmost Alaska Panhandle and B.C. to Sask. (Boivin 1967*a*; not known from Alta.), SE Man. (Falcon L.; Brereton L.), Ont. (N to Nikip L. at ca. 53°N, 92°W), Que. (N to Knob Lake, 54°48′N), s Labrador (Goose Bay), Nfld., N.B., P.E.I., and N.S., s to Calif., Colo., Wisc., Mich., Tenn., and N Ga.; NW Europe (the Outer Hebrides). MAPS and synonymy: *see* below.
1 Submersed leaves to 1 cm broad and with up to 13 nerves; [Ont. (N to near Ottawa), Que. (N to Anticosti Is. and the Gaspé Pen.), s N.B., and sw N.S.; MAPS: Fernald 1932: map 30, p. 114; Hultén 1958: map 191 (aggregate species, essentially the area of var. *nuttallii*), p. 211] . . . . .var. *epihydrus*
1 Submersed leaves not over 8 mm broad, 3–7-nerved; [*P. nuttallii* C. & S. and its var. *ramosus* Peck; *P. claytonii* Tuck.; *P. pensylvanicus* Willd.; transcontinental; MAPS: Hultén 1968*b*:72 (var. *ramosus*); Fernald 1932: map 31, p. 116] . . . . .var. *ramosus* (Peck) House

P. filiformis Pers.
/aST/X/GEA/ (HH) Calcareous or brackish waters, var. *borealis* N to the coasts of Alaska–Yukon–Dist. Mackenzie, s Baffin Is., N Ungava, Labrador (N to ca. 56°30′N), and Nfld., the aggregate species s to Calif., Ariz., Colo., Minn., Pa., Vt., and Maine; w and E Greenland N to ca. 75°N; Iceland; Eurasia; Africa; Australia. MAPS and synonymy: *see* below.
1 Leaves to 2 mm broad, obtuse or short-pointed; [*P. marinus* var. *mac.* Morong; *P. juncifolius sensu* Arthur Bennett, J. Bot. 46: 162. 1908, as to P.E.I. report, not Kern.; (Alta. (N to Wood Buffalo National Park at 59°41′N), Sask. (Ravenscrag; Breitung 1957*a*), Man. (Forrest, near Brandon; Gunisao L., NE of L. Winnipeg), Ont. (w James Bay at ca. 54°30′N), s James Bay (Charlton Is.), Que. (Côte-Nord; Gaspé Pen.; Magdalen Is.), P.E.I., and N.S. (Cape Breton Is.); MAPS (incomplete northwards); Fernald 1918*b*: map 7, p. 12, and 1925: map 27, p. 259] . . . . .var. *macounii* Morong
1 Leaves not over 0.5 mm broad.
    2 Spike at most about 2.5 cm long, its lower whorls less than 1 cm apart; leaves obtuse or shallowly notched; [*P. borealis* Raf.; transcontinental; MAP: Porsild 1957: map 11, p. 162] . . . . .var. *borealis* (Raf.) St. John

2 Spike to about 5 cm long, its lower whorls to 2.5 cm apart; leaves obtuse; [*P. interior*
Rydb.; St. Paul Is. and s Alaska to N Man. (Churchill); s Hudson Bay; E Que. (Gaspé
Pen. and Magdalen Is.); w Nfld.; MAPS (aggregate species): Hultén 1958: map 242, p.
261, and 1968*b*:78; Meusel Jaeger, and Weinert 1965:26] ....var. *filiformis*

P. foliosus Raf.

/sT/X/ (HH) Fresh or brackish water (ranges of Canadian taxa outlined below), s to Calif.,
Mexico, Tex., and Fla.; Guatamala; W.I.; Hawaii. MAPS and synonymy: *see* below.
1 Leaves to about 2.5 mm broad, 3–5-nerved; fruits obliquely suborbicular; [*P. niagarensis*
Tuck.; *P. pauciflorus* Pursh; sw Man. (N to about 20 mi N of Brandon) and s Ont. (N to
near Ottawa); MAPS: Hultén 1968*b*:77 (aggregate species); Fernald 1932: map 4, p. 44] ....
....var. *foliosus*
1 Leaves not over 1.5 mm broad, 1–3-nerved; fruits obliquely obovoid; [s Alaska (College);
Great Bear L.; N Alta., Sask. (Crane L.), Ont. (L. Nipigon, N of L. Superior), Que. (N to L.
Mistassini and the Gaspé Pen.), Labrador (N to Goose Bay), N.B., P.E.I., and N.S.; MAP:
Fernald 1932: map 5, p. 47] ....var. *macellus* Fern.

P. friesii Rupr.

/ST/X/EA/ (HH) Calcareous or brackish waters from s Alaska–Yukon to the Mackenzie R.
Delta, Great Bear L., s Dist. Keewatin, Ont. (N to Fort Severn, Hudson Bay, ca. 56°N), Que. (N to s
James Bay and the Côte-Nord), w Nfld., N.B. (Tidehead, near Campbellton), P.E.I., and N.S., s to
Wash., s B.C. (Bonaparte R., near Kamloops), s Alta., Sask. (Methy Portage, ca. 56°30′N; an early
collection by J. M. Macoun distributed as *P. zizii* Koch), S.Dak., Ind., and Va.; Eurasia. [*P. major*
(Fries) Morong; *P. mucronatus* Schrad.]. MAPS: Hultén 1968*b*:76, and 1958: map 251, p. 271;
Fernald 1932: map 7, p. 53; Porsild 1966: map 7, p. 67; Meusel, Jaeger and Weinert 1965:25.

P. gemmiparus Robins.

/T/E/ (HH) Quiet waters of s Que. (reported from L. St. Peter, about 50 mi NE of Montreal, by
Marie-Victorin 1935; also known from St-Tite, Champlain Co., about 30 mi N of L. St. Peter), Maine,
Mass., R.I., and Conn. MAP: Fernald 1932: map 13 (the Que. localities should be indicated), p. 69.

P. gramineus L.

/aST/X/GEA/ (HH) Lakes, ponds, and streams, the aggregate species from N Alaska and cent.
Yukon to the Mackenzie R. Delta, Great Bear L., s Dist Keewatin, Que. (N to the Larch R. at ca.
57°35′N), Labrador (N to ca. 55°N), Nfld., N.B., P.E.I., and N.S., s to Calif., Ariz., N.Mex., Nebr.,
Ind., Ohio, and N.J.; w Greenland N to ca. 67°30′N; Iceland; Eurasia. MAPS and synonymy: *see*
below.
1 Principal submersed leaves linear, 3-nerved, mostly less than 5 cm long and 3 mm broad
(up to 20 or 30 times as long as broad), their sides essentially parallel for most of their
length; [cent. Ont.: Baldwin 1958; MAP: Ogden 1943: map 12, p. 147] ....
....var. *myriophyllus* Robbins
1 Principal submersed leaves oblanceolate to narrowly elliptic, (3)5–9-nerved, often less
than 10 times as long as broad, their sides not parallel.
  2 Principal submersed leaves to 9(13) cm long and 1(1.5) cm broad, 7–9(11)-nerved;
    [transcontinental; MAP: Ogden 1943: map 11, p. 153] ....var. *maximus* Morong
  2 Principal submersed leaves to 4.5(6.5) cm long and 6(8) mm broad, 5–7-nerved ....
    ....var. *gramineus*
    3 Submersed leaves none, the emersed leaves crowded; [a form of muddy drying
      shores; L. Abitibi, Ont.: Baldwin 1958] ....f. *terrestris* (Schlecht.) Carpenter
    3 Submersed leaves present.
      4 Stem-internodes to 2 dm long, much longer than the leaves; [St. Helen's Is.,
        Montreal; Rouleau 1945] ....f. *longipedunculatus* (Mérat) House
      4 Stem-internodes mostly shorter than the leaves; [var. *graminifolius* Fries; *P.*
        heterophyllus of auth., not Schreb.; transcontinental; MAPS: Hultén 1968*b*:73;
        Ogden 1943: map 10, p. 147]. Genetic infiltration with *P. nodosus, P.*
        illinoensis, or the varieties of *P. alpinus* gives rise to a series of hybrids known
        as × *P. spathulaeformis* (Robbins) Morong (*P. gramineus* var. *spathulaefor-*

mis Robbins). Such plants are reported from Nfld., Anticosti Is., Magdalen Is., and N.S. by M. L. Fernald (Rhodora 23(272): 191. 1921) and from Que. and Ont. by Ogden (1943). × *P. subnitens* Hagstr. (*P. gramineus* × *P. perfoliatus* var. *bupleuroides*) is reported from Nfld. by R. B. Kennedy (Rhodora 32(373): 4. 1930) and from Ont., Que., Nfld., N.B., and N.S. by Ogden (1943). × *P. hagstromii* Bennett (*P. gramineus* × *P. richardsonii*) is reported from L. Nippissing, Ont., by Ogden (1943). × *P. methyensis* Benn., a purported hybrid between *P. gramineus* and another unknown species, is known from the type locality, Methy L., near Methy Portage, Saskatchewan . . . . .f. *gramineus*

P. illinoensis Morong
/T/X/ (HH) Lakes and streams, chiefly calcareous, from s B.C. (Sumas L., Chilliwack Valley; an isolated station in s Dist. Mackenzie) and N Mont. to s Man. (reported from Souris, about 20 mi sw of Brandon, by Ogden 1943; a 1936 collection by Marck in DAO from Bissett, near the Ont. boundary at ca. 51°N, may also belong here but requires verification; not known from Alta. or Sask.), Ont. (N to Golden L., Renfrew Co.), and sw Que. (N to Wakefield, Gatineau Co.), s to Calif., Colo., Tex., Ark., and N.C. (isolated stations in Fla.). [*P. lucens* and *P. zizii* of Canadian reports in part, not L. nor Koch, respectively; *P. ?angustifolius* of B.C. reports, not Berchtold & Presl; × *P. ?perplexus* Benn.]. MAPS: Hultén 1958: map 53, p. 73; Ogden 1943: map 13, p. 153; Meusel, Jaeger, and Weinert 1965: 24.

Hultén's map indicates a station on Anticosti Is., E Que., based upon the report of the Eurasian *P. lucens* L. from there by J. Adams (Can. Field-Nat. 49(8):138. 1935; DAO). Further study of this collection is required to verify its identity. The above Sumas L. collection from B.C. (J. M. Macoun, No. 26, 815; CAN) comprises the type material of × *P. perplexus* Benn., believed by Bennett to represent a hybrid between *P. natans* and *P. nodosus.* Ogden (1943), however, believes it to be typical *P. illinoensis.* He reports a hybrid between *P. illineonsis* and *P. nodosus* from Ile-Bizard, near Montreal, Que.

P. longiligulatus Fern.
/T/EE/ (HH) Calcareous waters of Minn., Mich., s Ont. (Sarnia Bay, Lambton Co., Dodge, no. 135, distributed as *P. hillii* Morong; reported from Stokes Bay, Bruce Pen., L. Huron, by Krotkov 1940), NW Nfld. (type from Flower Cove, Straits of Belle Isle), N.Y., and Conn. MAP: Fernald 1932: map 12, p. 67.

P. natans L.
/aST/X/GEA/ (HH) Lakes and quiet streams from s Alaska (*see* Hultén 1941: map 80, p. 125), NW Dist. Mackenzie (Porsild and Cody 1968), and B.C. to Alta. (near Lake Louise), Sask. (N to Windrum L. at ca. 56°N), Man. (N to near Flin Flon), Ont. (N to w James Bay at ca. 53°N), Que. (N to E James Bay at ca. 54°30'N and the Côte-Nord), Nfld., N.B., P.E.I., and N.S., s to Calif., N.Mex., Nebr., Ohio, and N.J.; southernmost Greenland; Iceland; Eurasia. [Var. *prolixus sensu* John Macoun 1888, perhaps not Koch]. MAPS: Hultén 1968b:71, and 1962: map 97, p. 107; Ogden 1943: map 8, p. 127; Meusel, Jaeger, and Weinert 1965:24.

P. nodosus Poir.
/sT/X/EA/ (HH) Ponds and streams from B.C. (Kamloops, the Chilcotin Valley about 100 mi NW of Kamloops, and Prince George, ca. 54°N; CAN) to Mont., S.Dak., Minn., Ont. (N to the Ottawa dist.), Que. (N to l'Islet Co.), s Labrador (Goose Bay, ca. 53°20'N), and s N.B. (Sunbury Co.; not known from P.E.I. or N.S.), s to s Calif., Mexico, Tex., Ala., and Va.; Europe and w Asia; E ?Asia; Africa. [*P. americanus* C. & S.; *P. fluitans* of auth., perhaps not Roth; *P. lonchites* Tuckerm. in part; *P. occidentalis* Sieber]. MAPS: Hultén 1962: map 136, p. 145; Ogden 1943: map 7, p. 127.

P. oakesianus Robbins
/sT/EE/ (HH) Peaty or sandy pools from Ont. (known only from the SE shore of L. Superior at Batchawana Bay) to Que. (N to Anticosti Is. and the Côte-Nord near the mouth of the Matamek R., Saguenay Co.), s Labrador (Battle Harbour, 52°19'N; Hustich and Pettersson 1943), Nfld., N.B. (Charlotte and Westmorland counties; not known from P.E.I.), and N.S., s to Wisc., Pa., and N.J. MAPS: Ogden 1943: map 9, p. 127; Meusel, Jaeger, and Weinert 1965:24.

P. obtusifolius Mert. & Koch
/T/X/EA/ (HH) Cold streams and lakes: s-cent. Yukon (Sheldon L.; CAN, distributed as *P. porsildiorum,* revised by Porsild) and sw B.C. (Cowichan L., Vancouver Is.; CAN) to Alta. (Glenevis, 45 mi NW of Edmonton; CAN); Sask. (L. Athabasca); Ont. (N to Sandy L. at ca. 53°N, 93°W) to Que. (N to the Côte-Nord and Gaspé Pen.), s Labrador (Labrador side of the Blanc Sablon R. at ca. 51°30′N; GH), Nfld., N.B. (Petit Rocher; Woodstock; not known from P.E.I.), and N.S., s to Kans., Minn., Mich., and N N.J.; Eurasia. MAPS: Hultén 1958: map 44, p. 63; Fernald 1932: map 16, p. 76; Meusel, Jaeger, and Weinert 1965:25.

P. pectinatus L.
/aST/X/EA/ (HH) Chiefly brackish or alkaline waters from N Alaska, s-cent. Yukon, and NW Dist. Mackenzie to L. Athabasca (Alta. and Sask.), Man. (N to The Pas, ca. 54°N), Ont. (N to w James Bay at ca. 53°N), Que. (N to the Larch R. at ca. 57°35′N), Nfld., N.B., P.E.I., and N.S., s to Baja Calif., Mexico, Tex., and Fla.; S. America; Eurasia; Africa. [Incl. f. *pseudomarinus* Benn., *P. flabellatus* Bab., and *P. interruptus* Kit.]. MAPS: Hultén 1968b:77, and 1962: map 119, p. 129; Porsild 1966: map 8, p. 67; Meusel, Jaeger, and Weinert 1965:25.

P. perfoliatus L.
/aST/X/GEA/ (HH) Chiefly brackish or calcareous waters, the aggregate species from N Alaska–Yukon–Dist. Mackenzie to s Dist. Keewatin, northernmost Ont., Que. (N to SE Hudson Bay at ca. 55°N and the Côte-Nord), Labrador (N to the Hamilton R. basin), Nlfd., N.B., P.E.I., and N.S., s to Calif., Colo., Nebr., Ohio, and N.C. (isolated stations from La. to Fla.); ssp. *bupleuroides* with an isolated station in E Greenland at ca. 71°N; Iceland; Eurasia; Africa; Australia. MAPS and synonymy: *see* below.
1 Stipules coarse, soon disintegrating into strong white fibres; leaves narrowly lanceolate to lance-ovate, to about 1 dm long, coarsely nerved; peduncles usually enlarged toward summit; beak of fruit to 1 mm long; [var. *lanceolatus* Robbins, not Blytt; *P. rich.* (Benn.) Rydb.; transcontinental, largely replacing the other varieties northwards; MAPS: Ogden 1943: map 15 (*P. rich.*), p. 175; Hultén 1962: map 172, p. 183, and 1968b:74; Porsild 1966: map 9 (*P. rich.*), p. 68; A. Löve, Sven. Bot. Tidskr. 48(1): fig. 2 (evidently incl. other varieties in the NE area), p. 217. 1954] ....ssp. *richardsonii* (Benn.) Hult.
1 Stipules delicate and soon disappearing; leaves ovate-lanceolate to rotund, at most about 6 cm long, delicately nerved; peduncles slender to summit; beak of fruit very short.
  2 Stem at most 1.5 mm thick; leaves at most 2 cm broad; [*P. bupleuroides* Fern., the type from Holyrood, Nfld.; Ont. (N to Cache L., Algonquin Park, about 65 mi SE of North Bay), Que. (Knob Lake, 54°48′N; Gaspé Pen.; Magdalen Is.), Nfld., N.B., P.E.I., and N.S.; E Greenland at ca. 71°N; MAPS: Ogden 1943: map 17, p. 175; also on the above-noted maps by Hultén and Löve] ....ssp. *bupleuroides* (Fern.) Hult.
  2 Stem to 2 mm thick; leaves to 3 cm broad; [s ?Yukon and NW Dist. ?Mackenzie, (Porsild 1951a); Que. (near Montreal, Gaspé Pen.; Côte-Nord), s Labrador (at ca. 51°30′N), and N.B. (Kings Co.); MAPS: Hultén 1962: map 172, p. 182; Ogden 1943: map 16 (E area), p. 175] ....ssp. *perfoliatus*

P. polygonifolius Pourret
/T/E/E/ (HH) Shallow pools and muddy shores: St-Pierre and Miquelon and E Nfld.; Sable Is., N.S.; Madeira and Azores; Europe; N Africa. [*P. oblongus* Viviani]. MAPS: Hultén 1958: map 150, p. 169 (citing other total-area maps by Taylor and Heslop-Harrison); Ogden 1943: map 4, p. 91; Meusel, Jaeger, and Weinert 1965:24.

P. praelongus Wulfen
/ST/X/EA/ (HH) Cool waters of lakes and streams from the Aleutian Is. (Atka Is.) and Alaska (N to near Fairbanks at ca. 65°N; V. L. Harms, Can. Field-Nat. 83(3): 255. 1969; not known from the Yukon) to L. Athabasca (Alta. and Sask.; an isolated station at Eskimo L., NW Dist. Mackenzie), Man. (N to ca. 55°N), Ont. (N to Fort Severn, Hudson Bay, ca. 56°N), Que. (N to E James Bay at ca. 53°45′N and the Gaspé Pen.), Labrador (N to Carol L. at ca. 53°N), Nfld., N.B., P.E.I., and N.S., s to Calif., Colo., Iowa, Ind., and N.J.; Iceland; Eurasia. MAPS: Hultén 1968b:73, and 1962: map 80, p. 89; Ogden 1943: map 14, p. 175.

Forma *elegans* Tiselius, the extremely flaccid phase, is reported from Nfld. by Ogden (1943; Humber R. system; GH; CAN).

P. pulcher Tuckerm.
/T/EE/ (HH) Peaty or muddy waters and shores from s Minn., s Ont. (Boivin 1967*a*), and N.S. (Digby, Queens, Lunenburg, and Halifax counties) to E Tex. and Fla. MAP: Ogden 1943: map 6, p. 127.

P. pusillus L.
/ST/X/EA/ (HH) Basic or alkaline waters, the aggregate species from cent. Alaska–Yukon–Dist. Mackenzie to s-cent. Sask., Man. (N to York Factory, Hudson Bay, ca. 57°N), Ont. (N to the Fawn R. at ca. 54°30′N), Que. (N to the Côte-Nord), ?Nfld. (Boivin 1967*a*), N.B., P.E.I., and N.S., s to Calif., Mexico, Tex., and Va. MAPS and synonymy: *see* below.
1   Leaves to 3 mm broad; [*P. panormitanus* Biv. and its var. *major* Fisch.; transcontinental;
     MAP: Hultén 1962: map 118 (*P. pan.*), p. 127]. Gleason (1958) unites *P. berchtoldii*
     (margins of stipules free) with *P. pusillus* (margins connate to middle) and expresses the
     opinion that, "While free stipules or united ones may be characteristic of some species,
     there is little reason to infer that both types or gradations between them may not occur
     within a single species." According to Löve and Bernard (1959), the plant with united
     stipules should be known as *P. panormitanus* Biv. (*P. pusillus sensu* Fernald, not L.),
     whereas the one with free stipules is the true *P. pusillus* L. (*P. berchtoldii* Fieb.) . . .var. *pusillus*
1   Leaves at most 1 mm broad; [*P. panormitanus* var. *minor* Biv.; s B.C. (Vancouver Is.;
     Similkameen R.), s Sask., Man. (N to York Factory, Hudson Bay, 57°N), and Ont. (s of L.
     Nipigon)] . . . . .var. *minor* (Biv.) Fern. & Schub.

P. robbinsii Oakes
/T/(X)/ (HH) Muddy waters: B.C. (N to Prince George, ca. 54°N) to Oreg. and Wyo.; w Ont. (N to the Kenora dist. and the N shore of L. Superior; TRT; reported N to Oba L. at ca. 49°N and the Missinaibi R. by John Macoun 1888) to Que. (N to L. Mistassini), N.B., and N.S. (not known from P.E.I.), s to Calif., Idaho, Wyo., Minn., Ind., Ala., and Del. MAP: Hultén 1968*b*:74.
     Forma *cultellatus* Fassett (leaf-margins entire rather than minutely serrulate; type from Great Cloche Is., N of Manitoulin Is., N L. Huron) is known from the type locality and from Renfrew Co. near Mattawa, Ont.

P. spirillus Tuckerm.
/T/EE/ (HH) Quiet waters from S.Dak. and Minn. to Ont. (N to the w and E shores of L. Superior and the Ottawa dist.), Que. (N to L. St. John; E to Rimouski Co.), Nfld. (Rushy Pond, near Grand Falls), N.B., and N.S. (not known from P.E.I.), s to Iowa, Wisc., Ohio, and Va. [*P. dimorphus* of auth., not Raf.; *P. ?hybridus sensu* John Macoun 1888, not Michx.]. MAP: Fernald 1932: map 26, p. 101.
     The inclusion of E Man. in the range given by Fernald *in* Gray (1950) is perhaps based upon the report of *P. diversifolius* Raf., a similar but more southern species, from Norway House, off the NE end of L. Winnipeg, by Hooker (1839; s B.C. is also included in the range by Hitchcock et al. 1969). The proper disposal of these reports remains in doubt. Another similar southern species is *P. capillaceus* Poir., reported from s Ont. by Gaiser and Moore (1966; Lambton Co.). It may be distinguished from *P. spirillus* as follows:
1   Submersed leaves obtuse or acute, not bristle-tipped, to 2 mm broad; floating leaves
     rounded at tip and minutely emarginate, 5–15-nerved; lowest spikes on ascending or
     arching peduncles to about 4 mm long . . . .*P. spirillus*
1   Submersed leaves setaceous, bristle-tipped, less than 1 mm broad; floating leaves (when
     present) mostly acute or short-pointed, 3–7-nerved; lowest spikes on divergent or
     recurved peduncles to 13 mm long; (Lambton Co., s Ont.; MAP: Fernald 1932: map 28,
     p. 109) . . . .*P. capillaceus* Poir.

P. strictifolius Benn.
/sT/X/ (HH) Calcareous waters (ranges of Canadian taxa outlined below), s to Utah, Nebr., Ind., Pa., and Vt. MAPS and synonymy: *see* below.

1   Leaves obtuse or abruptly mucronate; [Sask. (Cumberland House, ca. 54°N); Ont. (N to
James Bay at 54°22′N; RIM) and Que. (E James Bay at 52°37′N; RIM); MAP: Fernald
1932: map 8 (the James Bay stations should be indicated), p. 56] . . . . .var. *strictifolius*
1   Leaves tapering to a slender tip; [*P. pusillus* var. *rut.* (Fern.) Boivin; *P. pusillus* var.
vulgaris of auth., not Fries; *P. rutillus* of Canadian reports in large part, not Wolfg.; SE
Yukon, the Mackenzie R. Delta, and Great Bear L. to Man. (N to York Factory, Hudson
Bay, 57°N), Ont. (N to Hearst, 49°42′N), and sw Que. (North Wakefield, about 15 mi NW of
Ottawa, Ont.); MAP: Fernald 1932: map 9 (somewhat incomplete northwards), p. 58] . . . .
. . . . .var. *rutiloides* Fern.

P. subsibiricus Hagstr.
/S/X/A/   (HH)   Fresh or brackish waters: cent. Alaska (*see* Hultén 1941: map 83, p. 125;
P. porsildiorum), SE-cent. Yukon (Sheldon L.; CAN), and NW Dist. Mackenzie (type of *P. pors.* from
Eskimo L.); Ont. (w James Bay at ca. 54°N), James Bay (Twin Islands, ca. 53°N), and w-cent. Que.
(Great Whale R., Hudson Bay, ca. 55°20′N); N Asia. [*P. porsildiorum* Fern.; *P. rutilus* as to citations
of *P. "rutilans"* from James Bay by John Macoun 1888, not Wolfg.]. MAPS: Hultén 1968b:75;
Fernald 1932: map 3 (*P. pors.*), p. 41.

P. vaginatus Turcz.
/aST/X/EA/   (HH)   Deep fresh or brackish waters from the Aleutian Is., N Alaska, s-cent. Yukon,
and the Mackenzie R. Delta to Great Bear L., Alta. (N to L. Mamawi, near the w end of L.
Athabasca), Sask. (N to near Prince Albert), s Dist. Keewatin, Man. (N to Churchill), Ont. (N to Fort
Severn, Hudson Bay, ca. 56°N), Que. (N to s Ungava Bay and the Côte-Nord), Nfld., P.E.I., and
N.S. (not known from N.B.), s to Oreg., Wyo., N.Dak., Mich., and N.J.; Eurasia. [*P. moniliformis* St.
John]. MAPS: Hultén 1968b:78, and 1962: map 25, p. 33.

P. vaseyi Robbins
/T/EE/   (HH)   Quiet muddy or calcareous waters from Ont. (N to the Ottawa dist.) to Que. (N to
near Trois-Rivières; Marie-Victorin 1935) and s N.B. (Kings Co.; not known from P.E.I. or N.S.), s to
Minn., Ill., Ohio, Pa., and Mass. MAP: Fernald 1932: map 24, p. 96.
    The similar *P. lateralis* Morong, ranging, according to Fernald *in* Gray (1950) from Mich. to E
Mass. and Conn., is reported from s Ont. by Gaiser and Moore (1966; Lambton Co.). It may be
distinguished from *P. vaseyi* as follows:
1   Fruiting plants bearing dilated floating leaves to 8 mm broad, these with marginless
petioles; sterile plants bearing only linear-filiform bristle-tipped leaves to 0.5 mm broad;
fruits distinctly keeled on the back . . . .*P. vaseyi*
1   Fruiting plants bearing only linear acutish leaves to 1 mm broad; sterile (but flowering)
plants bearing floating leaves to 4 mm broad, these tapering to margined petioles; fruits
rounded on the back; (Lambton Co., s Ont.; MAP: Fernald 1932: map 25, p. 98) . . . .
. . . .*P. lateralis* Morong

P. zosterifolius Schum.
/ST/X/EA/   (HH)   Quiet waters from cent. Alaska and NW Dist. Mackenzie (not known from the
Yukon) to Alta. (N to Wood Buffalo National Park), Sask. (N to ca. 55°N), Man. (reported N to York
Factory, Hudson Bay, 57°N, by Hooker 1839), Ont. (N to Big Trout L. at ca. 54°N, 90°W), Que. (N to
Anticosti Is. and the Gaspé Pen.), N.B., and N.S. (not known from P.E.I.), s to N Calif., Utah, Nebr.,
Ind., and Va.; Eurasia. MAPS: Hultén 1968b:75, and 1962: map 159, p. 169; Fernald 1932: map 2
(*P. zosteriformis*), p. 37.
    The N. American plant may be distinguished as ssp. *zosteriformis* (Fern.) Hult. (*P. zosteriformis*
Fern.; *P. zosterifolius* var. *americanus* Benn.; *P. compressus* of many American auth., not L.),
differing from the typical Eurasian phase in its relatively large, truncate-based fruits with a thin
wing-like dorsal keel and a nearly erect marginal beak, the Old World plant having narrowly
obovate or oval fruits with a low, essentially wingless dorsal keel and a subcentral, depressed or
decurrent-recurved beak.

RUPPIA L.   [59]   Ditch-grass.   Ruppie

(Ref.: M. L. Fernald and K. M. Wiegand, Rhodora 16: 119–27. 1914)

1   Leaves scattered or but slightly tufted, at most about 12 cm long, their sheaths less than 2
    cm long; fruits 2 or 3 mm long, commonly brown to blackish . . . .*R. maritima*
1   Leaves often in fan-like groups, to about 2 dm long, their sheaths to about 7 cm long;
    fruits nearly symmetrical, olivaceous, dotted with red, 3 or 4 mm long; (saline or alkaline
    waters from B.C. to Man.) . . . .*R. occidentalis*

R. maritima L.   Ditch-grass
/sT/X/EA/   (HH)   Tidal waters and saline pools and ditches, the aggregate species from s
?Alaska (the map given for *R. spiralis* by Hultén 1941: map 87, p. 126, probably applies here; *see*
his discussion, p. 104) through B.C. to Baja Calif. and Mexico, and from E Que. (N to the Côte-Nord
and Anticosti Is.) to s ?Labrador (Boivin 1967a), Nfld., N.B., P.E.I., and N.S., s to Fla., with
intermediate stations on James Bay (Ont. and Que.); reports from Sask. require confirmation,
perhaps referring to *R. occidentalis*; S. America; W.I.; Eurasia. MAP (*R. spiralis*): Hultén 1968b:79.

1   Fruits only slightly oblique, bluntish or at least not distinctly beaked.
    2   Mature peduncles at least 1 dm long, strongly spiraling toward base; podogynes
        (fruit-stalks) to 3 cm long; [*R. ?spiralis* Dum.; Eurasia; ?Calif.] . . . . .var. *maritima*
    2   Mature peduncles at most 6 cm long, rarely spiraling.
        3   Peduncles to 6 cm long; podogynes at least 6 mm long; [*R. obliqua* Schur; E Que.,
            Nfld., N.B., P.E.I., and N.S.] .var. *obliqua* (Schur.) Asch. & Graebn.
        3   Peduncles at most 1 cm long; podogynes mostly less than 6 mm long.
            4   Podogyne longer than its fruit; [*R. intermedia* Thed.; E Que. and w Nfld.] . . . .
                . . . . .var. *intermedia* (Thed.) Asch. & Graebn.
            4   Podogyne commonly shorter than its fruit; [*R. brachypus* Gay; James Bay,
                Ont.; E Que.; w Nfld.] . . . . .var. *brevirostris* Agardh.
1   Fruits strongly eccentric or curved, distinctly beaked.
    5   Fruit at most 1.5 mm long, longer than its podogyne; [E Que. and Nfld. (type from
        Norris Arm)] . . . . .var. *exigua* Fern. & Wieg.
    5   Fruit 2 or 3 mm long, shorter than its podogyne.
        6   Podogynes less than 1 cm long; peduncles at most about 1.5 cm long; [James
            Bay, Que., E Que., N.B., P.E.I., and N.S.] . . . . .var. *subcapitata* Fern. & Wieg.
        6   Podogynes mostly 1 cm long or more.
            7   Peduncles commonly less than 3 cm long, not spiraling; [*R. rostellata* Koch;
                Vancouver Is.; ?Sask. (Breitung 1957a; *R. maritima*); Que., Nfld., N.B., P.E.I.,
                and N.S.] . . . . .var. *rostrata* Agardh.
            7   Peduncles usually at least 3 cm long (to 3 dm), flexuous or spiraling; [*R.
                spiralis* of auth., not Dumort.; Nfld., N.B., and N.S.] . . . . .var. *longipes* Hagstr.

R. occidentalis Wats.   Widgeon-grass
/T/W/   (HH)   Tidal waters and saline or alkaline ponds from B.C. (type from near Kamloops;
reports from Alaska require confirmation) to Alta. (Moss 1959), Sask. (common throughout the
prairie region; Fraser and Russell 1944), and s Man. (Pelican L., NE of Turtle Mt.; Virden;
St-Ambrose, near the s end of L. Manitoba), s to Nebr. and w Minn. [*R. maritima* var. *occ.* (Wats.)
Graebn.; *R. lacustris* Macoun].

ZANNICHELLIA L.   [62]   Horned Pondweed

Z. palustris L.
/sT/X/EA/   (HH)   Fresh, brackish, or alkaline waters, the aggregate species from B.C. (N to
Vanderhoof, ca. 53°30′N; an isolated station reported from the N coast of the Seward Pen., Alaska,
at 66°20′N, by A. E. Porsild, Rhodora 34(401): 94. 1932) to Alta.–Sask. (N to ca. 55°N according to
Hultén's map), Man. (N to 30 mi SE of Dauphin L., ca. 51°N), James Bay (Ont. and Que. coasts),
Que. (N to L. Mistassini and the Côte-Nord), Nfld., N.B., P.E.I., and N.S., s to Mexico, Tex., Mo.,
Tenn., and Fla.; S. America; Eurasia; Africa. MAPS and synonymy: *see* below.

1   Fruit sessile or short-stalked, nearly smooth on the back, the body rarely over 3 mm long;

[transcontinental; MAPS (aggregate species): Hultén 1962: map 124, p. 133, and 1968*b*:
79] . . . . .var. *palustris*
1   Fruit commonly distinctly stalked, more or less dentate on the back, the body at least 3
mm long; [var. *pedunculata* of auth., not Gay; cent. Ont. (w James Bay N to ca. 53°N);
Que. (L. Mistassini; Temiscouata Co. to the Côte-Nord, Anticosti Is., Magdalen Is. and
Gaspé Pen.), Nfld., N.B., P.E.I., and N.S.; MAP: Potter 1932: map 2, p. 71] . . . .
. . . . .var. *major* (Boenn.) Koch

ZOSTERA L.   [55]

Z. marina L.   Eelgrass, Grass-wrack.   Mousse de mer
/aST/(X)/GEeA/   (HH)   Shallow coastal waters: Aleutian Is. and Alaska (isolated stations along
the Seward Pen., w-cent. Alaska; *see* Hultén 1941: map 72, p. 123) through coastal B.C. to s Calif.;
SE Dist. Keewatin (Eskimo Point, ca. 61°N; A. E. Porsild, Rhodora 34(401): 91. 1932) and N Man.
(Churchill); James Bay (Charlton Is.; Ont. and Que. coasts); Que. (N to the Côte-Nord) to Labrador
(N to ca. 54°N), Nfld., N.B., P.E.I., and N.S., s to S.C.; sw Greenland at ca. 65°N; Iceland; Europe; E
Asia; N Africa. MAPS and synonymy: *see* below.
      Concerning the causes of the abrupt decline of this important waterfowl food along the Atlantic
coast in the early 1930's, *see* Clarence Cottam (Rhodora 41(487):257–60. 1939) and N. E.
Stevens (Rhodora 41(487):260–62. 1939).
1   Leaves to 12 mm broad, thinnish and subtransparent, mostly with 5–7 strong nerves (and
numerous finer ones); [Eurasia only; MAPS (aggregate species): Hultén 1962: map 186, p.
197, and 1968*b*:69; Meusel, Jaeger, and Weinert 1965:26; Meusel 1943:39; Ostenfeld,
Pflanzenareale 1: map 38, p. 4. 1927] . . . . .var. *marina*
1   Leaves to 6 mm broad, coriaceous and opaque, with 3(5) strong nerves; [var. *angustifolia*
Hornem.; N. America–Greenland range as given above for the aggregate species; MAPS:
the N. American area of the above-noted maps is referable here] . . . .
. . . . .var. *stenophylla* Asch. & Graebn.

NAJADACEAE (Naiad Family)

NAJAS L.   [64]   Naiad

Submersed aquatics of shallow, fresh or brackish waters. Stems bushy-branched. Leaves opposite, linear to lance-linear, spinulose-toothed, subtended by conspicuous sheathing stipules. Perianth none. Flowers unisexual, sessile in the leaf-axils and sheaths, the staminate a single stamen surrounded by a sac-like perianth enclosed in a bottle-like spathe, the pistillate a single naked pistil. Fruit a fusiform thin-walled 1-seeded achene.

1   Leaves filiform, abruptly dilated at the sheathing base, the upper edge of the broadly
    rounded or truncate basal auricles conspicuously toothed; each leaf-margin with at most
    20 minute spinules; staminate flowers commonly less than 1.5 mm long; seed with
    transversely elongate reticulation; (Ont. and N.S.) . . . .*N. gracillima*
1   Leaves linear, only gradually dilated at base, the expanded portion merely spinulose;
    each leaf-margin with up to 60 spinules; staminate flowers to about 3 mm long.
    2   Leaves acute to rounded at apex, flat, with straight tips; style at most 0.5 mm long;
        anthers 4-locular; seed dull, with rather coarse squarish reticulation; (s Ont. and
        s Que.) . . . .*N. guadalupensis*
    2   Leaves acuminate, with inrolling margins and recurving tips; style commonly over 1
        mm long; anthers 1-locular; seed shining, with obscure fine hexagonal reticulation;
        (essentially transcontinental) . . . .*N. flexilis*

N. flexilis (Willd.) Rostk. & Schmidt

/sT/X/EA/   (HH)   Shallow fresh to brackish waters from s Dist. Mackenzie (s of Great Slave L.) and B.C. (N to Cariboo, ca. 53°N) to N Alta. (Moose L., 59°36′N; not known from Sask.), SE Man. (N to Bissett, near the Ont. boundary at ca. 51°N), Ont. (N to the N shore of L. Superior), Que. (N to L. Mistassini and the Côte-Nord), Nfld., N.B., and N.S. (not known from P.E.I.), s to Calif., S.Dak., Iowa, and N.C.; Europe (often as a fossil); two widely separated stations in Asia. [*Caulinia* Willd.; *N. canadensis* Michx.; incl. var. *robusta* Morong]. MAPS: Hultén 1958: map 194, p. 213; Meusel, Jaeger and Weinert 1965: 27; Dansereau 1957: fig. 2A, p. 33; Fernald 1929: map 38, p. 1515.

N. gracillima (A. Br.) Magnus

/T/EE/   (HH)   Peaty or sandy ponds and shores from Minn. to E Ont. (Algonquin Park and Chalk River, Renfrew Co.; not known from Que.) and N.S. (Hants and Queens counties; not listed by Roland 1947; not known from P.E.I. or N.B.), s to Mo., Ky., and Va. MAPS: Meusel, Jaeger, and Weinert 1965 (generalized); R. T. Clausen, Rhodora 38(454), map 3 (U.S.A. area), p. 339. 1936.

N. guadalupensis (Spreng.) Magnus

/T/(X)/   (HH)   Shallow waters from Oreg. to Idaho, Minn., Wisc., s Ont. (Port Franks, Lambton Co.; OAC; several stations on the U.S.A. sides of lakes Erie and Ontario), and SW Que. (Montreal and other islands of the Hochelago Archipelago), s to Baja Calif., Colo., Tex., La., Ala., and Fla. MAP: R. T. Clausen, Rhodora 38(454), map 4 (U.S.A. area; *see* this paper for a discussion of the genus), p. 343. 1936

JUNCAGINACEAE (Arrow-grass Family)

Plants of fresh, brackish, or saline wet habitats. Leaves linear, entire, alternate or basal, sheathing at base. Flowers small, greenish, perfect, hypogynous, in racemes, the perianth-segments in 2 whorls of 3 each. Stamens 6. Pistils 3 or 6, finally distinct. Ovary superior. Fruit consisting of 3 or 6 follicles, each with 1 or 2 seeds. (Transcontinental species; incl. Scheuchzeriaceae).

1   Leaves both basal and alternate along the stem, each with a terminal pore; flowers 3 or
    4 mm long, few in a loose bracted raceme; mature follicles spreading, 2-seeded . . . .
    . . . .*Scheuchzeria*
1   Leaves all basal, closed at tip; flowers smaller, numerous in a long naked spike-like
    raceme; mature follicles connivent along the central axis, 1-seeded . . . .*Triglochin*

SCHEUCHZERIA L.   [67]

S. palustris L.
/ST/X/EA/   (Hsr)   Bogs and peaty shores from s Alaska, s-cent. Yukon, and cent. Dist. Mackenzie (Porsild and Cody 1968) to L. Athabasca (Alta. and Sask.), Man. (N to Reindeer L. at 57°52′N), Ont. (N to Hawley L. at 54°34′N), Que. (N to E James Bay at ca. 54°N and the Côte-Nord), Labrador (N to the Hamilton R. basin), Nfld., N.B., and N.S. (not known from P.E.I.), s to Calif., N.Mex., Nebr., Ohio, Pa., and N.J.; Eurasia. [Incl. var. *americana* Fern.]. MAPS: Hultén 1968*b*: 81, and 1962: map 149, p. 159; Meusel, Jaeger, and Weinert 1965: 27; W. A. Sledge, Watsonia 1(1): 30. 1949; Meusel 1943: fig. 27f.

TRIGLOCHIN L.   [66]   Arrow-grass.   Troscart

1   Fruit oblong or ovoid, commonly less than 6 mm long, rounded at base, the carpels
    usually 6, each with a recurving beak, the axis between them slender; scape to over
    8 dm tall . . . .*T. maritima*
1   Fruit linear, narrowly obclavate, commonly 7 or 8 mm long, blunt and beakless, the 3
    carpels separating from below upward, the axis between them broadly 3-winged, their
    awl-pointed bases spreading; scape usually less than 4 dm tall . . . .*T. palustris*

T. maritima L.
/aST/X/EA/   (Hr)   Saline, brackish, or fresh marshes and shores from N Alaska, s Yukon, and the coast of Dist. Mackenzie to s Dist. Keewatin, northernmost Ont., Que. (N to s Ungava Bay), Labrador (N to ca. 56°30′N), Nfld., N.B., P.E.I., and N.S., s to Baja Calif., N Mexico, Tex., Nebr., N Ill., N Ohio and Pa.; Patagonia; Iceland; Eurasia; N Africa. [Incl. *T. concinna* Davy, *T. elata* Nutt., and *T. gaspense* Löve & Lieth]. MAPS: Hultén, 1968*b*: 80, and 1962: map 112, P. 121.

   Forma *multifissa* Lepage, the fruits comprising up to 12 (rather than at most 6) carpels, is known from the type station at the mouth of the Opinaga R., N Ont., at 54°12′N.

T. palustris L.
/aST/X/GEA/   (Hrr (grh))   Bogs and brackish or alkaline marshes from the E Aleutian Is. and N Alaska to cent. Yukon, the coast of Dist. Mackenzie at Bathurst Inlet, Man. (N to Churchill), northernmost Ont., Que. (N to the George R. at 58°31′N), Labrador (N to 57°40′N), Nfld., N.B., P.E.I., and N.S., s to Calif., N.Mex., Nebr., Ill., N Ohio, and Pa.; s S. America; w Greenland N to 70°43′N, E Greenland N to 73°35′N; Iceland; Eurasia. MAPS: Hultén 1968*b*:80, and 1962: map 104, p. 113.

   Forma *fernaldiana* Rousseau, the fruits not constricted at apex, is known from the type station along the Jupiter R., Anticosti Is., E Que.

LILAEACEAE (Flowering Quillwort Family)

LILAEA H. & B. [69] Flowering Quillwort

Small scapose annual marsh plant with soft linear-subulate leaves to about 3.5 dm long, these tufted on a very short stem. Perianth none, the flowers in inflorescences of two types. Lower flowers enclosed by the sheathing leaf-bases, all pistillate, each consisting of a single pistil surmounted by a filiform style to about 1 dm long terminated by a capitate stigma. Other flowers born in short spikes, the lower ones pistillate (consisting of a single pistil), the upper ones staminate (consisting of a single stamen), those in the middle of the spike often perfect (with 1 pistil and 1 stamen); fruit dry, 1-seeded, that of the upper flowers ribbed and 2-winged, that of the others merely ribbed. (Sometimes merged with the Juncaginaceae).

L. scilloides (Poir.) Haum.
/T/W/ (T) Wet ground and mud-flats of sw B.C. (mouth of the Somas R., Alberni, Vancouver Is., where taken by John Macoun in 1887 and Carter in 1917; CAN) and the Cypress Hills region of sw Alta. and se Sask., s to Baja Calif. and Mexico; S. America [*Phalangium* Poir.; *L. subulata* Humb. & Bonpl.].

ALISMATACEAE (Water-plantain Family)

(Ref.: J. K. Small, North Am. Flora 17 (part 1): 43–62. 1909.

Scapose herbs of shallow water or wet places. Leaves all basal, long-petioled. Flowers perfect or unisexual or both, hypogynous, in peduncled whorls or heads borne in the axils of a bracted, racemose, umbellate, or paniculate terminal inflorescence. Green sepals and white or pinkish petals each 3. Stamens 6 to many. Ovary superior. Pistils numerous, distinct, in fruit forming a whorl or head of 1-seeded achenes. (Alismaceae).

1 Achenes borne in a single whorl on the small flat receptacle; stamens usually 6; leaf-blades not sagittate.
  2 Petals entire; achenes minutely beaked by the lateral style; inflorescence a panicle of whorled branches each bearing a simple or compound umbel . . . .*Alisma*
  2 Petals erose or incised; achenes long-beaked by the apical style; inflorescence a terminal umbel or, by proliferation, a succession of 2 or 3 umbels; leaves lanceolate to ovate; (?Vancouver Is.) . . . .*[Damasonium]*
1 Achenes in several series covering the surface of a relatively large convex or globose receptacle.
  3 Achenes plump, evenly ribbed but not winged, to 1.5 mm long, the lateral beak (when present) less than 0.3 mm long; whorls 3-bracted and with many secondary branchlets; flowers all perfect, in 1 or 2 whorls (up to 6 pedicels in each whorl), the pedicels commonly recurved near the apex; stamens 6 or 9; submersed leaves consisting of lance-linear phyllodia, the emersed leaves with a lanceolate blade to about 3 cm long; (?Ont.) . . . .*[Echinodorus]*
  3 Achenes flattened, distinctly winged, the style persisting as a distinct beak; whorls 3-bracted but lacking branchlets; upper flowers mostly staminate; stamens usually more than 9; leaf-blades often sagittate . . . .*Sagittaria*

ALISMA L. [70]   Water-plantain

(Ref.: Hendricks 1957)

1 Petals commonly pink or purplish; stamens barely equalling the ovaries, their subglobose anthers usually less than 0.5 mm long; styles hooked or coiled, at most 0.7 mm long; mature achenes usually with 3 ridges and 2 grooves down the back; lower panicle-branches mostly simple or with only 1 whorl of branchlets; leaves submersed and ribbon-like or, if emersed, with lanceolate to narrowly elliptic blades, these tapering at base and usually less than 1 dm long; (Alta. to sw Que.) . . . .*A. gramineum*
1 Petals usually white; stamens surpassing the ovaries; styles erect, commonly at least 1 mm long; mature achenes usually with 2 ridges and 1 groove down the back; panicle much compounded; leaves essentially all emersed and with elliptic to ovate blades, these rounded to subcordate at base, to about 2.5 dm long; (transcontinental) . . . .
. . . .*A. plantago-aquatica*

A. gramineum Gmel.

/T/(X)/EA/   (Hel)   Shallow water and muddy shores (ranges of Canadian taxa outlined below), s to N Calif., Idaho, Colo., S.Dak., Minn., and N.Y.; Eurasia; N Africa. MAPS and synonymy: *see* below.

1 Fruits commonly over 2 mm long, thick-walled; [*A. geyeri* Torr.; incl. vars. *graminifolia* (Wahl.) Hendricks and *angustissima* (DC.) Hendricks; sw B.C. (Hitchcock et al. 1969), Alta. (N to Fort Saskatchewan), Sask. (N to Battleford), and Man. (Brandon; Rossburn; Red R. near Winnipeg); s Ont. (St. Lawrence R. NE of L. Ontario) and s Que. (N to Montreal); MAPS: Hultén 1958: map 259, p. 279 (also noting two total-area maps by Samuelsson); Meusel, Jaeger, and Weinert 1965: 28; Meusel 1943: fig. 31e; Hendricks 1957: fig. 8 (vars. *ang.* and *gram.*), p. 486] . . . . .var. *gramineum*
1 Fruits at most 2 mm long, thin-walled; [*A. wahl.* (Holmb.) Juz.; reported by Frère Rolland-Germain, Ann. ACFAS 10: 93. 1944, from near Cascade Rapids, s Que., where probably introd.; MAP: Hultén 1958: map 259, p. 279] . . . .
. . . . .var. *wahlenbergii* (Holmb.) Raymond & Kucyniak

A. plantago-aquatica L.

/sT/X/EA/ (Hel) Shallow waters and muddy shores, the aggregate species from B.C. (N to Vanderhoof, ca. 53°30′N) to N Alta. (L. Athabasca), Sask. (N to Prince Albert), Man. (N to near York Factory, Hudson Bay, ca. 57°N), Ont. (N to s James Bay at 52°11′N), Que. (N to near Rimouski, Rimouski Co.), Nfld., N.B., P.E.I., and N.S., s to Baja Calif., Mexico, Tex., and Fla.; S. America; Eurasia; N Africa. MAPS and synonymy: *see* below.

1 Petals lilac or roseate; [var. *michaletii* Hendricks; generally conceded to be confined to Eurasia–Africa, but possibly introd. in N. America; MAPS: Hultén 1962: map 151, p. 161; Meusel 1943: fig. 31d; Meusel, Jaeger, and Weinert 1965: 28] . . . . .var. *plantago-aquatica*

1 Petals white.
  2 Petals to 6 mm long (flowers to about 13 mm broad); sepals 3 or 4 mm long, with broad scarious margins; styles about equalling the ovaries; stamens about twice as long as the ovaries, their ovoid anthers to 0.8 mm long; achenes to 3 mm long; [*A. brevipes* Greene; *A. trivialis* Pursh; *A. ?natans* L.; transcontinental; MAPS: American area of the above-noted maps by Hultén and Meusel; Hendricks 1957: fig. 4, p. 476] . . . . .var. *americanum* Schultes & Schultes
  2 Petals 1 or 2 mm long (flowers to 3.5 mm broad); sepals less than 3 mm long, not markedly scarious-margined; styles less than half as long as the ovaries; stamens only slightly surpassing the ovaries, their subglobose anthers at most 0.5 mm long; achenes to 2 mm long; [*A. parviflorum* Pursh; *A. subcordatum* Raf.; Sask (N to near Saskatoon), ?Man. (Lowe 1943), s Ont., and Que. (N to near Quebec City); MAPS: Hultén 1962: map 151 (*A. sub.*), p. 161; Hendricks 1957: fig. 4, p. 476] . . . .
  . . . . .var. *parviflorum* (Pursh) Farw.

[DAMASONIUM Juss.] [74]

[D. californicum Torr.]
[The tentative report of this species of the w U.S.A. (Oreg. and Idaho to cent. Calif.) from sw B.C. by John Macoun (1888; Somass R. near Alberni, Vancouver Is.) is based upon *Alisma plantago-aquatica,* the relevant collection in CAN. (*Machaerocarpus* Small).]

[ECHINODORUS Richard] [75] Burhead

[E. parvulus Engelm.]
[The inclusion of Ont. in the range of this species of the E U.S.A. (N to Ill. and Mass.) by Gleason (1958) is probably based upon the Agassiz report noted by John Macoun (1888; "North shore of Lake Superior"), this requiring clarification. (*E. tenellus* (Mart.) Buch.).]

SAGITTARIA L. [78] Arrowhead. Flèche d'eau

(Ref.: J. K. Small, North Am. Flora 17 (part 1): 48–62. 1909)
1 Leaves linear to ovate, sagittate, the basal lobes linear to triangular-ovate; flowers all slender-pedicelled; anther-filaments glabrous, slender.
  2 Fruiting heads less than 1.5 cm thick; achenes at most about 2.5 mm long, with broadly rounded dorsal and ventral keels, the erect or suberect beak borne well in from the margin, less than 0.5 mm long; bracts to 4 cm long, acute to acuminate; (transcontinental) . . . .*S. cuneata*
  2 Fruiting heads to 3 cm thick; achenes to 3.5 mm long, with broadly winged dorsal keel, the beak to 2 mm long.
    3 Sides of achene with 1 or 2 low ridges, the ventral margin wingless; beak of achene at most 1.5 mm long, obliquely ascending; bracts firm, lanceolate, attenuate to a prolonged tip, to 4 cm long; (s ?Ont.) . . . .[*S. engelmanniana*]
    3 Sides of achene plane, the ventral margin winged; beak of achene to 2 mm long, subhorizontal; bracts at most 1 cm long, thin, obtuse to acute; B.C. to N.S.) . . . .*S. latifolia*
1 Leaves tapering to an unlobed base or with one or two narrow arching basal appendages; anther-filaments dilated.

Sagittaria

4 Sepals accrescent, appressed to the fruiting head; lower flowers perfect; anther-filaments glabrous; leaves normally represented by thick spongy phyllodia; (E Que. and N.B.) . . . .*S. spatulata*
4 Sepals reflexed in fruit; lower flowers pistillate or plants unisexual; anther-filaments scaly-pubescent; leaves either phyllodial or with linear-lanceolate to broadly elliptic blades.
  5 Flowering scape usually bent near the lowest whorl; lower (pistillate) flowers subsessile, the staminate ones on long pedicels; beak of the slenderly or obscurely keeled achene at least 1 mm long, erect or arching, the fruiting head thus with a markedly prickly appearance; bracts roundish, obtuse; leaves linear to oval; (Ont. and sw Que.) . . . . *S. rigida*
  5 Flowering scape straight, the flowers all long-pedicelled; beak of the distinctly keeled achene not over 0.6 mm long; bracts ovate; leaves represented either by broadly linear-lanceolate phyllodia or by petioled linear to lance-elliptic blades; (Ont. to Nfld. and N.S.) . . . .*S. graminea*

S. cuneata Sheldon
/ST/X/ (Hel) Shallow water and muddy shores from N -cent. Alaska (Hultén 1950) and s Yukon to Great Slave L., L. Athabasca, Man. (N to Reindeer L. at 57°54′N), Ont. (N to Fort Severn, Hudson Bay, ca. 56°N), Que. (N to the Côte-Nord and Gaspé Pen.), s Labrador (Goose Bay, 53°18′N), ?Nfld. (Boivin 1967a), N.B., and N.S. (not known from P.E.I.), s to Calif., N.Mex., N Tex., Iowa, and N.Y. [*S. arifolia* Nutt.]. MAP: Hultén 1968b: 81.
Forma *hemicycla* Fern. (leaf-blades ovate and strongly rounded at summit, rather than lanceolate to ovate and acute or acutish) is known from tidal shores of the St. Lawrence R. estuary, E Que. (type from St-Augustin, Portneuf Co.; also known from the Gaspé Pen.).

[*S. engelmanniana* Sm.]
[*S. brevirostra* Mack. & Bush (*S. eng.* var. *br.* (M. & B.) Bogin; Kans. and Okla. to Ill. and Ind.) is reported from s Ont. by Soper (1949) and Boivin (1967a) and from N.S. by Fassett (1957). However, in view of the great plasticity of the genus and the close relationship of this species to *S. latifolia,* it is felt that the above reports require further confirmation.]

S. graminea Michx.
/T/EE/ (Hel) Shallow waters and muddy or sandy shores (ranges of Canadian taxa outlined below), s to Tex., Ill., Ohio, and Fla.
1 Fruiting head less than 1 cm thick; achenes at most 2 mm long, with a narrow dorsal keel and often 1 or 2 slender ridges on the sides, the beak minute; anthers about equalling their filaments; [*S. sagittifolia* var. *simplex* Hook. in part; Ont. (N to Algonquin Park, Renfrew Co.), E Que. (N to the Côte-Nord), SE Labrador (Forteau, ca. 51°20′N), Nfld., N.B., P.E.I., and N.S.; reports from Sask. require confirmation] . . . . .var. *graminea*
1 Fruiting head to 2 cm thick; achenes at least 2.5 mm long, with a broad whitish dorsal wing, the sides usually with a wing-like keel; achene-beak about 0.5 mm long; anthers shorter than their filaments; [*S. cristata* Engelm.; Ont.: N to the N shore of L. Superior] . . . .
. . . . .var. *cristata* (Engelm.) Bogin.

S. latifolia Willd.
/T/X/ (Hel) Shallow waters and wet shores, the aggregate species from B.C. (N to Williams Lake, ca. 52°N) to Alta., Sask. (N to Hudson Bay Junction, 52°52′N), Man. (N to near Flin Flon), Ont. (N to w James Bay at 54°12′N), Que. (N to the Côte-Nord and Gaspé Pen.; not known from Labrador or Nfld.), N.B., P.E.I., and N.S., s to Calif., Mexico, and Tex.; N S. America; W.I.; Hawaii.
1 Later leaf-blades acute or, if obtuse, strongly narrowed from base to apex, the basal lobes arching or divergent; achenes rounded at summit . . . . .var. *latifolia*
  2 Body of leaf at least 2/3 as broad as long, the basal lobes relatively broad; [*S. variabilis* Engelm. in part and its vars. *latifolia* (Willd.) Engelm. and ?*diversifolia* Engelm.; *S. sagittifolia sensu* Hooker 1838, in part, not L.; *S. sag.* var. *macrophylla* Hook.; range of the species] . . . . .f. *latifolia*
  2 Body of leaf at most 2/5 as broad as long; (essentially throughout the range).

3   Body of leaf at least 1/7 as broad as long; [*S. hastata* Pursh; *S. variabilis* var. *hastata* (Pursh) Engelm.; *S. sagittifolia (variabilis)* var. *angustifolia* Hook. in part] . . . . .f. *hastata* (Pursh) Robins.

3   Body of leaf at most 1/8 as broad as long; [*S. gracilis* Pursh; *S. variabilis* var. *gr.* (Pursh) Engelm.] . . . . .f. *gracilis* (pursh) Robins.

1   Later (or all) leaf-blades obtuse or rounded at summit, the terminal lobe about as broad as long, the broad subacute to obtuse basal lobes parallel or only slightly divergent; achenes subtruncate at summit.

4   Bracts and calyx glabrous; [*S. obtusa* Muhl.; *S. variabilis* var. *ob.* (Muhl.) Engelm.; Ont. to N.B. and N.S.] . . . . .var. *obtusa* (Muhl.) Wieg.

4   Bracts and calyx densely pubescent; [*S. pubescens* Muhl.; *S. variabilis* var. *pub.* (Muhl.) Engelm.; s Ont.: near Belleville, Hastings Co.; John Macoun 1888] . . . . .var. *pubescens* (Muhl.) Sm.

S. rigida Pursh
/T/EE/   (Hel)   Shallow waters and wet places from Ont. (N to the Ottawa dist.; reports from Man. require confirmation) to sw Que.. (N to the Montreal dist.) and s Maine, s to Nebr., Tenn., and Va. [Incl. var. *elliptica* Engelm.; *S. heterophylla* Pursh, not Schreb., and its var. *rig.* (Pursh) Engelm.]. MAP (*S. heter.*; E area): Fassett 1928: map 1, pl. 9.

Forma *fluitans* (Engelm.) Fern., the leaves reduced to slender bladeless phyllodia, is known from s Ont. (Dore and Gillett 1955: St. Lawrence Seaway region; Landon 1960: Norfolk Co.).

S. spatulata (Sm.) Buch.
/T/E/   (Hel)   Tidal mud of brackish estuaries from E Que. (mouth of the Matapedia R., Gaspé Pen.) and E N.B. (several estuaries; not known from P.E.I. or N.S.) to Va. [*Lophotocarpus calycinus* of N.B. reports, not *S. calycina* Engelm., basionym; *L. (S.) cal.* var. *spongiosus* Engelm.) Fassett (*L. spon.* (Engelm.) Sm.); *S. montevidensis* var. *spon.* (Engelm.) Boivin]. MAPS: Raymond 1950*b*: fig. 40, p. 105; Fassett 1928: map 2, pl. 11.

BUTOMACEAE (Flowering Rush Family)

BUTOMUS L.   [81]   Flowering Rush.   Jonc fleuri

Aquatic or marsh herb with linear basal leaves to about 1 m long and 1 cm broad. Scape to about 1.5 m tall, terminated by a simple umbel of numerous perfect, hypogynous, pink flowers about 2 cm broad. Sepals and petals each 3. Stamens 9, their anthers red. Pistils 6, united at the very base. Ovary superior. Fruit consisting of a whorl of 6 inflated long-beaked many-seeded follicles about 1 cm long. (Introd.).

B. umbellatus L.
Eurasian; introd. and rapidly spreading in shallow waters and on river flats in Ont., Que., P.E.I., and N.S. (as also in the Great Lakes region and Idaho.). For an account of this species in N. America (with MAP), see E. L. Core (Ohio J. Sci. 41: 79–85. 1941).

According to Marie-Victorin (1935), this attractive plant was first observed in Canada about 1897 at Laprairie, near Montreal, Que., and first reported on in 1905. C. H. Knowlton (Rhodora 25 (300): 220. 1923) reports that, by 1923, it had spread down the St. Lawrence R. to near the E end of L. St. Peter and Marie-Victorin notes that, by 1935, it had reached the brackish shores of the river estuary at St-Jean-Port-Joli, l'Islet Co., about 50 mi below Quebec City, and had mounted some of the tributaries (among them, the Richelieu R. as far as L. Champlain). It is now known down the St. Lawrence as far as Rivière-Ouelle, Kamouraska Co., and Ile-aux-Coudres, Charlevoix Co.

Knowlton notes an independent appearance of the plant as early as 1906 along the canal at Ottawa, Ont. L. O. Gaiser (Rhodora 51(612): 385–90. 1949) reports collections from the N shore of L. Erie in S Ont. (Kent, Essex, and Welland counties; see her map, p. 388, of the distribution in Ont. and Que.) and Dore and Gillett (1955) cite collections made in the St. Lawrence Seaway region NE of L. Ontario. A map of the distribution in Ont. is given by Montgomery (1956: fig. 2, p. 93). Groh (1944) reports an introduction from the St. Lawrence R. to a pond at Charlottetown, P.E.I., thence to Brackley Point. It is also known from Southport. I.V. Hall (Can. Field-Nat. 73(1): 53. 1959) reports a 1958 collection from Annapolis Royal, Annapolis Co., N.S.

Forma *vallisneriifolius* (Sag.) Glück, a sterile phase with very elongate and thin, submersed or finally floating leaves, is reported from SW Que. by Bernard Boivin (Ann. ACFAS 8: 94. 1942; Chateauguay, near Montreal).

HYDROCHARITACEAE (Frog's-bit Family)

Aquatic herbs with entire, submersed or floating, basal or cauline leaves. Flowers perfect or unisexual or both, sessile within or pedicelled from a subtending sheath. Sepals 3, distinct or united at base. Petals 3, white, commonly smaller than the sepals, or none. Stamens 1–12. Ovary inferior. Fruit ripening under water, indehiscent.

1 Leaves on an elongate stem, sessile, veinless, 1-ribbed, mostly less than 2 cm long, the median and upper commonly in whorls of 3 or 4, crowded and overlapping, the lower opposite; plants unisexual; pistillate flower solitary and sessile in the spathe, elevated to the surface by elongation of the perianth-tube to over 3 cm long; seeds few . . . .*Elodea*
1 Leaves basal or sub-basal; seeds numerous.
   2 Leaf-blades thin and ribbon-like, to over 2 m long and about 2 cm broad; pistillate flowers elevated to the surface on a greatly elongating pedicel that later coils spirally to submerge the fruit; stigmas 3, nearly sessile, 2-lobed; (Man. to N.S.) . . . .*Vallisneria*
   2 Leaf-blades broadly ovate or ovate-cordate to reniform, long-petioled and floating; flowers about 2 cm broad; styles 6–9, deeply 2-parted.
      3 Leaves to about 8 cm long and broad; pistillate flowers on thick pedicels at most 2.5 cm long; plant rooting or occasionally free-floating in deeper water . . . . .*[Limnobium]*
      3 Leaves at most about 3 cm broad; pistillate flowers mostly on longer pedicels; plant free-floating; (introd. in Ont. and sw Que.) . . . .*Hydrocharis*

ELODEA Michx. [87] Waterweed

1 Middle and upper leaves opposite, flaccid, to about 2.5 cm long and 2 mm broad; pistillate spathe usually at least 3 (up to 7) cm broad, its flower with sepals and petals each about 5 mm long; sepals and petals of staminate flowers about 2.5 and 4 mm long, respectively; capsule about 1 cm long, its seeds about 6 mm long; (s Alta. and sw Sask.) . . .*E. longivaginata*
1 . Middle and upper leaves in whorls of 3 (4); pistillate spathe rarely over 2 cm long; capsule at most 7 mm long, its seeds about 4.5 mm long.
   2 Leaves broadly lanceolate to oblong or ovate, minutely serrulate, usually obtuse, to about 17 mm long and 4 mm broad (averaging about 2 mm broad); staminate spathe over 1 cm long, its flower to 1 cm broad, reaching the surface by the great elongation of the calyx-tube to simulate a thread-like peduncle; (s B.C. to N.S.) . . . .*E. canadensis*
   2 Leaves linear or narrowly linear-lanceolate, essentially entire, to about 13 mm long and averaging less than 1.5 mm broad; staminate spathe subglobose, about 2 mm long, its flower 4 or 5 mm broad, sessile, separating and floating at anthesis; (Ont. and sw Que.) . . . .*E. nuttallii*

E. canadensis Michx.
/T/X/ (HH) Quiet waters (often calcareous) from s B.C. (Vancouver Is.; Penticton; Kootenay L.) to s Alta. (Moss 1959), Sask. (N to ca. 53°N), Man. (N to Reindeer L. at 57°54′N), Ont. (N to w James Bay at ca. 53°N), Que. (N to the Gaspé Pen.; type from near Montreal), N.B. (Boivin 1967a; not known from P.E.I.), and N.S., s to Calif., Colo., Okla., Iowa, Ala., and N.C.; introd. over large parts of Europe. [*Anacharis* Planch.; *Philotria* Britt.; *Udora* Nutt.; *U. verticillata* Spreng.; *E. planchonii* Casp.]. MAPS: Hultén 1962: map 227, p. 239; Harold St. John, Rhodora 67(769), fig. 5, p. 33. 1965.

E. longivaginata St. John
/T/WW/ (HH) Shallow waters from s Alta. (type from a lake at the northern edge of the Milk River Ridge, s of Lethbridge; also known from Hay, about 50 mi NE of Lethbridge), sw Sask. (basis of the report of *E. nuttallii* from Swift Current by Bernard Boivin, Nat. can. (Que.) 87: 26. 1960), and N.Dak to Mont., Wyo., Utah, Colo., and N.Mex. MAP: Harold St. John, Rhodora 67(769), fig. 4, p. 32. 1965.

E. nuttallii (Planch.) St. John
/T/X/ (HH) Shallow, fresh or slightly brackish waters from N Idaho ("perhaps in B.C.";

Hitchcock et al. 1969), N Colo., Nebr., Minn., Wisc., Ind., and Ohio to Ont. (N to the Kapuskasing R. system at ca. 49°N; Baldwin 1958), sw Que. (St-Jerôme, N of Montreal; Ile Ste-Thérèse, Richelieu R.), and Maine, s to Calif., Kans., Mo., Miss., s Ala., and N.C. [*Anacharis* Planch.; *Philotria* Rydb.; *Serpicula (A.) occidentalis* Pursh]. MAP: Harold St. John, Rhodora 67(769), fig. 4, p. 32. 1965.

HYDROCHARIS L.   [98]

H. morus-ranae L.   Frog's-bit (of Europe)
European; introd. in shallow water and along muddy shores of s Ont. (Rideau R. from near Smiths Falls to Ottawa; Ottawa R. from Ottawa to Oka, Que.) and s Que. (St. Lawrence R. from Montreal to L. St. Peter; *see* Que. map by Robert Joyal, Nat. can. (Que.) 97(5), map A, fig. 1, p. 562. 1970). MAP: W. G. Dore, Can. Field-Nat. 82(2): 77. 1968.

According to W. H. Minshall (Can. Field-Nat. 54(3): 44. 1940), this species was first noticed as established in N. America in 1936 at the arboretum of the Central Experimental Farm, Ottawa, Ont., records indicating that it had been admixed with a planting of other aquatics from Switzerland in 1932 in a trench connecting with the Rideau Canal. Dore's above-noted map illustrates its spread southwards and eastwards since that date. Père Louis-Marie (Inst. Agric. d'Oka. Rev. 35: 111–12. 1961) reports various stations for the plant around Montreal and notes that it had spread into several isolated basins of the Aquatic Garden following planting at the Montreal Botanical Garden. However, since there is no direct access to the St. Lawrence R., it seems probable that all of the stations in that system are derived from the original infestation at Ottawa.

[LIMNOBIUM Richard]   [97]

[*L. spongia* (Bosc) Steud.]   American Frog's-bit
[This species ranges in stagnant waters from Ill. and Del. to Tex., Fla., and tropical America. It is reported by Gray (1950) from the Lake Ontario shores of N.Y., where evidently now extinct, so that its discovery on the Canadian side of the lake is improbable.]

VALLISNERIA L.   [89]   Tapegrass, Eelgrass

V. americana Michx.   Water-celery.   Herbe aux anguilles
/T/EE/   (HH)   Quiet waters from N.Dak. to s Man. (N to Elphinstone, about 50 mi NW of Brandon), Ont. (N to L. Nipigon and L. Abitibi; Pritchard, Univ. Toronto Stud., Biol. Ser. 39: 80 and 95. 1935), Que. (N to L. Timiskaming and the Quebec City dist.), s N.B. (St. Stephen; near St. John), and N.S. (the report from P.E.I. by McSwain and Bain 1891, as *V. spiralis,* is probably based upon *Zostera marina,* as is certainly the case with an early report from Nfld. by Bell, according to M. L. Fernald, Rhodora 13(150): 111 (footnote). 1911), s to Tex. and Fla. [*V. spiralis* of Canadian reports, not L.; *V. spir.* var. *americana* (Michx.) Torr.]. MAPS: Fernald 1929: map 1, p. 1488; Frère Marie-Victorin, Contrib. Inst. Bot. Univ. Montréal 46: fig. 2, p. 11. 1943.

GRAMINEAE (Grass Family)

(Ref.: Hitchcock and Chase 1951; Dore 1959; Dore and Roland 1942; Hubbard 1955; North Am. Flora 17 (various authors): 77–638. 1909–39).
Herbs with terete or flattened stems (culms), these commonly hollow except at the nodes (evident as raised rings around the culm). Leaves narrow, usually 2-ranked, parallel-veined, their sheaths usually split open lengthwise on the side opposite the blade. Flowers perfect or unisexual, solitary to several in spikelets (the central axis of which is the rachilla). Spikelets disposed in spikes, racemes, or panicles (rarely in heads), each spikelet subtended by usually 2, opposite, mostly firm, more or less boat-shaped sterile bracts (glumes). Lower (outer) glume commonly smaller than the upper (inner) one or sometimes obsolete (both glumes obsolete in *Leersia* and in the pistillate flowers of *Zizania*). Flowers without normal perianth, each subtended by a lower (outer), more or less boat-shaped, awned or awnless bract (lemma) and commonly an additional upper (inner), usually softer bract (palea). Stamens commonly 3 or 6. Ovary superior, 1-locular. Styles or stigmas usually 2 (1 in *Nardus*), the stigmas papillate or plumose. Fruit a caryopsis (seed with adherent thin pericarp).

KEY TO TRIBES

1   Spikelets usually more or less dorsally compressed (the glumes and lemmas rounded or somewhat flattened on the back, never folded lengthwise and then more or less boat-shaped or keeled; *Milium* may be sought here), usually composed of 1 perfect terminal floret (but the staminate spikelets of the unisexual *Zea mays* comprising 2 staminate florets) subtended by a sterile or rudimentary floret; articulation of the rachilla always below the glumes, these falling with the spikelet at maturity; (Subfamily **PANICOIDEAE**).
    2   Spikelets all unisexual, the 2-flowered staminate spikelets in slender spike-like racemes of a large terminal panicle, the pistillate spikelets in short-peduncled several- or many-rowed spikes ("ears") enclosed in numerous spathes ("husks"), the long styles protruding as a mass of silky threads; (a single genus, *Zea*) . . . .
. . . .[Tribe **MAYDEAE**] (p. 227)
    2   Spikelets all with the central floret perfect.
        3   Lemmas with slender bent and twisted awns to about 1.5 cm long (or often nearly or quite awnless in *Miscanthus* and *Andropogon hallii*), the lemma of the fertile floret thin and translucent, merely a basal appendage to the awn; glumes subequal, indurated; rachis and pedicels long-villous with silky hairs; plants mostly tall stout perennials of dry or sandy habitats . . . . .Tribe **ANDROPOGONEAE** (p. 221)
        3   Lemmas awnless (sometimes awned in *Echinochloa*), the lemma of the fertile floret firm or indurated and covering the free grain; glumes unequal, herbaceous or membranous, the lower one much the smaller or obsolete; spikelets lacking long silky hairs (subtended by bristles in *Setaria*; bur-like in *Cenchrus*); annuals or perennials . . . . .Tribe **PANICEAE** (p. 227)
1   Spikelets usually more or less laterally compressed (sometimes terete; distinctly dorsally compressed only in *Milium*), composed of 1 or more functional florets, the sterile floret, if any, usually the uppermost (lowermost in *Hierochloë*; if only 1 floret is present, this is usually subtended by a bristle-like prolongation of the rachilla); rachilla usually articulated above the glumes, these persisting on the mature inflorescence as empty husks (articulation below the glumes in *Beckmannia, Holcus, Leersia* (glumes obsolete), *Spartina,* and *Sphenopholis*); (Subfamily **POACOIDEAE**).
    4   Spikelets sessile or nearly so in a 2-rowed spike.
        5   Spikelets or groups of spikelets in 2 rows on opposite sides of the rachis, forming a bilateral solitary elongate spike . . . . .Tribe **HORDEAE** (p. 226)
        5   Spikelets in 2 rows on one side of the flat rachis, forming 1-sided spikes, these head-like or elongate, solitary or in terminal heads, racemes, panicles, or digitate clusters . . . . .Tribe **CHLORIDEAE** (p. 223)
    4   Spikelets distinctly pedicelled in compact to open panicles (rarely simple racemes;

if in spike-like panicles, the spike more or less terete, its spikelets not distinctly 2-rowed).

6   Spikelets unisexual; pistillate spikelets erect, subulate or linear, terete, lacking glumes, their often firm lemmas closely clasping the 3-nerved palea by a pair of strong lateral nerves; staminate flowers pendulous on the lower panicle-branches, their thin lemmas acuminate or short-awned; stamens 6; leaves flat, to 5 cm broad; culms soft, to about 3 m tall; aquatic annual; (a single genus, *Zizania*) . . . . .Tribe **ZIZANIEAE** (p. 228)

6   Spikelets with 1 or more perfect florets; chiefly perennials.

7   Spikelets very flat, 1-flowered, the awnless lemma and palea subequal in length and of similar texture, both strongly keeled, the lemma clasping the much narrower, usually 3-nerved palea by a pair of strong marginal nerves; glumes obsolete or nearly so; stamens 6; (a single genus, *Leersia*) . . . . .Tribe **ORYZEAE** (p. 227)

7   Spikelets more or less compressed but not flat; both glumes usually present (glume solitary in all but the terminal floret of *Lolium,* solitary and minute in *Nardus*); stamens mostly 3 (sometimes 2 or 1).

8   Spikelet with 2 sterile or staminate lemmas below the single fertile floret and falling with it (these lemmas scale-like or bristly in *Phalaris*); palea 1-nerved . . . . .Tribe **PHALARIDEAE** (p. 228)

8   Spikelets with no sterile or staminate lemmas below the fertile floret or florets (but commonly terminating in a sterile floret or its pedicel); palea 1–2-nerved or sometimes wanting.

9   Spikelets 1-flowered; glumes often surpassing the lemma . . . . .Tribe **AGROSTIDEAE** (p. 219)

9   Spikelets with 2 or more perfect florets.

10 Glumes shorter than the lowest floret (except in *Dupontia* and *Sieglingia,* these with a tuft of hairs at base of lemma); lemmas awnless or awned at or near tip or between the pair of terminal teeth; spikelets commonly with 3 or more perfect florets . . . . .Tribe **FESTUCEAE** (p. 224)

10 Glumes usually equalling or surpassing the lowest floret or all of them; awn of lemma, when present, dorsal; spikelets mostly with not more than 3 perfect florets . . . . .Tribe **AVENEAE** (p. 221)

Tribe AGROSTIDEAE

1   Lemma indurated, thicker and much harder than the glumes.

2   Spikelets dorsally compressed, about 3.5 mm long, on slender pedicels, the obtuse glumes and shining awnless lemma rounded on the back; glumes falling with the floret; grain free; branches of the elongate-ovoid panicle with widely spreading or drooping branches; leaves to over 1.5 cm broad; (Man. to Nfld. and N.S.) . . . .*Milium*

2   Spikelets nearly terete or slightly laterally compressed, the acute or obtuse glumes and awned lemma folded or keeled; lemma firmly enclosing the palea and grain; rachilla jointed above the glumes, these persisting as empty husks on the mature inflorescence.

3   Florets plump, with a short oblique callus at base; glumes mostly ovate or oblong, obtuse to acuminate; awn of lemma weak and deciduous . . . .*Oryzopsis*

3   Florets slender, with the callus prolonged into a slender stipe; glumes narrow, acute to bristle-tipped; awn of lemma firm and persistent.

4   Lemma-awn simple, more or less distinctly twice-geniculate (therefore with 3 sections); (western species) . . . .*Stipa*

4   Lemma-awn 3-forked . . . .*Aristida*

1   Lemma as thin as or thinner than the glumes; grain free from the lemma.

5   Spikelets about 1 mm long, purplish, very short-pedicelled; inflorescence spike-like or racemose, narrowly linear, to about 2 cm long; leaves narrowly linear, to 6 cm long; culms to about 1.5 dm tall, surrounded at base by the sheaths of radical leaves;

annual or biennial (with somewhat the aspect of *Carex capillaris*; introd. on
Vancouver Is.) . . . .*[Mibora]*
5 Spikelets and leaves mostly longer.
 6 Floret raised on a short slender stipe above the glumes, these falling with it;
   stamen 1; palea 1-nerved, or 2-nerved with the nerves close together; glumes
   narrow, minutely hispid on the keel . . . .*Cinna*
 6 Floret not on a slender stipe; stamens mostly 3; palea 2–3-nerved or obsolete;
   lemma awned or awnless.
  7 Articulation below the glumes, these falling with the spikelet; inflorescence a
    very dense spike-like or lobed panicle.
   8 Glumes long-awned, scabrid or pilose; lemmas short-awned, much
     shorter than the glumes; (introd.) . . . .*Polypogon*
   8 Glumes awnless.
    9 Lemmas awnless, about half as long as the glumes, these merely
      somewhat scabrid; palea narrow and hyaline; (*P. semiverticillatus*) . . . .
      . . . .*Polypogon*
    9 Lemmas dorsally awned, about equalling the glumes, their margins
      usually connate at base; glumes ciliate or woolly at least at the base
      of the nerves; palea none . . . .*Alopecurus*
  7 Articulation above the glumes, these persisting as empty husks on the mature
    inflorescence.
   10 Lemma short-pointed or awned at tip, tightly embracing the grain; plants
     often with short or elongate scaly rhizomes.
    11 Rachilla not prolonged beyond the palea; body of the 3-nerved,
      awn-tipped or merely mucronate lemma less than 5 mm long; glumes
      well-developed (or the lower one minute or obsolete in *M. schreberi*) . . . .
      . . . .*Muhlenbergia*
    11 Rachilla prolonged beyond the palea as a slender bristle; body of the
      5-nerved lemma to 1 cm long, its awn to about 2.5 cm long; lower glume
      often obsolete; leaves to about 1.5 cm broad; (Ont. eastwards) . . . .
      . . . .*Brachyelytrum*
   10 Lemma awnless or dorsally awned, loosely embracing the grain.
    12 Glumes prominently keeled and folded, ciliate on the keels, awn-
      tipped, much longer than the awnless lemma and palea; inflorescence
      a very dense spike-like panicle . . . .*Phleum*
    12 Glumes not conspicuously keeled and tightly folded, commonly
      glabrous or merely scabrous on the nerves, awnless or very
      short-awned; inflorescence a spike-like to open panicle.
     13 Grain plump and bladdery to globose, readily falling from the
       spikelet, loosely invested by the outer membrane, this readily
       slipping away when moist; glumes 1-nerved; palea equalling or
       surpassing the awnless 1-nerved lemma, often splitting between
       the strong nerves at maturity; leaves tapering to long points . . . .
       . . . .*Sporobolus*
     13 Grain not plump and bladdery, its outer membrane tightly adherent.
      14 Palea definitely smaller and thinner than the awnless or dorsally
        awned lemma, or obsolete; glumes surpassing the lemma; floret
        beardless or very short-bearded at base.
       15 Rachilla prolonged behind the palea as a short bristle;
         leaves 1 or 2 mm broad; (B.C. and sw Alta.) . . . .*Podagrostis*
       15 Rachilla not prolonged; palea sometimes wanting . . . .*Agrostis*
      14 Palea nearly like the lemma and about as large, 2-nerved.
       16 Callus at base of lemma not bearded; panicle-branches
         erect or ascending; (arctic and subarctic regions).
        17 Culms stout, to over 1 m tall, strongly stoloniferous;
          leaves to nearly 1.5 cm broad; ligule to 5 mm long;
          panicle purplish, to about 3 dm long; spikelets about 5

mm long, about equalled by the upper glume; (transcontinental) . . . .*Arctagrostis*

    17 Culms slender, usually densely tufted, nonstoloniferous; leaves narrow, soft, with a somewhat boat-shaped tip; ligules less than 2 mm long; panicle less than 3 cm long, scarcely surpassing the leaves; spikelets not over 2 mm long; glumes minute; (transcontinental) . . . .*Phippsia*

16 Callus at base of lemma bearded; plants mostly stoloniferous or rhizomatous.

    18 Lemma awned from middle or near base, 3–5-nerved, the awn included or exserted from the side of the glumes, these surpassing the lemma; callus-hairs shorter than or surpassing the lemma; panicle-branches spreading or ascending . . . .*Calamagrostis*

    18 Lemma awnless; panicle to about 4 dm long, its branches strongly ascending; leaves to 8 mm broad; elongate rhizomes present.

        19 Panicle spike-like; spikelets at least 8 mm long; glumes subequal, longer than the floret; lemma 3–5-nerved; callus-hairs to 3 mm long; leaves elongate; (coastal sands) . . . .*Ammophila*

        19 Panicle open; spikelets about 7 mm long; glumes unequal, the lower one shorter than the floret; lemma 1-nerved; callus-hairs over half as long as the lemma; leaves tapering into a long thread-like point; ligule a zone of short hairs; (sandy prairies from B.C. to w Ont.) . . . .*Calamovilfa*

Tribe ANDROPOGONEAE (p. 218)

1 Lemmas awnless or with a short straight awn shorter than the spikelet (*Andropogon hallii* may often be sought here); racemes numerous in a broad fan-shaped inflorescence; spikelets all alike and perfect, about 5 mm long, surrounded by silky hairs about twice as long as the spikelets; leaves to over 1.5 cm broad; plant perennial, to about 2 m tall, with underground rhizomes; (introd. in s Ont.) . . . .*[Miscanthus]*

1 Lemmas of fertile florets bearing a long bent and twisted awn from the apex or between the apical teeth (or often awnless or nearly so in *Andropogon hallii*); spikelets dissimilar, the sessile ones perfect, the pedicelled ones sterile.

    2 Racemes solitary or digitately clustered, with at least several joints, flexuous or straight, greenish to purplish, the hairs of the rachises and pedicels white or yellowish; tall tufted perennials . . . .*Andropogon*

    2 Racemes very numerous, commonly reduced to only 2 or 3 joints, slender-peduncled and often drooping in a large open panicle.

        3 Pedicelled spikelet well developed and staminate, its glumes glabrous; glumes of the sessile perfect spikelet silky; lemma-awns early deciduous; leaves to about 2 cm broad; (introd.) . . . .*[Sorghum]*

        3 Pedicelled spikelet represented only by the pedicel, this 4 or 5 mm long and densely clothed with brownish-yellow hairs; lower glume of the sessile perfect spikelet hirsute, the upper glume glabrous; lemma-awns persistent; leaves to about 1 cm broad; perennial with scaly rhizomes; (s Man. to Que.) . . . .*Sorghastrum*

Tribe AVENEAE (p. 219)

1 Lemma awnless (or very short-awned or merely mucronate; *Holcus* with a short hooked awn), faintly nerved.

    2 Panicle spike-like; glumes unequal in length but similar in shape, acute, scabrous on the keel, slightly shorter than the florets, persisting on the rachilla after the

Gramineae

florets fall; lemma acute or mucronate, entire at apex . . . .*Koeleria*
2 Panicle dense (but scarcely spike-like) to open.
   3 Glumes dissimilar (subequal in length but the upper one upwardly dilated and
      shorter than the upper floret), falling with the floret; leaves to 5 mm broad . . . .
      . . . .*Sphenopholis*
   3 Glumes similar in shape, both tapering from base to apex, lanceolate to ovate;
      leaves to 1 cm broad.
      4 Glumes subequal in length, ovate, falling with the floret, the broader upper one
         much surpassing the florets; spikelets 2-flowered, to 6 mm long; panicle
         purplish, to about 1.5 dm long; lower perfect floret with a short hooked awn
         below the apex; (introd.) . . . .*Holcus*
      4 Glumes unequal in length, lanceolate, persisting on the rachilla after the florets
         fall, shorter than the column of florets; spikelets 2–4-flowered, to 9 mm long;
         panicle silvery-green, becoming whitish brown, to 3 dm long; plant glabrous or
         the leaf-sheaths pilose; (*T. melicoides*; Ont. to Nfld. and N.S.) . . . .*Trisetum*
1 Lemmas distinctly awned; spikelets disarticulating above the glumes, these persisting on
the rachilla as empty husks.
   5 Florets 2, the terminal one perfect, its scabrous lemma with a short straight awn; lower
      floret staminate, its scabrous lemma bearing a bent and twisted awn up to twice the
      length of the body; glumes minutely scabrous, the longer upper one about equalling
      the florets; spikelets 7 or 8 mm long; rachilla hairy; panicle to about 3 dm long,
      shining; leaves to 1 cm broad; (introd.) . . . .*Arrhenatherum*
   5 Florets 2–many, all alike or the uppermost one staminate; panicle usually smaller.
      6 Rachilla not prolonged behind the palea of the uppermost floret; spikelets
         2-flowered, yellowish and shining, about 4 mm long, both florets perfect; lemma
         bidentate, with a bent awn to 4 mm long inserted below the middle of the body;
         glumes surpassing the florets; leaves setaceous; annuals; (introd.) . . . .*Aira*
      6 Rachilla prolonged behind the palea of the uppermost floret; spikelets
         2–many-flowered, the florets all perfect or the uppermost one staminate or
         rudimentary.
         7 Awn borne from the notch between the apical teeth of the lemma, flattened and
            twisted; glumes subequal, usually surpassing the uppermost floret; inflores-
            cence a raceme or few-forked panicle, its stiff, ascending to appressed
            branches often with only a single terminal spikelet; ligule a tuft of long
            stiff hairs; perennials . . . .*Danthonia*
         7 Awn definitely dorsal, not flattened; inflorescence an open or spike-like panicle
            of many spikelets; ligule not a tuft of hairs.
            8 Lemma-awn jointed near the middle, the upper part clavate, the joint with a
               circle of stiff hairs; lemmas not keeled; glumes 1-nerved, subequal,
               awnless, about equalling or slightly surpassing the florets; spikelets to 4
               mm long; panicle dense and narrow, to 8 cm long; glaucous or purplish
               perennial to about 4 dm tall, with closely tufted convolute-acicular basal
               leaves; (introd. in sw B.C.) . . . .[*Corynephorus*]
            8 Lemma-awn not jointed, slender-tipped.
              9 Lemmas 4-nerved, 2–4-toothed at the subtruncate apex, the awn borne
                from near or below the middle of the body; glumes subequal; spikelets
                at most 7 mm long, usually 2-flowered; perennials . . . .*Deschampsia*
             9 Lemmas 5–9-nerved, 2-toothed at the tapering apex.
               10 Glumes 1–3-nerved, unequal, rarely over 6 mm long, sometimes
                  falling with the florets; lemmas 5-nerved, keeled, their awns borne
                  from well above the middle of the body; panicle-branches erect
                  or ascending; perennials . . . .*Trisetum*
               10 Glumes subequal, to 2.5 cm long, persisting as empty husks after
                  the florets fall; lemmas 5–9-nerved, rounded on the back, their
                  geniculate awns borne from near or below the middle of the body.
                 11 Glumes at least 2 cm long, smooth on the rounded back,
                    7–9-nerved; spikelets pendulous, 2-flowered or with a rudimen-

tary third floret, several on each drooping primary panicle-
branch; annuals; (introd.) . . . .*Avena*
11 Glumes not over 1.5 cm long, scabrous-keeled (at least above
middle), 3–5-nerved (rarely 1-nerved); spikelets 3–6-flowered,
mostly not more than 3 on each ascending or erect primary
panicle-branch; perennials with creeping rhizomes . . . . .*Helictotrichon*

Tribe CHLORIDEAE (p. 218)

1 Spikelets unisexual; pistillate spikelets in a bur-like cluster sessile in the dilated base of
the two uppermost leaves, their upper glumes thick and indurated, terminating in 3 stiff
spine-like lobes; staminate spikelets (often on separate plants) in 1–3 short spikes
terminating slender culms to about 3 dm tall; leaves 1 or 2 mm broad; stoloniferous
perennials; (s Sask. and sw Man.) . . . .*Buchloë*
1 Spikelets perfect.
   2 Inflorescence head-like, of 2 or 3 spikelets enclosed in the broad sheath of the
     uppermost of the short (to 3 cm long) rigid sharp-pointed fascicled leaves; low-
     spreading much-branched annual forming mats to about 5 dm across; (Alta.;
     introd. in Sask. and ?Man.) . . . .*Munroa*
   2 Inflorescence not head-like.
     3 Spikes terminal, solitary or in digitate clusters at the top of the culm; spikelets
       articulated above the glumes.
       4 Spike solitary, to about 5 cm long; spikelets 5 or 6 mm long, with 1 perfect floret
         and 1 or more sterile rudiments above this; lemmas densely long-villous at
         base, the fertile one 3-cleft, the divisions short-awned, the rudimentary ones
         stipitate and often reduced to 3 awns; leaves about 2 mm broad; (*B. gracilis*;
         B.C. to Man.) . . . .*Bouteloua*
       4 Spikes 2 or more; lemmas awnless, glabrous or merely ciliate on the keel;
         leaves to over 4 mm broad; culms to about 6 dm tall; (introd.).
         5 Culms coarse, tufted, branching at base; leaves rather soft; inflorescence
           with up to 10 spike-like branches to about 1.5 dm long; spikelets about 5
           mm long, with up to 6 florets; lemmas acute, with 3 strong green nerves
           close together forming a keel; fibrous-rooted annual . . . .*Eleusine*
         5 Culms wiry, diffusely branched; leaves stiff; ligule a conspicuous ring of
           white hairs; inflorescence with up to 8 spike-like branches to about 7 cm
           long; spikelets about 2 mm long, 1-flowered; lemmas obtuse, boat-shaped,
           ciliate on the keel; perennial from extensively creeping rhizomes . . . .*Cynodon*
     3 Spikes disposed in racemes or panicles.
       6 Spikelets about 3 mm long, suborbicular, laterally compressed, 1-flowered (in
         ours; or with the rudiment of a second floret), articulated below the glumes, in
         short erect or ascending spikes, these forming a narrow, more or less
         interrupted, nearly simple panicle or compound spike; glumes boat-shaped,
         equal, transversely wrinkled, falling with the floret; leaves to 8 mm broad;
         culms to over 1 m tall; tufted annual of moist or wet habitats; (trans-
         continental) . . . .*Beckmannia*
       6 Spikelets at least 4 mm long, lanceolate to ovate-lanceolate, the narrow
         glumes unequal; spikes solitary or racemose along the main axis of the
         inflorescence.
         7 Spikelets to 1 cm long, articulated above the glumes, each with a
           rudimentary sterile (rarely staminate) floret (commonly reduced to 3 awns)
           borne on the prolonged rachilla above the perfect floret; fertile lemma
           3-cleft at summit, each of the teeth usually short-awned by an excurrent
           nerve; perennials of dry prairies . . . .*Bouteloua*
         7 Spikelets 1-flowered, lacking sterile florets; lemma entire or with 2 rounded
           teeth at apex.
           8 Spikelets about 4 mm long, articulated above the lance-subulate
             glumes, appressed, alternate and distant along one side of the slender

3-angled rachis, forming very slender stiff spikes (these diverging from the curving main axis of the inflorescence); leaves at most about 5 cm long and 2 mm broad; low, diffusely branched annual of prairies or open alkaline habitats; (Alta. to sw Man.) .... *Schedonnardus*

8   Spikelets to 1.5 cm long, articulated below the acute to awned glumes, closely overlapping in dense, 1-sided, erect to ascending spikes; leaves to over 1 m long; erect perennials of fresh, saline, or alkaline moist or wet habitats, with spreading scaly rhizomes .... *Spartina*

Tribe FESTUCEAE (p. 219)

1   Glumes about equalling or surpassing the lowest floret; panicle-branches usually glabrous; leaves 2 or 3 mm broad; (essentially transcontinental in arctic and subarctic regions).
    2   Plant strongly stoloniferous, commonly less than 3 dm tall; panicle to about 1 dm long, open or contracted; spikelets shining, bronze-colour and often purple-tinged; lemmas with a tuft of hairs at base; anthers to over 1.5 mm long .... *Dupontia*
    2   Plants densely caespitose; spikelets brownish to dark purple, 2–4-flowered; lemmas lacking a tuft of hairs at base .... *Colpodium*
1   Glumes mostly shorter than the lowest floret.
    3   Panicle plumose, the rachilla-joints bearded with long silky hairs surpassing the lemmas at maturity; leaves to about 6 cm broad; culms stout and reed-like, to over 4 m tall (the tallest of our grasses); (transcontinental) .... *Phragmites*
    3   Panicle not plumose, the rachilla-joints beardless or, if bearded, the hairs shorter than the lemmas.
        4   Panicle dense, spike- or head-like, somewhat 1-sided; spikelets dimorphic, the terminal one of each cluster fertile, 2–3-flowered, sessile and nearly hidden by the lower sterile ones (these awn-like and in 2 fan-like vertical ranks); leaves soft, to about 3 mm broad; (introd.) .... *Cynosurus*
        4   Panicle more open and scarcely spike-like.
            5   Spikelets subsessile in dense 1-sided clusters at the ends of the stiff naked branches of the 1-sided panicles, the branches appressed-ascending after anthesis; leaves to 8 mm broad; plant coarse, glaucous, scabrous, to about 1.5 m tall; (introd.) .... *Dactylis*
            5   Spikelets not in 1-sided clusters; panicle-branches spirally arranged.
                6   Lemmas 3-nerved, awnless .... *GROUP A*
                6   Lemmas 5–11-nerved (sometimes only faintly so) .... *GROUP B*

GROUP A (Festuceae)

1   Glumes and lemmas rounded on the back (certain western species of *Poa* may be sought here); lemmas strongly 3-nerved; panicle-branches floriferous nearly to base.
    2   Lemmas rigid and coriaceous, smooth and shining, without scarious margins, acuminate or mucronate-pointed (their nerves convergent at summit into a short stout cusp); panicle narrow and few-flowered, drooping, to 2.5 dm long, its branches simple; leaves to 2 cm broad; culms to 1 m tall, from creeping rhizomes; (s ?Ont.) ...... [*Diarrhena*]
    2   Lemmas subcoriaceous or cartilaginous; panicle-branches mostly branching.
        3   Culms erect, wiry, hard-based, tufted, rarely over 7 dm tall; leaves rigid, usually sparingly hairy above, to 1 cm broad, attenuate to tip; panicle to about 3 dm long, its branches appressed-ascending; spikelets subterete, commonly dark purple; lemmas to 5 mm long, acute or acuminate, obscurely nerved; (introd.) .... *Molinia*
        3   Culms decumbent and soft, rooting at the lower nodes; leaves soft, glabrous, to 1.5 cm broad, rounded at tip; panicle rarely over 1.5 dm long, its branches spreading; spikelets strongly compressed; lemmas to 3 mm long, strongly nerved, rounded or subtruncate at apex; (transcontinental) .... *Catabrosa*
1   Glumes and lemmas keeled, the latter faintly to strongly nerved.
    4   Culm-nodes bearded, the wiry tufted culms to about 1 m tall, disintegrating at the

nodes when mature; leaves rigid, 1 or 2 mm broad, they and their sheaths scabrous; panicle to 7 cm long, the stiff branches finally divergent; spikelets remotely 3–6-flowered, short-pedicelled, usually rose-purple; lemma 2-cleft, its short awn barely surpassing the broadly rounded lobes, its nerves strongly ciliate; palea densely long-ciliate along the upper half of the margins; (s Ont.) . . . .*Triplasis*

4   Culm-nodes not bearded; lemma and palea not as above.
    5   Spikelets 3–many-flowered; glumes and lemmas thin and scarious, translucent, only the prominent broad nerves green; lemmas not cobwebby at base; grain free from the palea, not furrowed; annuals of gravelly or sandy shores and dry waste places . . . .*Eragrostis*
    5   Spikelets with usually not more than 5 (sometimes 6 or 7) florets; glumes and lemmas herbaceous, only the margins thin and translucent; grain closely embraced, commonly furrowed.
        6   Perennials; lemmas mostly distinctly nerved, often pubescent and often subtended by a tuft of cobwebby hairs on the callus just below the base . . . .*Poa*
        6   Annuals with fibrous roots.
            7   Lemmas obscurely nerved, glabrous and lacking cobwebby hairs on the callus; leaves narrowly linear, scabrous above; culms rarely over 3 dm tall; (introd.) . . . .*Eremopoa*
            7   Lemmas mostly more distinctly nerved, more or less pubescent at least on the nerves and more or less cobwebby on the callus; (*P. annua; P. bolanderi; P. howellii*) . . . .*Poa*

GROUP B *(Festuceae)*

1   Inflorescence a simple 1-sided raceme of finally reflexed or drooping spikelets on a slender flexuous axis; leaves flat or folded; stoloniferous aquatic perennials . . . . . .*Pleuropogon*
1   Inflorescence usually a panicle.
    2   Lemmas dissimilar, the 2 or 3 uppermost ones empty and convolute into a club-shaped mass or enclosing one another at apex; panicle with a few solitary branches (or sometimes reduced to a raceme) . . . .*Melica*
    2   Lemmas all alike, the upper ones free; panicle usually compound.
        3   Lemmas cordate at base, faintly nerved, about as broad as long, with spreading margins; spikelets not much longer then broad, the inflated florets horizontally divergent; (introd.) . . . .*Briza*
        3   Lemmas not cordate; spikelets definitely longer then broad, their florets ascending.
            4   Lemmas keeled, awnless, acute or subacute (blunt-tipped in *Poa compressa*).
                5   Plants with perfect florets in each spikelet, these 2–several-flowered; lemmas 5-nerved, herbaceous or membranaceous, mostly scarious-tipped, the nerves often pubescent or cobwebby at base . . . .*Poa*
                5   Plants unisexual; spikelets 5–16-flowered, crowded in small panicles; lemmas faintly 9–11-nerved, rigid and coriaceous, smooth; leaves involute to flat, stiffish; plants of saline or alkaline soils, with creeping rhizomes covered with papery whitish scales . . . .*Distichlis*
            4   Lemmas rounded on the back or keeled only near tip (certain western species of *Poa* may be sought here).
                6   Lemmas only obscurely nerved, obtuse.
                      7   Panicle loose, its branches drooping, 2 or 3 at a node; lemmas glabrous; leaves flat, often over 1 cm broad; ligules normal; coarse aquatic plant with rather soft culms from creeping rhizomes; (transcontinental in arctic and subarctic regions) . . . .*Arctophila*
                      7   Panicle simple, its branches appressed-ascending; lemmas hairy-tufted at base, their incurved margins ciliate; leaves firm, convolute, to 3 mm broad, the basal ones tufted; ligule a tuft of hairs; culms tufted, to about 5 dm tall; (introd.) . . . .*Sieglingia*
                6   Lemmas distinctly nerved.

8 Lemmas distinctly 2-lobed at apex, often awned from the notch or just below it.

  9 Grain pubescent at summit; stigmas sessile, plumose, borne laterally from below the summit of the ovary; lemmas 5–9-nerved; callus at base of lemma not fringed with hairs; ligule not sheathing the culm . . . .*Bromus*

  9 Grain glabrous; styles present; lemmas 7-nerved; callus with a ring of short stiff hairs; spikelets typically bronze to purplish; ligule encircling the culm and continuous with the front of the sheath; leaves to 5 mm broad; culms to about 1 m tall; (transcontinental) . . . .*Schizachne*

8 Lemmas entire or minutely 2-toothed at apex, the awn (when present) terminal.

  10 Callus at base of ovary hairy; ovary pubescent; lemmas somewhat toothed at apex, 7-nerved, 1 or more pairs of lateral nerves excurrent as sharp points; leaves to 1 cm broad; culms stout, to about 2 m tall; (plant of marshes and shallow waters from B.C. to Man.) . . . .*Scolochloa*

  10 Culms and ovary glabrous; lemmas entire or merely erose at summit, the nerves rarely excurrent.

    11 Lateral nerves of the 5-nerved lemma arched and converging to the midrib, a terminal awn often present; (plants of usually dry soil and rocky slopes).

      12 Annual; glumes subulate-lanceolate; lemma with a scabrous awn to 3 mm long; anther usually solitary, included in the cleistogamous floret; grain linear-cylindric; leaves narrowly linear, commonly involute; culms mostly less than 5 dm tall . . . .*Vulpia*

      12 Perennial; anthers 3; grain ellipsoid or ovoid; culms often taller . . . .*Festuca*

    11 Lateral nerves of lemma parallel and nearly straight, not conspicuously arching to the midrib; lemmas awnless; (plants of mostly wet or moist habitats).

      13 Nerves of lemma faint, generally 5; stigmas sessile; leaf-sheaths open; (plants of mostly saline or alkaline habitats) . . . .*Puccinellia*

      13 Nerves of lemma usually prominent, generally 7; (plants of fresh habitats).

        14 Leaf-sheaths open, their margins free and overlapping; upper glume 3-nerved; stigmas sessile; grain with an apical tuft of minute white hairs and an oblong hilum; panicle open and lax; culms slender, weak, loosely decumbent from creeping or floating bases . . . .*Torreyochloa*

        14 Leaf-sheaths (at least the upper) with margins united nearly to summit; upper glume 1-nerved; styles present; grain with 2 apical prongs and a long linear hilum . . . . .*Glyceria*

Tribe HORDEAE (p. 218)

1 Spikelets usually 2 or 3 at each joint of the spike-rachis, 1–6-flowered; lemmas obscurely 5-nerved.

  2 Spikelets commonly 1-flowered, in 3's at each joint of the flattened and disarticulating rachis, the lateral pair each reduced to 1–3 spreading scabrous awns to 8 cm long; glumes prolonged and commonly bristle-like; spike nodding, pale green to purple . . . .*Hordeum*

  2 Spikelets 2–6-flowered, in 2's at each joint of the rachis, all with perfect florets, awned or awnless; spike erect or more or less nodding.

  3 Glumes bristle-like, divided to base into 2 divergent awns up to 8 cm long; rachis of spike (to about 1 dm long) promptly disarticulating at the joints; leaves narrow, flat or involute, finely pubescent; culms tufted, to about 5 dm tall; (B.C. to s Sask.) . . . .*Sitanion*

  3 Glumes, if prolonged, not bristle-like; spike-rachis not readily disarticulating . . . . .*Elymus*

1 Spikelets solitary at each joint of the rachis; lemmas 5–7-nerved.

  4 Spikelets 1-flowered; lower glume obsolete; upper glume minute; lemmas narrow, subulate-tipped or short-awned, scabrous, their margins tightly inrolled; style 1 (nearly unique for the family); spike 1-sided, to about 1 dm long; leaves bristle-like, their whitish sheaths crowded at the bases of the densely tufted culms, these mostly less than 3 dm tall; (introd.) . . . .*Nardus*

  4 Spikelets with at least 2 florets; spike symmetrical, the 2 rows of spikelets on opposite sides of the rachis.

    5 Spikelets edgewise to the rachis, the glume on the inner edge usually wanting (only the terminal spikelet with 2 glumes); (introd.) . . . .*Lolium*

    5 Spikelets flatwise to the rachis, both glumes present.

      6 Lemmas rounded on the back or obscurely keeled near the middle, their nerves converging toward summit . . . .*Agropyron*

      6 Lemmas excentrically keeled, the keel much closer to the inner margin (adjacent to the rachis) of the lemma than to the outer margin; annuals; (introd.).

        7 Glumes relatively broad, 3-nerved; lemmas rounded at the awnless to long-awned apex, their nerves not converging at the summit . . . .[*Triticum*]

        7 Glumes linear-subulate, 1-nerved; lemmas gradually tapering to summit, their nerves converging to form an awn up to 5 cm long . . . .[*Secale*]

Tribe MAYDEAE (p. 218)

A single genus; (introd.) . . . .[*Zea*]

Tribe ORYZEAE (p. 219)

A single genus . . . .*Leersia*

Tribe PANICEAE (p. 218)

1 Spikelets subtended by an involucre of bristles or fused spines.

  2 Spikelets subtended by 1 or more scabrous bristles, these not falling with the spikelet; inflorescence a bristly, dense or somewhat interrupted, spike-like panicle; (introd.) . . . .*Setaria*

  2 Spikelets (1–5 together) enclosed in a subglobose hairy spiny bur dropping intact from the rachis; leaves to 8 mm broad, often becoming involute; culm branching from base; (sandy shores of s Ont.) . . . .*Cenchrus*

1 Spikelets not subtended by bristles or spines.

  3 Inflorescence a cluster of slender, simple, commonly purplish, often curved, spike-like racemes, these approximate or subdigitate on a short terminal axis; spikelets about 2 mm long, awnless, short-pedicelled, alternate in 2 rows on one side of the thin-winged rachis; (introd) . . . .*Digitaria*

  3 Inflorescence a panicle with spreading or ascending branches.

    4 Branches of the inflorescence bearing dense stout 1-sided racemes of almost sessile spikelets, these awnless or awned, often with bristly hairs; coarse annual chiefly of fields, roadsides, and waste places . . . .*Echinochloa*

    4 Branches of the symmetrical panicle spreading or ascending, bearing scattered long-pedicelled awnless dorsally compressed spikelets; annual, or if perennial, frequently with basal winter-rosettes of short leaves unlike the stem-leaves . . . .*Panicum*

Tribe PHALARIDEAE (p. 219)

1   Panicle open or contracted (if somewhat spike-like in *H. pauciflora,* the culms not over 2 dm tall and the leaves only 1 mm broad); sterile lemmas boat-shaped, nearly as long as the glumes, indurated and hairy; spikelets broad and shining, awnless or awned, comprising a central perfect floret with 2 stamens subtended by 2 staminate florets with 3 stamens; glumes subequal, broad and papery; (transcontinental species) . . . . *Hierochloë*
1   Panicle spike-like or with spike-like primary branches, green or yellowish; spikelets with 1 central perfect floret subtended by 2 sterile lemmas.
  2   Glumes subequal, boat-shaped, much surpassing the florets, glabrous; lemmas awnless, the sterile ones much reduced and scaly or bristly; fertile lemma indurated and shining in fruit; stamens 3; spikelets laterally flattened . . . . *Phalaris*
  2   Glumes very unequal; sterile lemmas dorsally awned, longer than the fertile floret; fertile lemma remaining hyaline or membranaceous; stamens 2; spikelets subterete; plant sweetly fragrant on drying; (introd.) . . . . *Anthoxanthum*

Tribe ZIZANIEAE (p. 219)

A single genus; (Sask. to N.B. and N.S., partly introd.) . . . . *Zizania*

AGROPYRON Gaertn.   [405]

(Ref.: Bowden 1965; hybrids are listed at the end of the following treatment)
1   Spikes more or less distinctly pectinate, strongly flattened and comb-like, the crowded spikelets usually strongly divergent in the two opposite rows; anthers usually over 4 mm long, their tips much surpassing the middle of the palea; leaves relatively short and rigid; tufted fibrous-rooted annuals; (introd.).
  2   Spikes small, ellipsoid-ovoid or subglobose, usually less than 2 cm long; spikelets glabrous, to 1 cm long; glumes lanceolate, subulate-tipped, thick-keeled; lemmas to 6 mm long, with an awn-point 1 mm long; leaves 2 or 3 mm broad, very scabrous and sparsely short-pubescent; (introd. in SE Alta., sw Sask., and s Man.) . . . . *A. triticeum*
  2   Spikes longer, linear to narrowly ovoid; spikelets to 1.5 cm long; glumes with an awn 1.5–3 mm long; leaves mostly broader.
    3   Spikes indistinctly pectinate, linear, to 1.5 dm long, nearly or quite glabrous; lemmas to 8.5 mm long, pointed or with an awn to over 3 mm long; leaves to 6 mm broad, scabrous above, they and their smooth or minutely scabrous sheaths otherwise glabrous or spreading-pubescent; (introd. from B.C. to Ont.) . . . . . *A. sibiricum*
    3   Spikes strongly pectinate, short-linear to narrowly ovoid, rarely over 6 cm long; lemmas to 6 or 7 mm long, with awns usually at least 2 mm long.
      4   Spikelets nearly or quite glabrous, separated by distinct gaps; glumes scabrous only on the keel at the base of the awn; culms smooth or scaberulous below the spike; (widely introd.) . . . . *A. pectiniforme*
      4   Spikelets copiously villous, crowded and without distinct gaps; glumes long-ciliate on the keel; culms usually puberulent below the spike . . . . [*A. cristatum*]
1   Spikes not pectinate, usually elongate and relatively narrow, the spikelets ascending to appressed; leaves mostly longer, flat or involute; perennials.
  5   Plants without elongate creeping rhizomes (short rhizomes may sometimes be present); glumes usually persistent on the rachis after the fall of the readily detached individual florets; anthers usually less than 4 mm long, their tips rarely surpassing the middle of the palea.
    6   Lemmas awnless or tipped with straight ascending awns; (transcontinental) . . . .*A. trachycaulum*
    6   Lemmas typically tipped with strongly arched-divergent awns.
      7   Culms prostrate or decumbent-spreading, often flexuous, to about 4 dm long, the 2 or 3 leaves usually less than 5 cm long, to 3 mm broad; spikes to 7 cm long, the rachis disarticulating at maturity; lemma-awns to 2.5 cm long; (sw Alta.) . . . .*A. scribneri*

7 Culms erect, straight, to about 1 m tall, the leaves mostly longer; spikes commonly longer, the rachis continuous.
    8 Spikelets relatively distant (each extending barely above the base of the next one above on the other side of the rachis); lemma-awns 1 or 2 cm long; leaves 1 or 2 mm broad; (B.C. to Sask.) . . . .*A. spicatum*
    8 Spikelets rather crowded and distinctly overlapping; lemma-awns to 4 cm long; leaves to 4 mm broad . . . . [*A. bakeri*]
5 Plants with elongate creeping rhizomes; mature spikelets usually falling intact (the glumes often falling with them), the individual florets not readily detached; anthers mostly over 4 mm long, their tips much surpassing the middle of the palea.
    9 Lemmas typically pubescent over the back (*A. smithii* var. *molle* may be sought here); rachilla pubescent between the spikelets; leaves pale to very glaucous, firm or rigid, commonly involute, rarely over 5 mm broad.
        10 Lemmas long-awned, their awns arched-divergent at maturity, to about 1.5 cm long; glumes about 1 cm long, awn-pointed; (B.C. to Sask.) . . . . *A. albicans*
        10 Lemmas awnless, at most merely mucronate.
            11 Spike to about 2.5 dm long; spikelets to 2.5 cm long, with up to 10 florets; glumes to 13 mm long; (?B.C.) . . . . [*A. elmeri*]
            11 Spike to about 12 cm long; spikelets to 1.5 cm long, with rarely more than 8 florets; (B.C. to Man.; introd. in Ont.) . . . . *A. dasystachyum*
    9 Lemmas typically glabrous or merely scabrous, or pubescent only along the margins near the base (*A. albicans* var. *griffithsii* will key out here, as also sometimes *A. elmeri*).
        12 Lemmas and glumes very blunt, awnless; leaves glaucous, to 7 mm broad; (introd. in Sask.) . . . . [*A. intermedium*]
        12 Lemmas and glumes acute, acuminate, or straight-awned.
            13 Glumes rigid, gradually tapering to a short awn, rather faintly nerved; spikelets to 3 cm long, with up to 13 florets; leaves firm or rigid, glaucous, mostly involute, rarely over 5 mm broad; cartilaginous band of upper nodes of culm shorter than thick; (B.C. to w Ont.; introd. eastwards) . . . . *A. smithii*
            13 Glumes not rigid, commonly strongly nerved, usually rather abruptly narrowed to the awnless or straight-awned apex; spikelets commonly less than 2 cm long; leaves flat or involute, to about 1 cm broad; cartilaginous band of upper nodes of culm about as long as thick.
                14 Top of culm pithy at anthesis; leaves hard, very glaucous, mostly involute, with remote coarse ribs; spike nearly square in cross-section; glumes leathery; spikelets with up to 11 florets; (introd. in N.S.) . . . .*A. pungens*
                14 Culm hollow at anthesis; leaves lax, with crowded fine nerves; spike usually not square; spikelets with up to 9 florets; glumes herbaceous; (introd., transcontinental) . . . .*A. repens*

A. albicans Scribn. & Sm.
/T/WW/ (Grh) Plains and dry hills (ranges of Canadian taxa outlined below), s to Utah, Colo., and S.Dak.
1 Lemmas more or less densely pubescent; [B.C. (Lillian L., Vancouver Is.), Alta. (N to near Fort Saskatchewan), and Sask. (N to Humboldt, 52°12′N; Breitung 1957a)] . . . . . . .var. *albicans*
1 Lemmas glabrous; [*A. griffithsii* Scribn. & Sm.; Alta. (Waterton Lakes; Pincher Creek; Calgary) and Sask. (Saskatoon; Hodgeville; Crestwynd)] . . . .
. . . . .var. *griffithsii* (Scribn. & Sm.) Beetle

[*A. bakeri* Nels.]
[This species of the w U.S.A. (Wash. to N.Mex.; N Mich.) is reported from Alta. by A. S. Hitchcock (1935; taken up by Boivin 1967a, under the name *A. trachycaulum* var. *bakeri* (Nels.) Boivin). Bowden (1965), however, states that, "Specimens from Canada that have been identified as *A. bakeri* proved to belong to other taxa".]

[A. cristatum (L.) Gaertn.]   Crested Wheat-Grass
[Eurasian; most or all reports of this species from Canada apparently refer to *A. pectiniforme* and *A. sibiricum*, Bowden (1965) stating that, "Although plants of this species have sometimes been grown in experimental plots in Canada, no escapes have so far been collected". (*Bromus* L.).]

A. dasystachyum (Hook.) Scribn.
/sT/WW/   (Grh)   Dry prairies, sandhills, and sandy shores, the aggregate species from s Yukon, N B.C., and N Alta. to Sask. and s Man., s to Calif., Colo., Nebr., and N Ill. (rarely introd. in Ont. according to Bowden 1965). Synonymy (together with a distinguishing key to the closely related *A. yukonense*): see below.

1   Spikelets distinctly overlapping the bases of the next ones above on the same side of the
        rachis, the median internodes of the spike commonly not over 6 mm long; callus at base
        of lemma glabrate in the middle, short-hirsute on the sides; [Alaska–Yukon–N B.C., the
        type from Fort Yukon, Alaska; MAPS: Hultén 1968*b*: 184, and 1942: map 191, p. 263] . . . .
        . . . .*A. yukonense* Scribn. & Merr.
1   Spikelets at most only slightly overlapping, the median internodes of the spike commonly
        7 or 8 (up to 13) mm long; callus at base of lemma short-villous or silky . . . . . .*A. dasystachyum*
    2   Glumes prominently 5-nerved, linear to linear-lanceolate; spikes to about 2.5 dm long, the
            rachis pubescent below the spikelets, its median internodes to 13 mm long; lemmas densely
            long-villous; paleas to about 11 mm long; [*A. psammophilum* Gillett & Senn, the type from
            near Kincardine, Bruce Co., s Ont.; also known from Manitoulin Is., L. Huron; MAP: J. M.
            Gillett and H. A. Senn, Can. J. Bot. 39(5), fig. 1, p. 1172. 1961] . . . .
            . . . . .var. *psammophilum* (Gillett & Senn) Voss
    2   Glumes 3-nerved, only the midnerve conspicuous; spikes to 11 cm long, the rachis
            glabrous below the spikelets, its median internodes 6 or 7 mm long; lemmas glabrous
            or villous; paleas to about 7 mm long; [B.C. to Man.].
        3   Lemmas glabrous or merely scabrous (sometimes with longer hairs at the base);
                [*A. riparium* Scribn. & Sm.] .var. *riparium* (Scribn. & Sm.) Bowden
        3   Lemmas typically densely villous to lanate; [*Triticum repens* var. *d.* Hook.,
                the type from Carlton House, Sask.; *A. subvillosum* (Hook.) Nels.; incl.
                A. elmeri Scribn.] . . . . .var. *dasystachyum*

[A. elmeri Scribn.]
[This species of Wash.–Oreg. is reported from B.C. by Hubbard (1955), probably on the basis of the inclusion of B.C. in the range given by A. S. Hitchcock (1935). However, as indicated in the key to the genus, it differs from *A. dasystachyum* chiefly in its larger floral dimensions, probably representing merely an extreme of that species, with which it is merged in the present treatment.]

[A. intermedium (Host) Beauv.]
[Eurasian; reported by Breitung (1957*a*) as planted in the prairie region of Sask. to rejuvenate range land, its long-creeping rhizomes serving to bind the soil against wind erosion. There appears to be no indication, however, of its escaping. The closely related *A. elongatum* (Host) Beauv. (densely tufted, lacking creeping rhizomes) is also reported by Breitung as cult. on wet alkaline meadows of the Saskatchewan prairies.]

A. pectiniforme R. & S.   Crested Wheat-Grass
Eurasian; dry fields and waste places of s Yukon (Porsild 1951*a*; as *A. cristatum*), w Dist. Mackenzie (N to Fort Simpson, ca. 62°N), B.C. (N to Williams Lake), Alta. (N to Waterways, 56°42′N), Sask. (N to Candle L., about 45 mi NE of Prince Albert), Man. (N to Lynn Lake), Ont. (Lake of the Woods Provincial Park; Carleton, Wellington, and Elgin counties), Que. (Montreal; Ste-Anne-de-la-Pocatière, Kamouraska Co.), and N.S. (Kentville). [*A. cristatum* of Canadian reports, at least in major part, not *Bromus cristatus* L., basionym; *A. cristatiforme* Sarkar]. MAP: Hultén 1968*b*: 182.

A. pungens (Pers.) R. & S.
European; introd. along beaches of the Atlantic coast from N.S. (Cumberland, Cape Breton,

Victoria, and Inverness counties; *see* N.S. map by Dore and Roland 1942: 218) to Cape Cod, Mass. [*A. acadiense* Hubbard].

A. repens (L.) Beauv.   Witch-Grass, Couch-Grass.   Chiendent
Eurasian (but considered by Fernald *in* Gray 1950, as perhaps native on gravelly and sandy shores of E N. America); an agressive weed throughout much of N. America, difficult to eradicate by reason of its very elongate stolons, from s Alaska and s Dist. Mackenzie to Man. (N to Churchill), James Bay, Que. (N to the Hamilton R. basin), Labrador (N to Cartwright, 53°42′N), and the Atlantic Provinces. MAPS and synonymy: *see* below.
1   Glumes broadly scarious-margined, rather abruptly narrowed at apex; (the following
    forms are known from our area) . . . . .var. *repens*
    2   Rachis glabrous except for ciliate edges.
        3   Lemmas and glumes at most subulate-tipped; [*Triticum* L., *Elymus* Gould;
            Elytrigia Nevski; *A. ?caesium* Presl; MAPS: Hultén 1968*b*: 183, and 1962: map
            199, p. 211] . . . . .f. *repens*
        3   Lemmas (and often the glumes) definitely awned . . . . .f. *aristatum* (Schum.) Holmb.
    2   Rachis hairy on the sides as well as on the edges.
        4   Lemmas and glumes at most subulate-tipped . . . . .f. *trichorrachis* Rohlena
        4   Lemmas (and often the glumes) definitely awned . . . . .f. *pilosum* (Scribn.) Fern.
1   Glumes narrowly margined, tapering gradually to the apex; (the following forms are
    known from our area) . . . . .var. *subulatum* (Schreb.) Rchb.
    5   Rachis glabrous except for ciliate edges.
        6   Lemmas and glumes at most subulate-tipped . . . . .f. *subulatum*
        6   Lemmas (and often the glumes) definitely awned . . . .
        . . . . .f. *vaillantianum* (Wulf. & Schreb.) Fern.
    5   Rachis hairy on the sides as well as on the edges.
        7   Lemmas and glumes at most subulate-tipped . . . . .f. *heberachis* Fern.
        7   Lemmas (and often the glumes) definitely awned . . . . .f. *setiferum* Fern.

A. scribneri Vasey
/T/WW/   (Hs)   Windswept alpine slopes from s ?B.C. (Hitchcock et al. 1969) and sw Alta. (Waterton Lakes; Breitung 1957*b*) to Oreg., N Ariz., and N N.Mex.

A. sibiricum (Willd.) Beauv.
Eurasian; introd. in Alaska (Fairbanks; Hultén 1950), s Dist. Mackenzie (Fort Smith, ca. 60°N; Porsild and Cody 1968), B.C. (between Lytton and Spences Bridge; near Dawson Creek), Alta. (Fort Saskatchewan), Sask. (Maple Creek; Moose Jaw; Eastend; Parkbeg), Man. (Carberry; Otterburne), and w Ont. (Thunder Bay). [*Triticum* Willd.; incl. *A. desertorum* (Fisch.) Schult., the relatively hairy phase with longer-awned glumes and lemmas].

A. smithii Rydb.
/T/WW/   (Grh)   Dry prairies and plains (ranges of Canadian taxa outlined below), s to Calif., Ariz., Tex., Ark., and Tenn. (largely or wholly introd. in the East). MAPS and synonymy: *see* below.
1   Lemmas glabrous or glabrate; [*A. glaucum* var. *occidentale* Scribn.; *A. occid.* Scribn.;
    B.C. (N to Pouce Coupe, ca. 55°45′N), Alta. (N to Fort Vermilion, 58°24′N), Sask. (N to
    Tisdale), Man. (N to Riding Mountain National Park), and Ont. (Ingolf; Thunder Bay;
    Nipigon; Michipicoten; Manitoulin Is.; Sudbury; Chalk River; Port Hope; (perhaps native
    in the w Ont. localities and introd. eastwards)); introd. at Wrightville, Que., and in s Alaska
    (Hultén 1942); MAP (aggregate species): Hultén 1968*b*: 183] .var. *smithii*
1   Lemmas more or less densely pubescent; [*A. spicatum molle* S. & S.; *A. molle* (S. & S.)
    Rydb.; essentially the range of var. *smithii*; MAP: Raup 1930: map 27, p. 203] . . . .
    . . . . .var. *molle* (Scribn. & Sm.) Jones

A. spicatum (Pursh) Scribn. & Sm.
/ST/WW/   (Hs)   Dry prairies, plains, and open woods (ranges of Canadian taxa outlined below), s to Calif., N.Mex., and w Nebr. MAP and synonymy: *see* below.
1   Lemmas with an arching awn commonly 1 or 2 cm long; [*Festuca* Pursh; *A. divergens*

Nees; reports of *A. bakeri* Nels. from B.C. and Alta. are probably referable here at least in part; cent. Alaska (Hultén 1942), cent. Yukon (N to near Dawson), B.C., sw Alta. (N to Banff), and Sask. (N to Scott and Sutherland); MAP (aggregate species): Hultén 1968b: 184] . . . . .var. *spicatum*
1   Lemmas awnless or with a short straight awn; [*A. divergens inerme* S. & S.; *A. inerme* (S. & S.) Rydb.; B.C. (N to Williams Lake) and sw Alta. (Waterton Lakes; Breitung 1957b)] . . . .
. . . . .var. *inerme* (Scribn. & Sm.) Heller

A. trachycaulum (Link) Malte
/aST/X/ (Hs) Prairies, gravelly shores, thickets, etc., the aggregate species from N Alaska–Yukon–Dist. Mackenzie to Banks Is., Devon Is., Ellesmere Is. (N to ca. 82°N), northernmost Labrador, and Nfld., s to Calif., N.Mex., Nebr., Ill., Ind., and Md. MAPS and synonymy (together with a distinguishing key to the closely related *A. macrourum*): see below.
1   Lemma-backs usually densely short-pubescent (sometimes glabrous or glabrate in var. *latiglume*); glumes 3–4(5)-nerved.
  2   Median rachis-internodes over 1 cm long; glumes with narrow membranous margins, 3–5-nerved, their bodies much shorter than the first lemma-bodies; first paleas about 2 mm long; [*Triticum* Turcz.; *A. sericeum* Hitchc. (*A. dasystachyum* var. *ser.* (Hitchc.) Boivin); incl. × *Agrohordeum jordalii* Meld.; Alaska–Yukon and w Dist. Mackenzie (Mackenzie R. Delta and Norman Wells; Bowden 1965); MAP: Hultén 1968b: 185] . . . .
. . . .*A. macrourum* (Turcz.) Drobov
  2   Median rachis internodes over 1 cm long; glumes with hyaline margins to 1 mm broad, 3–4-nerved, their bodies slightly shorter than the first lemma-bodies; first paleas about 9 mm long; [*A. latiglume* (Scribn. & Sm.) Rydb. and its vars. *alboviride* and *pilosiglume* Hult.; *A. violaceum* (Hornem.) Lange and its vars. *latiglume* Scribn. & Sm. and *hyperarcticum* Polunin; *A. alaskanum* Scribn. & Merr. and its var. *arcticum* Hult.; *A. ?angustiglume* Nevski; *A. ?boreale* (Turcz.) Drob. (incl. × *Agroelymus hultenii* Meld.); *A. caninum* var. *latiglume* (Scribn. & Sm.) Pease & Moore; transcontinental, the common form northwards extending N to Ellesmere Is. at ca. 82°N; MAPS: Porsild 1957: map 56, p. 167; Raup 1947: pl. 16; combine the maps by Hultén 1968b: 185 (*A. violaceum*) and p. 187 (*A. boreale*)] . . . .
. . . .*A. trachycaulum* var. *latiglume* (Scribn. & Sm.) Beetle
1   Lemma-backs essentially glabrous, merely minutely scabrous in the upper third or on the upper part of the nerves or short-hirsute on the sides at base; glumes (4,5)6–7-nerved . . . .
. . . .*A. trachycaulum*
  3   Awns nearly equalling or longer than their lemmas.
    4   Body of glumes to 12 mm long; lemma-awns to 2 cm long; spikes usually less than 8 mm thick; [vars. *caerulescens* and *pilosiglume* Malte; *A. caninum* f. *pubescens* of auth., not var. *pubescens* Scribn. & Sm.; transcontinental; MAP: Fernald 1933: map 18 (E area), p. 178] . . . . .var. *glaucum* (Pease & Moore) Malte
    4   Body of glumes to 18 mm long; lemma-awns to 3 or 4 cm long; spike commonly over 8 mm thick; [*A. caninoides* (Ramaley) Beal; *A. richardsonii* Schrad.; *A. subsecundum* (Link) Hitchc.; *A. caninum* of Canadian reports in large part, not *Triticum caninum* L.; *A. gmelinii* of Canadian reports in large part, not Scribn. & Sm.; transcontinental, N to cent. Yukon; MAPS: Raup 1947: pl. 16; Hultén 1968b: 186 (*A. sub.*)] . . . . .var. *unilaterale* (Cassidy) Malte
  3   Awns wanting or at most about half as long as their lemmas.
    5   Spikelets close, the tip of one overlapping the base of the next upper one on the same side; internodes of rachis mostly 2-angled.
      6   Body of glume to about 1.5 cm long; fruiting spike to 12 mm thick; [var. *glaucescens* Malte; var. *fernaldii* (Pease & Moore) Malte; *A. tenerum majus* Vasey; *A. caninum* var. *tenerum* f. *fernaldii* Pease & Moore and var. *hornemannii sensu* Pease & Moore in large part, not *Triticum biflorum* var. *hornemannii* Koch; *A. donianum* Buch.; MAPS: Hultén 1968b:190 (*A. pauciflorum* var. *majus*); Fernald 1933: map 16 (E area), p. 172] . . . .
. . . . .var. *majus* (Vasey) Fern.
      6   Body of glume at most 1 cm long; fruiting spike at most 6 mm thick; [var.

tricholeum (Piper) Malte; *A. caninum* var. *tenerum* f. *ciliatum* (Scribn. & Sm.)
Pease & Moore; *A. tenerum (pauciflorum)* var. *nov.* Scribn.; MAPS: Fernald
1933: map 17 (E area), p. 175; Hultén 1968b: 189; a map of the Que.–Labrador
distribution is given by Dutilly, Lepage, and Duman 1953: fig. 7, p. 29] . . . .
. . . . .var. *novae-angliae* (Scribn.) Fern.

5   Spikelets not crowded, the top of one scarcely reaching the base of the next upper
one on the same side; internodes of rachis often 4-angled; [var. *tenerum* (Vasey)
Malte; *Triticum* Link; *Roegneria* Nevski; *A. tenerum* Vasey; *A. teslinense* Porsild &
Senn; *A. pauciflorum* (Schwein.) Hitchc., not Schur; MAPS: Fernald 1933:
map 15 (E area), p. 170; Hultén 1968b: 188 (*A. pauciflorum*)] . . . . .var. *trachycaulum*

A. triticeum Gaertn.   Annual Wheat-Grass
European; introd. in s Alta. (Medicine Hat), s Sask. (Cypress Hills and Maple Creek; Breitung
1957a), and s Man. (Brandon; Winnipeg).

Agropyron Hybrids

Hybridization in *Agropyron* and other grasses has been intensively investigated experimentally
and in the field by Bowden (1965; 1966; 1967) and the following list of hybrids reported as
occurring naturally in Alaska–Canada (including the hybrid genera × *Agroelymus* Camus and ×
Agrohordeum Camus) is based upon these papers. Following present custom, the names of the
putative parents are given in alphabetical sequence.

Agropyron dasystachyum vars. *dasy.* and *riparium* × *A. trachycaulum* var. *tr.*: (× *A.
pseudorepens* Scribn. & Sm. nm. *pseudorepens*); Alta., Sask., Man., and s Ont. (Bruce Pen., L.
Huron).

A. das. var. *psammophilum* × *A. trachycaulum* var. *tr.*: (× *A. pseudorepens* nm. *sennii* Boivin, the
type from Bruce Co., s Ont.).

A. das. vars. *das., psamm.,* and *riparium* × *A. trachycaulum* var. *glaucum*: (× *A. pseudorepens*
nm. *vulpinum* (Rydb.) Boivin; *Elymus (A.) vulpinus* Rydb.); B.C., Alta., Sask., s Ont. (Manitoulin Is.,
N L. Huron), and E Que. (Bonaventure, Gaspé Pen.).

A. das. × *A. spicatum* var. *inerme*: sw Alta. (Boivin 1967a).

A. das. × *Elymus innovatus*: (× *Agroelymus turneri* Lepage, the type from Fort Saskatchewan,
Alta.); also known from Banff, Alta., and Saskatoon, Sask.

A. scribneri × *A. violaceum* (*trachycaulum* var. *latiglume*): (× *A. brevifolium* Scribn.; *A.
subsecundum* var. *andinum* (Scribn. & Sm.) Hitchc.; *A. andinum* (S. & S.) Rydb.); Mt. Richards,
Waterton Lakes, sw Alta.

A. sericeum (macrourum) × *Elymus sibiricus*: (× *Agroelymus palmerensis* Lepage, the type from
Palmer, Alaska (*E. canadensis* being originally named as the *Elymus* parent); × *Agroelymus
hodgsonii* Lepage); Alaska and sw Dist. Mackenzie

A. sericeum (macrourum) × *Hordeum jubatum*: (× *Agrohordeum pilosilemma* Mitchell &
Hodgson, the type from Galena, Alaska); also known from other Alaskan localities.

A. smithii × *Elymus innovatus*: (× *Agroelymus bowdenii* Boivin, the type from Beaverlodge, Alta.);
also known from Sinclair Hot Springs, B.C.

A. spicatum × *A. violaceum* (*trachycaulum* var. *latiglume*): Livengood, Alaska.

A. spic. × *A. trach.* var. *unilaterale*: Waterton Lakes, sw Alta.

A. spic. × *A. yukonense*: the Yukon.

A. trachycaulum var. *?trach.* × *Elymus canadensis*: (× *Agroelymus mossii* Lepage, the type from
Lake Louise, Alta.; × *Agroelymus cayouetteorum* Boivin, the type from s Que.).

A. trach. var. *trach.* or sometimes var. *novae-angliae* × *Elymus mollis*: (× *Agroelymus jamesensis*
Lepage, the type from Old Factory, James Bay, Que., and its vars. *anticostensis* and *stoloniferus*
Lepage; × *Agroelymus adamsii* Rousseau and its nothomorphs *jamesensis, longispica,* and
semiaelvus Lepage; *Agropyron repens* × *Elymus mollis* sensu J. Adams, Can. Field-Nat. 50(7):
117. 1936); also known from other localities in B.C. (Vancouver Is.) and Que.

A. trach. var. *trach.* × *Hordeum jubatum: (Elymus macounii* Vasey, the type from Kamloops, B.C.; × *Agrohordeum mac.* (Vasey) Lepage); cent. Alaska–Yukon–Dist. Mackenzie through B.C., Alta., Sask., and Man.

A. trach. var. *glaucum* × *A. yukonense*: the Yukon.

A. trach. var. *glaucum* × *Elymus hystrix* var. *bigeloviana:* (× *Agroelymus dorei* Boivin, the type from Breckenridge, Gatineau Co., Que.); reported from Ont. and Que. by W. G. Dore, Can. Field-Nat. 64(1): 39. 1950

A. trach. vars. × *Elymus innovatus:* (× *Agroelymus hirtiflorus* (Hitchc.) Bowden (*Elymus hirt.* Hitchc.); × *Agroelymus ontariensis* Lepage); B.C., Alta., Sask., and Ont.; MAP: Meusel, Jaeger, and Weinert 1965: 43.

A. violaceum (trach. var. *latiglume)* × *A. yukonense*: the Yukon.

A. violaceum × *Elymus innovatus:* (× *Agroelymus colvillensis* Lepage, the type from Umiat, Alaska, the only known locality).

A. violaceum × *Elymus mollis:* (× *Agroelymus ungavensis* (Louis-Marie) Lepage: *Agropyron ungavense* Louis-Marie (the type from near Fort Chimo, Ungava) and its f. *ramosum* Louis-Marie.

AGROSTIS L.   [242]   Bentgrass.   Agrostide

1 Palea at least half as long as the lemma.
  2 Dwarf alpine perennial species, the tufted culms mostly not over 1.5 dm tall (to about 3 dm), the leaves mostly basal, to about 1 dm long and 1 or 2 mm broad, flat or folded; panicle narrow, purple, to about 8 cm long, the branches appressed or somewhat spreading; spikelets about 2 mm long; lemmas awnless; anthers 0.6–0.7 mm long; (?B.C.) . . . .[*A. humilis*]
  2 Culms commonly taller, the leaves and panicles mostly longer; (introd.).
    3 Ligules of middle and lower leaves broader than long, at most 1 mm long; leaves at most 5 mm broad; panicle open, to 2 dm long, its branches spikelet-bearing chiefly above the middle; anthers to 1.5 mm long; perennial, often stoloniferous; (introd., transcontinental) . . . .*A. tenuis*
    3 Ligules of middle and lower leaves about as long as or longer than broad, to 6 mm long; panicle compact, with many of its spikelets on short branches near the rachis.
      4 Annual, the more or less tufted culms to about 6 dm tall; leaves convolute, smooth, narrow and short; ligules to 5 mm long; awns to about 1 cm long, about 3 times as long as the lemmas; anthers about 0.5 mm long; (introd. in s B.C.) . . . .*A. interrupta*
      4 Perennials with creeping stolons or subterranean rhizomes; awns typically wanting or short; anthers about 1.3 mm long; (introd., transcontinental) . . . .*A. stolonifera*
1 Palea minute and nerveless or wanting.
  5 Plants with creeping rhizomes; anthers to about 1.5 mm long; perennials.
    6 Panicle contracted and almost spike-like, to 1 dm long, its branches appressed, mostly floriferous to near base; lemmas awnless; culms to about 4 dm tall; (sw ?B.C.) . . . .[*A. pallens*]
    6 Panicle narrow but open, to about 1.5 dm long, its branches spreading-ascending, floriferous mostly toward top; lemmas awnless or awned; culms to about 1 m tall; (sw B.C.) . . . .*A. diegoensis*
  5 Plants tufted; rhizomes wanting or very short.
    7 Panicle very diffuse, the spikelets borne chiefly near the ends of the strongly scabrous branches; lemmas typically awnless or short-awned; anthers to about 0.5 mm long; spikelets to 2(3) mm long; perennial; (transcontinental) . . . .*A. hyemalis*
    7 Panicle narrow to open but not diffuse, the spikelets commonly borne from near or below the middle to the ends of the usually glabrous or only slightly scabrous panicle-branches.
      8 Panicle narrow, its branches mostly strongly ascending, usually some of the lower ones spikelet-bearing from near the base, commonly with short

branchlets in their axils; lemmas typically awnless; anthers to about 0.7 mm long.

> 9 Annual; spikelets to about 2.5 mm long; lemmas awnless; leaves 1 or 2 mm broad ....[*A. rossiae*]
>
> 9 Tufted perennials.
>> 10 Spikelets to 4 mm long; lemmas to 2 mm long; panicle to about 3 dm long; leaves to 1 cm broad; culms to over 1 m tall; (B.C. to Sask.) ....*A. exarata*
>>
>> 10 Spikelets at most 2.5 mm long; lemmas about 1.5 mm long; panicle less than 1 dm long; leaves 1 or 2 mm broad; culms mostly less than 2 dm tall; (B.C. and w Alta.) ....*A. variabilis*

8 Panicle open and broader, its branches spreading or loosely ascending, without short floriferous branches in their axils; perennials.

> 11 Lemmas typically awnless; panicle-branches minutely scabrous.
>> 12 Spikelets about 1.5 mm long; panicle to about 1 dm long; anthers about 0.3 mm long; leaves narrow, mostly basal; culms to about 3 dm tall ....[*A. idahoensis*]
>>
>> 12 Spikelets 2 or 3 mm long; panicle to over 3 dm long; anthers to 1.5 mm long; leaves rather numerous, to 6 mm broad; culms to over 1 m tall; (Ont. to Nfld. and N.S.) ....*A. perennans*
>
> 11 Lemmas typically awned, the awn bent; leaves mostly basal.
>> 13 Panicle-branches smooth, in 2's or 3's, the rather loose panicle to 7 cm long and 3 cm thick, usually brownish purple; spikelets mostly 3 or 4 mm long; anthers to 0.7 mm long; culms to 4 dm tall, tufted; (transcontinental) ....*A. borealis*
>>
>> 13 Panicle-branches scabrous, in clusters of up to 6, the panicle contracted before and after flowering, to 11 cm long and 5 cm thick, brownish-stramineous; spikelets less than 3 mm long; anthers to 1.5 mm long; culms to 6 dm tall, often finally producing long trailing stolons with leaves densely tufted at the nodes; (introd.) ....*A. canina*

A. borealis Hartm.

/aST/X/GEA/ (Hs) Gravelly or rocky open soil, the aggregate species from the Aleutian Is. and n-cent. Alaska to cent. Yukon, Great Bear L., L. Athabasca (Alta. and Sask.), Man. (n to Baralzon L., ca. 60°N), cent. Dist. Keewatin, Baffin Is. (n to the Arctic Circle), northernmost Ungava–Labrador, and Nfld. (not known from the Maritime Provinces), s to sw B.C. (Vancouver Is.), sw Alta., nw Sask., n Man., James Bay, and e Que. (Côte-Nord; Gaspé Pen.), and in the mts. to ?Wash., Utah, Colo., and n New Eng.; w and e Greenland n to ca. 72°30′N; n Eurasia. MAPS and synonymy: *see* below.

1 Lemmas awnless; spikelets to 4.5 mm long; branches of mature panicle strongly ascending; [*A. paludosa* Scribn., the type from Blanc-Sablon, Côte-Nord, e Que.; also known from Anticosti Is., e Que., and Nfld.]................var. *paludosa* (Scribn.) Fern.

1 Lemmas dorsally awned; branches of mature panicle spreading-ascending.

> 2 Leaves to 4 mm broad, flat; spikelets to 4 mm long; [*A. rubra* var. *amer.* Scribn.; Que. (n to Richmond Gulf, Hudson Bay, at ca. 56°20′N, Seal L. at ca. 56°40′N, and the Gaspé Pen.), Labrador (n to Goose Bay), and w Nfld.] ....var. *americana* (Scribn.) Fern.
>
> 2 Leaves mostly not over 2 mm broad; spikelets mostly 2 or 3 mm long ....var. *borealis*
>> 3 Lemmas much prolonged; [perhaps a pathological state; e Que. (Mingan Is.) and Nfld. (type of var. *mic.* Eames from the Blow-me-down Mts.)] ....f. *micrantha* (Eames) Fern.
>>
>> 3 Lemmas normal; [*A. canina* vars. *aenea* (*A. aenea* Trin.) and *melaleuca* Trin. (*A. melaleuca* (Trin.) Hitchc.); *A. ?longiligula sensu.* J. M. Macoun 1913, and of Alaskan reports, not Hitchc.; *Aira labradorica* Steud.; transcontinental; MAPS (aggregate species); Porsild 1957: map 18, p. 163; Raup 1930: map 11, p. 202; Hultén 1958: map 179, p. 199, and 1968*b*:98; Böcher 1954: fig. 13 (top), p. 56; Meusel, Jaeger, and Weinert 1965:47] ....f. *borealis*

A. canina L.   Brown or Velvet-Bent
/aST/-/GEA/   (Hs)   Ssp. *montana* probably native in Greenland, the typical form introd. in N. America. MAPS and synonymy: *see* below.
1   Panicle usually strongly contracted and spike-like after flowering; stolons wanting;
    rhizomes present, the culms tufted; [Greenland; Joergensen, Soerensen, and
    Westergaard 1958] ....ssp. *montana* Hartm.
1   Panicle little contracted after flowering; culms usually stoloniferous, decumbent and
    often rooting at the nodes; rhizomes wanting ....ssp. *canina*
    2   Panicle green; [var. *pallida* Rchb.; N.S.: near Glasgow, Pictou Co., where probably
        introd.; D. S. Erskine, Rhodora 53(635): 266. 1951] ....var. *varians* Asch. & Graebn.
    2   Panicle purplish ....var. *canina*
        3   Lemmas awnless; [Newcastle, N.B., and Glenwood, Nfld.] ....f. *mutica* (Gaud.) Döll
        3   Lemmas dorsally awned.
            4   Stolon-nodes bearing clusters of small leaves; [type from Cape Forchu, Cape
                Breton Is., N.S.; perhaps merely an autumnal phase] ....f. *fasciculata* Rousseau
            4   Stolon-nodes naked; [*A. ?macounii* Scribn.; *A. ?trinii* Turcz.; introd. in Ont.
                (Ottawa), Nfld., N.B., P.E.I., and N.S., and reported from Victoria, B.C., by
                Henry 1915; a collection in CAN from Cartwright, Labrador, 53°42′N, has been
                placed here by Malte but the report from Labrador by Stearns 1884, is thought
                by St. John 1922, to be based upon *A. borealis*; collections in CAN from P.E.I.
                (Borden and Wood Is.) have been named a hybrid between *A. canina* and *A.
                tenuis* by Malte; MAPS (aggregate species): Hultén 1962: map 197, p. 209;
                Meusel, Jaeger, and Weinert 1965: 47; the map by Hultén 1968b, for *A. trinii*
                indicates the occurrence of this doubtfully distinct taxon in Alaska] ....f. *canina*

A. diegoensis Vasey
/t/W/   (Hsr)   Meadows and open woods at low to moderate elevations from sw B.C. (Vancouver Is. and adjacent islands and mainland; CAN; V) and Mont. to s Calif. and Nev. [*A. foliosa* Vasey; *A. pallens foliosa* (Vasey) Hitchc.; *A. pallens sensu* Eastham 1947, not Trin.].

A. exarata Trin.   Spike Redtop
/sT/WW/   (Hs)   Moist or dryish open places at low to moderate elevations, the aggregate species from the Aleutian Is. (type from Unalaska) and s Alaska to NW Dist. Mackenzie (Porsild and Cody 1968), s through B.C.–Alta. and s Sask. to Calif., Mexico, and Tex. MAP and synonymy: *see* below.
1   Lemmas dorsally awned; [*A. inflata* Scribn.; *A. microphylla* var. *major* Vasey; *A. ?micro.*
    as to the Vancouver Is. report by Hitchcock et al. 1969, perhaps not Steud.; s B.C.:
    John Macoun 1888] ....var. *monolepis* (Torr.) Hitchc.
1   Lemmas awnless.
    2   Panicle relatively small; leaves relatively narrow; [Aleutian Is. (type locality) and
        s Alaska] ....var. *purpurascens* Hult.
    2   Panicle to 2.5 dm long; leaves to 8 mm broad; [var. *minor* Hook.; *A. alaskana* Hult. and
        its var. *breviflora* Hult.; *A. drummondii* Torr.; *A. asperifolia, A. grandis,* and *A. scouleri*
        Trin.; *A. canina* vars. *aenea* and *melaleuca sensu* Bongard 1833, in part, not Trin.; *A.
        oregonensis* Nutt., not Vasey (the report of which from Vancouver Is. by Hitchcock et
        al. 1969, requires confirmation); *A. glomerata sensu* Carter and Newcombe 1921, not
        (Presl) Kunth; Aleutian Is. (type from Unalaska) and s Alaska (*see* Hultén 1942: map
        109, p. 392), B.C., Alta., and Sask. (Cypress Hills, Swift Current, and Mortlach;
        Breitung 1957a; concerning reports from Man., *see* Scoggan 1957); MAP (aggregate
        species): Hultén 1968b: 101] ....var. *exarata*

[A. humilis Vasey]
[Alaska and B.C. are included in the range assigned this species of the w U.S.A. (Wash. and Mont. to Nev. and Colo.) by Henry (1915) and Hitchcock et al. (1969), respectively, and it is reported from B.C. by John Macoun (1890; Mt. Queest and Griffin L. in the Kamloops dist.) and Hubbard (1955; Mt. Garibaldi, N of Vancouver, and Windermere, Columbia Valley). None of the above collections have been located, however, and one by Macoun from Rogers Pass, B.C., originally placed here,

was later referred to *Podagrostis thurberiana* by Hitchcock. Further studies should be made before this species is admitted to our flora.]

A. hyemalis (Walt.) BSP.   Hairgrass, Ticklegrass.   Foin fou
/ST/X/eA/   (Hs)   Sterile wet or dry open soil, the aggregate species from the Aleutian Is. and N-cent. Alaska (*see* Hultén 1942: maps 111a, b, and c) to s Yukon, Great Bear L., L. Athabasca (Alta. and Sask.), s Dist. ?Keewatin (Boivin 1967a), Man. (N to Baralzon L. at ca. 60°N), northernmost Ont., Que. (N to the Larch R. at 57°35′N), Labrador (N to 58°06′N), Nfld., N.B., P.E.I., and N.S., s to Calif., Mexico, Tex., and Fla.; introd. in s Greenland; E Asia. MAPS and synonymy: *see* below.
1   Spikelets averaging about 1.5 mm long (to 2.4 mm), on pedicels averaging less than 1
       mm long (at most 2 mm long), aggregated into small terhinal clusters; anthers roundish,
       less than 0.5 mm long; [*Trichodium (A.) laxiflorum* Michx.; alternatively spelled
       "*hiemalis*"; U.S.A. only] . . . . .var. *hyemalis*
1   Spikelets 2 or 3 mm long, on pedicels averaging 2 mm long (to 5 mm), more loosely
       arranged.
    2   Panicle relatively compact, rarely 1/3 of the total height of the plant, to about 2 dm
           long; culm-leaves 1 or 2 mm broad . . . . .var. *geminata* (Trin.) Hitchc.
        3   Lemmas dorsally awned; [*A. geminata* Trin., the type from Unalaska, Alaska;
               A. scabra var. *gem.* (Trin.) Swallen; transcontinental; MAP: Hultén 1968b: 102
               (*A. gem.*)] . . . . .f. *geminata*
        3   Lemmas awnless; [*A. gem.* f. *ex.* Fern., the type from the Madeleine R., Gaspé
               Pen., E Que.; range of f. *geminata*] . . . . .f. *exaristata* (Fern.) Scoggan
    2   Panicle very diffuse at maturity, to over 2/3 the total height, to about 4 dm long;
           culm-leaves to 5 mm broad . . . . .var. *tenuis* (Tuck.) Gl.
        4   Lemmas dorsally awned; [*A. scabra* f. *tuck.* Fern.; *A. scabra* var. *septentrionalis* f.
               ?setigera Fern.; w and E James Bay.; Magdalen Is., E Que.; Nfld.] . . . .
               . . . . .f. *tuckermanii* (Fern.) Scoggan
        4   Lemmas awnless; [not *A. tenuis* Sibth.; *A. (Trichodium) scabra* Willd. and its var.
               septentrionalis Fern.; *A. nutkaensis* Kunth; *A. nootkaensis* Trin.; *Trichodium
               album* Presl; transcontinental; MAPS (*A. scabra*): Hultén 1968b: 102; Raup 1947:
               pl. 16] . . . . .f. *tenuis*

[*A. idahoensis* Nash]
[The report of this species of the w U.S.A. (N to Wash. and Mont.) from Fairbanks, Alaska, by Hultén (1942) was later referred by Hultén (1968b; with map) to *A. clavata* Trin. It is tentatively reported from Dist. Mackenzie by W. J. Cody (Can. Field-Nat. 75(2): 58. 1961; Fort Simpson). Collections in CAN from Jasper Park and Mountain Park, sw Alta., originally placed here, have been referred by Porsild and Boivin to *A. variabilis,* to which taxon the report from B.C. by Eastham (1947) may also refer. The report from the Gaspé Pen., E Que., by Fernald (1925) is based upon *A. hyemalis* f. *tenuis,* the relevant collection in GH.]

A. interrupta L.
European; reported from the Okanagan Valley, B.C., by A. S. Hitchcock (1935) and also from B.C. (Nanaimo, Vancouver Is.; Lumby, near Osoyoos; near Nelson) by Hubbard (1955) and Eastham (1947). [*Apera* Beauv.].

[*A. pallens* Trin.]
[The report of this species of the w U.S.A. (Wash. to Calif.) from sw B.C. by Eastham (1947; Lulu Is. and Vancouver) has been referred above to the scarcely distinct *A. diegoensis.*]

A. perennans (Walt.) Tuckerm.   Upland Bent
/T/EE/   (Hs)   Open woods, thickets, and dryish open soil (ranges of Canadian taxa outlined below), s to Tex. and Fla. (possible relict stations reported from Wash. and Oreg. by C. G. Carlbom, Madroño 20(4): 220. 1969).
1   Panicle-branches ascending to divergent; pedicels about equalling or shorter than the
       spikelets; ligules to 5 mm long; culms rather firm . . . . .var. *perennans*

2 Lemmas awnless; [*Cornucopiae* Walt.; *A. oreophila* Trin. in part; *Trichodium (A.)*
decumbens Michx.; Ont. (N to near L. Nipissing and Ottawa), Que. (N to the
Gaspé Pen.), Nfld. (Tompkins; ACAD; reported from Gander by Bowden 1960), N.B.,
P.E.I., and N.S.] . . . . .f. *perennans*
     2 Lemmas dorsally awned; [not yet known from Canada] . . . .[f. *chaetophora* Fern.]
1 Panicle-branches (and branchlets) divergent, the panicle very lax; many of the pedicels
1–3 times the length of the spikelets; ligules to 3 mm long; culms weak; (perhaps merely
a minor shade form) . . . . .var. *aestivalis* Vasey
     3 Lemmas awnless; [Que. and N.S.] . . . . .f. *aestivalis*
     3 Lemmas dorsally awned; [Lac Tremblant, Terrabonne Co., Que.; Fernald 1933] . . . .
     . . . . .f. *atherophora* Fern.

[*A. rossiae* Vasey]
[An 1890 collection from Rogers Pass, B.C., by J. M. Macoun, distributed as *A. alpina* but later
revised to *A. varians* by A. S. Hitchcock and to *A. rossiae* by Malte, is referred by Porsild to *A.
variabilis,* to which other reports from B.C. and Alta. probably refer. The report from Bonne Bay,
Nfld., by Fernald (1933) is referred by Eilif Dahl (Rhodora 64(758): 117–18. 1962) to the Eurasian
A. tenuis. A. rossiae is an endemic of Yellowstone National Park, Wyo.]

A. stolonifera L.   Redtop.   Foin follette
Eurasian (but considered by Gray 1950, under *A. alba,* to be indigenous northwards in N.
America), the Alaska–Canada ranges, MAPS, and synonymy given below.
1 Plant with prolonged subterranean scaly rhizomes and decumbent leafy shoots; panicle
usually purple, to 3 dm long, its branches spreading in fruit; leaves to 8(9) mm broad;
ligules to 4 mm long; culms to 1.5 dm tall; [*A. alba* var. *major* Gaud.; *A. gigantea* Roth and
its var. *dispar* (Michx.) Philipson (*A. dispar* Michx.); introd. in s Alaska, the Yukon
(Dawson), B.C., ?Alta., Man. (N to Grand Rapids), Ont. (Thunder Bay; Ottawa; Kingston),
E Que. (where considered native by Fernald *in* Gray 1950), N.B., P.E.I., and N.S.;
southernmost Greenland; MAP: Hultén 1968*b*: 100 (*A. gig.*)] . . . . .var. *major* (Gaud.) Farw.
1 Plant not rhizomatous, the creeping superficial stolons, when present, bearing erect leafy
shoots; panicle usually stramineous, its branches becoming suberect to closely
appressed in fruit; ligules to 6 mm long; [introd., transcontinental].
     2 Plant with more or less extensively creeping superficial stolons; panicle cylindric,
compact, to about 2 dm long, its strongly ascending to appressed branches to about 4
cm long; leaves to about 5 mm broad; [*A. alba* and its vars. *maritima* (Lam.) Mey. and
palustris (Huds.) Pers.; *A. maritima* Lam.; *A. palustris* Huds.; *A. stol.* var. *compacta*
Hartm.; Aleutian Is.–Alaska (Sitka; Seward; Popoff Is.) and B.C. to Alta., Sask. (N to
Hudson Bay Junction, 52°52′N), Man. (N to Churchill), Ont. (N to w James Bay), Que.
(N to E James Bay and the Côte-Nord), SE Labrador (N to Cartwright, 53°42′N),
Nfld., N.B., P.E.I., and N.S.; sw Greenland] . . . . .var. *palustris* (Huds.) Farw.
     2 Plant seldom with definite stolons, the culms ascending from a decumbent, often
rooting base; panicle often looser (but its relatively long branches becoming suberect
in fruit), to about 3 dm long; leaves to about 8 mm broad; [*A. alba* of auth., not L.,
which was based on a species of *Poa*; range of var. *palustris* but less northern; MAPS
(aggregate species): Hultén 1962: map 188, p. 199, and 1968*b*: 100). A hybrid
between *A. stolonifera* and *A. (stol.* var.) *tenuis* is reported from P.E.I. by Malte (1928)
and a collection in CAN from St. Paul Is., N.S., has been referred to it. An evident
hybrid with *Polypogon monspeliensis* (× *Agropogon littoralis* (With.) Hubbard;
Polypogon littoralis (With.) Sm.; *P. interruptus* HBK.) is reported from sw B.C. by John
Macoun (1888, *P. litt.*; Sidney and Victoria, Vancouver Is.] . . . . .var. *stolonifera*

A. tenuis Sibth.   Rhode Island Bent
Eurasian; Alaska–Canada distribution, MAPS, and synonymy: *see* below.
1 Culms less than 2 dm tall, densely tufted; panicle usually less than 4 cm long; stolons
short or none; [perhaps a mere pathological form, the fruits often blackened by smut;
E Que. (Gaspé Pen.), St-Pierre and Miquelon, and Nfld.] . . . . .var. *pumila* (L.) Druce

1 Culms to about 8 dm tall, loosely to densely tufted; panicle to about 2 dm long; plant
   often stoloniferous . . . . .var. *tenuis*
      2 Lemmas dorsally awned; [incl. var. *setulosa* (Murb.) Holmb.; *A. stricta* Willd., not
         Gmel.; range of f. *tenuis*] . . . . .f. *aristata* (Sincl.) Wieg.
      2 Lemmas awnless; [*A. sylvatica* Huds.; *A. vulgaris* With.; *A. alba* vars. *sylv.* (Huds.)
         Sm. and *vulg.* (With.) Thurb.; introd. in s Alaska, B.C. (N to 57°25′N), Ont. (N to
         Ottawa), Que. (N to the Côte-Nord), SE Labrador (Goose Bay), Nfld., N.B., P.E.I., and
         N.S.; southernmost Greenland; MAPS (aggregate species): Hultén 1962: map 204,
         p. 215, and 1968*b*: 99] . . . . .f. *tenuis*

A. variabilis Rydb.
/T/W/ (Hs) Rocky creeks and mountain slopes at high altitudes from B.C. (reported N to Queen
Charlotte Is. by John Macoun 1888, as *A. varians*, but his report from Sitka, SE Alaska, is referred
by Hultén 1942, to *A. exarata; A. rossiae* reported from Garibaldi, Kokanee, Nelson, McBride,
Manning Park, and Hope–Princeton by Eastham 1947; *see* above under this species) and sw Alta.
(N to Jasper) to Calif. and Colo. [*A. rossiae* of B.C. and Alta. reports, not Vasey; *A. ?varians* Trin.,
not Thuill.].

AIRA L.   [265]   Hairgrass

1 Panicle open, to about 7 cm long, the silvery shining spikelets about 3 mm long, clustered
   near the ends of the spreading capillary branches; (introd. in s Yukon and sw B.C.) . . . .
   . . . .*A. caryophyllea*
1 Panicle dense and spike-like, to about 3 cm long; spikelets yellowish and shining, to
   4 mm long; (introd. in sw B.C. and N.S.) . . . .*A. praecox*

A. caryophyllea L.   Silver Hairgrass
European; introd. in s-cent. Yukon (Hultén 1950; between Bear Creek and Champagne) and sw
B.C. (Queen Charlotte Is.; Vancouver Is. and adjacent islands and mainland). MAPS: Hultén 1968*b*:
109; Meusel, Jaeger, and Weinert 1965: 50.

A. praecox L.
European; introd. in sw B.C. (Vancouver Is. and adjacent mainland at New Westminster; CAN;
John Macoun 1888) and in N.S. (Seal Is. and Mud. Is., Yarmouth Co.; NSPM).

ALOPECURUS L.   [225]   Foxtail.   Vulpin

1 Spikelets 5 or 6 mm long; glumes acute; (introd.).
      2 Lemma much shorter than the divergent-tipped glumes, its short awn included;
         spikelets more or less urn-shaped; culms mostly solitary from the tips of long stolons;
         perennial . . . .*A. ventricosus*
      2 Lemmas and glumes subequal, 5 or 6 mm long, the awn exserted at least 2 mm;
         spikelets elliptic or oblong-ovate; culms tufted, arising from among tussocks of
         leaves.
            3 Panicle slender, tapering at each end; glumes scabrous on the keel; lemma-awn
               exserted up to 8 mm; annual . . . .[*A. myosuroides*]
            3 Panicle cylindric, dense; glumes long-ciliate on the keel and midnerves; lemma-
               awn exserted at most about 5 mm; perennial with creeping rhizomes . . . .*A. pratensis*
1 Spikelets at most 4 mm long (rarely 5 mm in the annual *A. saccatus*).
      4 Glumes densely woolly all over, to 4 mm long; panicle oblong, to about 5 cm long and
         1 cm thick; lemma-awn inserted near base, exserted only slightly or up to 5 mm;
         leaves to 5 mm broad; perennial; (transcontinental in arctic, subarctic, and alpine
         regions) . . . .*A. alpinus*
      4 Glumes merely long-ciliate or pilose.
            5 Awn included or exserted not over 1.5 mm beyond the glumes, attached at or near
               the middle of the lemma; spikelets about 2 mm long; anthers about 1 mm long;

panicle slender, to about 8 cm long and 5.5 mm thick; perennial; (transcontinental) . . . .*A. aequalis*

5   Awn exserted at least 2 mm, attached toward the base of the lemma.

   6   Spikelets 4 or 5 mm long; lemma-awn exserted to 8 mm; anthers about 1 mm long; panicle to about 4 cm long, relatively loose; annual; (s ?B.C.) . . . .[*A. saccatus*]

   6   Spikelets 2 or 3 mm long; panicle narrow and very dense, to about 7.5 cm long.

      7   Anthers about 0.5 mm long; lemma-awn exserted 2 or 3 mm; annual, the tufted culms much branched at base; (s B.C. and s Sask.) . . . .*A. carolinianus*

      7   Anthers to about 2 mm long; plants perennial, the decumbent culms rooting at the nodes; (introd., transcontinental) . . . .*A. geniculatus*

A. aequalis Sobol.
/aST/X/GEA/ (Hs) Shallow water, shores, ditches, etc., the aggregate species from the Aleutian Is. and N-cent. Alaska to cent. Yukon, Great Bear L., L. Athabasca (Alta. and Sask.), Man. (N to Reindeer L. at 57°48′N), Ont. (N to the Winisk R. at ca. 55°N), Que. (N to Ungava Bay), Labrador (N to Saglek Bay, 58°34′N), Nfld., N.B., P.E.I., and N.S., s to Calif., N.Mex., Kans., Ohio, and Md.; w Greenland N to ca. 70°30′N; isolated in E Greenland near the Arctic Circle; Iceland; Eurasia; N Africa. MAPS and synonymy: *see* below.

1   Leaf-sheaths only slightly inflated; panicle to about 8 cm long, finally long-exserted; [*A. aristulatus* Michx,; *A. geniculatus* vars. *arist.* (Michx.) Torr. and *robustus* Vasey; *A. caespitosus* Trin.; *A. subaristatus* Pers.; transcontinental; MAPS (aggregate species): Hultén 1968b: 91, and 1962: map 100, p. 109] . . . . .var. *aequalis*

1   Leaf-sheaths inflated; panicle usually not over 3.5 cm long, its base often included in the upper leaf-sheath; [*A. aristulatus* var. *merriamii sensu* St. John 1922, not *A. howellii* var. *merriamii* Beal, basionym; Alaska; E Que. (Côte-Nord), Labrador (Saglek Bay, 58°34′N; CAN; reported from the Hamilton R. basin by Abbe 1955), and NW Nfld.; w Greenland N to ca. 70°N] . . . . .var. *natans* (Wahl.) Fern.

A. alpinus Sm.   Alpine foxtail
/AST/X/GEA/ (Grh) Meadows and margins of streams (often alpine) from the coasts of Alaska–Yukon–Dist. Mackenzie throughout the Canadian Arctic Archipelago to northernmost Ellesmere Is. and northernmost Ungava–Labrador, s to s-cent. Yukon, SE Dist. Keewatin, NE Man. (Churchill), and northernmost Ont. (mouth of the Black Duck R., Hudson Bay, ca. 56°50′N), and in the mts. through B.C. to Utah and Colo.; isolated in the Cypress Hills of sw Sask.; circumgreenlandic (but a large gap in SE Greenland); Eurasia. [Incl. *A. behringianus* Gand., *A. ?borealis* Trin., *A. glaucus* Less., *A. occidentalis* Scribn. & Tweedy, and *A. stejnegeri* Vasey]. MAPS: Hultén 1968b: 90; Porsild 1957: map 15, p. 162; Savile 1961: map A, p. 928.

A. carolinianus Walt.
/T/X/ (T) Shores, ditches, fields, and low grounds (probably introd. in the West) from sw B.C. (Vancouver Is. and Yale, Fraser Valley; CAN; not known from Alta.) to s Sask. (Wood Mountain Trail; CAN), Wisc., Ind., Ohio, Pa., and N.J., s to Calif., N.Mex., Tex., and Fla. [*A. macounii* Vasey; *A. geniculatus* var. *caespitosus* Scribn.].

A. *geniculatus* L.
Eurasian; introd. in ditches, pools, and wet clearings from s Alaska–B.C. to Alta. (reported from Sask. by Boivin 1967a, but not listed by Breitung 1957a; reports from s Man. by Lowe 1943, probably refer to *A. aequalis*), Ont. (Essex and Northumberland counties), Que. (N to the Côte-Nord), s ?Labrador (Boivin 1967a), Nfld. (Rouleau 1956), N.B., P.E.I., and N.S.; s Greenland. [Incl. var. *microstachyus* Uechtr.; *A. pallescens* Piper & Beattie]. MAPS: Hultén 1968b: 92, and 1962: map 203, p. 215.

[A. *myosuroides* Huds.]
[Eurasian; introd., probably in seed of other grasses, at Experimental Stations in sw B.C. (Saanichton, Vancouver Is.; Hubbard 1955) and sw Man. (Brandon; G. A. Stevenson, Can. Field-Nat. 79(3): 174. 1965), but not established.]

A. pratensis L.   Meadow Foxtail
Eurasian; commonly cult, and escaped to meadows, pastures, and damp clearings from s Alaska (Skagway) and B.C. (Vancouver Is.) to Alta. (Edson; not known from Sask.), s Man. (Brandon; Winnipeg), Ont. (N to New Liskeard, 47°31′N), Que. (N to Anticosti Is. and the Gaspé Pen.), Labrador (N to Venison Tickle, 52°57′N), Nfld., N.B., P.E.I., and N.S. MAPS: Hultén 1968b: 89, and 1962: map 201, p. 213.

Forma *breviaristatus* Beck (the short-awned extreme) is reported from Nfld. by W. M. Bowden (Can. J. Bot. 38(4): 546. 1960; Stephenville).

[A. saccatus Vasey]
[The report of this species of the w U.S.A. (Wash. to Calif.) from s B.C. by Henry (1915; Vancouver Is. and Yale) is probably based upon *A. carolinianus* (see above). (*A. californicus* and *A. howellii* Vasey).]

A. ventricosus Pers.   Creeping Foxtail
Eurasian; known in N. America only from s Nfld. (Upper Ferry, Codroy Valley, and Mt. Pearl, near St. John's; I.J. Green, Rhodora 46(550): 386–87. 1944).

AMMOPHILA Host   [249]   Sand-reed, Psamma, Marram

1   Leaves puberulent above, their ligules thin, to about 3 cm long; panicle to about 2 dm
    long; callus-hairs to about 3 mm long; (introd. in B.C.) . . . .*A. arenaria*
1   Leaves merely scabrous on the nerves above, their ligules firm and chartaceous, at most
    3 mm long; panicle to 4 dm long; callus-hairs mostly not over 2 mm long; (sandy
    shores of the Great Lakes and Atlantic Ocean) . . . .*A. breviligulata*

A. arenaria (L.) Link   Marram Grass
European; this valuable sand-binder is reported by Eastham (1947) as introd. for that purpose in B.C. at Clayoquot, Vancouver Is., and apparently established following spontaneous introduction on Sidney Is., near Victoria. [*Arundo* L.].

A. breviligulata Fern.   Beachgrass.   Gourbet or Oyat
/T/EE/   (Grh)   Coastal sands and dunes of Ont. (shores of the Great Lakes) and from E Que. (St. Lawrence R. estuary from l'Islet Co. to the Côte-Nord, Gaspé Pen. and Magdalen Is.; also known from L. St. John and L. Champlain) to w Nfld., N.B. (Buctouche; St. John; Miscou Is.), P.E.I., and N.S. (the report from s Labrador by Fernald *in* Gray 1950, may refer to the Côte-Nord, E Que.). [*A. (Calamagrostis) arenaria* of E Canadian reports, not (L.) Link; *A. arundinacea* of E Canadian reports, not Host].

ANDROPOGON L.   [134]   Beardgrass, Bluestem.   Bardon

1   Racemes solitary at the tip of each peduncle; pedicels and rachis-joints upwardly
    enlarged; awns to 14 mm long; leaves to 7 mm broad, their sheaths strongly keeled;
    (B.C. to N.S.) . . . .*A. scoparius*
1   Racemes in digitate clusters of 2 or more at the tip of each peduncle; pedicels and
    rachis-joints not upwardly enlarged; leaves to 1 cm broad, their sheaths not strongly
    keeled.
    2   Awn of sessile (fertile) spikelet wanting or at most 6 mm long, pointed straight
        forward; rachis-joint and sterile pedicel densely long-villous; rhizome elongate;
        plant glaucous; (sw Man.; introd. in Ont. and Que.) . . . .*A. hallii*
    2   Awn of sessile spikelet at least 8 mm long, geniculate and twisted toward base;
        rachis-joint and sterile pedicel ciliate, the joints short-hispid at base; rhizomes short;
        plant scarcely glaucous; (SE Sask. to Que.) . . . .*A. gerardii*

A. gerardii Vitman   Big Bluestem
/T/(X)/   (Hs)   Prairies, shores, and dry open ground from SE Sask. (Carlyle, St. Hubert, Wapella, and Yorkton; Breitung 1957a) to Man. (N to Grand Rapids, near the NW end of L. Winnipeg), Ont. (N

to the Albany R. s of James Bay at 51°34′N), and Que. (N to St-Vallier, l'Islet Co.; MT), s to Ariz., N Mexico, Tex., and Fla. [*A. furcatus* Muhl.; *A. provincialis* Lam., not Retz.].

A. hallii Hack.   Sand Bluestem
/T/WW/   (Hs)   Prairies, sandhills, ravines, and dry ground from E Mont. to sw Man. (N to St. Lazare, about 75 mi NW of Brandon), s to Ariz., Tex., Kans., and Iowa; introd. farther eastwards, as in s Ont. (Point Pelee, Essex Co.; GH) and s Que. (N to Grosse-Ile, Montmagny; MT; GH). [*A. chrysocomus* Nash; incl. var. *incanescens* Hack. (*A. gerardii* var. *inc.* (Hack.) Boivin)].

A. scoparius Michx.   Broom-Beardgrass
/T/X/   (Hs)   Dry to moist prairies, clearings, and open woods (ranges of Canadian taxa outline below), s to Ariz., Mexico, Tex., and Fla.
1   Basal third of the rachis-joints beardless; [*Schizachyrium* Nash; SE B.C. (Fairmont Hot Springs), s Alta. (Calgary and Morley; John Macoun 1888), Sask. (Breitung 1957*a*), and Man. (N to Duck Mt.)] . . . . .var. *scoparius*
1   Rachis-joints bearded nearly to base.
    2   Glumes of fertile spikelets at most 6 mm long; beard of rachis-joints comparatively sparse and short; [s Alta. (Bow River; CAN); s Ont. (Welland Co.: reported from Point Abino by House 1930, and from Port Colborne by Zenkert 1934)] . . . .var. *frequens* Hubbard
    2   Glumes of fertile spikelets to 11 mm long; beard of rachis-joints abundant and long.
        3   Racemes with not more than 10 fertile spikelets, usually very flexuous; [Ont., the type from Shirleys Bay, near Ottawa), Que. (N to Pontiac Bay, Pontiac Co.), N.B. (Victoria and York counties), and N.S. (Wolfville, Kings Co.)] . . . . .var. *septentrionalis* Fern.
        3   Raceme with at least 11 fertile florets, rarely flexuous.
            4   Inflorescence elongate and racemiform; sterile rudiment beyond the upper-most floret less than 5 mm long; [s Ont. (Long Point, Norfolk Co.) and sw Que. (Ironside, Gatineau Co.; Fernald 1935)] . . . . .var. *neomexicanus* (Nash) Hitchc.
            4   Inflorescence shorter and with numerous appressed branches; sterile rudiment to about 8 mm long; [*A. littoralis* Nash; reported from the Great Lakes region of s Ont. by Boivin 1967*a*] . . . . .var. *littoralis* (Nash) Hitchc.

ANTHOXANTHUM L.   [205]   Vernal Grass

1   Panicle whitish green, to about 4 cm long; spikelets about 6 mm long; glumes essentially glabrous; awns of both sterile lemmas more or less geniculate near the blackish base and protruding from the glumes; leaves to 3 mm broad; culms less than 4 dm tall, strongly bushy-branched; annual . . . .*A. aristatum*
1   Panicle brownish green, to about 7 cm long; spikelets about 1 cm long; glumes scabrous on the keel to villous throughout; awn of lower sterile lemma straight, not protruding from the glumes; leaves to 7 mm broad; culms to about 1 m tall, unbranched or weakly branched; perennial, sweetly fragrant on drying . . . .*A. odoratum*

A. aristatum Boiss.   Annual Vernal Grass
European; introd. in sw B.C. (Vancouver Is. and Barnston Is.; Eastham 1947) and reported from Ont. by Fernald *in* Gray (1950). [*A. puelii* Lec. & Lam.; *A. odoratum* var. *puelii* (Lec. & Lam.) Coss. & Dur.].

A. odoratum L.   Sweet Vernal Grass.   Flouve or Foin d'odeur
Eurasian; natzd. in fields, pastures, and waste places from s Alaska (Unalaska and Sitka) and B.C. (Boivin 1967*a*) to Calif. and from Ont. (N to Algonquin Park and the Ottawa dist.) to Que. (N to the Gaspé Pen.; reported N to the Côte-Nord by Saint-Cyr 1887), Nfld., N.B., P.E.I., and N.S.; s Greenland. [*A. alpinum* Löve & Löve]. MAPS: Hultén 1968*b*: 83, and 1962: map 194, p. 205.
    Forma *giganteum* Junge (panicle to 14 cm long, rather than at most about 7 cm; spikelets to 12 mm long, rather than at most 10 mm) is known from N.S. (near New Glasgow, Pictou Co.; DAO; GH).

ARCTAGROSTIS Griseb. [240]

A. latifolia (R. Br.) Griseb.
/AS/X/GEA/ (Grh) Damp to marshy tundra from the coasts of Alaska–Yukon–Dist. Mackenzie, throughout the Canadian Arctic Archipelago (type from Melville Is.) to northernmost Ellesmere Is., Baffin Is., and northernmost Ungava–Labrador, s to n B.C. (s to the Wicked R. at 56°03′N), n Alta. (s to Wood Buffalo National Park at 58°54′N), n ?Sask. (Hultén 1942; *A. poaeoides*), n Man. (s to the mouth of the Black Duck R., Hudson Bay, ca. 56°50′N), and n Que. (s to s Ungava Bay); w and e Greenland n to ca. 69°N; Iceland; n Eurasia. maps and synonymy: *see* below.

1   Outer glume 4 or 5 mm long; [type from Foxe Is., s Baffin Is.; also known from Akpatok Is., Ungava Bay, n Que.] . . . . .var. *longiglumis* Polunin
1   Outer glume to 2.5 mm long.
  2   Panicle to 2 dm long and 5 cm thick, pale green to purplish; culms to 1.5 dm long; [incl. vars. *crassispica* Bowden, *nahannensis* Porsild, and *?alaskensis* Vasey; *Vilfa (Arctagrostis; Colpodium) arund.* Trin., the type from Kotzebue Sound, Alaska; *A. macrophylla* Nash; Alaska–Yukon–w Dist. Mackenzie, n B.C. (s to the Wicked R. at 56°03′N), and n Alta. (s to Wood Buffalo National Park at 58°54′N); maps: Hultén 1968*b*: 94; Raup 1947: pl. 16 (*A. arund.*)] . . . . .var. *arundinacea* (Trin.) Griseb.
  2   Panicle to 12 cm long and 2.5 cm thick, pale green or purplish to dark brown-purple; culms mostly not over 5 dm tall . . . . .var. *latifolia*
    3   Lemmas tipped with a short but distinct dark awn; [e Dist. Keewatin, Southampton Is., and Baffin Is.; Polunin 1940] . . . . .f. *aristata* Holmb.
    3   Lemmas awnless; [*Colpodium* R. Br.; incl. var. *angustifolia* (Nash) Hult. (*A. angust.* Nash) and *A. poaeoides* Nash; transcontinental; maps: Porsild 1957: map 17, p. 163, and 1951*b*: fig. 2 (aggregate species), p. 142; Hultén 1962: map 12 (aggregate species), p. 19 and 1968*b*: 94] . . . . .f. *latifolia*

ARCTOPHILA Rupr. [379]

A. fulva (Trin.) Rupr.
/ASs/X/GEA/ (Hel (Hsr)) Marshy tundra and margins of tundra pools from the coasts of Alaska–Yukon–Dist. Mackenzie, to Prince Patrick Is., Boothia Pen., and s-cent. Baffin Is., s to Alaska–Yukon–Dist. Mackenzie, n Man. (s to Churchill), n Ont. (s to Cape Henrietta Maria, nw James Bay), and n Que. (s to e Hudson Bay at ca. 58°N); an isolated station reported by Abbe 1955, on L. Melville, Hamilton R. basin, s-cent. Labrador; another isolated station possible in cent. Ont., a collection by Bell in 1882 (type of *A. gracilis* Holm; CAN) being labelled "N. of Lake Superior"; w Greenland between ca. 63° and 66°N; n Eurasia. [*Poa (Colpodium) fulva* Trin., the type from Eschscholtz Bay, nw Alaska; *A. brizoides, A. chrysantha,* and *A. gracilis* Holm; *A. effusa* Lange and its f. *depauperata* Nath.; *A. laestadii* Rupr.; *A. mucronata* of American auth., not Hack.]. maps: Hultén 1968*b*: 149, and 1962: map 9, p. 17; Porsild 1957: map 41, p. 166; Tolmachev 1952: map 9 (very incomplete for N. America).

Forma *aristata* (Polunin) Scoggan (*Colpodium fulvum* f. *arist.* Polunin; *A. trichopoda* Holm, in part; lemmas distinctly short-awned rather than awnless) is known from the type locality, Lake Harbour, s Baffin Is., and from Mansel Is. and Nottingham Is., n Hudson Bay.

ARISTIDA L. [208] Triple-awned Grass, Needlegrass

1   Lemma-awns unequally divergent, the central one (at least when dry) coiled near base; annuals.
  2   Central awn at most 1 cm long, loosely coiled (usually in a half to a full turn); lateral awns rarely over 3 mm long; glumes subequal, to 1 cm long; leaves to 2 mm broad; leaf-sheaths borne to summit of culm; (s Ont.) . . . .*A. dichotoma*
  2   Central awn to over 1.5 cm long, usually coiled in 1 to 3 full turns near base; lateral awns to over 1 cm long; upper glume to 15 mm long; lower glume to 12 mm long; leaves about 1 mm broad, usually involute; (s ?Man.) . . . .[*A. basiramea*]
1   Lemma-awns about equally divergent at base, not coiled; perennials.
  3   Awns essentially equal, to 7 or 8 cm long; panicle to about 1 dm long; lower glume to

13 mm long; upper glume at least 1.5 cm long; leaves filiform, usually involute; (s B.C.
to sw Man.) . . . . *A. longiseta*
3   Awns unequal, the central one longer than the lateral pair, not much over 3 cm long;
panicle to 2 or 3 dm long, often up to half the total height of the plant.
    4   Panicle to 3 dm long; lower glume to 14 mm long, slightly surpassing the upper
       one; central awn to over 3 cm long; lateral awns to about 2.5 cm long; leaves flat,
       to 4 mm broad; culms to about 7 dm tall; (s Ont.) . . . .*A. purpurascens*
    4   Panicle to 2 dm long; glumes at most 1 cm long, the upper one slightly surpassing
       the lower one; central awn to about 2.5 cm long; lateral awns less than 2 cm long;
       leaves often involute, at most 2 mm broad; culms to about 5 dm tall . . . . . .[*A. intermedia*]

[A. basiramea Engelm.]
[Reports of this species of the U.S.A. (N to N.Dak. and Mich.) from Man. by John Macoun (1888; near Brandon) and Shimek (1927; near MacGregor) require confirmation, being probably based upon some other species, perhaps *Stipa viridula*.]

A. dichotoma Michx.   Poverty-Grass
/T/EE/   (T)   Dry sterile soil from s Ont. (Port Colborne, Welland Co.; John Macoun 1888; Boivin 1967a) to Ohio, N.Y., and New Eng., s to Tex. and Fla.

[A. intermedia Scribn. & Ball]
[Dry soil from Nebr. to s Mich., s to Tex. and Miss. Collections in MT and GH from Long Point, "Leeds" (?Norfolk) Co. (Marie-Victorin et al., No. 45,831, in 1932) have been placed here but require further study.]

A. longiseta Steud.
/T/WW/   (Hs)   Dry plains and foothills from s B.C. (N to Kamloops) to s Alta. (Lethbridge), s Sask. (Val Marie), and sw Man. (Aweme, about 30 mi SE of Brandon), s to N Mexico, Tex., Iowa, and Minn.; possibly introd. in Alaska according to Hultén (1942; *A. purpurea*).
1   Culms mostly not over 3 dm tall; leaves to 1.5 dm long; panicle loose; lemma-awns to
8 cm long; (w U.S.A. only) . . . . .var. *longiseta*
1   Culms to about 5 dm tall; leaves averaging somewhat longer; panicle relatively stiff and
compact; lemma-awns to 6 cm long; (*A. purpurea sensu* Hultén 1942, and John Macoun 1888, not Nutt.; *A. fasciculata* (*A. adscensionis* L.) *sensu* John Macoun 1890, not Torr.; Canadian range as outlined above) . . . . .var. *robusta* Merr.

A. purpurascens Poir.
/T/EE/   (Hs)   Sandy or gravelly soil from Kans. to Mich. and s Ont. (Squirrel Is., at the mouth of the St. Clair R., Lambton Co.; Dodge 1915; Boivin 1967a), s to Tex. and Fla.

ARRHENATHERUM Beauv.   Oat-Grass

A. elatius (L.) Beauv.   Tall Oat-Grass
Eurasian; fields and roadsides in Alaska (Petersburg) and B.C. (N to Queen Charlotte Is.) and from Ont. (N to Kapuskasing, 49°24′N) to Que. (N to Anticosti Is. and the Gaspé Pen.), Nfld., and N.S. [*Avena elatior* L.; *Arr. avenaceum* (Scop.) Beauv.]. MAPS: Hultén 1968b: 121, and 1962: map 206, p. 217.
    Forma *flavescens* (Niels.) Holmb. (panicle pale yellowish or greenish rather than purplish) is known from Que. (Kapuskasing; CAN; MT).

AVENA L.   [273]   Oat.   Avoine

1   Lemmas with 2 awn-like scabrous bristles at apex in addition to the dorsal awn (this stout, geniculate, more than twice as long as the lemma), typically more or less pilose on the back; spikelets about 2.5 cm long, mostly 2–3-flowered, all of the florets lacking joints and falling together at maturity; panicle less than 2 dm long, rather open or 1-sided; leaves and sheaths glabrous . . . .[*A. strigosa*]

1   Lemmas merely toothed at apex; panicle to over 2.5 dm long.
   2   All of the florets (2 or 3) of the spikelet jointed and readily falling separately at
     maturity; lemmas pubescent over the back, about 2 cm long, with a stout twisted
     geniculate awn to 4 cm long; glumes to 2.5 cm long; leaves hairy at base and
     marginally ciliate; sheaths of lower leaves pubescent ....*A. fatua*
   2   None of the florets or only the lowest one jointed, the florets all falling together at
     maturity.
      3   Lemmas densely bristly-hairy over the back in the lower half, the basal hairs about
        5 mm long; lemma-awn twisted and flexuous, to 7 cm long; glumes to 5 cm long;
        spikelets with up to 5 florets, the lowest floret jointed; leaves marginally ciliate ....
        ....*A. sterilis*
      3   Lemmas glabrous or sparingly hairy toward base, awnless or with a nearly straight
        awn; glumes to 2.5 cm long; spikelets with 2 or 3 florets, none of these jointed;
        leaves and sheaths glabrous ....*A. sativa*

A. fatua L.   Wild Oat
Eurasian; fields and waste places of SE Alaska (Juneau), S Dist. Mackenzie (Boivin 1967a), B.C., Alta. (N to Wood Buffalo National Park), Sask. (N to ca. 55°N), Man. (N to Churchill), Ont. (N to Longlac, N of L. Superior at 49°47′N), Que. (N to L. St. John and the Côte-Nord), Nfld., N.B., P.E.I., and N.S. [Forma *pilosissima* S. F. Gray]. MAPS: Hultén 1968b: 119, and 1962: map 211, p. 223; B. R. Baum, Can. J. Bot. 46(8): figs. 2 and 4, p. 1014. 1968 (Baum's fig. 3 also indicating the occurrence of putative hybrids with *A. sativa* throughout Canada (except Man. and P.E.I.) and in the Yukon).

A. sativa L.   Oat.   Avoine
Eurasian; persisting after cultivation or spontaneous along roadsides and railways in Alaska–Yukon and all of the provinces; SW Greenland [Incl. *A. annua* L.]. MAPS: Hultén 1968b: 120; B. R. Baum, Can. J. Bot. 46(8): fig. 1 (record of escapes), p. 1014. 1968 (Baum's fig. 5, p. 1015, also outlining the areas of cultivation in Canada).

A. sterilis L.   Animated Oat
Eurasian; reported from Waterloo and adjacent counties in S Ont. by F. H. Montgomery (Can. Field-Nat. 62(2): 80. 1948).
   The common name expresses the action of the fruit in creeping on moist surfaces through hygroscopic movements of the awns, resulting in burying the fruit beneath the soil surface.

[A. strigosa Schreb.]
[The report of this Eurasian species as a weed in cultivated fields near Sooke, Vancouver Is., B.C., by J. M. Macoun (1894) requires confirmation.]

BECKMANNIA Host   [303]

B. syzigachne (Steud.) Fern.   Slough-Grass
/ST/X/GEA/   (T)   Wet ground and shallow water, ssp. *baicalensis* from cent. Alaska–Yukon –Dist. Mackenzie to L. Athabasca (Alta. and Sask.), Man. (N to Churchill), Ont. (N to Fort Severn, Hudson Bay, ca. 56°N), Que. (N to Fort George, E James Bay, ca. 54°N, L. Mistassini, the Côte-Nord, and Gaspé Pen.), P.E.I. (Charlottetown), and N.S. (Bedford, near Halifax; not known from N.B., Labrador, or Nfld.), S to Calif., N.Mex., Kans., Mo., and Pa. (perhaps introd. in the eastern part of the N. American area); introd. in southernmost Greenland; Eurasia. MAPS and synonymy (together with treatments of the typical form of Asia and the Eurasian *B. erucaeformis*, with which it has been united as var. *uniflora*): see below.
1   Perennial with creeping stolons, the culms usually solitary at the rhizome-nodes,
    tuberous-thickened at base; glumes less than 2 mm long, broadest above the middle,
    sparingly short-hispid at the base of the keel, broadly hyaline-margined, short-pointed;
    anthers about 1.5 mm long; panicle-branches simple, erect, spikelet-bearing from base;
    spikelets plump; [Eurasia only; MAP: Hultén 1962: map 164, p. 173] ....
    ....*B. erucaeformis* (L.) Host

1 Annual or biennial, lacking stolons, the fibrous-rooted culms tufted, not tuberous-thickened; glumes about 2.5 mm long; anthers at most 0.7 mm long; panicle-branches naked toward base . . . .*B. syzigachne*

    2 Ligules to 3 mm long, subacute; panicle-branches usually simple (the lower ones occasionally divided), erect; spikelets subrotund, cinereous, slightly compressed; glumes broadest above the middle, smooth, narrowly hyaline-margined, short-pointed; [*Panicum (B.) syz.* Steud.; Asia only; MAPS: Hultén 1962: map 164 (the Asiatic area referred by him to ssp. *baicalensis*), p. 173; Tetsuo Koyama and Shoichi Kawano, Can. J. Bot. 42(7): fig. 12, p. 876. 1964] . . . . .ssp. *syzigachne*

    2 Ligules at least 4 mm long, acute; panicle-branches divided, suberect; spikelets obovate, pale green, compressed; glumes broadest at the middle, short-hispid at the base of the keel, broadly hyaline-margined, acuminate; [*B. eruc.* var. *uniflora* Scribn.; Alaska–Canada range outlined above; MAPS: Hultén 1968*b*: 122, and on the above-noted maps by Hultén and by Koyama and Kawano] . . . .
. . . . .ssp. *baicalensis* (Kusnez.) Koyama & Kawano

BOUTELOUA Lag.   [295]   Grama-Grass, Mesquite-Grass

1 Spikes numerous, at most 2 cm long, spreading or reflexed, each with at most 8 spikelets 7 mm long or more; leaves to 5 mm broad; culms to about 1 m tall; (SE Sask. to s Ont.) . . . .
. . . .*B. curtipendula*

1 Spikes to over 3.5 cm long, usually 2 per culm (sometimes solitary, rarely more than 3 or 4); spikelets 5 or 6 mm long, mostly 35 or more in a dense comb-like group on the outer side of the curved rachis; leaves 1 or 2 mm broad; culms mostly less than 6 dm tall.

    2 Rachis of spike extending beyond the spikelets as a slender bristle to 8 mm long; spikes to about 2 cm long; upper glume hirsute with long spreading tuberculate-based hairs (the tubercles black) . . . .[*B. hirsuta*]

    2 Rachis not prolonged beyond the spikelets; spikes to 4 cm long; glumes only slightly tuberculate-hairy; (B.C. to Man.) . . . .*B. gracilis*

B. curtipendula (Michx.) Torr.   Tall Grama-Grass
/T/X/ (Hsr) Dry prairies and sandhills from Mont. to SE Sask. (Oxbow and Hitchcock; Breitung 1957*a*), s to Stony Mountain, about 25 mi N of Winnipeg), and s Ont. (probably introd.; reported from Heely Falls, Northumberland Co., and from Shannonville Station, Hastings Co., by John Macoun 1888; also known from Spottiswood Lake, Brant Co., and about 10 mi s of Cambridge, Waterloo Co.), s to Calif., Mexico, Tex., Ala., and Ga.; S. America [*Chloris* Michx.; *B. racemosa* Lag.]. MAP: F. W. Gould and Z. J. Kapadia, Brittonia 16: fig. 2, p. 184. 1964).

Concerning the Ont. stations, Montgomery (1956) notes that the Spottiswood Lake plants "seem to be highly sterile . . . and just surviving in the area . . . there is the possibility that it is a relict stand of the prairie intrusion into Ontario".

B. gracilis (HBK.) Lag.   Blue Grama-Grass
/T/WW/ (Hsr) Dry prairies and sandhills from SE B.C. to s Alta., Sask (N to McKague, ca. 52°45′N), and s Man. (N to Bield, s of Duck Mt.), s to Calif., Mexico, Tex., Mo., and N Ill. [*Chondrosium (Eutriana) gracile* HBK.; *Atheropogon (B.; Eutriana) oligostachyus* Nutt.; *B. hirsuta* of Canadian reports, not Lag.]. MAPS: *Atlas of Canada* 1957: map 11, sheet 38.

Forma *pallida* (Scribn.) Boivin (*B. olig.* var. *pall.* Scribn.; spikelets yellowish rather than purplish) is reported from Man. by Boivin (1967*a*; Rathwell, sw of Portage la Prairie).

[*B. hirsuta* Lag.]   Hairy Grama
[Reports of this species of the w U.S.A. (N to S.Dak. and Wisc.) from Canada are apparently all based upon *B. gracilis,* relevant collections in several herbaria.]

BRACHYELYTRUM Beauv.   [216]

B. erectum (Schreb.) Beauv.   Bearded Short-husk
/T/EE/ (Hsr) Woods and thickets from Minn. to Ont. (N to Renfrew, Carleton, and Stormont

counties; reported by John Macoun 1888, from near Port Arthur (Thunder Bay)), Que. (N to the Bell R. at 49°39′N, Anticosti Is., and the Gaspé Pen.), Nfld., N.B., P.E.I., and N.S., s to La., Miss., Ala., and Ga. [Incl. var. *glabratum* (Vasey) Koyama & Kawano (var. *septentrionale* Babel); *Muhlenbergia* Schreb.; *Dilepyrum* Farw.; *D. (M.; B.) ?aristosum* Michx.]. MAP: Tetsuo Koyama and Shoichi Kawano, Can. J. Bot. 42(7): fig. 4 (A), p. 863. 1964.

BRIZA L. [367] Quaking Grass

1 Spikelets rarely more than 8, mostly 1 or 2 cm long and at least 1 cm broad, with up to 20 florets, tinged with golden-pink (becoming brownish); glumes to 6 mm long; leaves to 7 mm broad; ligules to 4 mm long; annual; (introd.) . . . .[*B. maxima*]
1 Spikelets usually more numerous, at most 7 mm long and with less than 10 florets; (introd.).
    2 Spikelets about 3 mm long and with up to 7 florets, whitish green; leaves to 8 mm broad; ligules to 1 cm long; annual . . . .[*B. minor*]
    2 Spikelets to 7 mm long and with up to 9 florets, whitish or tinged with purple; leaves to 4 mm broad; ligules at most about 1 mm long; perennial . . . .*B. media*

[B. maxima L.] Big Quaking Grass
[European; known from s Ont. (London; Boivin 1967a), s Que. (Lotbinière, about 60 mi SE of Quebec City, where taken by Saint-Cyr in 1883; MT) and N.B. (Norton, Kings Co., where taken in a garden by Hay in 1877; NBM). The Ont. and Que. collections may have also been garden specimens or, if escaped, probably not established.]

B. media L. Doddering Dillies
Eurasian; cult. in N. America and known as a garden-escape or casual waif in B.C. (Alberni and Nanaimo, Vancouver Is.), s Ont (near London, Middlesex Co.; Thornbury, Grey Co.), and N.S. (Digby, Yarmouth, and Hants counties); reported from Nfld. by Reeks (1871; 1873).

[B. minor L.] Little Quaking Grass
[European; reported from Canada by A. S. Hitchcock (1935), from B.C. by Jepson (1951), and from sw B.C. by Boivin (1967a), where probably not established.]

BROMUS L. [389] Brome-Grass. Brome
(Ref.: Wagnon 1952)
1 Tip of lemma with its two teeth prolonged to over 2 mm beyond the base of the awn (or *B. brizaeformis* with a very broad obtuse inflated awnless lemma); annuals; (introd.).
    2 Spikelets clavate, dilated toward the apex; lower glume 1-nerved; upper glume 3-nerved; lemmas narrow, lanceolate, tapering to tip, the apical teeth to over 2 mm long; panicle-branches rarely bearing more than 3 spikelets; leaves and sheaths pubescent, the leaves to 4 or 5 mm broad.
        3 Panicle contracted, ovoid, erect, usually purplish, to about 1 dm long, its short branches (much shorter than the subsessile spikelets) and the summit of the culm short-pilose; lemmas about 1.5 cm long, their awns somewhat longer; upper glume about 1 cm long; spikelets to about 4 cm long (including awns); (introd. on Vancouver Is.) . . . .*B. rubens*
        3 Panicle open, its long branches spreading or drooping.
            4 Panicle-branches and summit of culms scabrous but otherwise glabrous; lemmas linear-lanceolate, about 2 cm long, their awns about 3 cm long; upper glume about 1.5 cm long; spikelets to about 6 cm long (including awns); panicle to 2 dm long; (introd. in s B.C. and s Ont.) . . . .*B. sterilis*
            4 Panicle-branches and summit of culm short-pilose; lemmas broadly lanceolate.
                5 Upper glume about 2 cm long; lemmas to 2.5 cm long, their awns to about 5 cm long; spikelets glabrous or more or less puberulent, to about 1 dm long (including awns); panicle to 2 dm long; (introd. in sw B.C.) . . . .*B. rigidus*
                5 Upper glume, lemmas, and lemma-awns each to about 12 mm long; spikelets villous or pilose, 2 or 3 cm long (including awns); panicle to 1.5 dm long; (widely introd.) . . . .*B. tectorum*

2 Spikelets attenuate toward apex; lower glume 3-nerved; upper glume 5–9-nerved; lemmas broad, oblong, elliptic, or oval, the apical teeth mostly less than 1 mm long.

  6 Anthers about 4 mm long; panicle usually narrow, to 3 dm long, finally nodding, more or less purple-tinged, its longer slender scabrous branches mostly bearing about 7 glabrous spikelets (these to 2.5 cm long); lower glume about 4 mm long, the upper one somewhat longer; lemmas to about 8 mm long, the straight awn about as long; leaves to 6 mm broad, they and their sheaths pilose; (introd. in B.C., s Ont., and Que.) . . . . *B. arvensis*

  6 Anthers less than 2.5 mm long; panicle usually not purple-tinged.

    7 Lemmas awnless or with a very short awn less than 3 mm long, broadly obovate-rhombic, obtuse, strongly angular at the margin, to 9 mm long and about as broad; spikelets to about 2.5 cm long; anthers about 1 mm long; panicle to 1 dm long, loose, 1-sided, nodding, few-flowered; leaves to 5 mm broad, they and their sheaths pilose; (introd. in Alaska, B.C., and s Ont.) . . . . *B. brizaeformis*

    7 Lemmas with elongate awns over 2 mm long, narrower in outline.

      8 Pedicels nearly all shorter than their spikelets; lemmas soft-pubescent, prominently nerved, they and their awns each to 9 mm long; panicle erect or contracted, to 1 dm long, its pilose branches short, the lower ones bearing several spikelets; anthers about 2 mm long; leaves to 5 mm broad; plant soft-hairy throughout; (introd., transcontinental) . . . . *B. mollis*

      8 Pedicels nearly all as long as or longer than their spikelets; lemmas with less prominent nerves, usually nearly or quite glabrous; panicle-branches scabrous, not pilose.

        9 Leaf-sheaths glabrous (or the lower ones sometimes puberulent); leaves to 6 mm broad, glabrous (more rarely pilose); lemmas at maturity with their margins somewhat inrolled (so that the lemmas of one row do not overlap those of the other row and the pedicels and rachis are exposed to view); anthers 2 mm long; (introd., transcontinental) . . . . *B. secalinus*

        9 Leaf-sheaths and their blades (to 4 or 5 mm broad) pilose; lemmas at maturity overlapping those of the adjacent row, the pedicels and rachis not exposed.

          10 Awns of the lower florets in the spikelet shorter than those of the upper florets, twisted and finally spreading or divergent; anthers about 1 mm long.

            11 Panicle 1-sided, its branches usually straight; spikelets ovate-oblong, to 5 cm long; lemmas prominently angular at the margins; (introd. in B.C., Alta., and Man.) . . . . *B. squarrosus*

            11 Panicle open, its lower branches usually bearing several lanceolate spikelets to 2.5 cm long; lemmas not angular or only faintly so at the margins; (widely introd.) . . . . *B. japonicus*

          10 Awns of all florets in the spikelets approximately subequal, straight at maturity; anthers to 2 mm long; (widely introd.).

            12 Lemmas angular at the margins, slightly longer than the palea; spikelets to 2 cm long; panicle to 1.5 dm long . . . . *B. commutatus*

            12 Lemmas not angular, equalled by the paleas; spikelets to 1.5 cm long, mostly purplish; panicle to 7 cm long . . . . *B. racemosus*

1 Tip of lemma only minutely indented, its teeth minute or not over 0.5 mm long.

  13 Spikelets strongly flattened, the lemmas distinctly compressed-keeled; lower glume 3–5-nerved; upper glume 5–9-nerved; leaves glabrous or sparingly pilose above.

    14 Lemmas awnless (or the awn at most about 2 mm long), scabrous to almost glabrous; palea distinctly shorter than the lemma; culms glabrous; annuals or biennials; (introd. in Ont. and s Labrador) . . . . *B. catharticus*

    14 Lemmas with an awn to about 1 cm long; perennials.

      15 Leaves generally lacking basal auricles, to 1.5 cm broad, their veins relatively

fine; spikelets mostly 1 or 2 at the ends of each panicle-branch; (Alaska–
B.C.) . . . .*B. sitchensis*
   15 Leaves mostly with basal auricles, averaging less than 1 cm broad,
coarse-veined; panicle-branches usually bearing several spikelets (some-
times to near the base); (B.C. to Sask.; introd. in Ont.) . . . .*B. carinatus*
13 Spikelets plump or moderately flattened but the lemmas not compressed-keeled;
perennials.
   16 Creeping rhizomes present; lower glume 1-nerved; upper glume 3-nerved;
lemmas awnless or with awns to 6 mm long; anthers to 7 mm long (trans-
continental, introd.; ssp. *pumpellianus* native) . . . .*B. inermis*
   16 Creeping rhizomes wanting; lemmas awned.
      17 Lower glume usually 3-nerved, sometimes 5-nerved (usually 1-nerved in *B.
nottowayanus*); upper glume usually 5–7-nerved (3-nerved in *B. porteri*);
leaf-auricles wanting.
         18 Upper glume 3-nerved; lemma-awn commonly less than 3.5 mm long;
anthers commonly not over 3.5 mm long; leaves to 5(6) mm broad; culms
with commonly 3 or 4 nodes; (B.C. to Man.) . . . .*B. porteri*
         18 Upper glume 5–7-nerved.
            19 Lemma-awn at least 5 mm long; upper glume 5–7-nerved; anthers 3–5
mm long; spikelets to 4 cm long; leaves to 13 mm broad, pilose above,
their sheaths mostly covering the nodes, these up to 9 in number;
(s ?Ont.) . . . .[*B. nottowayanus*]
            19 Lemma-awn at most 3 mm long; upper glume 5-nerved; anthers about 2
mm long; spikelets at most about 2.5 cm long; leaves to 1 cm broad,
pilose on both surfaces or glabrous, the tips of at least the lower ones
boat-shaped; nodes at most 5, mostly exserted; (SE Man. to Que.) . . . .
. . . .*B. purgans*
      17 Lower glume usually 1-nerved (often 3-nerved in *B. orcuttianus*; sometimes so
in other species); upper glume usually 3-nerved (sometimes 5-nerved in *B.
pubescens*).
         20 Panicle contracted, with short erect branches; lemmas glabrous or
minutely scabrous or puberulent, their awns commonly 6 or 7 mm long;
anthers to 6.5 mm long; leaf-auricles wanting; sterile tufts of slender plicate
leaves borne at base of culm, these with only 2 or 3 nodes; (introd. in B.C.,
Ont., and Que.) . . . .*B. erectus*
         20 Panicle open, its branches spreading or drooping to reflexed; anthers
rarely over 5 mm long; culm-nodes usually more numerous (except in *B.
orcuttianus*).
            21 Sheaths overlapping, their summits densely pubescent and nearly
closed, with two horizontal flanges usually prolonged into auricles;
lemmas almost glabrous or minutely silky toward base, their awns to 6
mm long; anthers at most 2.5 mm long; (range obscure; see text) . . . .
. . . .*B. ciliatus*
            21 Sheaths mostly not overlapping, usually neither auricled nor densely
pubescent at the V-shaped orifice (lower leaves often auricled in *B.
anomalus*).
               22 Culm-nodes rarely more than 3; leaves glabrous, to 12 mm broad;
lemmas to about 1.5 cm long, pubescent across the back or merely
along the margins and toward the base, their awns to over 7 mm
long; anthers to 5.5 mm long; (?Vancouver Is.) . . . .[*B. orcuttianus*]
               22 Culm-nodes commonly at least 4; anthers rarely over 4 mm long.
                  23 Lemmas sparingly to rather copiously pubescent along the
margins and toward base but usually glabrous across the back
(sometimes pubescent across the back in *B. vulgaris*), to about
1.5 cm long; leaves more or less pilose above.
                    24 Leaves commonly over 1 cm broad, their ligules to 5 mm

long; lemmas to 1 cm long, the awn to 8 mm long; anthers
to over 3 mm long; (B.C. and Alta.) . . . .*B. vulgaris*
24 Leaves usually less than 1 cm broad, their ligules about 1
mm long; lemmas to 1.5 cm long, the awn to 5 mm long;
anthers usually less than 1.5 mm long; (transcontinental) . . . .
. . . .*B. canadensis*
23 Lemmas pubescent across the back as well as along the
margins.
25 Leaves rarely more than 5 mm broad, glabrous to pubescent
or pilose, the lower ones commonly auricled; culm-nodes
rarely more than 7, pubescent, the internodes glabrous;
ligules to 1 mm long; lemmas to 1 cm long, their awns to 3(5)
mm long; anthers to 4 mm long; (B.C. and sw Alta.) . . . .
. . . .*B. anomalous*
25 Leaves often over 1 cm broad, mostly lacking conspicuous
auricles.
26 Culm-nodes commonly more than 10, usually glabrous,
the internodes glabrous; leaves usually glabrous; ligules
to 1 mm long; lemmas to about 1.5 cm long, their awns to
5(7) mm long; anthers less than 3 mm long; (Alta. to
Que.) . . . .*B. pubescens*
26 Culm-nodes at most 8, pubescent, the internodes
glabrous or puberulent just below the nodes; ligules 2 or 3
mm long; lemmas to 12 mm long, their awns to 7 mm
long; anthers to 4 mm long; (coastal Alaska–B.C.) . . . .
. . . .*B. pacificus*

B. anomalus Rupr.   Nodding Brome
/T/WW/ (Hs) Open woods and thickets from B.C. (N to François Lake, ca. 54°N; Eastham
1947) to sw Alta. (Waterton Lakes; Breitung 1957*b*) and Sask. (A. S. Hitchcock 1935), s to s Calif.,
Mexico, and Tex. MAP: Wagnon 1952: fig. 33 (B), p. 451.

B. arvensis L.
Eurasian; known from fields and along roadsides in s B.C. (Invermere, Columbia Valley), s Ont.
(Guelph; Ottawa), and Que. (St-Alexandre, Kamouraska Co.; QSA; RIM).

B. brizaeformis Fisch. & Mey.
Eurasian; waste places and roadsides of Alaska (Seward and Nome; Hultén 1942), B.C. (Trail;
Kootenay), and s Ont. (Lambton, Middlesex, Haldimand, Waterloo, and Wellington counties). MAP:
Hultén 1968*b*: 176.

B. canadensis Michx.
/ST/X/ (Hs) Open woods, thickets, and fields from cent. Alaska–Yukon to Great Slave L., Sask.
(N to the Methy R. at ca. 56°N), Man. (N to Gillam, about 165 mi s of Churchill), Ont. (N to the Winisk
R. at ca. 55°N), Que. (N to NE James Bay; type from L. St. John), Nfld., N.B., P.E.I., and N.S., s to
Baja Calif., N Mexico, Tex., and N.C. [*B. ciliatus* of most Canadian reports, not L.; *B. dudleyi* Fern.;
B. richardsonii Link; *B. purgans* vars. *longispicata* and *pallidus* Hook.; *B. ?segetum* C. & S.]. MAPS
(as *B. ciliatus*): Hultén 1968*b*: 173; Wagnon 1952: fig. 30, p. 451. (*See* discussion under *B.
ciliatus*).

B. carinatus Hook. & Arn.   California Brome
/T/WW/ (Hs) Open woods, thickets, fields, and waste places (ranges of Canadian taxa
outlined below), s to Baja Calif., Mexico, and Tex. (introd. farther eastwards). MAP and synonymy:
see below.
1 Panicle-branches often deflexed; lemma-awns to about 1.5 cm long; [*B. hookerianus*
Thurb. and its var. *minor* Scribn.; *Ceratochloa breviaristata* Hook. (*B. brev.* (Hook.)

Thurb.), not *B. brev.* Buckl.; B.C. (N to the Nass R. at ca. 54°N; CAN) and sw Alta. (Castle Mt., near Banff; CAN); introd. in s Ont. (Milton, Halton Co.; TRT)] . . . . .var. *carinatus*

1 Panicle-branches spreading or drooping but not deflexed; lemma-awns mostly less than 7 mm long; [*B. marg.* Nees; *B. sitchensis* var. *marg.* (Nees) Boivin; Alaska (probably introd.; Hultén 1942), B.C. (N to Hazelton, 55°15′N), sw Alta. (Waterton Lakes; Breitung 1957*b*), and s Sask. (Cypress Hills, Maple Creek, and Wood Mountain; Breitung 1957*a*); reported from Ont. by Boivin 1967*a*, where probably introd.; MAP (*B. marg.*): Hultén 1968*b*:177] . . . . .var. *marginatus* (Nees) Hitchc.

B. catharticus Vahl   Rescue-Grass

Tropical America; collections from Ont. (grounds of the Central Experimental Farm, Ottawa; DAO) and s-cent. Labrador (Goose Bay, 53°19′N; ACAD; DAO) have been placed here by Dore. [*Ceratochloa* Henrard; *B. unioloides* (Willd.) HBK.].

B. ciliatus L.

/T/EE/   (Hs)   Open woods, thickets, and clearings, the type material grown from seed collected by Kalm somewhere in Canada but the range obscure through past confusion with *B. canadensis.* According to B. R. Baum (Can. J. Bot. 45(10): 1849. 1967), *B. latiglumis* (Shear) Hitchc. (*B. purgans* var. *lat.* Shear; *B. altissimus* Pursh) is a superceded synonym of *B. ciliatus* and the following range is based upon records of that species: Sask. (N to Tisdale, ca. 52°50′N) to s Man. (Winnipeg; Westbourne; Otterburne), Ont. (N to the Ottawa dist.), Que. (N to L. St. Peter at Nicolet), and N.B. (Kings and Victoria counties; not known from P.E.I. or N.S.), s to N.Mex., Tex., Mo., N.C., and Md.

B. commutatus Schrad.

European; dry roadsides and waste places of Alaska (Hultén 1942), B.C. (Vancouver Is. and Saltspring Is.; CAN), sw Alta. (Waterton Lakes; Breitung 1957*a*), Ont. (N to Carleton Co.; TRT; not listed by Gillett 1958), Que. (N to York, Gaspé Pen.; GH), and N.S. (not known from N.B. or P.E.I.). [*B. pratensis* Ehrh., not Lam.]. MAP: Hultén 1968*b*: 179.

B. erectus Huds.

European; fields and waste places of sw B.C. (Cedar Hill, Vancouver Is., the type locality of *B. macounii* Vasey, the report of which from the Lewes R., Yukon Territory, by John Macoun 1888, requires confirmation), s Ont. (London, Guelph, and Kingston; Montgomery 1956), and sw Que. (Lanoraie, about 45 mi NE of Montreal; MT). [*Zerna* Gray; *B. macounii* Vasey].

B. inermis Leyss.   Awnless or Hungarian Brome

/aST/X/EA/   (Hsr)   Alaska–Yukon–Dist. Mackenzie and all provinces, the Canadian ranges of the typical form of Eurasia (introd. in fields and waste places of N. America) and the native ssp. *pumpellianus* (tundra and gravelly or sandy slopes and shores; s to Calif., N.Mex., Ohio, and Vt.) outlined below, together with MAPS and synonymy.

1 Lemmas glabrous or minutely scabrous, awnless or with awns to 3 mm long; leaves and culms essentially glabrous . . . . .ssp. *inermis*

  2 Panicle-branches widely spreading to reflexed; [introd. near Montreal, Que., and Grand Falls, Nfld.] . . . . .var. *divaricatus* Rohlena

  2 Panicle-branches spreading-ascending to erect . . . . .var. *inermis*

    3 Lemmas blunt and awnless.

      4 Spikelets normal; [*Zerna inermis* (L.) Lindm.; transcontinental, introd.; MAPS: Wagnon 1952: fig. 35, p. 453; Meusel, Jaeger and Weinert 1965: 42; Hultén 1962: map 202, p. 213, and 1968*b*: 174] . . . . .f. *inermis*

      4 Spikelets transformed to leafy shoots commonly about 4 cm long; [SE Man. and La Trappe, Que., the type locality] .f. *proliferus* Louis-Marie

    3 Lemmas with awns to 3 mm long; [Argenteuil Co., Que., and Goose Bay, Labrador] . . . . .f. *aristatus* (Schur) Fern.

1 Lemmas pubescent at least near margins, awnless or with awns to 6 mm long; upper surface of leaves usually pilose, the lower surface usually glabrous; culms pubescent near the nodes . . . . .ssp. *pumpellianus* (Scribn.) Wagnon

Grameneae

5   Lemmas pubescent on margins only, or also on the main nerves and base of back; glumes glabrous; [*B. pumpellianus* and *B. polyanthus* Scribn.; Alaska to Great Bear L. and N Man.; MAPS: Hultén 1968*b*: 174 (*B. pump.*); Wagnon 1952: fig. 36 (*B. inermis* ssp. *pump.* var. *purp.*), p. 453] . . . . var. *pumpellianus*

5   Lemmas densely hairy across the back; glumes sparsely to densely pubescent; [*B. arcticus* Shear; incl. *B. pump.* var. *villosissimus* Hult. and ssp. *dicksonii* Mitchell & Wilton; Alaska–Yukon–Dist. Mackenzie; MAPS: Hultén 1968*b*: 175 (*B. pump.* vars. *arct.* and *vill.*); Wagnon 1952: fig. 36 (*B. inermis* ssp. *pump.* var. *arct.*), p. 453] . . . . .var. *tweedyi* (Scribn.) Hitchc.

B. japonicus Thunb.   Japanese Chess
Eurasian; fields and waste places of s B.C. (Kamloops; Kelowna), s Alta. (Pincher Creek; Medicine Hat), Sask. (Breitung 1957*a*), Man. (Brandon; Treesbank; Otterburne), Ont. (N to the Ottawa dist.; *see* s Ont. map by Montgomery 1956: fig. 3, p. 94), and sw Que. (Iberville, Iberville Co.). MAP: Hultén 1962: map 218, p. 229.

B. mollis L.   Soft Chess
Eurasian; fields and waste places of s Alaska (Nome; Hultén 1942), B.C. (N to Hazelton, ca. 55°15′N), sw Alta. (Rocky Mountain Park; CAN), s Ont. (N to Wellington and Prince Edward counties), Que. (N to Ste-Anne-de-la-Pocatière, Kamouraska Co.; QSA), St-Pierre and Miquelon (Rouleau 1956), N.B., and N.S.; Greenland. [*B. hordeaceus* ssp. *mollis* (L.) Hyl.]. MAPS: Hultén 1968*b*: 178 (*B. hordeaceus*, incl. *B. mollis*), and 1962: map 198, p. 209.

The similar *B. thominii* Hard. (lower lemmas at most 7.5 mm long rather than at least 8 mm; lower glumes at most 6 mm long rather than usually at least 7 mm; upper glumes at most about 6.5 mm long rather than usually at least 8 mm) is reported by F. C. Seymour (Rhodora 68 (774): 171. 1966) as having been taken on Vancouver Is., B.C., by Rosendahl in 1907, where, however, probably not established.

[*B. nottowayanus* Fern.]
[The report of this species of SE Va. from s Ont. by Wagnon (1952; *see* his MAP, fig 32, p. 451) requires further confirmation.]

[*B. orcuttianus* Vasey]
[The report of this species of the w U.S.A. (Wash. to Calif. and Ariz.) from near Victoria, Vancouver Is., B.C., by John Macoun (1888; this taken up by Henry 1915; *see* Wagnon 1952: MAP, fig. 38, p. 462) requires confirmation.]

B. pacificus Shear
/sT/W/   (Hs)   Moist thickets near the coast from SE Alaska (*see* Hultén 1942: map 181, p. 399) through coastal B.C. (CAN; V) to w Oreg. MAP: Hultén 1968*b*: 176.

B. porteri (Coult.) Nash
/T/WW/   (Hs)   Dry or moist thickets and clearings from s B.C. (Hanceville, about 150 mi NW of Kamloops) to s Alta. (Red Deer; Milk River Ridge; Porcupine Hills; Sweet Grass Hills), Sask. (N to Battleford), and Man. (N to Riding Mt.), s to s Calif., Ariz., and N.Mex. [Scarcely distinct from *B. anomalus*, with which it is often merged; *B. kalmii* of auth. in part, not Gray]. MAP: Wagnon 1952: fig. 32, p. 451.

B. pubescens Muhl.
/T/(X)/   (Hs)   Open woods, thickets, and rocky slopes from Alta. (N to Edmonton) to Sask. (N to near Saskatoon), Man. (N to Lac du Bonnet, about 50 mi NE of Winnipeg), Ont. (N to the Ottawa dist.), and Que. (N to Cabano,Temiscouata Co.), s to Nebr., Okla., and Va. [*B. purgans* of Canadian reports in large part, not L. (*see* B. R. Baum, Can. J. Bot. 45(10): 1849. 1967) and its f. *laevivaginatus* Wieg.; *B. ciliatus* f. *laeviglumis* (Scribn.) Wieg.]. MAPS: the map for *B. pubescens* by Wagnon 1952: fig. 31, p. 451, indicates its absence in Canada, but his map for *B. purgans*, fig. 37, p. 453, probably applies here. However, in view of the past confusion between these two species, the above outline of the range can only be considered tentative.

B. purgans L.
/T/EE/ (Hs)   Dry or moist open soil and thickets from SE Man. (near Otterburne) to S Ont. (N to Waterloo and Wentworth counties) and SW Que. (N to the Montreal dist.), S to Tex. and Ga. [*B. kalmii* Gray]. *See* discussion under *B. pubescens*.

B. racemosus L.
Eurasian; introd. in cent. Alaska–Yukon (Hultén 1942; CAN), B.C. (N to Mons and Spences Bridge), Sask. (Regina), Ont. (N to Ottawa; John Macoun 1888; not listed by Gillett 1958, nor Boivin 1967*a*), Que. (reported N to Quebec City by John Macoun 1888), Nfld. (a collection from St. John's has been placed here by Dore; not listed by Rouleau 1956), ?N.B. (Fowler 1885; John Macoun 1888), P.E.I. (Montague, Kings Co.; DAO), and N.S. MAP: Hultén 1968*b*: 179.

B. rigidus Roth
European; reported from Vancouver Is., B.C. by Henry (1915; *B. maximus*) and there are collections in CAN from Victoria, Nanaimo, and Sidney. [*Anisantha* Hyl.; *B. maximus* Desf., not Gilib.; *B. ?rigens* L.].

B. rubens L.   Foxtail Chess
European; reported from SW B.C. by J. M. Macoun (1913; Nanaimo, Vancouver Is.), this taken up by Boivin (1967*a*).

B. secalinus L.   Cheat, Chess.   Brome des seigles
Eurasian; fields and waste places from Alaska (Hultén 1942) and B.C. to Alta. (Boivin 1967*a*; not listed for Sask. by Breitung 1957*a*; the report from S Man. by Shimek 1927, requires confirmation), Ont. (N to the N shore of L. Superior), Que. (N to Rimouski, Rimouski Co.), Nfld. (Boivin 1967*a*; not listed by Rouleau 1956), N.B., and N.S. MAPS: Hultén 1968*b*: 178, and 1962: map 208, p. 218.

B. sitchensis Trin.
/sT/W/ (Hs)   Woods and banks near the coast from the Aleutian Is. and S Alaska (*see* Hultén 1942: map 183, p. 400; type from Sitka) through coastal B.C. to Oreg. [*B. aleutensis* Trin.]. MAP: Hultén 1968*b*: 1779 The reports from Alta. to Ont. by Boivin (1967*a*) may be referable to *B. carinatus* of the present treatment.

B. squarrosus L.
Eurasian; fields and waste places of B.C. (reported from Kamloops by Eastham 1947, and from Roosville by Hubbard 1955), Alta. (Moss 1959), and S-cent. Man. (Boivin 1967*a*).

B. sterilis L.
Eurasian; roadsides and waste places of B.C. (N to Quesnel) and S Ont. (Lambton, Wentworth, and Lincoln counties). [*Anisantha* Nevski].

B. tectorum L.   Downy Chess.   Brome des toits
Eurasian; roadsides and waste places of S Alaska–Yukon and B.C. to Alta., Sask. (N to Prince Albert), Man. (Brandon; Emerson), Ont. (N to Ottawa), Que. (N to St-Joseph-de-la-Rive, Charlevoix Co.; Marcel Raymond and James Kucyniak, Rhodora 50(595): 177. 1948), N.B. (St. Andrews; Fredericton), and N.S. (Kings and Halifax counties); S Greenland. [*Anisantha* Nevski]. MAPS: Hultén 1968*b*: 172, and 1962: map 217, p. 229.

   Var. *glabratus* Spenner (spikes glabrous rather than villous) is known from B.C. (Vancouver Is.; Pemberton; Trail; Hazelton).

B. vulgaris (Hook.) Shear
/T/W/ (Hs)   Rocky woods and shady ravines from B.C. (N to Hazelton, 55°15'N) and SW Alta. (Waterton Lakes; Porcupine Hills; Livingstone Valley) to Calif. and Wyo. [*B. purgans* var. *vulgaris* Hook., the type presumably from w Canada; *B. ciliatus* vars. *ligulatus* and *pauciflorus* Vasey; *B. eximius* (Shear) Piper]. MAP: Wagnon 1952: fig. 31 (B), p. 451.

BUCHLOË Engelm.   [308]

B. dactyloides (Nutt.) Engelm.   Buffalo-Grass
/T/WW/   (Hsr)   Dry plains and prairies from SE Sask. (Estevan, about 115 mi SE of Regina; DAO) and SW Man. (Coulter, about 70 mi SW of Brandon; CAN) to N Mexico, Tex., Minn., and W La. [*Sesleria* Nutt.; *Bulbilis* Raf.].

CALAMAGROSTIS Adans.   [247]   Reed-Bentgrass

(Ref.: G. L. Stebbins, Rhodora 32(375): 35–57. 1930)
1   Glumes linear-subulate; lemmas about 3 mm long, 3-nerved below, 2-nerved above the insertion of the awn, much surpassed by the callus-hairs; panicle slender, to 3.5 dm long, often becoming bronze or purplish, at maturity silky from the protruding callus-hairs; leaves pale, firm, often becoming involute, to 8(13) mm broad; (introd. in S Ont.) ....*C. epigejos*
1   Glumes lanceolate to ovate; lemmas 3–5-nerved below, more or less strongly 4-nerved above the insertion of the awn, surpassing or only slightly shorter than the callus-hairs.
  2   Lemma-awn more or less geniculate; palea nearly or quite equalling the lemma.
    3   Awn surpassing the glumes and protruding from the top of the spikelet (or sometimes shorter than the glumes in *C. deschampsioides*); callus-hairs in 2 lateral tufts united by a zone of shorter hairs across the front; culms commonly less than 5 dm tall.
      4   Panicle open, pyramidal, to 9 cm long, purplish, the spreading branches spikelet-bearing toward tip; glumes acuminate, glabrous, to 6 mm long; awn borne near or below middle of lemma; callus-hairs about half as long as the lemma; lemma-awns weakly geniculate; culms often decumbent at base, with slender elongate rhizomes; (essentially transcontinental in arctic and subarctic brackish marshes) ....*C. deschampsioides*
      4   Panicle contracted, stiff and spike-like, the short appressed-erect branches spikelet-bearing from near base; glumes acute, minutely scabrous, to 8 mm long; awn borne near base of lemma; callus-hairs about 1/3 as long as the lemma; lemma-awns strongly geniculate; leaves scabrous; culms tufted (sometimes with short rhizomes), covered at base with dry whitish marcescent leaves; (transcontinental) ....*C. purpurascens*
    3   Awn shorter than the glumes, sometimes projecting at the side of the spikelet; panicle compact; rhizome usually elongate (short in *C. nutkaensis*).
      5   Callus-hairs less than 1 mm long and at most 1/4 as long as the lemma, in 2 lateral tufts separated by a naked frontal zone (under the median lemma-nerve); panicle greenish or purplish, glaucous, to over 1.5 dm long; (?Ont., St-Pierre and Miquelon, Nfld., and N.S.) ....*C. pickeringii*
      5   Callus-hairs mostly at least 1/3 as long as the lemma.
        6   Callus-hairs about 3 mm long and about 2/3 as long as the lemma, in 2 lateral tufts separated by a naked frontal zone; lemma-awn about 4 mm long; (Ont., Que., and SW Nfld.) ....*C. lacustris*
        6   Callus-hairs at most about 1/2 as long as the lemma.
          7   Panicle rather loose, to over 3 dm long, usually purplish, the branches rather stiffly ascending; lemma-awns only weakly geniculate; callus-hairs scarcely 1/2 as long as the lemma; leaves to 12 mm broad, their sheaths glabrous on the collar (junction of sheath and blade); culms stout, to over 1.5 m tall; (Alaska–W B.C.) ....*C. nutkaensis*
          7   Panicle more spikelike, rarely over 1.5 dm long; lemma-awns strongly geniculate.
            8   Callus-hairs about 1/2 as long as the lemma, the 2 lateral tufts separated by a frontal zone of shorter hairs; panicles usually pale, at most about 1 dm long; glumes scabrous; leaves mostly less than 2 mm broad; culms at most about 4 dm tall, scabrous below the panicle; plants tufted, often with short rhizomes; (B.C. to S Man.) ....
            ....*C. montanensis*

        8   Callus-hairs about 1/3 as long as the lemma, the 2 lateral tufts separated by a naked frontal zone; panicles pale or purplish, to 1.5 dm long; glumes glabrous; leaves to 4 mm broad, flat, drying involute near tip; culms to 1 m tall, glabrous; plant with slender deep-seated rhizomes; (B.C. to s Sask.) . . . .*C. rubescens*

2   Lemma-awn straight or barely arched, fine and inconspicuous, not exserted; palea distinctly shorter than the lemma; rhizomes elongate.

      9   Leaf-sheaths retrorsely pubescent on the collar (junction of sheath and blade); panicle pale or purplish, narrow but rather lax, rarely over 1.5 dm long; glumes about 4 mm long; leaves to 7 mm broad; (s B.C.–Alta.) . . . .*C. scribneri*

      9   Leaf-sheaths glabrous.

        10   Panicle mostly loose and open, broadly lanceolate to ovoid, to 3 dm long, its branches spreading during anthesis, ascending in fruit; callus-hairs equalling or surpassing the lemma, of uniform length except for an outer short ring; (transcontinental) . . . .*C. canadensis*

        10   Panicle narrow and contracted to somewhat open, its rigid branches appressed or ascending even at anthesis; callus-hairs shorter than the lemma (except sometimes in *C. lapponica*) and of unequal length, those at the sides distinctly longer and tufted.

          11   Spikelets 6 or 7 mm long; glumes awn-tipped; grain pubescent, particularly at summit; lemma awned above middle; prolongation of rachilla bearded only at summit; panicle to over 2 dm long and 4 cm thick; plant glaucous, to over 1.5 m tall; (N.S.) . . . .*C. cinnoides*

          11   Spikelets less than 6 mm long; glumes not awn-tipped; grain glabrous; prolongation of rachilla bearded throughout.

            12   Leaves harsh and scabrous, mostly flat and up to 8 mm broad or narrower and involute; ligules of upper leaves averaging about 4 mm long, their margins erose or lacerate; glumes opaque; top of culm scabrous; (transcontinental) . . . .*C. inexpansa*

            12   Leaves smooth or sometimes scabrous at tip and on the margins and upper surface, at most 5 mm broad and commonly involute; ligules of upper leaves averaging 2 mm long, their margins essentially entire; culms smooth except just below the panicle.

              13   Glumes thick and firm or almost indurate; panicle narrow and dense, its branches mostly not more than 1/5 the length of the panicle (or the panicle not over about 5 cm long).

                14   Leaves 4 or 5 mm broad; panicle commonly not over 5 cm long; (?Vancouver Is.) . . . .[*C. crassiglumis*]

                14   Leaves at most 2.5 mm broad; panicle to about 12 cm long; (E Que. and SE Labrador) . . . .*C. labradorica*

              13   Glumes thin and translucent at least at tip, about 1 mm broad; panicle-branches to about 1/3 the length of the panicle; (transcontinental).

                15   Panicle lax, narrow to somewhat open, strongly suffused with purple, commonly less than 1 dm long; longer callus-hairs about equalling the lemma, the awn of the latter inserted about 1/3 above the base; leaves about 2 mm broad, strongly involute; culms commonly less than 6 dm tall . . . .*C. lapponica*

                15   Panicle strict and commonly narrow, usually greenish or bronze, sometimes purplish, to 1.5 dm long; leaves to 4 mm broad, commonly involute . . . .*C. neglecta*

C. canadensis (Michx.) Nutt.   Blue-joint.   Foin bleu
/aST/X/GEA/   (Hsr)   Meadows, bogs, wet thickets, and open woods, the aggregate species from the Aleutian Is. and N Alaska–Yukon–Dist. Mackenzie to s Baffin Is., northernmost Ungava–Labrador, and Nfld., s to Calif., N.Mex., Nebr., Ohio, and Del.; w Greenland N to ca. 70°N, E Greenland N to 63°20′N; Eurasia. MAPS and synonymy: *see* below.

A hybrid between var. *langsdorfii* and *C. nutkaensis* is reported from the Aleutian Is. and Alaska by Hultén (1942). Reports of the Eurasian *C. lanceolata* Roth from Labrador by E. Meyer (1830) and Schlechtendal (1836), of *Arundo calamagrostis* L. from Labrador by Schranck (1818), and of *C. (Deyeuxia) strigosa* Bong. from Alaska by Hooker (1840) are all possibly based upon var. *scabra*.

1   Spikelets less than 4 mm long; glumes rounded on the back, weakly keeled, acute or acuminate; lemma at most 3 mm long, awn-bearing near the middle.

    2   Panicle loosely flowered; spikelets to 3.8 mm long; glumes distinctly surpassing the lemma, acute or acuminate; [*Arundo* Michx.; *Deyeuxia* Munro; transcontinental; MAP: Hultén 1968*b*: 104] . . . . .var. *canadensis*

    2   Panicle densely flowered; spikelets less than 3 mm long; glumes nearly or quite equalled by the lemma, obtuse or acutish; [*Deyeuxia (C.) macouniana* Vasey, the type from "Souris Plains, Assiniboia", probably in present-day Sask.; var. *campestris* Kearney; incl. var. *pallida* (Vasey & Scribn.) Stebbins; Alta. to Man.; w Nfld.; P.E.I.] . . . . .var. *macouniana* (Vasey) Stebbins

1   Spikelets to 6 mm long; glumes narrow, strongly keeled, distinctly acuminate; lemma to about 4 mm long, its awn inserted on the lower third.

    3   Spikelets at most 4.5 mm long; glumes often hyaline at tip and on margins, short-scabrous on the keel, elsewhere minutely scabrous; lemma at most 3.5 mm long; [*C. atropurpurea* Nash; transcontinental; MAP: Raup 1947: pl. 16 (incomplete northwards)] . . . . .var. *robusta* Vasey

    3   Spikelets to 6 mm long; glumes usually thick and opaque to tip, scabrous or ciliate on the keel, elsewhere minutely pilose; lemma to about 4 mm long.

        4   Culm-leaves broad and flat, their ligules to 1 cm long; panicle expanded at anthesis; glumes ciliate on the keel, elsewhere rather densely pubescent; [*Deyeuxia (C.) scabra* Kunth; *Arundo (C.) langsdorfii* Link; *C. hirtigluma* Steud.; transcontinental, the common form northwards; MAPS: Hultén 1962: map 177, p. 189, and 1968*b*: 104; Porsild 1957: map 19, p. 163; Raup 1947: pl. 16] . . . . .var. *scabra* (Kunth) Hitchc.

        4   Culm-leaves involute at least on drying, their ligules at most about 3 mm long; panicle compact even at anthesis; glumes short-scabrous on the keel, elsewhere minutely and sparingly pubescent; [N Ungava–Labrador (type from Nachvak Bay, Labrador, ca. 59°N) and w Greenland; MAP: Dutilly, Lepage, and Duman 1953: fig. 8, p. 31] . . . . .var. *arcta* Stebbins

C. cinnoides (Muhl.) Bart.
/T/E/   (Hsr)   Damp, sandy or peaty ground from N.S. (Halifax; GH) to Ala. and Ga. [*Arundo* Muhl.; *A. (C.) coarctata* Torr., not *sensu* Hooker 1839].

[*C. crassiglumis* Thurb.]
[The report of *Deyeuxia crassiglumis* (Thurb.) Vasey from Vancouver Is., B.C., by John Macoun (1888; this taken up by Henry 1915, and A. S. Hitchcock 1935) appears to be based upon *C. inexpansa* (a relevant collection in CAN).]

C. deschampsioides Trin.
/aST/(X)/EA/   (Hsr)   Brackish coastal marshes and shores: NW Alaska and the Mackenzie R. Delta to the E Aleutian Is. (Unalaska Is.) and S Alaska (see Hultén 1942: map 113, p. 393); w Hudson Bay, (var. *churchilliana* Polunin, the type from Churchill, Man., N to Chesterfield Inlet); N Que. (coast of James Bay N to ca. 56°N, SE Hudson Bay; E Ungava Bay); NE Europe; E Asia. [*Deyeuxia* Vasey; incl. var. *macrantha* Piper and *C. kolymaënis* Kom.]. MAPS: Hultén 1968*b*: 108, and 1962: map 10, p. 17; Schofield 1959: map 4, p. 111.

C. epigejos (L.) Roth   Feathertop, Bush-Grass
Eurasian; roadsides and waste places of Ont. (collections in CAN, DAO, and TRT from Flamboro, Wentworth Co., and from between Espanola and Little Current along the N shore of L. Huron; reported from near St. Thomas, Elgin Co., by Montgomery 1956). [*Arundo* L.].

C. inexpansa Gray
/aST/X/GeA/ (Hsr) Rocky, gravelly, or peaty ground from the E Aleutian Is. and N Alaska –Yukon–Dist. Mackenzie to Prince Patrick Is., cent. Dist. Keewatin, Man. (N to Churchill), Ont. (N to the Fawn R. at ca. 55°30′N), Que. (N to E Ungava Bay, L. Mistassini, and Anticosti Is.), Nfld., and N.S. (Cape Breton Is.; not known from N.B. or P.E.I.; reports from Labrador may refer to the Côte-Nord, E Que.), s to Calif., N.Mex., Nebr., Mo., Mich., N.Y., and Vt.; Greenland; E Asia. [*C. ?chordorrhiza* Porsild; *C. hyperborea elongata* Kearney (*C. el.* (Kearney) Rydb.); *Deyeuxia glomerata* Vasey; *D. neglecta* vars. *americana* and *robusta* Vasey]. MAP: Hultén 1968*b*: 105.

C. labradorica Kearney
/sT/EE/ (Hsr) Marshes and gravelly shores of E Que. (type from Bonne-Espérance, E Saguenay Co., Côte-Nord; also known from Anticosti Is.) and SE Labrador (Fox Harbour, 52°22′N). [*C. hyperborea* of Canadian reports in part, not Lange].

C. lacustris (Kearney) Nash
/T/EE/ (Hsr) Damp, rocky, gravelly, or peaty ground from Ont. (Pagwa, ca. 50°N; Wasaga Beach, L. Huron, Simcoe Co.) to Que. (region between E James Bay and L. Mistassini) and SW Nfld., s to N Minn., N Mich., N.Y., Vt., and N.H. [*C. breviseta (pickeringii)* var. *lac.* Kearney].

C. lapponica (Wahl.) Hartm.
/aST/X/GEA/ (Hsr) Moist, sandy or gravelly ground from N Alaska–Yukon and the coast of Dist. Mackenzie to NE Dist. Keewatin, N Ungava–Labrador (reported N to Port Burwell, 60°28′N), and s Baffin Is., s to B.C. (s to Lucerne, ca. 52°N), SW Alta., s-cent. Sask., Man. (Reindeer L. at 57°54′N), and Que. (s to ca. 55°N); w Greenland between ca. 67° and 70°N; Eurasia. [*Arundo* Wahl.; *Deyeuxia* Kunth; *C. alaskana* Kearney; *C. canadensis* var. *acuminata* Vasey; incl. vars. *groenlandica* Lange and *nearctica* Porsild]. MAPS: Hultén 1968*b*: 106, and 1962: map 26, p. 33; Young 1971: fig. 16, p. 89.

C. montanensis Scribn.
/T/WW/ (Hsr) Dry prairies and hillsides from SE B.C. (reported by Eastham 1947, from a roadside at Invermere, Columbia Valley, where possibly introd.) to Alta. (N to Peace Point, ca. 59°N; CAN), Sask. (N to McKague, 52°37′N), and Man. (N to Duck Mt.), s to Idaho, Colo., S.Dak., and Minn. [*C. neglecta* var. *candidula* Kearney].

C. neglecta (Ehrh.) Gaertn., Mey. & Scherb.
/aST/X/GEA/ (Hsr) Moist or wet ground, the aggregate species from the coasts of Alaska and Dist. Mackenzie to Devon Is., cent. Baffin Is., northernmost Ungava–Labrador, Nfld., N.B., P.E.I., and N.S., s to Oreg., Utah, Colo., Wisc., Mich., and N Maine; w Greenland N to 71°22′N, E Greenland N to 74°10′N; Iceland; Eurasia. MAPS and synonymy: *see* below.
1 Spikelets usually less than 2.5 cm long; glumes obtuse or merely acute; [*C. micrantha* Kearney, the type from Prince Albert, Sask.; reported from the Yukon (Whitehorse), Alta. (Athabasca Landing), and Sask. by G. L. Stebbins, Rhodora 32(375): 55. 1930, and from w and E James Bay by Dutilly, Lepage, and Duman 1954 and 1958] . . . .
. . . . .var. *micrantha* (Kearney) Stebbins
1 Spikelets to 5 mm long; glumes sharply acute or acuminate.
　2 Callus-hairs rarely over half as long as the lemma; awn inserted near the middle of the lemma; panicle commonly less than 5 cm long; culms usually less than 5 dm tall; [*C. borealis* Laest., not *Deyeuxia bor.* Macoun; *C. ?holmii* Lange; *Arundo (C.) groenlandica* Schrank; *C. lapponica* var. *brevipilis* Stebbins; reported from Nfld., Labrador, and Greenland by Fernald *in* Gray 1950, but Hultén's map indicates a transcontinental distribution; MAP: Hultén 1962: map 63, p. 73] . . . .
. . . . .var. *borealis* (Laest.) Kearney
　2 Callus-hairs to 3/4 as long as the lemma; awn inserted on the lower third of the lemma; panicle to 1.5 dm long; culms to 1 m tall; [*Arundo* Ehrh.; *Deyeuxia* Kunth; *D. borealis* Macoun; *D. vancouverensis* and *D. neglecta* var. *brevifolia* Vasey; transcontinental; MAPS: Hultén 1962: map 63, p. 73, and 1968*b*: 106; Porsild 1957:

map 22 (aggregate species), p. 163; Raup 1947: pl. 16 (agg. species); Meusel, Jaeger, and Weinert 1965: 49] . . . . .var. *neglecta*

C. nutkaensis (Presl) Steud.
/sT/W/ (Hsr) Coastal dunes and beaches from the Aleutian Is. and cent. Alaska (*see* Hultén 1942: map 118, p. 393) through w B.C. to cent. Calif. [*Deyeuxia nutkaensis* Presl, the type from Nootka Sound, Vancouver Is.; *C. (D.) aleutica* Trin.; *D. breviaristata* Vasey; *D. ?columbiana* Macoun]. MAP: Hultén 1968*b*: 105.

C. pickeringii Gray
/T/E/ (Hsr) Bogs and wet shores (ranges of Canadian taxa outlined below), s to Vt. and Mass.
1 Spikelets less than 4 mm long; culms to about 6 dm tall; [*C. breviseta debilis* Kearney, the type from Nfld.; also known from many counties in N.S.] . . . .
. . . . .var. *debilis* (Kearney) Fern. & Wieg.
1 Spikelets 4 or 5 mm long; culms to about 6 dm tall . . . . .var. *pickeringii*
  2 Spikelets viviparous; [type from near Tiddville, Digby Co., N.S.] . . . . .f. *vivipara* Louis-Marie
  2 Spikelets normal; [*Deyeuxia* Vasey; ?Ont. (a collection in OAC from Dryden, about 70
    mi E of Kenora, has been placed here by Dore but may belong to the closely related *C.
    lacustris*); Nfld. (Quirpon Is. and Bonavista North), St-Pierre and Miquelon (Boivin
    1967*a*), and N.S.; the report from Tabletop Mt., Gaspé Pen., E Que., by M. L. Fernald,
    Rhodora 9(105): 158. 1907, is now discredited; MAP: Fernald, 1933: map 20, p. 212] . . . .
    . . . . .f. *pickeringii*

C. purpurascens R. Br.
/AST/X/GeA/ (Hs(r)) Rocky soil and cliffs (often calcareous), the aggregate species from the coasts of Alaska and Dist. Mackenzie to Banks Is., Victoria Is., N Baffin Is., cent. Ellesmere Is., and N Que. (N to 57°42′N), s to Calif., Colo., S.Dak., L. Athabasca (Alta. and Sask.), and N Man. (s to the Seal R. at ca. 59°N); s-cent. Que. (L. Mistassini; isolated on the mts. near Bic, Rimouski Co.), and Nfld.; isolated on the N shore of L. Superior (Ont. and Minn.); w Greenland N to 76°36′N, E Greenland N to 77°40′N; NE Asia. MAPS and synonymy (together with the scarcely separable *C. lepageana*): see below.
1 Panicle commonly greenish, rarely over 9 cm long; spikelets less than 5 mm long;
  lemma-awn at most 5 mm long; fruit olivaceous; culms to about 7 dm tall, scattered, the
  plant weakly stoloniferous; [E Que., the type from Mont-Commis, Rimouski Co.] . . . .
  . . . .[*C. lepageana* Louis-Marie]
1 Panicle commonly pinkish or purplish, to 12 cm long; spikelets at least 5.5 mm long;
  lemma-awn to 1 cm long; fruit reddish; culms at most about 4 dm tall . . . .*C. purpurascens*
  2 Lateral tufts of callus-hairs separated by a zone of frontal hairs less than 1 mm long;
    [*Deyeuxia* Kunth; incl. ssp. *arctica* (Vasey) Hult. (*C. arctica* Vasey) and ssp.
    tasuensis Calder & Taylor; *C. poluninii* Soerensen; *C. yukonensis* Nash; *C. sylvatica*
    var. *americana* Vasey; incl. *Trisetum bongardii* Louis-Marie and *T. sesquiflorum* Trin.;
    transcontinental, the type from between Point L. and the coast of Dist. Mackenzie;
    MAPS: Porsild 1957: map 23, p. 163; Raup 1947: pl. 16 (incomplete eastwards);
    Meusel, Jaeger, and Weinert 1965: 49; Böcher 1954: fig. 51 (bottom), p. 189; Hultén
    1968*b*: 107] . . . . .var. *purpurascens*
  2 Lateral tufts of callus-hairs separated by a zone of frontal hairs at least 1 mm long.
    3 Glumes smooth; lemma-awn strongly twisted at base, exserted 3 or 4 mm beyond
      the tip of the glume; [type from Pond Inlet, N Baffin Is.; also known from Sussex L.,
      E-cent. Dist. Mackenzie] . . . . .var. *maltei* Polunin
    3 Glumes scabrous and often pubescent; lemma-awn less strongly twisted,
      exserted at most 1 or 2 mm . . . . .var. *laricina* Louis-Marie
      4 Panicle rarely over 6 cm long, rather dense; [type from the Larch R., Ungava,
        at ca. 57°30′N] . . . . .f. *compacta* Louis-Marie
      4 Panicle longer and more open; [Kaniapiskau, Larch, and Koksoak rivers,
        Ungava, the type from the Larch R. at ca. 57°35′N] . . . . .f. *laricina*

C. rubescens Buckl.
/T/WW/ (Hsr) Open woods, prairies, and banks from B.C. (N to Vanderhoof, ca. 54°N) to Alta. (N to Calgary), s Sask. (Cypress Hills; Maple Creek), and sw Man. (Forest, about 10 mi N of Brandon; Fort Ellice, about 70 mi NW of Brandon), s to s Calif. and Colo. [*Deyeuxia* Vasey; *D. (C.) suksdorfii* Scribn.; *C. luxurians* (Kearney) Rydb.; *C. sylvatica* of auth., not DC.; *C. (D.) porteri sensu* John Macoun 1888, not Gray, the relevant collection in CAN].

C. scribneri Beal
/T/W/ (Hsr) Moist meadows from s B.C. (reported from the Cascade region by Henry 1915, and from Kootenay National Park by Eastham 1947) and sw Alta. (Kicking Horse L.; Stebbins, loc. cit., as var. *imberbis* Stebbins (*C. canadensis* var. *imb.* (Stebbins) Hitchc.)) to Oreg. and Colo.

CALAMOVILFA (Gray) Hack. [250]

C. longifolia (Hook.) Scribn. Sand-Reed
/T/WW/ (Grh) Dry sandy prairies and sandhills (ranges of Canadian taxa outlined below), s to Idaho, Colo., Kans., Mo., and s Mich. MAP and synonymy: *see* below.
1 Sheaths glabrous except at the throat and sometimes along the margins; panicle usually oblong or subcylindrical; [*Calamagrostis longifolia* Hook., the type from Sask.; *Ammophila* Benth.; s B.C. (Tobacco Plains), Alta. (Moss 1959), Sask. (type locality; N to near Meadow L. at 54°20′N), Man. (N to ca. 54°N), and w Ont. (Ingolf, near the Man. boundary; evidently introd. farther eastwards at Pickle Lake); MAP: J. W. Thieret, Am. Midl. Nat. 63(1): fig. 2, p. 173. 1960] . . . . .var. *longifolia*
1 Sheaths softly villous; panicle-branches often widely spreading; [s Ont. (N to Cockburn Is., Manitoulin Is., and the Bruce Pen. of N L. Huron); MAP: on the above-noted map by Thieret] . . . . .var *magna* Scribn. & Merr.

CATABROSA Beauv. [347] Water Hairgrass

C. aquatica (L.) Beauv.
/aST/X/GEA/ (Hel) Marshes and shallow, fresh to brackish waters (ranges of Canadian taxa outlined below), s to E Oreg., Nev., N Ariz., Colo., Nebr., and Mich.; w Greenland; Iceland; Eurasia. MAPS and synonymy: *see* below.
1 Spikelets 2–4-flowered; panicle to 2.5 dm long; culms finely furrowed, their leaves broadly rounded at tip; [*Aira* L.; *Glyceria* Presl; *Poa (Glyceria) airoides* Koel.; sw Alaska; w Greenland N to ca. 70°N; Eurasia; MAPS (aggregate species): Hultén 1958: map 52, p. 71, and 1968b: 124] . . . . .var. *aquatica*
1 Spikelets 1 (rarely 2)-flowered; panicle to about 1.5 dm long; culms coarsely furrowed, their leaves subacute to obtuse at tip.
   2 Panicle-branches spreading horizontally, flowering nearly to base; [Ont. (N to Fort Severn, w Hudson Bay, ca. 56°N), Que. (N to Fort George, E James Bay, 53°50′N, L. Mistassini, and the Côte-Nord; type from Capucins, Gaspé Pen.; SE Labrador (Hamilton R. basin and Gready Is., ca. 53°45′N), Nfld., N.B., and P.E.I.; MAP: on the above-noted 1958 map by Hultén] . . . . .var. *laurentiana* Fern.
   2 Panicle-branches subascending, flowering chiefly from well above the base; [s B.C., Alta., Sask. (N to the Qu'Appelle R.), Man. (N to Churchill), Ont. (N to near Winisk, s Hudson Bay, ca. 55°20′N; *see* James Bay map by Dutilly and Lepage 1945: fig. 3, p. 196), and w Ungava (N to Richmond Gulf, Hudson Bay, ca. 56°10′N); (*Colpodium ?pauciflorum* Hook.); MAP: on the above-noted 1958 map by Hultén] . . . . . .var. *uniflora* Gray

CENCHRUS L. [174] Sandbur, Burgrass

C. longispinus (Hack.) Fern. Field Sandbur
/t/X/ (T) Sandy shores and clearings from Oreg. to s Ont. (*see* s Ont. map by Soper 1962: fig. 3, p. 9; considered by Soper to be native N to N Lambton, SE Waterloo, and s York counties but introd. near a railway station at Kincardine, sw Bruce Co.), s to Calif., Mexico, Tex., and Fla.; s S.

America; introd. in Eurasia. [*C. carolinianus, C. pauciflorus,* and *C. tribuloides* of Ont. reports, not Walt., Benth., nor L., respectively].

C. pauciflorus Benth., a native of Mexico and Tex., is reported by Lionel Cinq-Mars et al. (Nat. can. (Que.) 98(2): 194. 1971) as introd. at Quebec City, Que. *C. longispinus* may be the species involved but there remains the possibility that all of the Ont. material may eventually prove referable to the more southern, introduced species.

CINNA L.   [241]   Wood Reedgrass

1 Spikelets to 6 mm long; glumes strongly unequal; anthers over 1 mm long; leaves to about 4 dm long and 12 mm broad; (s Ont. and s Que.) . . . .*C. arundinacea*
1 Spikelets to 4.5 mm long; glumes subequal; anthers less than 1 mm long; leaves to 1.5 cm broad but rarely over 2.5 dm long; (transcontinental) . . . .*C. latifolia*

C. arundinacea L.

/T/EE/ (Hs) Swamps and moist woods from Minn. to s Ont. (N to Carleton, Russell, and Stormont counties), Que. (N to Mt. Tremblant Park, N of Montreal), and cent. Maine, s to Tex., Ark., Tenn., and Ga.; reports from P.E.I. are discounted by D. S. Erskine 1960. [*Agrostis (Muhlenbergia) cinna* Retz.].

C. latifolia (Trev.) Griseb.

/sT/X/EA/ (Hs) Woods, thickets, and clearings from the E Aleutian Is. and s Alaska (*see* Hultén 1942: map 104, p. 392) to sw Dist. Mackenzie, L. Athabasca, Man. (N to Flin Flon), Ont. (N to Big Trout L. at ca. 53°45′N), Que. (N to Richmond Gulf, Hudson Bay, the Koksoak R. at ca. 58°N, and the Côte-Nord), Labrador (N to the Hebron R. at ca. 56°N), Nfld., N.B., P.E.I., and N.S., s to Calif., N N.Mex., Tenn., and N.C.; Eurasia. [*Agrostis* Trev.; *C. expansa* Link; *C. (Muhlenbergia) pendula* Trin. and its vars. *acutiflora* Vasey and *glomerata* Scribn.; *C. arundinacea* var. *pend.* (Trin.) Gray]. MAPS: Hultén 1968*b*: 95, and 1962: map 75, p. 85; J. M. Gillett, Can. Field-Nat. 74(1): fig. 4, p. 20. 1960.

COLPODIUM Trin.   [379]

1 Panicle commonly less than 5 cm long; spikelets to about 7 mm long; anthers about 1 mm long; culms commonly less than 2 dm tall; (widespread arctic species) . . . .*C. vahlianum*
1 Panicle to about 1 dm long, its branches smooth or somewhat scabrous, spreading or ascending (or the lowest ones sometimes deflexed); spikelets to 9 mm long; anthers about 2.5 mm long; culms to about 5 dm tall; (Alaska) . . . .*C. wrightii*

C. vahlianum (Liebm.) Nevski

/AS/X/GEA/ (Hs) Moist fresh soils (often in clays by streams) from N Banks Is. to SE Boothia Pen. and northernmost Ellesmere Is., s to sw Yukon, w Dist. Mackenzie, Somerset Is., and northernmost Ungava (known only from Port Burwell, 60°25′N); w and E Greenland N of ca. 70°N; Iceland; Spitsbergen; Novaya Zemlya; arctic Asia. [*Poa* Liebm.; *Puccinellia* Scribn. & Merr.]. MAPS: Hultén 1968*b*: 148, and 1962: map 1, p. 9; Porsild 1957: map 42, p. 166.

A hybrid with *Phippsia algida* is reported from Ellesmere Is. by Porsild (1955; × *Puccinellia vacillans* (Fries) Schol.).

C. wrightii Scribn. & Merr.

/aS/W/eA/ (Hs) Known only from the Seward Pen., NW Alaska, and from the type locality on Arakamtchatch Is., on the Siberian side of Bering Strait. According to Hultén (1968*b*), the report from the Yukon by Porsild (1951*a*) is based upon *Poa vaseyochloa*. [*Poa* Hitchc.]. MAPS: Hultén 1968*b*: 148, and 1942: map 161, p. 397.

[CORYNEPHORUS Beauv.]   [269]

[C. canescens (L.) Beauv.]
[Eurasian; at one time well established at Jericho Air Station, Vancouver, B.C., but apparently extinct at the date of its report by Eastham (1947). *(Aira* L.).]

CYNODON Richard   [282]

C. dactylon (L.) Pers.   Bermuda Grass, Scutch-Grass
Eurasian; introd. in grasslands and waste places of sw B.C. (Vancouver Is.), Ont. (Toronto; Hamilton; Ottawa), and St-Pierre and Miquelon (Louis Arsène, Rhodora 29(346): 207. 1927). [*Panicum* L.].

CYNOSURUS L.   [373]

1   Panicle narrow and spike-like, to about 1 dm long and 1 cm thick; lemma-awns mostly not
    over 1 mm long; glumes of sterile spikelets merely mucronate; ligules barely 1.5 mm long;
    culms to about 8 dm tall; perennial; (introd., transcontinental) . . . .*C. cristatus*
1   Panicle ellipsoid or subglobose, head-like, 2 or 3 cm thick; lemma-awns to about 1 cm
    long; glumes of sterile spikelets long-awned; ligules to about 1 cm long; culms to about
    4 dm tall; annual; (introd. in sw B.C.) . . . .*C. echinatus*

C. cristatus L.   Crested Dog's-tail
Eurasian; introd. in grasslands and waste places of sw B.C. (Vancouver Is.), Ont. (Middlesex, Huron, Waterloo, Peel, York, and Carleton counties), Que. (Montreal; Lanoraie, Berthier Co.; Duchesnay, Portneuf Co.; Quebec City; Gaspé Basin), St-Pierre and Miquelon, Nfld. (Avalon Pen. and Bay of Islands), and N.S. (Kings, Cape Breton, and Inverness counties).

C. echinatus L.   Hedgehog Dog's-tail
European; introd. along sandy shores in sw B.C. (Victoria and Nanaimo, Vancouver Is.; CAN).

DACTYLIS L.   [372]

D. glomerata L.   Orchard-Grass.   Dactyle
Eurasian; introd. along roadsides and in dry fields and waste places in N. America (ranges of Canadian taxa outlined below).
1   Keels of glumes and lemmas merely scabrous or short-hispid; [introd. at the mouth of
    the Main R., Bonne Bay, Nfld., and in N.S.] . . . . .var. *detonsa* Fries
1   Keels of glumes and lemmas long-ciliate.
    2   Glumes and lemmas pubescent on the back; [introd. in N.B. and N.S.]. . . . .
        . . . . .var. *ciliata* Peterm.
    2   Glumes and lemmas essentially glabrous on the back; [introd. in Alaska (N to
        Middleton Is. at 59°29′N), B.C., Alta., Sask. (N to Waskesiu Lake, 53°55′N), Man. (N to
        Brandon and Winnipeg), Ont. (N to the N shore of L. Superior), Que. (N to Anticosti Is.),
        Nfld., N.B., P.E.I., and N.S.; MAPS (aggregate species): Hultén 1962: map 189,
        p. 201, and 1968b: 126] . . . . .var. *glomerata*

DANTHONIA DC.   [280]   Wild Oat-Grass

1   Panicle reduced to a single spikelet (or an abortive spikelet, rarely a perfect one, below),
    the pedicel commonly about 12 mm long; leaf-blades and sheaths pilose; (B.C. to s
    Sask.) . . . .*D. unispicata*
1   Panicle with usually at least 3 spikelets.
    2   Longer glumes to over 2 cm long; longer lemmas about 1 cm long; (s B.C. and sw
        Alta.).
        3   Lemmas glabrous on the back, pilose on the margins toward base; panicle with

mostly not more than 5 spikelets, open, the slender pedicels spreading or some-
what reflexed, more or less flexuous; (s B.C. to sw Sask.) . . . .*D. californica*

3 Lemmas sparingly to rather densely pilose over the back; panicle often with up
to 8 spikelets, compact, the pedicels appressed or ascending; (sw Alta. and sw
Sask.) . . . .*D. parryi*

2 Longer glumes commonly not over 1.5 cm long; longer lemmas at most about 8 mm
long.

4 Lemmas to 8 mm long, glabrous on the back, hairy only on the margins, their
awn-tipped teeth to about 3 mm long; panicle purplish or bronze, dense, to 7.5 cm
long, simple or the lowest appressed branches bearing 2 spikelets; leaves to 3 mm
broad; (essentially transcontinental) . . . .*D. intermedia*

4 Lemmas to 6.5 mm long, commonly pilose on the back.

5 Lemma-teeth about 3 mm long; panicle to 1 dm long, open, the loosely
ascending or spreading lower branches bearing 2 or 3 spikelets; leaves to 4
mm broad, the basal ones half to quite as long as the culm, this usually
geniculate at base; (Ont. to N.S.) . . . .*D. compressa*

5 Lemma-teeth less than 2 mm long; panicle usually less than 5 cm long,
subsimple, its branches finally tightly appressed, the lower ones mostly with
only 1 spikelet; leaves 1 or 2 mm broad, the basal ones much shorter than the
culm; (transcontinental) . . . .*D. spicata*

D. californica Boland.   California Oat-Grass

/T/WW/   (Hs)   Meadows and open woods from sw B.C. (Vancouver Is. and adjacent islands and
mainland; Chilliwack L.; Cascade; Kootenay R.), s Alta. (Moss 1959), and s Sask. (Cypress Hills;
Dinsmore; Mortlach; Old Wives L.; Red Deer Lakes) to Calif. and N.Mex.; Chile.

Most of our material appears to belong to var. *americana* (Scribn.) Hitchc. (*D. americana*
Scribn.; *D. macounii* Hitchc.; leaves more or less densely spreading-pilose) but Boivin (1967a)
reports the typical form (leaves glabrous) from Vancouver Is., B.C.

D. compressa Aust.

/T/EE/   (Hs)   Woodlands and clearings from Ont. (N to the Ottawa dist.) to Que. (N to
Cap-Rouge, near Quebec City; Groh and Frankton 1949), P.E.I. (D. S. Erskine 1960; not known
from N.B.), and N.S., s to Ohio, Tenn., and Ga.

This species is scarcely separable from *D. spicata* and is not listed by Boivin (1967a).

D. intermedia Vasey   Timber Oat-Grass

/sT/X/eA/   (Hs)   Meadows, peats, and gravels from s Alaska and s-cent. Yukon to L. Athabasca
(Alta. and Sask.), Man. (Duck Mt.; Riding Mt.), Ont. (N to the Fawn R. at ca. 55°30'N; CAN), Que. (N
to Great Whale R., SE Hudson Bay, at ca. 55°15'N, Carol L. at 53°04'N, and Anticosti Is.; type from
Mt. Albert, Gaspé Pen.), Labrador (N to Makkovik, 55°10'N; CAN), and Nfld. (not known from the
Maritime Provinces), s to N Calif., Ariz., N.Mex., S.Dak., and N Mich.; E Asia. [*Trisetum ?williamsii*
Louis-Marie; *see* Hultén 1942]. MAPS: Hultén 1968b: 121; Porsild 1966: map 10, p. 68; Fernald
1924: map 1, p. 560, and 1925: map 6, p. 251.

D. parryi Scribn.

/T/WW/   (Hs)   Grassland, open woods, and rocky slopes from sw Alta. (Waterton Lakes;
Pincher Creek; Crowsnest Pass; Elbow R.; Cardston; The Gap, Livingstone Valley) and sw Sask.
(Walsh; CAN) to Calif. and N.Mex.

D. spicata (L.) Beauv.   Poverty-Grass

/aST/X/   (Hs)   Sterile rocky or sandy open soil and thin woodlands (ranges of Canadian taxa
outlined below), s to Oreg., N.Mex., Tex., S.Dak., Ala., and N Fla.; introd. in southernmost
Greenland. MAP and synonymy: *see* below.

1 Glumes lance-attenuate, strongly nerved, their overlapping portion confined to the basal
quarter of the longer one; basal marcescent leaves strongly curving and twisting; [*Avena*
L.; *D. allenii* Aust.; Que. to N.S.); MAP (aggregate species): Hultén 1968b: 122] . . . .var. *spicata*

1 Glumes relatively broader, weakly nerved (apart from the midrib), their overlapping

portion up to half the length of the longer one; basal leaves scarcely curved or twisted; [*D. thermalis* Scribn.; SE Alaska–B.C. to L. Athabasca (Alta. and Sask.), Man. (N to Norway House, off the NE end of L. Winnipeg), Ont. (N to Sandy L. and w James Bay, both at ca. 53°N), Que. (N to Fort George, E James Bay, ca. 53°45′N, L. Mistassini, and the Côte-Nord), Labrador (N to 55°27′N; Hustich and Pettersson 1943), Nfld., N.B., P.E.I., and N.S.] . . . . .var. *pinetorum* Piper

D. unispicata (Thurb.) Munro
/T/WW/ (Hs) Open, often rocky or sandy ground from B.C. (N to François Lake, ca. 54°N), S Alta. (Crowsnest Pass and Pincher Creek; CAN), and sw Sask. (Cypress Hills; CAN) to Calif. and Colo. [*D. californica* var. *uni.* Thurb.].

DESCHAMPSIA Beauv.   [270]   Hairgrass.   Canche

1   Panicle narrow, to 3 dm long, the remote capillary branches appressed-erect; glumes to 6 mm long, equalling or slightly surpassing the florets; lemma-awn straight or nearly so; leaves soft, those of the stem to 1.5 mm broad, those of the basal tuft filiform; perennial; (B.C.–Alta.) . . . .*D. elongata*
1   Panicle usually open (narrow in *D. caespitosa* ssp. *beringensis*), the branches spreading or spreading-ascending, or the lower ones sometimes drooping.
  2   Annual, the slender culms to about 6 dm tall; leaves few, short and narrow; glumes to about 7 mm long, distinctly surpassing the florets; lemmas 2 or 3 mm long, their geniculate awns to over 6 mm long, long-exserted; panicle-branches commonly in pairs, stiffly ascending; (B.C.–Alta.) . . . .*D. danthonioides*
  2   Perennials, the usually more numerous leaves mostly at least 1/3 the length of the culm.
    3   Glumes much surpassing the florets, mostly about 5 mm long; lemmas 3 or 4 mm long, their geniculate awns inserted at about the middle of the lemma, included or those of the upper florets barely exserted; (transcontinental) . . . .*D. atropurpurea*
    3   Glumes shorter than or barely surpassing the florets.
      4   Spikelets commonly viviparous and lacking normal florets; glumes acuminate; lemma usually with long-acuminate apical teeth, awnless or awned above the middle (sometimes from near the apex in non-viviparous spikelets); leaves short and stiff; culms densely tufted, to about 4 dm tall; (NE Canada) . . . .*D. alpina*
      4   Spikelets with normally perfect florets.
        5   Lemma-awns strongly geniculate and exserted, to 7 mm long, inserted toward the base of the minutely scabrous, 4-toothed lemma; spikelets to 5 mm long; panicle-branches nearly smooth; leaves involute-setaceous; (introd. in Alaska–B.C.; native from Ont. to Labrador and Nfld.) . . . . . .*D. flexuosa*
        5   Lemma-awns straight or somewhat curved but not geniculate, from straight and included in the glumes to weakly geniculate and twice as long as the spikelet, usually inserted near the base of the glabrous, erose-truncate lemma but sometimes (var. *mackenzieana*) near or above its middle; spikelets to 7 mm long; panicle-branches usually distinctly scabrous; leaves typically flat or folded and only tardily involute, to 4 mm broad; (transcontinental) . . . .*D. caespitosa*

D. alpina (L.) R. & S.
/aS/E/GEA/ (Hs) Wet, gravelly or sandy freshwater shores and alpine meadows of Resolute Is. (SE of Baffin Is. in Hudson Strait) and N Labrador (S to ca. 57°N); w Greenland N to 67°44′N, E Greenland N to 66°36′N; Iceland; N Eurasia. [*Aira* L.]. MAPS: Hultén 1958: map 203, p. 223; Porsild 1957: map 24, p. 163.

D. atropurpurea (Wahl.) Scheele   Mountain Hairgrass
/aST/(X)/GEeA/ (Hs (Grh)) Woods and wet meadows: Aleutian Is.–S Alaska–S Yukon–sw Dist. Mackenzie through B.C. and sw Alta. to Calif. and Colo.; an isolated area in SE Dist. Keewatin (Eskimo Point) and N Man. (Baralzon L., ca. 60°N); northernmost Ungava–Labrador S to E Que.

(Laurentide National Park, Montmorency Co.; L. Mistassini; Knob Lake; Côte-Nord; Gaspé Pen.), w Nfld., and mts. of N N.Y. and N New Eng.; w and E Greenland N to ca. 65°N; N Europe; E Asia. [*Aira* Wahl.; *Vahlodea* Fries; incl. var. *minor* Vasey, var. *latifolia* (Hook.) Scribn. (*Aira lat.* Hook.; *D. lat.* (Hook.) Vasey), and ssp. *paramushirensis* (Kudo) Hult. and its var. *patentissima* Hult., all completely intergrading with the typical form in floral dimensions]. MAPS: Hultén 1968*b*: 115, and 1958: map 186, p. 205; Tolmachev 1952: map 10 (requiring extensive revision).

D. caespitosa (L.) Beauv.   Tufted Hairgrass
/AST/X/GEA/ (Hs) Damp (often calcareous) soils and shores, the aggregate species from the coasts of Alaska–Dist. Mackenzie to Prince Patrick Is., northernmost Ellesmere Is., northernmost Ungava–Labrador, Nfld., N.B., P.E.I., and N.S., s to Calif., Mexico, N.Mex., Minn., Ill., and Va.; w Greenland N to ca. 83°N, E Greenland between ca. 65° and 79°N; Iceland; Eurasia. MAPS and synonymy: *see* below.
1 Awn inserted near or above the middle of the lemma; [*D. mackenzieana* Raup, the type
    from L. Athabasca, Sask., the only known locality] .... var. *mackenzieana* (Raup) Boivin
1 Awn inserted toward the base of the lemma.
    2 Panicles diffuse, to over 4 dm long; leaves flat or only tardily involute, the basal ones
        to about 6 dm long, with ligules to over 1 cm long; lower culm-leaves to 4 dm long;
        culms to over 1.5 m tall and up to 6 mm thick at base.
        3 Spikelets mostly at least 3.5 mm long; [*Aira* L.; *D. (A.) pungens* Rydb.; *A. ?vivipara*
            Steud.; transcontinental; MAPS (aggregate species); Porsild 1957: map 25, p. 164;
            Hultén 1968*b*: 111, and 1962: map 58A, p. 67; Shoichi Kawano, Can. J. Bot.
            41(5): fig. 9, p. 733. 1963] .... var. *caespitosa*
        3 Spikelets 2 or 3 mm long; [introd. in N.B. and N.S., *see* D. S. Erskine, Rhodora
            53(635): 266. 1951] .... var. *parviflora* (Thuill.) Richter
    2 Panicles rarely much over 2 dm long; leaves often involute, the basal ones commonly
        not over 3 dm long, their ligules at most 7 mm long; lower culm-leaves less than 1 dm
        long; culms less than 8 dm tall and at most 2.5 mm thick at base; [transcontinental].
        4 Spikelets less than 5 mm long; panicle lax, rather diffuse in anthesis; [incl. vars.
            intercotidalis and *abbei* Boivin; *D. glauca* Hartm.; *Aira ambigua* Michx.,
            basionym, the type probably from L. Mistassini, Que.; MAP: Hultén 1968*b*: 112] ....
            ....var. *glauca* (Hartm.) Lindm. f.
        4 Spikelets to 7 mm long; panicle commonly contracted in anthesis; [vars. *alpina*
            and *?maritima* Vasey and *?longiflora* Beal; ssp. *orientalis* Hult. in large part; *D.
            bottnica sensu* John Macoun 1888, not Wahl.; incl. the reduced arctic extremes,
            D. beringensis Hult., *D. ?holciformis* Presl, *D. pumila* (Ledeb.) Ostenf., and *D.
            brevifolia* R. Br. *(Aira (D.) arctica* Spreng.; *A. curtiflora* Rydb.; *D. brachyphylla*
            Nash, not Phil.); MAPS: Hultén 1968*b*: 112, and 1962: map 58B (ssp. *orientalis*),
            p. 67] ....var. *littoralis* (Reut.) Richter

D. danthonioides (Trin.) Munro   Annual Hairgrass
/aST/W/ (T) Open ground from cent. Alaska–Yukon (probably introd.; *see* Hultén 1942: map 122 (*D. danth.*), p. 394) through B.C. and sw Alta. (Lake Louise; CAN) to Baja Calif. and Mexico; Chile. [*Aira* Trin.; *D. calycina* Presl]. MAP: Hulten 1968*b*: 111.

D. elongata (Hook.) Munro   Slender Hairgrass
/aST/W/ (Hs) Open ground from cent. Alaska and s Yukon (*see* Hultén 1942: map 123, p. 394) through B.C. and sw Alta. (Waterton Lakes; Castle Mt., near Banff; Porcupine Hills) to Baja Calif., N.Mex., and Mexico; Chile. [*Aira* Hook.]. MAP: Hultén 1968*b*: 110.
    The larger-dimensioned extreme, var. *ciliata* Vasey (*D. ciliata* (Vasey) Rydb.) is reported from B.C. by Rydberg (1922; this confirmed by a report from Nanaimo, Vancouver Is., by Bernard Boivin, Nat. can. (Que.) 75: 83. 1948).

D. flexuosa (L.) Trin.   Common Hairgrass
/aST/EE/GEeA/ (Hs) Dry open soil and thickets (ranges of Canadian taxa outlined below), s to Okla., Tenn., and Ga.; Greenland; Europe; an isolated station (?introd.) in w Siberia; E Asia. MAPS and synonymy: *see* below.

1	Spikelets to 7 mm long; panicle contracted; [*D. alba* R. & S.; Que. (E James Bay; cent. Ungava at ca. 55°N; Côte-Nord; Anticosti Is.; Gaspé Pen.), Labrador (N to Port Burwell, 60°25′N), Nfld., and N.S. (St. Paul Is. and Sable Is.); w and E Greenland N to ca. 67°N] . . . . .var. *montana* (L.) Ledeb.
1	Spikelets 4 or 5 mm long; panicle very loose and open . . . . .var. *flexusoa*
    2	Spikelets pale; [Moose R., Cumberland Co., N.S.; CAN] . . . . .f. *flavescens* Slyvén
    2	Spikelets bronze or purplish; [*Aira (Avena; Avenella) flex.* L.; Ont. (N to Renison, near James Bay at ca. 51°N), Que. (N to Fort George, E James Bay, 53°50′N, N Ungava at ca. 59°N, L. Mistassini, and the Côte-Nord), N Labrador (Komaktorvik, 59°22′N), Nfld., N.B., P.E.I., and N.S.; introd. in the Aleutian Is. (Attu), s Alaska (Sitka; Hultén 1942), and s B.C. (Vancouver; Eastham 1947); MAPS: Hultén 1958: map 232, p. 251, and 1968*b*: 114; Meusel, Jaeger and Weinert 1965: 51] . . . . .f. *flexuosa*

[DIARRHENA Beauv.]	[356]

[D. americana Beauv.]
[The outline-map of the area of this species of the E U.S.A. (N to S.Dak., Mich., Ohio, and W.Va.) by Tetsuo Koyama and Shoichi Kawano (Can. J. Bot. 42(7): fig. 9 (C), p. 872. 1964) includes the s tip of the Niagara Pen. along the shores of L. Erie, s Ont., probably erroneously, no Canadian reports having been found by the present writer. It should be searched for in s Ont.]

DIGITARIA Heist.	[166]	Finger-Grass, Crab-Grass

1	Upper glume about as long as the dark-brown or purple-black fertile lemma; lower glume rudimentary or obsolete; sterile lemma (and upper glume) more or less short-glandular-villous, especially between the nerves; racemes at most 6, commonly purplish, rather remote, less than 1 dm long; leaves (to 6 mm broad), sheaths, and culms glabrous, the culms to about 4 dm tall . . . . *D. ischaemum*
1	Upper glume rarely more than half as long as the pale greenish-brown fertile lemma; lower glume small and often deciduous; sterile lemma usually scabrous on the 5 strong nerves and minutely ciliate on the margins, otherwise glabrous; racemes up to 10 or more in a crowded subdigitate cluster at the top of the culm, to about 2 dm long; leaves (to 1 cm broad), sheaths, and culm-nodes more or less pilose, the culms to over 1 m tall . . . .
    . . . .*D. sanguinalis*

D. ischaemum (Schreb.) Muhl.	Small Crab-Grass
Eurasian; fields, roadsides, and waste places of s B.C. (N to Kamloops), sw Alta., s Man. (Winnipeg; not known from Sask.), Ont. (N to Cobalt, 47°25′N), Que. (N to Charlevoix and Rimouski counties), N.B. (Hampton; Fredericton; Sussex), P.E.I. (Dundee, Kings Co.; Grand Tracadie, Queens Co.), and N.S. [*Panicum* Schreb.; *Bothriochloa* Keng; *Syntherisma* Nash; *D. (S.) humifusa* Pers.; *P. glabrum* Gaud.]. MAP: Hultén 1962: map 213, p. 225.

D. sanguinalis (L.) Scop.	Large Crab-Grass
Eurasian; fields, roadsides, and waste places of s B.C. (Agassiz; Kamloops), s Alta. (Moss 1959), s Sask. (Breitung 1957*a*), s Man. (Brandon), Ont. (N to Ottawa), Que. (Montreal; Granby; ?St-Pierre and Miquelon (Boivin 1967*a*), P.E.I. (Charlottetown; not known from N.B.), and N.S. (Boivin 1967*a*). [*Panicum* L.; *Syntherisma* Dulac]. MAP: Hultén 1962: map 224, p. 235.

DISTICHLIS Raf.	[366]	Spike-Grass, Alkali-Grass

1	Leaves with mostly smooth margins and bluntish tips; spikelets less than 1.5 cm long; lower glume to 3.5 mm long; upper glume to 4 mm long; grain to 2 mm long, scarcely beaked; panicle compact, the spikelet-pedicels mostly not evident; (s B.C. to N.S.) . . . .
    . . . .*D. spicata*
1	Leaves usually scabrous on the margins and the sharp tip; spikelets to 2.5 cm long; lower glume to 8 mm long; upper glume to 7 mm long; grain at least 3 mm long, distinctly beaked; panicle looser, the spikelet-pedicels mostly discernible; (B.C. to Man.) . . . . . .*D. stricta*

Gramineae

D. spicata (L.) Greene
/T/D (coastal)/ (Grh) Coastal salt marshes: sw B.C. (Vancouver Is. and adjacent islands and mainland) to Baja Calif.; s N.B. (Shediac, Westmorland Co.; Sussex, Kings Co.), P.E.I. (Queens Co.), and N.S. to Fla. and Tex.; W.I.; S. America. [*Uniola* L.; *Brizopyrum* Hook. & Arn.; *B. (Poa) boreale* Presl]. MAP: J. A. Steyermark, Rhodora 42(493): 23. 1940.

D. stricta (Torr.) Rydb.
/T/WW/ (Grh) Coastal salt marshes and damp or wet alkaline flats in the interior from B.C. (introd. in SE Alaska) and sw Dist. Mackenzie (near Fort Smith at ca. 60°03′N; W. J. Cody, Can. Field-Nat. 70(3): 104. 1956) through Alta. to Sask. (N to Saskatoon) and Man. (N to Dawson Bay, L. Winnipegosis), s to Baja Calif., Mexico, Tex., and Mo. [*Uniola* Torr.; *D. dentata* Rydb. at least in part; *D. maritima (spicata)* var. *str.* (Torr.) Thurb.].

DUPONTIA R. Br. [380]

D. fisheri R. Br.
/AS/X/GEA/ (Grh (Hel)) Brackish or saline coastal marshes (ranges of N. American taxa outlined below; not known in the U.S.A.); N Eurasia. MAPS and synonymy: *see* below.
1   Panicle relatively loose, its usually elongate spreading branches bearing 1–2-flowered spikelets; glumes acute or acuminate, equalling the uppermost lemma; lemmas glabrous or nearly so; anthers at most 2 mm long; [*D. psilosantha* Rupr.; *D. micrantha* Holm; E Aleutian Is. and coasts of Alaska, the Yukon, and Dist. Mackenzie to N-cent. Baffin Is. at ca. 70°N and northernmost Ungava–Labrador, s to cent. Alaska, s Dist. Keewatin, N-cent. Man. (s to York Factory, Hudson Bay), w and E James Bay (s to ca. 52°N), and s-cent. Labrador (s to Holton Harbour, ca. 54°30′N); w Greenland between ca. 68° and 72°N, E Greenland between ca. 73° and 77°N; MAPS: Hultén 1968b: 150, and 1962: map 3, p. 11; Porsild 1957: map 44, p. 166] . . . . .ssp. *psilosantha* (Rupr.) Hult.
1   Panicle more or less contracted, its short ascending branches bearing 2–4-flowered spikelets; glumes often obtuse and shorter than the spikelet; lemmas appressed-hirsute along the keel and margins; anthers to over 2.5 mm long . . . . .ssp. *fisheri*
  2   Lemmas tipped with a short blackish awn; [NE Ont. (20 mi s of Cape Henrietta Maria, NW James Bay, at ca. 55°N), Baffin Is. (type from Pond Inlet), and northernmost Que. (Cape Smith, NE Hudson Bay, ca. 61°N, to Wolstenholme, 62°35′N, and Wakeham Bay, Hudson Strait, ca. 61°30′N] . . . . .var. *aristata* Malte
  2   Lemmas awnless; [coasts of Alaska–Yukon–Dist. Mackenzie, throughout the Canadian Arctic Archipelago to Ellesmere Is. at 81°25′N, s to the islands of N Hudson Bay (Southampton, Coats, Mansel) and s Baffin Is.; w Greenland at ca. 79°N; MAPS: on the above-noted map by Hultén; Hultén 1968b: 149; Porsild 1957: map 43, p. 166; Tolmachev 1952: map 8 (aggregate species)] . . . . .var. *fisheri*

ECHINOCHLOA Beauv. [166]

1   Leaves at most 6 mm broad; panicle-branches simple, usually less than 3 cm long, the lower ones distant as much as 1 cm; panicle pale green, its nodes and rachises with few or no bristles; spikelets awnless (awn of sterile lemma reduced to a short point), more or less 4-rowed; (introd. in B.C.) . . . .[*E. colonum*]
1   Leaves to 2 or 3 cm broad; panicle darker green to purple, its lower branches commonly more or less compound (of crowded glomerules); spikelets usually not arranged in rows.
  2   Upper glume with an awn to about 1 cm long; spikelets ellipsoid, about 1/3 as broad as long, strigose with appressed non-pustular-based trichomes; grain lanceolate to lance-ellipsoid; leaf-blades harshly scabrous; lower leaf-sheaths normally spreading-hispid; (Ont. and sw Que.) . . . .*E. walteri*
  2   Upper glume awnless or short-awned; spikelets ovoid, about 1/2 as broad as long; grain ovoid or oval; leaf-blades smooth or slightly scabrous, their sheaths glabrous; (introd., transcontinental) . . . .*E. crusgalli*

[E. colonum (L.) Link]   Jungle-rice
[Eurasian; reported from B.C. by Hubbard (1955) with, however, no localities nor indication of its being other than a casual ephemeral.]

E. crusgalli (L.) Beauv.   Barnyard-Grass.   Pied-de-coq
Eurasian; Alaska–Canada ranges, MAP, and synonymy: see below.
1   Panicle very dense, chocolate-purple, the ascending branches closely overlapping and
      with incurved tips; spikelets very turgid, awnless; sterile lemma soft-tipped, minutely and
      sparsely pubescent; lower palea purple; [Panicum (E.) frumentaceum Link; Alta. (Moss
      1959), s Man. (Brokenhead, about 30 mi NE of Winnipeg), s Ont. (Wellington Co.; Stroud
      1941), and N.S. (Dore and Roland 1942)] ....var. frumentacea (Link) Wight
1   Panicle relatively open, green to purple, the straight branches loosely ascending or
      spreading; spikelets scarcely turgid; sterile lemma apiculate and firm-tipped, it and the
      glumes more or less appressed-hispid; lower palea whitish; [transcontinental] ....var. crusgalli
      2   Spikelets with awns to about 4 mm long ....f. longiseta (Trin.) Farw.
      2   Spikelets awnless or very short-awned; [incl. var. mitis (Pursh) Peterm.; Panicum
            (Milium; Oplismenus; Orthopogon) L.; E. microstachya (Wieg.) Rydb.; E. (Panicum)
            muricata (Michx.) Fern. and vars.; E. (P.) pungens (Poir.) Rydb. and vars.; s Alaska
            Panhandle; B.C. to Alta. (N to Edmonton), Sask. (N to Prince Albert), Man. (N to
            Dauphin, N of Riding Mt.), Ont. (N to near Sault Ste. Marie), Que. (N to the Gaspé Pen.
            at Paspébiac), Nfld., N.B., P.E.I., and N.S.; MAP: Hultén 1962: map 220, p. 231] ....
            ....f. crusgalli

E. walteri (Pursh) Nash
/T/EE/   (T)   Saline or alkaline marshes, swamps, and shallow water from Minn. to Ont. (N to the
Ottawa dist.), Que. (N to Sorel, Richelieu Co.; MT), N.Y., and N.H., s to Tex and Fla. [Panicum
Pursh; P. crusgalli var. hispidum Ell.]. MAP: McLaughlin 1932: fig. 4 (requiring considerable
expansion), 341.

ELEUSINE Gaertn.   [304]

E. indica (L.) Gaertn.   Wiregrass
Eurasian; sandy yards and waste places of s Ont. (shores of L. Erie and L. Ontario in Essex, Kent,
Wentworth, and York counties) and sw Que. (Montreal). [Cynosurus L.]. MAP: Hultén 1962: map
226, p. 237.

ELYMUS L.   [411]   Wild Rye, Lyme-Grass

(Ref.: Church 1967; hybrids are listed at the end of the following treatment)
1   Lemmas awnless or merely mucronate, or with an awn much shorter than the lemma-
      body; anthers commonly at least 5 mm long.
      2   Plants with long extensively creeping stout rhizomes.
            3   Glumes very narrow, rigidly setaceous (one or both of a pair sometimes reduced
                  or wanting), the body to about 1 cm long and 0.4 mm broad, the narrow margins
                  thickened and hard, nerved; lemmas densely purplish-greyish-villous over the
                  back (rarely glabrate); leaves to 8 mm broad, mostly finely nerved above; culms
                  and rhizomes relatively slender; (B.C. to cent. Ont.) ....E. innovatus
            3   Glumes linear to linear-lanceolate, not setaceous, the body to over 1.5 cm long
                  and 3 mm broad, the thin and more or less membranaceous, narrow or broad
                  margins nerveless; lemmas silky-villous to near or above the middle; leaves to 1.5
                  cm broad, coarsely nerved, they and the stout culms often glaucous; rhizomes
                  coarse; (plants of sandy shores).
                  4   Summit of culm usually glabrous; rachis of spike usually glabrous except for
                        the long-hirsute margins and nodes; glumes rather firm (the prominent
                        midnerve stiff), long-acuminate, usually glabrous at base, their awns to about
                        3 mm long; (introd. in s B.C., Que., and Nfld.) ....E. arenarius
                  4   Summit of culm usually puberulent to villous; rachis puberulent to villous or

silky or partly glabrate; glumes thin and papery, flexible, acuminate, hirsute at base, their awns usually not over 1 mm long; (transcontinental on sandy shores) . . . .*E. mollis*

2   Plants tufted or with very short rhizomes; glumes subulate or rigidly setaceous.
    5   Spike very brittle and with a strongly jointed rachis, to 11 cm long; spikelets (1)2(3)-flowered; rachis-internodes about 3 mm long; glumes scabrous or short-hairy; lemmas to 8 mm long, short-hairy to densely villous; leaves scabrous on both sides, glaucous- or greyish-green, nearly all in a large dense basal tuft, their margins inrolled, their sheaths becoming conspicuously fibrillose; culms rarely over 8 dm tall; (introd. in Sask. and Man.) . . . .*E. junceus*
    5   Spike with a continuous flexible rachis, to 2 dm long; spikelets with up to 7 florets; rachis-internodes at least 5 mm long; glumes scabrous; lemmas usually glabrous in the centre, short-hairy toward the margins (rarely glabrate or completely short-hairy); leaves thick and hard, to 13 mm broad; culms to 3 m tall, their nodes usually puberulent; (B.C. to sw Sask.) . . . .*E. piperi*

1   Lemmas long-awned, the awn usually 1–4 cm long and often much longer than the lemma-body; anthers to about 3.5 mm long; leaves finely nerved above; plants tufted or with very short slender rhizomes.
    6   Glumes linear to linear-lanceolate, not setaceous, the body to about 2 cm long and usually less than 1.5 mm broad, the narrow or broad margins thin and more or less membranaceous, not nerved at the edges; leaves to about 1 cm broad.
        7   Lemmas sparingly to densely hirsute on or near the margins above the middle; glume-bodies to about 1 cm long, usually much shorter than the body of the lowest lemma; (Alaska–B.C.) . . . .*E. hirsutus*
        7   Lemmas glabrous or merely scabrous on the margins above the middle.
            8   Glume-bodies less than 1 cm long and usually not over 1 mm broad, much shorter than the body of the lowest lemma; lemmas densely scabrous on the back; (Alaska and Dist. Mackenzie) . . . .*E. sibiricus*
            8   Glume-bodies to over 1.5 cm long and 1.5 mm broad, about equalling the body of the lowest lemma; lemmas scabrous above middle, otherwise glabrous or sparingly scabrous throughout; (B.C. to Ont.) . . . .*E. glaucus*
    6   Glumes setaceous, subulate, or linear (or one or both of a pair sometimes reduced or wanting), the body commonly less than 1 mm broad, the margins usually thickened and hard, narrow, nerved.
        9   Glumes linear to subsetaceous, their bodies commonly at least 1 cm long and 1 or 2 mm broad; leaves glabrous or merely scabrous (those of *E. virginicus* sometimes pilose).
          10   Glume-bodies shorter than the lemma-bodies, thickened and indurated only near the straight bases; lemma-awns divergently curved at maturity; palea of lowest flower to 1.5 cm long; leaves to 2.5 cm broad; (B.C. to N.S.) . . .*E. canadensis*
          10   Glume-bodies nearly as long as, to much longer than, the lemma-bodies; lemma-awns straight at maturity; palea of lowest flower rarely over 8 mm long.
              11   Glume-bodies to 2 mm broad, often much surpassing the lemma-bodies, very rigid (thickened and indurated from the curved base to near the apex); paleas about equalling the lemma-bodies; leaves to over 1.5 cm broad; (B.C. to Nfld. and N.S.) . . . .*E. virginicus*
              11   Glume-bodies to 1 mm broad, about equalling the lemma-bodies, thickened and indurated only near the straight base; paleas much shorter than the lemma-bodies; leaves to 12 mm broad; (Ont.–Que.) . . . .*E. riparius*
        9   Glumes setaceous or subsetaceous (or reduced, rudimentary, or even wanting), their bodies rarely over 6 mm long (excluding the awned tips) and usually less than 0.5 mm broad.
          12   Glumes of each spikelet nearly equal, both present, long-awned, the midnerve prominent; lemmas commonly villous, their awns straight; palea of lowest flower usually less than 7 mm long; leaves commonly soft-villous above, to 1 cm broad, their usually villous sheaths expanded at the summit into a broad firm horizontal flange; (Ont.–Que.) . . . .*E. villosus*

12 Glumes of each spikelet mostly unequal in length (at least one of the pair reduced, rudimentary, or even absent); palea of lowest flower at least 7.5 mm long; leaves merely pilose on the veins above, to 1.5 cm broad.

    13 Spikelets soon widely divergent to horizontally spreading; spikes erect at maturity; glumes usually reduced to short or minute awns, the lower one usually obsolete, both often wanting in the upper spikelets; lemmas glabrous or pubescent, their awns straight; (s Man. to N.B.) . . . .*E. hystrix*

    13 Spikelets closely appressed; spikes arching at maturity; both glumes of a pair usually present, one often reduced; lemmas strigose to silky, their awns finally divergently curved; (SE Sask. to w Ont.) . . . .*E. diversiglumis*

E. arenarius L.   Sea Lyme-Grass, Strand-Wheat.   Seigle de mer

European; introd. in N. America and Greenland, either accidentally in sand ballast from ships or purposefully as a sand-binder; known from s B.C. (Vancouver Is.; near Trail), sw Que. (clones derived from Scottish seed planted at Lachute, near Montreal), Nfld. (Alexander Bay, Bonavista North dist.), and w Greenland (at ca. 60°40′ and 64°10′N). MAPS: Hultén 1962: map 185, p. 197; Meusel, Jaeger, and Weinert 1965: 42; A. Löve 1950: fig. 4, p. 32 (indicating a major extension into w Asia, contrary to Bowden's opinion); Bowden 1957: fig. 1 (native European area) and fig. 2 (introd. E N. America–Greenland area), p. 957.

E. canadensis L.

/sT/X/ (Hs) Dry, sandy or gravelly soil, the aggregate species from B.C. to Great Slave L. (?introd. in s Alaska), Alta. (N to McMurray, 56°44′N), Sask. (N to Prince Albert), Man. (N to Elk Is., L. Winnipeg, ca. 51°N), Ont. (N to the Albany R. sw of James Bay at ca. 52°10′N), Que. (N to the Gaspé Pen.), N.B., and N.S. (not known from P.E.I.), s to Calif., Tex., Mo., Ohio, and N.C.

1 Leaves firm, usually glabrous or merely scabrous (rarely pilose) above; spikes arching; glumes commonly 1 or 2 mm long; palea to 11 mm long.

    2 Lemma-backs villous-hirsute; [var. *?albanensis* Lepage; *Clinelymus* Nevski; *Hordeum* Asch. & Graebn.; *E. interruptus* Buckl.; *E. philadelphicus* L.; *E. robustus* Scribn. & Sm. and its var. *vestitus* Wieg.; *H. patulum* Moench; Dist. Mackenzie to N.B.; ?introd. in Alaska] . . . . .var. *canadensis*

    2 Lemma-backs glabrous or merely scabrous; [f. *?glaucifolius* (Muhl.) Fern.; *E. brachystachys* Scribn. & Ball; B.C., Alta., Man., and Que.] . . . . .var. *brachystachys* (Scribn. & Ball) Farw.

1 Leaves thin, usually pilose-villous above; spikes pendulous; glumes mostly less than 1 mm broad; palea to 1.5 cm long . . . . .var. *wiegandii* (Fern.) Bowden

    3 Lemma-backs villous-hirsute; [*E. wiegandii* Fern.; Sask. to N.B. and N.S.] . . . . .f. *wiegandii*

    3 Lemma-backs glabrous or merely scabrous; [*E. wiegandii* f. *calvescens* Fern.; Man., Que., and N.B.] . . . . .f. *calvescens* (Fern.) Boivin

E. diversiglumis Scribn. & Ball

/T/WW/ (Hs) Rich moist soil and alluvial thickets and woods from Wyo. and N.Dak. to Sask. (Saskatoon; Oxbow; Moose Mountain), Man. (N to Riding Mt.), w Ont. (Pigeon R., NW shore of L. Superior; collection in MT from railway-ballast at Cochrane), and Minn. [*E. interruptus* of American auth. in part, not Buckl.].

E. glaucus Buckl.   Blue Wild Rye

/sT/X/ (Hs) Moist or dry open thickets, hillsides, and shores, the aggregate species from SE Alaska–B.C. to Alta. (N to Yellowhead Pass, near Jasper; CAN), sw Sask. (Cypress Hills), s Man. (N to Oak Point, near the s end of L. Manitoba), and Ont. (N to Pine Portage on the Nipigon R. N. of L. Superior; CAN; reported from near trading posts along w James Bay where probably introd., by Dutilly, Lepage, and Duman 1954), s to Calif., N.Mex., Ark., Mich., and w N.Y. MAPS and synonymy: see below.

1 Leaves and their sheaths pubescent; [reported from B.C. by A. S. Hitchcock 1935, and from Alta. by Moss 1959] . . . . .var. *jepsonii* Davy

1 Leaves and their sheaths glabrous or merely scabrous.

2 Lemma-awns mostly not over 5 mm long and much shorter than the lemma-body; glume-awns at most 2 mm long; [*E. virescens* Piper; *E. howellii* Scribn. & Merr.; *E. petersonii* Rydb.; SE Alaska (*see* Hultén 1942: map 200, p. 401) and B.C. (Vancouver Is.; Chilliwack R.; Glacier); MAP: Hultén 1968*b*: 196] . . . . var. *virescens* (Piper) Bowden

2 Lemma-awns mostly over 1 cm long (to over 2.5 cm) and usually about as long as or much surpassing the lemma-body; glume-awns to over 5 mm long; [*Clinelymus* Nevski; *E. americanus* Vasey & Scribn.; *E. marginalis* Rydb.; *E. nitidus* Vasey; incl. var. *hirsutus* Malte; SE Alaska–B.C. to Ont.; MAP: Hultén 1968*b*: 195] . . . . .var. *glaucus*

E. hirsutus Presl
/sT/W/ (Hs) Moist woodlands and open ground from the Aleutian Is. and s Alaska (*see* Hultén 1942: map 197, p. 401; Hultén believes that the type locality is probably Yukatat Bay, Alaska, rather than Nootka Sound, Vancouver Is., as indicated by Presl) through B.C. to Oreg. [*E. borealis* Scribn.; *E. ciliatus* Scribn., not Muhl.]. MAP: Hultén 1968*b*: 196.

E. hystrix L. Bottle-brush Grass
/T/EE/ (Hs) Rich moist woods (ranges of Canadian taxa outlined below), s to Okla., Ark., Tenn., and Ga.

1 Lemmas pubescent; [*Hystrix patula* f. *big.* (Fern.) Gl.; Ont. (N to Ottawa), Que, (N to Hull and Montreal), and N.B. (Woodstock and Eel R., Carleton Co.)] . . . . .f. *bigeloviana* (Fern.) Bowden

1 Lemmas glabrous; [*Asprella (Asperella; Hystrix) hystrix* Humb.; *Hordeum hystrix* Schrenck, not Roth; *Hystrix patula* Moench; s Man. (Roseisle, about 30 mi s of Portage la Prairie; reports from Sask. require confirmation), Ont. (N to the Ottawa dist.), and Que. (Pontiac, Gatineau, Vaudreuil, Iberville, Rouville, and Missisquoi counties; the report from Nfld. by Reeks 1873, is probably erroneous] . . . . .f. *hystrix*

E. innovatus Beal
/ST/WW/ (Grh) Rocky or sandy ground and open woodlands (ranges of Canadian taxa outlined below), s to Mont., Wyo., and S.Dak. MAPS and synonymy; *see* below.

1 Lemmas velvety-pubescent on the back; [Alaska; type from Mi 264, Richardson Highway, the only known locality] . . . . .ssp. *velutinus* Bowden

1 Lemmas more or less densely villous (rarely glabrate) on the back . . . . .ssp. *innovatus*

2 Lemmas-backs glabrate except for marginal and apical scabridity; [Alta. (type from Beaverlodge) and Sask.] . . . . .var. *glabratus* Bowden

2 Lemma-backs villous or densely villous . . . . .var. *innovatus*

3 Spike lax; some of the spikelets subtended by 4 glumes of unequal length; [type from 40 mi up the Attawapiskat R. near James Bay, Ont., at 52°57′N] . . . . .f. *laxatus* Lepage

3 Spike compact; spikes subtended by 2 glumes; [*E. brownii* Scribn. & Sm.; *E. mollis* R. Br., not Trin.; Alaska–B.C. to Alta. (N to Wood Buffalo National Park), Sask. (N to Methy Portage, on the Clearwater R. at ca. 56°30′N), Man. (N to Churchill), and Ont. (known only from the mouth of the Black Duck R., Hudson Bay, ca. 56°40′N, and the w James Bay region); MAPS (aggregate species): Hultén 1968*b*: 195; Raup 1930: map 18 (requiring expansion), p. 202] . . . . .f. *innovatus*

E. junceus Fisch.
Eurasian; Breitung (1957*a*) notes that this is "A valuable forage grass adapted to the drier conditions found on the prairies and plains of southern Saskatchewan" but he does not indicate it as an escape. However, it was taken on wasteland at Brandon, Man., by G. Stevenson in 1952, where definitely an escape.

E. mollis Trin. Sea Lyme-Grass, Strand-Wheat. Seigle de mer
/aST/X/GeA/ (Grh) Beaches and sands, the aggregate species from the coasts of Alaska–Yukon–Dist. Mackenzie–B.C. to Banks Is., N Baffin Is., and northernmost Ungava–Labrador, s to L. Athabasca (Alta. and Sask.), N Man. (s to York Factory, Hudson Bay, 57°N), Ont. (s to the N shore of L. Superior), Que. (s to SE James Bay and the St. Lawrence R. estuary from l'Islet Co. to

the Côte-Nord, Anticosti Is., Gaspé Pen., and Magdalen Is.), s Labrador, Nfld., N.B., P.E.I., and N.S., and along the coasts to cent. Calif. and Mass.; w Greenland N to ca. 71°30′N, E Greenland N to 63°32′N; Iceland; E Asia (the closely related *E. arenarius* in Europe; see Bowden 1957). MAPS and synonymy: see below.

1 Spikes to about 1 dm long; glumes less than 1.5 cm long, densely hirsute on the back; lemmas densely fine-pubescent; [*E. villosissimus* Scribn.; *E. arenarius* vars. *vill.* (Scribn.) Polunin and *brevispicus* (Scribn. & Sm.) Boivin; Alaska (type from St. Paul Is.) to Banks Is., N Baffin Is., Dist. Keewatin, and northernmost Que.; MAPS: Bowden 1957: fig. 5, p. 965; Hultén 1968*b*: 194] . . . . .ssp. *villosissimus* (Scribn.) Löve
1 Spikes to about 2.5 dm long; glumes to over 3 cm long, usually soft-hirsute on the back (occasionally subglabrous); lemmas soft-hirsute . . . . .ssp. *mollis* var. *mollis*
  2 Culm glabrous below the spike; [*E. arenarius* f. *sim.* (Bowden) Boivin; sw B.C., the type from near Vancouver] . . . . .f. *simulans* Bowden
  2 Culm with long areas of dense short pubescence below the spike.
    3 Glumes glabrous on the back except for short hairs on the nerves, especially toward the tips; [*E. arenarius* f. *scab.* (Bowden) Boivin; E Que., N.B., and P.E.I. (type from near Alberton)] . . . . .f. *scabrinervis* Bowden
    3 Glumes sparsely to densely soft-hirsute on the back, usually with long hairs; [*E. arenarius* var. *villosus* Mey. and ssp. *mollis* (Trin.) Hult.; *E. ampliculmis* Prov.; *E. capitatus* Scribn.; *E. dives* Presl; *E. ?interior* Hult.; transcontinental; MAPS: Hultén 1968*b*: 193, and 1962: map 185, p. 197; Bowden 1957: fig. 5, p. 965; Porsild 1957: map 57, p. 168; A. Löve 1950: fig. 4, p. 32; Meusel, Jaeger, and Weinert 1965: 43] . . . . .f. *mollis*

E. piperi Bowden
/T/W/ (Hs) Dry prairies, plains, sand-hills, and in the vicinity of hot springs from B.C. (N to Kamloops), s Alta. (N to Calgary and Medicine Hat), and s Sask. (Cypress Hills; Vanguard; Indian Head) to Calif., Colo., and Minn. [*E. condensatus pubens* Piper, basionym; *E. ambiguus, E. cinereus, E. condensatus, E. triticoides,* and *E. villiflorus* of most or all Canadian reports, not Vasey & Scribn., Scribn. & Merr., Presl, Buckl., nor Rydb., respectively].

E. riparius Wieg.
/T/EE/ (Hs) Moist thickets and streambanks from Wisc. to s Ont. (Kent, Elgin, and Waterloo counties), s Que. (St-Augustin, Portneuf Co.; MT), and cent. Maine, s to Ark., Ky., and N Fla. [*E. canadensis* var. *rip.* (Wieg.) Boivin].

E. sibiricus L.
/ST/W/eA/ (Hs) Dry to wet meadows, open slopes, and coastal beaches and dunes of Alaska and sw Dist. Mackenzie (see Bowden 1964) and N B.C. (near Fort Nelson, 58°47′N; W. J. Cody, Can. Field-Nat. 81(4): 275. 1967); E Asia [*Clinelymus* Nevski; *E. pendulosus* Hodgson]. MAP: Hultén 1968*b*: 197.

E. villosus Muhl.
/T/EE/ (Hs) Thickets, rocky woods, and shores from Wyo. to s Ont. (N to the Ottawa dist.) and sw Que. (N to Montmagny, Montmagny Co.; Ernest Lepage, Ann. ACFAS 7: 95. 1941; see Que. map by Rouleau 1945: fig. 11, p. 173), s to N.Mex., Tex., Okla., and N.C. [*E. striatus* var. *vill.* (Muhl.) Gray; *E. ?europaeus sensu* Hooker 1840, not L.].

Forma *arkansanus* (Scribn. & Ball) Fern. (*E. ark.* S. & B.) is reported from s Ont. by Bowden (1964; Kettle Point, Lambton Co.).

E. virginicus L.
/T/X/ (Hs) Rich thickets and shores, the aggregate species from B.C. (Boivin 1967*a*) and ?Alta. (Moss 1959) to Sask. (N to Tisdale, 52°51′N), Man. (N to Flin Flon; Lowe 1943), Ont. (N to Renison, s James Bay, ca. 51°N; Hustich 1955), Que. (N to the Gaspé Pen.; reported from Anticosti Is. by Saint-Cyr 1887), Nfld., N.B., P.E.I., and N.S., s to Wash., Ariz., Tex., and Fla.

1 Glumes and lemmas long-awned, the lemma-awns commonly over 6 mm long (to about 1.5 cm) . . . . .var. *virginicus*

2 Glumes and lemmas glabrous or merely scabrous along the margins and toward the tips; [incl. vars. *halophilus* (Bickn.) Wieg. and *jejunus* (Ramaley) Bush; transcontinental] . . . . .f. *virginicus*

2 Glumes and lemmas hirsute; [*E. hirsutiglumis* Scribn.; *E. intermedius* (Vasey) Scribn. & Sm.; Ont., Que., N.B., and N.S.] . . . . .f. *hirsutiglumis* (Scribn.) Fern.

1 Glumes and lemmas mucronate to short-awned, the lemma-awns usually less than 2 mm long.

3 Glumes and lemmas glabrous or merely scabrous along the margins and toward the tips; [B.C. to Que.; type from Cumberland House, Sask.] . . . . .var. *submuticus* Hook.

3 Glumes and lemmas hirsute; [Ont. and Que. (type from the Gatineau R. near Wakefield)] . . . . .var. *jenkensii* Bowden

Elymus Hybrids

The following list of *Elymus* hybrids (including the hybrid genera × *Elyhordeum (Elymordeum)* Mansf. and × *Elysitanion* Bowden) occurring naturally in Alaska–Canada is based largely upon three papers by Bowden (1957; 1958; 1964; *see* also the listing of *Agropyron* hybrids).

Elymus canadensis × *E. virginicus:* (× *E. maltei* Bowden, the type an artificial hybrid, both parents of U.S.A. origin); the natural hybrid taken at Ottawa, Ont., and Nicolet, Nicolet Co., Que.

E. canadensis × *E. virginicus* var. *submuticus*: (× *E. maltei* nm. *churchii* Bowden, the type an artificial hybrid (*E. can.* from Wenatchee, Wash.; *E. virg.* var. *sub.* from Fort Francis, Ont.); the natural hybrid taken at Winnipeg, Man.

E. canadensis var. *wiegandi* × *E. virginicus*: (× *E. maltei* nm. *simulans* Bowden, the type an artificial hybrid (*E. can.* var. *wieg.* from Vt.; *E. virg.* from Metcalfe, Carleton Co., Ont.); the natural hybrid taken at Ottawa, Ont.

E. glaucus × *E. innovatus*: reported from the Ashnola R., B.C., by Bowden (1967).

E. glaucus × *E. mollis*: (× *E. uclueletensis* Bowden, the type from Ucluelet, Vancouver Is., B.C., the only known locality).

E. glaucus × *Sitanion hystrix*: (× *Elysitanion hansenii* (Scribn.) Bowden; *Elymus (S.) hansenii* Scribn.); reported by Bowden (1967) from s B.C. (Goodchap Mts., near Penticton; Mt. Baldy, near Bridesville).

E. hirsutus × *Hordeum brachyantherum (jubatum* ssp. *breviaristatum)*: (× *Elyhordeum (Elymordeum) schaackianum* Bowden, the type from Attu Is., Aleutian Is., the only known locality).

E. mollis × *E. hirsutus*: (× *Elymus aleuticus* Hult., the type from Atka Is., Aleutian Is., the only known locality; *see* Hultén 1942: map 194, p. 401).

E. mollis × *E. triticoides*: (× *E. vancouverensis* Vasey nm. *vancouverensis,* the type from Oak Bay, Vancouver Is., B.C.; MAP: Meusel, Jaeger, and Weinert 1965: 43).

Note: Bowden 1957 states that *E. triticoides* has not yet been collected in Canada but that it does occur in NW Wash., so that the presence of its hybrid in southernmost B.C. is not illogical. Two other nothomorphs of × *E. vancouverensis* are also known from B.C.: nm. *californicus* Bowden (Vancouver Is. and West Vancouver) and nm. *crescentianus* Bowden (Vancouver Is.; type from Crescent Beach, Vancouver).

E. mollis × *Hordeum jubatum*: (× *Elyhordeum dutillyanum* (Lepage) Bowden; ×*Elymordeum dut.* Lepage, the type from Old Factory, James Bay, Que., the only known locality).

E. mollis × *Hordeum jubatum* ssp. *breviaristatum*: (× *Elyhordeum dutillyanum* nm. *littorale* (Hodgson & Mitchell) Bowden; × *Elymordeum lit.* H. & M., the type from 18 mi sw of Palmer, Alaska, the only known locality).

E. virginicus × *Hordeum jubatum*: (× *Elyhordeum (Elymordeum) montanense* (Scribn.) Bowden; *Hordeum pammelii* Scribn. & Ball); Sask. (Tisdale), Que. (St-Jovite, Terrebone Co.), and N.S. (Lower Onslow and near Truro, Colchester Co.).

ERAGROSTIS Beauv. [341] Love-Grass

1 Culms capillary, creeping and often rooting at the nodes, forming mats; leaves to 3 mm

broad; panicle to 8 cm long; spikelets to 1.5 cm long; lemmas about 2 mm long; annual; (Sask. to Que.) . . . .*E. hypnoides*
1 Culms decumbent-ascending to erect; panicle mostly larger.
  2 Leaves bordered by wart-like glands; lemmas and glumes glandular on the keel; lateral nerves of lemma prominent; spikelets with up to about 40 flowers; annuals; (introd.).
    3 Spikelets densely 10–40-flowered, to over 1.5 cm long; upper glume about 2 mm long; lemmas at least 2 mm long, scabrous . . . .*E. megastachya*
    3 Spikelets less densely 8–20-flowered, at most about 1 cm long; upper glume less than 2 mm long; lemmas less than 2 mm long, nearly smooth . . . .*E. poaeoides*
  2 Leaves not glandular-warty; lemmas and glumes not glandular on the keel; spikelets with rarely more than 15 flowers.
    4 Stoloniferous perennial; culms rigid, forming dense stools, much shorter than the purple panicle; leaves to 6 mm broad; panicle-branches becoming stiffly divergent to reflexed, pilose in their axils; sheaths densely bearded at throat, often overlapping the lower panicle-branches; spikelets 3–15-flowered, to 8 mm long, on stiff pedicels; lemmas about 2 mm long, prominently nerved, scabrous on the keel; (s Ont. and s Que.) . . . .*E. spectabilis*
    4 Annuals with softer, often tufted culms; lemmas mostly about 1.5 mm long.
      5 Leaves to 6 mm broad; panicle to 3 dm long; spikelets 6–10-flowered, to about 7 mm long and 1 mm broad; lemmas faintly nerved; (introd. in Ont.) . . .[*E. orcuttiana*]
      5 Leaves rarely over 3 mm broad; panicle relatively short.
        6 Auricles of upper leaf-sheaths naked; panicle rather dense, its branches spikelet-bearing from near base; spikelets with less than 10 flowers, less than 5 mm long, about 1.5 mm broad, very short-pedicelled; lemmas faintly nerved; (introd.) . . . .*E. multicaulis*
        6 Auricles of upper leaf-sheaths long-ciliate; panicle looser.
          7 Lemmas strongly nerved; palea very persistent; glumes scabrous on the keel; spikelets to 8 mm long and with up to about 15 flowers, on pedicels rarely over 5 mm long, borne from near the base of the panicle-branches; (B.C.; s Ont. to N.S.) . . . .*E. pectinacea*
          7 Lemmas faintly nerved; palea soon deciduous; glumes smooth; spikelets borne along the upper 2/3 of the panicle-branches.
            8 Spikelets 2–6-flowered, to 4 mm long, mostly shorter than their pedicels; panicle slenderly ellipsoid; (introd.) . . . .*E. frankii*
            8 Spikelets 3–9-flowered, to 5 mm long, on pedicels of about the same length; panicle ovoid or pyramidal; (s ?Ont.) . . . .[*E. pilosa*]

E. frankii Meyer   Sandbar Love-Grass
Native in the ᴇ U.S.A.; introd. along sandy or gravelly roadsides and railways or in dry sandy waste places of Ont. (ɴ to Ottawa), sw Que. (Montreal; St-Jean) and N.B. (Boivin 1967a).

E. hypnoides (Lam.) BSP.   Teal Love-Grass
/T/X/   (T)   Mud-flats and sandy shores from Wash. to sᴇ Sask. (Glen Ewan, about 135 mi sᴇ of Regina; DAO; not listed by Breitung 1957a; the inclusion of B.C. in the range by Rydberg 1922, requires clarification), s Man. (ɴ to Grand Beach, near the s end of L. Winnipeg), Ont. (ɴ to the Ottawa dist.), sw Que. (ɴ to the shores of L. St. Peter near Sorel), and cent. Maine, s to Mont., Mexico and S. America; W.I. [*Poa* Lam.; *E. reptans sensu* John Macoun 1888, not *Poa reptans* Michx.].

E. megastachya (Koel.) Link   Stink-, Snake-, or Skunk-Grass
Eurasian; roadsides and waste places of sᴇ Sask. (Boivin 1967a; *E. cilianensis),* s Man. (Brandon; Morris; Otterburne), Ont. (ɴ to the Rainy River and Ottawa districts), Que. (ɴ to Masson, Papineau Co., and Montreal), and N.S. (Kings and Halifax counties). [*Poa* Koel.; *E. cilianensis* of Canadian reports, apparently not *Poa cil.* All.; *E. major* Host].

E. multicaulis Steud.
Asiatic; roadsides and waste places of s Ont. (Point Abino, Welland Co.; House 1930), sw Que. (La Trappe, Two Mountains Co.; St-Jean, St-Jean Co.; Bellerive, Labelle Co.), and N.S. (Boivin 1967a), [E. peregrina Wieg.].

[E. orcuttiana Vasey]
[Native in the w U.S.A. (Calif., Nev., and Ariz.); reported by Boivin (1967a) as introd. at Ottawa, Ont., where probably not established.]

E. pectinacea (Michx.) Nees
/T/X/ (T) Sandy shores, roadsides, etc., from ?B.C. (Fernald in Gray 1950; ?introd.) and Wash. to N.Dak., Ont. (N to Renfrew and Carleton counties), Que. (N to the shores of L. St. Peter near Sorel), and N.S. (Pictou, Pictou Co.; P.E.I.), s to s Calif., Mexico, Tex., and Fla. [Poa Michx.; P. (E.) caroliniana Spreng.; E. purshii Schrad.].

[E. pilosa (L.) Beauv.] India Love-Grass
[Reports of this Eurasian species from s Ont. by Dodge (1914; 1915), Zenkert (1934), and Core (1948) all appear to be referable to other species, particularly E. pectinacea. (Poa L.).]

E. poaeoides Beauv. Little Love-Grass
Eurasian; known from along roadsides and railways and in waste ground of Ont. (N to Renfrew and Carleton counties), Que. (N to near Quebec City; see Que. map by C. Rousseau 1968: map 9, p. 65), N.B. (St. Stephen, Charlotte Co.), ?P.E.I. (Boivin 1967a), and N.S. (Hants and Kings counties). [Poa (E.) eragrostis L.; E. minor Host].

E. spectabilis (Pursh) Steud. Tumble-Grass, Petticoat-climber
Native in the E U.S.A.; introd. in s Ont. (sandy ground at Woodstock, Oxford Co.; OAC; reported from Pelee Is., Essex Co., by Core 1948) and sw Que. (railway ballast at Stanbridge Station, Missisquoi Co.; Herb. M. Raymond). [Poa Pursh].

EREMOPOA Roshev. [378]

E. persica (Trin.) Roshev.
Asiatic; taken by G. A. Stevenson in 1955 in an area of over half an acre of ballast and wasteland at the Canadian National Railway freight yards in Brandon, sw Man. (DAO; GH), the first record of the plant in N. America. [Poa Trin.; P. heptantha Steud.].

FESTUCA L. [385] Fescue-Grass. Fétuque

1   Lemmas acute to cuspidate or short-awned (the awn rarely over 1.5 mm long); culms to over 1 m tall.
  2   Leaves chiefly basal and less than 2 mm broad, often involute; lemmas distinctly keeled toward apex.
    3   Leaves stiff and scabrous, early deciduous, their ligules ciliate, their chartaceous sheaths persistent for several years; panicle contracted in fruit; lemmas to 1 cm long; (Alaska–B.C. to Man.; Ont. (introd.); E Que. and Nfld.) . . . . .F. altaica
    3   Leaves soft, persistent, their ligules minutely ciliate; panicle slender but open, its branches spreading or ascending; lemmas to 8 mm long; (s B.C.) . . . . .F. viridula
  2   Stems relatively leafy; lemmas scarcely keeled.
    4   Spikelets mostly about 5 mm long and about 3-flowered, borne near the ends of the branches of the diffuse panicle; lemmas firm, less than 5 mm long; leaf-blades not auricled; (s Man. to N.S.) . . . . .F. obtusa
    4   Spikelets to 12 mm long, usually about 8-flowered; branches of the contracted panicle spikelet-bearing nearly to base; lemmas thin, to 8 mm long; leaf-blades auricled at base; (introd.) . . . . .F. elatior
1   Lemmas typically terminating in an awn at least 2 mm long.
  5   Leaves flat, to 1 cm broad; panicle loose and open, drooping, to 4 dm long, its

branches spreading or reflexed (the lower ones to 1.5 dm long); flowers not stipitate; lemma-awns to 2 cm long; (B.C. and sw Alta.) . . . .*F. subulata*
5 Leaves commonly involute (if flat, at most 3 or 4 mm broad).
  6 Flowers long-stipitate (the rachilla appearing to be jointed a short distance below the flower); glumes to 5 mm long; lemmas to 8 mm long; panicle to 2 dm long, loose, nodding, at least its lower branches drooping; leaves to 4 mm broad, their ligules strongly ciliate with long coarse hairs; (sw B.C.) . . . .*F. subuliflora*
  6 Flowers not stipitate; leaves rarely over 3 mm broad, their ligules minutely ciliate.
    7 Culms from a decumbent or somewhat creeping base; leaves to 3 mm broad, the lower sheaths reddish, soon becoming fibrous; lemmas to 8 mm long (excluding awns); anthers to 4 mm long; (transcontinental) . . . .*F. rubra*
    7 Culms erect or at least not with decumbent rhizome-like bases; leaves usually less than 1 mm broad, their sheaths commonly whitish or drab, firm and only tardily fibrous; lemmas (excluding awns) at most 6 mm long; anthers to 3 mm long.
      8 Panicle narrow and usually not over 5 cm long; lemma-body to 5 mm long, with an awn to 3 mm long; culms commonly less than 3 dm tall (at most 6 dm); (transcontinental) . . . .*F. ovina*
      8 Panicle somewhat open and commonly 1 or 2 dm long; lemma-body to 6 mm long; culms to over 1 m tall; (B.C. to sw Sask.; Ont.) . . . .*F. occidentalis*

F. altaica Trin.   Rough Fescue

/AST/D/eA/   (Hs)   Peaty or rocky meadows and barrens: N-cent. Alaska, the coast of the Yukon (King Point), the Mackenzie R. Delta, and Great Bear L., through B.C. and sw Alta. (N to Jasper and Fort Saskatchewan) to Sask. (N to Prince Albert) and s Man. (N to Duck Mt.), s to Oreg., Idaho, Colo., and S.Dak.; isolated in ?Mich. (A. S. Hitchcock 1935); Que. (Kaniapiscau and Larch rivers at ca. 57°35′N; Sawbill L. at 53°27′N, 66°42′W; serpentine outcrops of Megantic Co., s of Quebec City; serpentine tableland and gravel outwash plains of Mt. Albert, Gaspé Pen.; subalpine meadows of Tabletop Mt., Gaspé Pen.; see Que. map by Dutilly, Lepage, and Duman 1953: fig. 9, p. 36); w Nfld. (serpentine tableland of the Blomidon Mts., Bay of Islands); E Asia. [Incl. var. *major* (Vasey) Gl. (*F. scabrella* var. *major* Vasey; *F. campestris* Rydb.) and *F. scabrella* Torr. (*F. hallii* (Vasey) Piper; *F. ?macounii* Vasey), for varying interpretations of which see Hultén 1942, and Gleason 1958]. MAPS (the last two as *F. scabrella*): Hultén 1968b: 167; Raup 1947: pl. 15 (NW area); Fernald 1925: map 7, p. 251; Alexander Johnston, Ecology 39: fig. 1, p. 536. 1958.

According to Johnston, the migration of this and certain other disjunct species to their present isolated outposts in the East can be explained by their having been forced southward in the West during the last glaciation to a position where they could use the great "prairie peninsula" south of the Great Lakes as a route of migration northeastward during a postglacial xerothermic period; (see Scoggan 1950: 3–14).

In addition to the above native stations in the East, the species is reported from railway ballast at Cambridge, Waterloo Co., s Ont., by W. Herriot (Ont. Nat. Sci. Bull. 1: 30. 1905) and was taken by Soper in 1961 on a sandy hillside of a cemetery at Stanley, Huron Co. Forma *pallida* Jordal (spikelets pale yellowish-green rather than purplish) and f. *vivipara* Jordal (most or all of the florets replaced by bulblets) are reported from their type localities in the Brooks Range, Alaska, by L. H. Jordal (Rhodora 54(638): 36. 1952).

F. elatior L.   Meadow Fescue

Eurasian; introd. along roadsides and in meadows and waste places in N. America, as in s Alaska (Talkeetna), s Yukon (Whitehorse), B.C., Alta. (N to near Edmonton), Man. (N to Riding Mt.), Ont. (N to Moosonee, James Bay, 51°16′N), Que. (N to Rupert House, James Bay, 51°29′N), Nfld., N.B., P.E.I., and N.S.; sw Greenland. MAPS and synonymy: see below.
1 Spikelets 4–5-flowered; panicle open; lemmas usually short-awned; plant relatively stout; [*F. arundinacea* Schreb., not Liljebl.; B.C. (Boivin 1967a), Man. (Winnipeg), Ont., Que., and N.S. (Cape Breton Is.); MAP (*F. arund.*): Hultén 1968b: 166] . . . .
. . . . .var. *arundinacea* (Schreb.) Wimm.
1 Spikelets mostly 6–8-flowered; panicle contracted after flowering, much branched to nearly simple . . . . .var. *elatior*

> 2 Lemmas short-awned; [not yet reported from Canada but some of our material, particularly of var. *arundinacea,* may belong here] .f. *aristata* Holm.
> 2 Lemmas awnless; [*F. poaeoides* Michx.; incl. *F. pratensis* Huds.; range of the species] . . . . .f. *elatior*

F. obtusa Biehler   Nodding Fescue
/T/EE/   (Hs)   Moist woods and thickets from s Man. (N to Winnipeg) to Ont. (N to near Ottawa), Que. (N to near Montreal in Joliette Co.), ?N.B. (Fowler 1885; *F. nutans*), and N.S. (Kings, Hants, and Cumberland counties; not known from P.E.I.), s to Tex., Miss., and Fla. [*F. nutans* of Canadian reports, not Moench nor Biehler].

F. occidentalis Hook.   Western Fescue
/T/WW/   (Hs)   Dry rocky wooded slopes (ranges of Canadian taxa outlined below), s to Calif. and Wyo., with an isolated area in the Great Lakes region of N Mich. and Ont.
1 Lemma-awn to 7 mm long, equalling or slightly longer than the lemma-body, this to 6.5 mm long; [*F. ovina* var. *polyphylla* Vasey; *F. rubra* var. *?longearistata* Hack.; B.C. (N to Stuart L., ca. 54°30′N) and sw Alta.; Ont. (Bruce Pen. and Manitoulin Is., L. Huron; Carp L., near Sault Ste. Marie)] . . . . .var. *occidentalis*
1 Lemma-awn to about 4 mm long, distinctly shorter than the lemma-body, this to 7 mm long; [*F. ovina* var. *ingrata* Hack.; *F. idahoensis* Elmer; s B.C. (Vancouver Is. and adjacent islands and mainland; Lower Arrow L.), sw Alta. (Waterton Lakes; Breitung 1957b), and sw Sask. (Cypress Hills)] . . . . .var. *ingrata* (Hack.) Boivin

F. ovina L.   Sheep's-Fescue
/AST/X/GEA/   (Hs)   Dry open ground, sands, and rocky places, the aggregate species from the coasts of Alaska–Yukon–Dist. Mackenzie, throughout the Canadian Arctic Archipelago to northernmost Ellesmere Is. and northernmost Ungava–Labrador, s through all of the provinces to Calif., N.Mex., Kans., Ill., and S.C.; S. America; circumgreenlandic; Iceland; Eurasia. MAPS and synonymy (together with a distinguishing key to the scarcely separable *F. baffinensis*): *see* below.
1 Lemmas awnless or with awns rarely over 0.5 mm long.
    2 Flowers replaced by leafy shoots; stamens wanting; upper glume at least 3 mm long; lemmas soft, about 5 mm long; [*F. vivipara* (L.) Sm. and its var. *hirsuta* (Lge.) Schol.; Alaska, sw Yukon, cent. B.C. (Mt. Selwyn), and mts. of sw Alta.; E Que., s Labrador, and N Nfld.; MAPS: Porsild 1966: map 12A, p. 68; Hultén 1958: map 17, p. 37, and 1968b: 168] . . . . .var. *vivipara* L.
    2 Flowers normal, with anthers about 2 mm long; upper glume less than 3 mm long; lemmas coriaceous, tightly inrolled, at most 3.5 mm long; [*F. capillata* Lam.; introd. in s B.C. (Vancouver) and from Ont. to Nfld.] . . . . .var. *capillata* (Lam.) Alef.
1 Lemmas with awns over 1 mm long.
    3 Panicle loose and open, to about 1 dm long; anthers to over 2.5 mm long; lemmas coriaceous and strongly involute; [introd.].
        4 Leaves capillary or strongly involute, about 0.5 mm broad; panicles to about 5 cm long; spikelets to 7 mm long, with at most 5 flowers . . . . .var. *ovina*
            5 Lemmas glabrous; [introd., essentially transcontinental] . . . . .f. *ovina*
            5 Lemmas hispid; [introd. from Man. to N.S.] . . . . .f. *hispidula* (Hack.) Holmb.
        4 Leaves to about 1 mm broad; panicles to 1 dm long; spikelets to 1 cm long, with up to 9 flowers; [*F. duriuscula* L.; introd. in ?B.C. (Henry 1915) and in Ont., Que., Nfld., and Cape Breton Is.] . . . . .var. *duriuscula* (L.) Koch
    3 Panicle contracted, spikelike to loosely lanceolate; anthers at most about 1.5 mm long.
        6 Anthers to about 1.5 mm long; lemmas greenish, coriaceous and strongly involute; panicle to 1 dm long; [*F. saximontana* Rydb., the type from Banff, Alta.; *F. brachyphylla* ssp. *sax.* (Rydb.) Hult.; *F. ovina* ssp. *?alaskensis* Holmen and ssp. *sax.* var. *rydbergii* St-Yves; transcontinental; MAPS: Hultén 1968b: 169, and 1962: map 17, p. 25] . . . . .var. *saximontana* (Rydb.) Gl.
        6 Anthers at most about 1 mm long; lemmas bronze or purplish to dark

purplish-brown or blackish, membranaceous and usually not strongly involute; panicle rarely over 3 cm long; [transcontinental].
  7   Panicle narrowly cylindrical or lance-ovoid; lemma-awns 2 or 3 mm long; anthers to 1 mm long; culms glabrous, to about 2.5 dm tall, much overtopping the more or less involute leaves; [var. *brevifolia* (R. Br.) Wats. (*F. brevifolia* R. Br., not Muhl.); *F. brachyphylla* Schultes, the type from Melville Is.; *F. hyperborea* Holmen; *F. supina* Schur; *F. ?halleri sensu* Schlechtendal 1836, not All.; *F. ovina* var. *?alpina sensu* Hantzsch 1931, not *F. alpina* Suter; MAPS (*F. brach.*): Hultén 1968*b*: 168, and 1962: map 17, p. 25; Porsild 1957: map 53, p. 167; Raup 1947: pl. 15] . . . . .var. *brachyphylla* (Schultes) Piper
    8   Panicle purple to dark purplish-brown . . . . .f. *brachyphylla*
    8   Panicle light greenish-yellow; [type from Lake Harbour, Baffin Is.] . . . .
        . . . . .f. *flavida* Polunin
  7   Panicle ovoid; lemma-awns relatively short; anthers to 0.5 mm long; culms minutely hirsute at least above, not much overtopping the strongly involute leaves; [type from Pond Inlet, Baffin Is.; reported from Beartooth Plateau, on the Mont.–Wyo. boundary, and from Greenland by Hitchcock et al. (1969; also noting reports from the Rocky Mts. of Canada and Colo.); MAPS: Hultén 1968*b*: 169, and 1962: map 17, p. 25; Porsild 1957: map 54, p. 167, and 1955: fig. 17, p. 83] . . . .*F. baffinensis* Polunin

F. rubra L.   Red Fescue
/aST/X/GEA/   (Hsr (Grh))   Sandy, rocky, or peaty soils, the aggregate species from the coasts of Alaska–Yukon–Dist. Mackenzie to Banks Is., Victoria Is., and northernmost Ungava–Labrador, s through all of the provinces to s Calif., N.Mex., Tex., Mo., Tenn., and N.C.; w Greenland N to ca. 71°N, E Greenland N to ca. 74°N; Iceland; Eurasia. MAPS and synonymy: *see* below.
1   Flowers largely replaced by leafy tufts, the rare fertile spikelets with anthers at most 2 mm long; [f. *prolifera* (Piper) Hyl.; *F. prolifera* (Piper) Fern. and its var. *lasiolepis* Fern.; *F. rubra* var. *subvillosa* f. *?vivipara* Eames; E Dist. Mackenzie, Dist. Keewatin, and Sask. to Nfld.] . . . . .var. *prolifera* Piper
1   Flowers normal; anthers to about 4 mm long.
  2   Leaves stiff and somewhat wiry, strongly whitened; spikelets to 2 cm long, with up to 10 flowers; lemmas to 7.5 mm long; [E Que. to Nfld. and N.S.; introd. at L. Attawapiskat, N Ont.] . . . . .var. *juncea* (Hack.) Richter
  2   Leaves comparatively soft, not strongly whitened (except in var. *rubra* f. *glaucescens*).
    3   Lemmas densely villous to lanate; panicle often spike-like or with short branches, to 1.5 dm long; spikelets to 1.5 cm long; [var. *?lanuginosa* Mert. & Koch; var. *?villosa* Vasey; ssp. *?cryophila* (Krecz.) Hyl.; ssp. *?richardsonii* (Hook.) Hult.; *F. aren.* Osbeck; *F. richardsonii* Hook.; *Bromus ?secundus* Presl; transcontinental; MAPS: Hultén 1962: map 57B (ssp. *cry.*), p. 65; Porsild 1957: map 55, p. 167] . . . . .var. *arenaria* (Osbeck) Fries
    3   Lemmas typically glabrous or merely scabrous; panicle to over 2 dm long, often with elongate branches.
      4   Basal and cauline leaves all flat, to 3 mm broad; spikelets to over 1.5 cm long, with up to 10 flowers; lemmas to 7 mm long; [ssp. *?aucta* (Kretch. & Bobr.) Hult.; ?Aleutian Is. and s ?Alaska; E Que. to Nfld. and N.S.; MAP (ssp. *aucta*): Hultén 1968*b*: 171] . . . . .var. *multiflora* (Hoffm.) Asch. & Graebn.
      4   Basal leaves setaceous or linear-involute, to 1 mm broad.
        5   Basal offshoots all erect, bearing mostly erect elongate leaves to 6 dm long; culms to 9 dm tall; [var. *fallax* (Thuill.) Hack.; Que. to Nfld. and N.S.] . . . . .var. *commutata* Gaud.
        5   Basal offshoots all or mostly divergent or decumbent, forming loose mats.
          6   Lemmas awnless or merely short-pointed; [Que. (N to Hudson Strait), Nfld., and Greenland] . . . . .var. *mutica* Hartm.
          6   Lemmas distinctly awned . . . . .var. *rubra*
            7   Lemmas puberulent or strigose-hirsute; [var. *subvillosa* Mert. &

Koch; James Bay, Ont., to N Labrador and Nfld.] . . . .
. . . . .f. *squarrosa* (Fries) Holmb.
7 Lemmas glabrous.
  8 Spikelets to 17 mm long, with up to 10 flowers; upper glume to 6
     mm long; lemmas to 8 mm long; [James Bay, Ont., to Labrador
     and Nfld.] . . . . .f. *megastachys* (Gaud.) Holmb.
  8 Spikelets to 10 mm long, with up to 7 flowers; upper glume to 4.5
     mm long; lemmas to 6 mm long.
    9 Leaves glaucous; [Man., Que., Nfld., and N.S.] . . . .
       . . . . .f. *glaucescens* (Hartm.) Holmb.
    9 Leaves green; [vars. *?longiseta* and *pruinosa* Hack.;
       F. tenella sensu Saint-Cyr 1887, not Willd.; *F. ?borealis
       sensu* Hantzsch 1931, not Lange; transcontinental; MAPS:
       Hultén 1968*b*: 170 (excluding ssp. *aucta,* this being var.
       multiflora of the present treatment), and 1962: map 57A
       (aggregate species), p. 65] . . . . .f. *rubra*

F. subulata Trin. Bearded Fescue
/T/WW/ (Hs) Moist thickets and shaded banks from the Alaska Panhandle (type from Sitka)
through B.C. and sw Alta. (Waterton Lakes; CAN) to N Calif., Nev., Utah, and Wyo. MAPS: Hultén
1968*b*: 167; Koyama and Kawano 1964: fig. 9 (B), p. 872.

F. subuliflora Scribn. Crinkle-awn Fescue
/t/W/ (Hs) Moist shaded places from sw B.C. (Vancouver Is., the type from Coldstream;
adjacent mainland) to N Calif., chiefly along the coast.

F. viridula Vasey Green Fescue
/T/W/ (Hs) Mountain meadows and open slopes (chiefly between 1,000 and 2,000 ft altitude)
from B.C. (reported from about 6,000 ft on mts. near Nelson by Eastham 1947 who also cites Hope
Mt., Manning Park, and between Hope and Princeton; reports from Alta. require confirmation) to N
Calif., Idaho, and Mont.

GLYCERIA R. Br. [383] Manna-Grass

1 Spikelets linear, nearly terete, commonly over 1 cm long (to about 4 cm), the
   rachilla-internodes to 2.5 mm long; palea wing-margined; stamens 3; leaf-sheaths
   usually strongly compressed and 2-edged.
  2 Lemmas glabrous between the slightly scabrous nerves, thin, 3 or 4 mm long; upper
     (second) glume usually 2 or 3 mm long; anthers scarcely 1 mm long; spikelets to 18
     mm long, on slender pedicels up to 2/3 as long as the spikelets; leaf-blades at most
     about 5 mm broad; (transcontinental) . . . .*G. borealis*
  2 Lemmas minutely scabrous or hirsute between the usually distinctly scabrous nerves;
     spikelets commonly sessile, subsessile, or short-pedicelled.
    3 Lemmas about 3 mm long, firm, broadly rounded at summit; upper glume about
       1.5 mm long; anthers less than 1 mm long; spikelets 1 or 2 cm long, often purplish;
       leaves commonly less than 7 mm broad; (B.C.) . . . .*G. leptostachya*
    3 Lemmas to over 5 mm long; anthers at least 1 mm long.
      4 Lemmas to 8 mm long, thin, narrowed to an obtuse, slightly purple-tinged apex;
         tip of palea usually surpassing the lemma; anthers to about 3 mm long; upper
         glume 3 or 4 mm long; spikelets to 4 cm long; leaves to 8 mm broad; (introd.) . . . .
         . . . .*G. fluitans*
      4 Lemmas at most 5.5 mm long, firm; anthers at most 1.5 mm long; spikelets to
         about 2.5 cm long; leaves to 12 mm broad.
        5 Lemmas pale or green, not purple-tinged, commonly not over 4 mm long,
           the apex rounded or only slightly narrowed; palea usually surpassing the tip
           of the lemma; upper glume commonly over 3 mm long; (s Ont. to N.S.) . . . .
           . . . .*G. septentrionalis*

     5   Lemmas usually purple-tinged near the obtuse tip, to about 6 mm long; palea rarely surpassing the tip of the lemma; upper glume less than 3 mm long; (s B.C.) . . . .*G. occidentalis*

1   Spikelets lanceolate to ovate, commonly less than 1 cm long, more or less laterally compressed, the rachilla-internodes rarely over 1 mm long; palea wingless; stamens usually 2 (3 in *G. grandis* and *G. maxima*); leaf-sheaths scarcely compressed.

    6   Panicle contracted and narrow, its lower branches strongly ascending; upper (second) glume to 3 mm long.

       7   Panicle linear-cylindric, often nodding at summit, its branches closely appressed, the two lowest internodes at least 2 cm long; spikelets about 4 mm long, 3–4-flowered; lemmas less than 3 mm long; leaves to 5 mm broad; (Ont. to N.S.) . . . .*G. melicaria*

       7   Panicle ovoid or oblong, dense and compact, its two lowest internodes usually less than 2 cm long; spikelets to 9 mm long, 3–7-flowered; lemmas at least 3 mm long; leaves to 8 mm broad; (N.B. and N.S.) . . . .*G. obtusa*

    6   Panicle open and lax, its branches loosely ascending or often drooping.

       8   Nerves of lemma not prominently raised; upper glume 2 or 3 mm long; spikelets commonly 3 or 4 mm broad; leaves to 1 cm broad; (Sask.; Ont. to Nfld. and N.S.) . . . .*G. canadensis*

       8   Nerves of lemma prominently raised; spikelets at most about 2.5 mm broad.

          9   Lower glume less than 1 mm long, the upper glume at most 1.3 mm long, about twice as long as the lower one; lemmas firm; leaves to 7 mm broad; culms usually not over 1 m tall; (transcontinental) . . . .*G. striata*

          9   Glumes larger, the upper one usually less than twice as long as the lower one; culms commonly taller.

          10 Stamens 3; apex of lemma firm, only moderately scarious; panicle to over 4 dm long; culms stout.

            11 Culms to over 2 m tall; leaves to about 2 cm broad; upper glume about 3 mm long; lemmas to about 3.5 mm long; anthers to 2 mm long; (introd.) . . . .*G. maxima*

            11 Culms rarely over 1.5 m tall; leaves at most about 1.5 cm broad; upper glume less than 2.5 mm long; lemmas at most 2.5 mm long; anthers less than 1 mm long; (transcontinental) . . . .*G. grandis*

          10 Stamens 2; panicle less than 2 dm long.

            12 Apex of lemma with a broad hyaline margin; leaves to about 6 mm broad; culms slender, commonly less than 6 dm tall; (Alaska–B.C. to Man.) . . . .*G. pulchella*

            12 Apex of lemma firm, only moderately scarious; leaves to 12 mm broad; culms commonly 1 or 2 m tall; (B.C. and sw Alta.) . . . .*G. elata*

G. borealis (Nash) Batch.   Small Floating Manna-Grass

/sT/X/   (Hs)   Wet ground and shallow water from cent. Alaska–Yukon to L. Athabasca (Alta. and Sask.), Ont. (N to Lake River, NW James Bay, 54°22′N; *see* NE Canada map by Lepage 1966: map 4, p. 214), Que. (N to Fort George, E James Bay, 53°50′N, and the Côte-Nord), Labrador (N to the Hamilton R. basin), Nfld., N.B., P.E.I., and N.S., s to Calif., N.Mex., S.Dak., Ind., Pa., and N New Eng. [*Panicularia* Nash]. MAP: Hultén 1968b: 151.

G. canadensis (Michx.) Trin.   Rattlesnake-Grass

/T/EE/   (Hs)   Damp or wet ground and shores (ranges of Canadian taxa outlined below), s to N Ill., Ind., Tenn., and N.C.

1   Lemmas 3 or 4 mm long, distinctly nerved; spikelets to 7 mm long, with up to 10 florets; culm-nodes mostly exserted from the leaf-sheaths; [*Briza can.* Michx., not Nutt.; *G. laxa sensu* Robinson and Schrenk 1896, not Scribn.; Sask. (Boivin 1967a; not known from Man.) to Ont. (N to L. Nipigon), Que. (N to Fort George, E James Bay, ca. 53°45′N, and the Côte-Nord), Nfld., N.B., P.E.I., and N.S.]. Three sterile hybrid nothomorphs between *G. canadensis* and probably *G. striata* are reported by W. M. Bowden (Can. J. Bot. 38(2):

128. 1960) from the type locality, Mer Bleue, Ottawa, Ont., under the name × *G. ottawensis* Bowden . . . . .var. *canadensis*

1 Lemmas less than 3 mm long, obscurely nerved; spikelets about 4 mm long, with rarely more than 6 florets; culm-nodes mostly included in the leaf-sheaths; [*G. laxa* Scribn.; Ont. (N to the N shore of L. Superior) to Que. (Labelle and Terrebonne counties), N.B., P.E.I., and N.S.] . . . . .var. *laxa* (Scribn.) Hitchc.

G. elata (Nash) Hitchc.   Tall Manna-Grass
/T/W/   (Hs)   Wet ground and moist woodlands from s B.C. (N to Glacier, in Rogers Pass) and sw Alta. (Waterton Lakes; Crowsnest Pass) to s Calif. and N.Mex. [*Panicularia* Nash].

G. fluitans (L.) R. Br.   Floating Manna-Grass
?Eurasian (according to Hultén 1958, probably entirely introd. in N. America; considered native there by Fernald *in* Gray 1950); known in shallow waters in B.C. (Boivin 1967a), E Que. (Gaspé Pen.), Nfld., and N.S. (Colchester and Inverness counties), reports from elsewhere in Canada probably referring to *G. borealis*. [*Festuca* L.]. MAPS: Hultén 1958: map 113, p. 133; Meusel, Jaeger, and Weinert 1965: 40.
Two other closely related Old World species are reported from Canada by Boivin (1967a) and may be distinguished as follows:

1 Anthers about 0.5 mm long; teeth of lemma acute, prominent; [introd. in B.C.] . . . .*G. declinata* Bréb.

1 Anthers to over 1 mm long; teeth of lemma obtuse, rather obscure.
  2 Anthers to 1 mm long; lemmas less than 5 mm long; [introd. in s Ont. and N.S.] . . . .*G. plicata* Fries
  2 Anthers to 2 mm long; lemmas 6 or 7 mm long . . . .*G. fluitans*

G. grandis Wats.   Reed-meadow Grass
/ST/X/   (Hs)   Wet meadows, streambanks, etc., the aggregate species from cent. Alaska–Yukon to Great Slave L., N Alta., Sask. (N to Waddy L. at ca. 56°N), Man. (N to Reindeer L. at 57°37′N), Ont. (N to Sandy L. and w James Bay, both at ca. 53°N), Que. (N to L. St. John and Anticosti Is.; type from Que.), Nfld., N.B., P.E.I., and N.S., s to Oreg., Ariz., N.Mex., Iowa, Tenn., and Va. MAPS and synonymy: *see* below.

1 Spikelets to 1 cm long and with up to 10 florets; [reported by Leon Kelso, Rhodora 36(427): 266. 1934, from Alaska and from the type locality, Dawson, Yukon Territory] . . . . .var. *komarovii* Kelso

1 Spikelets rarely over 6 mm long, with at most 7(8) florets . . . . .var. *grandis*
  2 Spikelets yellowish or pale; [Ont. (near Ottawa; Dore 1959), N.B., and N.S. (type from Yarmouth)] . . . . .f. *pallescens* Fern.
  2 Spikelets purple; [*G. aquatica* Sm., not (L.) Presl; *Poa aquatica* var. *americana* Torr.; *G. maxima* var. *amer.* (Torr.) Boivin; *G. arundinacea* of auth., not Kunth; *G. max.* ssp. *grandis* (Wats.) Hult.; transcontinental; MAPS: Hultén 1962: map 173, p. 185, and 1968b: 153 (*G. max.* ssp. *gr.*); Meusel, Jaeger, and Weinert 1965: 40; Raup 1930: map 37, p. 203] . . . . .f. *grandis*

G. leptostachya Buckl.   Slim-head Manna-Grass
/T/W/   (Hel)   Shallow water from the s Alaska Panhandle (Wrangell; Hultén 1950) through coastal B.C. (Alberni, Vancouver Is.; J.K. Henry, Ottawa Nat. 31(5–6): 55. 1917) to cent. Calif. MAP: Hultén 1968b: 151.

G. maxima (Hartm.) Holmb.
Eurasian; introd. in Alaska (Fairbanks; L. H. Jordal, Rhodora 53(630): 156. 1951), N Alta. (near L. Athabasca; CAN), Ont. (N to the Ottawa dist. and the N shore of L. Huron; *see* s Ont. map by Montgomery 1956: fig. 4, p. 96), and Nfld. (near St. John's). [*Molinia* Hartm.; *G. spectabilis* Mert. & Koch]. MAPS: Hultén 1962: map 173, p. 185; Meusel, Jaeger, and Weinert 1965: 40.

G. melicaria (Michx.) Hubbard   Melic Manna-Grass
/T/EE/   (Hs)   Swamps and wet open or wooded ground from Ont. (N to the Petawawa R. in

Renfrew Co.) to Que. (N to Senneterre, 48°24'N), N.B., and N.S. (not known from P.E.I.), s to Tenn. and N.C. [*Panicum* Michx.; *G. torreyana* (Spreng.) Hitchc.; *Poa (G.; Panicularia) elongata* Torr.].

A hybrid with *G. striata* (× *G. gatineauensis* Bowden) is reported by W. M. Bowden (Can. J. Bot. 38(2): 126. 1960) from the type locality near Eardley, Gatineau Co., sw Que.

G. obtusa (Muhl.) Trin.
/T/E/   (Hs)   Peaty and wet sandy soils from N.B. (several localities in Charlotte Co.; CAN; DAO) and N.S. to E Pa. and N.C.

G. occidentalis (Piper) Nels.
/T/W/   (Hs)   Marshes, wet places, and shallow water from s B.C. (Vancouver Is.; Chilliwack; Columbia Valley) to N Calif. and Idaho. [*Panicularia* Piper; *G. septentrionalis sensu* Henry 1915, not Hitchc.].

G. pulchella (Nash) Schum.
/ST/WW/   (Hs)   Wet soil and shallow water from cent. Alaska–Yukon–Dist. Mackenzie (type from White River, the Yukon) to L. Athabasca (Alta. and Sask.) and Man. (N to Norway House, off the NE end of L. Winnipeg), the s limits uncertain but perhaps extending to N.Mex. and Mexico. [*Panicularia* Nash]. MAP: Hultén 1968b: 154.

G. septentrionalis Hitchc.   Floating Manna-Grass
/T/EE/   (Hs)   Wet soil or shallow water from Minn. to s Ont. (N to Hastings and Waterloo counties), sw Que. (N to Montreal), and N.S. (Digby, Digby Co.; G. L. Church, Am. J. Bot. 36: 156. 1949), s to Tex., Mo., Ky., and Ga. [*Panicularia* Hitchc.].

G. striata (Lam.) Hitchc.   Fowl-meadow Grass
/ST/X/   (Hs)   Moist ground (ranges of Canadian taxa outlined below), s to Calif., N Mexico, Tex., S.Dak., Iowa, Ala., and N Fla. MAP and synonymy: see below.
1   Spikelets to 4 mm long; tip of lemmas barely scarious; panicle to 3 dm long; leaves to over 7 mm broad, flat; [*Poa striata* Lam.; *P. (G.) nervata* Willd.; area uncertain through gradual intergrading with the more northern var. *stricta* but reported from Alta. to Nfld. by Fernald *in* Gray 1950; MAP (aggregate species): Hultén 1968b: 153] . . . . .var. *striata*
1   Spikelets to 4.5 mm long; lemmas with broad scarious tips; panicle at most about 1.5 dm long; leaves to about 5 mm broad, usually folded; [*G. nervata* var. *stricta* Scribn.; Alaska–Yukon–Dist. Mackenzie to Sask. (N to L. Athabasca), Man. (N to Churchill), Ont. (N to the Fawn R. at ca. 55°40'N, 88°W), Que. (N to E Ungava Bay), Labrador (N to the Hamilton R. basin), Nfld., N.B., P.E.I., and N.S.] . . . . .var. *stricta* (Scribn.) Fern.

HELICTOTRICHON Bess.   [273]

1   Panicle commonly less than 1 dm long, its short branches erect or appressed-ascending; spikelets 3–6-flowered; leaf-blades and sheaths glabrous; culms usually less than 4 dm tall; (B.C. to Man.) . . . .*H. hookeri*
1   Panicle to about 1.5 dm long, its flexuous branches ascending; spikelets mostly 2–3-flowered; leaf-blades and at least the lower sheaths typically short-villous; culms to over 1 m tall; (introd. on Anticosti Is., Que.) . . . .*H. pubescens*

H. hookeri (Scribn.) Henrard   Spike Oat
/T/WW/   (Hs)   Dry slopes, prairies, and plains from sw Yukon and sw Dist. Mackenzie to B.C., Alta. (the main area N to ca. 54°N; an isolated station near the w end of L. Athabasca; type a Drummond collection from the Rocky Mts., probably in Alta.), Sask. (N to McKague, ca. 52°45'N), and s Man. (N to Duck Mt.; introd. in sand-hills near Notre-Dame-du-Laus, Papineau Co., sw Que.), s to Mont., N.Mex., N.Dak., and Minn. [*Avena* Scribn.; *A. americana* Scribn.; *A. pratensis* var. *am.* Scribn.; *A. versicolor sensu* Hooker 1840, not Vill.]. MAPS: Hultén 1968b: 120; *Atlas of Canada* 1957: map 11, sheet 302.

H. pubescens (Huds.) Pilger   Hairy Oat
Eurasian; introd. in s Ont. (on a campus at London; C. Rousseau 1968) and E Que. (Port Menier, Anticosti Is.; DAO; see J. Adams, Can. Field-Nat. 50(7): 117. 1936). [Avena Huds.].

HIEROCHLOË R. Br.   [206]   Holy Grass

1   Lower staminate lemma (spikelets in this genus with 2 staminate florets and a terminal fertile one) always with a straight awn or a small point, the upper one with a rather long geniculate awn; panicle contracted, to 5 cm long; leaves to 3 mm broad, more or less revolute; plants loosely to densely caespitose or with short-creeping rhizomes . . . . . . .H. alpina
1   Both staminate lemmas awnless or merely short-pointed; plants with slender creeping rhizomes.
  2   Spikelets to 8 mm long, brownish or purplish; panicle pyramidal, to over 1 dm long; lower leaves flat, to 5 mm broad; culms to over 6 dm tall . . . . .H. odorata
  2   Spikelets to 6 mm long, pale; panicle linear-cylindric, contracted and subsimple, usually less than 2.5 cm long; leaves involute, about 1 mm broad; culms mostly less than 2 dm tall; (arctic regions) . . . . .H. pauciflora

H. alpina (Sw.) R. & S.
/AST/X/GEA/   (Hsr (Grh))   Dry tundra and rocky ground (chiefly acidic), the aggregate species from the coasts of Alaska–Yukon–Dist. Mackenzie to Ellesmere Is. at ca. 80°N, northernmost Ungava–Labrador, and Nfld., s to northernmost B.C., Great Slave L., N Man. (Baralzon L. at 60°N; reports from Churchill require confirmation), ?Ont. (the report from Moose Factory s of James Bay, by John Macoun 1888, requires confirmation), James Bay (an island at 53°27′N), Que. (s to Great Whale R., SE Hudson Bay, ca. 55°20′N, and Knob Lake, 54°48′N; isolated on mts. near St-Urbain, Charlevoix Co., and on the Shickshock Mts. of the Gaspé Pen.), and in the mts. to N Mont., N N.Y., and N New Eng.; circumgreenlandic; Eurasia. MAPS and synonymy: see below.
1   Awn of second staminate lemma inserted above the middle, straight, little or not at all twisted at base; lemma of fertile floret commonly 3-nerved or with only obscure intermediate nerves, with a bristle-like awn inserted at the entire apex; branches 1 or 2 at the lowest panicle-node, the solitary branch or the larger one of the pair bearing 3 or 4 spikelets; plant loosely caespitose or with short-creeping rhizomes; [*H. orthantha* Soer., the type from s Greenland; *Holcus (Hier.) monticola* Bigel.; E Que. (Mt. Albert, Gaspé Pen.), Nfld., Labrador (between 55°20′ and 58°13′N), and Greenland. The MAP by Hultén 1962: map 13, p. 21, should be expanded to indicate the occurrence in E Canada] . . . . .var. *orthantha* (Soer.) Hult.
1   Awn of second staminate lemma inserted below the middle, geniculate, strongly twisted; lemma of fertile floret 5-nerved, with a firm awn inserted below the 2-toothed apex; lowest panicle-node bearing 2 branches, these usually with only 1 or 2 spikelets; plant densely caespitose . . . . .var. *alpina*
  2   Stem-leaves to 7 mm broad; spikelets relatively broad; [type from Nettilling L., cent. Baffin Is.; Polunin 1940] . . . . .f. *soperi* Polunin
  2   Stem-leaves 1 or 2 mm broad; [*Holcus alpinus* Sw.; transcontinental; MAPS: Hultén 1968b: 84, and 1962: map 13, p. 21; Porsild 1957: map 12, p. 162; Tolmachev 1952: map 7 (incomplete); Raup, 1947: pl. 17] . . . . .f. *alpina*

H. odorata (L.) Beauv.   Vanilla, Indian, or Sweet Grass
/aST/X/GEA/   (Hsr)   Meadows, wet ground, and fresh to brackish shores from N Alaska–Yukon–Dist. Mackenzie to w Victoria Is., Great Slave L., L. Athabasca (Alta. and Sask.), Man. (N to Churchill), northernmost Ont.–Que.–Labrador, Nfld., N.B., P.E.I., and N.S., s to Oreg., Ariz., N.Mex., S.Dak., Pa., and N N.J.; southernmost Greenland; Iceland; Eurasia. [*Holcus (Torresia; Savastana) L.; Hol. borealis* Schrad.; *Hol. fragrans* Willd.; *Hierochloa arctica* Presl]. MAPS: Hultén 1968b: 84, and 1962: map 106, p. 115; Porsild 1957: map 13, p. 162; Meusel, Jaeger and Weinert 1965: 53.

H. pauciflora R. Br.
/ASs/X/EA/   (Hsr)   Wet mossy tundra (chiefly near the coast) from the coasts of Alaska–

Yukon–Dist. Mackenzie to Prince Patrick Is., Melville Is. (type locality), Devon Is., Baffin Is., and northernmost Que., s along the w coast of Hudson Bay to NE Man. (Churchill; Schofield 1959); isolated stations in N Ont. (NW James Bay at Cape Henrietta Maria and 34 mi southwards); Novaya Zemlya; Asia. MAPS: Hultén 1968b: 85, and 1962: map 5, p. 13 (noting three other total-area maps); Porsild 1957: map 14, p. 162; Tolmachev 1952: map 6 (requiring revision).

Forma *setigera* Lepage (Nat. can. (Que.) 81: 255. 1954; both of the staminate florets with awns exserted up to 1 mm rather than awnless) is known from the type locality, Cape Henrietta Maria, NW James Bay, Ont.

HOLCUS L.   [257]   Velvet-Grass

1 Lemmas about 2 mm long, obtuse, with a few silky hairs at base, the awn of the upper one hooked at apex, 1 or 2 mm long, scarcely exserted from the glumes; glumes obtusely acuminate, to 5 mm long, pubescent on the back and ciliate on the nerves; leaf-sheaths pubescent; culms tufted, softly villous below the nodes, these not bearded; (introd., transcontinental) . . . .*H. lanatus*
1 Lemmas about 2.5 mm long, acute, subtended by a conspicuous tuft of silky hairs, the awn of the upper one geniculate, to 4 mm long, exserted from the glumes; glumes long-acuminate, to 6 mm long, glabrous or merely minutely scabrous; leaf-sheaths glabrous or the lower ones more or less pubescent; culms glabrous, from slender rhizomes, the nodes with a tuft of retrorse hairs; (introd. in sw B.C.) . . . .*H. mollis*

H. lanatus L.   Common Velvet-Grass, Yorkshire Fog
Eurasian; sterile fields and grassy banks of s Alaska (Sitka), B.C. (Victoria; Glacier), Ont. (Middlesex, Wellington, and Wentworth counties), Que. (Sherbrooke and Stanstead counties), Nfld. (Whitbourne and vicinity), N.B. (Fredericton), and N.S. (Digby, Yarmouth, Shelburne, Annapolis, Queens, Kings, Lunenburg, and Cape Breton counties). [*Notholcus* Nash]. MAPS: Hultén 1968b: 109, and 1962: map 205, p. 217.

H. mollis L.   German Velvet-Grass
European; known in Canada only through a collection by Groh at Langley Prairie (DAO), near Vancouver, B.C.

HORDEUM L.   [410]   Barley.   Orge

(Ref.: Bowden 1962)
1 Leaf-blades with prominent auricles at base; glumes of central spikelets somewhat broadened above base; annuals; (introd.).
  2 Rachis of inflorescence continuous, flexuous and not disarticulating; glumes of central spikelets not ciliate . . . .[*H. vulgare*]
  2 Rachis disarticulating; glumes of central spikelets long-ciliate along the broadened basal part; (B.C. and Alta.).
    3 Anthers of central spikelets mostly less than 0.5 mm long; rachillae of lateral spikelets prolonged not more than 2.5 mm behind the palea; florets of central spikelets on pedicels to 1.5 mm long; inner glumes of the lateral spikelets about as broad as those of the central ones; rachis of inflorescence with longer hairs on the margins than on the sides . . . .*H. glaucum*
    3 Anthers of central spikelets commonly about 1 mm long; rachillae of lateral spikelets commonly prolonged about 4 mm behind the palea; inner glumes of the lateral spikelets usually narrower than those of the central ones; rachis with shorter hairs on the margins.
      4 Florets of central spikelets on pedicels commonly 1 or 2 mm long; paleas of lateral spikelets pubescent (as well as scabrous) between the nerves . . . .
       . . . .*H. leporinum*
      4 Florets of central spikelets usually subsessile (rarely sessile), the pedicel commonly not over 0.5 mm long; paleas of lateral spikelets merely scabrous on the nerves . . . .*H. murinum*

1   Leaf-blades not auricled at base or the auricles rudimentary.
   5   Glumes linear-lanceolate, about 1 mm broad above base (except the bristleform
       upper glume of the pedicelled sterile lateral spikelets), to about 1.5 cm long; spike to
       about 6 cm long; leaves to 6 mm broad; annual; (s ?B.C., sw Alta., and s Ont.) . . . .
       . . . .*H. pusillum*
   5   Glumes bristleform throughout (slightly broadened above base).
       6   Central spikelet 2-flowered; lateral spikelets perfect, with well-formed lemmas,
           nearly sessile; spike nodding, to about 1.5 dm long; leaves to 8 mm broad . . . .
           . . . . × *H. montanense*
       6   Central spikelet 1-flowered; lateral spikelets usually imperfect or rudimentary
           (their lemmas much reduced), on pedicels to 2 mm long.
           7   Tufted perennial; lemmas of lateral spikelets awned when present; spikes to
               12 cm long, soft, the rachis very brittle and readily disarticulating at maturity;
               anthers to 2 mm long; (transcontinental) . . . .*H. jubatum*
           7   Annuals; spikes to 7 cm long, stiff, the rachis not readily disarticulating; anthers
               commonly about 1 mm long.
               8   Lemmas of lateral spikelets awned; spikes ovate to ovate-oblong, rarely
                   over 5 cm long; glumes rigid; (introd. in s B.C. and s Ont.) . . . .*H. geniculatum*
               8   Lemmas of lateral spikelets acute, awnless; spikes linear-oblong, to 7 cm
                   long; glumes not rigid; (Vancouver Is.) . . . .*H. depressum*

H. depressum (Scribn. & Sm.) Rydb.
/t/W/   (T)   Meadows and moist open ground from sw B.C. (near Victoria, Vancouver Is.; Bowden
1962) to s Calif. and w Idaho. [*H. nodosum* var. *dep.* Scribn. & Sm.].

H. geniculatum All.
European; on ballast and in waste places of sw B.C. (near Victoria, Sidney, and Nanaimo,
Vancouver Is.) and s Ont. (Toronto; TRT). [*H. gussonianum* Parl.; *H. hystrix* Roth, not Schenck;
incl. *H. marinum* Huds.].

H. glaucum Steud.
Eurasian; known in Canada only from along a dry roadside in sw Alta. (Jasper National Park, where
taken by G. H. Turner in 1949; CAN, detd. Bowden). [*H. stebbinsii* Covas].

H. jubatum L.   Squirrel-tail Grass.   Queue d'écureuil
/aST/X/EA/   (Hs (T))   Coasts, shores, open ground, and roadsides and waste places, the
aggregate species from N-cent. Alaska–Yukon and w-cent. Dist. Mackenzie (an isolated station
also at the s end of Bathurst Inlet) to s Dist. Keewatin, Que. (N to Chimo, Ungava Bay), Labrador (N
to the Hamilton R. basin), Nfld., N.B., P.E.I., and N.S., s through most of the U.S.A. to Mexico and
S. America; Eurasia. MAPS and synonymy: see below.
1   Glumes commonly less than 1.5 cm long, straight (rarely arching from middle or near
    apex), ascending (sometimes slightly spreading); lemma-awns of central spikelets
    commonly less than 1 cm long; [*H. brachyantherum* Nevski; *H. boreale* Scribn. & Sm., not
    Gand.; *H. nodosum* of Canadian reports, not L.; *H. ?pratense* Huds.; Aleutian Is.–cent.
    Alaska–cent. Yukon–B.C.–Alta.; isolated in the Cypress Hills of SE Alta. and sw Sask.; s
    Labrador (along the Blanc-Sablon R.); NW Nfld. (Pistolet Bay); MAPS: Hultén 1968b:191
    (*H. brach.*); Meusel, Jaeger, and Weinert 1965:43 (*H. bor.*); Fernald 1925: map 24
    (*H. bor.*), p. 257] . . . . .ssp. *breviaristatum* Bowden
1   Glumes to over 7 cm long, straight or arching from base (rarely bent near apex),
    sometimes spreading and strongly recurved (at least when ripe); lemma-awns of central
    spikelets to over 7 cm long.
    2   Glumes and the lemma-awns of the central spikelets to 2.5 (3.5) cm long; [var.
        caespitosum (Scribn.) Hitchc. (*H. caesp.* Scribn.); cent. Alaska–Yukon–N Dist.
        Mackenzie–B.C.–Alta. to Sask. (N to Candle L., Saskatoon, and Sutherland) and Man.
        (N to Playgreen L., off the NE end of L. Winnipeg; MAP (var. *caesp.*): Hultén 1962:
        map 228, p. 239] . . . . .ssp. *intermedium* Bowden
    2   Glumes and the lemma-awns of the central florets to 7(9) cm long; [*Critesion* Nevski;

introd., transcontinental; MAPS: on the above-noted 1962 map by Hultén; Hultén 1968b:192]. Concerning hybridization with *Agropyron* and *Elymus, see* the listings of hybrids following the treatments of those genera . . . . .ssp. *jubatum*

H. leporinum Link
Eurasian; known in Canada from beaches and waste ground of sw B.C. (near Victoria, Sidney, and Nanaimo, Vancouver Is., and the adjacent mainland at Kitsilano Beach, Vancouver; Bowden 1962).

H. murinum L.   Mouse barley
European; known in Canada from beaches, ballast, and waste ground of sw B.C. (Vancouver Is. and adjacent islands, and Penticton; Bowden 1962).

H. pusillum Nutt.   Little Barley
/T/WW/   (T)   Open (often alkaline) ground from Wash. to s Alta. (Onefour, about 60 mi s of Medicine Hat; DAO; reported as introd. near Victoria, Vancouver Is., B.C., by John Macoun 1888; not known from Sask. or Man.) and s Ont. (along railways at Amherstburg and vicinity, Essex Co., where undoubtedly introd.; *see* Bowden 1962), s to s Calif., Mexico, Tex., La., Fla., and S. America.

[H. vulgare L.]   Barley
[Eurasian; much cult. in N. America and escaping to roadsides, railway ballast, and other waste places but not established, as in the Aleutian Is., s Alaska, all of the provinces, and s Greenland. MAP: Hultén 1968b:192.
    For a key to cultivars escaped in Canada (incl. *H. distichon* L., introd. in s Greenland), *see* Bowden (1962:1698, citing localities).]

KOELERIA Pers.   [346]

1   Panicle purplish, rarely over 4 cm long, compact; lemmas silky-pilose, to 5 mm long;
    leaves 1 or 2 mm broad, strongly involute; culms commonly less than 2 dm tall;
    (Alaska–Yukon–Dist. Mackenzie) . . . .*K. asiatica*
1   Panicle pale and shining, to over 1.5 dm long, often interrupted at base; lemmas
    scabrous, 3 or 4 mm long; leaves to 3 mm broad, flat or involute; culms to about 6 dm tall;
    (transcontinental) . . . .*K. cristata*

K. asiatica Domin
/aS/W/A/   (Hsr)   Sandy shores and dunes of N Alaska (Chipp R. at 70°41′N; Wiggins and Thomas 1962; *see* their Alaska map, fig. 16, p. 365), s-cent. Yukon (*see* Hultén 1942: map 132 (*K. cairn.*), p. 395), and NW Dist. Mackenzie (Porsild and Cody 1968); N Siberia. [*K. cairnesiana* Hult.]. MAPS: Hultén 1968b:123; Johnson and Viereck 1962: map 4, p. 22.

K. cristata (L.) Pers.   June-Grass
/sT/X/EA/   (Hs)   Dry prairies, sandy ground, and open woods from sw Dist. Mackenzie (J. W. Thieret, Can. Field-Nat. 75(3):112. 1961; reported from Alaska by Boivin 1967a; an isolated station in s-cent. Yukon indicated on the map by Hultén 1942: fig. 133 (*K. yuk.*), p. 395) and B.C. to Alta., Sask. (N to Prince Albert), Man. (N to the Nelson R. about 150 mi s of Churchill), s Ont. (N to Huron and Waterloo counties), Que. (s shore of L. St. John; MT; GH), and s Labrador (Goose Bay, Hamilton R. basin; CAN; not known from the Atlantic Provinces), s to Calif., Mexico, Tex., Mo., Ohio, and Del., and locally adventive to New Eng. (as possibly the case with some of the above E Canadian citations); Eurasia. [?*Aira (Poa) cristata* L.; *K. gracilis* Pers.; *K. latifrons* (Domin) Rydb.; *K. ?macrantha* (Ledeb.) Spreng.; *K. nitida* Nutt.; *K. pyramidata* (Lam.) Beauv.; *K. yukonensis* Hult.; incl. var. *major* Vasey]. MAPS (*K. grac.*): Hultén 1968b:124, and 1962: map 129, p. 139; Meusel, Jaeger, and Weinert 1965:46; Meusel 1943: fig. 30e.

Gramineae

LEERSIA Sw.   [194]   Cutgrass, Whitegrass

1   Lower panicle-branches usually whorled; spikelets to 6 mm long, strongly overlapping;
    lemmas stiffly ciliate on the keel, pilose on the sides; stamens 3; leaves to 12 mm broad,
    very rough on the margins with stiff colourless spinules, their sheaths roughly
    retrorse-scabrous; culms terete, from slender elongate rhizomes; (B.C.; Man. to N.S.) . . . .
    . . . . .L. oryzoides
1   Lower panicle-branches solitary; spikelets mostly less than 4 mm long, less strongly
    overlapping; lemmas sparsely ciliate on the keel and margins, glabrous to pilose on the
    sides; stamens 2; leaves usually not over 8 mm broad, they and their sheaths smoothish
    to minutely scabrous; culms slender and weak, erect to decumbent and rooting at base,
    from short thick scaly rhizomes; (Ont. to N.B.) . . . . .L. virginica

L. oryzoides (L.) Sw.   Rice-Cutgrass
/T/X/EA/   (Grh (Hsr))   Swamps, shores, and ditches (ranges of Canadian taxa outlined below),
s to Calif., Mexico, Tex., and Fla.; S. America; W. I.; Europe; w and E Asia. MAP and synonymy: *see*
below.
1   Leaves smooth and glabrous; [Que. (Lotbinière, Bellechasse, and Portneuf counties) and
    N.S. (Shelburne and Yarmouth counties]. Dore and Roland (1942) and Dore (1959) point
    out that in submersed plants the submersed leaves may be smooth, the emersed ones
    harsh, this form probably being merely an ecological phase of little or no taxonomic
    significance . . . . .f. *glabra* Eat.
1   Leaves harsh, their margins and often the nerves beneath scabrous-hispid.
    2   Panicle wholly included in its subtending leaf-sheath; [f. *clandestina* Eames; Que.,
        N.B., and N.S.] . . . . .f. *inclusa* (Wiesb.) Dorfl.
    2   Panicle exserted from the leaf-sheath or only partly included; [*Phalaris* L.;
        Homalocenchrus Poll.; swamps, ditches, and shores in B.C. (N to Kamloops; not
        known from Alta.; reports from Sask. require confirmation), s Man., Ont. (N to
        Batchawana Bay, E shore of L. Superior), Que. (N to the Gaspé Pen.), N.B., P.E.I.,
        and N.S. (the report from Nfld. by Reeks 1873, requires confirmation); E Asia
        (var. *japonica* Hack.); MAP: Hultén 1958: map 246, p. 265] . . . . .f. *oryzoides*

L. virginica Willd.   White Grass
/T/EE/   (Grh (Hsr))   Damp woods and thickets from Nebr. to Minn., Ont. (N to the Ottawa dist.),
Que. (N to Grondines, Portneuf Co., and Beauport, about 10 mi N of Quebec City; *see* Que. map by
Dominique Doyon and W. G. Dore, Can. Field-Nat. 81(1): fig. 2, p. 31. 1967), and N.B. (Boivin
1967a; not known from P.E.I. or N.S.; the report from Nfld. by Reeks 1873, requires clarification), s
to Tex. and Fla.
    Some of our s Ont. and sw Que. material is referable to var. *ovata* (Poir.) Fern. (*L. ovata* Poir.;
L. lenticularis Michx.; margins and often the keel of the lemmas bristly-ciliate rather than smooth or
only minutely ciliate).

LOLIUM L.   [395]   Darnel.   Ivraie

1   Lemmas typically long-awned, the awn to over 1 cm long; spikelets with at most 9 florets;
    rachis of spike scabrous; leaves to over 6 mm broad; culms to about 9 dm tall; annuals;
    (introd.).
    2   Glume rarely over 1.5 cm long (note that the glume is singlein *Lolium* and located
        on the side of the spikelet away from the rachis); lemmas oblong-lanceolate;
        spikes to 3 dm long . . . . .L. *persicum*
    2   Glume equalling or longer than the spikelet, to 2 cm long; lemmas elliptic-ovate;
        spikes commonly less than 2 dm long . . . . .L. *temulentum*
1   Lemmas awnless or with an awn at most 0.5 mm long; spikelets with up to over 10 florets;
    glume not surpassing the spikelet; leaves at most 4 mm broad; culms to about 6.5 dm tall;
    (introd.).
    3   Lemmas (at least in the upper florets) with a short awn to 0.8 mm long; glume shorter

than the spikelet (usually half the length), 5-nerved; culms scabrous near summit;
annual . . . . *L. multiflorum*
3 Lemmas awnless.
    4 Leaves, sheaths, culm, and rachis glabrous and smooth; glume shorter than the
       spikelet (usually not much more than half as long), 7–9-nerved; perennial, often
       stoloniferous . . . . *L. perenne*
    4 Leaves scabrous on the upper surface, their sheaths also minutely scabrous; culm
       usually scabrous below the very slender spike (the spikelets scarcely broader than
       the very scabrous rachis); glume shorter than or equalling the spikelet, 5–7-
       nerved; annual . . . . [*L. rigidum*]

L. multiflorum Lam.   Italian Rye-Grass
European; (ranges of Canadian taxa outlined below).
1 Annual; spikelets with usually at least 10 florets; glumes of the lowest spikelets often
   surpassing the contiguous floret; [*L. italicum* A. Br.; *L. perenne* ssp. *mult.* (Lam.) Husnot;
   s Alaska (Juneau), s-cent. Yukon (Dawson), B.C., Sask. (Tisdale), s Man. (near Carberry
   and Winnipeg), Ont. (N to the Ottawa dist.), Nfld., N.B., and N.S.; MAP (aggregate
   species): Hultén 1968*b*:180] . . . . var. *multiflorum*
1 Perennial, commonly lower; spikelets with less than 10 florets; glumes of the lowest
   spikelets shorter than the contiguous floret; [Glenwood, Nfld., where taken by Robinson
   and Schrenk in 1894; GH; CAN] . . . . var. *diminutum* Mutel

L. perenne L.   Common Darnel, Perennial Rye-Grass
Eurasian; fields, roadsides, and waste places of SE Alaska (Wrangell), B.C. (N to Prince Rupert),
Alta. (N to near Edmonton), Sask. (Indian Head), Man. (Delta; Winnipeg), Ont. (N to Ottawa dist.),
Que. (N to Rimouski, Rimouski Co.), Nfld. (St. John's; Bonavista Bay), N.B. (near St. John), P.E.I.
(Charlottetown), and N.S. (Kings, Halifax, Pictou, Inverness, and Victoria counties); sw Greenland.
MAP: Hultén 1968*b*:180.

L. persicum Boiss. & Hoh.   Persian Darnel
Eurasian; fields, roadsides, railway ballast, etc., of the Peace River dist. of B.C. and Alta., Sask. (N
to near North Battleford), Man. (N to Churchill), and Ont. (between L. Superior and L. Nipigon).
[*L. dorei* Boivin and its var. *laeve* Boivin]. MAP (Canadian stations): W. G. Dore, Sci. Agric. 30: pl. 2,
p. 161. 1950.

[L. rigidum Gaud.]   Wimmera Rye-Grass
[European; reported by G. A. Stevenson (Can. Field-Nat. 79 (3):174. 1965) as having been taken
at Brandon, Man., in a stand of sweet clover grown from seed imported from Australia, but not
persisting.]

L. temulentum L.   Bearded or Poison Darnel.   Ivraie
Eurasian; (ranges of N. American taxa outlined below):
1 Lemmas awned; [incl. var. *macrochaeton* A. Br.; s Alaska (St. Michael), the Yukon
   (Dawson), B.C. (Vancouver Is.; Henry 1915), Alta. (Altario, ca. 52°N, and Northmark,
   55°36′N; by Groh 1946; not listed by Moss 1959), Man. (Dominion City; Winnipeg), Ont.
   (N to Ottawa), Que. (Fernald *in* Gray 1950), and N.B. (Carleton, York Co.; CAN); MAP
   (aggregate species): Hultén 1968*b*:181] . . . . var. *temulentum*
1 Lemmas awnless; [the Yukon (Hunker Creek, near Dawson; CAN), sw B.C. (Vancouver
   Is.; CAN), s Alta. (Lacombe; CAN), s Man. (Winnipeg; WIN), and s Ont. (Lambton,
   Wellington, Peel and Frontenac counties)] . . . . var. *leptochaeton* A. Br.

MELICA L.   [355]   Melic-Grass

1 Spikelets relatively narrow (commonly more than 3 times as long as broad); glumes
   usually narrow, scarious-margined; sterile lemmas similar to the fertile ones.
    2 Lemmas long-awned from a 2-toothed apex; culms not bulbous at base.
       3 Panicle narrow, its short branches mostly paired and appressed; lemma-awn to 1

cm long; spikelets to 1.5 cm long (excluding awns); leaves to 6 mm broad, closely veined, strongly scabrous, their ligules usually pubescent to tip; (SE ?B.C.) . . . .
. . . .[M. aristata]

3 Panicle open, its few short branches mostly solitary and spreading; lemma-awn rarely over 5 mm long; spikelets to 2 cm long (excluding awns); leaves to 1.5 cm broad, weakly scabrous, their prominent veins wide-spaced, their ligules usually glabrous except at base; (s B.C.–sw Alta.; Ont.) . . . . .M. smithii

2 Lemmas awnless or tipped with an awn less than 2 mm long; leaves to about 6 mm broad; (B.C.–Alta.)

4 Lemmas mucronate or tipped with an awn less than 2 mm long; spikelets to 1.5 cm long; panicle narrow, its branches appressed; culms not bulbous at base; (sw B.C.) . . . . .M. harfordii

4 Lemmas acuminate or obtuse, awnless; spikelets to 2 cm long; culms bulbous at base.

5 Lemmas acuminate, to 13 mm long, usually pilose; panicle narrow, its short branches usually appressed, sometimes spreading; (B.C.–Alta.) . . . . . . .M. subulata

5 Lemmas narrowed to an obtuse or obscurely 2-toothed apex, to 10 mm long, not pilose; panicle broad, its branches long and spreading; (s ?B.C.) . . . . .[M. geyeri]

1 Spikelets relatively broad (at most about 2 or 3 times as long as broad); glumes broad and papery; lemmas awnless, the sterile ones small and convolute, usually hidden in the upper ones.

6 Leaves to 2 cm broad; panicle narrow and dense, its very short branches ascending; (introd. in s Ont.) . . . . .M. altissima

6 Leaves at most about 5 mm broad.

7 Culms from knotted rhizomes, not distinctly bulbous at base; panicle open, its long branches spreading; spikelets to 11 mm long, often pendulous, usually 2-flowered, about equalling the glumes; (s ?Ont.) . . . .[M. mutica]

7 Culms bulbous at base; panicle narrow, its short branches appressed; spikelets to 1.5 cm long, with often more than 2 flowers, distinctly surpassing the glumes; (B.C. and s Alta.).

8 Pedicels capillary, flexuous or recurved; lower glume averaging about 5 mm long; lemmas to 8 mm long; corm stipitate on the rhizome . . . . .M. spectabilis

8 Pedicels stouter, appressed; lower glume averaging about 7 mm long; lemmas to over 12 mm long; corm sessile on the rhizome . . . . .M. bulbosa

M. altissima L.
Eurasian; sometimes cult. as an ornamental in N. America and reported as a garden-escape in s Ont. by Landon (1960; Norfolk Co.).

[M. aristata Thurb.]
[This species of the w U.S.A. (Wash. to Calif.) is reported from B.C. by John Macoun (1888; "abundant in the valleys of the Selkirk Mountains"; this taken up by Henry 1915) and by Hubbard (1955; Field, SE B.C.). Voucher-specimens have not been seen, however, and these reports require confirmation.]

M. bulbosa Geyer   Onion-Grass
/T/WW/   (Gst)   Rocky woods and hills from B.C. (Lillooet; Merritt; Okanagan; Kokanee; reported N to Fort McLeod, ca. 55°N, by John Macoun 1888) and sw Alta. (Waterton Lakes; Breitung 1957b) to Calif. and Colo. [M. bella Piper].

[M. geyeri Munro]
[This species of the w U.S.A. (Oreg. to Calif.) is reported from sw B.C. by J. K. Henry (Ott. Nat. 31 (5–6):55. 1917; Alberni, Vancouver Is.; this taken up by Eastham 1947, and Hubbard 1955). In the apparent absence of voucher-specimens, however, these reports require confirmation.]

M. harfordii Boland.
/t/W/ (Hs) Open dry woods and slopes from sw B.C. (Vancouver Is. and Mayne Is.; CAN) to Calif.

[M. mutica Walt.]
[This species of the E U.S.A. (N to Iowa and Md.) is reported from s Ont. by Stroud (1941; Wellington Co.) but no voucher-specimens have been seen nor is it listed by Soper (1949).]

M. smithii (Porter) Vasey
/T/WW/ (Hs) Moist woodlands: s B.C. (Vancouver Is.; Agassiz; Fernie; Yoho Valley; Nelson; Popkum) and sw Alta. (Waterton Lakes; Castle Mt., near Banff) to Oreg. and Wyo.; Great Lakes region of N Mich. and s Ont. (Bruce Pen., L. Huron; MT; reported from Port Franks, Lambton Co., by Dodge 1915). [*Avena* Porter; *Bromelica* Farw.]. MAP: Fernald 1935: map 11, p. 217.

M. spectabilis Scribn.   Purple Onion-Grass
/T/W/ (Gst) Rocky open woods and thickets from B.C. (near Spences Bridge, Sophie Mt. sw of Rossland at 5,500 ft elevation, the west summit of the North Kootenay Pass, and McLeod L. at ca. 54°N; CAN) and sw Alta. (Castle Mt., near Banff; CAN) to N Calif. and Colo. [Scarcely separable from *M. bulbosa,* with which it is apparently merged by Boivin (1967a)].

M. subulata (Griseb.) Scribn.
/sT/W/ (Gst) Meadows and shaded slopes from the E Aleutian Is. (Unalaska) and SE Alaska (Sitka; Hultén 1942) through B.C. and sw Alta. (Waterton Lakes; Breitung 1957b) to Calif. and w Wyo.; Chile. [*Bromus subulatus* Griseb., the type from Unalaska, Aleutian Is.; *M. acuminata* Boland.]. MAP: Hultén 1968b:125.

[MIBORA Adans.]   [227]

[M. minima (L.) Desv.]
[European; this delicate annual (with the general aspect and stature of *Carex capillaris*) is known in N. America only in nurseries in N.Y. and Mass. (Fernald *in* Gray 1950) and in sw B.C. (grounds of the Experimental Station near Sidney, s Vancouver Is., where taken by Macoun in 1914 but apparently not established, no later collections being known; CAN). (*Agrostis* L.; *M. verna* Beauv.).]

MILIUM L.   [213]

M. effusum L.   Millet-Grass.   Millet
/T/EE/ (Hsr) Damp or rocky woods and thickets from s Man. (N to Norgate, E of Riding Mt.) to Ont. (N to the N shore of L. Superior), Que. (N to the Côte-Nord, Anticosti Is., and Gaspé Pen.), Nfld., N.B., and N.S. (not known from P.E.I.), s to NE S.Dak., N Ill., N Ind., Ohio, and Md.; Iceland; Eurasia. [Incl. var. *cisatlanticum* Fern.]. MAPS: Hultén 1958: map 128, p. 147, and 1937: fig. 12, p. 126; Meusel, Jaeger, and Weinert 1965:56.

[MISCANTHUS Anderss.]   [110]   Silver Grass

[M. sacchariflorus (Maxim.) Hack.]
[Asiatic; sparingly cult. as an ornamental and known, presumably as a garden-escape, from s Ont. (Elgin, Oxford, Wentworth, York, Ontario, and Stormont counties) and sw Que. (Victoriaville; Clarenceville).]

MOLINIA Schrank   [340]

M. caerulea (L.) Moench   Moor-Grass
Eurasian; fields and roadsides of Ont. (Frontenac and Carleton counties), N.S. (Louisburg, Cape Breton Co.; CAN), St-Pierre and Miquelon (GH), and ?Nfld. (Boivin 1967a). [*Aira* L.]. MAP: Hultén 1958: map 142, p. 161.

MUHLENBERGIA Schreb. [215]

1 Panicle typically open and diffuse; spikelets at most 2 mm long, on capillary pedicels longer than the awnless glabrous lemma; leaves at most 2 mm broad.
  2 Upper glume about half as long as the lemma; panicle long-exserted, up to 4 times as long as thick; rhizomes wanting; (Ont. to Nfld. and N.S.) .... *M. uniflora*
  2 Upper glume about equalling the lemma; panicle nearly as broad as long, exserted only at maturity; culms from elongate slender rhizomes; (B.C. to Man.) .... *M. asperifolia*
1 Panicle slender, contracted, often dense; spikelets sessile or very short-pedicelled, often closely overlapping.
  3 Lemmas neither pilose nor bearded at base (the basal hairs inconspicuous); leaves 1 or 2 mm broad.
    4 Lemma with a delicate straight awn about 1 cm long, the lemma-body about 4 mm long; lower glume mostly 1 or 2 mm long, the upper one a little longer; panicle usually more than half the entire height of the plant; densely tufted annual, the culms rarely to 1.5 dm tall; (s ?B.C.) .... [*M. depauperata*]
    4 Lemma awnless or merely minutely awn-tipped.
      5 Lemmas glabrous, about 3 mm long, the ovate glumes about half as long; anthers about 1.5 mm long; panicle to about 1 dm long; leaf-ligules 2 or 3 mm long; culms wiry, to 6 dm tall, arising singly or in tufts from the bases of old culms (more rarely from slender stolons); perennial; (essentially trans-continental) .... *M. richardsonis*
      5 Lemmas minutely pubescent.
        6 Lemma 2 mm long, the ovate glumes about half as long; anthers about 0.5 (to 0.8) mm long; panicle usually less than 5 cm long; leaf-ligules to 3 mm long; tufted annual with delicate filiform culms commonly not over 1.5 dm tall (but occasionally to 3 dm); (s B.C.) .... *M. filiformis*
        6 Lemma 3 or 4 mm long, not much surpassing the acuminate-cuspidate glumes; anthers to 1.2 mm long; panicle to about 1 dm long; ligules barely 1 mm long; culms wiry, to about 4 dm tall, in dense tufts with hard bulb-like scaly perennial bases but lacking rhizomes; (Alta. to Man.) .... *M. cuspidata*
  3 Lemmas pilose or bearded at base.
    7 Glumes rounded or subtruncate, the lower one minute or obsolete, the upper one less than 1 mm long; lemma about 2 mm long, its awn to 5 mm long; leaves to 4 mm broad; culms weak, flattened, often rooting at the nodes, lacking basal rhizomes; (s Ont.) .... *M. schreberi*
    7 Glumes linear-attenuate to awned, to 8 mm long; culms firm, from scaly rhizomes.
      8 Glumes tapering to long awns, to 8 mm long (including awns), much surpassing the lemma; panicle-branches densely flowered to base; leaves to 8 mm broad.
        9 Culms usually with several stiffly erect branches from the middle nodes, the internodes glabrous and shining; leaf-sheaths keeled; ligule about 4 mm long; anthers less than 1 mm long; grain about 2 mm long; (B.C. to Man. and w Ont.) .... *M. racemosa*
        9 Culms simple, rarely with erect basal branches, the internodes puberulent nearly to base; leaf-sheaths scarcely keeled; ligule minute; anthers to 1.5 mm long; grain to 1.5 mm long; transcontinental) .... *M. glomerata*
      8 Glumes merely attenuate or subulate at tip, typically awnless, the upper (larger) one about 3 mm long, rarely surpassing the lemma; longer panicle-branches devoid of spikelets at base.
        10 Anthers over 1 mm long; glumes ovate-lanceolate, abruptly narrowed to a short subulate tip, prominently green-keeled, much shorter than the lemma; lemma pruinose-puberulent, its awn to 1 cm long; leaves to 1.5 cm broad; ligule obsolete; internodes minutely retrorse-pilose below the pubescent nodes; (s Ont. and s Que.) .... *M. tenuiflora*
        10 Anthers less than 1 mm long; glumes linear to lance-attenuate, at least the

upper one nearly as long as the lemma; leaves at most 8 mm broad; ligule usually evident.
11 Culms glabrous throughout; lemma typically awnless; (Ont., Que., and N.B.) . . . .*M. frondosa*
11 Culm definitely puberulent below the glabrous nodes.
12 Lemma tapering to a delicate awn to 1.5 cm long; principal panicle-branches to 3 cm apart; spikelets silvery-green or whitish, often slenderly pedicelled; glumes distinctly unequal; (sw Que.) . . . .
. . . .*M. sylvatica*
12 Lemma typically awnless; principal panicle-branches at most 1 cm apart; spikelets green or purplish, subsessile; glumes subequal; (B.C. to N.S.) . . . .*M. mexicana*

M. asperifolia (Nees & Meyen) Parodi   Scratchgrass
/T/WW/   (Hsr)   Damp sandy ground from B.C. (N to Kamloops) to Alta. (N to Grande Prairie, ca. 55°10′N), Sask. (N to Saskatoon), and Man. (N to Vista, 60 mi NW of Brandon), s to Calif., Mexico, and Tex. [*Sporobolus* N. & M.].

M. cuspidata (Torr.) Rydb.
/T/WW/   (Hs(r))   Prairies and sandy or gravelly soil from Alta. (N to Grande Prairie, ca 55°N), Sask. (N to Saskatoon), and Man. (N to Grand Rapids, near the NW end of L. Winnipeg), s to N.Mex., Okla., Mo., and Ky. [*Vilfa cuspidata* Torr., the type a Drummond collection from "Saskatchewan River, Rocky Mountains", presumably in w Alta.; *Sporobolus cusp.* (Torr.) Wood; *V. gracilis* Trin. in part].

[M. depauperata Scribn.]
[The apparent report of this species of the sw U.S.A. (Ariz. and Colo. to Mexico) from B.C. by Henry (1915; *Sporobolus dep.*) would probably be found referable to *M. filiformis* if voucher-specimens were available.]

M. filiformis (Thurb.) Rydb.
/T/WW/   (T)   Open woods and alpine meadows and rocks from sw B.C. (Ucleulet, Vancouver Is.; cliffs at E end of Chilliwack L., alt. 3,500 ft; CAN) to S.Dak., s to Calif. and N.Mex. [*Sporobolus* Rydb.; *Vilfa depauperata* var. *fil.* Thurb.].

M. frondosa (Poir.) Fern.
/T/EE/   (Hsr)   Damp open woods, clearings, shores, and waste places from Ont. (N to the NW shore of L. Superior; doubtfully reported from B.C. by Boivin 1967a) to Que. (N to L. St. Peter; reports from farther north may refer to *M. mexicana*), and N.B. (Victoria and York counties; not known from P.E.I. or N.S.), s to E Tex., Mo., Tenn., and N Ga.
Forma *commutata* (Scribn.) Fern. (lemma bearing a slender awn rather than awnless) is known from Ont. (N to the Ottawa dist., Dore 1959) and Que. (N to L. St. Peter at Sorel; MT).

M. glomerata (Willd.) Trin.
/sT/X/   (Hsr)   Meadows, bogs, wet rocks, and shores (ranges of Canadian taxa outlined below), s to Oreg., Nev., Wyo., Minn., Ohio, Pa., and Conn. MAP and synonymy: see below.
1 Keel and awn of glumes copiously hispid; panicle-branches crowded; [*Polypogon glom.* Willd.; *M. setosa* (Biehler) Trin., not (HBK.) Kunth; s Ont. to N.S. (Fernald *in* Gray 1950)] . . . . . .
. . . . .var. *glomerata*
1 Keel and awn of glumes merely scabrous; panicle often interrupted below; [*M. racemosa* var. *cinn.* (Link) Boivin; *M. setosa* var. *cinn.* (Link) Fern.; s Yukon–Dist. Mackenzie–B.C.–Alta.–Sask. to Man. (N to Hill L., N of L. Winnipeg), Ont. (N to Hawley L., 54°24′N), Que. (N to the E James Bay watershed at ca. 54°N and the Côte-Nord; the northernmost Ont.–Que. stations are indicated in a map by Lepage 1966: map 5, p. 214), Nfld., N.B., P.E.I., and N.S.; MAP: Hultén 1968b:88] . . . . .var. *cinnoides* (Link) Herm.

M. mexicana (L.) Trin.
/T/X/ (Hsr) Thickets, damp clearings, and shores from B.C. (New Westminster, Agassiz, and near a hot spring on the Laird R. at 59°23′N; CAN) to Sask. (Boivin 1967*a*; not known from Alta.), Man. (N to Bellhampton, about 75 mi NE of Brandon), Ont. (N to the Albany R. sw of James Bay at ca. 52°N), Que. (N to the Gaspé Pen.), N.B., and N.S. (Cumberland, Hants, Halifax, and Kings counties; not known from P.E.I.), s to Calif., N.Mex., Kans., Ohio, and N.C. [*Agrostis* L.; *M. foliosa* of Canadian reports in large part, not (R. & S.) Trin.].

Forma *ambigua* (Torr.) Fern. (lemmas with a delicate awn to 1 cm long rather than awnless) and f. *setiglumis* (Wats.) Fern. (glumes awned rather than awnless) are known from Ont. and Que., f. *setiglumis* also from N.S.

M. racemosa (Michx.) BSP.
/T/WW/ (Hsr) Open woods, thickets, and clearings from B.C. (Liard Hotsprings, ca. 59°N, Agassiz, and about 70 mi w of Golden; CAN) to Alta. (Banff, Red Deer, and Edmonton; CAN), Sask. (N to Prince Albert), Man. (N to Duck Mt. and Rossburn, about 70 mi NW of Brandon), and Ont. (ledges by Kakagi L., near Kenora, at 49°10′N; CAN; a collection from a wet meadow near Kingston has also been placed here, as well as one from railway tracks at Schreiber, N shore of L. Superior), s to Oreg., N Mexico, N.Mex., Kans., Mo., and Ill.; introd. along railways eastwards to N N.H. [*Agrostis* Michx.].

Because of differing interpretations of this species and *M. glomerata* by various authors, their treatment here can only be considered tentative.

M. richardsonis (Trin.) Rydb.
/sT/X/ (Hsr) Damp thickets, shores, and rocky slopes from s Yukon, Great Bear L., and sw Dist. Mackenzie to B.C., Alta. (N to L. Athabasca), Sask. (N to ca. 55°N), s Man. (Brandon; Duck Mt.; Baldur), Ont. (w James Bay watershed s to ca. 50°N), Que. (known only from Anticosti Is. and the Gaspé Pen.; CAN; MT), and N.B. (York and Carleton counties; CAN; GH; ACAD; not known from P.E.I. or N.S.), s to Baja Calif., Ariz., Mexico, Nebr., Mich., Ohio, and Maine. [*Vilfa* Trin.; *Sporobolus* Merr.; *V. (S.) depauperata* Torr., not *M. dep.* Scribn.; incl. *V. (M.) squarrosa* Trin.]. MAPS: Porsild 1966: map 14, p. 68; Hultén 1968*b*:87.

M. schreberi Gmel.   Drop-seed, Nimble Will
/t/EE/ (Hs) Woodlands, thickets, roadsides, etc., from Nebr. to Mich., s Ont. (Essex, Kent, Elgin, Lambton, Middlesex, Welland, Lincoln, and Wentworth counties), Vt., and N.H., s to E Mexico, E Tex., and Fla. [*M. diffusa* Willd.].

M. sylvatica Torr.
/T/EE (Hsr) Damp thickets, rocky woods, clearings, and shores from Que. (a collection in CAN from Leamy·L., near Hull, has been placed here by Dore; collections in MT from Argenteuil and Missiquoi counties have also been referred here) to N.Y. and Maine, s to NE Tex., Ark., Ala., and N.C.

A collection in CAN from Griffin L., near Kamloops, B.C., has been placed here by Swallen but, having been taken along a roadside, was doubtless introd. at that locality. Reports from Ont. and N.B. by John Macoun (1888; relevant collections in CAN) are mostly based upon *M. frondosa* or *M. mexicana*. Collections from sw Que. have been referred to f. *attenuata* (Scribn.) Palmer & Steyerm. (lemmas awnless rather than with a delicate awn to 1.5 cm long; Gatineau and St-Jean counties; MT).

M. tenuiflora (Willd.) BSP.
/T/EE/ (Hsr) Rocky woods and shaded slopes and cliffs from Iowa to Wisc., s Ont. (Waterloo, Lincoln, Welland, and Hastings counties; CAN), sw Que. (Gatineau and Missiquoi counties; MT), and s Vt., s to Okla., Ark., Tenn., and Ga. [*M. willdenowii* Trin.]. MAP: Tetsuo Koyama and Shoichi Kawano, Can. J. Bot. 42(7): fig. 4 (C), p. 863. 1964.

M. uniflora (Muhl.) Fern.
/T/EE/ (Hs) Meadows and sandy or peaty soil from Ont. (N to the N shore of L. Superior) to Que. (N to E James Bay at 52°26′N, L. Mistassini, and the Gaspé Pen.), Nfld., St-Pierre and

Miquelon, N.B., and N.S. (not known from P.E.I.), s to Wisc., Mich., and N.J. [*Poa* Muhl.; *Sporobolus* Scribn. & Merr.; *S. serotinus* Gray]. MAPS: McLaughlin 1932: fig. 10 (incomplete), p. 345; Dore, p. 4 of a pamphlet (Dorset, s Ont., field trip) prepared for the 1959 Ninth International Botanical Congress).

The report from B.C. by Boivin (1967a) may be referable to *M. filiformis*. Depauperate but completely intergrading forms have been separated as var. *terrae-novae* Fern. (type from St. John's, Nfld.).

MUNROA Torr. [319]

M. squarrosa (Nutt.) Torr. False Buffalo-Grass
/T/WW/ (T) Dry plains, prairies, and hills from Mont. to s Alta. (Red Deer; Hardisty; Medicine Hat) and N.Dak., s to Calif., Ariz., and Tex.

E. W. Tisdale and A. C. Budd (Can Field-Nat. 62(6):174. 1948) report a small patch growing on a garden path of a ranch in the valley of the South Saskatchewan R. in sw Sask. They note that the plant, being an annual and known at that station for several years, had been able during that period to produce viable seed but gave no sign of having spread into the surrounding native prairie sod. Concerning reports from Man., *see* Scoggan (1957).

NARDUS L. [394]

N. stricta L. Matgrass
Eurasian; rocky banks, sandy fields, and roadsides of s Ont. (reported by Montgomery 1956, as extending for about 100 yds along a sandy roadside near Ilfracombe, about 30 mi E of Parry Sound), Que. (Megantic, Terrebonne, Frontenac, and Wolfe counties), Nfld. (near St. John's and Salmonier), and N.S. (Yarmouth and Shelburne counties); also introd. in s Greenland. MAPS: Hultén 1958: map 101, p. 121; Meusel, Jaeger, and Weinert 1965:57.

According to Fernald *in* Gray (1950), the plant appears to be indigenous in sandy or peaty soil of SE Nfld., but Hultén (1958) notes Raymond's opinion that it is adventive there (*see* note under *Luzula campestris*).

ORYZOPSIS Michx. [210] Mountain-Rice

1   Panicle diffuse, the branches forking regularly in pairs; glumes tipped with a sharp firm point, much exceeding the densely long-silky lemma, this with an awn about 5 mm long; leaves linear-filiform, involute; culms densely tufted; (B.C. to s Man.) . . . .*O. hymenoides*
1   Panicle with erect to spreading branches, not diffuse; glumes blunt to merely soft-pointed.
   2   Leaves flat, to over 1 cm broad; glumes about 8 mm long, usually distinctly 7-nerved; lemmas pubescent.
      3   Culms with sheaths usually crowded at the base, the leaf-blades usually less than 1 cm broad, glaucous above; basal leaves at most about 1 cm broad; panicle mostly less than 1 dm long, its simple erect paired branches each bearing a single spikelet; lemma pale green or yellowish, its awn less than 1.5 cm long; (trans-continental) . . . .*O. asperifolia*
      3   Culms leafy to summit, the blades to over 1.5 cm broad; panicle to 3 dm long, many of its branches forking; lemma becoming black in fruit, its awn to 2.5 cm long; (Ont. and sw Que.) . . . .*O. racemosa*
   2   Leaves involute, at most about 2 mm broad; glumes mostly obscurely 5-nerved, to about 5 mm long.
      4   Lemmas usually glabrous (rarely sparingly appressed-pubescent), at most 2.5 mm long, their straightish awns to 1 cm long; glumes acuminate, to 4 mm long; panicle open, to about 1.5 dm long; (B.C. to sw Man.) . . . .*O. micrantha*
      4   Lemmas distinctly appressed-pilose.
         5   Glumes obtuse; lemmas grey or pale green, rather densely pubescent, their

awns usually 1 or 2 mm long; panicle to about 8 cm long, its branches erect
or ascending; (essentially transcontinental) ....*O. pungens*
5　Glumes abruptly acute; lemmas yellow or brown, their flexuous awns longer.
　　6　Panicle to about 6 cm long, narrow, its branches appressed; lemma-awn
　　　　once-bent, about 5 mm long; (s B.C. and sw Alta.) ....*O. exigua*
　　6　Panicle to about 1.5 dm long, open, its branches spreading; lemma-awn
　　　　1 or 2 cm long, bent near the middle and recurved toward summit;
　　　　(transcontinental) ....*O. canadensis*

O. asperifolia Michx.
/T/X/　(Grh)　Woods, thickets, and peaty openings from s Dist. Mackenzie (sw of Great Slave L.; CAN) and B.C. to Alta. (N to Wood Buffalo National Park at 59°34′N), Sask. (N to Prince Albert), Man. (N to Gillam, about 165 mi s of Churchill), Ont. (N to L. Nipigon and w James Bay at ca. 52°N), Que. (N to L. Mistassini, Anticosti Is., and the Gaspé Pen.; type locality: "Hudson Bay to Quebec"), Nfld., N.B., and N.S. (not known from P.E.I.), s to Utah, N.Mex., S.Dak., and Va.

O. canadensis (Poir.) Torr.
/T/X/　(Grh)　Open woods, hillsides, and peaty soil from B.C. (Pouce Coupe, near the Alta. boundary at ca. 55°30′N; CAN) to Alta. (N to Wood Buffalo National Park), Sask. (N to Cochin, ca. 52°N), SE ?Man. (Lowe 1943), Ont. (N to the N shore of L. Superior), Que. (N to the Larch R. at ca. 56°40′N, L. Mistassini, and the Côte-Nord), sw Labrador (L. Ashuanipi at ca. 53°N, 66°10′W), Nfld., N.B. (Westmorland Co.), P.E.I. (Kings Co.), and N.S. (Shelburne, Halifax, Colchester, and Cumberland counties), s to Calif., N.Mex., Minn., Mich., and N.H. [*Stipa* Poir.; *S. juncea* Michx., basionym, the type from Hudson Bay; *S. (O.) macounii* Scribn.]. MAP (Que. eastwards): Marcel Raymond, Ann. ACFAS 19: fig. 1, p. 89. 1953.

O. exigua Thurb.
/T/W/　(Hs)　Dry open ground or open woods from s B.C. (Kingsvale and Merritt; Eastham 1947) and sw Alta. (Waterton Lakes; Breitung 1957*b*) to Oreg., Nev., and Colo.

O. hymenoides (R. & S.) Ricker　　Indian Rice-Grass, Silkgrass
/T/WW/　(Hs)　Dry prairies, sand-hills, and sandy blow-outs from s B.C. (N to Lillooet and Kamloops) to Alta. (N to Lesser Slave L. at ca. 55°20′N), s Sask., and sw Man. (N to St. Lazare, about 75 mi NW of Brandon), s to s Calif., Mexico, and Tex. [*Stipa* R. & S.; *Eriocoma* Rydb.; *E. (O.) cuspidata* Nutt.; *S. membranacea* Pursh, not L.].

O. micrantha (Trin. & Rupr.) Thurb.　　Little-seed Rice-Grass
/T/WW/　(Hs)　Dry open woods, sandy prairies, and rocky slopes from B.C. (N to Cariboo, ca. 53°N) to Alta. (N to the Peace R. at ca. 57°45′N; CAN), Sask. (N to Saskatoon), and s Man. (Routledge; Shilo; Onah), s to Calif., N.Mex., and Okla. [*Urachne mic.* T. & R., the type from Sask.].

O. pungens (Torr.) Hitchc.
/sT/X/　(Grh)　Rocky, sandy, or peaty soils from SE Yukon to Great Bear L., L. Athabasca (Alta. and Sask.), Man. (N to Gillam, about 165 mi s of Churchill), Ont. (N to Hudson Bay at ca. 56°30′N), Que. (N to the Ungava R. at ca. 58°N), N.B. (Pabineau Falls; CAN), and N.S. (not known from P.E.I.), s to Colo., S.Dak., Ind., and N.J. [*Milium* Torr.; *O. parviflora* Nutt.; *Urachne brevicaudata* Trin.]. MAP: Hultén 1968*b*:87.

O. racemosa (Sm.) Ricker
/T/EE/　(Grh)　Rich rocky woods (often calcareous) from Ont. (N to the Ottawa dist. and w to Lake of the Woods) to Que. (N to St-Joachim, NE of Quebec City in Montmorency Co.; *see* Que. map by Doyon and Lavoie 1966: fig. 14, p. 818) and sw Maine, s to E Mo., Ky., and Va. [*O. melanocarpa* Muhl.].

PANICUM L.   [166]   Panic-Grass.   Panic

1   Basal leaves elongate and similar to the culm-leaves, not forming a winter-rosette;
    spikelets glabrous, acute or acuminate; panicles essentially uniform ....*GROUP 1*
1   Basal leaves short, unlike the culm-leaves, often forming winter-rosettes; vernal panicles
    borne on simple culms, their spikelets not seed-forming; culms later freely branching and
    bearing numerous reduced fertile panicles (often concealed by the leaves) of ellipsoid to
    obovoid spikelets; perennials without elongate rhizomes.
    2   Lower internodes of the stem greatly shortened, the leaves all sub-basal ....*GROUP 2*
    2   Lower internodes not greatly shortened, the leaves well distributed on the culm.
        3   Ligule consisting of a zone of hairs 2–5 mm long, conspicuously protruding from
            the leaf-sheaths ....*GROUP 3*
        3   Ligule none or minute, or a zone of hairs less than 2 mm long.
            4   Principal culm-leaves distinctly cordate at base, mostly 1–3 cm broad;
                spikelets hairy ....*GROUP 4*
            4   Principal culm-leaves narrowed or rounded at base; spikelets hairy or
                glabrous.
                5   Leaf-blades distinctly hairy above; spikelets hairy ....*GROUP 5*
                5   Leaf-blades glabrous above or merely with a few widely scattered hairs
                    (sometimes marginally ciliate, or hairy at the very base); spikelets
                    glabrous or hairy ....*GROUP 6*

GROUP 1

1   Perennials; culms simple at base.
    2   Culms hard, terete, from hard, closely scaly rhizomes; leaf-sheaths scarcely
        compressed; spikelets ovoid, not all 1-sided; anthers about 2 mm long; (SE Sask.
        to N.S.) ....*P. virgatum*
    2   Culms soft and easily compressed, from a hardened persistent-leafy knotty crown;
        rhizomes not developed; leaf-sheaths laterally compressed; spikelets lanceolate,
        more or less 1-sided along the smaller panicle-branches; anthers very small.
        3   Leaves to 12 mm broad, their merely erose ligules at most 1 mm long, their
            sheaths keeled; spikelets to 2.2 mm long; panicles present at most of the
            nodes, the terminal one finally exserted; (sw B.C. and s Ont.) ....*P. rigidulum*
        3   Leaves at most 8 mm broad, their long-ciliate ligules to 3.5 mm long, their sheaths
            scarcely keeled; spikelets to 3.5 mm long; panicles long-exserted; (N.S.) ....
            ....*P. longifolium*
1   Annuals with a tuft of fibrous roots at the soft base; culms solitary, commonly branching
    from the lower nodes.
    4   Leaf-sheaths glabrous; leaves glabrous (rarely pilose), to 2.5 cm broad; culms
        compressed, the nodes glabrous; lower glume 1/5–1/4 the length of the spikelet,
        broadly rounded or truncate at apex; (Ont. to N.S.) ....*P. dichotomiflorum*
    4   Leaf-sheaths, and commonly the blades, papillose-hispid; nodes mostly bearded;
        lower glume at least 1/3 the length of the spikelet, acute or subacute.
        5   Spikelets ovoid, about 5 mm long; grain about 2 mm thick; mature panicle arching
            or nodding; leaves 1 or 2 cm broad; (introd.) ....*P. miliaceum*
        5   Spikelets at most 4 mm long; grain less than 1 mm thick; panicles erect.
            6   Spikelets ellipsoid to ovoid, acute or short-acuminate, about 2 mm long; grain
                blackish; panicle less than twice as long as thick; leaves to 8 mm broad; (Ont.
                and Que.) ....*P. philadelphicum*
            6   Spikelets lance-acuminate or lance-ovoid; grain stramineous.
                7   Panicle 2 or 3 times as long as thick; spikelets to 3.5 mm long; pulvini at
                    base of panicle-branches glabrous or merely puberulent; leaves at most
                    7 mm broad; (Ont. and sw Que.) ....*P. flexile*
                7   Panicle less than twice as long as thick; basal pulvini copiously long-
                    hispid; leaves to 2 cm broad; (B.C. to N.S.) ....*P. capillare*

GROUP 2

1   Spikelets acutely beaked (the tips of the upper glume and sterile lemma extending at least 0.5 mm beyond the fertile lemma), to about 4 mm long, glabrous or puberulent; leaves and sheaths glabrous to long-pilose; (SE Sask. to N.S.) . . . .*P. depauperatum*
1   Spikelets rounded or merely subacute at the beakless tip, the upper glume and sterile lemma equalling the fertile lemma.
   2   Culms solitary or few in a tuft; spikelets to 3.5 mm long, minutely hairy; leaves and sheaths long-papillose-pilose; (s Man. and s Ont.) . . . .*P. perlongum*
   2   Culms numerous in dense tufts; spikelets to about 3 mm long, glabrous or pilose; leaves and sheaths glabrous to sparingly long-papillose-pilose; (SE Man. to N.S.) . . . .
. . . .*P. linearifolium*

GROUP 3

1   Leaf-sheaths and culm glabrous or minutely puberulent between the nerves (or the sheaths sometimes ciliate).
   2   Panicle-branches widely spreading or ascending, the panicle ovoid; spikelets to 2.1 mm long; leaves to 12 mm broad, their ligule-hairs to 5 mm long; (var. *lindheimeri*; Ont. to N.S.) . . . .*P. lanuginosum*
   2   Panicle-branches strongly ascending, the panicle about 3 times as long as thick; spikelets to 1.8 mm long; leaves to 6 mm broad, their ligule-hairs 2 or 3 mm long; (N.S.) . . . .*P. spretum*
1   Leaf-sheaths and culm distinctly pubescent.
   3   Spikelets to 2.5 mm long, finely hairy; leaves to 13 mm broad, more or less papillose-pilose on both surfaces, the ligule-hairs to 5 mm long; (s Ont.) . . . .*P. villosissimum*
   3   Spikelets at most 2.1 mm long.
      4   Plants typically greyish-velvety-villous; leaves to about 1 cm broad.
         5   Autumnal phase ascending or spreading, branching from the middle and upper nodes, the reduced fascicled leaf-blades strongly ciliate; (transcontinental) . . . .
. . . .*P. lanuginosum*
         5   Autumnal phase prostrate, branching from the base and lower nodes, forming close mats; leaf-blades not ciliate; (B.C. and sw Alta.) . . . .*P. thermale*
      4   Plants not velvety.
         6   Culms copiously pilose with long, horizontally spreading hairs to 4 mm long; leaves to 6 mm broad, long-pilose on both surfaces; (s Ont.) . . . .*P. praecocius*
         6   Culms variously pubescent (if pilose, the hairs shorter and not horizontally spreading).
            7   Leaves of vernal culms glabrous or nearly so above, appressed-pubescent beneath, firm, to 7 mm broad; (s B.C.) . . . .*P. occidentale*
            7   Leaves of vernal culms pilose above, pilose or appressed-pubescent beneath.
               8   Spikelets rarely over 1.5 mm long, finely hairy; autumnal phase erect or leaning, never forming a mat; (s ?Ont. and sw N.S.) . . . .*P. meridionale*
               8   Spikelets to 2 mm long; autumnal phase prostrate or widely spreading and forming a mat.
                  9   Culms leafy below, sparingly branching from the base and lower nodes; (Sask.; Ont. to N.S.) . . . .*P. subvillosum*
                  9   Culms evenly leafy, branching from the upper nodes; (s B.C.) . . . .
. . . .*P. pacificum*

GROUP 4

1   Spikelets at most 1.8 mm long, nearly spherical; leaves to 1.5 cm broad, the primary nerves scarcely stronger than the sets of 3–5 intermediate ones; culms glabrous except at the appressed-hairy nodes; (s Ont.) . . . .*P. sphaerocarpon*
1   Spikelets 2.1 mm long or more, ellipsoid to ovoid or obovoid.

2   Primary nerves of the leaf-blades (excluding the midnerve) scarcely differentiated from the secondary nerves, the blades to 2.5 cm broad; leaf-sheaths much shorter than the lower culm-internodes, the culms to 7 dm tall, glabrous to sparsely villous or crisp-puberulent, often purple-tinged; spikelets to 3.2 mm long, sparsely short-hairy; (s ?Ont.) . . . .[*P. commutatum*]

2   Primary nerves of the leaf-blades (including the midnerve) raised and sharply differentiated from the secondary ones, the blades to 3 or 4 cm broad.

    3   Culm-nodes densely retrorse-bearded, the glabrous to puberulent culms to 7 dm tall; leaves rarely more than 5 times as long as broad; spikelets 3.8–5.2 mm long, sparsely hairy . . . .[*P. latifolium*]

    3   Culm-nodes not more densely pubescent than the sheaths or internodes; leaves often more than 5 times as long as broad; spikelets 2.5–3.7 mm long.

        4   Leaf-sheaths (or some of them) strongly papillose-hirsute; spikelets to 3.2 mm long, sparsely hairy; culms to 1.5 m tall; (s Ont., s Que., and N.S.) . . . .*P. clandestinum*

        4   Leaf-sheaths ciliate or hairy at the summit, otherwise glabrous to sparsely short-pilose or sparsely papillose; spikelets to 3.7 mm long, softly short-villous; culms to 1 m tall; (Ont. and Que.) . . . .*P. macrocarpon*

GROUP 5

1   Leaf-sheaths viscid and glabrous along the middle near summit, soft-hairy elsewhere and densely reflexed-villous at base; leaf-blades to 1.5 cm broad, softly hairy on both sides; culms to over 1 m tall, from a knotted crown, softly pubescent except for a glabrous viscid band just below the bearded nodes; spikelets abruptly apiculate, soft-hairy, to 2.7 mm long . . . .[*P. scoparium*]

1   Leaf-sheaths not viscid; spikelets subacute to rounded at tip.

    2   Leaves to 12 mm broad, papillose-hirsute on both sides, varying to subglabrous above, their sheaths papillose-hirsute with spreading hairs; culms to 6 dm tall, minutely puberulent; spikelets to 4 mm long, papillose-hirsute with hairs to 1 mm long; (Alta. to Ont.) . . . .*P. leibergii*

    2   Leaves rarely over 6 mm broad; culms to 3 or 4 dm tall.

        3   Spikelets about 1.5 mm long, finely hairy; leaves to 4 mm broad, pilose above with ascending hairs, softly pubescent to merely puberulent beneath; sheaths pilose; (s ?Ont. and N.S.) . . . .*P. meridionale*

        3   Spikelets to about 3 mm long, softly short-villous; leaves to 6 mm broad, papillose-hirsute on both sides, varying to subglabrous above; sheaths papillose-hirsute; (Alta. to Man.) . . . .*P. wilcoxianum*

GROUP 6

1   Spikelets 2.6–4 mm long.

    2   Leaves softly pubescent or more or less papillose-hirsute beneath, to 12 mm broad; culms to 6 or 7 dm tall.

        3   Spikelets papillose-hirsute with hairs to 1 mm long; (Alta. to Ont.) . . . .*P. leibergii*

        3   Spikelets glabrous to softly short-hairy, the hairs not over 0.3 mm long; (B.C. to Ont.) . . . .*P. oligosanthes*

    2   Leaves glabrous beneath.

        4   Ligule obsolete; leaves to about 1.5 cm broad, typically about 15 times as long as broad; spikelets to 3 mm long, short-hairy to subglabrous; culms to 5 dm tall, smooth to minutely canescent, usually short-villous at the nodes; (Ont. and s Que.) . . . .*P. bicknellii*

        4   Ligule a zone of hairs to 2 mm long; leaves at most about 10 times as long as broad; spikelets to about 4 mm long.

            5   Panicle narrow, its branches strictly erect; leaves erect or nearly so, to 2 cm broad, glabrous on both sides except for the papillose-ciliate base; culms to 5 dm tall; (Sask. to N.S.) . . . .*P. xanthophysum*

      5 Panicle ovoid, its branches spreading or ascending; leaves spreading, to 12
         mm broad, glabrous to softly hairy or sparsely papillose-pilose beneath; culms
         to 7 dm tall, often purplish, glabrous or pubescent especially below; (B.C. to
         Ont.) . . . .*P. oligosanthes*

1   Spikelets at most 2.5 mm long.
   6 Spikelets glabrous.
      7 Spikelets about 1.5 mm long, usually puberulent; culms to 5 dm tall, their nodes
         beardless; leaves to 5 mm broad, cartilaginous-margined, puberulent beneath;
         (?Vancouver Is.) . . . .[*P. ensifolium*]
      7 Spikelets to about 2 mm long; leaves glabrous or merely ciliate at base.
         8 Panicle-branches or leaf-sheaths or both mottled with pale spots; first (lower)
            glume subrotund, to 1/3 as long as the spikelet; culms to 9 dm tall, bearded at
            the nodes, otherwise glabrous; leaves to 13 mm broad . . . .[*P. nitidum*]
         8 Panicle-branches and sheaths not mottled; first glume deltoid-ovate, to 1/2 as
            long as the spikelet; culms to 7 dm tall, glabrous or the nodes (especially the
            lower ones) bearded; leaves to 8 mm broad; (s Ont.) . . . .*P. dichotomum*
   6 Spikelets pubescent.
      9 Culms to 5 dm tall, densely short-pubescent; leaves to 7 mm broad, minutely
         puberulent beneath; ligule a zone of hairs to 1.5 mm long; spikelets to 1.9 mm
         long, finely pubescent; (Ont. and s Que.) . . . .*P. columbianum*
      9 Culms glabrous (except the nodes).
         10 Uppermost leaf-blade erect or strongly ascending, the leaves to 14 mm broad,
            usually glabrous beneath; culms to 6 dm tall; spikelets to 2.3 mm long,
            puberulent; (Ont. to Nfld. and N.S.) . . . .*P. boreale*
         10 Uppermost leaf-blade widely spreading or deflexed.
           11 Principal leaves 3–5 mm broad, glabrous to puberulent beneath; culms to 4
              dm tall, their nodes not bearded; panicle-branches and leaf-sheaths not
              mottled; (?Vancouver Is.) . . . .[*P. ensifolium*]
           11 Principal leaves 5–13 mm broad, ciliate at base, otherwise glabrous; culms
              to 9 dm tall, glabrous except the bearded nodes; panicle-branches or
              leaf-sheaths or both mottled with pale spots . . . .[*P. nitidum*]

P. bicknellii Nash
/T/EE/ (Hs) Dry open woods, thickets, and clearings (ranges of Canadian taxa outlined below),
s to Mo. and Ga.
1   Lower glume to 1.2 mm long; spikelets less than 3 mm long; leaf-blades less than
   1 cm broad, taper-pointed; [s ?Ont. and sw Que. (Longueuil, near Montreal; near Mt.
   Johnson, about 20 mi SE of Montreal; Contrecoeur); this and var. *calliphyllum* are
   possibly hybrids between *P. dichotomum* and *P. linearifolium* or related species] . . . .
   . . . . .var. *bicknellii*
1   Lower glume to 2.5 mm long; spikelets to 3.2 mm long; leaf-blades to over 1.5 cm broad;
   [*P. calliphyllum* Ashe; s Ont. (Galt and Waterloo counties)] . . . . .var. *calliphyllum* (Ashe) Gl.

P. boreale Nash
/T/EE/ (Hs) Thickets, fields, meadows, and shores (ranges of Canadian taxa outlined below), s
to N Ind., N Ohio, and N N.J.
1   Leaves glabrous, their sheaths (except the very lowest) also glabrous; (Ont. (N to near
   Ottawa), Que. (N to the Nottaway R. at 50°21′N, L. St. John, and Trois-Pistoles,
   Temiscouata Co.), Nfld., N.B., P.E.I., and N.S.; [*P. dichotomum sensu* Fowler 1885, in
   part, Reeks 1871 and 1873 (Nfld.), and John Macoun 1888, as to the Glenelg, N.S.,
   citation] . . . . .var. *boreale*
1   Leaf-blades (at least beneath) and their sheaths pubescent; (Ont. (N to Chat Falls on the
   Ottawa R.) and sw Que. (along the Gatineau R. near Ottawa and the Ottawa R.
   about 60 mi E of Ottawa)) . . . . .var. *michiganense* Farw.

P. capillare L.   Old-witch Grass.   Mousseline
/T/X/ (T) Open sandy fields, clearings, and waste places, the aggregate species from B.C. (N

to Kamloops) to Alta., Sask. (N to Indian Head), Man. (N to Washow Bay, L. Winnipeg), Ont. (N to thc Albany R. sw of James Bay at 52°11′N), Que. (N to the Gaspé Pen.; not known from Labrador or Nfld.), N.B., P.E.I., and N.S., s to Calif., Tex., and Fla.

1   Pulvini in the axils of the leaf-petioles densely hirsute with hairs 1 or 2 mm long; spike-
    lets to about 4 mm long, long-acuminate; panicle to 2/3 the height of the entire plant;
    [*P. barbipulvinatum* Nash; B.C. to N.S.] . . . . .var. *occidentale* Rydb.
1   Axillary pulvini glabrous to obscurely short-villous; spikelets 2 or 3 mm long.
    2   Panicle usually less than 1/2 the height of the entire plant, soon exserted from the
        uppermost leaf-sheath; leaves to 1 cm broad; culms usually geniculate or decumbent
        at base, commonly with numerous secondary panicles below the terminal one and
        often repeatedly forking; [*P. gattingeri* Nash; s Ont. and sw Que.] . . . . .var. *campestre* Gatt.
    2   Panicle often 2/3 the height of the entire plant, its base included in the uppermost
        leaf-sheath (its lower branches therefore appressed); leaves to over 1.5 cm broad;
        culms commonly erect; [var. *agreste* Gatt.; transcontinental] . . . . .var. *capillare*

P. clandestinum L.
/T/EE/ (Hs) Thickets, shores, and alluvial ground from E Kans. to Iowa, s Ont. (Essex, Lambton, and Norfolk counties), sw Que. (N to Cap Rouge, near Quebec City), and N.S. (Yarmouth, Shelburne, Kings, Lunenburg, Halifax, and Guysborough counties; not known from N.B. or P.E.I.), s Tex. and N Fla.

P. columbianum Scribn.
/T/EE/ (Hs) Dry or sandy open ground and thin woods from Wisc. to Ont. (N to Quetico Park, about 100 mi w of Thunder Bay, the Sudbury dist. and Renfrew and Carleton counties), Que. (N to Taschereau, 48°40′N), and s Maine, s to Ill., Tenn., and Ga. [Incl. *P. tsugetorum* Bosc].

[*P. commutatum* Schultes]
[The report of this species (N to Ill. and Mass.) from the delta of the St. Clair R. in Lambton Co., s Ont., by Dodge (1915) requires confirmation.]

P. depauperatum Muhl.
/T/EE/ (Hs) Dry open soil or thin woods (ranges of Canadian taxa outlined below), s to E Tex., Tenn., and S.C.

1   Leaf-sheaths copiously pilose; [*P. rectum* R. & S.; SE Sask. (Moosomin; Breitung 1957a),
    s Man. (N to Lac du Bonnet, about 40 mi NE of Winnipeg), Ont. (N to the NW shore of L.
    Superior at 48°20′N and Renfrew and Carleton counties), and Que. (N to L. St. Peter)] . . . .
    . . . . .var. *depauperatum*
1   Leaf-sheaths glabrous or nearly so . . . . .var. *psilophyllum* Fern.
    2   Panicle reduced to 1–few spikelets in the lowest axils, often hidden by the basal
        leaf-sheaths; [Ont., Que., and N.S.] . . . . .f. *cryptostachys* Fern.
    2   Primary panicle finally exserted (later panicles as in f. *cryptostachys*); [var. *involutum*
        (Torr.) Wood; s Man. (N to Victoria Beach, near the SE end of L. Winnipeg), Ont. (N to
        Kenora, Pie Is., L. Superior, and Matheson), Que. (N to Taschereau, 48°40′N, and L.
        St. John), N.B., P.E.I. (Queens Co.), and N.S.] . . . . .f. *psilophyllum*

P. dichotomiflorum Michx.
/T/EE/ (T) Moist ground and waste places, the aggregate species from Minn. to Ont. (N to the Ottawa dist.), Que. (N to the Montreal dist.), and w N.S. (Yarmouth and Shelburne counties; not known from P.E.I. or N.B.), s to Mexico, Tex., and Fla.

1   Spikelets ovoid to slenderly ellipsoid, obtuse or abruptly short-pointed, about 2 mm long;
    leaf-blades at most about 5 mm broad; [reported from N.S.] . . . . .var. *puritanorum* Svenson
1   Spikelets oblong-lanceolate, acuminate, to about 3.5 mm long.
    2   Culms geniculate, the lower nodes enlarged; lower and primary sheaths inflated;
        leaf-blades to 2 cm broad; base of panicle included or short-exserted; [s Ont. and N.S.] . . . .
        . . . . .var. *geniculatum* (Wood) Fern.
    2   Culms nearly erect, the lower nodes only slightly enlarged; sheaths scarcely inflated;

leaf-blades to about 1 cm broad; terminal panicle becoming long-exserted; [s Ont. and s Que.] ....var. *dichotomiflorum*

P. dichotomum L.
/t/EE/ (Hs) Dry open woods, thickets, and clearings from Mo., Ill., and Mich. to s Ont. (Lambton, Waterloo, Brant, Lincoln, and Welland counties; not known from Que. or the Atlantic Provinces) and Maine, s to E Tex. and Fla.

Reports of this species from elsewhere in Canada other than s Ont. (as by Reeks 1871 and 1873; Fowler 1885; John Macoun 1888; Fernald *in* Gray 1950) are mostly based upon *P. boreale*, *P. lanuginosum* and vars., and *P. subvillosum* (relevant collections in CAN, GH, and NBM).

[P. ensifolium Baldw.]
[*P. tenue* Muhl. (*P. unciphyllum* Trin.) is included by Gleason and Cronquist (1963) in their treatment of this species. The report of *P. tenue* from Vancouver Is., B.C., by Carter and Newcombe (1921) is probably erroneous, it being a species of the E U.S.A. (Va. to Fla.).]

P. flexile (Gatt.) Scribn.
/T/EE/ (T) Dry or moist (chiefly calcareous) open woods, meadows, ledges, and sands from S.Dak., Ill., and Mich. to s Ont. (N to Renfrew and Carleton counties, s Que. (N to Pontiac and Gatineau counties), and w New Eng., s to E Tex. and Fla.; also reported as a probable introduction near Otterburne, s Man., by Löve and Bernard (1959) and from L. St. John, Que., by Frère Marie-Victorin (Contrib. Inst. Bot. Univ. Montréal 4:114. 1925).

P. lanuginosum Ell.
/T/X/ (Hs) Moist or dry sandy open soil or thin woods, the aggregate species from B.C. (N to Kamloops and Sicamous) to s Alta. (Medicine Hat), ?Sask. (Boivin 1967a), Man. (N to Deerhorn, about 70 mi NW of Winnipeg), Ont. (N to the N shore of L. Superior and the Albany R. s of James Bay at ca. 52°10′N), Que. (N to Amos, 48°34′N, and L. St. Peter), Nfld., N.B., P.E.I., and N.S.
1 Culms and sheaths glabrous or merely minutely puberulent in the internerves; leaves glabrous (except for basal cilia), not papillose; [*P. lindheimeri* Nash; Ont. to N.S.] ....
....var. *lindheimeri* (Nash) Fern.
1 Culms and sheaths more or less hairy.
  2 Culms and sheaths soft-villous, scarcely papillose; leaves soft-hairy on both surfaces, non-papillose ....var. *lanuginosum*
  2 Culms and sheaths distinctly papillose-pilose on both surfaces.
    3 Leaves glabrous above; [*P. tennesseense* Ashe; reported from Essex Co., s Ont., by Core 1948] ....var. *tennesseense* (Ashe) Gl.
    3 Leaves distinctly papillose-pilose above.
      4 Panicle-axis glabrate or sparsely pilose only in the axils of its branches; spikelets to 2 mm long; [Man. to N.S.] ....var. *septentrionale* Fern.
      4 Panicle-axis pilose.
        5 Spikelets to 2 mm long; leaves short-pilose above; [incl. var. *siccanum* Hitchc. & Chase (*P. columbianum* var. *sic.* (H. & C.) Boivin); *P. huachucae* Ashe; transcontinental] ....var. *fasciculatum* (Torr.) Fern.
        5 Spikelets to 1.5 mm long; leaves long-pilose above with hairs 3–6 mm long; [*P. implicatum* Scribn.; *P. laxiflorum sensu* John Macoun 1888, not Lam.; Ont. to N.S.] ....var. *implicatum* (Scribn.) Fern.

[P. latifolium L.]
[According to B. R. Baum (Can. J. Bot. 45: 1847–48. 1967), the plant known as *P. latifolium* L. by American authors should now be called *P. macrocarpon* Le Conte, true *P. latifolium* L. (*P. boscii* Poir.) being a species of the E U.S.A.]

P. leibergii (Vasey) Scribn.
/T/(X)/ (Hs) Prairies, meadows, and open woods from Alta. and ?Sask. (Boivin 1967a) to Man. (N to Sasaginnigak L., about 110 mi NE of Winnipeg) and Ont. (reported from Lambton Co., s Ont.,

by Dodge 1915; elsewhere known in cent. Ont. N to the Missinaibi R. at 49°41'N), s to Kans., Pa., and N.Y. [Incl. var. *baldwinii* Lepage].

P. linearifolium Britt.
/T/EE/ (Hs) Dry woods and sandy open places from Man. (N to Lac du Bonnet, about 50 mi NE of Winnipeg; DAO) to Ont. (N to Quetico Park, about 100 mi E of Thunder Bay, Sudbury, and Renfrew and Carleton counties), Que. (N to N of Mont-Laurier, Labelle Co., about 100 mi N of Hull, and L. St. Peter), and N.S. (not known from N.B. or P.E.I.), s to Tex. and Ga.

Some of our material is referable to var. *werneri* (Scribn.) Fern. (*P. werneri* Scribn.; leaf-sheaths essentially glabrous rather than copiously pilose).

P. longifolium Torr.
/T/EE/ (Hs) Peaty or sandy ground and shores from Ohio and sw N.S. (Yarmouth and Queens counties; ACAD; CAN; GH; see N.S. map by Dore and Roland 1942:281) to Tex. and Fla.

The N.S. plant is referable to var. *tusketense* Fern. (spikelets to 3.5 mm long rather than at most 3 mm, the upper glume usually slightly shorter than the sterile lemma rather than equalling or surpassing it, the panicle-branches closely ascending or appressed rather than spreading-ascending; type from Gavelton, Yarmouth Co.).

P. macrocarpon Le Conte
/T/EE/ (Hs) Dry open woods from Wisc. to Ont. (N to Renfrew and Carleton counties) and sw Que. (N to Pontiac Co. across the Ottawa R. from Deep River, Renfrew Co.), s to Kans. and N.C. [*P. latifolium* of most or all Canadian reports, not L.; see discussion under that species].

P. meridionale Ashe
/T/EE/ (Hs) Dry open ground and thin woods of s ?Ont. (Sarnia, Lambton Co.; Dodge 1915) and sw N.S. (Gavelton, Yarmouth Co.; CAN; GH), s to N Ala. and N Ga. MAPS: M. L. Fernald, Rhodora 39(468): map 46, p. 478. 1937; McLaughlin 1932: fig. 8 (incomplete), p. 343.

P. miliaceum L.   Millet, Broom-corn Millet
Eurasian; waste places of B.C. (New Westminster; Henry 1915), Alta. (Fort Saskatchewan; CAN), Man. (N to Stonewall, about 30 mi N of Winnipeg), Ont. (N to Sault Ste. Marie and Ottawa), Que. (N to Ste-Anne-de-la-Pocatière, Kamouraska Co.; DAO), P.E.I., and N.S. MAP: Hultén 1962: map 221, p. 233.

[*P. nitidum* Lam.]
[Reports of this taxon and its vars. *barbulatum* (Michx.) Chapm. (*P. barb.* Michx.) and *ramulosum* Torr. (*P. microcarpon* Muhl.) from B.C., Ont., and N.B. by John Macoun (1888; 1890) and from E Que. by Leon Provancher (Nat. can. (Que.) 19:246. 1890; Magdalen Is.) are largely referable to *P. lanuginosum* var. *fasciculatum* and *P. pacificum* (relevant collections in CAN).]

P. occidentale Scribn.
/T/W/ (Hs) Peat bogs and moist sandy ground from sw B.C. (type from Nootka Sound, Vancouver Is.; also known from the s mainland E to Nelson and Kootenay L.) to s Calif. and Ariz. [*P. dichotomum* var. *pubescens* Munro].

P. oligosanthes Shultes
/T/X/ (Hs) Sandy woods and open ground (the range of the Canadian taxon outlined below), s to Calif., Tex., and Fla.
1   Longer panicle-branches with rarely more than 6 remote spikelets (to 4 mm long) on
    pedicels to 1.5 cm long; leaves to about 1 cm broad, their sheaths appressed-pubescent;
    [*P. pauciflorum* Ell.; E U.S.A. only] . . . . .var. *oligosanthes*
1   Longer panicle-branches with up to 12 spikelets (about 3.5 mm long) on pedicels mostly
    less than 5 mm long; leaves to 1.5 cm broad, their sheaths glabrous or spreading-hirsute;
    [*P. scribnerianum* Nash; *P. scoparium sensu* John Macoun 1888, not Lam. (relevant
    collections in CAN); s B.C. (N to Spences Bridge), s Man. (N to Carberry and Winnipeg),

and s Ont. (Essex, Kent, Lambton, Huron, Middlesex, Elgin, Norfolk, and Welland counties)] . . . . .var. *scribnerianum* (Nash) Fern.

P. pacificum Hitch. & Chase
/T/W/ (Hs) Sandy shores and slopes and moist rock-crevices from s B.C. (Vancouver Is.; Agassiz; Yale-Lytton; Manning Park; Keremeos; Osoyoos; Nakusp; Shuswap L.) to s Calif. and Ariz. [*P. nitidum* var. *barbulatum sensu* John Macoun 1890, not *P. barb.* Michx., relevant collections in CAN].

P. perlongum Nash
/T/EE/ (Hs) Dry prairies and thin woods from s Man. (N to Sidney, about 40 mi E of Brandon) and Ont. (a collection from Pontypool, Durham Co., has been placed here by Dore; MT; TRT) to Tex., Ark., and Ind. [*P. depauperatum* var. *per.* (Nash) Boivin].

P. philadelphicum Bernh.
/T/EE/ (T) Rocky or sandy open soil and thin woodlands (ranges of Canadian taxa outlined below), s to E Tex. and Ga.
1 Mature terminal panicle long-exserted; spikelets about 2 mm long, mostly in pairs at the tips of the branchlets; pulvini (swollen bases of the leaves) short-hispid or glabrous; [Ont. (N to Renfrew and Carleton counties) and Que. (a collection in MT from Duparquet, ca. 48°30′N, has been placed here by Dore, who annotates it as probably introd. there)] . . . .
. . . . .var. *philadelphicum*
1 Mature terminal panicle included at base or short-exserted, the peduncle rarely 1/4 as long as the panicle; pulvini glabrous; [*P. tuckermanii* Fern.; Ont. (N to Renfrew and Carleton counties), Que. (N to near Mont-Laurier, about 80 mi N of Hull, and Montmagny, Montmagny Co.), ?N.B., and ?N.S.] . . . . .var. *tuckermanii* (Fern.) Stey. & Schmoll

P. praecocius Hitchc. & Chase
/t/EE/ (Hs) Dry prairies, open woods, and clearings from Nebr. to Minn. and s Ont. (Squirrel Is., Lambton Co.; GH, detd. A. S. Hitchcock), s to Tex., Mo., and Ind.

P. rigidulum Bosc
/t/(X)/ (Hs) Wet meadows and shores: sw B.C. (Sproat L., near Alberni, Vancouver Is., where "no doubt of its being indigenous"; John Macoun 1888) and cent. Calif.; the main area in the East from Kans. to Ill., Mich., s Ont. (Tweed, Hastings Co.; TRT, detd. Dore; reported from Wellington Co. by Stroud 1941; not listed by Soper 1949), and Maine (the report from P.E.I. by McSwain and Bain 1891, requires confirmation), s to Tex. and Fla. [*P. agrostoides* Spreng. in part].

[*P. scoparium* Lam.]
[Reports of this species of the E U.S.A. (N to Okla. and Mass.) from s B.C. and s Ont. by John Macoun (1888) are referable to *P. oligosanthes* var. *scribnerianum* (relevant collections in CAN).]

P. sphaerocarpon Ell.
/t/EE/ (Hs) Dry open soil and thin woodlands from Kans. and Mich. to s Ont. (Essex, Lambton, Huron, and Kent counties) and s Vt., s to Mexico, Tex., Fla., and N S. America. [*P. microcarpon* var. *sp.* (Ell.) Vasey; *P. dichotomum* var. *sp.* (Ell.) Wood].

P. spretum Schultes
/T/EE/ (Hs) Wet peats and sands from Mo. to Mich., N.Y., New Eng., and N.S. (Annapolis, Digby, Yarmouth, Queens, Halifax, and Lunenburg counties; ACAD; CAN; GH; *see* N.S. map by Dore and Roland 1942:276), s to Tex. and Fla.

P. subvillosum Ashe
/T/EE/ (Hs) Dry woods and sandy ground from Sask. (a collection in CAN from the s side of L. Athabasca has been placed here by Raup; reported from Creighton and Amisk L. by Breitung 1957a; not known from Man.) and Ont. (N to Thunder Bay; MT) to Que. (N to Anticosti Is.), N.B., P.E.I., and N.S., s to Mo., Ind., N Pa., and Long Island.

Phalaris

P. thermale Boland.
/T/W/ (Hs) Wet saline soils (commonly in the vicinity of hot springs) from s B.C. (Hope and Fairmont, Columbia Valley; CAN) and sw Alta. (Banff; CAN, detd. Hitchcock and Chase) to Calif. and Wyo.

P. villosissimum Nash
/t/EE/ (Hs) Sandy ground and dry thin woodlands from Minn. and Mich. to s Ont. (Squirrel Is., Lambton Co; TRT; Dodge 1915), N.Y., and Mass., s to Tex. and Fla.

P. virgatum L. Switchgrass
/T/EE/ (Grh) Dry or moist sandy soils (ranges of Canadian taxa outlined below), s to Ariz., Mexico, the Gulf States, and Cent. America; Bermuda; S. America.
1 Spikelets to 6 mm long; rhizomes elongate and creeping, the culms solitary to many in a loose tussock; [SE Sask. (near Gainsborough, about 160 mi SE of Regina; Breitung 1957a), Man. (N to the Red Deer R. N of Porcupine Mt.; John Macoun 1888), Ont. (N to the Albany R., sw James Bay, 51°24′N; RIM), and Que. (N to Contrecoeur, Verchères Co.; MT)] ....var. *virgatum*
1 Spikelets at most 4 mm long; rhizomes very short and closely interlocking, forming dense crowns with many close culms; [var. *?cubense* of N.S. reports, not Griseb.; w N.S. (Yarmouth, Shellburne, Lunenburg, Kings, and Queens counties; type from Great Pubnico L., Yarmouth Co.)] ....var. *spissum* Linder

P. wilcoxianum Vasey
/T/EE/ (Hs) Dry prairies, sand-hills, and thin woods from s-cent. Alta. (known only from Fort Saskatchewan; CAN, detd, Chase and Swallen), to Sask. (N to Prince Albert; Breitung 1957a), and sw Man. (N to St. Lazare, about 75 mi NW of Brandon; CAN), s to Colo., N.Mex., Kans., Ind., and Wisc.

P. xanthophysum Gray
/T/EE/ (Hs) Dry sandy or rocky ground and open woods from E Sask. (Amisk L. region at ca. 54°45′N; Breitung 1957a) to SE Man. (N to Victoria Beach, near the SE end of L. Winnipeg), Ont. (N to Quetico Park, about 100 mi w of Thunder Bay, and Renfrew and Carleton counties), Que. (N to L. St. Peter), N.B. (Pabineau Falls, Victoria Co.; CAN), and N.S. (Lunenburg Co.; E. C. Smith and J. S. Erskine, Rhodora 56(671): 246. 1954; not known from P.E.I.), s to Minn., Mich., Pa., and New Eng.

PHALARIS L. [204] Canary-Grass. Phalaride

1 Inflorescence to over 2 dm long, with many dense clavate branches, these spreading in anthesis, appressed-ascending in fruit; spikelets lanceolate; glumes 5 or 6 mm long, strongly laterally compressed, essentially wingless; leaves to 2 cm broad; perennial of moist places, to about 2 m tall; (transcontinental) ....*P. arundinacea*
1 Inflorescence a single ovate spike-like panicle at most about 4 cm long; spikelets broadly obovate; glumes broadly boat-shaped, whitish with green veins, the keel broadly winged, each face of the strongly laterally compressed glume 2-nerved; leaves to 7 mm broad; annuals to about 7 dm tall; (introd.).
  2 Glumes commonly 7 or 8 mm long; the winged keel not fringe-toothed; fertile lemma about 6 mm long; sterile lemmas 2 ....*P. canariensis*
  2 Glumes about 5 mm long, the winged keel with a fringe-toothed margin; fertile lemma about 3 mm long; sterile lemma 1 ....*P. minor*

P. arundinacea L. Reed-Canary-Grass. Roseau
/ST/X/EA/ (Grh) Shores, meadows, and wet places from cent. Alaska and s Yukon to Great Bear L., Alta. (N to L. Athabasca), Sask. (N to Prince Albert), Man. (N to Bear L. at ca. 55°20′N, 96°W), Ont. (N to the Shamattawa R. at 54°13′N), Que. (N to E James Bay at 53°40′N, L. Mistassini, and Anticosti Is.), Nfld., N.B., P.E.I., and N.S., s to Calif., N.Mex., ?Tex., and N.C.; Eurasia. [*Digraphis* Trin.; *Typhoides* Moench]. MAPS: Hultén 1968b:82, and 1962: map 113, p. 123; Meusel,

Jaeger, and Weinert 1965:52; the northeasternmost stations in Canada are indicated in a map by Lepage 1966: map 6, p. 216.

Forma *variegata* (Parnell) Druce (var. *picta* L.; leaves broadly white-striped), the ornamental ribbon-grass, is known as a casual adventive or garden-escape in Ont., Que., N.B., P.E.I., and N.S.

P. canariensis L.   Canary- or Birdseed-Grass.   Graines d'oiseaux
A native of N. Africa and the Canary Is.; roadsides and waste places of Alaska (Juneau; Fairbanks), the Yukon (Fort Selkirk), Dist. Mackenzie (Boivin 1967a), B.C. (Vancouver Is.; New Westminster), Alta. (N to Fort Saskatchewan), Sask. (N to Saskatoon), Man. (N to Strathclair, about 40 mi NW of Brandon), Ont. (N to Sault Ste. Marie and Ottawa), Que. (N to Jonquière, near L. St. John), N.B., P.E.I., and N.S. MAP: Hultén 1968b:82.

P. minor Retz.
European; introd. in sw B.C. (taken by John Macoun in 1887 and 1893 on ballast heaps at Nanaimo, Vancouver Is.; CAN) and N.B. (Gleason 1958). MAP: Hultén 1968b:83.

PHIPPSIA R. Br.   [229]

P. algida (Soland.) R. Br.
/AS/X/GEA/   (Hs)   Open damp soil (particularly in well-manured areas near bird-cliffs or rock-perches) from the coasts of Alaska and Dist. Mackenzie (an isolated area in cent. Alaska; not known from the Yukon) throughout the Canadian Arctic Archipelago to northernmost Ellesmere Is., s to s-cent. Dist. Keewatin, Coats Is. in N Hudson Bay, N Que. (Ungava Bay), and Labrador (s to ca. 55°N); mts. of Wyo.; circumgreenlandic; Iceland; N Eurasia. [*Agrostis* Soland.; *Catabrosa* Fries; *P. (Vilfa) monandra* H. & A.; incl. var. *algidiformis* (Sm.) Boivin, the taller and more upright f. *vestita* Holmb., and *P. concinna* (Fries) Lindeb. (see Theodore Mosquin and D. E. Hayley, Can. J. Bot. 44 (9):1214. 1966)]. MAPS: Hultén 1968b:92, and 1962: map 2, p. 9; Tolmachev 1952: map 3 (incomplete northwards).

PHLEUM L.   [223]   Timothy.   Phléole or Fléole

1   Panicle ovoid or short-cylindric, mostly less than 3 cm long but up to nearly 1.5 cm thick, green or drab to purplish; glume-awns over 1.5 mm long, about as long as or longer than the glume-body; anthers less than 2 mm long; leaves commonly less than 6 mm broad, the upper sheaths more or less inflated near the middle; culms from prolonged or creeping non-bulbous bases or caespitose, smooth throughout; (transcontinental) . . . .
. . . .*P. alpinum*
1   Panicle slender-cylindric, to over 1 dm long, green or drab; glume-awns at most 1.5 mm long, mostly about half as long as the body; anthers to 2 mm long; leaves to 1 cm broad; sheaths close; culms solitary or tufted, from bulbous bases, roughish at summit; (introd., transcontinental) . . . .*P. pratense*

P. alpinum L.   Mountain-Timothy
/aST/(X)/GEA/   (Hsr)   Damp or wet meadows, shores, and slopes: Aleutian Is. and cent. Alaska–Yukon–w Dist. Mackenzie through B.C. and sw Alta. (isolated stations in the L. Athabasca region) to Calif. and N.Mex.; Ont. (N shore of L. Superior; NW James Bay) and N Mich.; Que. (coasts of E James Bay and SE Hudson Bay to Ungava Bay, s to the Côte-Nord, Anticosti Is., and Gaspé Pen.) to Labrador (N to Ramah, 58°54'N), Nfld., N N.B. (Restigouche R.), and N.S. (Inverness Co., Cape Breton Is.; not known from P.E.I.), s in the mts. to N.H. and Maine; Mexico; S. America; w and E Greenland N to ca. 70°N; Iceland; Eurasia. [Incl. var. *americanum* Fourn., *P. commutatum* Gaud. (see W. M. Bowden, Rev. Can. Biol. 19:286. 1960; ssp. *comm.* (Gaud.) Hult.),and *P. haenkeanum* Presl]. MAPS: Hultén 1968b:88 (*P. comm.* var. *amer.*), and 1958: map 216, p. 235; Meusel, Jaeger, and Weinert 1965:55; Böcher 1954: fig. 21 (top and bottom; *P. comm.*), p. 79.

Forma *bracteolatum* Dansereau (the normally naked spike subtended by a foliaceous bract) is known from the w and E coasts of James Bay and reported from the type locality along the shores of Rivière-du-Brick, Anticosti Is., E Que., by Pierre Dansereau (Nat. can. (Que.) 72: 142. 1945).

P. pratense L.   Common Timothy.   Mil
Eurasian; widely cult. for fodder and freely escaping in Alaska–Yukon–Dist. Mackenzie and all the
provinces (in Man. N to Churchill; in Labrador N to the Hamilton R. basin); sw Greenland. MAP and
synonymy: see below.
1   Panicle rarely over 1 dm long and 6 mm thick; leaves commonly less than 5 mm broad;
    culms relatively decumbent, often with leafy tufts at anthesis; [P. nodosum L.; James Bay
    (Fernald in Gray 1950) and Nfld. (CAN; GH)] ....var. nodosum (L.) Huds.
1   Panicle to over 2 dm long and 1 cm thick; leaves to 1 cm broad; culms usually erect, from
    slightly arching bases, with few or no leafy tufts ....var. pratense
    2   Some or all of the spikelets changed into leafy tufts; [s Man. (near Otterburne, about
        30 mi s of Winnipeg) and Que. (Portneuf Co. and near Montreal)] ....
        ....f. viviparum (Gray) Louis-Marie
    2   Spikelets all normal.
        3   Spike bearing a basal bract; [Man. and Que.] ....f. bracteatum A. Br.
        3   Spike naked at base; [Introd., transcontinental; MAP: Hultén 1968b:89].......f. pratense

PHRAGMITES Trin.   [333]

P. australis (Cav.) Trin.   Reed.   Roseau
/sT/X/EA/   (Grh (Hel))   Marshes, shallow water, ditches, etc., from sw Dist. Mackenzie (Yohin L.
at 61°12′N; W. J. Cody, Can. Field-Nat. 77(2):111. 1963) and s B.C. to Alta. (N to L. Athabasca),
Sask. (N to Windrum L. at ca. 56°N), Man. (N to Cross L. at ca. 54°30′N), Ont. (N to Sandy L. at ca.
53°N, 93°W), Que. (N to L. St. John, Anticosti Is., and the Gaspé Pen.; not known from the
Côte-Nord, Labrador, or Nfld.), N.B., P.E.I., and N.S., s to Baja Calif., Mexico, Tex., La., Ind., and
Md.; Eurasia. [Arundo Cav.; A. phragmites L.; P. communis Trin. and its var. berlandieri (Fourn.)
Fern.; See W. D. Clayton, Taxon 17(2):168. 1968]. MAP: Hultén 1962: map 166 (P. comm.), p. 175.

PLEUROPOGON R. Br.   [363]   Semaphore Grass

1   Spikelets green, to 3(3.5) cm long; lemmas to about 8 mm long, their awns to over 1 cm
    long; palea not subtended by bristles; leaves to 7 mm broad; culms to 1.5 m tall;
    (sw B.C.) ....P. refractus
1   Spikelets dark purple or purplish black, 1 or 2 cm long; lemmas to 5 mm long, mucronate
    or short-awned; palea subtended by a pair of conspicuous hairy bristles; leaves 2 or
    3 mm broad; culms 1 or 2 dm tall; (transcontinental in arctic regions) ....P. sabinei

P. refractus (Gray) Benth.   Nodding Semaphore-Grass
/T/W/   (Hs (?r))   Bogs, wet meadows, and mountain streams from sw B.C. (Vancouver Is. and
adjacent mainland up to ca. 3,500 ft alt.) to N Calif. [Lophochlaena Gray].

P. sabinei R. Br.
/Aa/(X)/GEA/   (Grh (Hel))   Shallow water and muddy shores of pools and streams throughout
the Canadian Arctic Archipelago to northernmost Ellesmere Is. (type from Melville Is.), s to the
coast of Dist. Mackenzie, N Hudson Bay (Coats Is.), and northernmost Ungava–Labrador; w
Greenland s to ca. 76°N, E Greenland s to ca. 70°N; Spitsbergen; Novaya Zemlya; arctic Asia (an
isolated station in the Altai Mts. of w-cent. Asia). MAPS: Hultén 1968b:126, and 1958: map 4, p. 23;
Porsild 1957: map 40, p. 165, 1955: fig. 15, p. 51, and 1951b: fig. 7, p. 143; Atlas of Canada 1957:
map 4, sheet 38.

POA L.   [378]
Bluegrass, Meadow-Grass, Speargrass.   Pâturin

1   Spikelets little compressed, much longer than broad; lemmas more or less rounded on
    the back, the keel and intermediate nerves obscure; panicle usually narrow, its relatively
    short branches mostly appressed-ascending (or the lowest ones sometimes spreading;
    panicle open and pyramidal only in P. gracillima); leaves 1–2(3) mm broad, mostly folded
    or involute (flat in P. ampla and sometimes so in P. canbyi and P. gracillima), chiefly

basal; tufted perennials without rhizomes; ("bunchgrasses"; B.C. to Sask.; *P. canbyi* also in E Que.) . . . .*GROUP 1*
1   Spikelets distinctly compressed laterally, the glumes and lemmas keeled.
   2   Plants with distinct, more or less extensively creeping rhizomes; perennials . . . .
     . . . .*GROUP 2*
   2   Plants more or less tufted or caespitose, normally lacking rhizomes (but *P. leptocoma* often becoming stoloniferous).
      3   Lemmas with a tuft of long cobwebby hairs at the base . . . .*GROUP 3*
      3   Lemmas mostly lacking a tuft of long cobwebby hairs at the base (sometimes sparsely webbed in *P. laxa* and *P. pattersonii*; a few webby hairs normally present in *P. nemoralis*); perennials . . . .*GROUP 4*

GROUP 1

1   Lemmas glabrous or minutely scabrous but not crisp-puberulent; panicle contracted; culms to over 1 m tall.
   2   Leaf-sheaths minutely scabrous; ligules of principal leaves about 4 mm long, decurrent-based; (s B.C. and s Alta.) . . . .*P. nevadensis*
   2   Leaf-sheaths smooth; ligules short; (B.C. to Sask.).
      3   Leaves involute; panicle to 2 dm long; lemmas about 4 mm long . . . .*P. juncifolia*
      3   Leaves flat; panicle at most about 1.5 dm long; lemmas to 6 mm long . . . .*P. ampla*
1   Lemmas usually distinctly crisp-puberulent on the back toward the base.
   4   Leaves and their sheaths more or less scabrous; panicle contracted . . . .[*P. scabrella*]
   4   Leaves and their sheaths nearly or quite glabrous.
      5   Panicle rather open, the lower branches naked at base, ascending or somewhat spreading; culms usually decumbent at base, to about 6 dm tall; (?Alaska) . . . .
       . . . .[*P. gracillima*]
      5   Panicle contracted, the branches appressed or somewhat divergent only at anthesis).
         6   Culms slender, commonly not over 3 dm tall (to 5 or 6 dm), with numerous short offshoots at base; leaves usually folded; panicle commonly not over 1 dm long; (B.C. to Sask.) . . . .*P. sandbergii*
         6   Culms stouter, to over 1 m tall, with usually relatively few basal offshoots; leaves flat or folded; panicle to about 2 dm long; (B.C. to Sask.; E Que.) . . . .*P. canbyi*

GROUP 2

1   Upper glume to about 8 mm long or sometimes longer, not much surpassed by the pubescent, distinctly nerved, uppermost lemma; panicle contracted; plants of coastal sands.
   2   Panicle to about 3 dm long; spikelets to about 12 mm long; upper glume to about 11 mm long; anthers to 2.5 mm long; leaves glaucous, to about 12 mm broad; culms to about 1 m tall and 9 mm thick at base; (Alaska; Hudson Bay–James Bay; E Que., Labrador, and Nfld.) . . . .*P. eminens*
   2   Panicle usually not over 12 cm long; upper glume to about 8 mm long; anthers to about 2 mm long; leaves green, rarely over 6 mm broad; culms rarely over 4 dm tall and 4 mm thick at base.
      3   Spikelets to about 1.5 cm long; plants unisexual, the pistillate plants with abortive stamens; (B.C.) . . . .*P. macrantha*
      3   Spikelets to 9 mm long; plants with perfect flowers; (Que. and Labrador) . . . .
       . . . .*P. labradorica*
1   Upper glume at most 5 or 6 mm long, usually considerably surpassed by the uppermost lemma; spikelets usually less than 8 mm long.
   4   Lemmas lacking a tuft of long cobwebby hairs at base (*P. compressa* sometimes with a scant web).
      5   Panicle open-pyramidal, the slender branches spikelet-bearing near the tips, mostly in 2's or 3's, the lower ones spreading or somewhat reflexed; intermediate

nerves of lemma distinct, the marginal nerves and the keel either glabrous or pubescent; plant generally dioecious, mostly pistillate; (B.C. to s Man.) . . . . . .*P. nervosa*

5 Panicle more or less contracted (or open but narrow in *P. glaucifolia*), the rather short branches ascending to suberect; intermediate nerves of lemma obscure; plants mostly perfect-flowered (*P. confinis* dioecious).

  6 Culms solitary or few together, from long slender entangled rhizomes, strongly flattened, slender, wiry, blue-green, geniculate at base; leaves to 4 mm broad; panicle dense, its ascending branches solitary or paired, spikelet-bearing nearly to base; spikelets to 8 mm long; lemmas blunt-tipped; (introd. in dry soils; transcontinental) . . . .*P. compressa*

  6 Culms more or less caespitose or tufted, not strongly flattened; leaves 2 or 3 mm broad; (B.C. to Man.).

    7 Lower glume at most 3 mm long, 1-nerved; spikelets with up to 8(9) florets, the lowest lemma about 3 mm long; anthers about 1.5 mm long; panicle-branches mostly solitary or in pairs . . . .*P. arida*

    7 Lower glume 4 or 5 mm long, 3-nerved; spikelets with at most 4 florets, the lowest lemma about 5 mm long; panicle-branches mostly in whorls; plant glaucous . . . .*[P. glaucifolia]*

4 Lemmas with a tuft of long cobwebby hairs at base (but the web sometimes rather scant).

  8 Panicle contracted, its branches usually strongly ascending or suberect.

    9 Plants unisexual; panicle to about 3 cm long; lemmas sparingly webbed at base, the nerves faint; culms often geniculate, usually less than 2 dm tall; (Vancouver Is.) . . . .*P. confinis*

    9 Plants not unisexual, the flowers perfect; lemmas usually distinctly 5-nerved; culms usually over 2 dm tall; (transcontinental; see under *P. pratensis*) . . . .*P. alpigena*

  8 Panicle open-pyramidal, at least its lower branches usually horizontally divergent; (transcontinental).

    10 Lemmas profusely webby-lanate over the lower half, obtuse, with broad hyaline margins, 6 or 7 mm long; panicle purplish, violet, or greyish violet; leaves mostly 3 or 4 mm broad, the short median culm-leaf usually solitary; leaf-sheaths rather loose, the lower ones marcescent and overlapping, scabrous; culms to about 4 dm tall . . . .*P. arctica*

    10 Lemmas cobwebby only near base; panicle green; culm-leaves often 2 or more; culms often taller . . . .*P. pratensis*

GROUP 3

1 Panicle up to half the entire height of the plant; lemmas scantily webbed at base; annuals.

  2 Lemmas glabrous except on the scabrous keel, the marginal nerves rather indistinct, the intermediate nerves obsolete; panicle to 1.5 dm long, at first contracted, the distant branches few, glabrous, stiffly spreading; leaves to 5 mm broad, their sheaths glabrous; (?Vancouver Is.) . . . .*[P. bolanderi]*

  2 Lemmas pubescent on the lower half, the nerves all distinct; panicle to 2.5 dm long, open, the spreading branches in rather distant fascicles, scabrous; leaves relatively narrow, their sheaths glabrous or retrorsely scabrous; (sw B.C.) . . . .*P. howellii*

1 Panicle commonly much less than half the height of the plant; perennials.

  3 Culms more or less bulbous at base; spikelets very often forming proliferous buds (var. *vivipara* Koeler), these with a dark-purple base, their subtending bracts extended into slender green tips to 1.5 cm long; intermediate nerves of lemma obscure; panicle somewhat contracted, its branches mostly floriferous to base; leaves 1 or 2 mm broad; (introd. in s B.C.) . . . .*P. bulbosa*

  3 Culms not bulbous-based; spikelets normal, not proliferous.

    4 Lemmas glabrous (except for the basal web) or the keel sometimes pubescent at base.

      5 Leaf-sheaths retrorsely scabrous; ligules of upper leaves commonly at least 5

mm long; lemmas distinctly 5-nerved; panicle pyramidal, its lower nodes with mostly 3 or more spreading or ascending branches; spikelets with 2 or 3 florets, often borne from below middle of panicle-branches to apex; (introd., transcontinental) . . . .*P. trivialis*

    5   Leaf-sheaths glabrous; panicle open and lax, rather 1-sided; spikelets with up to 5 florets, borne near the ends of the panicle-branches.

        6   Panicle narrow, often drooping, its capillary branches appressed or ascending, solitary or in pairs; lemmas glabrous, long-webbed at base; leaves to 3 mm broad, their ligules very short; culms to about 1 m tall; (sw B.C.) . . . .*P. marcida*

        6   Panicle very open, its few slender branches spreading or drooping; leaves to 5 mm broad; (eastern species).

            7   Lemmas villous toward the base of the keel, the intermediate nerves obscure; panicle diffuse, its branches mostly in 4's or 5's, its base often included in the upper leaf-sheaths; anthers at most 0.7 mm long; ligules barely 1 mm long; (Ont. to N.S.) . . . .*P. alsodes*

            7   Lemmas glabrous on the keel, the intermediate nerves usually distinct; panicle-branches mostly in 2's or 3's; ligules often longer; (Ont. to Nfld. and N.S.) . . . .*P. saltuensis*

    4   Lemmas copiously villous on the keel and marginal nerves.

        8   Culms relatively slender, to about 7 dm tall; leaves to 4 mm broad, their sheaths glabrous or nearly so; lemmas to 4.5 mm long, their intermediate nerves distinct; stolons often developed; (B.C.–Alta.) . . . .*P. leptocoma*

        8   Culms relatively stout; leaves to 4 or 5 mm broad; stolons wanting.

            9   Spikelets bronzed at tip, the panicle relatively dark, to 3 dm long; (transcontinental) . . . .*P. palustris*

            9   Spikelets pale.

               10 Leaf-sheaths usually distinctly retrorse-scabrous; panicle to 3 dm long; intermediate nerves of lemma distinct; (B.C.) . . . .*P. occidentalis*

               10 Leaf-sheaths smooth; panicle to 1.5 dm long; intermediate nerves of lemma obscure; (transcontinental) . . . .*P. nemoralis*

GROUP 4

1   Lemmas glabrous or merely minutely scabrous; perennials.

    2   Leaves scabrous; lemmas distinctly 5-nerved.

        3   Leaves filiform, mostly basal; sheaths glabrous; panicle contracted or somewhat open, usually not over 8 cm long, its branches commonly solitary or in pairs; culms tufted, commonly not over 6 dm tall; (s B.C. to sw Man.) . . . .*P. cusickii*

        3   Leaves flat, to about 1 cm broad, their sheaths scabrous; panicle open and nodding, to over 1.5 dm long, its branches commonly in 4's or 5's, spikelet-bearing toward their ends; culms to over 1 m tall, solitary or few in a tuft; (introd. in sw Que.) . . . . . .
          . . . .*P. chaixii*

    2   Leaves smooth or smoothish, to about 3 mm broad; panicle rarely over 6 cm long, its branches commonly solitary or in pairs; culms tufted, to about 4 dm tall; intermediate nerves of lemma rather obscure; (s B.C.; *P. vaseyochloa* also reported from Alta.).

        4   Culm-leaves 2 or 3 mm broad, the leaves of the basal innovations more slender or filiform; panicle contracted, to about 6 cm long; culms to about 4 dm tall . . . .*P. epilis*

        4   Culm-leaves (and leaves of the innovations) involute-filiform, 1 or 2 mm broad.

            5   Panicle open (the branches spreading), to about 5 cm long; culms to about 2 dm tall . . . .*P. vaseyochloa*

            5   Panicle contracted, to about 3 cm long; culms usually less than 1 dm tall, usually scarcely surpassing the leaves . . . .*P. lettermanii*

1   Lemmas pubescent, at least on the lower part of the keel or nerves or both.

    6   Winter-annual with soft, weakly rooting base, the weak culms to about 5 dm tall, often rooting at the lower nodes; lemmas usually distinctly 5-nerved, obtuse; anthers at

most 1 mm long; panicle open-pyramidal, its branches commonly solitary or in pairs; leaves soft, to about 3 mm broad; (introd., transcontinental) . . . .*P. annua*

6 Perennials; intermediate nerves of lemma usually obscure; anthers usually over 1 mm long.

    7 Leaves to about 6 mm broad, the lower culm-leaves subacute to obtuse at apex; ligules 2 mm long or more; glumes ovate; panicle open.

        8 Caudex to 1.5 cm thick, densely covered with whitish papery sheaths; glumes round-ovate, abruptly pointed; lemmas villous on keel and margins; anthers at least 1.5 mm long; (transcontinental) . . . .*P. alpina*

        8 Caudex less than 5 mm thick, loosely brown-sheathed; glumes ovate, acuminate; lemmas long-pilose on keel and margins to above the middle and at the base of the intermediate nerves; anthers less than 1.5 mm long; culms loosely caespitose; (E Que.) . . . .*P. gaspensis*

    7 Leaves mostly not over 3 mm broad, often more or less involute, attenuate to tip; glumes lanceolate to ovate, acute or acuminate.

        9 Leaves folded or involute, firm and stiff, their ligules less than 1 mm long; panicle contracted, to about 7 cm long; culms tufted, to about 5 dm tall; plant generally dioecious, mostly pistillate . . . .[*P. fendleriana*]

        9 Leaves flat or, if involute, rather lax or soft; plants perfect-flowered.

          10 Culms wiry; panicle to about 12 cm long, its more or less scabrous branches mostly 1 or 2 at a node; lemmas pubescent on nerves and margins and over the back at base.

            11 Basal leaves involute; panicle commonly not over 5 cm long; lemmas and glumes with very broad membranous shining margins giving the spikelets a glassy lustre; spikelets sometimes viviparous and then deformed; ligules to 5 mm long; culms rarely over 2.5 dm tall, the basal leaf-sheaths with a whitish lustre; (high-arctic regions) . . . .*P. hartzii*

            11 Basal leaves normally flat except sometimes toward apex; spikelets neither viviparous nor with distinct lustre; ligules commonly about 2 mm long; culms often taller; plant more or less blue-glaucous; (transcontinental in arctic, subarctic, and alpine regions) . . . .*P. glauca*

          10 Culms scarcely wiry and rigid.

            12 Panicle nodding, to 1.5 dm long, its branches in 2's or 3's, the lower ones arcuate-drooping; leaves 1 or 2 mm broad, their ligules to 5 mm long; culms to about 5 dm tall; (B.C. and sw Alta.; E ?Que.) . . . . .*P. stenantha*

            12 Panicle erect.

              13 Culms densely tufted, commonly less than 2 dm tall; leaves about 1 mm broad; panicle contracted.

                14 Panicle to about 2.5 cm long (often not over 1.5 cm long); (arctic regions) . . . .*P. abbreviata*

                14 Panicle to about 4 cm long; (sw Alta.) . . . .*P. pattersonii*

              13 Culms solitary or few, scarcely tufted.

                15 Culms filiform, flexuous, to about 3 dm tall; panicle rarely over 6 cm long, its smooth flexuous branches solitary or in pairs, spikelet-bearing near tip; leaves commonly 1 or 2 mm broad; ligules to 5 mm long; (cent. Baffin Is. and N Labrador to Nfld. and E Que.) . . . .*P. laxa*

                15 Culms stouter and taller; panicle to about 1.5 dm long, its branches mostly in whorls of 3–5, spikelet-bearing from near the middle to tip; leaves to over 3 mm broad; ligules less than 2 mm long; (transcontinental) . . . .*P. nemoralis*

P. abbreviata R. Br.

/AS/X/GEA/ (Hs) Dry gravelly slopes and exposed ridges (often calcareous) from N Alaska–?Yukon–Dist. Mackenzie to Banks Is., Melville Is. (type locality), and Ellesmere Is., s to s Victoria Is., Melville Pen., and Prince Charles Is. in Foxe Basin (reported from Digges Is. in Hudson

Gramineae

Strait and from Port Bowen, Labrador, by John Macoun 1888); w and E Greenland s to ca. 69°N; Spitsbergen; Novaya Zemlya; isolated stations in arctic Asia. MAPS: Hultén 1968b:147, and 1958: map 2, p. 21; Porsild 1957: map 36, p. 165; Savile 1961: map C, p. 928; H. Steffen, Bot. Zentralbl. Beih. 57(B):393. 1937.

P. alpina L.
/aST/X/GEA/ (Hs) Rocky shores, outcrops, and crevices (often calcareous) from N Alaska–Dist. Mackenzie to cent. Dist Keewatin, cent. Baffin Is., northernmost Ungava–Labrador, and Nfld., s through B.C.–Alta. to Oreg., Utah, and Colo., and from N Sask. (L. Athabasca; Hasbala L. at ca. 59°45′N) to s Man., N Mich., s Ont. (Bruce Pen., L. Huron), Que. (s to the Laurentide National Park and Rimouski Co.), and N.S. (Victoria Co., Cape Breton Is.; not known from N.B. or P.E.I.); w and E Greenland N to ca. 75°N; Iceland; Eurasia. [Incl. vars. *bivonae* (Parl.) St. John, *brevifolia* Gaud., and *frigida* (Gaud.) Rchb., these all apparently mere ecological extremes]. MAPS: Hultén 1968b:129, and 1958: map 212, p. 231 (noting, also, a 1928 total-area map by Pawlowski); Porsild 1957: map 34, p. 165; Meusel, Jaeger, and Weinert 1965:31.
    The viviparous f. *vivipara* (Willd.) Boivin is reported from Greenland by Boivin (1967a).

P. alsodes Gray
/T/EE/ (Hs) Rich woods and thickets from Ont. (N to the Ottawa dist.) to Que. (N to Mont Tremblant Park N of Montreal; MT), N.B. (Boivin 1967a; not known from P.E.I.), St-Pierre and Miquelon, and N.S. (Cumberland, Colchester, Inverness, and Victoria counties), s to Tenn., N.C., and New Eng. [*P. paludigena* Fern. & Wieg.].

P. ampla Merr.   Big Bluegrass
/sT/WW/ (Hs(r)) Meadows and moist open ground or dry rocky slopes from SE Alaska and s-cent. Yukon (*see* Hultén 1942: map 138, p. 395) through B.C., Alta., and s Sask. (Cypress Hills and Gull Lake, about 30 mi sw of Swift Current) to Calif., N.Mex., and Nebr. [*P. confusa* and *P. truncata* Rydb.; merged with *P. juncifolia* by Hitchcock et al. 1969]. MAP: Hultén 1968b:144.

P. annua L.   Annual Bluegrass
Eurasian; cult. fields, roadsides, and waste ground. MAPS: *see* below.
1 Plant an annual or winter-annual, sometimes rooting at the lower nodes; [Aleutian Is.–Alaska–Yukon (*see* Hultén 1942: map 140, p. 395) and B.C.–Alta. to Sask. (N to Montreal L. at ca. 54°30′N), Man. (N to the Cochrane R. at ca. 58°N), Ont. (N to L. Attawapiskat, 52°14′N, 87°53′W), Que. (N to Eastmain, E James Bay, ca. 52°15′N, L. Mistassini, and the Côte-Nord), Labrador (N to ca. 53°30′N), Nfld., N.B., P.E.I., and N.S.; sw Greenland; MAPS (aggregate species): Hultén 1968b:145, and 1962: map 187, p. 199] . . . . .
. . . . .var. *annua*
1 Plant becoming perennial, the shoots creeping and rooting, producing leafy and flowering tufts at their tips; [St-Pierre and Miquelon; Fernald *in* Gray 1950] . . . . .var. *reptans* Haussk.

P. arctica R. Br.
/AST/X/GEA/ (Grh) Moist to dryish tundra and shores of ponds and streams, the aggregate species from the Aleutian Is. and coasts of Alaska–Yukon–Dist. Mackenzie throughout the Canadian Arctic Archipelago (type from Melville Is.) to Ellesmere Is. at ca. 80°N and northernmost Ungava–Labrador, s to s Alaska–Yukon–Dist. Mackenzie–Dist. Keewatin, N Man. (Churchill), N Ont. (Fort Severn, Hudson Bay, ca. 56°N), Que. (s to E James Bay and s Ungava Bay), s Labrador, and Nfld., and in the mts. through B.C. and Alta. to Oreg. and N.Mex.; circumgreenlandic; Iceland; Spitsbergen; Eurasia. MAPS and synonymy: *see* below.
1 Lemmas lacking lanate hairs between the nerves; [St. Lawrence Is., Alaska; Hultén 1942] . . . . .
. . . . .var. *glabriflora* Roshew.
1 Lemmas copiously lanate between the nerves.
    2 Ligules about 4 mm long; spikelets purplish grey or brownish; lemmas strigose or minutely scabrous above . . . . .var. *lanata* (Scribn. & Merr.) Boivin
        3 Spikelets viviparous; [*P. lanata* var. *viv.* Hult., the type from Umiat Is., Aleutian Is.; *P. arctica* f. *neophora* Boivin] . . . . .f. *neophora* (Boivin) Scoggan
        3 Spikelets normal; [*P. lanata* Scribn. & Merr., the type from Unalaska Is., Aleutian

310

Is.; incl. *P. williamsii* Nash; Aleutian Is.–Alaska–Yukon (*see* Hultén 1942: map 150, p. 396), sw Dist. Mackenzie, and L. Athabasca, Sask.; MAP: Hultén 1968*b*:138 (*P. lan.*)] . . . . .f. *lanata*

2 Ligules normally not over 2 mm long; spikelets usually purplish or green; lemmas glabrous above.

    4 Rhizomes rarely present; leaves soft and flat; spikelets normal; [Prince Patrick Is. to Ellesmere Is. at ca. 80°N, s to N Man. (Churchill) and N Que. (s to Port Harrison, E Hudson Bay); w and E Greenland s to ca. 69°N; *P. cenisia* f. *caespitans* Simm. in sched., the type from Ellesmere Is.; MAP: Porsild 1957: map 32, p. 164] . . . .
. . . . .var. *caespitans* (Nannf.) Boivin

    4 Rhizomes present . . . . .var. *arctica*

        5 Spikelets viviparous; [var. *vivipara* Hook., the type from the Canadian eastern Arctic; coast of E Dist. Keewatin near the Arctic Circle; Ellesmere Is. N to ca. 80°N; Devon Is.; Baffin Is.; w and E Greenland; MAPS: Porsild 1957: map 33, p. 165; Savile 1961: map F, p. 928] . . . . .f. *vivipara* (Hook.) Scoggan

        5 Spikelets normal; [incl. ssp. *longiculmis* Hult., *P. cenisia* All., *P. grayana* Vasey, *P. ?groenlandica* Steud., *P. longipila* Nash, and *P. rigens* Hartm.; range of the species, the type from Melville Is.; MAPS: Hultén 1968*b*:129, and 1958: map 6, p. 25; Porsild 1957: map 31, p. 164; Tolmachev 1952: map 1 (incomplete northwards); Raup 1947: pl 15] . . . . .f. *arctica*

For an exhaustive treatment of the *P. arctica* complex, *see* J.A. Nannfeldt (Symb. Bot. Ups. 4(4):1–85. 1940). Several other more or less closely related species (?microspecies) have been reported from the Aleutian Is. and Alaska. They may be distinguished from *P. arctica* as follows:

1 Lemmas glabrous or merely scabrous, lacking prolonged or lanate hairs.

    2 Glumes long and narrow, very acute, reaching nearly to the top of the upper flower in the often 2-flowered spikelet; [Alaska; E Asia; MAP: Hultén 1968*b*:132] . . . .
. . . .[*P. macrocalyx* Trautv. & Mey.]

    2 Glumes relatively short with respect to the upper florets; [type from Kodiak Is., Alaska; MAP: Hultén 1968*b*:133] . . . .[*P. eyerdamii* Hult.]

1 Lemmas with prolonged or lanate hairs at least toward the base of the intermediate nerves.

    3 Spikelets purplish; lemmas commonly copiously lanate toward base.

        4 Glumes lanceolate, not pruinose; [var. *lanata*] . . . .*P. arctica*

        4 Glumes relatively broader, pruinose with a waxy bloom; [type from Hoona, Alaska; MAP: Hultén 1968*b*:138] . . . .[*P. norbergii* Hult.]

    3 Spikelets green or brownish.

        5 Panicle nodding; glumes and lemmas both about 7 mm long; marginal nerves of lemma prominent, reaching almost to the tip of the lemma; [type presumably from Atka Is., Aleutian Is.; MAP: Hultén 1968*b*:139] . . . .[*P. turneri* Scribn.]

        5 Panicle erect or nearly so; glumes and lemmas shorter; marginal nerves of lemma distinct but shorter and less prominent.

            6 Base of culm surrounded by a cylinder of old leaf-sheaths; plant less than 3 dm tall, short-stoloniferous; [*P. komarovii* Roshev. and its var. *vivipara* Roshev. (*P. ?nascopieana* Polunin); Aleutian Is.–Alaska; MAP: Hultén 1968*b*:139] . . . .
. . . .[*P. malacantha* Kom.]

            6 Base of culm not surrounded by a cylinder of old leaf-sheaths.

                7 Slender plant with narrow leaves, stoloniferous; [transcontinental] . . . . .*P. arctica*

                7 Stouter plant with broader leaves, nonstoloniferous; [type from Shumigan Is., Alaska; incl. var. *aleutica* Hult. and its f. *vivipara* (Hult.) Boivin; MAP: Hultén 1968*b*:140] . . . .[*P. hispidula* Vasey]

P. arida Vasey

/T/WW/ (Grh) Prairies, plains, and alkaline flats from s B.C. (N to Spences Bridge) to Alta. (N to Wood Buffalo National Park at 59°41′N; CAN; GH), Sask. (N to McKague, ca. 52°45′N), and s Man. (N to Forrest, about 10 mi N of Brandon), s to Ariz., Tex., and Iowa. [*P. andina* Nutt., not Trin., and its var. *purpurea* Vasey; *P. overi, P. pratensiformis,* and *P. pratericola* Rydb.]. MAP: Raup 1930: map 24, p. 203.

[P. bolanderi Vasey]
[The report of this annual species of the w U.S.A. (Wash. and Idaho to Calif.) from sw B.C. by John Macoun (1888; Sooke, Vancouver Is.; taken up by Henry 1915) is based upon *P. howellii,* the relevant collection in CAN.]

P. bulbosa L.
Eurasian; introd. in ?Alaska (Hitchcock et al. 1969) and s B.C. (Vancouver Is.; Kamloops; Armstrong; and Okanagan L.). MAP: Hultén 1962: map 216, p. 226.

P. canbyi (Scribn.) Piper
/sT/(X)/ (Hs) Sandy or dry ground and dry calcareous ledges from cent. Yukon through B.C., sw Alta., and s Sask. (Moose Jaw; CAN) to N Calif., N.Mex., and Nebr.; isolated in the Great Lakes region of Minn. and Mich. and in E Que. (Kamouraska, Rimouski, Bonaventure, and Gaspé counties). [*P. laevigata* Scribn.; *P. lucida* Vasey; *P. sandbergii sensu* Fernald 1925, the relevant collection in GH]. MAPS Hultén 1968*b*:142; Porsild 1966: map 15, p. 68. See *P. scabrella.*

P. chaixii Vill.
European; introd. on the grounds of the Mackenzie King Estate at Kingsmere, sw Que., near Ottawa, Ont. (Gillett 1958; Dore 1959).

P. compressa L.   Wiregrass
Eurasian; dry soils of s Alaska–Yukon–Dist. Mackenzie and all the provinces (N to Fort Severn, Hudson Bay, ca. 56°N, and Goose Bay, Labrador, 53°18′N). [Incl. var. *langeana* (Rchb.) Koch]. MAPS: Hultén 1968*b*:140, and 1962: map 207, p. 219.

P. confinis Vasey
/t/W/ (Hs) Coastal meadows and sand dunes from sw B.C. (several localities on Vancouver Is.; CAN) to Calif. (According to Hultén (1942), reports from Alaska are probably referable to some other species).

P. cusickii Vasey
/T/WW/ (Hs) Dry prairies and sand-hills or rocky slopes from s Yukon (Alaska Highway at 60°47′N; CAN) and B.C. to s Alta. (Kananaskis, Manyberries, Medicine Hat, and the Cypress Hills; CAN), s Sask. (N to Wood Mountain and Moose Jaw), and sw Man. (Brandon; Oak Lake), s to Calif., Colo., and N.Dak. [*P. filifolia* Vasey, not Schur; *P. subaristata* Scribn., not Phil. MAP: Hultén 1968*b*:141.

P. eminens Presl
/aST/D (coastal)/neA/ (Grh) Gravelly or sandy seashores: Aleutian Is. and NW Alaska to sw B.C. (Vancouver Is., the type from Nootka Sound); James Bay and Hudson Bay N to ca. 52°45′N (concerning a report from Churchill, Man., *see* Scoggan 1957); Que. (St. Lawrence R. estuary from l'Islet Co. to the Côte-Nord, Anticosti Is., and Gaspé Pen.; an isolated station at Port Burwell, 60°28′N) and Nfld.; NE Asia. [*P. (Glyceria) glumaris* Trin.]. MAPS: Hultén 1968*b*:131; Potter 1932: map 3 (very incomplete), p. 72.

P. epilis Scribn.
/T/W/ (Hs) Montane meadows (mostly above timberline) from B.C. (N to Griffin L., near Kamloops, and Rogers Pass; CAN) and sw Alta. (N to Jasper; CAN) to Calif., Nev., Utah, and Colo. [*P. paddensis* Williams; *P. subpurpurea* Rydb.; *P. purpurascens* Vasey, not Spreng.; merged with *P. cusickii* by Hitchcock et al. 1969].

[P. fendleriana (Steud.) Vasey]   Mutton-Grass
[Reports of this species of the w U.S.A. (N to Wash. and S.Dak.) from s B.C. by John Macoun (1890; Spences Bridge, this taken up by later auth.), s Alta. by John Macoun (1888; Cypress Hills, this taken up by Moss 1959), and from Man. by A.S. Hitchcock (1935) require clarification. (*Eragrostis* Steud.; *P. eatonii* Wats.; *P. californica* (Munro) Scribn., not Steud.).]

P. gaspensis Fern.
/aST/E/ (Hs) Rocky and gravelly shores and slopes of Labrador (Mugford Tickle, 57°47′N; CAN) and E Que. (Bic Mt., Rimouski Co.; Bonaventure Is., Tabletop Mt., and the Ste-Anne-des-Monts R. (type locality), Gaspé Pen.); tentatively reported from the Kokrines Mts. of Alaska by A.E. Porsild (Rhodora 41(485):180. 1939).

P. glauca Vahl
/AST/X/GEA/ (Hs) Gravelly or rocky places (often calcareous), the aggregate species from the coasts of Alaska–Yukon–Dist. Mackenzie–Dist. Keewatin throughout the Canadian Arctic Archipelago to Ellesmere Is. (N to ca. 80°N), northernmost Ungava–Labrador, and Nfld., s to Wash., Utah, Colo., N Minn.–Mich., s Ont. (Bruce Pen., L. Huron), Que. (s to Kamouraska Co. and the Gaspé Pen.), N.B., N.S. (Cape Breton Is.; not known from P.E.I.), and New Eng.; circumgreenlandic; Iceland; Eurasia. MAPS and synonymy: see below.
1   Spikelets greenish or only slightly purple-tinged; glumes distinctly unequal; culms rather
    soft, to about 8 dm tall; [*P. glaucantha* Gaud.; Ont. (shores of L. Superior), Que. (Bic,
    Rimouski Co.; Ste-Anne-des-Monts R., Gaspé Co.; George R. at ca. 57°N), NW Nfld., and
    N.S. (Kings, Cumberland, Inverness, and Victoria counties); w Greenland N to 66°50′N] . . . .
    . . . . .ssp. *glaucantha* (Gaud.) Lindm.
1   Spikelets mostly strongly purple-tinged; glumes subequal; culms stiffly erect, mostly less
    than 3 dm tall.
    2   Panicle dense and almost spike-like; spikelets very dark, commonly 4 or 5 mm long;
        [*P. conferta* Blytt; Ont. (Pigeon Bay, NW shore of L. Superior); NW Nfld. (Pistolet Bay);
        w Greenland N to 66°50′N] . . . . .ssp. *conferta* (Blytt) Lindm.
    2   Panicle rather loose; spikelets 3 or 4 mm long; [*P. caesia* Sm.; *P. rupicola* Nash;
        transcontinental; MAPS (aggregate species): Hultén 1968b:136; Porsild 1957: map
        35, p. 165; Raup 1947: pl. 15; Meusel, Jaeger, and Weinert 1965:32] . . . . .ssp. *glauca*

[*P. glaucifolia* Scribn. & Williams]
[Reports of this species of the w U.S.A. (N to Wash. and Minn.) from B.C. by Hubbard (1955; Tetana L.), from Alta. by Moss (1959) and Raup (1935), from Sask. by Breitung (1957a), and from Man. by Scoggan (1957), appear generally referable to *P. arida*, of which *P. glaucifolia* may be merely the larger-dimensioned extreme.]

[*P. gracillima* Vasey]
[According to Hultén (1942), reports of this species of the w U.S.A. (Wash. and Mont. to Calif. and Colo. from Alaska are based upon the closely related (?identical) *P. stenantha* Trin. The inclusion of B.C.–Alta. in the range by Hitchcock et al. (1969) also requires clarification.]

P. hartzii Gand.
/Aa/(X)/GE/ (Hs) Exposed sands, gravels, and alluvial clays from Great Bear L. to Ellesmere Is. at ca. 80°N, s to Baffin Is. at ca. 70°N; w and E Greenland between ca. 69° and 77°30′N; Spitsbergen. [Incl. *P. ammophila* Porsild; see Porsild 1943, and Polunin 1959]. MAPS: Hultén 1958: map 159, p. 179; Porsild 1957: map 37, p. 165.
    Nannfeldt (1935) notes Scholander's opinion that the anthers of this species are always sterile and that it may be a hybrid between *P. abbreviata* and *P. glauca*. He agrees that, if a hybrid, *P. glauca* may be one of the parents, but believes that the second parent is more likely *P. arctica*. Forma *prolifera* (Simm.) Boivin (var. *vivipara* Polunin; *P. glauca* var. *atroviolacea* f. *pro.* Simm.; spikelets viviparous) is known from Ellesmere Is. (type locality) and a small island near Devon Is.

P. howellii Vasey & Scribn.
/T/W/ (T) Rocky banks and shaded slopes from s B.C. (several collections in CAN from Vancouver Is., Mayne Is., and Eagle Pass, w of Revelstoke, have been placed here by Macoun and Malte) to s Calif. [*P. bolanderi sensu* John Macoun 1888, not Vasey, the relevant collection in CAN].

P. juncifolia Scribn.   Alkali Bluegrass
/T/WW/ (Hs) Alkaline prairies and meadows from B.C. (N to Williams Lake), s Alta. (Moss

1959), and Sask. (N to McKague, 52°37'N; CAN: DAO) to N Calif., Colo., and S.Dak. [*P. brachyglossa* Piper].

P. labradorica Steud.
/aS/E/ (Grh) Sandy or muddy brackish shores of NW Que. (E James Bay and E Hudson Bay N to Richmond Gulf at ca. 56°30'N) and Labrador (type locality; N to Bowdoin Harbour, 60°24'N). [*P. lab.* Fern.; *P. glumaris sensu* Fernald and Sornborger 1899, not Trin.; scarcely distinct from *P. eminens* Presl, with which it is merged by Polunin 1959]. MAP: Dutilly, Lepage, and Duman 1958: fig. 5 (incomplete for Labrador), p. 59.

P. laxa Haenke
/aST/E/GE/ (Hs) Rocky ground, cliffs, and alpine slopes (ranges of Canadian taxa outlined below), s in the mts. to N.Y. and New Eng. MAPS and synonymy: *see* below.
1 Lemmas of the lower florets with short lines of hairs along the keel and marginal nerves (confined to about the basal 1/3 of the lemma), the hairs tapering toward tip, acuminate; [*P. flexuosa* Sm., not Muhl.; N-cent. Baffin Is. at ca. 70°N, s along the coast to N Ungava (Wolstenholme; Sugluk; Wakeham Bay) and Labrador (s to ca. 58°N); w Greenland at ca. 70°N and 72°N; MAPS: Hultén 1958: map 48, p. 67; Porsild 1957: map 38 (*P. flex.*) p. 165; Meusel, Jaeger, and Weinert 1965:32] . . . . .ssp. *flexuosa* (Sm.) Hyl.
1 Lemmas of the lower florets with longer lines of hairs (the keel hairy to about 2/3 its length from the base), the hairs cylindrical, with rounded tips.
  2 Leaves thin, to about 1.5 mm broad, with prominent nerves; ligules mostly about 2.5 mm long; panicle-branches elongate, usually bearing a single spikelet; [*P. fernaldiana* Nannf.; E Que. (Bic, Rimouski Co.; mts. of the Tabletop group, Gaspé Co.) and Nfld.; MAPS: on the above-noted maps by Hultén and Meusel, Jaeger, and Weinert] . . . .
  . . . . .ssp. *fernaldiana* (Nannf.) Hyl.
  2 Leaves thicker, to over 2 mm broad, less prominently nerved; ligules mostly about 4 mm long; lowermost panicle-branches relatively short and often themselves branched; [Europe only, reports from Canada referring to the above taxa; MAPS: on the above-noted maps by Hultén and Meusel, Jaeger, and Weinert] . . . . .ssp. *laxa*
For an exhaustive treatment of the *P. laxa* group, *see* Nannfeldt (1935). The following Alaskan species (?microspecies) appear to be more or less closely related to *P. laxa* or *P. leptocoma.*
1 Panicle-branches long and slender; leaves about 0.5 mm broad; [type from Popof Is., Alaska; MAP: Hultén 1968*b*:146] . . . .[*P. brachyanthera* Hult.]
1 Panicle-branches short.
  2 Leaves about 0.5 mm broad; panicle-branches capillary; [*P. ?jordalii* Porsild; Aleutian Is.–Alaska and E Asia; MAP: Hultén 1968*b*:147] . . . .[*P. pseudoabbreviata* Roshev.]
  2 Leaves about 2 mm broad; panicle-branches relatively stout; [*P. glacialis* Scribn. & Merr., not Stapf.; type from Yakutak Bay, Alaska; MAP: Hultén 1968*b*:143] . . . .
  . . . .[*P. merrilliana* Hitchc.]

P. leptocoma Trin.
/aST/W/eA/ (Hs(r)) Bogs and wet ground from N-cent. Alaska (type from Sitka), sw Yukon (*see* Hultén 1942: map 152, p. 397), and w Dist. Mackenzie through the mts. of B.C. and sw Alta. to Calif. and N.Mex.; E Asia (the Kuriles, Kamchatka, and Penshina). [*P. paucispicula* Scribn. & Merr.; *P. reflexa* Vasey & Scribn.; incl. *P. laxiflora* Buckl.]. MAP: Hultén 1968*b*:145.
   Plants with the palea-nerves merely scabrous (rather than pilose) have been named var. *scabrinervis* Hultén (1942; type from Skagway, Alaska).

P. lettermanii Vasey
/T/W/ (Hs) Rocky alpine or subalpine summits and slopes from B.C. (near Summit Pass, 58°31'N; Mt. Selwyn, ca. 56°N; Antimony Mt. and the Dunn Range, Kamloops dist.; Mt. Garibaldi, NE of Vancouver) and sw ?Alta. (Boivin 1967*a*) to Calif. and Colo.

P. macrantha Vasey
/t/W/ (Grh) Sand dunes along the coast from sw B.C. (Vancouver Is. and adjacent islands; CAN; V) to N Calif. [*P. douglasii* var. *mac.* (Vasey) Boivin].

P. marcida Hitchc.
/t/W/ (Hs) Bogs and wet shady places from sw B.C. (Vancouver Is. and Vancouver; A.S. Hitchcock 1935; Eastham 1947; Hubbard 1955; Boivin 1967*a*) to NW Oreg. [*P. saltuensis* var. *mar.* (Hitchc.) Boivin].

P. nemoralis L.   Foin à vaches
/sT/X/GEA/ (Hs) Thickets, open woods, rocky slopes, sands, and shores (ranges of Canadian taxa outlined below), s to Calif., Mexico, Tex., Nebr., Minn., Ohio, and Va.; w Greenland N to ca. 62°N, E Greenland N to 66°18′N; Iceland; Eurasia. MAPS and synonymy: *see* below.
1   Ligules very short, truncate; glumes acuminate, about equalling the lowest lemma; lower panicle-nodes commonly with 3 or more branches; [according to Hultén 1962, and his map 99, p. 109, var. *nemoralis* is introd. in the Aleutian Is., s Alaska, and w-cent. Yukon, the map apparently indicating a native area in the West (where much more southern than var. *interior*) comprising the southern half of B.C.–Alta.–Sask. and a small section of sw Man., in the East comprising Ont. (N to the James Bay watershed at ca. 55°N), Que. (reported N to Fort McKenzie, s of Ungava Bay at 56°50′N by Lepage 1966; *see* his NE Canada map 7, p. 216), s Labrador (N to the Hamilton R. basin; a collection in GH from Ramah, 58°54′N, has also been placed here, perhaps erroneously), Nfld., N.B., and N.S. (a report from P.E.I. by McSwain and Bain 1891, requires confirmation); Greenland; MAPS: Hultén 1968*b*:137, and 1962: map 99, p. 109; Raup 1947: pl. 15 (aggregate species; accepting N Labrador reports); Meusel, Jaeger, and Weinert 1965:33] . . . .var. *nemoralis*
1   Ligules to over 1 mm long; glumes merely acute, shorter than the lowest lemma; lower panicle-nodes commonly with only 2 branches; [*P. interior* Rydb.; *P. caesia* var. *strictior* Gray; *P. rupicola* of Sask. reports, not Nash; s-cent. Dist. Mackenzie, sw Dist. Keewatin, and B.C.–Alta–Sask. to Man. (N to Reindeer L.; the dot for Churchill on Hultén's map probably refers to an introd. specimen of var. *nemoralis*) and extreme sw Ont. near the Man. boundary; reported from Que. by Boivin 1967*a*, where perhaps introd.; MAPS: on the above-noted maps by Hultén and Meusel, Jaeger, and Weinert] . . . .
. . . . .var. *interior* (Rydb.) Butters & Abbe

P. nervosa (Hook.) Vasey
/sT/WW/ (Grh) Thickets, open woods, and dry slopes from B.C. (N to Hudson Hope, ca. 56°N) to Alta. (N to Beaverlodge, ca. 54°30′N), ?Sask. (the reports from the Cypress Hills and Donovan by Breitung 1957*a*, were later referred by him (Am. Midl. Nat. 61(2):510. 1959) to *P. (sandbergii) secunda)*, and sw Man. (Virden and Melita; W.M. Bowden, Can. J. Bot. 39(1):129. 1961), s to Calif. and N.Mex. [*Festuca nervosa* Hook., the type from Vancouver Is., B.C.; *P. cuspidata* Vasey, not Nutt.; *P. olneyae* Piper; *P. wheeleri* Vasey].

P. nevadensis Vasey   Nevada Bluegrass
/sT/W/ (Hs) Moist meadows and wet places from southernmost ?Yukon (*see* Hultén 1968*a*) through B.C. and w Alta. to Calif. and Colo.; (introd. on wool waste in Maine). MAP: Hultén 1968*b*:143. See *P. scabrella.*

P. occidentalis Vasey
/sT/W/ (Hs) Open woods and moist banks from s Alaska through B.C. (reported from Rogers Pass, between Revelstoke and Golden, by Henry 1915, and from Allison Pass in Manning Provincial Park, SE of Hope, by Hubbard 1955) to Colo. and N.Mex. MAP: Hultén 1968*b*:136.

P. palustris L.   Fowl Meadow-Grass
/sT/X/EA/ (Hs) Wet meadows and damp soil from the E Aleutian Is. and s Alaska–Yukon–Dist. Mackenzie to L. Athabasca (Alta. and Sask.), Man. (N to Churchill), Ont. (N to the Shamattawa R. at ca. 55°N), Que. (N to s Ungava Bay, L. Mistassini, and the Côte-Nord), Labrador (N to Cartwright, 53°42′N), Nfld., N.B., P.E.I., and N.S., s to Calif., Ariz., Mo., Tenn., and N.C.; introd. in s Greenland; Eurasia. [*P. crocata* Michx.; *P. rotundata* Trin.; *P. serotina* Ehrh.; *P. triflora* Gilib.]. MAPS: Hultén 1968*b*:137, and 1962: map 114, p. 123; Meusel, Jaeger, and Weinert 1965:33.

P. pattersonii Vasey
/T/W/ (Hs) Mts. of sw Alta. (collections in CAN from the E end of Simpson Pass, sw of Banff (detd. Nannfeldt) and from Crowsnest Pass and Moose Mt. on the Elbow R. (the last two tentatively placed here by Porsild); reported from Waterton Lakes by Breitung 1957*b*; not listed by Boivin 1967*a*) to ?Oreg., Nev., Utah, and Colo.

P. pratensis L.   Kentucky Bluegrass.   Foin à vaches
/AST/X/GEA/ (Grh) Moist to dry soil of meadows, fields, tundra, and shores, the aggregate species from the coasts of Alaska–Yukon–Dist. Mackenzie throughout the Canadian Arctic Archipelago to northernmost Ellesmere Is., northernmost Ungava–Labrador, and Nfld., s throughout most of the U.S.A. (the typical form largely introd. in N. America; the closely related *P. alpigena* circumgreenlandic except for a large gap in SE Greenland); Iceland; Eurasia. MAPS and synonymy (together with the scarcely separable *P. alpigena* (Fries) Lindm.): *see* below.
1   Both glumes lanceolate to ovate-lanceolate; intermediate nerves of lemma glabrous;
    basal leaves to 3 mm broad, new basal leaf-tufts borne from the basal sheaths; culms
    to 3 mm thick at the compressed and often geniculate base; [incl. *P. brintnellii* Raup,
    P. irrigata Lindm., and *P. subcaerulea* Sm.; transcontinental but less northern than the
    following taxa, largely introd. and cult., the native N limits very uncertain] ....var. *pratensis*
1   Lowest glume narrowly lanceolate; basal leaves generally filiform or involute; culms
    subterete, 1 or 2 mm thick at base.
    2   Intermediate nerves of lemma glabrous; new basal leaf-tufts borne from the basal
        sheaths; [*P. angustifolia* L.; *P. agassizensis* Boivin & Löve in part (*see* W.M. Bowden,
        Can. J. Bot. 39: 125. 1961); s Dist. Mackenzie–B.C. to Sask. (N to Hudson Bay
        Junction, ca. 53°N), Man. (N to Riding Mt.), Ont. (N to the NW and NE shores of L.
        Superior), Que. (N to Anticosti Is.), Nfld., N.B., and N.S.; MAP: Hultén 1968*b*:135
        (*P. angust.*)] ....var. *angustifolia* (L.) Gaud.
    2   Intermediate nerves of lemma soft-pilose; new basal leaf-tufts nearly all on separate
        stolons or offsets ....*P. alpigena* (Fries) Lindm. f.
        3   Spikelets viviparous; [incl. var. *vivipara* (Fr.) Schol.; *P. pratensis* f. ?*prolifera* Simm.;
            Banks Is. to northernmost Ellesmere Is., s to Southampton Is.; Greenland; MAPS:
            Porsild 1957: map 30, p. 164; Hultén 1962: map 7, p. 15] ....
            ....var. *colpodea* (Fries) Schol.
        3   Spikelets normal; [transcontinental in the northern part of the range; a probable
            hybrid with *P. (arctica* var.) *lanata* is reported from Alaska by Hultén 1942; MAPS:
            Hultén 1968*b*:135, and 1962: map 7, p. 15; Porsild 1957: map 29, p. 164] ....
            ....var. *alpigena*

P. saltuensis Fern. & Wieg.
/T/EE/ (Hs) Open woods, thickets, and clearings (ranges of Canadian taxa outlined below), s to Iowa, N Ill., Ky., Pa., and R.I.
1   Spikelets less than 4 mm long; lemmas to 3 mm long; [Ont. (N to Kapuskasing, 49°24'N),
    Que. (N to the Nottaway R. at ca. 51°N and the Gaspé Pen.), Nfld. (type from Glenwood),
    N.B., P.E.I., and N.S.] ....var. *microlepis* Fern. & Wieg.
1   Spikelets to 5.5 mm long; lemmas 3 or 4 mm long; [incl. *P. languida* Hitchc. (*P. debilis*
    var. *acutiflora* Vasey); Ont. (N to Nipigon, N shore of L. Superior), Que. (N to Anticosti Is.
    and the Gaspé Pen.), N.B., P.E.I., and N.S.] ....var. *saltuensis*

P. sandbergii Vasey
/T/WW/ (Hs) Dry open woods, prairies, and rocky slopes from B.C. (concerning a report from the Yukon, *see* Hultén 1968*a*) to Alta. (N to near Fort Saskatchewan; CAN) and Sask. (N to L. Athabasca), s to Calif., Colo., Minn., and Mich. [*P. buckleyana* Nash; *P. secunda* of auth., not Presl; *P. tenuifolia* (Thurb.) Buckl., not Rich.]. MAP (*P. buckl.*): Raup 1947: pl. 15. (See *P. scabrella*.).

[*P. scabrella* (Thurb.) Benth.]   Pine Bluegrass
[According to Hitchcock et al. (1969:680), this species of the w U.S.A. (N to Wash. and Idaho) is completely replaced in Canada (and largely in E Wash. and Oreg.) by *P. canbyi* Scribn., both,

however, being completely transitional to *P. sandbergii* (and to *P. nevadensis* as the lemmas become less crisp-puberulent and more scabrous). Reports and collections from our area should thus probably be referred to these last three species, as also, the MAP by Hultén (1968b:142). (*Atropis* Thurb.; *P. acutiglumis* Scribn.).]

P. stenantha Trin.
/sT/W/eA/ (Hs) Moist open ground from the Aleutian Is., N Alaska, and Great Slave L. (W.J. Cody, Can. Field-Nat. 70(3):104. 1956) through B.C. and Alta. to Oreg., Idaho, and Colo.; E Asia. [*P. flavicans* Griseb.]. MAP: Hultén 1968b:144. The type material is from Kamchatka and s Alaska.

The report from E Que. by Ernest Lepage (Ann. ACFAS 18:78. 1952; Bic, Rimouski Co.; RIM, detd. Swallen) may finally prove referable to some other species such as *P. canbyi,* already known from that region. Viviparous material (f. *vivipara* (Hult.) Boivin) is known from the Aleutian Is. and Alaska (Hultén 1942).

P. trivialis L.   Rough-stalked Meadow-Grass
Eurasian (but considered native northwards in N. America by Fernald *in* Gray 1950): moist ground from the Aleutian Is. and s Alaska–s Yukon–B.C.–Alta. (not known from Sask.) to ?Man. (Riding Mt.; Lowe 1943), Ont. (N to the Ottawa dist.), Que. (N to the Gaspé Pen.; reports from Labrador by Schranck 1818, E. Meyer 1830, and Schlechtendal 1836, probably refer to some other species), Nfld., N.B., P.E.I., and N.S. [Incl. var. *filiculmis* Scribn.]. MAPS: Hultén 1968b:132, and 1962: map 193, p. 205.

P. vaseyochloa Scribn.
/sT/W/ (Hs) Mts. of s-cent. Yukon (MacMillan Pass, ca. 63°20′N; CAN) through ?B.C. (Victoria, Vancouver Is.; Henry 1915) and ?Alta. (included in the range by Rydberg 1922) to Oreg. and Idaho. [*P. porsildii* Gjaerevoll; *Colpodium wrightii sensu* Porsild 1951a, not Scribn. & Merr.; according to Hitchcock et al. 1969, identical with *P. leibergii* Scribn., described two years earlier than *P. vaseyochloa*]. MAP: Hultén 1968b:141.

PODAGROSTIS Scribn. & Merr.   [242A (*Agrostis*)]

1   Spikelets to over 3 mm long; palea nearly as long as the lemma; prolongation of rachilla up to 1/2 as long as the flower; panicle usually purple, to about 1.5 dm long; culms to about 6 dm tall; (Alaska–B.C.) . . . .*P. aequivalvis*
1   Spikelets about 2 mm long; palea distinctly shorter than the lemma; prolongation of rachilla less than 1/3 as long as the flower; panicle usually pale green or pale purple, to about 7 cm long; culms to about 4 dm tall; (Alaska–B.C.–Alta.) . . . .*P. thurberiana*

P. aequivalvis (Trin.) Scribn. & Merr.
/sT/W/ (Hs) Wet meadows and bogs from the E Aleutian Is. and s Alaska (*see* Hultén 1942: map 105, p. 392) through B.C. to Oreg. [*Agrostis (Deyeuxia) aeq.* Trin., the type from Sitka, Alaska; *A. canina* var. *aeq.* Trin.]. MAP: Hultén 1968b:96.

P. thurberiana (Hitchc.) Hult.
/sT/W/ (Hs) Bogs and moist places from the E Aleutian Is. and SE Alaska (*see* Hultén 1942: map 106, p. 392) through B.C. and sw Alta. (Waterton Lakes; Breitung 1957b) to cent. Calif. and Colo. [*Agrostis* Hitchc.; *A. atrata* Rydb.]. MAP: Hultén 1968b:96.

POLYPOGON Desf.   [233]   Beardgrass

1   Glumes and lemmas awnless, the glumes scabrous at least near the keel; tufted or stoloniferous perennial . . . .[*P. semiverticillatus*]
1   Glumes and lemmas awned.
    2   Perennial, the tufted culms to over 8 dm tall; glumes to 3 mm long, scabrous, their awns 3–5 mm long; (introd. on Vancouver Is.) . . . .*P. interruptus*
    2   Annual; glumes about 2 mm long, short-pilose, their awns 6–8 mm long; (widely introd.) . . . .
    . . . .*P. monspeliensis*

P. interruptus HBK.
European; introd. in salt marshes of sw B.C. (Victoria and Sidney, Vancouver Is.; CAN). [*Alopecurus* Poir.; *Agrostis (P. (Vilfa) lutosa* of auth., not Poir.; *A. (P.) littoralis* With., not Lam.; intermediate in characters between *A. stolonifera* and *P. monspeliensis* and perhaps a hybrid (× *Agropogon litt.* (With.) Hubbard) between these two species].

P. monspeliensis (L.) Desf.
European; damp soil and waste places of Alaska (Loring, Sitka, and Nome; Hultén 1942), the Yukon (Dawson and Fairbanks; Hultén 1942), B.C. (Vancouver Is., Vancouver, and the Kamloops dist.; CAN), Alta. (Medicine Hat; CAN), Man. (Delta, s end of L. Manitoba; CAN), Ont. (Elgin and York counties), Que. (St-Roche-des-Aulnets, l'Islet Co.), and ?N.B. (A.S. Hitchcock 1935). [*Alopecurus* L.]. MAP: Hultén 1968b:93.

[P. semiverticillatus (Forsk.) Hyl.]
[The report of this species of the w U.S.A. (Wash. and Mont. to Calif. and Colo.) from sw B.C. by John Macoun (1888, as *Agrostis verticillata;* "Apparently introduced at Victoria and Nanaimo, Vancouver Island") is based upon *P. interruptus* (relevant collections from both localities in CAN). (*Phalaris* Forsk.; *Agrostis* Chr.; *A. verticillata* Vill.).]

PUCCINELLIA Parl.   [384]   Alkali-Grass, Goose-Grass

(Ref.: Fernald and Weatherby 1916; Swallen 1944; Soerensen 1953)
1   Anthers at least 1.2 mm long (*P. lucida* and *P. nutkaensis* may sometimes be sought here).
  2   Plants markedly stoloniferous, with overground runners, the culm-leaves to 2 mm broad; panicle to 6 cm long, tinged with red or purple, its essentially smooth branches commonly 2 or 3 from the lowest nodes and bearing up to 3 spikelets; spikelets to 11 mm long and with up to 6 flowers; lemmas entire or erose-denticulate (but not erose-ciliolate) at summit.
    3   Palea-keel strongly spinulose-ciliate; lemmas to 3.5 mm long, hairy at the very base of the nerves (short hairs also present on the callus below the lemma); anthers to 2.5 mm long, fertile and dehiscent; pedicels upwardly thickened; panicle-branches stiffly ascending; ligules about 1.5 mm long, rounded at apex; flowering culms to 2.5 dm long, decumbent-based, commonly 3-leaved; bulbil-like lateral offshoots of the runners usually arising from the axils of 1-year-old leaves; (Greenland) . . . .*P. maritima*
    3   Palea-keel glabrous; lemmas to 4.5 mm long, completely glabrous (hairs also absent on the callus); anthers to 2 mm long, neither fertile nor dehiscent; pedicels not thickened; panicle-branches at first stiffly ascending, later reflexed; ligules less than 1 mm long, truncate; flowering culms to 1.5 dm long, procumbent, 2-leaved; bulbil-like lateral offshoots of the runners arising opposite the point of insertion of the leaf; (transcontinental) . . . .*P. phryganodes*
  2   Plants densely to rather loosely tufted, lacking overground runners.
    4   Panicle greenish, open and pyramidal, to about 1 dm long, its slender, flexuous, lower branches horizontally spreading; spikelets to 6 mm long, 3–5 -flowered; lemmas glabrous, about 3.5 mm long, acutish; leaves mostly in an erect basal tuft; (B.C. to Sask.) . . . .*P. lemmonii*
    4   Panicle usually more or less reddish or purplish, contracted, its branches appressed or ascending (or finally stiffly spreading in *P. grandis*); lemmas more or less pubescent on the lower half, rounded or obtuse to subacute.
      5   Panicle-branches smooth or essentially so for most of their length below the minutely scabrous pedicels of the spikelets.
        6   Spikelets to 5 mm long, with rarely more than 4 flowers; lemmas to 3 mm long, the nerves slightly hairy at base; palea-keel sparingly spinulose or even glabrous; anthers to 1.5 mm long; panicle to 6 cm long, its branches bearing up to 6 spikelets; leaves to 1.3 mm broad; culms to about 2 dm tall; (western Arctic) . . . .*P. agrostidea*

      6   Spikelets to 12 mm long, with up to 9 flowers; lemmas to 4 mm long, pubescent on the lower half; palea-keel spinulose-ciliate; anthers to 1.7 mm long; panicle to about 1.5 dm long, its branches bearing up to 10 spikelets; leaves to 3.5 mm broad; culms to over 5 dm tall; (Atlantic coast) . . . .
. . . .*P. americana*

   5 Panicle-branches distinctly scabrous for a considerable distance below the pedicels of the spikelets.

      7   Lemmas erose-ciliolate at summit; palea-keel spinulose toward apex, long-hairy toward base; anthers to 1.5 mm long; panicle to 3 dm long; leaves to 3 mm broad; culms to over 6 dm tall; (*P. groenlandica* of Greenland) . . . .*P. arctica*

      7   Lemmas entire or erose-denticulate (but not erose-ciliolate) at summit.

         8   Palea-keel strongly ciliate; lemmas to 3.2 mm long; anthers to 2 mm long; spikelets to 11 mm long, with up to 9 flowers; panicle rarely over 9 cm long; culms to 2.5 dm tall, their leaves to 2 mm broad; (Alaska to Ellesmere Is.) . . . .*P. arctica*

         8   Palea-keel obscurely ciliate; lemmas to 4 mm long; anthers to 1.5 mm long; spikelets to 1.5 cm long, with up to 12 flowers; panicle to 2 dm long; culms to 9 dm tall, their leaves to 3.5 mm broad; (Alaska–Yukon– B.C.) . . . .*P. grandis*

1 Anthers usually not over 1 mm long (sometimes to 1.2 mm in *P. lucida* and *P. nutkaensis*).

   9 Plants markedly stoloniferous, with overground runners, these developing lateral bulbil-like offshoots between the nodes; culms erect or prostrate, to about 12 cm tall, usually bearing 2 leaves about 1 mm broad; panicle rarely over 4 or 5 cm long, with up to 3 smooth branches from the lowest node; lemmas to 3.2 mm long, glabrous; palea-keel glabrous or with a few spinules toward tip; anthers about 0.6 mm long; (Hudson Bay–James Bay and Atlantic coast) . . . .*P. ambigua*

   9 Plants densely to rather loosely tufted, lacking overground runners.

    10 Panicle-branches essentially smooth for most of their length below the often minutely scabrous pedicels, the panicle to about 13 cm long.

      11 Lemmas erose-ciliolate at apex, more or less hairy at the base of the nerves; palea-keel more or less spinulose on the upper half, glabrous or with a few long hairs toward base; spikelets to 7 mm long and with up to 5 flowers; panicle greenish or tinged with red or light purple, to about 13 cm long, its branches appressed or ascending, each bearing up to 15 or more spikelets; culms to 3 dm tall, bearing 2 or 3 leaves.

        12 Lemmas acute or acutish, to 3.5 mm long; anthers to 1 mm long; panicle-branches mostly paired; leaves involute, gradually tapering toward tip, they and the culms stiffly erect and conspicuously glaucous; (James Bay, Ont., to N.B. and P.E.I.) . . . .*P. laurentiana*

        12 Lemmas obtuse or truncate, to about 2.5 mm long; anthers to 0.8 mm long; panicle-branches up to 4 from the lowest nodes; leaves flat and rather lax, to 3 mm broad, abruptly pointed at apex, they and the geniculate culm less glaucous; (s ?James Bay; Que. to Labrador–Nfld.) . . . .*P. coarctata*

      11 Lemmas entire or merely erose-denticulate (but not ciliolate) at apex; culms mostly not over 2 dm tall, bearing 1 or 2 leaves to about 2 mm broad; (chiefly arctic and subarctic regions; *P. langeana* s to N.S.).

        13 Panicle greenish to yellowish or brownish (or tinged with pinkish or light purple), to 6(8) cm long, the lowest nodes with 2 or 3 ascending branches each bearing up to 5 spikelets; spikelets to 9 mm long, with up to 7 flowers; lemmas to 4 mm long, their nerves faintly pilose at base; palea-keel faintly spinulose-ciliate near tip; anthers to 1 mm long; culms to about 2 dm tall, 2-leaved . . . .*P. andersonii*

        13 Panicle purplish to dark purple, rarely over 5 cm long, its branches in pairs from the lowest nodes; culms commonly about 1 dm tall, usually bearing a solitary leaf.

14 Palea-keel completely glabrous; panicle to 5 cm long, its lowest
branches finally more or less spreading, bearing up to 6 spikelets with
up to 7 flowers each; lemmas to 2.3 mm long, usually glabrous; anthers
about 0.5 mm long . . . . .P. langeana

14 Palea-keel strongly spinulose in the upper half, long-hairy toward base;
panicle usually less than 2.5 cm long, its branches appressed, each
with 1 (rarely 2) spikelet with at most 4 flowers; lemmas to over 3 mm
long, strongly hairy on the lower half of the nerves; anthers to 0.8 mm
long . . . . .P. bruggemannii

10 Panicle-branches distinctly scabrous for a considerable distance below the
pedicels of the spikelets.

15 Leaves to over 6 mm broad, flat, commonly 3 or 4 on a culm, this to about 8 dm
tall; panicle green or light purple, with several branches at the lower nodes;
lemmas to 3 mm long, obtuse to subtruncate, lightly pubescent at base; keels
of palea ciliate nearly to base; anthers to about 1 mm long.

16 Panicle finally broadly pyramidal, lax, commonly 2 or 3 dm long and about
2 dm broad at base, the branches soon divergent and finally deflexed;
spikelets 4 or 5 mm long; lower glume about 1 mm long; upper glume
about 2 mm long; lemmas obtuse or subtruncate, erose-ciliolate at
summit; (introd.) . . . . .P. distans

16 Panicle narrower, its branches ascending or spreading but not deflexed;
lemmas erose-serrulate (but scarcely ciliolate) at the obtuse or bluntish
(but scarcely subtruncate) summit.

17 Panicle long-exserted, ellipsoid to ovoid, commonly not over 1.5 dm
long, the ascending or slightly spreading branches spikelet-bearing to
near base; spikelets commonly 4 or 5 mm long; lower glume less than 1
mm long; upper glume about 1.5 mm long; lemmas to 2.5 mm long,
their midribs commonly excurrent as short sharp points; culms to about
8 dm tall; (P.E.I. and N.S.) . . . . .P. fasciculata

17 Panicle linear-cylindric, to about 3 dm long, its base included in the
upper leaf-sheath; spikelets to 7 mm long; lower glume about 2 mm
long; upper glume about 2.5 mm long; lemmas to 3 mm long, their
midribs not excurrent; culms rarely over 6 dm tall; (James Bay
and E Que.) . . . . .P. macra

15 Leaves at most about 3 mm broad, commonly involute at least toward tip.

18 Anthers at most 0.6 mm long; lemmas to 2.3 mm long, glabrous except at
the very base, obtuse; palea-keel sparsely spinulose above the middle;
panicle to 2 dm long, whitish or yellowish to bronze or purplish, its
branches spreading or reflexed; (western species).

19 Panicle open and pyramidal; pedicels short and thickened, the
spikelets subsessile and crowded; leaves narrow and involute,
glaucous, those of the innovations bristle-like; ligules relatively firm,
pubescent . . . . .P. interior

19 Panicle oblong; pedicels more or less elongated and slender; leaves
mostly flatter, green; ligules thin, glabrous . . . . .P. hauptiana

18 Anthers commonly over 0.6 mm long; lemmas mostly over 2.3 mm long,
usually more copiously pubescent at least toward base.

20 Lemmas mostly 3 or 4 mm long, erose-ciliolate at apex.

21 Panicle dark purple, usually not over 1 dm long, dense, its lower
nodes with fascicles of longer and shorter ascending branches;
lemmas copiously pubescent below the middle, acute or acuminate; palea-keel spinulose above the middle, long-hairy toward
base; anthers to 0.8 mm long; culms to about 3 dm tall; (arctic and
subarctic regions) . . . . .P. angustata

21 Panicle pale green or faintly purple-tinged, to over 1.5 dm long,
becoming open and somewhat pyramidal; anthers to 1.2 mm long;
culms to over 6 dm tall; (subarctic and temperate regions).

22 Lemmas copiously long-hairy below the middle, acute; palea-keel spinulose above the middle, long-hairy toward base; (Man. to ?Labrador and E Que.) ....*P. lucida*

22 Lemmas glabrous or merely slightly pubescent at the very base, obtuse to subacute; palea-keel spinulose above the middle, glabrous or with a few long hairs toward base; (Alaska–B.C.) ....
....*P. nutkaensis*

20 Lemmas rarely as much as 3 mm long, usually obtuse to rounded at apex (or sometimes subacute in *P. nuttalliana*); anthers to 0.9 mm long.

23 Palea-keel spinulose above the middle, long-hairy toward base; lemmas erose-ciliolate at the obtuse or subtruncate apex; panicle to about 2.5 dm long, usually pale but tinged with purple, its branches ascending to horizontally spreading or the lowest ones deflexed; culms to about 8 dm tall; (western species) ....*P. sibirica*

23 Palea-keel spinulose above the middle, glabrous or merely with a few long hairs toward base.

24 Panicle to about 3 dm long, yellowish green or slightly purple-tinged, its branches spreading to ascending, the lower ones fascicled; lemmas erose-denticulate, blunt to subacute, pubescent below the middle; leaves to 3 mm broad, flat or becoming involute; culms to 9 dm tall; (Alaska–B.C. to Man.) ....
....*P. nuttalliana*

24 Panicle rarely as much as 1.5 dm long, its lowest branches mostly in pairs; lemmas sparingly pilose near base; (arctic and subarctic regions).

25 Culms to about 2 dm tall, leafy to above the middle (the upper sheath reaching the panicle); leaves rather lax, flat (sometimes involute), to 3 mm broad, green or slightly glaucous; old leaf-sheaths not scarious, decaying; panicle greenish or faintly purple-tinged; glumes thin and trans-lucent, lustrous; lemmas rounded or truncate at summit ....
....*P. vaginata*

25 Culms to about 5 dm tall, leafy only toward the base; leaves rigid, involute, to 2 mm broad, glaucous; old basal leaf-sheaths whitish or reddish, scarious and shining; panicle purplish; glumes rather firm, opaque; lemmas abruptly acuminate ....*P. deschampsioides*

P. agrostidea Soer.

/aS/W/ (Hs) Dry tundra of sw Banks Is. (type from De Salis Bay), s Victoria Is. (Cambridge Bay; CAN), ?Ellesmere Is. (a collection in CAN from Lake Hazen has been placed here tentatively by Porsild), and s-cent. Yukon (Alaska Highway at Mi 945; CAN). MAPS: Hultén 1968*b*:157; Porsild 1957: map 45, p. 166.

P. ambigua Soer.

/sT/EE/ (Hsr) Brackish coastal sands and salt marshes of w Hudson Bay (Churchill), SE James Bay (Old Factory, 52°37′N), and the Gulf of St. Lawrence region in E Que. (St-Omer, Bonaventure Co., Gaspé Pen.; MTJB), SE Nfld., N.B. (Grande Anse, Gloucester Co.), and P.E.I. (type from Alberton, Prince Co.). [*P. distans* f. *amb.* (Soer.) Boivin].

Soerensen (1953) concludes that this species includes the type of *P. alaskana* Scribn. & Merr. (*P. paupercula* var. *al.* (S. & M.) Fern. & Weath.) but that *P. paupercula* as interpreted by Fernald and Weatherby (1916) includes both stoloniferous and nonstoloniferous forms; the stoloniferous forms perhaps all to be referred to *P. ambigua,* the nonstoloniferous forms, at least in part, to *P. langeana* (the true *Glyceria paupercula* Holm). *See* A.E. Porsild (Can. Field-Nat. 83(2):163–64. 1969).

P. americana Soer.

/T/E/ (Hs) Coastal sands and salt marshes from E Que. (Paspébiac, Bonaventure Co., Gaspé Pen.; GH) to N.B. (Charlotte, Kent, and Westmorland counties; type from Anlac), P.E.I. (Bunbury, Queens Co.), N.S. (Digby, Yarmouth, Shelburne, Annapolis, Kings, Lunenburg, Victoria, and Cape Breton counties), St-Pierre and Miquelon, R.I., and Del. [*P. maritima* of E Canadian reports in large part, not (Huds.) Parl.].

P. andersonii Swallen

/AS/X/GeA/ (Hs) Brackish shores and coastal salt marshes from Prince Patrick Is. to southernmost Ellesmere L., s to the NW coast of Alaska (type from Lay Point) and the N coast of Dist. Keewatin; an isolated area in N ?Que. (E James Bay between ca. 53°and 54°20′N; Dutilly, Lepage, and Duman 1958); w and E Greenland N of ca. 69°N; Novaya Zemlya. MAPS: Hultén 1968*b*:160; Porsild 1957: map 47, p. 166.

Keyed out below to distinguish them from *P. andersonii* are four species (?microspecies) known in N. America only from Alaska:

1 Lemmas not over 3 mm long; anthers less than 1 mm long.
  2 Palea surpassing the lemma, this obtuse, glabrous, to 2 mm long; panicle-branches ascending or, at maturity, spreading or reflexed; culms rather soft, to 2.5 dm tall; (var. *sublaevis* Holmb.; Alaska and E Asia; MAP: Hultén 1968*b*:160) . . . .[*P. kamtschatica* Holmb.]
  2 Palea a little shorter than the lemma, this acutish, obscurely pubescent at the base of the lateral nerves, to 2.8 mm long; culms rigid, to 4 dm tall; (type from Port Hobron, Alaska; MAP: Hultén 1968*b*:159) . . . .[*P. hultenii* Swallen]
1 Lemmas 3.5–4 mm long; anthers to 1.5 mm long.
  3 Panicle-branches ascending, elongate; spikelets 5–7-flowered, to 1 cm long, the florets spreading; culms to 4 dm tall; (type from the Kenai Peninsula, Alaska; MAP: Hultén 1968*b*:162) . . . .[*P. glabra* Swallen]
  3 Panicle-branches stiffly spreading or reflexed.
    4 Spikelets 2–3-flowered, to 7 mm long; lemmas to 4 mm long, obtuse; culms densely tufted, erect, to 6 dm tall; (type from Cook Inlet, Alaska; MAP: Hultén 1968*b*:162) . . . .[*P. triflora* Swallen]
    4 Spikelets 5–7-flowered, to 8 mm long; lemmas to 3.5 mm long, acute or subobtuse, sometimes irregularly toothed; culms to 3 dm tall, erect from a rather long decumbent base . . . .*P. andersonii*

P. angustata (R. Br.) Rand & Redf.

/Aa/(X)/GEwA/ (Hs) Fresh soils and exposed areas from s Alaska (Hultén 1968*a*) and Prince Patrick Is. to northernmost Ellesmere Is., s to s Banks Is. and s Baffin Is.; w Greenland N of ca. 68°N, E Greenland N of ca. 70°N; arctic Eurasia. [*Poa (Glyceria) ang.* R. Br., the type from Melville Is.; *P. pumila* of American auth. in part, not *Glyceria pum.* Vasey]. MAPS: Hultén 1968*b*:159; Porsild 1957: map 46, p. 166.

Reports from N Que. by Polunin (1940) and Dutilly and Lepage (1951*a*) require confirmation. The report from Annapolis, N.S., by John Macoun (1888; *Gly. ang.*) is based upon *P. americana,* the relevant collection in CAN. His report from the Gaspé Pen., E Que., probably refers to *P. langeana.*

P. arctica (Hook.) Fern. & Weath.

/Aa/(X)/G/ (Hs) Sandy river banks, terraces, and clayey flood plains of N Yukon (Shingle Point and Herschel Is.; CAN), NW Dist. Mackenzie (Atkinson Point and Cape Dalhousie; CAN), s Victoria Is., Axel Heiberg Is., and Ellesmere Is. N to ca. 80°N; w Greenland N to ca. 72°N. [*Glyceria arctica* Hook., the type from arctic w Canada; incl. *P. groenlandica* and *P. poacea* Soer.]. MAPS: Hultén 1968*b*:156; Porsild 1957 (1964 revision), map 333, p. 202.

P. bruggemannii Soer.

/Aa/(X)/ (Hs) Damp sandy tundra (particularly below owl perches and around lemming mounds) from Prince Patrick Is. (type locality) to s Ellesmere Is., s to King William Is. MAP: Porsild 1957: map 48, p. 166.

P. coarctata Fern. & Weath.
/aST/E/GE/ (Hs)   Brackish or saline shores and calcareous ledges of ?Ont. (s James Bay at 51°33'N; Dutilly and Lepage 1963), E Que. (Côte-Nord), Labrador (N to Eclipse Harbour, 59°50'N), and Nfld. (type from Notre Dame Bay); w Greenland N to ca. 71°N, E Greenland N to 73°10'N; Iceland; arctic Europe. [*P. borreri* (Bab.) Hitchc.; *P. retroflexa* ssp. *borealis* var. *virescens* Lange]. MAPS: Hultén 1958: map 262, p. 281; Löve and Löve 1956: fig. 4, p. 131.

P. deschampsioides Soer.
/aS/(X)/G/ (Hs)   Dry clayey soils and tundra of sw Yukon, Dist. Mackenzie (NW coast; Great Bear L.), N Man. (Churchill), and northernmost Ungava (Suglej Bay, 62°15'N); w Greenland between ca. 66°20'and 72°N; type locality). [Incl. *P. rosenkrantzii* Soer.]. MAPS: Porsild 1966: map 17, p. 69; Hultén 1968b:163.

P. distans (L.) Parl.
Eurasian; saline, alkaline, or calcareous soils (often in waste ground and along roadsides) of the Yukon (Dawson; CAN, verified by Soerensen), N B.C. (Bennett, ca. 59°N; CAN), sw Alta. (Banff and Nordegg; CAN), Sask. (Weyburn and Tisdale; CAN), Man. (N to The Pas), Ont. (N to Fort Severn, Hudson Bay, ca. 56°N), Que. (near asbestos mines in Beauce and Megantic counties and along streets in Montreal; the report from the Gaspé Pen. by John Macoun, Proc. Trans R. Soc. Can. 1 (Sect. 4): 128. 1883, requires confirmation), St-Pierre and Miquelon, N.B., and N.S. [*Poa* L.; *Glyceria* Wahl.; *P. ?pumila* of American auth. in part, not *Glyceria pum.* Vasey, basionym; *P. pum.* var. *?minor* Wats.]. MAP: Hultén 1968b:165.

P. fasciculata (Torr.) Bickn.
/T/(X)/E/ (Hs)   Nev., Utah, and Ariz. (Hitchcock and Chase 1951); salt marshes along the coast from P.E.I. (Charlottetown; ACAD) and N.S. (Kings Co.) to Va., and inland along shores of w N.Y.; Europe. [*Poa* Torr.].

P. grandis Swallen
/ST/W/ (Hs)   Salt marshes and sandy or rocky seashores of the Alaska Panhandle (Swallen 1944) and the Yukon (Whitehorse; Swallen 1944) through coastal B.C. (Vancouver Is. and adjacent islands; Vancouver) to Calif. [Merged with *P. lucida* by Hitchcock et al. 1969]. MAP: Hultén 1968b:161.

P. hauptiana (Krecz.) Kitagawa
/aST/(X)/ (Hs)   Wet ground and river banks from N Alaska (*see* N Alaska map by Wiggins and Thomas 1962:363; Hultén 1950, refers his 1942 listings of *P. distans* and *P. tenuiflora* to this taxon) and the Yukon (Dawson; Swallen 1944) to the mts. of sw Alta. (Banff; Swallen 1944); N Ont. (Winisk R. near s Hudson Bay at 55°12'N; Lepage 1966, detd. Swallen). [*Atropis* Krecz.]. MAP: Hultén 1968b:164.

P. interior Soer.
/Ss/W/ (Hs)   Open ground (chiefly away from the coast) of Alaska–Yukon (Hultén 1950; type from the Alaska Range), w Dist. Mackenzie (near Fort Simpson, ca. 62°N; CAN), N B.C. (Bennett, ca. 59°N; CAN), and N Alta. (sw end of L. Athabasca; CAN). MAPS: Hultén 1968b:165; Porsild 1966: map 18, p. 69.

P. langeana (Berl.) Soer.
/AST/X/GeA/ (Hs)   Coastal sands and salt marshes: St. Lawrence Is., Bering Strait, and N Alaska; Devon Is. and Baffin Is. s to N Man. (Churchill), Que. (s to NE James Bay, the Côte-Nord, Gaspé Pen., and Bic, Rimouski Co.; not known from Ont.), s Labrador, Nfld., N.B., P.E.I., and N.S.; w Greenland between ca. 65° and 71°N; arctic E Asia. [*Glyceria lang.* Berl., the type from w Greenland; *G. (P.) tenella* Lange in part; *P. ?longiglumis* (Fern. & Weath.) Raymond; *G. paupercula* Holm, not *P. paup. sensu* Fernald and Weatherby 1916 (see *P. ambigua*)]. MAPS: Hultén 1968b:157; Porsild 1957: map 49, p. 167; Soerensen 1953: fig. 114, p. 179.

P. laurentiana Fern. & Weath.
/aST/EE/G/ (Hs) Sandy coasts and salt marshes of Ont. (w James Bay between ca. 52° and
53°N; Dutilly, Lepage, and Duman 1954), E Que. (Rivière-du-Loup, Temiscouata Co., to Bic,
Rimouski Co., and the s coast of the Gaspé Pen.; type from Tracadigash Mt., Carleton), P.E.I.
(Charlottetown), and N N.B. (Restigouche and Gloucester counties); sw Greenland at 64°03′N
(somewhat atypical according to Soerensen 1953).

P. lemmonii (Vasey) Scribn.
/T/W/ (Hs) Moist alkaline soil from sw B.C. (Victoria and Sidney, Vancouver Is.; CAN), Alta.
(Pend d'Oreille; CAN), and s Sask. (Crane Lake, Moose Jaw, and Boulder L., SE of Watrous; CAN,
all detd. Fernald and Weatherby) to Calif. and Utah. [*Poa* and *Glyceria* Vasey].

P. lucida Fern. & Weath.
/ST/EE/ (Hs) Salt marshes and coastal sands of NE Man. (Churchill; CAN), N Ont. (w James
Bay N to 54°52′N; CAN), and Que. (James Bay–Hudson Bay N to Great Whale R., ca. 55°20′N; St.
Lawrence R. estuary from Charlevoix and Kamouraska counties to the s coast of the Gaspé Pen.,
the type from Cacouna, Temiscouata Co.).
    The report from Labrador by Rouleau 1956, requires confirmation. The citation from Newcastle
Is., sw B.C., by Fernald and Weatherby 1916, is based upon *P. nuttalliana,* the relevant collection
in CAN.

P. macra Fern. & Weath.
/sT/EE/ (Hs) Coastal sands and ledges of N Ont. (w James Bay N to ca. 55°N; CAN) and Que.
(E James Bay N to 54°37′N; Saguenay Co. of the Côte-Nord; Gaspé Pen., the type from
Bonaventure Is.). MAPS: Dutilly, Lepage, and Duman 1954: fig. 4 (E James Bay stations should be
added), p. 45; the maps by Fernald 1924: map 7, p. 569, and 1925: map 26, p. 257, indicate only
the type station.

P. maritima (Huds.) Parl.
/aST/–/GE/ (Hsr) Coastal salt marshes of sw Greenland (N to ca. 61°30′N; *see* Greenland map
by Soerensen 1953: fig. 113, p. 178) and w Europe. [*Poa* Huds.; *Pucc. ?porsildii* Soer.; reports
from Canada refer chiefly to *P. americana*].

P. nutkaensis (Presl) Fern. & Weath.
/sT/W/ (Hs) Coastal sands and salt-marshes from the Aleutian Is. and s Alaska (*see* Hultén
1942: map 172, p. 398) through coastal B.C. (type from Nootka, Vancouver Is.) to Wash. and cent.
?Calif. [*Poa* Presl; *Poa (Glyceria; Pucc.) festucaeformis sensu* Hooker 1840, and John Macoun
1888, not Host]. MAP: Hultén 1968*b*:158.

P. nuttalliana (Schultes) Hitchc.
/sT/WW/ (Hs) Moist prairies, saline shores, and alkaline flats from s Alaska–Yukon (CAN;
Hultén 1950) and B.C.–Alta. to Sask. (N to McKague, ca. 52°45′N) and Man. (N to Nejanilini L. at
59°22′N; CAN; introd. eastwards, as at Longlac, Ont., where taken along a railway), s to Calif., N
Mexico, Tex., and Minn. [*Poa* Schultes; *Poa (Glyceria; Pucc.) airoides* Nutt.; *Pucc. cusickii*
Weath.; *P. tenuiflora* (Turcz.) Scribn. & Merr.; *P. fasciculata sensu* Hooker 1840, not (Torr.)
Bickn.]. MAP: Hultén 1968*b*:163.

P. phryganodes (Trin.) Scribn. & Merr.
/ASs/X/GEA/ (Hsr) Clayey seashores (often flooded at high tide) from the coasts of
Alaska–Yukon–Dist. Mackenzie to Prince Patrick Is., Ellesmere Is. (N to ca. 79°30′N), and
northernmost Ungava–Labrador, s along the coasts to NE Man. (Churchill), w and E James Bay,
and Labrador (s to Cartwright, 53°42′N); circumgreenlandic; Spitsbergen; arctic Eurasia. [*Poa
phryganodes* Trin., the type from Kotzebue Sound, Alaska; *Catabrosa (Glyceria) vilfoidea* of
Alaska–Canada reports, not Anderson; incl. *Pucc. geniculata* (Turcz.) Krecz.]. MAPS: Hultén
1968*b*:155, and 1962: map 182, p. 193; Porsild 1957: map 50, p. 167.
    Because this species has never been observed fruiting and pollen is not normally formed, the

anthers never dehiscing, W.M. Bowden (Can. J. Bot. 39(1):136. 1961) believes that it should be regarded as a hybrid of uncertain parentage.

P. sibirica Holmb.
/ASs/WW/A/ (Hs) Beaches, shores, and alkaline flats from the coasts of Alaska–Yukon–Dist. Mackenzie (E to Bathurst Inlet) and s Victoria Is. to Great Bear L. and N Alta. (delta of the Athabasca R. at L. Athabasca); N Asia. [Incl. *P. borealis* Swallen]. MAPS (*P. borealis*): Hultén 1968*b*:164; Porsild 1957 (1964 revision), map 334, p. 202.

P. vaginata (Lange) Fern. & Weath.
/AS/(X)/G/ (Hs) Coastal sands and clays, the ranges of taxa outlined below (not known in the U.S.A.), together with MAPS and synonymy.
1 Panicle long-exserted, its branches spikelet-bearing only at apex; anthers to 1.2 mm long; [known only from northernmost Que. (Wakeham Bay, 61°40′N; type locality) and northernmost Labrador (Eclipse Harbour, 59°50′N)] . . . . .var. *elegans* Soer.
1 Panicle scarcely exserted, its branches mostly spikelet-bearing to below the middle; anthers rarely over 0.8 mm long.
  2 Panicle-branches up to 6 from the lower nodes, stiff; spikelets in part subsessile, to about 6 mm long; glumes acute or acutish, relatively strongly nerved, the lower one to 2 mm long, the upper one to 2.6 mm long; [NE Man. (Churchill), s Baffin Is., and N Labrador (Hebron, 58°13′N); w Greenland between ca. 69° and 73°N (type locality), E Greenland between ca. 70°30′ and 74°N; MAP: Porsild 1957: map 52, p. 167] . . . . .var. *paradoxa* Soer.
  2 Panicle-branches commonly in pairs from the lower nodes, slender and more or less drooping; spikelets short-pedicelled, to 8 mm long; glumes obtuse, very obscurely nerved, the lower one at most 1 mm long, the upper one at most 1.8 mm long; [*Glyceria vaginata* Lange, the type from Greenland; incl. *P. contracta* (Lange) Soer.; coasts of Alaska–Yukon–NW Dist. Mackenzie to Victoria Is., Baffin Is., and Ellesmere Is. at 81°25′N, s to NE Man. (Churchill), Southampton Is., and Digges Is., Hudson Strait; w Greenland between ca. 67° and 79°N, E Greenland between ca. 70° and 77°N; MAPS: Hultén 1968*b*:161; Porsild 1957: map 52, p. 167] . . . . .var. *vaginata*

SCHEDONNARDUS Steud. [292]

S. paniculatus (Nutt.) Trel. Tumble-Grass
/T/WW/ (T) Dry alkaline prairies (particularly around salt-licks) from s Alta. (Milk River to Medicine Hat) to Sask. (N to Swift Current and Moose Jaw) and sw Man. (Melita; Medora; near Turtle Mt.), s to Ariz., Tex., and La.; S. America. [*Lepturus* Nutt.; *S. texanus* Steud.].

SCHIZACHNE Hack. [355A (*Melica*)]

S. purpurascens (Torr.) Swallen False Melic-Grass
/sT/X/EA/ (Hs) Thickets and woods from s Alaska–Yukon and Great Bear L. to L. Athabasca (Alta. and Sask.), Man. (N to Tod L. at ca. 56°45′N), Ont. (N to the Severn R. s of Hudson Bay at ca. 55°40′N), Que. (N to s Ungava Bay, L. Mistassini, and Anticosti Is.; not known from the Côte-Nord), Labrador (N to the Hebron R. at 58°06′N), Nfld., N.B., P.E.I., and N.S., s to Mexico, N.Mex., S.Dak., Wisc., and Pa.; E Europe; Asia. [*Trisetum* Torr.; *Melica* Hitchc.; *S. komarovii* Roshev.; *Avena (S.) callosa* Turcz.; *A. (M.; Bromelica) striata* Michx., not Lam.; *Bromus subulatus* of Alaskan reports, not Griseb.]. MAPS: Hultén 1968*b*:125; Tetsuo Koyama and Shoichi Kawano, Can. J. Bot. 42(7): fig. 3, p. 861. 1964; J.M. Gillett, Can. Field-Nat. 74: fig. 4 (top), p. 20. 1960.

Forma *albicans* Fern. (panicle pale rather than bronze or purplish; type from Mt. Albert, Gaspé Pen., E Que.) is known from Sask. (Boivin 1967*a*), Que., N.B., P.E.I., and N.S.

SCOLOCHLOA Link [381]

S. festucacea (Willd.) Link Sprangle-top
/sT/WW/EA/ (Grh (Hel)) Marshes and shallow water from B.C. (N to Hudson Hope, ca. 56°N)

and s Dist. Mackenzie (Yellowknife) to Alta. (N to Wood Buffalo National Park at 59°41′N), Sask. (N to Prince Albert), and Man. (N to Cross L., NE of L. Winnipeg), s to Oreg., Nebr., and Iowa; Eurasia. [*Arundo* Willd.; *Fluminia* Hitchc.; *Festuca (Flu.) arundinacea* Liljebl., not Schreb.; *Fest. borealis* Mert. & Koch]. MAPS: Hultén 1968*b*:150, 1958: map 257, p. 277, and 1937*b*: fig. 14, p. 129; Meusel, Jaeger, and Weinert 1965:40; Raup 1930: map 26 (now requiring considerable expansion), p. 203.

[SECALE L.] [407] Rye

[S. cereale L.] Rye
[Eurasian; the cultivated rye, occasionally escaped to roadsides and waste places in s Alaska (Juneau, where introd. in packing straw; Hultén 1942) and Canada (N to s Yukon–Dist. Mackenzie; reported from all provinces and sw Greenland but not established. MAP: Hultén 1968*b*:190].

SETARIA Beauv. [171] Bristly Foxtail

1　Bristles downwardly barbed; panicle often interrupted, to 12 mm thick; leaves blue-green, to about 1 cm broad; (introd.) . . . .*S. verticillata*
1　Bristles upwardly barbed; panicle more compact (sometimes interrupted at base); (introd.).
　2　Bristles below each spikelet 5 or more, upwardly barbed, yellowish to orange; fertile lemma strongly cross-wrinkled; panicle to 14 mm thick; margins of leaf-sheaths not ciliate; leaves to 12 mm broad, often twisted . . . .*S. glauca*
　2　Bristles not more than 3, stramineous or purplish; fertile lemma not distinctly cross-wrinkled; margins of leaf-sheaths ciliate.
　　3　Spikelets 2 or 3 mm long, the grain dropping free from the glumes; panicle to 3.5 cm thick; leaves to over 3 cm broad . . . .*S. italica*
　　3　Spikelets at most 2.5 mm long, falling with the grain; panicle less than 2.5 cm thick; leaves to 1.5 cm broad . . . .*S. viridis*

S. glauca (L.) Beauv.　Foxtail, Pigeon-Grass.　Foin sauvage
Eurasian; known from roadsides, fields, and waste places in all of the provinces except Nfld. (but in St-Pierre and Miquelon; N to Regina, Sask., and Goudbout, Saguenay Co., E Que.). [*Panicum* L.; *Chaetochloa* Scribn.; *S. lutescens* (Weigel) Hubbard]. MAP: Hultén 1962: map 223, p. 235.

S. italica (L.) Beauv.　German or Hungarian Millet.　Millet des oiseaux
Eurasian; fields, roadsides, and waste places of B.C. (Comox, Vancouver Is.; Henry 1915), Ont. (N to the Ottawa dist.), Que. (N to Rimouski, Rimouski Co.), and N.S. [*Panicum* L.; *Chaetochloa* Scribn.].

S. verticillata (L.) Beauv.
Eurasian; fields, roadsides, and waste places of sw Dist. Mackenzie (Fort Simpson, ca. 62°N; W.J. Cody, Can. Field-Nat. 75(2):59. 1961), B.C. (Kamloops; CAN), Man. (Brandon), Ont. (N to the Ottawa dist.), and Que. (N to Montreal). [*Panicum* L.].
　Forma *ambigua* (Guss.) Boivin (axis of panicle merely scabrous rather than pilose; spikelet-bristles upwardly barbed rather than retrorse-barbed) is reported from s B.C. and s Ont. by Boivin (1967*a*).

S. viridis (L.) Beauv.　Green Foxtail, Bottle-Grass
Eurasian; fields, roadsides, and waste places in N. America (ranges of Canadian taxa outlined below). MAP and synonymy: *see* below.
1　Culms decumbent at base, mostly less than 3 dm tall; leaves mostly less than 5 mm broad; panicle usually lobed, rarely over 5 cm long but often 2 or 3 cm thick; bristles mostly 3 or 4 mm long and less than twice as long as the spikelets; [Que. and N.B.; Gleason 1958, vol. 1:236] . . . . .var. *breviseta* (Döll) Hitchc.
1　Culms commonly over 3 dm tall; leaves to over 5 mm broad; panicle not lobed, to about 8 cm long and 1 cm thick; bristles mostly more than twice as long as the spikelets . . . . .var. *viridis*

2 Panicle somewhat open, its branches evident; [Otterburne, s Man., about 30 mi s of Winnipeg; Löve and Bernard 1959] . . . . .f. *ramuliflora* Chr.
2 Panicle dense and spike-like, its branches scarcely evident; [*Panicum* L.; *Chaetochloa* Scribn.; incl. var. *weinmannii* (R. & S.) Brand, intermediate between the two above varieties; transcontinental, introd. from s Alaska (Juneau), s Dist. Mackenzie (Fort Simpson), and B.C. to Nfld. and N.S.; MAP: Hultén 1962: map 214, p. 225] . . . . .f. *viridis*

SIEGLINGIA Bernh. [335]

S. decumbens (L.) Bernh. Heather-Grass
Eurasian; known from thickets, peaty pastures, and turfy acidic soils of sw N.S. (Digby, Yarmouth, and Shelburne counties; see N.S. map by Dore and Roland 1942) and s Nfld. (Rennie's R.; Robinson and Schrenk 1896; considered native there by Fernald *in* Gray 1950, but *see* discussion under *Luzula campestris*). [*Festuca* L.; *Triodia* Beauv.]. MAPS: Hultén 1958: map 111, p. 131; Fernald 1929: map 30, p. 1502, and 1918b: map 13, pl. 13.

SITANION Raf. [411] Squirrel-tail

S. hystrix (Nutt.) Sm.
/T/WW/ (Hs) Dry open woods, plains, and rocky slopes from B.C. (N to Vanderhoof, ca. 54°N) to s Alta. (Crowsnest Pass; Waterton Lakes; Manyberries; Medicine Hat) and s Sask. (Val Marie; Beechy), s to Calif., Mexico, Tex., and w Mo. [*Aegilops* Nutt.; *S. longifolium* Sm.; *S. (Elymus) elymoides* Raf.]. (See *Elymus glaucus* in the treatment of *Elymus* hybrids).

SORGHASTRUM Nash [134]

S. nutans (L.) Nash Indian Grass, Wood-Grass. Faux-sorgho
/T/(X)/ (Hsr (Grh)) Prairies, dry slopes, and open woods from Wyo. to s Man. (N to Brokenhead, about 30 mi NE of Winnipeg; the report from Sask. by Rydberg 1922, requires confirmation), Ont. (N to the Nipissing Dist. and Renfrew and Carleton counties), and Que. (N to 22 mi N of Mont-Laurier, Labelle Co.; Marie-Victorin and Rolland-Germain 1942; MT), s to Ariz., Tex., and Fla. [*Andropogon* L.; *Chrysopogon* Benth.; *Sorghum* Gray; *Trichachne* Buam; *A. avenaceus* Michx.].

[SORGHUM (*Sorgum*) Moench] [134]

1 Stoloniferous perennial with extensively creeping scaly rhizomes . . . .[*S. halapense*]
1 Fibrous-rooted annual . . . .[*S. vulgare*]

[S. halapense (L.) Pers.] Johnson Grass, Egyptian Millet
[European; reported from s Ont. by Boivin (1967a), where doubtless a casual waif and not established. (*Holcus* L.).]

[S. vulgare Pers.] Sorghum or Broom-corn
[Eurasian; Sudan Grass (var. *sudanense* (Piper) Hitchc.; *S. sud.* (Piper) Stapf), native in The Sudan, Africa, is known as a casual waif in Ont. (Lambton and Carleton counties) and sw Que. (Laprairie and Shefford counties).]

SPARTINA Schreb. [283] Cord- or Marsh-Grass

1 Culms stout, to about 2.5 m tall, often over 1 cm thick at base; fresh leaves flat, to 1.5 cm broad.
2 Leaves smooth; spikes to 1.5 dm long; glumes awnless, glabrous or only sparingly scabrous on the lower part of the keel; rhizome flaccid; (E Que. to Nfld. and N.S.) . . . .*S. alterniflora*
2 Leaves with scabrous margins and tips; glumes scabrous on the keel, the upper one

tapering to an awn to 7 mm long; rachis often prolonged but not surpassing the
terminal spikelet; rhizome hard; (transcontinental) . . . .*S. pectinata*
1   Culms slender, at most 6 mm thick at base; spikes rarely over 5 cm long.
    3   Leaves very scabrous above, flat when fresh, to 5 mm broad; spikes several,
        appressed; spikelets less than 9 mm long; glumes ciliate on the keel, the lower one
        about half as long as the upper; lemmas ciliate on the keel; (B.C. to s Ont.) . . . .*S. gracilis*
    3   Leaves smooth, involute, at most about 2 mm broad; spikes ascending or divergent;
        glumes scabrous on the keel, the lower one less than half as long as the upper;
        lemmas scabrous on the keel; (E Que. to Nfld. and N.S.) . . . .*S. patens*

S. alterniflora Loisel.   Salt-water Cord-Grass.   Herbe salée
/T/E/ (Hsr (Hel)) Saline shores and salt marshes, the aggregate species from E Que. (St.
Lawrence R. estuary from St-Jean-Port-Joli, l'Islet Co., to the Côte-Nord, Anticosti Is., and Gaspé
Pen.) to Nfld., N.B., P.E.I., N.S., and N.J.; introd. in Wash. and locally introd. in Europe, where it
produces with the European *S. maritima* (Curtis) Fern. the aggressive sand-binding hybrid, ×
S. townsendii H. & J. Groves. MAP and synonymy: see below.
1   Spikelets not crowded; rachis often prolonged far beyond the terminal spikelets; lemmas
    glabrous; [*S. stricta* var. *alt.* (Loisel.) Gray; E Que. to Nfld. and N.S.; MAP: Fernald 1929:
    map 35, p. 1505] . . . . .var. *alterniflora*
1   Spikelets strongly overlapping.
    2   Lemmas glabrous; rachis rarely prolonged beyond the terminal spikelet; [*S. glabra*
        Muhl.; *S. stricta* var. *gl.* (Muhl.) Gray; E Que.; reported from N.B. by G.F. Matthew,
        Nat. Hist. Soc. N.B. (April), 1869, and from N.S. by John Macoun 1888] . . . .
        . . . . .var. *glabra* (Muhl.) Fern.
    2   Lemmas sparingly pilose; rachis often prolonged beyond the terminal spikelet;
        [*S. glabra* var. *pilosa* Merr.; N.B. and N.S.] . . . . .var. *pilosa* (Merr.) Fern.

S. gracilis Trin.   Alkali Cord-Grass
/sT/WW/ (Hsr) Meadows, marshes, and prairies (particularly in alkaline flats and around
salt-springs) from s Dist. Mackenzie (N to Fort Simpson, ca. 62°N) and B.C.–Alta. to Sask. (N to
Hoosier, ca. 51°40′N) and Man. (N to Dawson Bay, N L. Winnipegosis, and the N end of L.
Winnipeg; reported from sw Ont. by Boivin 1967a), s to Calif., N.Mex., and Kans. MAP: Raup 1930:
map 25, p. 203.

S. patens (Ait.) Muhl.   Salt-meadow Grass.   Musotte
/T/EE/E/ (Hsr) Salt-marshes and brackish or saline shores from SE Que. (St. Lawrence R.
estuary from St-Jean-Port-Joli, l'Islet Co., to the Gaspé Pen. and Magdalen Is.) to Nfld., N.B.,
P.E.I., and N.S. (inland in SE Mich. and w N.Y.), s along the coast to Fla. and Tex.; Europe; N Africa.
Var. *monogyna* (Curtis) Fern. (var. *juncea* (Michx.) Hitchc.; *S. juncea* (Michx.) Willd.; plant
relatively coarse, with up to 9 spikes rather than at most 4, the spikelets at most 10 mm long rather
than to 13 mm, the upper glume acute or even blunt rather than acuminate) is reported from Nfld.
by Reeks (1873), from N.B. by Fowler (1885), and from E Que., Nfld., N.B., and N.S. by John
Macoun (1888). Collections in PEI and NSPM from Charlottetown, P.E.I., have been referred to the
series of hybrids between *S. patens* and *S. pectinata* (× *S. cespitosa* Eat.), this also reported from
N.S. by W.G. Dore and Christopher Marchant (Can. Field-Nat. 82(3):184. 1968; Lower Onslow,
Colchester Co.).

S. pectinata Link   Fresh-water Cord-Grass, Slough-Grass.   Chaume
/sT/X/ (Hsr (Grh)) Fresh, brackish, or saline shores and marshes (ranges of Canadian taxa
outlined below), s to Oreg., Utah, N.Mex., Tex., Mo., and N.C.
1   Spikes to 1.5 dm long but usually not over 5 mm broad; awns appressed; [s Ont. (Soper
    1949), Que. (St. Helen's Is., Montreal; Rouleau 1945), P.E.I. (Tignish), and N.S. (Truro)] . . . . . .
    . . . . .var. *suttiei* (Farw.) Fern.
1   Spikes to 11 cm long and 8 mm broad; awns divergent . . . . .var. *pectinata*
    2   Leaves marked with white along the margins and midrib; [known only from Ile-Perrot,
        near Montreal, Que., the type locality] . . . . .f. *variegata* Vict.
    2   Leaves uniformly green; [*Dactylis (S.; Trachynotia) cynosuroides* and *Trachynotia*

(Limnetis; S.) polystachya of Canadian reports, not L. nor Michx., respectively; *S. cyn.* var. *michauxiana* f. *major* St. Ives; Dist. Mackenzie (Great Slave L.) and s B.C. (Sea Is. at the mouth of the Fraser R.; Eastham 1947) to Alta. (N to Moose Jaw), Man. (N to Oak Point, about 60 mi NW of Winnipeg), Ont. (N to the Attawapiskat R., w James Bay, at ca. 53°N), Que. (N to the Nottaway R. at 50°52′N, L. St. John, the Côte-Nord, Anticosti Is., and Gaspé Pen.), Nfld., N.B., P.E.I., and N.S.] . . . . .f. *pectinata*

SPHENOPHOLIS Scribn. [344A (*Eatonia*)]

1 Upper glume broadly obovate, rounded and somewhat hooded at summit, its margins firm and shining; lower glume linear-oblong; anthers less than 1 mm long; spikelets about 3 mm long; panicle lance-cylindric and rather dense; [B.C. to ?Ont.] . . . .*S. obtusata*
1 Upper glume broadly oblanceolate to obovate, acute to rounded but not hooded at summit, its margins relatively thin; panicle more open.
  2 Lower glume linear-attenuate; upper glume broadly oblanceolate, acute or subacute; lemmas lanceolate, glabrous except near apex of keel; anthers 0.5 mm long; spikelets to about 4 mm long; [transcontinental] . . . .*S. intermedia*
  2 Lower glume narrowly oblong; upper glume obovate, rounded and often abruptly pointed at summit; lemmas oblong, the upper one scabrous; anthers 1 or 2 mm long; spikelets about 3 mm long; [s Ont.] . . . .*S. nitida*

S. intermedia Rydb.
/sT/X/ (Hs) Meadows, prairies, shores, and damp slopes from cent. Alaska (Tanana Hot Springs; *see* Hultén 1942: map 131, p. 395) and Great Slave L. to B.C.–Alta., Sask. (N to Meadow Lake, 54°08′N), Man. (N to the Hayes R. at about 80 mi SW of York Factory), Ont.(N to the Fawn R. at ca. 55°30′N; *see* James Bay basin map by Lepage 1966: map 8, p. 218), Que. (N to Rupert House, SE James Bay, 51°29′N), Nfld. (Bonne Bay and Port au Port Bay; CAN; MT), N.B., P.E.I., (Central Badeque, Prince Co.; DAO), and N.S. (Hants, Colchester, Kings, and Victoria counties), s to Oreg., Ariz., Tex., and Fla. [*Eatonia* Rydb.; *S. pallens* var. *major* (Torr.) Scribn.; *E. pensylvanica* of most or all reports from Canada other than from s Ont., not (Spreng.) Gray]. MAP: Hultén 1968*b*:123.

S. nitida (Biehler) Scribn.
/t/EE/ (Hs) Dry or moist woods and hillsides from Mo. to Mich., s Ont. (Niagara Falls, Welland Co.; CAN, detd. Dore; reported from Waterloo Co. by Montgomery 1945), N.Y., and Mass., s to Tex. and Fla. [*Aira* Biehler; *Eatonia* Nash; *A. (E.) pensylvanica* Spreng.].

S. obtusata (Michx.) Scribn.
/T/X/ (Hs) Dry to wet soils, open woods, and shores (ranges of Canadian taxa outlined below), s to Calif., Mexico, Tex., and Fla.
1 Panicle to about 2 dm long and 3 cm thick, its branches not strongly appressed; leaf-sheaths glabrous or merely scabrous; [*Aira* L.; *Eatonia* and *Reboulea* Gray; *S. gracilis* Kunth; incl. var. *koelerioides* Scribn.; s B.C. (N to Kamloops), s Alta. (Moss 1959), s Sask. (Cypress Hills; Manitou L.), Man. (N to Ochre River, s of Dauphin L.), and ?Ont. (reported N to w James Bay by Dutilly, Lepage, and Duman 1954, but this and reports from s Ont. by Core 1948, Dodge 1915, and John Macoun 1888, are perhaps based upon *S. intermedia;* the report from s Que. by Marie-Victorin 1935, also requires clarification)] . . . . . .
. . . . .var. *obtusata*
1 Panicle usually less than 1 dm long and 1.5 cm thick, its branches tightly appressed and subequal; leaf-sheaths puberulent; [*Trisetum lobatum* Trin.; B.C. (N to Kamloops), ?Alta., and s Man. (Otterburne, about 30 mi s of Winnipeg; Löve and Bernard 1959)] . . . .
. . . . .var. *lobata* (Trin.) Scribn.

SPOROBOLUS R. Br. [230] Drop-seed, Rush-Grass

1 Panicles to 2 or 3 dm long, relatively open, their branches spreading or ascending; perennials.
  2 Leaf-sheaths densely villous-bearded dorsally on both sides at summit, otherwise glabrous; leaves flat or drying involute, to 6 mm broad; glumes acute, the upper one usually not more than 3 mm long and about equalling the lemma; base of panicle usually included in the upper leaf-sheath; (B.C. to sw Que.) . . . .*S. cryptandrus*
  2 Leaf-sheaths beardless at the glabrous or merely somewhat pilose summit; leaves involute-setaceous; lower glume subulate above a broader base, the upper one acuminate into a keeled or involute tip, to 6 mm long and usually slightly surpassing the lemma; panicle becoming long-exserted; (Sask. to sw Que.) . . . .*S. heterolepis*
1 Panicles usually shorter, contracted and more or less spike-like, their branches erect or appressed; terminal panicle included in the upper leaf-sheath or partly exserted, lateral ones (when present) partly or wholly included; leaf-sheaths glabrous or merely somewhat pilose at summit.
  3 Hard-based tufted perennial; leaves to 4 mm broad, pilose above near base, the involute-filiform top scabrous; spikelets to 6.5 mm long; glumes unequal, obtuse or subacute, the lower one about half as long as the glabrous lemma; (s ?Ont., s Que., and N.B.) . . . .*S. asper*
  3 Soft-based tufted annuals; leaves about 2 mm broad, essentially glabrous (margins of sheaths often long-ciliate), involute at tip; glumes subequal, acuminate, about equalling the lemma.
    4 Lemmas glabrous; spikelets 2 or 3 mm long; (Alta.; Man. to w N.B.) . . . .*S. neglectus*
    4 Lemmas pubescent; spikelets to 6 mm long; (Ont. to N.B. and N.S.) . . . . . .*S. vaginiflorus*

S. asper (Michx.) Kunth   Tall Drop-seed
/T/EE/   (Hs)   Dry open soil from N.Dak. and Ohio to s ?Ont. (the report from the Niagara R. gorge in Welland Co. by Zenkert 1934, requires confirmation), sw Que. (Ile-St-Paul, near Montreal; MT; CAN), and N.B. (St. John R. near Fredericton; ACAD; CAN), s to Oreg., Ariz., Tex., La., Tenn., and Va. [*S. longifolius* Wood].

S. cryptandrus (Torr.) Gray   Sand-Drop-seed
/T/X/   (Hs)   Dry sandy soil from s B.C. (N to Redstone, about 65 mi w of Williams Lake at ca. 52°N; CAN) to s Alta. (Milk River, Lethbridge, and Medicine Hat; CAN), Sask. (N to Prince Albert and Nipawin), Man. (N to St. Lazare, about 75 mi NW of Brandon), Ont. (N to Renfrew and Carleton counties), and sw Que. (Pontiac and Deux-Montagnes counties), s to Calif., N Mexico, Tex., Mo., Iowa, and Minn. [*Agrostis* Torr.; *Vilfa* Torr.; *V. triniana* Steud.; incl. ssp. *fuscicolus* (Hook.) Jones & Fassett and var. *occidentalis* Jones & Fassett].
  [*S. airoides* Torr. is ascribed to SE B.C. by Hitchcock et al. (1969). It differs from *S. cryptandrus* in its terete and pithy (rather than grooved and solid) stems and its more open panicle, the spreading branches spikelet-bearing chiefly near the tips (rather than spikelet-bearing to near the base). The report requires confirmation.]

S. heterolepis Gray   Northern or Prairie Drop-seed
/T/(X)/   (Hs)   Rocky ground and prairies from Wyo. and N.Dak. to Sask. (N to McKague, 52°37′N), s Man. (N to Brandon and Selkirk), Ont. (N to the Ottawa dist.), and sw Que. (N to the Montreal dist.; see map of NE limits by Rouleau 1945: fig. 12, p. 54; not known from the Atlantic Provinces), s to Tex., Ark., Ill., and Pa. [*Vilfa* Gray].

S. neglectus Nash
/T/X/   (T)   Dry sterile or sandy soil from Wash. to Alta. (Medicine Hat; Moss 1959), s Man. (reported from Baldur and Winnipeg by Lowe 1943, and from railway ballast at Otterburne, about 30 mi s of Winnipeg, by Löve and Bernard 1959), Que. (N to Grondines, Portneuf Co.; see Que. map by Dominique Doyon and W.G. Dore, Can Field-Nat. 81(1): fig. 1, p. 31. 1967), and w N.B. (Fernald in Gray 1950; GH: not known from P.E.I. or N.S.), s to Mont., Ariz., Tex., La., and Va. [*S. vaginiflorus* var. *negl.* (Nash) Scribn.].

S. vaginiflorus (Torr.) Wood   Poverty-Grass
/T/X/   (T)   Dry sterile or sandy soil from Nebr. and Minn. to Ont. (N to the Ottawa dist.), sw Que. (N to Pontiac Co. and the Montreal dist.), N.B. (Keswick, near Fredericton; CAN), and N.S. (Annapolis and Kings counties; not known from P.E.I.), s to Ariz., N.Mex., Tex., and Ga. [*Vilfa* Torr.].

Var. *inequalis* Fern. (palea prolonged into a slender beak much surpassing the lemma) is known from Ont. (N to Renfrew and Carleton counties) and sw Que. (N to Rougemont, Rouville Co.).

STIPA L.   [209]   Feathergrass, Speargrass

1   Glumes at least 1.5 cm long; lemma at least 8 mm long, its awn usually 1 or 2 dm long; panicle to about 2 dm long, narrow, its branches strongly ascending.
    2   Floret pale brown when mature; lemma to 12 mm long, dorsally pubescent all over (usually sparsely so above); glumes mostly about 2 cm long; callus to 4 mm long; leaves to 2(3) mm broad, their ligules mostly 3–6 mm long, rounded to acute or acuminate; (B.C. to sw Man.) . . . .*S. comata*
    2   Floret dark brown at maturity; lemma to 2.5 cm long, dorsally pubescent only marginally; glumes to 4 cm long; callus to 7 mm long; leaves to 5 mm broad, their ligules to 2.5(5) mm long, rounded to emarginate; (B.C. to Ont.) . . . .*S. spartea*
1   Glumes at most about 1.5 cm long; lemma mostly less than 8 mm long, its awn rarely over 6 cm long.
    3   Panicle open, its branches spreading or drooping, spikelet-bearing only near the tip.
        4   Leaves to 3 mm broad, their ligules barely 0.5 mm long; glumes to 1 cm long; lemma to 6 mm long, its awn to 2.5 cm long; (B.C. to Man.) . . . .*S. richardsonii*
        4   Leaves about 1 mm broad, their ligules to 3 mm long; glumes to 1.5 cm long; lemma to 1 cm long, its awn to 7 cm long; (?Ont.) . . . .[*S. avenacea*]
    3   Panicle narrow, its branches ascending to erect, often spikelet-bearing to near the base; lemma indurate.
        5   Awn (to about 3 cm long) sabterminal, the margins of the lemma prolonged and usually thickened above the point of insertion of the awn; glumes to 1 cm long; lemmas to 7.5 mm long; callus barely 0.5 mm long; leaves to 2.5(3) mm broad; (sw B.C.) . . . .*S. lemmonii*
        5   Awn terminal, the margins of the lemma not prolonged beyond its base; leaves to 5(6) mm broad.
            6   Palea glabrous; callus about 0.5 mm long, not very sharp; glumes to 12 mm long; lemma to 6 mm long, its awn to about 4 cm long; ligules to 3 mm long; (B.C. to Man.) . . . .*S. viridula*
            6   Palea pubescent; callus to about 1.5 mm long, sharp; glumes to over 15 mm long; lemma to 8 mm long, its awn to about 6 mm long; ligules rarely over 1 mm long; (B.C. to sw Sask.) . . . .*S. occidentalis*

[*S. avenacea* L.]   Black Oat-Grass
[This species of the E U.S.A. (N to Wisc. and Mass.) is reported from Pointe-aux-Pins, near Sault Ste. Marie, Ont., by John Macoun (1888) but no voucher-specimen has been located. A 1901 collection in CAN taken by Macoun at Sarnia, Lambton Co., s Ont., proves to be *S. spartea*. It is not listed by Soper (1949).]

S. comata Trin. & Rupr.   Needle-and-thread, Speargrass
/sT/WW/   (Hs)   Dry plains and prairies (ranges of Canadian taxa outlined below), s to Calif., Tex., Iowa, and Ind. MAPS and synonymy: *see* below.
1   Panicle-base commonly enclosed in the upper lead-sheaths; glumes to about 2 cm long; lemmas to 12 mm long; upper section of awn flexuous; [s Yukon (Whitehorse) and B.C. to Alta. (N to Peace Point, ca. 59°N), Sask. (N to Prince Albert; type, as first collection cited, from Carleton House Fort, about 40 mi sw of Prince Albert), and sw Man. (N to near Cranberry Portage, ca. 55°N); introd. in s Ont. (Squirrel Is., Lambton Co.; Dodge 1915); MAPS (aggregate species): Hultén 1968*b*:85; Raup 1930: map 17, p. 202] . . . . .var. *comata*

1 Panicle-base commonly exserted; glumes and lemmas averaging a little longer; upper
   section of awn straightish and shorter; [*S. tweedyi* Trin; B.C. (N to Pouce Coupe, near
   Dawson Creek, ca. 55°45′N), Alta. (N to L. Mamawi, off the sw end of L. Athabasca; Raup
   1936), and Sask. (Saskatoon; St. Gregor)] . . . . .var. *intermedia* Scribn. & Tweedy

S. lemmonii (Vasey) Scribn.
/t/W/   (Hs)   Dry open ground and open woods from s B.C. (Victoria, Vancouver Is.; CAN, detd.
Hitchcock; reported from Comox Spit, Vancouver Is., by Eastham 1947) to Calif. and Ariz.
[*S. pringlei* var. *lemmonii* Vasey].

S. occidentalis Thurb.
/sT/WW/   (Hs)   Grasslands and sagebrush plains up to subalpine forest and ridges, the
aggregate species from s Yukon (Whitehorse and near Macintosh; CAN) and B.C. to Alta. (N to the
Peace River Dist.) and w Sask. (N to Carlton House, about 40 mi sw of Prince Albert), s to Baja
Calif., Mexico, Tex., and S.Dak. MAPS and synonymy: see below.
   Stipa bloomeri Boland. (*Oryzopsis bloomeri* (Boland.) Ricker) is apparently represented from
our area by collections in CAN from Alta. (near Edmonton), Sask. (Breakmore; Ribstone Creek),
and sw Man. (mouth of the Qu'Appelle R.). It is now generally accepted as being a sterile hybrid
between *Oryzopsis hymenoides* and *Stipa occidentalis* (*Stiporyzopsis bloomeri* (Boland.)
Johnson).
1 Awn subplumose over the first segment and usually over the second one; [s B.C.:
   reported from Merritt and Manning Provincial Park by Eastham 1947, and from
   Summerland and Trail by Hubbard 1955] . . . . .var *occidentalis*
1 Awn glabrous to scabrous (but not subplumose even on the first segment).
   2 Awns at most 3.5 cm long; callus relatively blunt, scarcely 1 mm long, pubescent
      except at the short tip; [*S. viridula* var. *minor* Vasey; *S. columbiana* Macoun;
      S. williamsii Scribn.; range of the species; MAPS (*S. columb.*): Hultén 1968*b*:86;
      Porsild 1966: map 19, p. 69] . . . . .var. *minor* (Vasey) Hitchc.
   2 Awns to 6 mm long; callus sharp, its inner curve with a glabrous area extending from
      the naked tip; [*S. nelsonii* Scribn.; reported from Alta. by A.S. Hitchcock 1935, and
      from Sask. by Rydberg 1922] . . . . .var. *nelsonii* (Scribn.) Hitchc.

S. richardsonii Link
/ST/WW/   (Hs)   Dry to moist ground and open woods from cent. Alaska (Hultén 1950), s Yukon
(Whitehorse; CAN), and B.C. to Alta. (N to Wood Buffalo National Park at ca. 59°N), Sask. (N to
Meadow Lake, 54°08′N), and s Man. (Riding Mt. and Duck Mt.; reports from Ont., Que., and N.B.
by John Macoun 1888, and from N.B. by Fowler 1885, are based upon *Oryzopsis canadensis*,
relevant collections in CAN), s to Wash., Idaho, Colo., and Iowa. The type material consists of
plants grown at Berlin from seed sent by Richardson, presumably from w Canada. MAP: Hultén
1968*b*:86.

S. spartea Trin.   Porcupine-Grass
/sT/WW/   (Hs)   Dry prairies and plains (ranges of Canadian taxa outlined below), s to N.Mex.,
Okla., and Pa. MAP: see below.
1 Lemma-body over 1.5 cm long, its awn over 1 dm long; glumes to 5 cm long; [s Dist.
   Mackenzie (near Fort Simpson) and B.C. to Alta. (N to Fort Saskatchewan), Sask. (N to
   Yorkton), and Man. (N to Duck Mt.); isolated stations in sw Yukon, N Alta., and s Ont.
   (Kent, Lambton, Huron, Bruce, and Brant counties and the Algoma Dist. N of L. Huron);
   MAP: *Atlas of Canada* 1958: map 11, sheet 38] . . . . .var. *spartea*
1 Lemma-body usually less than 1.5 cm long, its awn less than 1 dm long; glumes 2 or 3 cm
   long; [s Dist. Mackenzie (near Fort Simpson) and B.C. to Alta. (N to McMurray, ca.
   56°45′N), Sask. (N to Prince Albert), and Man. (N to Duck Mt.)] . . . . .var. *curtiseta* Hitchc.

S. viridula Trin.
/sT/(X)/   (Hs)   Plains and dry slopes from sw Dist. Mackenzie (Mackenzie R. at 61°47′N; W.J.
Cody, Can. Field-Nat. 77(2):113. 1963) and B.C. to Alta. (N to Wood Buffalo National Park at

57°45'N), Sask. (presumed type locality; N to Prince Albert), and Man. (N to Rossburn, s of Riding Mt.), s to Ariz., N.Mex., Kans., and Minn.; an isolated station in NW N.Y.

TORREYOCHLOA Church [383A (*Glyceria*)]

1 Lemmas with 7 distinct nerves, to 3.5 mm long; spikelets to 7 mm long; (essentially transcontinental) . . . .*T. pallida*
1 Lemmas with 5 prominent nerves (an additional pair of small marginal ones may be present), about 2 mm long; spikelets 4 or 5 mm long; (B.C.–Alta.) . . . .*T. pauciflora*

T. pallida (Torr.) Church
/sT/X/eA/ (Hs (Hel)) Pools, swampy ground, and wet shores (ranges of Canadian taxa outlined below), s to Mo., Tenn., and Va.; Asia. MAPS and synonymy: *see* below.
1 Leaves to over 1 cm broad, their ligules commonly at least 5 mm long; spikelets to 7 mm long; lemmas to 3.5 mm long; anthers to 1.5 mm long; grain about 1.5 mm long; [*Windsoria* Torr.; *Glyceria* Trin.; *Panicularia* Ktze.; *Puccinellia* Clausen; (Ont. (N to near Ottawa), Que. (N to near Montreal), and N.S.; MAP: Tetsuo Koyama and Shoichi Kawano, Can. J. Bot. 42(7): fig. 9(A), p. 872. 1964] . . . . .var. *pallida*
1 Leaves mostly less than 4 mm broad, their ligules commonly less than 4 mm long; spikelets about 4 mm long; lemmas less than 3 mm long; anthers at most about 0.5 mm long; grain less than 1 mm long; [var. *fern.* Hitchc.; *Glyceria (Puccinellia) fern.* (Hitchc.) St. John; *G. neogaea* of Canadian reports, not Steud., and its f. *natans* Vict. & Rolland-Germain; N-cent. B.C.; L. Athabasca, Sask.; ?Man.; Ont. (N to Hearst), Que. (N to E James Bay at ca. 52°N, the Côte-Nord, and Gaspé Pen.), Nfld., N.B., P.E.I., and N.S.; MAPS: on the above-noted map by Koyama and Kawano; Meusel, Jaeger, and Weinert 1965:40] . . . . .var. *fernaldii* (Hitchc.) Dore

T. pauciflora (Presl) Church
/sT/WW/ (Hs (Hel)) Shallow water, marshes, and wet meadows from s Alaska (*see* Hultén 1942: map 166 (*Gly. p.*), p. 398) and s Yukon (Watson L. at 60°05'N; CAN) through B.C. and Alta. (N to Lesser Slave L.) to Calif., N.Mex., and S.Dak. [*Glyceria pauciflora* Presl, the type from Nootka Sound, Vancouver Is., B.C.; *Panicularia* Ktze.]. MAP: Hultén 1968b:152 (*Gly. pauc.*).

TRIPLASIS Beauv. [335]

T. purpurea (Walt.) Chapm. Sand-Grass
/t/EE/ (T) Dry sandy ground from Minn. to s Ont. (Essex, Kent, Lambton, Huron, Norfolk, Haldimand, Welland, and Lincoln counties; *see* s Ont. map by Soper 1962: fig. 2, p. 8), N.Y., and s Maine, s to E Colo., Tex., and Fla.

TRISETUM L. [271]

1 Lemmas entire at apex, awnless or with a minute included awn just below the tip; glumes about 5 mm long; culms to about 1 m tall.
  2 Panicle rather lax, nodding, to about 3 dm long; lemmas acute, 5 or 6 mm long; leaves to 8 mm broad; (Ont. to Nfld. and N.B.) . . . .*T. melicoides*
  2 Panicle rather dense, erect, at most about 1.5 dm long; lemmas blunter, 4 or 5 mm long; leaves mostly not over 4 mm broad; sheaths scabrous, the lower ones rarely pilose; (Cypress Hills of SE Alta. and sw Sask.) . . . .*T. wolfii*
1 Lemmas 2-toothed at apex, with an exserted, slender, bent and twisted, dorsal awn.
  3 Panicle narrow and spike-like (the spikelets subsessile or very short-pedicelled), dense or interrupted toward base, usually less than 1 dm long; upper (inner) glume not much longer than the lower one; leaves flat or involute, commonly less than 3 mm broad; culms densely tufted, rarely more than 5 dm tall, typically hairy at least near the spike; (transcontinental) . . . .*T. spicatum*
  3 Panicle rather open; upper glume considerably longer than the lower one; leaves flat,

to over 5 mm broad; culms commonly taller, glabrous (leaf-sheaths glabrous or
pubescent).
4   Panicle relatively few-flowered, loose, lax or drooping, the filiform branches naked
     below, the florets distant; leaf-blades to 12 mm broad; (s Alaska, B.C., and sw Alta.) . . . .
     . . . .*T. cernuum*
4   Panicle many-flowered, from rather loose to dense and interrupted, the florets
     more crowded; leaves commonly less than 1 cm broad, their sheaths glabrous or
     the lower ones sparingly pilose.
     5   Spikelets to nearly 1 cm long, the glumes persistent on the rachis; (Alaska-
          Yukon) . . . .*T. sibiricum*
     5   Spikelets about 5 mm long, the glumes falling with the spikelet; (introd.) . . . .
          . . . .*T. flavescens*

T. cernuum Trin.
/sT/W/   (Hs)   Moist woods from s Alaska (*see* Hultén 1942: map 125, p. 394; type from Sitka)
through B.C. and sw Alta. (Waterton Lakes; Breitung 1957b) to Calif. and w Mont. [*Avena* Kunth; *A.
(T.) nutkaensis* Presl; incl. *T. canescens* Buckl. and the reduced alpine extreme, *T. montanum*
Vasey]. MAP Hultén 1968b:116.

T. flavescens (L.) Beauv.   Yellow Oats
Eurasian; occasionally introd. along roadsides and in fields in N. America, as in the Yukon (Boivin
1967a), B.C. (Hubbard 1955), Alta. (Waterton Lakes), Ont. (N to Ottawa), Que. (Rimouski,
Rimouski Co.; RIM; CAN), and N.S. (Meteghan, Digby Co.; DAO; MT). [*Avena* L.].

T. melicoides (Michx.) Scribn.
/T/EE/   (Hs)   Gravelly or sandy soil, river banks, and lake shores (ranges of Canadian taxa
outlined below), s to Wisc., Mich., N.Y., and Maine.
1   Leaf-sheaths glabrous; [*Aira* Michx.; *Graphephorum* Desv.; Ont. (N to the Fawn R. at ca.
     55°30'N; *see* James Bay region map by Dutilly, Lepage, and Duman 1958: fig. 6, p. 62),
     Que. (N to E James Bay at 52°37'N, L. St. John, Anticosti Is., and the Gaspé Pen.; not
     known from the Côte-Nord but reports from Labrador by Schranck, 1818, E. Meyer 1830,
     and Schlechtendal 1836, may possibly refer to that region), Nfld., N.B., and N.S.
     (not known from P.E.I.)] . . . . .var. *melicoides*
1   Leaf-sheaths pilose; [var. *cooleyi* (Gray) Scribn. (*Dupontia cool.* Gray); *Graph. mel.* var.
     major Gray; Ont. (N to the N shore of L. Superior and Mattice, 49°37'N), Que. (N to
     Anticosti Is. and the Gaspé Pen.), Nfld., and N.B. (Victoria Co.)] . . . . . .var. *majus* (Gray) Scribn.

T. sibiricum Rupr.
/aS/W/EA/   (Hs)   Moist tundra from the coasts of Alaska–Yukon to s-cent. Yukon and the
Alaska Panhandle; E Europe; Asia. [Incl. the reduced arctic extreme, var. *litorale* Rupr.]. MAPS:
Hultén 1968b:119; Johnson and Viereck 1962: map 3, p. 22.

T. spicatum (L.) Richter
/AST/X/GEA/   (Hs)   Tundra, alpine meadows, rocky slopes, and shores, the aggregate species
from the coasts of Alaska–Yukon–Dist. Mackenzie throughout the Canadian Arctic Archipelago to
northernmost Ellesmere Is., northernmost Ungava–Labrador, Nfld., N.B., and N.S. (not known from
P.E.I.), s in the mts. of the West to Calif. and Mexico, farther eastwards s to L. Athabasca (Alta. and
Sask.), cent. Man., L. Superior (Wisc. and Ont.), James Bay, s Que., and N New Eng., and mts. of
N.C.; S. America; circumgreenlandic; Iceland; Eurasia. MAPS and synonymy: *see* below.
1   Glumes pilose, the upper one to 4.5 mm long; panicle to about 7 cm long; [Hudson Bay
     and James Bay, Ont., to N Labrador and Nfld.; MAPS: Hultén 1962: map 53, p. 61, and
     Sven. Bot. Tidskr. 53(2): fig. 7, p. 219, and fig. 9, p. 223. 1959; Meusel, Jaeger, and
     Weinert 1965:47] . . . . .var. *pilosiglume* Fern.
1   Glumes glabrous or merely scabrous on the keel; [transcontinental; MAPS: Hultén 1962:
     maps 53 and 54, p. 61, the Canadian or total areas of the aggregate species also being
     shown in these maps and in maps by Porsild 1957: map 28, p. 164, Raup 1947: pl. 16,
     Meusel 1943: fig. 2a, and Meusel, Jaeger, and Weinert 1965:47].

2 Culm rather densely pubescent with hairs equalling its diameter; (type from s
Greenland, the only known locality) ....var. *villosissimum* Lange
2 Culm less densely pubescent; [transcontinental].
  3 Panicle very dense, rarely over 3 cm long, bronze; lemmas 3 or 4 mm long; [*Aira
spicata* and *A. subspicata* L.; *T. subspicatum* (L.) Beauv.; *T. groenlandicum* and
T. labradoricum Steud.; *T. (Calamagrostis) sesquiflorum* Trin. in part; *T. triflorum*
(Bigel.) Löve & Löve in part; *T. airoides* (Koel.) Beauv.; *Avena ?squarrosa*
Schranck; MAP: Hultén 1968b:117] ....var. *spicatum*
  3 Panicle looser; lemmas to 6 mm long.
    4 Panicle less than 7 cm long, very dense except at base, green, bronze, or
purple; upper glume 4 or 5 mm long, acute to short-pointed; [*T. maidenii* Gand.;
T. alaskanum Nash] ....var. *maidenii* (Gand.) Fern.
    4 Panicle to over 12 cm long, interrupted, silvery green, becoming whitish brown;
upper glume to 6.5 mm long, attenuate to tip; [incl. the nearly glabrous
extreme, var. *majus* (Vasey) Farw. (*T. americanum* Gand.); *Avena mollis*
Michx.; *Rupestrina pubescens* Prov.; MAP: Hultén 1968b:118] ....
....var. *molle* (Michx.) Beal

T. wolfii Vasey
/T/WW/ (Hs) Meadows and moist to dry ground and open woods from Wash. to Alta. (Waterton
Lakes; Cypress Hills) and sw Sask. (Cypress Hills; Breitung 1957a), s to Calif. and N.Mex.
[*Graphephorum* Vasey; *T. brandegei* Scribn.].

[TRITICUM L.] [408] Wheat. Blé

1 Adjacent spikelets on same side of rachis distinctly separated; glumes broadly rounded
dorsally, the keel sharply winged only toward apex (sometimes narrowly winged toward
base); lemmas long-awned or awnless, the awns usually divergent ....[*T. aestivum*]
1 Adjacent spikelets on same side of rachis crowded and closely overlapping; glumes less
rounded dorsally, the keel broadly winged from apex to base; lemmas usually long-
awned, the awns parallel, to 12 cm long ....[*T. turgidum*]

[T. aestivum L.] Wheat
[Eurasian; extensively cult. in N. America and escaped (but not established) to roadsides, railway
ballast, and other waste places in Alaska–Yukon and probably all of the provinces (apparently not
yet recorded from N.B.; see the listing of Canadian localities by Bowden 1962:1708–09, the
northernmost one being Churchill, Man.); sw Greenland. (*T. sativum* Lam.; *T. vulgare* Vill.). MAP:
Hultén 1968b:191.].

[T. turgidum L.] Poulard Wheat
[Eurasian; Bowden 1962, notes that "Durum or macaroni wheat, an Old World cultivar, has been
collected only occasionally as an escape in Western Canada". He cites collections from Sask.
(near Demaine) and s Man. (Morris, about 30 mi s of Winnipeg), where, "well-established and
common weeds along gravelled highways" and possibly self-perpetuating.]

VULPIA K.C. Gmel. [385A (*Festuca*)]

1 First (lower) glume at most 2.5 mm long, less than half as long as the second glume, this
2.5–5 mm long (averaging about 4 mm); lemma-awns averaging about 13 mm long.
  2 Lemmas strongly ciliate above the middle; (sw B.C.; introd. in Alaska–Yukon) ....
....*V. megaleura*
  2 Lemmas minutely scabrous, not ciliate; (introd. in sw B.C.) ....*V. myuros*
1 First (lower) glume rarely less than 3 mm long, at least half as long as the second glume.
  3 Awn usually shorter than the lemma; spikelets with (6)7–12(15) florets; (s B.C. to s
Ont.; introd. in sw Que.) ....*V. octoflora*
  3 Awn equalling or surpassing the lemma; spikelets with usually less than 6 florets.

4   Spikelets and branches of the inflorescence erect; lemmas minutely scabrous but
never pubescent; plant glabrous; (introd. in B.C.) . . . .*V. bromoides*
4   Spikelets and/or branches of the inflorescence spreading to reflexed; lemmas
sometimes hairy; leaf-sheaths often hairy; (sw B.C.) . . . .*V. microstachys*

V. bromoides (L.) Gray
European; dry sandy, gravelly, or rocky ground and waste places of B.C. (Vancouver Is.,
Vancouver, Powell River, and Prince Rupert, ca. 54°20′N; CAN; reported from Ladner and Gibsons
Landing, near New Westminster, and from Penticton by Eastham 1947). [*Festuca* L.; *Bromus (F.;
V.) dertonensis* All.]. MAP: Meusel, Jaeger, and Weinert 1965:34.

V. megaleura (Nutt.) Rydb.   Foxtail Fescue
/t/W/   (T)   Open sterile ground from s B.C. (Vancouver Is. and adjacent islands and the sw
mainland at Yale; introd. in Alaska (Tanana Hot Springs) and the Yukon (Dawson); some of the s
B.C. material also annotated as introd.) to Baja Calif. and Ariz. (?introd. at a few localities
eastwards); S. America. [*Festuca* Nutt.]. MAP: Hultén 1968*b*:171.

V. microstachys (Nutt.) Munro
/t/W/   (T)   Coastal strand, open woods, and lower montane slopes from sw B.C. (Vancouver Is.,
Mayne Is., and Cascade; CAN) to Baja Calif. and N.Mex.; S. America. [*Festuca* Nutt.; *F. (V.)
pacifica* Piper; *F. (V.) reflexa* Buckl.; incl var. *pauciflora* Scribn.]

V. myuros Gmel.
European; introd. in sw B.C. (Cowichan and Sidney, Vancouver Is.; CAN). [*Festuca* L.]. MAPS:
Hultén 1962: map 222, p. 233; Meusel, Jaeger, and Weinert 1965:34.

V. octoflora (Walt.) Rydb.
/T/X/   (T)   Dry sterile soil from B.C. (N to Telegraph Trail, ca. 54°N; CAN) to s Alta. (near
Manyberries and Medicine Hat; CAN), s Sask. (Piapot; Breitung 1957*a*), s ?Man. (the report from
Carberry by Shimek 1927, requires confirmation), s Ont. (N to Waterloo and Northumberland
counties; introd. along railway ballast in sw Que. near Shawville, Pontiac Co., and near
Buckingham, Papineau Co.), and N.J., s to Baja Calif., Mexico, Tex., and Fla. [*Festuca* Walt.].
    Canadian material appears to be referable to var. *tenella* (Willd.) Fern. (*F. ten.* Willd.;
inflorescence spike-like rather than an open raceme, the lemma-awns to 3 mm (rather than 7 mm)
long).

[ZEA L.]   [102]

[Z. mays L.]   Maize, Indian Corn.   Blé d'Inde
[A native of Mexico and S. America; extensively cult. in N. America and sometimes escaped to
roadsides and waste places, as in s B.C. (Hubbard 1955), s Ont., and sw Que., but not naturalized.]

ZIZANIA L.   [190]

Z. aquatica L.   Wild Rice.   Riz sauvage or Folle Avoine
/T/EE/   (T)   Quiet fresh or brackish waters and muddy shores (ranges of Canadian taxa outlined
below), s to Tex. and Fla. MAP and synonymy: *see* below.
1   Pistillate lemma thin and slenderly ribbed, opaque; aborted spikelets less than 1 mm
    thick.
    2   Mature pistillate lemma to 2 cm long, usually strigose, its awn to 7 cm long; leaves to 5
        cm broad, their ligules to 2.5 cm long; plant to about 3 m tall; [reported from the N
        shore of L. Superior, Ont., by John Macoun 1888, and from Mont-Laurier, Que., about
        80 mi N of Hull, by Marie-Victorin and Rolland-Germain 1942; Macoun's citation from
        the Rosseau R., SE Man., probably refers to var. *interior*; MAP: Fassett 1928: map 2,
        pl. 9] . . . . .var. *aquatica*
    2   Mature pistillate lemma at most 1 cm long, glabrous or scabrous, its awn to 3 cm long;
        leaves to about 12 mm broad, their ligules about 3 mm long; plant mostly less than

1 m tall; [incl. var. *subbrevis* Boivin; Ont. (Casselman, about 20 mi E of Ottawa) and Que. (N to Charlevoix and l'Islet counties; type from Lévis, opposite Quebec City)] . . . .
. . . . .var. *brevis* Fassett

1 Pistillate lemma firm and coarsely corrugated, lustrous, strigose only in the slender furrows or at the summit; abortive spikelets at least 1.5 mm thick.

   3 Lower pistillate branches of inflorescence with up to over 25 spikelets; leaves to 3 cm broad, their ligules to 1.5 cm long; plant to about 3 m tall; [Sask. (Boivin 1967*a*), Man. (Whiteshell Forest Reserve E of Winnipeg; E coast of L. Winnipeg N to 52°20′N, where abundant and harvested annually by the Indians; *see* A.I. Hallowell, Rhodora 37(440): 302–04. 1935), ?N.B., and N.S.]. A collection in TRT from Hamilton, Wentworth Co., s Ont., has been placed here, as well as a collection in OAC from the Muskoka Dist. E of Georgian Bay, L. Huron. They probably represent introduced colonies. Dore (1959) is of the opinion that stands near Ottawa are the result of introduction. As noted below, var. *angustifolia* is considered native in the St. John R. valley of E N.B. and T.E. Steeves (Rhodora 52(614):34. 1950) points out the occurrence there, also, of larger plants evidently representing var. *interior,* making the significant observation that, "it appeared that the size was correlated with the ecological conditions, smaller plants occurring on the more sandy mud." Further studies are necessary to confirm the viewpoint that the various forms occupy fairly distinct geographical areas. D.S. Erskine (1960) reports var. *interior* as introd. at Southport, P.E.I., and (Rhodora 53(635):267. 1951) in N.S. (Kings, Cumberland, and Inverness counties, evidently to attract wildfowl) . . . . .var. *interior* Fassett

   3 Lower pistillate branches with rarely over 6 spikelets; leaves commonly less than 1.5 cm broad, their ligules rarely over 1 cm long; plant usually less than 1.5 m tall; [*Z. palustris* L.; Ont. (N to Hail L., ca. 52°N, 88°15′W), Que. (N to near Quebec City), N.B., P.E.I., and N.S.]. The map of the northeastern part of the area given by Fassett (1928: map 3, pl. 9) does not indicate the occurrence in either P.E.I. or N.S. Baldwin (1958) is of the opinion that the plant has been introduced into the Clay Belt of Ont. between Geraldton and Timmins, probably by duck hunters. Concerning the Hail L., Ont., station noted above, Dutilly, Lepage, and Duman (1954) reserve decision as to whether the plant is native or introduced, pointing out that it has been known for a long time in the Sioux Lookout area of w-cent. Ont. and that the Hail L. station may eventually prove to be connected through intervening stations with the main area. D.S. Erskine (1960) notes that the P.E.I. plant was reputedly introduced as duck food, although it is native in the St. John R. valley of N.B. *See* Dore (1959) . . . .
. . . . .var. *angustifolia* Hitchc.

CYPERACEAE (Sedge Family)

Herbs with jointed, usually solid, terete or often 3-angled stems (culms). Leaves narrow, parallel-veined, entire, commonly 3-ranked, their sheaths closed (or becoming ruptured in age), the stem-leaves sometimes nearly bladeless and reduced to their sheaths. Flowers perfect or unisexual, solitary in the axils of usually spirally overlapping bracts (scales) disposed in spikes or spikelets, these often subtended by sterile scales. Spikelets solitary (as in *Eleocharis*) or variously disposed in simple or compound inflorescences. Perianth none or represented by hypogynous bristles or scales. Stamens commonly 3. Ovary superior. Fruit a flattish (lenticular) achene and styles 2, or a 3-angled (trigonous) achene and styles 3.

1   Flowers unisexual (plants monoecious or rarely dioecious), the staminate and pistillate
     spikes or portions of them dissimilar in appearance; perianth-bristles none.
     2   Achene naked, whitish, bony, subglobose, apiculate, usually borne on a low 3-angled
          disk (hypogonium); style 3-cleft; spikelets in solitary or few small compact lateral
          clusters; pistillate spikelets 1-flowered, intermixed with clusters of few-flowered
          staminate spikelets; lower spirally imbricate scales empty; culms 3-angled;
          (s Ont.) . . . . *Scleria*
     2   Achene enclosed in a sac or spathe-like glume, sessile or sometimes stipitate.
          3   Achene enclosed in a spathe open on one side above the middle; spikelets
               unisexual and 1-flowered, or with 1 staminate and 1 pistillate flower, in short
               spikes disposed in elongate heads or spike-like panicles; leaves filiform or
               involute, shorter than the culms, these usually less than 3.5 dm tall, stiff, obscurely
               angled, densely tufted, leafy only near base; (chiefly arctic and subarctic
               regions) . . . .*Kobresia*
          3   Achene enclosed in a sac (perigynium) open only at the entire or bidentate
               summit, through the orifice of which the 2 or 3 stigmas (usually 4 in *C.
               concinnoides*) are exserted; staminate and pistillate flowers borne in different
               parts of the same spike or in separate spikes on the same plant (or the plant rarely
               dioecious); culms 3-angled; perennials . . . .*Carex*
1   Flowers all perfect; achene naked; spikelets essentially uniform.
     4   Scales of spikelets strictly 2-ranked, keeled, the many-flowered spikelets disposed in
          spikes or head-like clusters.
          5   Inflorescences axillary, from the leaf-sheaths; spikes to about 3 cm long,
               composed of 2 ranks of linear flattened spikelets up to 2.5 cm long; perianth of up
               to 9 downwardly barbed bristles; achene flattened, linear-oblong; style 2-cleft;
               leaves flat, to about 1 dm long; culms to about 1 m tall, terete, hollow, jointed;
               perennial (transcontinental) . . . .*Dulichium*
          5   Inflorescences terminal, of head-like clusters or simple or compound umbels;
               spikelets smaller; perianth-bristles none; style 2–3-cleft; culms mostly 3-angled,
               solid, not jointed . . . .*Cyperus*
     4   Scales spirally arranged (or more or less 2-ranked in *Dichromena,* with whitish
          petaloid scales, and sometimes so in *Eleocharis,* with solitary terminal spikelets).
          6   Spikelets usually with only 1 fertile flower, the lower scales either empty or
               staminate; perennials.
               7   Achene terete, ovoid, broadly truncate at base, about 3 mm long,
                    conspicuously pointed but lacking a distinct terminal tubercle; style 2–3-cleft;
                    perianth-bristles none; inflorescence a panicle of up to 4 umbel-like cymes,
                    each peduncle terminating in a head-like cluster of up to 10 dark-brown
                    spikelets about 4 mm long; leaves to 3 mm broad, channelled; culms to about
                    1 m tall, obscurely 3-angled, stoloniferous; (Sask. to Nfld. and N.S.) . . . .*Cladium*
               7   Achene lenticular or flattened, strongly narrowed toward base, crowned by a
                    distinct, usually paler tubercle (enlarged style-base); style 2-cleft; culms more
                    or less 3-angled.
                    8   Spikelets flattened, to 7 mm long, with whitish keeled scales, sessile in a
                         single dense terminal head (to 1.5 cm thick) subtended by an involucre of
                         elongate, mostly white-based, leaf-like bracts; achene transversely

wrinkled, obovate, about 1 mm long; perianth-bristles none; leaves 2 or 3 mm broad; culms to over 6 dm tall; perennial from a creeping rhizome; (s ?Ont.) . . . .*[Dichromena]*
   8   Spikelets terete, the ultimate loose to head-like clusters on axillary and terminal peduncles; perianth-bristles normally present; culms caespitose or stoloniferous . . . .*Rhynchospora*
6.   Spikelets mostly many-flowered.
   9.   Base of style persistent as a tubercle at summit of achene.
     10   Spikelets few to several in a head-like or umbel-like bracted cyme to about 1 cm long, 4-angled, purple-brown, the central one sessile, the lateral ones usually long-peduncled; achene 3-angled, barely 1 mm long, transversely corrugated, capped with a minute depressed tubercle; style 3-cleft; perianth-bristles none; leaves involute-filiform, mostly basal, minutely ciliate, much shorter than the culm; culms filiform, to 4 dm tall; densely tufted annual; (Ont. and sw Que.) . . . .*Bulbostylis*
     10   Spikelet solitary and terminal, terete, the subtending bract or bracts scale-like; achene lenticular or trigonous; style 2–3-cleft; perianth-bristles often present; culms flattened or variously angled to nearly terete, the basal coloured sheaths usually bladeless; chiefly tufted perennials from matted or creeping rhizomes . . . .*Eleocharis*
   9   Base of style not persistent as a tubercle.
     11   Flower subtended by 1 or more inner scales (bracteoles) in addition to the normal larger scale or with dilated sepal-like bristles; spikelets 1–3 in a terminal head subtended by 2 or 3 foliaceous bracts of unequal length; tufted annuals to about 3 dm tall.
       12   Perianth of 3 slender retrosely barbed bristles (mostly surpassing the achene) alternating with 3 dilated perianth-scales; basal bractlets none; culms commonly bearing 2 or 3 well-developed leaf-blades, the sheaths and spikelets often pilose . . . .*[Fuirena]*
       12   Perianth wanting; flower subtended by a solitary blunt basal bractlet; culms bladeless except for a single leaf subtending the inflorescence and simulating a continuation of the culm (or this leaf often accompanied by 1 or 2 smaller divergent bracts); plants glabrous; (s Ont. and sw Que.) . . . .*Hemicarpha*
     11   Flower not subtended by inner basal scales.
       13   Style dilated below, the narrowly triangular base tubercle-like but not persistent; spikelets 1–4 at the tips of the rays of the terminal, simple or compound, umbel-like cyme; perianth-bristles none; leaves chiefly basal, to 3 mm broad, shorter than the culms; annuals or perennials; (Ont. and sw Que.) . . . .*Fimbristylis*
       13   Style slender, not dilated at base; perianth-bristles usually present, sometimes deciduous; chiefly tufted or stoloniferous perennials (only *Scirpus juncoides* and *S. smithii* annual).
         14   Perianth-bristles not more than 8 (sometimes none or rudimentary, or deciduous), often barbed, shorter than to rarely over 3 times as long as the lenticular or trigonous achene; style 2–3-cleft; floral-scales firm or rigid; culms terete or 3-angled . . . .*Scirpus*
         14   Perianth-bristles very numerous, not barbed, soft and silky, many times longer than the compressed-trigonous achene; style 3-cleft; floral-scales membranaceous; culms nearly terete . . . .*Eriophorum*

BULBOSTYLIS (Kunth) Clarke [471]

B. capillaris (L.) Nees
/T/EE/ (T) Dry to muddy open soil from Minn. to Ont. (N to Renfrew and Carleton counties), sw Que. (Montreal dist. and Pontiac, Labelle, Argenteuil, and Missisquoi counties), and Maine, s to

Tex. and Fla. (according to Gleason 1958, also in the Pacific States and Cuba). [*Stenophyllus* Britt.].

CAREX L. [525] Sedge. Laiche

(Ref.: Mackenzie 1931, 1935; 1940; *see* listing of hybrids at end)

Key to Sections

1 Spike solitary, usually bractless but sometimes subtended by an inconspicuous setaceous bract rarely much longer than the spike.
  2 Achenes lenticular or plano-convex (or plump, but not trigonous); stigmas 2; leaves filiform or setaceous.
    3 Culms solitary or few from filiform creeping rhizomes or stolons; spike unisexual or androgynous (staminate above, pistillate below); perigynia plump, shining, many-nerved dorsally, obscurely many-nerved ventrally, about 3 mm long, with a short-conic subentire beak, divergent or reflexed in age .....Section *DIOICAE* (p. 357)
    3 Culms densely tufted.
      4 Spike unisexual or gynaecandrous (pistillate above, staminate below); perigynia finely several-nerved dorsally, nerveless or obscurely nerved ventrally; achene naked at base.
        5 Culms at most about 1 dm tall, scarcely surpassing the leaves; spike gynaecandrous, subglobose, less than 1 cm long; perigynia about 2 mm long, densely white-puncticulate, faintly nerved, nearly beakless, surpassing their obtuse scales; (*C. ursina;* transcontinental in arctic regions) ....
        ....Section *HELEONASTES* (p. 360)
        5 Culms to about 7 dm tall, usually considerably surpassing the leaves; spike unisexual or gynaecandrous, to 2.5 cm long, the serrulate beak about half as long as the body; pistillate scales usually acute, about equalling the body of the perigynium; (*C. exilis;* Ont. to Labrador, Nfld., and N.S.) ....
        ....Section *STELLULATAE* (p. 375)
      4 Spike androgynous (staminate above, pistillate below), at most 1 cm long; perigynia short-beaked; pistillate scales obtuse; achene with a slender bristle (rudimentary rachilla) at base; (chiefly arctic and subarctic regions).
        6 Perigynia lanceolate, ascending, tapering to a substipitate base and a short beak; leaves overtopping the culm, this at most about 1.5 dm tall ....
        ....Section *NARDINAE* (p. 365)
        6 Perigynia subglobose, soon spreading, rounded to the sessile base, to 3 mm long, with a smooth slender beak less than 1 mm long, finely few-nerved dorsally, nerveless ventrally; spikes subglobose; leaves shorter than the culms; (a single species, *C. capitata*; transcontinental) ....
        ....Section *CAPITATAE* (p. 355)
  2 Achenes trigonous (triangular in cross-section); stigmas 3.
    7 Perigynia distinctly inflated; pistillate scales persistent.
      8 Culms to 9 dm tall, sharply 3-angled, scabrous above, in dense or loose clumps from short thick blackish rhizomes; leaves clustered at base, much surpassing the culm; spike gynaecandrous, to over 3 cm long and 2 cm broad, subtended by a non-sheathing long-attenuate bract; perigynia strongly several-ribbed above, squarrose, their beaks (to 3.5 mm long) radiating in all directions; styles continuous with the achene; (*C. squarrosa;* s Ont. and s Que.) ....
      ....Section *SQUARROSAE* (p. 374)
      8 Culms to about 2 dm tall, obtusely angled and smooth or nearly so, numerous from slender but tough brown scaly rhizomes; leaves involute-filiform, to about 1.5 dm long, stiff, clustered toward the base of the culm and shorter than or surpassing it; spike androgynous, to about 1.5 cm long and 1 cm broad, bractless; perigynia nerveless, smooth and shining, their beaks about 0.5 mm

long; style jointed with the achene; (a single species, *C. engelmannii*; s B.C.) . . . .
. . . . .Section *INFLATAE* (p. 362)

7 Perigynia scarcely inflated; culms commonly less than 2 (rarely over 4) dm tall.

  9 Plants dioecious (unisexual), with either wholly staminate or wholly pistillate spikes; perigynia densely pubescent, short-pointed, substipitate, obscurely nerved at base, to 3 mm long, their beaks at most 0.5 mm long, their subtending scales with purplish-brown to blackish centres; culms stiff, mostly not clothed at base with the dried-up leaves of the previous year . . . .
. . . . .Section *SCIRPINAE* (p. 374)

  9 Plants monoecious, the spike androgynous (staminate above, pistillate below).

    10 Pistillate scales soon deciduous; perigynia glabrous; leaves to 2 mm broad; culms stiff or wiry.

      11 Perigynia subulate to linear-lanceolate (the beak very slender), to over 6 mm long, stramineous, finely many-nerved, stipitate, at least the lower ones reflexed at maturity; style continuous with the achene; culms from slender elongate rhizomes, the dried-up leaves of the previous year few; (transcontinental species) . . . .
. . . . .Section *ORTHOCERATES* (p. 366)

      11 Perigynia broader in outline, at most 4 mm long, the beak rarely over 1 mm long; style jointed with the achene and deciduous; culms to 2 or 3 dm tall, the dried-up leaves of the previous year conspicuous.

        12 Perigynia coriaceous and shining, finely many-grooved, dark brown to blackish, sessile, spreading-ascending; culms roughened above, from slender tough blackish scaly rhizomes; (a single species, *C. obtusata*; B.C. to Man.) . . . . .Section *OBTUSATAE* (p. 365)

        12 Perigynia membranaceous and shining, finely many-nerved, yellowish-brown to brown, strongly slender-stipitate, the lower ones often reflexed at maturity; culms smooth or scabrous above; (western species) . . . . .Section *CALLISTACHYS* (p. 355)

    10 Pistillate scales persistent; perigynia not becoming reflexed, sessile or short-stipitate, beakless or short-beaked; style jointed with the achene and deciduous.

      13 Staminate scales with margins united nearly to the middle; perigynia glabrous, subalternate, substipitate, rounded at the beakless apex, finely many-nerved, their scales yellowish-green to pale brown; leaves very soft and lax, less than 1.5 mm broad, the dried-up ones of the previous year conspicuous; culms flaccid, smooth or slightly scabrous just below the spike; (a single species, *C. leptalea*, of shaded or damp habitats; transcontinental) . . . . .Section *POLYTRICHOIDEA* (p. 374)

      13 Staminate scales loose, their margins free to the base; leaves and culms firm; (plants of relatively dry habitats).

        14 Perigynia lanceolate, finely many-nerved, their beaks to 1 or 1.5 mm long; culms smooth or somewhat scabrous above, the dried-up leaves of the previous year conspicuous; (Alaska–Yukon–Dist. Mackenzie–B.C.) . . . . .Section *CIRCINATAE* (p. 355)

        14 Perigynia oblong-obovoid, nerveless or nearly so, their beaks at most 0.5 mm long.

          15 Perigynia minutely puberulent at summit; pistillate scales with very broad white-hyaline margins (the whole scale sometimes largely white-hyaline), longer and wider than and entirely concealing the perigynia until maturity; achene subtended by a slender bristle-like rachilla; leaves filiform or involute-acicular, fibrillose at base, their old brown sheaths persistent, the blades usually broken off; culms to about 3 dm tall, smooth or nearly so, forming dense tussocks, the rhizomes very short; (a single species, *C. filifolia;* B.C. to sw Man.) . . . .Section *FILIFOLIAE* (p. 359)

          15 Perigynia glabrous; plants with elongate rhizomes.

16 Perigynia to 6 mm long, their tawny scales with broad white-hyaline margins, usually surpassing and wider than the perigynia and half enveloping them; staminate part of the spike to 2.5 cm long; leaves of the previous year reduced to bladeless sheaths; culms very scabrous above, to about 4 dm tall, loosely caespitose, from thick brown scaly rhizomes; (a single species, *C. geyeri*; s B.C. and sw Alta.) . . . . .Section *FIRMICULMES* (p. 359)

16 Perigynia to about 4 mm long, surpassing their dark brown scales at maturity; staminate part of the spike less than 1.5 cm long; leaves of the previous year very conspicuous; culms rarely over 2 dm tall, usually somewhat scabrous above, from slender cord-like rhizomes; (a single species, *C. rupestris;* transcontinental in arctic, subarctic, and alpine regions) . . . . .Section *RUPESTRES* (p. 374)

1 Spikes 2 or more (sometimes not readily distinguishable).
  17 Achenes lenticular or plano-convex; stigmas 2.
    18 Spikes of 2 kinds, the terminal staminate (at least below), the lower entirely or mostly pistillate and peduncled (if sessile, more or less elongate).
      19 Style continuous with the achene and of the same bony texture, persistent; perigynia nearly nerveless, their short beaks entire or shallowly bidentate; bracts leaf-like, much surpassing the inflorescence; (*C. physocarpa; C. saxatilis*) . . . . .Section *VESICARIAE* (p. 376)
      19 Style jointed with the achene, soon withering; perigynia nerved or nerveless, beakless or with a short entire beak.
        20 Bract at base of inflorescence distinctly sheathing the slender to nearly filiform culm.
          21 Leaves wiry, folded, to about 1.5 mm broad, whitish to greyish-green, marcescent; perigynia lustrous, 5 or 6 mm long, lightly several-nerved, about equalling their dark-brown to purplish-black scales; blades of bracts rudimentary or short; culms to about 4.5 dm tall, from hard scaly rhizomes; (*C. petricosa* var. *misandroides*; Alaska, Que., and Nfld.) . . . . .Section *FERRUGINEAE* (p. 358)
          21 Leaves soft, flat or channelled toward base, green or yellowish-green, to 4 or 5 mm broad; perigynia dull, at most 3 mm long, plump, nerved or nerveless, exceeding their scales; blades of bracts elongate; culms to about 6 dm tall, loosely caespitose or substoloniferous . . . . .Section *BICOLORES* (p. 353)
        20 Bracts nearly or quite sheathless, their blades well developed; perigynia nerved or nerveless; culms mostly stouter.
          22 Pistillate scales acuminate to long-awned, mostly much longer than the perigynia; pistillate spikes nearly all peduncled; achenes usually with a deep indentation in one side near the middle . . . . .Section *CRYPTOCARPAE* (p. 356)
          22 Pistillate scales obtuse to acute (if acuminate, the upper spikes mostly sessile), shorter than to exceeding the perigynia; achenes not indented . . . . .Section *ACUTAE* (p. 348)
    18 Spikes essentially uniform, the lateral ones sessile.
      23 Culms branching, the long prostrate ones of the previous year bearing erect flowering branches from the axils of the old dried-up leaves; spikes 3–5, androgynous, in a compact head to 1.5 cm long; bracts inconspicuous or wanting; perigynia strongly nerved on both faces, short-beaked; (a single species, *C. chordorrhiza;* transcontinental in bogs) . . . . .Section *CHORDORRHIZAE* (p. 355)
      23 Culms simple, erect or ascending.
        24 Plant usually dioecious, with either completely staminate or completely pistillate heads; perigynia ovate-lanceolate, about 4 mm long, more or less

strongly nerved dorsally, obscurely nerved ventrally, the bidentate beak
more than half as long as the body; anthers to 5 mm long; lowest bract
often conspicuous; leaves to 2 mm broad; (*C. douglasii*; B.C. to s Man.) . . . .
. . . . .Section *DIVISAE* (p. 357)

24 Plants monoecious, the spikes either androgynous or gynaecandrous.
  25 Some (especially the terminal one) or all of the spikes gynaecandrous
     (androgynous only in *C. disperma*; the section *Arenariae* may
     sometimes key out here).
     26 Perigynia narrowly to broadly wing-margined, distinctly beaked, not
        thickened at base; lower bracts inconspicuous to prolonged; plants
        more or less densely tufted . . . . .Section *OVALES* (p. 366)
     26 Perigynia at most thin-edged, corky-thickened at base.
        27 Perigynia thin-edged, ascending to spreading or reflexed at
           maturity, not puncticulate, very spongy at base and distinctly
           beaked; bracts inconspicuous or wanting; plants more or less
           densely caespitose . . . . .Section *STELLULATAE* (p. 375)
        27 Perigynia with rounded margins, ascending or spreading-
           ascending.
           28 Perigynia at most about 4 mm long, finely but densely
              puncticulate, nearly beakless or short-beaked; bracts incon-
              spicuous or wanting (except in *C. trisperma*); plants more or
              less rhizomatous or stoloniferous . . . .
              . . . . .Section *HELEONASTES* (p. 360)
           28 Perigynia at least 4 mm long, not puncticulate, with long
              slender bidentate beaks; bracts usually conspicuous (except
              in *C. bromoides*); plants caespitose . . . .
              . . . . .Section *DEWEYANAE* (p. 356)
  25 Some or all of the spikes androgynous.
     29 Rhizomes cord-like and extensively creeping; culms mostly solitary.
        30 Perigynia narrowly wing-margined above the middle, slenderly
           but strongly nerved on both faces, their beaks sharply
           bidentate; spikes often dissimilar, staminate or gynaecandrous
           ones often mixed in the head with androgynous or pistillate
           ones; at least the lowest bract prolonged . . . .
           . . . . .Section *ARENARIAE* (p. 350)
        30 Perigynia nearly wingless, their beaks obscurely bidentate.
           31 Spikes densely aggregated in a compact bractless head less
              than 1.5 cm long and appearing like a single spike; perigynia
              faintly nerved at base dorsally, nerveless ventrally; pistillate
              scales obtuse; bracts wanting; leaves strongly involute;
              culms obtusely angled, at most 2.5 dm tall . . . .
              . . . . .Section *FOETIDAE* (p. 359)
           31 Spikes more or less separated and distinguishable at
              maturity; bracts wanting or the lower 1 or 2 short-
              prolonged . . . . .Section *DIVISAE* (p. 357)
     29 Rhizome short, dark brown to black, fibrillose; perigynia with
        bidentate beaks; bracts setaceous but often well developed.
        32 Spikes in mostly simple, compact or interrupted heads;
           perigynia few-nerved dorsally or nearly nerveless; bracts often
           well developed .Section *BRACTEOSAE* (p. 354)
        32 Spikes in branching spike-like heads, usually 2 or more on each
           lateral branch; perigynia often distinctly nerved, at least dorsally.
           33 Leaf-sheaths loose, commonly cross-puckered or dotted
              ventrally; culms soft and spongy (flattened under pressure),
              thick, sharply angled, their sides concave; perigynia spongy
              or corky at base, thin and soft, somewhat inflated, usually at
              least 4 mm long .Section *VULPINAE* (p. 378)

33 Leaf-sheaths close; culms slender and firm; perigynia firm, not inflated, at most about 3.5 mm long.

 34 Inner band of leaf-sheath cross-puckered; leaves to 6 mm broad; bracts setaceous, mostly overtopping the spikes and branches of the head; scales mostly awned; perigynia dorsally nerved or nerveless . . . . Section *MULTIFLORAE* (p. 365)

 34 Inner band of leaf-sheath not cross-puckered; leaves commonly not over 3 mm broad (to 6 mm in *C. cusickii* of B.C.); bracts mostly short or wanting; scales acute or short-acuminate but scarcely awned; perigynia few-nerved dorsally, obscurely few-nerved at base ventrally . . . . Section *PANICULATAE* (p. 373)

17 Achenes trigonous (triangular in cross-section); stigmas 3; spikes differentiated, some of them strictly pistillate.

 35 Lateral spikes very numerous, sessile, short; heads generally dioecious, the pistillate ones very large (to 6 cm long and 5 cm thick), the staminate ones to 4 cm long and 1 cm thick; perigynia to 1.5 cm long (including beak), very coriaceous, smooth and shining, strongly many-nerved on both faces, strongly lacerate-wing-margined ventrally nearly to base, the beak sharply bidentate, nearly as long as the perigynium-body; pistillate scales acuminate or cuspidate, shorter than the perigynia; bracts from little to strongly developed; leaves to 8 mm broad, firm, yellowish green, sharply serrulate; culms to about 3.5 dm tall, strongly black-fibrillose at base, stiff and stout, from elongate rhizomes; (a single species, *C. macrocephala*; coastal sands of sw B.C.) . . . . . Section *MACROCEPHALAE* (p. 364)

 35 Lateral spikes usually not very numerous, mostly more or less peduncled; heads usually smaller; perigynia not lacerate-winged.

  36 Staminate scales with margins united at base, very tight and somewhat tubular; at least the lowest pistillate scale green and bract-like, its base more or less hiding the 2-edged perigynium; perigynia plump, firm, glabrous, long-beaked; plants densely cespitose . . . . Section *PHYLLOSTACHYA* (p. 373)

  36 Staminate scales loosely ascending, their margins free to base; pistillate scales not bract-like.

   37 Style jointed with the achene, not indurated, soon disarticulating and withering; perigynia at most about 1 cm long.

    38 Achene only obscurely 3-angled, with rounded or convex sides, closely filling the body of the beaked, nerveless perigynium; bract at base of inflorescence sheathless, short-sheathing, or wanting.

     39 Perigynia glabrous or nearly so; culms sharply angled, scabrous at least above.

      40 Perigynia long-beaked, stipitate, much shorter than to about equalling their lance-ovate, attenuate scales; bracts at base of the non-basal spikes setaceous; leaves to 5 mm broad, firm, harsh; culms rarely over 1 dm tall; plant densely caespitose; (*C. tonsa*; Alta. to s Labrador and N.S.) . . . . Section *MONTANAE* (p. 364)

      40 Perigynia short-beaked, sessile, slightly surpassing their ovate scales; leaves scabrous along the revolute margins; plants strongly stoloniferous.

       41 Staminate spike 1; pistillate spikes 1 or 2, sessile in a compact head; perigynia coriaceous, lustrous; bract scale-like, shorter than the inflorescence; leaves at most 1.5 mm broad, attenuate; culm less than 2 dm tall; plants densely tufted; (a single species, *C. supina*; transcontinental) . . . . Section *LAMPROCHLAENAE* (p. 362)

       41 Staminate spikes 1 or 2; pistillate spikes 1–3, slenderly cylindric and up to 7 cm long, all peduncled; perigynia

granulose-roughened, minutely hispidulous above; foliaceous bract equalling or surpassing the inflorescence;
leaves to 6 mm broad; culms to 6 dm tall; plant loosely tufted;
(a single species, *C. flacca*; introd. in s Ont., sw Que., and N.S.) ...
..............................Section *PENDULINAE* (p. 373)

  39 Perigynia more or less pubescent, stipitate.

     42 Pistillate scales tipped with a short rough awn; beak of
perigynium conic, not over 0.5 mm long; achene capped by a
dark ring about 0.5 mm in diameter; inflorescence head-like,
only the lowest pistillate spike sometimes short-peduncled;
bracts scale-like to setaceous-prolonged; leaves all near the
base of the culm, 2 or 3 mm broad; plant slightly stoloniferous;
(a single species, *C. caryophyllea;* introd. in s ?Ont.) ....
...............................[Section *PRAECOCES*] (p. 374)

     42 Pistillate scales awnless; beak of perigynium at least 0.5 mm
long; achene not capped; lower spikes often more or less
remote; bracts scale-like to leaf-like; plants caespitose or
stoloniferous ....Section *MONTANAE* (p. 364)

  38 Achenes more sharply 3-angled; perigynia sometimes beakless.

    43 Leaves (and often sheaths and culms) pubescent at least beneath;
at least the lower bracts leaf-like.

      44 Bract at base of inflorescence with a prolonged closed sheath;
perigynia glabrous; plants loosely caespitose.

       45 Culms pilose; terminal spike staminate throughout;
perigynia finely several-nerved, their bidentate beaks about
half as long as the body; pistillate scales acute to short-
cuspidate; (*C. castanea*; sw Man. to Nfld. and N.S.) ....
...............................Section *SYLVATICAE* (p. 376)

       45 Culms glabrous; terminal spike pistillate above the clavate
staminate base; perigynia inflated, nerved or nerveless,
their short beaks entire or slightly bidentate ....
...............................Section *GRACILLIMAE* (p. 359)

      44 Bract at base of inflorescence sheathless or short-sheathing.

       46 Perigynia beakless or abruptly short-beaked, nerved or
nerveless, glabrous or pubescent; terminal spike sometimes
pistillate above; culms, leaves, and sheaths pilose; plants
caespitose ....Section *VIRESCENTES* (p. 377)

       46 Perigynia long-beaked, pubescent; terminal spike staminate
throughout.

        47 Perigynia nerveless, tightly filled by the achene; pistillate
scales serrulate along the rounded summit and cuspidate
tip, otherwise glabrous; leaves, sheaths, and culms
soft-pilose, the leaves to 1 cm broad; plant loosely
caespitose; (a single species, *C. hirtifolia*; Ont. to N.S.) ....
...............................Section *TRIQUETRAE* (p. 376)

        47 Perigynia strongly 15–20-ribbed, at least the summit
usually empty; scales pilose on back; leaves generally
sparsely hairy, at most about 4 mm broad, their sheaths
densely hirsute; culms nearly or quite glabrous; plant
loosely caespitose and rhizomatous; (*C. hirta*; introd. in
Que., P.E.I., and N.S.) ....Section *HIRTAE* (p. 361)

    43 Leaves and culms glabrous (sheaths pubescent in *C. hitchcockiana*).

      48 Perigynium closely filled to tip by the achene, abruptly very
short-beaked; bracts sheathing or short-sheathing, their blades
wanting or rudimentary.

       49 Staminate spike at most 8 mm long, sessile or nearly so,
overtopped by two or more of the upper pistillate spikes;

perigynia finely few-nerved, glabrous; pistillate scales whitish to pale brown; leaves involute-filiform; culms capillary, wiry, to 4 dm tall, densely tufted from elongate stolons; (a single species, *C. eburnea*; transcontinental) . . . . .Section *ALBAE* (p. 350)

49 Staminate spike overtopping the pistillate ones; pistillate scales straw-colour to purple or brown; plants caespitose or stoloniferous . . . . .Section *DIGITATAE* (p. 357)

48 Perigynium not closely filled by the achene, at least the tip usually empty; blade of at least the lower bract usually well developed (sometimes rudimentary in the Section *Ferrugineae*).

50 Bract at base of inflorescence sheathless or short-sheathing (long-sheathing only in *C. laxa* of the Section *Limosae*) . . . .*GROUP A* (p. 346)

50 Bract at base of inflorescence with a prolonged tubular sheath . . . .*GROUP B* (p. 347)

37 Style continuous with the achene and of the same bony texture; perigynia often inflated, beaked, to 2 cm long and over 1.5 mm thick, the beak often emarginate or bidentate; at least the lowest spike leafy-bracted.

51 Perigynia broadly obovoid, nerved or essentially nerveless, truncate or abruptly rounded at summit to a subulate bidentate beak with teeth at most 0.5 mm long; plants loosely to densely caespitose; (Ont. and sw Que.) . . . . .Section *SQUARROSAE* (p. 374)

51 Perigynia subulate to ovoid or subglobose, rather gradually tapering to a mostly longer-toothed beak.

52 Perigynia more or less leathery, scarcely inflated, glabrous or pubescent, nerved, in linear- to oblong-cylindric spikes; pistillate scales obtuse to acuminate or short-awned; culms 1–few from strongly stoloniferous bases . . . . .Section *PALUDOSAE* (p. 372)

52 Perigynia mostly thin and papery, usually strongly inflated, glabrous (if minutely pubescent, then in subglobose spikes); culms mostly in dense tussocks (if 1–few, from elongate stolons and with turgid perigynia).

53 Pistillate scales with scabrous awns equalling or longer than the blades; pistillate spikes elongate and densely flowered; perigynia prominently nerved, with firm-toothed beaks at least 1/3 as long as the body . . . . . . .Section *PSEUDO-CYPERAE* (p. 374)

53 Pistillate scales blunt to cuspidate or with a smooth awn less than half as long as the body; pistillate spikes cylindric to subglobose, mostly fewer-flowered; perigynium-beak mostly shorter.

54 Perigynia at most 1 cm long, obscurely to strongly nerved; staminate spikes often more than 1 . . . . .Section *VESICARIAE* (p. 376)

54 Perigynia mostly 1 or 2 cm long (if less, then in subglobose spikes); staminate spike usually solitary).

55 Perigynia lance-subulate to -conic, at most 3 mm thick, delicately many-nerved, barely inflated, soon deciduous; (eastern species) . . . . .Section *FOLLICULATAE* (p. 359)

55 Perigynia broader, to 8 mm thick, rather coarsely nerved, usually much inflated, persistent . . . . .Section *LUPULINAE* (p. 363)

GROUP A

1 Perigynia pubescent, beaked; pistillate spikes sessile or the lowest ones short-peduncled; plants with long horizontal rhizomes and stolons . . . . .Section *HIRTAE* (p. 361)

1   Perigynia glabrous or merely with scabrous margins; pistillate spikes often long-peduncled.
    2   Perigynium with a long, more or less curving beak; culm sharply angled, to about 8 dm tall.
        3   Leaves to over 1.5 cm broad, the dried-up ones of the previous year conspicuous; perigynia abruptly narrowed to a minutely bidentate beak; pistillate scales acute or acuminate; culms from long horizontal stout scaly rhizomes ....Section *ANOMALAE* (p. 350)
        3   Leaves rarely over 5 mm broad, pale green, the basal sheaths slightly filamentose; perigynia glabrous, gradually tapering into a subentire beak; pistillate scales mucronate or short-cuspidate; culms caespitose, from very short rootstocks; (*C. prasina*; Ont and s Que.) ....Section *GRACILLIMAE* (p. 359)
    2   Perigynia beakless or very short-beaked, glabrous.
        4   Perigynia plump, subglobose to ellipsoid, spreading-ascending, nerveless; culms sharply angled, nearly smooth.
            5   Terminal spike staminate throughout; perigynia longer than broad, granular-roughened; pistillate spikes to 7 cm long, mostly peduncled; leaves glaucous, firm, to 6 mm broad, with scabrous revolute margins; culms at most 6 dm tall, loosely tufted from very long glaucous stolons; (a single species, *C. flacca*; introd. in s Ont., s Que., and N.S.) ....Section *PENDULINAE* (p. 373)
            5   Terminal spike clavate at a staminate base below a long pistillate summit; perigynia as broad as long, transversely rugose; pistillate spikes at most 3.5 cm long, short-peduncled; leaves flat, rather thin, to 1 cm broad; culms in clumps from a thick rhizome; (a single species, *C. shortiana*; s Ont.) ....Section *SHORTIANAE* (p. 374)
        4   Perigynia usually distinctly compressed, strongly appressed-ascending; plants caespitose to long-stoloniferous, the culms usually strongly purple-tinged (sometimes brownish-tinged) at base.
            6   Roots glabrous (except for rootlets); terminal spike commonly sessile or short-peduncled, usually pistillate or monoecious, sometimes staminate; bracts short; perigynia nerved or nerveless ....Section *ATRATAE* (p. 351)
            6   Roots closely covered with a yellowish felt; terminal spike usually staminate throughout, long-peduncled; bracts relatively long; perigynia more or less nerved ....Section *LIMOSAE* (p. 363)

GROUP B (see p. 346)

1   Perigynia soon divergent to reflexed, nerved, the bidentate beak with erect teeth; pistillate spikes subglobose to thick-cylindric, at least the upper ones usually sessile, rarely to 2.5 cm long; plants caespitose or short-stoloniferous ....Section *EXTENSAE* (p. 358)
1   Perigynia ascending or at least not strongly divergent, the beak (when present) entire or but slightly notched (if deeply notched, the perigynia not strongly nerved and the bracts ascending).
    2   Pistillate scales blackish; perigynia flattened-triangular in cross-section, nerved or nerveless; plants caespitose or short-stoloniferous ....Section *FERRUGINEAE* (p. 358)
    2   Pistillate scales green to light reddish-brown; perigynia triangular to suborbicular in cross-section.
        3   Terminal spike regularly pistillate except at base; lateral spikes 1 cm long or more, loosely ascending to drooping; perigynia nerved or nerveless, beaked or beakless, their scales pale; culms to over 6 dm tall; plants caespitose ....Section *GRACILLIMAE* (p. 359)
        3   Terminal spike regularly staminate throughout.
            4   Perigynium-beak nearly half to quite as long as the body.
               5   Base of plant strongly and coarsely fibrillose; scales acute or awned; perigynium-beak equalling the globose-ovoid nerveless body; plant caespitose; (a single species, *C. sprengelii*; B.C. to Que. and N.B.) ....Section *LONGIROSTRES* (p. 363)

    5 Base of plant not fibrillose; perigynium-beak shorter than the body.
       6 Pistillate scales brownish purple, acute; perigynia glaucous, faintly nerved at maturity, the beak about 1/3 as long as the body; plant long-stoloniferous; (*C. vaginata*; transcontinental) . . . .
        . . . . .Section *PANICEAE* (p. 373)
       6 Pistillate scales pale brown to green or white; perigynia not glaucous; plants caespitose, lacking elongate stolons.
          7 Plants with reddish-purple to dark-brown bases; lower sheaths lacking green blades; pistillate spikes linear-cylindric, to over 6 cm long; perigynia more or less distinctly nerved at least toward base . . . .
            . . . . .Section *SYLVATICAE* (p. 376)
          7 Plants with drab or brown bases; lowest sheaths with elongate green blades; pistillate spikes at most 1.5 cm long; perigynia few-nerved or nerveless; (transcontinental) . . . .
            . . . . .Section *CAPILLARES* (p. 355)
   4 Perigynia beakless or with a short essentially entire beak not more than 1/4 the length of the body.
       8 Pistillate scales with purplish-brown or -black margins; plants loosely stoloniferous . . . . .Section *PANICEAE* (p. 373)
       8 Pistillate scales whitish or green to reddish-brown.
          9 Plant loosely stoloniferous; perigynia at most 3.5 mm long, often resinous-dotted, obscurely nerved; scales reddish brown; leaves stiff, glaucous, often folded, to 4 mm broad; (*C. crawei*; transcontinental) . . . .
            . . . . .Section *GRANULARES* (p. 360)
          9 Plants nonstoloniferous or nearly so; pistillate scales whitish to pale brown.
            10 Perigynia nerveless or with elevated nerves, glabrous or hispidulous, the spongy stipe to 1 mm long; perigynium-beak often somewhat oblique; pistillate scales often short-awned . . . .
              . . . . .Section *LAXIFLORAE* (p. 362)
            10 Perigynia sessile, glabrous, nearly beakless.
              11 Perigynia at most 4 mm long, with elevated nerves, their scales acuminate or cuspidate . . . . .Section *GRANULARES* (p. 360)
              11 Perigynia to 6 mm long, their nerves impressed, their scales often rough-awned; (eastern species) . . . .
             . . . . .Section *OLIGOCARPAE* (p. 365)

Section *ACUTAE* (see p. 342)

1 Perigynia distinctly nerved at least dorsally; terminal spike or spikes normally entirely staminate; fertile culms arising from the center of the previous year's tufts of dried-up leaves.
   2 Lower leaf-sheaths of the flowering-culms breaking ventrally and becoming fibrillose; pistillate spikes erect or suberect (their scales mostly rough-cuspidate or rough-awned), the upper ones often staminate at apex; perigynia to 4.5 mm long, rather obscurely to strongly several-nerved on both faces, abruptly short-beaked, the beak strongly bidentate; leaves to about 1 cm broad; culms to about 1 m tall, from long horizontal stolons; (?B.C.) . . . .[*C. barbarae*]
   2 Lower sheaths not becoming noticeably fibrillose.
      3 Perigynia coriaceous, strongly ribbed, to 3.5 mm long, their beaks bidentate; lowest bract commonly about equalling the inflorescence; culms to over 1 m tall, from long horizontal stout stolons; (?B.C.–?Alta.) . . . .[*C. nebrascensis*]
      3 Perigynia softer, slenderly nerved, their beaks entire.
         4 Plants strongly stoloniferous, with horizontal stolons; lowest bract shorter than or only slightly surpassing the inflorescence; (Ont. to Nfld. and N.S.) . . . .*C. acuta*
         4 Plants loosely to densely caespitose or with ascending stolons; lowest bract usually considerable surpassing the inflorescence.

     5   Lowest bract short-sheathing; perigynia short-stipitate, few-ribbed on both faces; pistillate scales tinged with brown-red; culms densely caespitose; sterile shoots phyllopodic; (transcontiental) ....*C. lenticularis*

     5   Lowest bract sheathless or nearly so; perigynia more strongly stipitate; pistillate scales blackish; culms less densely caespitose; sterile shoots aphyllopodic; (western species).

          6   Perigynia light green (or glaucous-green in age), nerved, very minutely granular; scales long-persistent; (s Alaska–B.C.–w Alta.) ....*C. kelloggii*

          6   Perigynia yellowish green, ribbed, papillate-roughened; scales soon deciduous; (Alaska–B.C.) ....*C. hindsii*

1 Perigynia nerveless ventrally, nerveless or lightly few-nerved at the base dorsally.

   7 Fertile culms all or mostly arising laterally and not enveloped at base by the dried-up previous year's tufts of leaves; perigynium-beak entire or merely emarginate; (eastern species).

     8 Pistillate spikes (at least the lower) strongly curved or nodding; perigynia nerveless or nearly so, their beaks about 0.5 mm long, bent (and twisted when mature); culms stout-based, in open clumps from stout forking rhizomes and cord-like roots; (s Ont. to N.B. and N.S.) ....*C. torta*

     8 Pistillate spikes erect or suberect; perigynia usually lightly few-nerved at least dorsally, their beaks straight; culms generally more slender-based.

          9 Perigynia inflated, brownish at maturity; pistillate spikes cylindric, rarely much attenuate at base, rarely staminate at tip, to 4 cm long, mostly sessile, their scales soon divergent; achenes suborbicular; lower leaf-sheaths rarely fibrillose; prolonged stolons wanting; (Ont. to N.B. and St-Pierre and Miquelon) ....*C. haydenii*

          9 Perigynia not inflated, unequally biconvex, green or straw-coloured; pistillate spikes often clavate or tapering at base, often with staminate tips, to about 1 dm long, the lower ones frequently peduncled; scales appressed-ascending; achenes somewhat narrower in outline; lower leaf-sheaths fibrillose ventrally; (Man. to N.B., N.S., and ?Nfld.) ....*C. stricta*

   7 Fertile culms arising from the centre of the previous year's tufts of dried-up leaves.

    10 Terminal spike gynaecandrous (pistillate above and staminate below, like the lower ones; or the spikes sometimes with a few staminate flowers at the apex), the spikes to 3 cm long, short-peduncled and forming a more or less digitate cluster; scales blunt or obtuse, rusty-blackish-brown; perigynia to 2.5 mm long, abruptly contracted into a barely distinct, short, slightly emarginate beak; (Alaska–Yukon–B.C.–Alta.) ....*C. eleusinoides*

    10 Terminal 1–3 spikes staminate, the lower ones pistillate (or sometimes staminate at apex).

      11 Lowest bract leaf-like, equalling or surpassing the inflorescence; pistillate scales with a slender midvein or broader pale centre; culms commonly over 6 dm tall and in relatively large clumps.

         12 Perigynia turgid; scales of spikes spreading; (B.C. and sw Alta.) ......*C. aperta*

         12 Perigynia not turgid; scales appressed.

            13 Perigynia ovate-orbicular, olive-green, scarcely 2 mm long; (?B.C.) ....
....[*C. interrupta*]

            13 Perigynia narrower in outline, light green to straw-colour, to 3.5 mm long.

               14 Leaf-sheaths not coloured ventrally at the mouth; lower pistillate spikes not nodding; pistillate scales not whitened at tip; culms smooth or more or less roughened above; long horizontal stolons present; (transcontinental) ....*C. aquatilis*

               14 Leaf-sheaths usually strongly dark-coloured ventrally at the mouth; lower pistillate spikes erect to spreading or drooping; pistillate scales whitened at tip, especially in age; long horizontal stolons wanting; (s Alaska–w B.C.) ....*C. sitchensis*

      11 Lowest bract more or less setaceous, shorter than the inflorescence; pistillate

scales with an obsolete or slender midvein; culms stiff, arising solitary or few together, commonly less than 6 dm tall.

15 Plants densely caespitose, from short stout obliquely ascending rootstocks; pistillate spikes usually 2 or 3, dark purple, often staminate at apex, to 2.5 cm long, their mainly blackish scales obtuse to acutish; perigynia to 2.5 mm long; leaves less than 3 mm broad; culms to 5 dm tall; (western Arctic) . . . .*C. lugens*

15 Plants strongly stoloniferous.

16 Dried first-year leaves at base of fertile culm rigid and conspicuous, concealing the base of the culm; leaves of fertile culms all blade-bearing, the lower sheaths neither purplish nor hispidulous dorsally; culms smooth or roughened above.

17 Perigynia plano-convex, minutely punctate, appressed, the straight beak to 0.3 mm long; (transcontinental) . . . .*C. bigelowii*

17 Perigynia soon turgid, papillose, squarrose-spreading, the often abruptly bent beak to 0.5 mm long; (Yukon–B.C.–sw Alta.) . . . .
. . . .*C. scopulorum*

16 Dried first-year leaves of fertile culms much desiccated, not rigid or conspicuous and not concealing the base of the culms; lowest leaves (of the season) of the fertile culms bladeless, the lower sheaths purplish and more or less hispidulous dorsally; culms usually strongly roughened above.

18 Lower bladeless sheaths of fertile culms conspicuous; sterile shoots strongly aphyllopodic; (?B.C.) . . . .[*C. miserabilis*]

18 Lower bladeless sheaths of fertile culms inconspicuous and largely hidden by old dead leaves.

19 Perigynia plano-convex or flattened-biconvex, appressed-ascending; sterile shoots phyllopodic; (s B.C.–Alta.) . . . .
. . . .*C. gymnoclada*

19 Perigynia deeply concave ventrally, convex dorsally, out-curving and spreading; sterile shoots aphyllopodic; (s ?B.C.) . . . .
. . . .[*C. campylocarpa*]

Section *ALBAE* (see p. 346)

A single species; (transcontinental) . . . .*C. eburnea*

Section *ANOMALAE* (see p. 347)

1 Perigynia 2-ribbed, otherwise nerveless, smooth or somewhat rugose, to about 3 mm long; leaves light green or glaucous-green, smooth except on the veins and margins toward apex; culms scabrous on the angles above, purplish-tinged toward base; (s B.C.) . . . . . .
. . . .*C. amplifolia*

1 Perigynia 2-ribbed and also strongly several-nerved, scabrous-papillate, to 4 mm long; leaves dark green, very scabrous above; culms harshly scabrous above, brownish at base; (Ont. to N.S.) . . . .*C. scabrata*

Section *ARENARIAE* (see p. 343)

1 Inner band of leaf-sheaths nerveless; culm smoothish or slightly scabrous; perigynia 5 or 6 mm long, the beak 2/3 as long as the body; head to 3.5 cm long; spikes less than 10, the lowest small and pistillate, the middle largely staminate, the terminal one usually pistillate above and with a prolonged staminate base; pistillate scales acute; rhizomes covered with brown scales; (B.C. to Que.) . . . .*C. siccata*

1 Inner band of leaf-sheaths green-nerved nearly to summit; culms scabrous; perigynia less than 5 mm long, the beak not more than 1/2 as long as the body; head to 7 cm long;

spikes to about 25 in number, the crowded upper ones chiefly staminate or pistillate, the lower ones pistillate; rhizomes covered with blackish scales.
  2 Leaf-sheaths shorter than the upper nodes; pistillate spikes subglobose or short-ovoid, less than 1 cm long, their pale-brown scales obtuse to cuspidate; (B.C. to James Bay) ....*C. sartwellii*
  2 Leaf-sheaths covering the upper nodes; lower pistillate spikes ovoid or ellipsoid, to 1.5 cm long, their chestnut-brown scales acute; (introd. in s Ont. and sw Que.) ....
....*C. disticha*

Section *ATRATAE* (*see* p. 347)

1 Terminal spike usually entirely staminate (*C. adelostoma* may often key out here, and sometimes *C. parryana*; in some plants of *C. hallii* often entirely pistillate, but not gynaecandrous); culms arising centrally from the midst of leaves of the previous year and clothed at base with their dried-up remains.
  2 Lateral (pistillate) spikes all sessile or short-peduncled.
    3 Terminal spike sometimes staminate, sometimes entirely pistillate (when staminate, usually short-peduncled and up to 2.5 cm long); pistillate spikes (often solitary) to 3 cm long, short-peduncled or the middle one sessile; perigynia to about 3 mm long, 2-ribbed laterally, otherwise nerveless or lightly few-nerved dorsally; leaves to 4 mm broad, very long-attenuate; culms to about 6 dm tall; (s Man.) ....*C. hallii*
    3 Terminal spike always entirely staminate.
      4 Terminal (staminate) spike inconspicuous, to 8 mm long, overtopped and largely hidden by the contiguous pistillate spikes; perigynia less than 3 mm long, straw-colour below, purplish black near summit, rather obscurely few-nerved; pistillate scales purplish black with obscure lighter midvein and narrow white hyaline apex, obtuse to acute; leaves to 2 mm broad; culms to about 2.5 dm tall, long-stoloniferous; (Alaska to Baffin Is. and N Que.) ....
....*C. holostoma*
      4 Terminal (staminate) spike to over 2 cm long; perigynia to 4.5 mm long, finely many-nerved; pistillate scales orbicular, abruptly mucronate, purplish red with hyaline margins and 3-nerved lighter centre, about as wide as but only half the length of the perigynium; leaves to 8 mm broad; culms to about 9 dm tall; (B.C. to sw Sask.) ....*C. raynoldsii*
  2 Lateral spikes (at least the lower ones) long-peduncled.
    5 Basal leaf-sheaths bladeless, scarcely fibrillose, the leaves of the previous year not persistent or much desiccated at anthesis, only the upper 2–4 leaves of the fertile culms blade-bearing; culms purplish-tinged at base; perigynia strongly flattened, to 4 or 5 mm long; (B.C. and Alta.).
      6 Scales obtuse or acute, the midvein inconspicuous; perigynia 2-ribbed marginally, otherwise nerveless; achenes long-stipitate ....*C. montanensis*
      6 Scales with a conspicuous midvein usually more or less excurrent as a short cusp; perigynia 2-ribbed and very obscurely nerved; achenes short-stipitate ....
....*C. spectabilis*
    5 Basal leaf-sheaths blade-bearing, often fibrillose, the culms clothed at base with the dried-up leaves of the previous year and bearing most of their leaves along their lower third of length.
      7 Perigynia much inflated, to about 3.5 mm long, 2-ribbed marginally, otherwise nerveless or nearly so, the style prominently exserted to about 1 mm; leaves to 3 mm broad; culms purplish-tinged at base; (chiefly coasts of Alaska, B.C., Que., Labrador, and Nfld.) ....*C. stylosa*
      7 Perigynia strongly flattened, to 4 or 5 mm long, 2-ribbed marginally and also faintly nerved on the faces; leaves to 5 or 6 mm broad; culms brownish-tinged at base; (western species).
        8 Pistillate spikes linear, their scales with a prominent midvein, more or less

concealing the smooth light-green perigynia; (Alaska–Yukon–w Dist.
Mackenzie). . . . .*C. nesophila*

    8 Pistillate spikes cylindric to short-oblong, their scales with an inconspicu-
ous midvein, scarcely concealing the relatively dark granular-roughened
perigynia; (Alaska–Yukon–Dist. Mackenzie–B.C.–Alta.) . . . .*C. podocarpa*

1 Terminal spike normally gynaecandrous (pistillate above, staminate below; sometimes
atypically staminate or nearly so, especially in *C. adelostoma,* and sometimes entirely
pistillate in *C. media* and *C. norvegica*).

  9 Lateral (pistillate) spikes all sessile or short-peduncled; culms tinged with reddish
purple at base.

    10 Plants scarcely caespitose, the solitary or loosely tufted culms arising from small
tufts of leaves terminating long horizontal stolons and not clothed at base by the
dried-up leaves of the previous year; basal sheaths bladeless, finally becoming
fibrillose; perigynia obscurely several-nerved, markedly granulose, their subtending scales ferruginous to purplish brown or near black; (transcontinental).

      11 Perigynia to 3 mm long, greyish green, the beak obsolescent; pistillate scales
acuminate, shorter than to as long as the perigynia; terminal spike often
entirely staminate; leaves 2 or 3 mm broad; culms to 3.5 dm tall . . . . .*C. adelostoma*

      11 Perigynia to 4 mm long, bluish grey, with a short but distinct beak; pistillate
scales often long-awned, equalling or surpassing the perigynia; terminal spike
typically gynaecandrous (staminate in f. *heterostachya* Anderss.); leaves to
4 mm broad; culms to about 1 m tall . . . .*C. buxbaumii*

    10 Plants loosely to densely caespitose, arising centrally from the midst of leaves of
the previous year and strongly clothed at base with their dried-up remains; leaves
to 3 mm broad; pistillate scales blunt to subacute, shorter than, to about equalling,
the perigynia, these rarely over 3 mm long, 2-ribbed marginally but otherwise
nerveless or faintly nerved.

      12 Inflorescence relatively loose, the terminal spike (sometimes nearly entirely
staminate) to about 2 cm long, more or less strongly separated from the lower
ones and usually surpassing the lowest bract; pistillate scales mostly about
equalling the perigynia, dark reddish-brown, the green midvein prominent to
apex; culms stiff, to about 4 dm tall, from long slender horizontal stolons;
(B.C. to Man.) . . . .*C. parryana*

      12 Inflorescence compact, the terminal spike (sometimes staminate only at the
very base or pistillate throughout) closely crowded by the lateral spikes, rarely
over 8 mm long; lowest bract usually very short but frequently a long one
developed that surpasses the head; pistillate scales much shorter than the
perigynia, reddish brown to purplish black, their midveins obscure or obsolete;
(chiefly arctic and subarctic regions).

        13 Culms stiff, to 3 dm tall, caespitose and forming dense tussocks; perigynia
to 2.5 mm long, becoming reddish or purple, strongly granulose, abruptly
short-beaked, the beak erect; (Dist. Keewatin–Man. to Ungava–
Labrador–Nfld.) . . . .*C. norvegica*

        13 Culms weaker and flexuous, to 6 dm tall, more loosely caespitose;
perigynia to 3.5 mm long, whitish to pale brown, less granulose, tapering to
spreading or recurving beaks; (transcontinental) . . . .*C. media*

  9 Lateral spikes (at least the lower ones) long-peduncled; culms strongly purple-tinged
at base.

    14 Midnerve of the purplish-black pistillate scales excurrent from the obtuse apex as
a prominent rough awn or cusp; perigynia 4 or 5 mm long, strongly compressed,
nerveless or nearly so ventrally, finely many-nerved dorsally; leaves to 4 mm
broad, the lower sheaths bladeless and becoming very sparingly fibrillose; culms
to 6 dm tall, in dense clumps from short stout scaly rootstocks; (Alaska–B.C.) . . . .
. . . .*C. gmelinii*

    14 Midnerve of pistillate scales not excurrent.

      15 Spikes commonly at least 6 per head, oblong-cylindric; perigynia very
chartaceous (papery), much longer than the dark purplish-brown scales;

leaves to 7 mm broad, the lower sheaths bladeless; culms to about 1 m tall, in large dense clumps from very short rootstocks; (B.C.) ....*C. mertensii*
   15 Spikes commonly less than 6 per head, not oblong-cylindric.
      16 Perigynia more or less leathery in texture, suborbicular in section and somewhat inflated (the flattish side turned outward), few-nerved, to about 5 mm long; scales ferruginous, equalling or surpassing the perigynia; leaves to 3.5 mm broad, very smooth (even toward tip), stiff and thick, with a prominent midrib, the lower marcescent sheaths bladeless and strongly fibrillose; culms to about 3.5 dm tall, from long creeping rhizomes; (s Yukon) ....
      ....*C. sabulosa*
      16 Perigynia membranaceous, to 4 mm long, typically strongly compressed, nerveless or nearly so; leaves roughened toward apex; lower sheaths blade-bearing and mostly not very strongly fibrillose.
         17 Perigynia not granular-roughened (under a lens), at least the margins green or whitish green; leaves to 7 mm broad; (western species).
            18 Lateral spikes linear, gynaecandrous, the lower ones nodding on long slender peduncles; scales purplish brown, shorter than the perigynia; culms to 9 dm tall; (?B.C.) ....[*C. bella*]
            18 Lateral spikes oblong, pistillate, erect on stiff peduncles; scales blackish, about equalling the perigynia; culms to 6 dm tall; (s B.C.–Alta.) ....*C. epapillosa*
         17 Perigynia granular-roughened (under a lens) especially on the upper margins, yellowish brown or dark-tinged.
            19 Head compact, only the lowest spike slightly separate and short-peduncled; scales purplish black, with very conspicuous white-hyaline apex and upper margins, about equalling the purplish-black perigynia; leaves to 5 mm broad; culms stiff, to 3 dm tall; (mts. of B.C. and Alta.) ....*C. albonigra*
            19 Head loose, at least the lowest spike usually remote and on a peduncle about equalling to about twice as long as the spike; culms mostly taller, slender and flexuous or nodding above.
               20 Spikes to 1 cm thick, their obtusish to acute scales black or fading brownish-black; perigynia to 4 mm long; leaves to 8 mm broad; (Alaska–B.C.–Alta; Hudson Bay–James Bay) ....*C. atrata*
               20 Spikes about 5 mm thick, their acute to acuminate scales light to dark purplish-red; perigynia at most 3.5 mm long; leaves mostly less than 5 mm broad; (essentially transcontinental) ....
               ....*C. atratiformis*

Section *BICOLORES* (*see* p. 342)

1 Perigynia minutely serrulate below the minute purple-tinged beak, about 2.5 mm long, yellowish- to glaucous-green, obscurely nerved, their obovate scales purplish brown; spikelets 3–5, crowded and sessile or the lowest ones more or less remote and peduncled; terminal spike normally gynaecandrous; culms less than 1 dm tall, over-topped by the leaves, these less than 2 mm broad; (s Dist. Keewatin–N Man.) ....*C. rufina*
1 Perigynia not serrulate, beakless or nearly so; culms mostly surpassing the leaves.
   2 Lowest bract short-sheathing, shorter than to little surpassing the inflorescence, with broad black basal auricles; perigynia white-granulose, essentially nerveless; scales purplish black, obtuse or short-pointed; spikes crowded, the terminal one gynaecandrous; leaves to 2.5 mm broad; culms to 1.5 dm tall; (transcontinental in arctic and subarctic regions) ....*C. bicolor*
   2 Lowest bract long-sheathing, leaf-like and usually surpassing the inflorescence, without basal auricles; spikes distant to approximate.
      3 Pistillate spikes rather loosely flowered, the perigynia golden-yellow to brown at maturity, fleshy and translucent, coarsely ribbed, their scales whitish to

orange-brown, obtusish to short-cuspidate, spreading; terminal spike usually staminate throughout; (transcontinental) . . . .*C. aurea*

  3 Pistillate spikes more densely flowered, the perigynia whitish-pulverulent, dry, rather obscurely nerved, their reddish-brown to purplish scales ascending or appressed.

      4 Spikes mostly remote, the terminal one usually staminate throughout; scales of pistillate spikes firm, mostly about equalling or slightly surpassing the perigynia, their tips acuminate or awned; scales of staminate spike narrowed to blunt tips, not white-margined; upper sheaths truncate at orifice . . . .[*C. hassei*]

      4 Spikes usually closely crowded and overlapping, the terminal one usually staminate only at base or entirely pistillate; scales of pistillate spikes membranaceous, mostly rounded at summit or merely short-pointed, mostly shorter than the perigynia; principal staminate scales broadly rounded at summit, with pale scarious margins; upper sheaths with a V-shaped orifice; (B.C. to Que.) . . . .*C. garberi*

Section BRACTEOSAE (see p. 343)

(see p. 343)

1 Leaf-sheaths loose, septate-nodulose and usually mottled dorsally with green and white, the nerveless ventral (inner) surface thin and friable; leaves to 1 cm broad; perigynia plano-convex, nerveless on the inner face, faintly nerved or nerveless dorsally.

  2 Perigynia drab or pale brown, to 5.5 mm long; scales long-acuminate to awn-tipped, about as long as the perigynia; achenes suborbicular, tipped by an ovoid style-base about 0.8 mm long; head dense, normally not over 3 cm long; (SE Sask. to s Ont.) . . . .*C. gravida*

  2 Perigynia pale green, at most 4.5 mm long; achenes broadly ovoid, the style-base not conspicuously expanded; (Ont. to Que. and N.B.) . . . .*C. sparganioides*

1 Leaf-sheaths close, usually smoothish; leaves at most about 4.5 mm broad.

  3 Inflorescence ovoid or subglobose, at most about 2 cm long, setaceous-bracted; perigynia plano-convex, not spongy-thickened at base, their margins scarcely incurved.

      4 Perigynia broadly deltoid, subtruncate at base, rarely over 3.5 mm long and 2.5 mm broad, nerveless ventrally, obscurely nerved dorsally; bracts bristle-form, at most 2 cm long; leaves to 3 mm broad; (s Ont.) . . . .*C. leavenworthii*

      4 Perigynia lance- to elliptic-ovate, tapering to base, nerveless or with a few nerves at the base of the dorsal face.

         5 Perigynia 2 or 3 mm long and about 1 mm broad; lowest bract setaceous but usually at least 1 cm long; leaves to 4.5 mm broad; culms to about 8 dm tall; (s Ont. and s Que.) . . . .*C. cephalophora*

         5 Perigynia 4 or 5 mm long and about 2 mm broad; bracts usually wanting; leaves at most about 3.5 mm broad; culms rarely over 6 dm tall; (B.C. to sw Sask.) . . . .*C. hoodii*

  3 Inflorescence elongate and looser, the heads bracted or bractless.

      6 Perigynia conspicuously spongy-thickened below the middle, their nerve-like margins inflexed or incurved.

         7 Summit of the finely striate perigynia smooth; perigynia biconvex, about 3 mm long, green to brown, corrugated below; scales brownish, acute, soon deciduous, nearly equalling the perigynia; leaves flat, to about 3 mm broad, shorter than to about equalling the stiff erect culm; (s Ont.) . . . .*C. retroflexa*

         7 Summit of the essentially nerveless plano-convex perigynia minutely serrulate-ciliate; scales whitish, obtuse or rounded at apex; (s Man. to N.S.).

            8 Culms stiffish and suberect; perigynia to 4.5 mm long, deep green; stigmas strongly coiled; leaves to 3 mm broad . . . .*C. convoluta*

            8 Culms weak and lax; stigmas not normally strongly coiled; perigynia at most 3.5 mm long; leaves at most 2 mm broad . . . .*C. rosea*

      6 Perigynia not spongy-thickened below middle (except in *C. spicata*), their minutely serrulate-ciliate margins scarcely incurved.

9 Perigynia lance-ovate, to 3.5 mm long, obscurely nerved dorsally, nerveless ventrally, completely covered by the pale-brown, prominently awned scales; leaves flat, thin, at most 2.5 mm broad; head linear-cylindric, to 5 cm long, subtended by a bract with a broad brown scarious base; ligules dark-margined, broader than long; (B.C. and Alta. to sw Man.) . . . .*C. hookerana*

9 Perigynia narrowly ovate to suborbicular, not completely covered by the scales.

  10 Ligule prolonged, much longer than broad; leaves to 3 mm broad; perigynia narrowly ovate, about 5 mm long, lustrous, essentially nerveless, spongy-thickened below the middle; sclaes green to tawny, acuminate, scarcely as long as the perigynia; (introd. in s Ont. and N.S.) . . . .*C. spicata*

  10 Ligule rarely longer than broad; perigynia at most 3.5 mm long, not spongy-thickened below middle.

    11 Scales reddish brown, merely short-pointed, much shorter than the broadly ovate, essentially nerveless perigynia; inflorescence rather loose, the lower spikes remote; leaves to 3 mm broad; (introd. in N.B.) . . . . . .
. . . .*C. pairaei*

    11 Scales pale green, becoming pale brown, rough-awned, nearly or quite equalling the broadly ovate to suborbicular perigynia; inflorescence compact; leaves harsh, often folded, to 4 mm broad; (s Ont. and sw Que.) . . . .*C. muhlenbergii*

Section *CALLISTACHYS* (*see* p. 341)

1 Culms smooth, dark brown at base, with up to 9 leaves, from stout creeping brownish-black scaly rootstocks; leaves to 3 mm broad, flat or channelled; staminate flowers many, conspicuous; perigynia early reflexed; (s Alaska–B.C.–sw Alta.) . . . .*C. nigricans*

1 Culms often scabrous above, yellowish brown at base, with mostly not more than 4 leaves, very densely tufted; leaves to 2 mm broad, very strongly channelled; staminate flowers few; perigynia tardily reflexed; (Alaska–Yukon–Dist. Mackenzie–B.C.–Alta.) . . . .
. . . .*C. pyrenaica*

Section *CAPILLARES* (*see* p. 348)

1 Leaves flat or slightly channelled, to 2.5 mm broad; perigynia contracted to beak, 2-ribbed, otherwise nerveless; staminate spike several-flowered, to about 8 mm long . . . .*C. capillaris*

1 Leaves strongly channelled, less than 1 mm broad; perigynia tapering to beak, 2-ribbed and slenderly few-nerved; staminate spike few-flowered, to 6 mm long . . . .*C. williamsii*

Section *CAPITATAE* (*see* p. 340)

A single species; (transcontinental) . . . .*C. capitata*

Section *CHORDORRHIZEAE* (*see* p. 342)

A single species; (transcontinental) . . . .*C. chordorrhiza*

Section *CIRCINATAE* (*see* p. 341)

1 Plants with tough, slender, long-creeping, very scaly, light yellowish-brown rootstocks; culms to about 4 dm tall; leaves flat, to 2.5 mm broad; perigynia yellowish green, to 4 mm long, the beak to 1 mm long; pistillate scales tinged with chestnut-brown; (Alaska–s Yukon–s Dist. Mackenzie–n B.C.) . . . .*C. anthoxanthea*

1 Plants densely caespitose, the rootstocks very short; culms commonly less than 2 dm tall; leaves involute, about 0.5 mm broad; perigynia straw-colour, to 6 mm long, the beak to 1.5 mm long; pistillate scales red-tinged; (s Alaska–Yukon–B.C.) . . . .*C. circinata*

Section *CRYPTOCARPAE* (see p. 342)

1 Culms in stools, the rootstocks short; perigynia membranaceous, essentially nerveless; pistillate scales tinged with reddish-brown, at least the lower ones with long rough awns; leaves to about 12 mm broad, their ligules much longer than broad; lower leaf-sheaths usually prominently filamentose; (fresh habitats from Man. to Nfld. and N.S.) . . . .*C. crinita*
1 Culms solitary or few from each crown, from long horizontal leafless stolons or long-creeping rootstocks.
  2 Perigynia shining, nerveless, brown in age, coriaceous; lower leaf-sheaths of sterile culms strongly filamentose ventrally; ligules longer than broad; (coastal B.C.) . . . .
  . . . .*C. obnupta*
  2 Perigynia dull green to light brown, smooth or minutely granular to strongly papillate; lower leaf-sheaths of sterile culms rarely noticeably filamentose; ligules as broad as or broader than long; (plants of salt-marshes or tidal flats).
    3 Pistillate scales long-awned; perigynia essentially nerveless, submembra- naceous; pistillate spikes drooping; leaves to about 12 mm broad; (coasts of Hudson Bay–James Bay and the Atlantic Ocean) . . . .*C. paleacea*
    3 Pistillate scales obtuse to merely subulate-tipped (rarely short-awned); perigynia firmer in texture.
      4 Pistillate spikes slender-peduncled, loosely spreading or drooping (a variety of *C. salina* will also key out here), to 8 cm long, their scales castaneous or dark brown to dark purple or blackish, 2 or 3 times as long as the perigynia; perigynia more or less strongly nerved on both faces; leaves firm, to 12 mm broad, commonly about half as long as the culms; (ssp. *cryptocarpa* in the Aleutian Is., Alaska, and B.C.; the typical form in ᴇ Que., s Labrador, and s Greenland) . . . .*C. lyngbyei*
      4 Pistillate spikes subsessile or strongly ascending to erect on short straight peduncles, their scales straw-colour or ferrugineous to (more rarely) dark purple; perigynia with only a few faint nerves at base.
        5 Spikes cylindric to clavate-cylindric, to 7 cm long and up to 7 in number; pistillate scales lanceolate, subulate-tipped, with a 3-nerved centre, 2 or 3 times as long as the perigynia; leaves to 9 mm broad, usually equalling the culm; (transcontinental) . . . .*C. salina*
        5 Spikes oblong, at most 2.5 cm long and 5 in number; pistillate scales ovate, obtuse to acute, less than twice as long as the perigynia; leaves at most 3 mm broad.
          6 Spikes few-flowered, to 1 cm long, their scales 3-nerved at centre, the lowest one enlarged at base and spathiform, equalling the inflores- cence; leaves 1 or 2 mm broad, equalling the culm; culms rarely over 1.5 dm tall, arching (transcontinental) . . . .*C. subspathacea*
          6 Spikes dense and many-flowered, to 2.5 cm long, their scales 1-nerved, the lowest one foliaceous and overtopping the inflorescence; leaves 2 or 3 mm broad, shorter than the culm; culms to 5 dm tall, erect; (coastal Alaska) . . . .*C. ramenskii*

Section *DEWEYANAE* (see p. 343)

1 Leaves to about 2.5 mm broad, scarcely glaucous; culms very rough above; perigynia narrowly lanceolate, at most 1.3 mm broad, slenderly but strongly few-nerved on both faces; pistillate scales obtusish to acuminate; spikes lance-cylindric, bractless or very short-bracted; (s Ont. to N.S.) . . . .*C. bromoides*
1 Leaves to 5 mm broad, light green or yellowish green and more or less glaucous; culms smooth or somewhat roughened above; perigynia lanceolate to ovate; pistillate scales obtuse to awned; spikes thicker, the lowest one subtended by a bract sometimes equalling or surpassing the inflorescence.
  2 Rootstocks rarely elongate, the culms densely caespitose; perigynia rounded at base, to 5.5 mm long, many-nerved dorsally at base, nerveless ventrally, shallowly

bidentate, the upper part of the body covered by the translucent scale; achenes to 2.5 mm long; (transcontinental) . . . .*C. deweyana*
2   Rootstocks slender, elongate; perigynia tapering to a stipitate base, rarely over 4.5 mm long, several-nerved dorsally on the lower half, nerveless or nearly so ventrally; achenes less than 2 mm long; (B.C.).
    3   Perigynia shallowly bidentate, the upper part of their bodies usually not covered by the scales, these never tinged with red-brown; spikes ovoid or oblong . . . . .*C. leptopoda*
    3   Perigynia deeply bidentate, the upper part of their bodies covered by the scales, these usually tinged with red-brown; spikes linear-oblong . . . .*C. bolanderi*

Section *DIGITATAE* (see p. 346)

1   Terminal spike pistillate at base, staminate above; capillary peduncles of lower (and often basal) spikes much longer than the spike; pistillate scales cuspidate or awned; perigynia nerveless, glabrous or slightly pubescent; leaves mostly basal, to 5 mm broad, firm; culms to 3 dm tall, from stout branching rootstocks; plant forming purple-based lax mats; (transcontinental) . . . .*C. pedunculata*
1   Terminal spike staminate throughout; basal or prolonged peduncles wanting, the pistillate spikes sessile or short-peduncled; pistillate scales blunt.
    2   Perigynia glabrous, 2-keeled, otherwise nerveless, distinctly beaked; leaves channelled above, at most 1.5 mm broad; densely tufted plant with smooth obtusely angled culms, these at most 1.5 dm tall; (transcontinental) . . . .*C. glacialis*
    2   Perigynia pubescent; leaves flat or somewhat channelled, to 4 mm broad; plants loosely caespitose, the slender rootstocks often very elongate.
        3   Staminate spike to 6 mm long, often sessile; pistillate spikes less than 1 cm long, their scales about half as long as the 2-ribbed and several-nerved perigynia; inflorescence subtended by a greenish or brownish tubular sheath; leaves to 3 mm broad; culms smooth except at summit, to 2 dm tall; (transcontinental) . . . . .*C. concinna*
        3   Staminate spike to over 2 cm long, sessile or short-peduncled; perigynia 2-ribbed, otherwise nerveless or obscurely nerved; leaves to 4 mm broad; culms to 3.5 dm tall, with purplish-brown sheaths at base, these with or without short green blades.
            4   Perigynia loosely pubescent, broader and longer than the ciliate scales; staminate spike sessile or nearly so; pistillate spikes rarely over 1 cm long or with more than 10 flowers; stigmas normally 4; bracts very short-sheathing; culms smooth or nearly so; (s B.C. and sw Alta.) . . . .*C. concinnoides*
            4   Perigynia appressed-pubescent, narrower and shorter than the smooth scales; staminate spike usually noticeably peduncled; pistillate spikes to about 2 cm long and with up to about 25 flowers; stigmas 3; bracts long-sheathing and spathe-like; culms strongly scabrous above; (B.C. to James Bay) . . . .*C. richardsonii*

Section *DIOICAE* (see p. 340)

1   Stems usually filiform and more or less terete, obscurely furrowed; perigynia finally widely divergent to reflexed, abruptly contracted to a short scabrous beak; (transcontinental) . . . .*C. gynocrates*
1   Stems evidently 3-angled and distinctly furrowed; perigynia ascending to weakly spreading, gradually tapering to a smooth beak; (E Greenland) . . . .*C. parallela*

Section *DIVISAE* (see p. 343)

1   Culms obtusely triangular, smooth, to about 3 dm tall, from slender brownish-scaly rhizomes.
    2   Plant usually unisexual, to 3 dm tall, the slender but stiff, smooth, obtusely triangular culms from slender elongate brownish-scaly rhizomes; staminate heads to 4 cm long, the anthers to 5 mm long; pistillate heads to 5 cm long, the many spikes closely aggregated but usually readily distinguishable; perigynia ovate-lanceolate, about 4 mm long, rather strongly nerved dorsally, lightly nerved ventrally, the beak more than

half as long as the body; leaves to 2.5 mm broad, usually involute; (B.C. to s Man.) . . . .
. . . .*C. douglasii*
2 Plants commonly monoecious, the spikes androgynous; perigynia finely nerved
dorsally, essentially nerveless ventrally; heads compact, at most 2 cm long, the few
to several spikes not readily distinguishable; (B.C. to Man.) . . . .*C. stenophylla*
1 Culms sharply triangular, scabrous above; heads relatively loose, the numerous spikes
rather readily distinguishable, usually androgynous (or individual plants sometimes
unisexual); perigynia finely nerved dorsally, essentially nerveless ventrally; leaf-blades
flat or channelled.
3 Perigynia chestnut-coloured, less than 3 mm long, unequally biconvex, the beak at
most 1/3 as long as the body; head rarely over 2.5 cm long; leaves to 4 mm broad;
lower leaf-sheaths light brown; culms to about 5 dm tall, from slender rhizomes;
(Alta. and Sask.) . . . .*C. simulata*
3 Perigynia becoming blackish, to 4.5 mm long, plano-convex, the beak to about half as
long as the body; lower leaf-sheaths dark brown to black; leaves rarely over 3 mm
broad; rhizomes stout.
4 Scales very dark chestnut-brown, shining; perigynia glossy, scarcely hyaline at
the orifice; culms to about 3 dm tall; (w B.C.) . . . .*C. arenicola*
4 Scales lighter and dull; perigynia dull, strongly hyaline at the orifice; culms to over
7 dm tall; (Yukon–B.C. to Ont.) . . . .*C. praegracilis*

Section *EXTENSAE* (see p. 347)

1 Plants loosely tufted, with short decumbent leafy-tipped stolons; all bracts long-
sheathing, with erect blades; perigynia ascending; staminate spike peduncled; (E Que.,
St-Pierre and Miquelon, and Nfld.) . . . .*C. hostiana*
1 Plants densely cespitose; all but the lowest bracts sheathless or nearly so, with divergent
to reflexed blades; perigynia soon spreading or reflexed.
2 Perigynia to 7 mm long, their beaks half as long as to equalling the body, commonly
strongly recurved; culms to about 7 dm tall, usually much surpassing the leaves.
3 Staminate spike usually peduncled; lowest bract less than twice as long as the
head; perigynium-beak minutely serrulate; pistillate scales blunt; leaves glaucous
or blue-green; culms obtusely angled; (Ont. to Nfld. and N.S.) . . . .*C. lepidocarpa*
3 Staminate spike sessile or short-peduncled; lowest bract to 4 times as long as the
head; pistillate scales acuminate; perigynium-beak smooth; leaves yellow-green;
culms acutely angled; (transcontinental) . . . .*C. flava*
2 Perigynia less than 4 mm long, their straightish beaks shorter than the body; culms to
about 3.5 dm tall, obtusely angled.
4 Plant clear green; leaves to 3 mm broad; terminal spike entirely staminate;
pistillate spikes either crowded at summit of culm or in two distinct groups;
(introd. in E ?Canada) . . . .[*C. serotina*]
4 Plant yellowish green; leaves rarely over 2 mm broad (sometimes broader
following a second flowering); terminal spike sessile and gynaecandrous, the
basal staminate flowers sometimes scarcely visible; pistillate spikes clustered
around the terminal one.
5 Leaves to 4.5 mm broad, about as long as the arching or somewhat decumbent
culm; upper leaves and lowest bract with prolonged and convex summits;
perigynium-beak not much shorter than the body; (Ont. to Nfld. and N.S.) . . . .
. . . .*C. demissa*
5 Leaves at most about 2 mm broad, commonly surpassing the erect culm; upper
leaves and lowest bract with concave or truncate summits; perigynium-beak
about one-third as long as the body; (transcontinental) . . . .*C. viridula*

Section *FERRUGINEAE* (see pp. 342, 347)

1 Terminal spike normally androgynous, the upper lateral spike sessile and either
staminate, androgynous, or pistillate, the lower 1 or 2 spikes pistillate or androgynous,

drooping on capillary peduncles; slender horizontal stolons present; perigynia finely many-nerved; (Alaska–Yukon–Dist. Mackenzie and mts. of B.C. and Alta.; E Que. and Nfld.) . . . . .C. petricosa
1 Terminal spike staminate or gynaecandrous, the lateral ones pistillate; plants densely caespitose or with short ascending stolons, the culms much surpassing the leaves.
    2 Perigynia narrowly lanceolate, about 1 mm broad, the upper half very long-tapering; terminal spike gynaecandrous; lowest bract with a well-developed blade; (transcontinental) . . . . .C. misandra
    2 Perigynia broader in outline and over 1 mm broad, rather abruptly short-beaked; terminal spike gynaecandrous or staminate.
        3 Lateral spikes ovoid or short-oblong; lowest bract with a short or rudimentary blade; plants loosely caespitose; (transcontinental) . . . . .C. atrofusca
        3 Lateral spikes linear-oblong to oblong; lowest bract often with a well-developed blade; plants densely caespitose.
            4 Pistillate spikes linear-oblong to narrowly oblong, to about 3 cm long, their scales tinged with purplish black, the midvein not extending to the apex; (B.C.) . . . . . .
            . . . . .C. ablata
            4 Pistillate spikes oblong, to about 2 cm long, their scales reddish brown, the midvein extending nearly or quite to the apex; (?B.C.) . . . .[C. luzulina]

Section FILIFOLIAE (see p. 341)

A single species; (B.C. to Man.) . . . . .C. filifolia

Section FIRMICULMES (see p. 342)

A single species; (s B.C. and sw Alta.) . . . . .C. geyeri

Section FOETIDAE (see p. 343)

1 Heads ellipsoid to thick-cylindric; scales and flattish perigynia appressed-ascending; leaves scabrous toward tip . . . . .C. langeana
1 Heads globose-ovoid to subglobose; scales and slightly inflated perigynia soon divergent; leaves smooth or barely scabrous near tip; (transcontinental along coasts; mts. of B.C. and sw Alta.) . . . . .C. maritima

Section FOLLICULATAE (see p. 346)

1 Leaves to over 1.5 cm broad; sheaths of bracts prolonged at summit; perigynia lance-conic, to 1.5 cm long, at most only slightly surpassing their commonly cuspidate or awned scales; culms robust, to over 1 m tall; (Ont. to Nfld. and N.S.) . . . . .C. folliculata
1 Leaves mostly not over 3.5 mm broad; sheaths of bracts concave at summit; perigynia lance-subulate, at most about 13 mm long, more than twice as long as their awnless scales; culms slender, at most 6 dm tall; (Ont. to s Labrador, Nfld., and N.S.) . . . .
. . . . .C. michauxiana

Section GRACILLIMAE (see pp. 345, 347)

1 Leaves pilose beneath, to about 8 mm broad, their sheaths pilose; terminal spike nearly always pistillate at summit; perigynia sharply several-nerved.
    2 Pistillate scales shorter than the obscurely nerved perigynia, obtuse, acute, or merely short-pointed; lateral spikes with a few sterile scales at base, not over 3 cm long, mostly much shorter than the recurving filiform peduncles; (s Ont. and sw Que.) . . . .
    . . . . .C. formosa
    2 Pistillate scales about equalling or surpassing the several-nerved perigynia, narrowed into a prominent awn; lateral spikes fertile to base, to 4.5 cm long, subsessile or short-peduncled; (sw Que.) . . . . .C. davisii

1   Leaves and sheaths glabrous, or the latter with a few scattered hairs.
    3   Pistillate spikes 3 mm thick or more, their scales acute or short-pointed; terminal spike pistillate at summit or staminate throughout; perigynia 3 or 4 mm long, acutely trigonous, otherwise nerveless, gradually tapering into a definite curved beak; lowest bract with a short pale sheath; leaves pale green, at most 6 mm broad; plant brown at base; (s Ont. and s Que.) . . . .*C. prasina*
    3   Pistillate spikes not over 3 mm thick, their scales obtuse or short-pointed; terminal spike nearly always pistillate at summit; perigynia sharply several-nerved, beakless; lowest bract with a tubular sheath; leaves deep green; plant reddish purple at base; (SE Man. to Nfld. and N.S.) . . . .*C. gracillima*

Section *GRANULARES* (see p. 348)

1   Peduncle of staminate spike much surpassing the uppermost pistillate spike; perigynia nearly nerveless or few-nerved; leaves glaucous, often folded, to 4 mm broad; plant loosely stoloniferous, the culms mostly solitary; (transcontinental) . . . .*C. crawei*
1   Peduncle of staminate spike none or shorter than the uppermost pistillate spike; perigynia usually strongly nerved; leaves warm green to glaucous, flat, to over 1 cm broad; culms in tussocks from very short rhizomes; (SE Sask. to N.S.) . . . .*C. granularis*

Section *HELEONASTES* (see pp. 340, 343)

1   Spike solitary, globose, commonly 5 or 6 mm long, staminate at base; perigynia about 2 mm long, obscurely nerved, nearly beakless, surpassing their ovate obtuse scales; culms densely caespitose, rarely over 6 dm tall, only slightly longer than the involute-filiform leaves; (transcontinental along arctic and subarctic coasts) . . . .*C. ursina*
1   Spikes 2 or more; culms mostly taller.
    2   Spikes staminate at summit, their bases with at most 3 finely nerved, biconvex, minutely beaked perigynia less than 3 mm long (or the terminal spike with up to 6 perigynia); heads much interrupted, to about 2.5 cm long; leaves mostly less than 2 mm broad; plant loosely tufted or slightiy stoloniferous; (transcontinental) . . . . .*C. disperma*
    2   Spikes pistillate at summit, staminate at base; perigynia plano-convex.
        3   Inflorescence compact, usually even the lowest spikes more or less overlapping; (transcontinental).
            4   Spikes very crowded, to 10 or more, even the lowest ones very slightly separated and hiding the rachis; perigynia finely nerved on both faces, about equalling or slightly longer than their subtending scales; leaves pale green, soft and flat.
                5   Head silvery, commonly about 1 cm long (but up to 2 cm); perigynia elliptic-oval, broadest near the middle, to 3.5 mm long, nearly beakless, densely white-puncticulate; leaves to 2 mm broad; plant loosely tufted in large clumps, from very slender elongate rhizomes, the culms to about 6 dm tall . . . .*C. tenuiflora*
                5   Head green, becoming brown, to 3 cm long; leaves to 4 mm broad; culms very soft, in loose stools from very short rhizomes . . . .*C. arcta*
            4   Spikes less crowded, rarely more than 4, at least the lowest ones well separated (but usually somewhat overlapping), the rachis plainly visible between the spikes.
                6   Perigynia ellipsoid, faintly nerved, densely punctate, yellowish brown, tapering to a conical but scarcely beaked apex; plant loosely caespitose; (arctic and subarctic regions) . . . .*C. amblyorhyncha*
                6   Perigynia lanceolate, distinctly (but often minutely) beaked.
                     7   Perigynia essentially nerveless, golden- to reddish-brown; leaves to 3 mm broad; culms obtusely angled and smooth except at summit; (fresh habitats) . . . .*C. lachenalii*
                     7   Perigynia finely but distinctly nerved, greyish- to yellowish-brown;

leaves at most 2 mm broad; culms rather sharply angled and more or less scabrous except below.

8 Culms very scabrous at least below the head; leaves to 2 mm broad; perigynia mostly pale yellowish-brown, about 3 mm long; usually all of the spikes gynaecandrous; (fresh habitats) ....*C. heleonastes*

8 Culms generally smooth (or slightly scabrous below the head); leaves rarely over 1.5 mm broad; perigynia drab to pale brown; usually only the terminal spike gynaecandrous; (saline or brackish shores) ....*C. glareosa*

3 Inflorescence interrupted, only the upper spikes overlapping, the lower ones well separated or remote, the spikes usually more than 4 in number.

9 Lowest bract bristle-form, many times surpassing the subtended spike; spikes 1–3, all distant; perigynia finely many-nerved, beaked, to 4 mm long; plant long-stoloniferous; (transcontinental) ....*C. trisperma*

9 Lowest bract much shorter or wanting; upper spikes contiguous.

10 Perigynia beakless or minutely beaked, strongly many-nerved dorsally, more lightly nerved ventrally, about twice as long as their hyaline scales; plants long-stoloniferous.

11 Perigynia less than 2 mm long; spikes to about 7 in number in a head to 4 cm long; lowest bract to 3 cm long; leaves to 3 mm broad; (Alaska to Dist. Mackenzie) ....*C. bonanzensis*

11 Perigynia to 3 mm long; spikes rarely more than 5 in a head to 2.5 cm long; lowest bract less than 1 cm long; leaves to 2 mm broad; (B.C. to Ont.) ....*C. loliacea*

10 Perigynia distinctly beaked; (transcontinental).

12 Culms smooth, at most 4.5 dm tall, from slender rhizomes and stolons; perigynia faintly nerved, to 3.3 mm long, substipitate, about equalled and closely covered by their reddish-brown scales; spikes up to 6 in a head to about 5 cm long, the terminal spike with a prolonged staminate base; leaves yellowish- or glaucous-green, rarely over 2.5 mm broad; (plant of saline or brackish habitats) ....*C. mackenziei*

12 Culms scabrous above, to over 1 m tall, densely caespitose, the rhizomes usually very short; perigynia longer than the white-hyaline to silvery-brown scales; terminal spike not much prolonged at base.

13 Spikes greenish to brown, 5–10-flowered; perigynia finally loosely spreading, faintly nerved dorsally, essentially nerveless ventrally, to 2.7 mm long, the distinct beak minutely serrulate; leaves green, at most 2.5 mm broad ....*C. brunnescens*

13 Spikes whitish to silvery-brown, 10–30-flowered; perigynia ascending, obscurely to distinctly nerved on both faces; leaves glaucous, to 4 mm broad ....*C. canescens*

Section *HIRTAE* (see pp. 345, 346)

1 Leaf-sheaths and backs of scales pilose; perigynia strongly nerved, their teeth 1 or 2 mm long, rigid, the beak more than half as long as the strongly 15–20-ribbed body; leaves soft and flat, to 4 mm broad, mostly sparingly hairy; culm obtusely angled, nearly smooth, to 6 dm tall; (introd. in Que., P.E.I., and N.S.).....................................*C. hirta*

1 Leaf-sheaths (and blades) and backs of scales glabrous; teeth of perigynium-beak less than 1 mm long, the beak less than half as long as the body; culms to over 1 m tall.

2 Perigynia to 7 mm long, strongly 15–20-nerved, their beaks about half as long as the body; leaves to about 8 mm broad; culms sharply angled and stiff, scabrous; (dry sandy habitats from Alta. to Nfld. and N.S.) ....*C. houghtonii*

2 Perigynia mostly not over 5 mm long, their obscure nerves largely concealed beneath the dense pubescence.

3 Beak of perigynium soft and hyaline-tipped, marked with purple on the back, only obscurely bidentate or finally split; lower bract rarely reaching the usually solitary,

Cyperaceae

clavate, subsessile or very short-peduncled staminate spike; leaves firm, harsh, to 5 mm broad, with revolute margins; culms sharply angled; plant of dry sandy habitats; (s ?Ont.) . . . .[C. vestita]

3 Beak of perigynium firm, with stiff sharp teeth; lower bract often overtopping the 1–3, long-peduncled staminate spikes; (transcontinental in damp to wet habitats).
4 Leaves filiform-convolute except at base, rarely over 2 mm broad, smooth and wiry; culms obtusely angled and smooth except sometimes at tip; (var. americana) . . . . .C. lasiocarpa
4 Leaves flat, to 5 mm broad, scabrous, their margins revolute; culms sharply triangular and scabrous above . . . . .C. lanuginosa

Section *INFLATAE* (see p. 341)

A single species; (sw B.C.) . . . . .C. engelmannii

Section *LAMPROCHLAENAE* (see p. 344)

A single species; (transcontinental) . . . . .C. supina

Section *LAXIFLORAE* (see p. 348)

1 Sheaths of culm spathiform and bladeless or with rudimentary blades, reddish purple; leaves of the sterile shoots to over 2.5 cm broad, their sheaths reddish purple; staminate spike reddish purple; perigynia sharply angled, lightly many-nerved, to 5 mm long, sparsely hispidulous, the beak erect or outwardly curving; scales white-hyaline, pointed, somewhat shorter than the perigynia; (Ont. to N.S.) . . . . .C. plantaginea
1 Sheaths blade-bearing; staminate spike whitish to purplish-brown or deep brown.
2 Upper cauline leaf and lowest bract shorter than to at most 2(3) times as long as their sheaths; leaves of sterile shoots to 3 cm broad, those of the culms much shorter and rarely over 5 mm broad; perigynia sharply angled, finely many-nerved, more or less hispidulous, short-stipitate, short-beaked, their scales acute to acuminate; (Ont. and s Que.).
3 Perigynia olivaceous, 5 mm long or more; peduncles of lower pistillate spikes to 5 cm long; staminate spike deep brown; basal leaves bright green, purplish at base, less than 2 cm broad; culms ascending, to 8 dm tall . . . . .C. careyana
3 Perigynia green, less than 5 mm long; peduncles of lower pistillate spikes shorter than the spikes; staminate spike pale brown; basal leaves glaucous, pale-based, to 3 cm broad; culms weak, flexuous or loosely spreading, mostly less than 3 dm tall . . . . .C. platyphylla
2 Upper cauline leaf and lowest bract many times longer than their sheaths; leaves of sterile shoots not much more than twice as broad as the culm-leaves; bract-sheaths green (leaf-sheaths sometimes purplish).
4 Perigynia acutely or subacutely angled, with nearly flat sides, hispidulous, finely many-nerved, to 4 mm long, short-stipitate, short-beaked, their pale-hyaline scales acute to acuminate; (s Ont. and sw Que.).
5 Leaves green, at most 5 mm broad; lateral spikes pistillate to base, to 3 cm long and 4 mm thick, ascending, the upper one often subsessile . . . . .C. digitalis
5 Leaves pale green or glaucous, to 12 mm broad; lateral spikes mostly with 1 or 2 staminate flowers or empty scales at base, to 2 cm long and 6 mm thick, all slender-peduncled, the lower ones sometimes loosely spreading or drooping on long filiform peduncles . . . . .C. laxiculmis
4 Perigynia obtusely angled at least below the middle or barely angled (the sides convex), glabrous, usually more prominently stipitate, their obtuse to rounded or truncate scales often cuspidate or awned.
6 Perigynia nerveless or obscurely nerved, to 4 mm long, nearly symmetrical, the straightish beak nearly as long as the slender stipe; staminate spike sessile or short-peduncled, partly hidden by the rather crowded upper pistillate spikes;

362

pistillate spikes to 3 cm long and 4 mm thick; leaves at most 1 cm broad; (Ont. to s Labrador, Nfld., and N.S.) ....*C. leptonervia*

6 Perigynia strongly many-nerved.

    7 Staminate spike usually subsessile or short-peduncled, partly hidden by the rather crowded upper pistillate spikes; (s Man. to N.S.) ....*C. laxiflora*

    7 Staminate spike usually long-peduncled, its base usually overtopping the uppermost pistillate spike; perigynia strongly outward-curving; upper bract rarely overtopping the inflorescence.

        8 Perigynia about 3 mm long, their short beaks very abruptly bent; sterile shoots forming conspicuous culms, their leaves at most about 8 mm broad; fertile culms purple-tinged at base, their leaves at most about 5 mm broad; (var. *gracillima*; s Ont. and sw Que.) ....*C. laxiflora*

        8 Perigynia 4 mm long or more, their relatively long beaks straight or slightly curved; sterile shoots forming mere tufts of leaves, these to 1.5 cm broad; fertile culms green or brown-tinged at base.

            9 Perigynia rather sharply angled above, 5 or 6 mm long, substipitate, up to 12 in a spike; culm-leaves to 1 cm broad; (s B.C.) ....*C. hendersonii*

            9 Perigynia more obtusely angled above, 4 or 5 mm long, more strongly stipitate, up to 20 in a spike; culm-leaves to 7 mm broad; (var. *angustifolia*; s Ont. and N.S.) ....*C. laxiflora*

Section *LIMOSAE* (see p. 347)

1 Lowest bract with a tubiform green sheath to 2 cm long; scales obtuse (or the lower ones subacute); perigynia to 3.8 mm long, slender-nerved; leaves 1 or 2 mm broad; culms to 4 dm tall; (Alaska and NW Dist. Mackenzie) ....*C. laxa*

1 Lowest bract sheathless or very short-sheathing and spathiform.

    2 Pistillate scales with the whitish midrib excurrent as a very slender rough bristle-like awn to 12 mm long, the ovate-oblong body black, with slightly hyaline margins; perigynia to 6 mm long, rather obscurely several-nerved; leaves to 5 mm broad; culms from densely matted tough scaly rhizomes; (Alaska–Yukon–B.C.) ....*C. macrochaeta*

    2 Pistillate scales at most long-acuminate; perigynia usually more distinctly nerved, their scales ovate to suborbicular.

        3 Pistillate scales long-acuminate; leaves to 3 mm broad; plants loosely caespitose, from much branching, slender, short to elongate rhizomes; (transcontinental) ....
....*C. paupercula*

        3 Pistillate scales obtuse to acute or abruptly short-cuspidate; plants from loosely forking, slender, elongate rhizomes; (transcontinental).

            4 Culms obtusely triangular, smooth, to 3.5 dm tall; leaves green, to 2 mm broad, flat; perigynia inflated-triangular in cross-section, to 3.5 mm long, their scales purple-brown to blackish, obtuse to abruptly mucronate, soon deciduous ....
....*C. rariflora*

            4 Culms sharply triangular, to about 6 dm tall; perigynia to 4.5 mm long, their scales acutish to abruptly short-cuspidate, persistent ....*C. limosa*

Section *LONGIROSTRES* (see p. 347)

A single species; (B.C. to Que. and N.B.) ....*C. sprengelii*

Section *LUPULINAE* (see p. 346)

1 Achene subsessile, its style straight or loosely contorted above the middle; pistillate spikes subglobose to globose; perigynium-beak much shorter than the body; plants densely caespitose, lacking elongate rhizomes.

    2 Perigynia lance-rhomboid, cuneate at base, dull, firm, usually minutely hispid below the middle; pistillate spikes globose, 3 or 4 cm thick, with up to about 30 flowers;

leaves pale green or grey-green, to about 1.5 cm broad, firm and scabrous; (s Ont. and s Que.) ....*C. grayii*
  2 Perigynia lanceolate to ovoid, rounded at base, lustrous, softer, glabrous; pistillate spikes subglobose, to 3 cm thick, with at most about 15 flowers; leaves dark green, less than 1 cm broad, soft and lax; (Man. to Nfld. and N.S.) ....*C. intumescens*
1 Achene stipitate, its style strongly contorted just above the base; pistillate spikes short-cylindric, to 8 cm long and 3.5 cm thick; perigynium-beak equalling the body; plants loosely to densely caespitose, from creeping rhizomes.
  3 Achene ovoid, longer than wide, each side oval in shape and nearly flat; perigynia pale green to drab-brown, rather firm; leaves to about 12 mm broad; (Ont. to N.S.) ....*C. lupulina*
  3 Achene rhomboid, as wide as long, each side broadly diamond-shaped and distinctly concave; perigynia straw-colour to tawny, thin; staminate spike often peduncled; pistillate spikes sessile or short-peduncled; leaves to about 1.5 cm broad; (s Ont. and s Que.) ....*C. lupuliformis*

Section *MACROCEPHALAE* (see p. 344)

A single species; (s Alaska and coastal B.C.) ....*C. macrocephala*

Section *MONTANAE* (see pp. 344, 345)

1 Perigynia glabrous or essentially so, about 3 mm long; pistillate scales mostly long-tapering, about equalling to much longer than the perigynia; most of the spikes crowded about the bases of the firm and hard scabrous leaves, these to 5 mm broad; (Alta. to s Labrador and N.S.) ....*C. tonsa*
1 Perigynia distinctly pubescent, at least above.
  2 Culms of various lengths, the shorter ones crowded among the leaf-bases (elongate ones sometimes lacking), their spikes often entirely pistillate; leaves at most about 3.5 mm broad.
    3 Bract of lowest pistillate spike (on elongate culms) leaflet-like and normally exceeding the head (the staminate spike sessile or short-peduncled); remnants of old leaves soft and scarcely shredded; pistillate scales obtuse to acute or short-cuspidate, much shorter than the perigynia.
      4 Rhizomes slender, freely branching; culms normally loosely caespitose, curved or spreading, smooth except at tip; leaves soft and thin, to 2 mm broad; staminate spike to about 5 mm long, inconspicuous; (transcontinental) ...*C. deflexa*
      4 Rhizomes stout; culms mostly more densely caespitose; leaves thin but firm, to 2.5 mm broad; staminate spike to over 12 mm long; (B.C. to w Ont.) ....*C. rossii*
    3 Bract of lowest pistillate spike (on elongate culms) scale-like and shorter than the head (the staminate spike usually long-peduncled and to over 12 mm long); leaves thin but firm; culms scabrous above; (transcontinental) ....*C. umbellata*
  2 Culms all elongate, none of the spikes hidden among the leaf-bases nor entirely pistillate.
    5 Perigynium-body distinctly longer than thick, tightly investing the achene; plants loosely to densely caespitose.
      6 Lower pistillate spikes remote, short-peduncled, usually leafy-bracted, the head to 6 cm long; perigynia pale, less than 3 mm long; leaves soft, pale green, at most 2 mm broad; culms weak, to about 4 dm tall, shorter than or somewhat surpassing the leaves; (Ont. to Nfld. and N.S.) ....*C. novae-angliae*
      6 Lower pistillate spikes crowded or overlapping, short-bracted; (B.C. to N.S.) ....*C. nigromarginata*
    5 Perigynium-body about as thick as long, somewhat loose over the achene; staminate spike to 2 cm long.
      7 Plant densely caespitose, nonstoloniferous, the leaf-bases scarcely or only slightly fibrillose; perigynia to 4 mm long; leaves to about 7 mm broad; culms to about 6 dm tall; (s Man. to N.S.) ....*C. communis*

      7   Plant with elongate cord-like fibrillose stolons; leaves rarely over 3 mm broad, their bases reddish and usually with persistent brush-like tufts of fibres; culms mostly less than 4 dm tall.

         8   Mature perigynia about 1.5 mm broad, the body obtusely triangular in cross-section; leaves rather soft and only slightly scabrous; (B.C. to N.S.) . . . .*C. pensylvanica*

         8   Mature perigynia to 2 mm broad, the body suborbicular in cross-section.

            9   Pistillate spikes suborbicular, sessile or nearly so, their scales tawny or reddish brown; perigynia puberulent or short-pubescent; leaves relatively stiff; (var. *digyna*; B.C. to Ont.) . . . .*C. pensylvanica*

            9   Pistillate spikes oblong to suborbicular, the lowest ones peduncled, their scales brownish to dark reddish-brown; perigynia short-hispid; leaves thin but firm; (B.C.) . . . .*C. inops*

Section *MULTIFLORAE* (see p. 344)

1   Leaves mostly equalling or surpassing the culm; perigynia green to straw-colour, several-nerved dorsally, essentially nerveless ventrally, to 3 mm long, tapering to a distinctly bidentate beak up to 2/3 as long as the body, the upper ones surpassed by the long-awned scales; (transcontinental) . . . .*C. vulpinoidea*

1   Leaves mostly longer than the culm; perigynia not surpassed by their subtending scales, their obscurely notched beaks at most about 1/3 as long as the body.

    2   Scales strongly white-hyaline-margined, straw-colour to brownish, obtusish to awned; perigynia blackish at maturity, to 4 mm long, obscurely striate on both faces; (?B.C.) . . . .[*C. alma*]

    2   Scales not at all or but little hyaline-margined, reddish brown, rough-awned; perigynia yellowish, less than 3 mm long, nerveless or faintly (sometimes plainly) few-nerved dorsally, nerveless ventrally; (sw Que.) . . . .*C. annectens*

Section *NARDINAE* (see p. 340)

1   Perigynia about 2.5 mm long, with smooth beaks; spike subglobose to ellipsoid; leaves flattish and relatively coarse, their sheaths dull brown; (Alaska) . . . .*C. jacobi-peteri*

1   Perigynia 3 or 4 mm long, with scabrous beaks; spike ellipsoid or ovoid; leaves filiform, their sheaths pale reddish-brown; (transcontinental) . . . .*C. nardina*

Section *OBTUSATAE* (see p. 341)

A single species; (Alaska–Yukon–Dist. Mackenzie–B.C. to Man.) . . . .*C. obtusata*

Section *OLIGOCARPAE* (see p. 348)

1   Perigynia tight over the achene, definitely angled and more or less beaked, narrowed to base; pistillate spikes loose, 1–9-flowered; leaves at most 7 mm broad; (Ont. and s Que.).

    2   Sheaths normally glabrous; leaves mostly 2–4 mm broad; perigynia less than 4.5 mm long; achenes with a minute straight beak . . . .*C. oligocarpa*

    2   Sheaths minutely hispid; leaves to 7 mm broad; perigynia to nearly 6 mm long; achenes with a minute sharply bent beak . . . .*C. hitchcockiana*

1   Perigynia loose to close over the achene, empty above, obscurely angled, scarcely beaked, rounded at base; pistillate spikes dense and mostly many-flowered; leaf-sheaths glabrous.

    3   Peduncles, axis of inflorescence, margins of bract-sheaths, and usually the midvein of the pistillate scales scabrous; perigynia at most 4 mm long, sublustrous; leaves to 5 mm broad; (Ont. to Nfld. and N.S.) . . . .*C. conoidea*

    3   Peduncles, axis of inflorescence, margins of bract-sheaths, and usually the midvein of the pistillate scales smooth; perigynia opaque, to 5 mm long; staminate spike sessile or short-peduncled; leaves to 1 cm broad.

4 Leaves very glaucous, coriaceous and thick; bracts linear-lanceolate, their loose sheaths enlarged upward; pistillate spikes with up to about 60 flowers; (s Ont.) . . . .*C. flaccosperma*

4 Leaves warm green to slightly glaucous, weaker; bracts elongate-linear, with cylindric tight sheaths; pistillate spikes with up to about 20 flowers; (s Ont. to w N.B.) . . . .*C. amphibola*

Section *ORTHOCERATES* (see p. 341)

1 Perigynia 4 or 5 mm long, obscurely many-striate, with a slender process (the rachilla) projecting from the orifice and surpassing the stigmas; lowest leaf-sheaths blade-bearing; culms smooth, to about 2.5 dm tall . . . .*C. microglochin*

1 Perigynia about 7 mm long, finely many-striate, lacking a rachilla; lowest sheaths bladeless; culms scabrous above, to about 6 dm tall . . . .*C. pauciflora*

Section *OVALES* (see p. 343)

1 Head (the inflorescence) to about 8 cm long, more or less flexuous and often moniliform, at least the lowest spike not reaching the base of the next one above; bracts setaceous or wanting, rarely slightly surpassing the head.
 2 Pistillate scales equalling or surpassing the tips of the perigynia and nearly or completely covering their bodies; leaf-sheaths tight.
  3 Leaves stiff and very glaucous, to 4.5 mm broad, with a pair of rounded small knob-like auricles at base, the inner band of their sheaths very strongly prolonged; spikes distinctly clavate, silvery to pale brown; perigynia to 5 mm long, strongly nerved on both faces; (eastern coastal sands) . . . .*C. silicea*
  3 Leaves neither very stiff and glaucous nor auricled at base, the inner band of their sheaths less prolonged.
   4 Perigynia sharply nerved on both faces, about 4 mm long; spikes silvery-green to whitish brown; leaves to 5 mm broad; (Man. to s Labrador and N.S.) . . . .*C. argyrantha*
   4 Perigynia finely nerved dorsally, essentially nerveless ventrally; spikes brown to reddish brown; (transcontinental).
    5 Perigynia ovate, 4 or 5 mm long, their flattened beaks serrulate to tip; leaves to about 4 mm broad . . . .*C. aenea*
    5 Perigynia lanceolate, to 6 or 7 mm long, their subterete beaks smooth toward tip; leaves mostly less than 3.5 mm broad . . . .*C. praticola*
 2 Pistillate scales shorter than the perigynia and not concealing them above.
  6 Leaves stiff and very glaucous, to 4.5 mm broad, with a pair of small rounded knob-like auricles at base, the inner band of their sheaths strongly prolonged; spikes distinctly clavate, silvery to pale brown; perigynia to 5 mm long, strongly nerved on both faces; (eastern coastal sands) . . . .*C. silicea*
  6 Leaves neither very stiff and glaucous nor auricled at base, the inner band of their sheaths less prolonged.
   7 Perigynia less than 2 mm broad.
    8 Perigynia thin and scale-like except where slightly distended over the achene, to 5 mm long, sharply nerved dorsally, obscurely nerved ventrally; leaves to 8 mm broad, their sheaths loose.
     9 Inner band of leaf-sheaths white-hyaline, weak and friable; tips of perigynia loosely ascending to recurved; head interrupted; (Man. to Labrador and Nfld.) . . . .*C. projecta*
     9 Inner band of leaf-sheaths firm, green-striate; tips of perigynia closely appressed; head relatively compact; (Ont. to N.S.) . . . .*C. tribuloides*
    8 Perigynia thin but firm, clearly distended over the achene and scarcely scale-like; leaf-sheaths tight.
     10 Body of perigynium obovate to obovate-rotund, broadest near the abruptly short-beaked summit; perigynia to 4.5 mm long, nerved on

both faces; leaf-sheaths green-striate on the inner band; (N.B. and N.S.) . . . .
. . . .*C. albolutescens*

    10 Body of perigynium ovate, broadest near or below the middle;
          leaf-sheaths veinless on the inner band.

          11 Head nearly straight to arching, its spikes distinctly clavate;
               perigynia to 3.5 mm long, greenish to stramineous, finely nerved
               over the achene on both faces; leaves to 3.5 mm broad; (Ont.) . . . .
               . . . .*C. festucacea*

          11 Head commonly nodding above the lowest spike, its spikes not
               markedly clavate; perigynia finely nerved on both faces or nerveless
               ventrally; (B.C. to N.S.) . . . .*C. tenera*

  7 Perigynia 2 mm broad or more; leaf-sheaths close.

    12 Pistillate scales awn-tipped.

          13 Leaves scabrous-margined, to 5.5 mm broad; perigynia obovate to
               obovate-rotund, broadly winged, abruptly contracted to the beak,
               obscurely nerved on both faces or nerveless ventrally; (s ?Ont.) . . . .[*C. alata*]

          13 Leaves smooth except at tip, at most about 2.5 mm broad; perigynia
               distinctly finely many-nerved on both faces.

               14 Perigynia lance-ovate, tapering gradually to the beak; spikes
                   rhomboid-elliptic, tapering gradually to summit and base, the
                   terminal one tapering gradually to the short staminate base; (E Que.
                   to Nfld. and N.S.) . . . .*C. hormathodes*

               14 Perigynia broadly ovate to subrotund, abruptly contracted to the
                   beak; spikes more rounded, the terminal one rounded to the pro-
                   longed staminate base . . . .[*C. straminea*]

    12 Pistillate scales blunt to acute or short-cuspidate but scarcely awned.

          15 Perigynia narrowly ovate, to 2.5 mm broad, lightly nerved dorsally,
               the inner face nerveless or barely nerved at the golden-yellow base;
               head to 5 cm long, the brownish rhomboid spikes acutish at both
               ends; leaves stiffish, light green, to 3 mm broad; culms wiry, to 6 dm
               tall; (s B.C. to s Man.) . . . .*C. xerantica*

          15 Perigynia broadly ovate or obovate to subrotund; head often longer, its
               spikes more rounded at both ends; leaves softer; culms weaker, to over
               1 m tall.

               16 Perigynia pale green to dull brown, to 2.5 mm broad, nerved on both
                 faces, their obovate to obovate-rotund bodies broadest above the
                 middle; leaves to 3.5 mm broad; (N.B. and N.S.) . . . .*C. albolutescens*

               16 Perigynia stramineous, to 3.5 mm broad, strongly nerved dorsally,
                 obscurely nerved ventrally (the translucent wing often with 1 or 2
                 nerves), its broadly ovate to ovate-rotund body broadest below the
                 middle; leaves to 4.5 mm broad; (B.C. to sw Que.) . . . . .*C. merritt-fernaldii*

1 Head more compact, usually less than 4 cm long, not moniliform, all of the spikes (except
sometimes the lowest) more or less overlapping.

  17 Bracts much surpassing the very dense head; leaf-sheaths tight.

    18 Bracts leaf-like, the lower ones to about 2 dm long; perigynia lance-subulate, thin,
        5 or 6 mm long, barely 1 mm broad, obscurely nerved on both faces, the slender
        beak longer than the body; scales greenish white, lance-acuminate, about half as
        long as the perigynia; leaves to 4 mm broad; culm smooth; (s Yukon–Dist.
        Mackenzie–B.C. to sw Que.) . . . .*C. sychnocephala*

    18 Bracts conspicuously surpassing the head but whitish-hyaline-margined at base
        and scarcely leaf-like; perigynia narrowly ovate, at most 5 mm long, the tawny- or
        reddish-brown-tipped beak shorter than the body; scales reddish brown, with
        hyaline margins; culms roughened below the head.

          19 Perigynia lightly nerved on both faces, the bidentate beak not hyaline at the
               orifice; pistillate scales narrowly ovate, acuminate to awned; head to 3 cm
               long; (B.C.) . . . .*C. unilateralis*

          19 Perigynia lightly nerved dorsally, nerveless ventrally, the shallowly bidentate

beak hyaline at the orifice; pistillate scales oblong-ovate, acute to short-cuspidate; head to 2 cm long; (B.C. to Sask.) . . . .*C. athrostachya*
17　Bracts setaceous or wanting, rarely somewhat surpassing the dense to rather loose head.
　　20　Head very dense, all of the spikes strongly overlapping; leaf-sheaths close.
　　　　21　Perigynium-beak flattened and margined at tip, serrulate to apex; perigynia finely nerved dorsally, essentially nerveless ventrally, their subtending scales brown or light reddish-brown; leaves to 5 mm broad; culms to about 9 dm tall; (transcontinental).
　　　　　　22　Pistillate scales shorter and narrower than the perigynia, these to 4 mm long and 2 mm broad, dull green or brownish; spikes to 9 mm long; head to 12 mm thick, the bracts all short-setaceous . . . .*C. bebbii*
　　　　　　22　Pistillate scales equalling the length and breadth of the perigynia and largely concealing them, the perigynia to 5 mm long and 3 mm broad, olive green to blackish, shining; spikes to 12 mm long; head to 1.5 cm thick, the lowest bract dilated at base and cuspidate-prolonged . . . .*C. adusta*
　　　　21　Perigynium-beak slender and nearly terete, scarcely margined at tip, the upper 1 or 2 mm little, if at all, serrulate; pistillate scales shorter and narrower than the perigynia, usually strongly dark-tinged; bracts mostly scale-like or the lowest setaceous and sometimes awned.
　　　　　　23　Perigynia much flattened, thin and scale-like except where distended by the achene.
　　　　　　　　24　Scales copper-brown to purple-black; perigynia copper-brown, to 4 mm long and 2 mm broad, lightly several-nerved dorsally, obscurely few-nerved toward base ventrally; leaves to 4.5 mm broad; rhizomes brownish; (B.C. to sw Sask.; Que. and Labrador) . . . .*C. macloviana*
　　　　　　　　24　Scales and perigynia not copper-tinged; rhizomes blackish.
　　　　　　　　　　25　Perigynia to 6 mm long and 2.75 mm broad, obscurely nerved dorsally, nerveless or nerved at base ventrally; scales brownish black or blackish; leaves to 4 mm broad; culms to about 4 dm tall; (B.C. and Alta.) . . . .*C. haydeniana*
　　　　　　　　　　25　Perigynia at most 5 mm long and 2 mm broad; culms to 1 m tall.
　　　　　　　　　　　　26　Perigynia ovate, strongly margined, appressed, lightly nerved dorsally, few-nerved toward base ventrally; scales dark-chestnut to brownish-black; leaves to 6 mm broad; (var. *microptera*; B.C. to sw Sask.) . . . .*C. macloviana*
　　　　　　　　　　　　26　Perigynia lance-ovate, very narrowly margined, spreading-ascending, lightly several-nerved on both faces; scales dull brown; leaves to about 4.5 mm broad; (var. *microptera*; Yukon–B.C. to Sask.) . . . .*C. macloviana*
　　　　　　23　Perigynia plano-convex, thicker.
　　　　　　　　27　Perigynia at most 3.5 mm long and 1.5 mm broad.
　　　　　　　　　　28　Perigynia (and scales) blackish, very obscurely nerved, the margins of the beak entire or nearly so; leaves to 3 mm broad; culms to about 3.5 dm tall; (s B.C.) . . . .*C. illota*
　　　　　　　　　　28　Perigynia greenish (becoming stramineous), finely many-nerved dorsally, several-nerved toward base or nearly nerveless ventrally, the beak-margins strongly serrulate; scales reddish brown; leaves to 3.5 mm broad; culms to about 7 dm tall; (sw ?B.C.) . . . .*C. subfusca*
　　　　　　　　27　Perigynia to 4 mm long and 2 mm broad; leaves to 4.5 mm broad.
　　　　　　　　　　29　Perigynia copper-brown at maturity, submembranaceous, lightly several-nerved dorsally, obscurely few-nerved toward base ventrally; (B.C. to sw Sask.; Que. and Labrador) . . . .*C. macloviana*
　　　　　　　　　　29　Perigynia deep green in anthesis, tinged with yellowish brown in age, obscurely several-nerved dorsally, nerveless or sometimes obscurely few nerved toward base ventrally; culms to about 8 dm tall; (B.C.–Alta.) . . . .*C. preslii*

20 Head more open, the lowest spike sometimes not reaching the base of the next one above.

30 Pistillate scales nearly equalling or slightly surpassing the perigynia, about as broad as the perigynia and more or less concealing them; leaf-sheaths tight.

31 Perigynium-beak flattened and margined at tip, serrulate to apex; perigynia ovate, to about 5 mm long and 3 mm broad; scales light reddish-brown.

32 Perigynia membranaceous, drab, lightly nerved dorsally, the inner face nerveless or barely nerved at the golden-yellow base; head to 1 cm thick; lowest bracts often prolonged but setaceous; leaves to 3 mm broad; (s B.C. to s Man.) .... *C. xerantica*

32 Perigynia coriaceous, olive-green or blackish, shining, strongly nerved dorsally, essentially nerveless ventrally; head to 1.5 cm thick; lowest bract dilated at base and cuspidate-prolonged; leaves to 5 mm broad; (transcontinental) .... *C. adusta*

31 Perigynium-beak slender and nearly terete, scarcely margined at tip, the upper 1 or 2 mm little if at all serrulate; perigynia at most 2.5 mm broad.

33 Perigynia to 8 mm long, oblong-lanceolate (broadest near the top of the achene), strongly but finely many-nerved on both faces; scales light reddish-brown; leaves to 3 mm broad; (B.C. to Sask.) .... *C. petasata*

33 Perigynia at most 6.5 mm long.

34 Perigynium-beak not white-hyaline at apex; perigynia to 5 mm long, strongly slender-nerved dorsally, their scales dull reddish-brown; head to 4 cm long; leaves to 4 mm broad.

35 Perigynia with bidentate beaks, light reddish-brown at tip, nerveless or lightly nerved ventrally; (introd. in the Atlantic Provinces) .... *C. leporina*

35 Perigynia with obliquely cut, minutely bidentate beaks, deep reddish-brown at tip, strongly nerved ventrally; (B.C.) .... *C. tracyi*

34 Perigynium-beak strongly white-hyaline at apex; perigynia finely nerved dorsally, nerveless or basally nerved ventrally.

36 Leaves to about 2 mm broad; head stiff, to about 2.5 cm long; perigynia to 6 mm long, their scales brownish or reddish-brown to brownish-black; culms stiff, to about 3 dm tall, in large stools, the rhizomes densely matted; (B.C.–Alta.) .... *C. phaeocephala*

36 Leaves to about 3.5 mm broad; heads and culms weaker.

37 Pistillate scales shining, dark chestnut-brown with silvery white hyaline margins; perigynia to over 6 mm long; head rather flexuous, to 4.5 cm long; (s B.C.) .... *C. piperi*

37 Pistillate scales dull reddish-brown; heads mostly shorter and more compact.

38 Perigynia to 6.5 mm long and 2 mm broad, subcoria-ceous, their scales with broad silvery white-hyaline margins; (transcontinental) .... *C. praticola*

38 Perigynia to 5 mm long and 1.5 mm broad, their scales with narrow white-hyaline margins; (B.C.–Alta.) . . .*C. platylepis*

30 Pistillate scales shorter than the perigynia and not completely concealing them.

39 Perigynium-beak slender and nearly terete, scarcely margined at the entire or only slightly serrulate tip; pistillate scales reddish brown with lighter centre and narrow hyaline margins; leaves to 4 mm broad, their sheaths tight; culms to 8 dm tall.

40 Perigynia to 4 mm long and 2 mm broad, obscurely several-nerved dorsally, nerveless or sometimes obscurely nerved toward base ventrally; (B.C.–Alta.) .... *C. preslii*

40 Perigynia to 3.5 mm long and 1.5 mm broad, finely many-nerved

dorsally, several-nerved toward base or nearly nerveless ventrally;
(sw ?B.C.) . . . .[C. subfusca]
39 Perigynium-beak flattened and margined at the serrulate tip; scales mostly
greenish white to light brown or reddish brown.
  41 Perigynia at most 2 mm broad.
    42 Leaf-sheaths loose (C. normalis may key out here).
      43 Leaves to 4 mm broad; sterile culms poorly developed, their
        erect or ascending leaves usually clustered near the top;
        perigynia to 3.5 mm long and 2 mm broad, several-nerved
        dorsally, obscurely few-nerved ventrally . . . .[C. straminea]
      43 Leaves to 8 mm broad; sterile culms strongly developed, their
        numerous leaves divergent, not clustered near the top;
        perigynia finely nerved on both faces, at most 1.5 mm broad.
        44 Perigynia lanceolate, thin and scale-like, barely distended
          over the achene, their tips closely appressed; (Ont. to N.S.) . . . . . .
          . . . . .C. tribuloides
        44 Perigynia ovate, firm and distinctly distended over the
          achene, to 4 mm long, their tips spreading or somewhat
          recurved at maturity; (Alta. to E Que.) . . . . .C. cristatella
    42 Leaf-sheaths tight (or rather loose in C. normalis); perigynia nerved
      on both faces; sterile culms often poorly developed, their erect or
      ascending leaves usually clustered near the top.
      45 Perigynia narrowly lanceolate to narrowly ovate; leaves to 4 mm
        broad; (transcontinental).
        46 Perigynia firm, relatively thick and obviously distended over
          the achene, narrowly lanceolate, about 1 mm broad . . . .
          . . . . .C. crawfordii
        46 Perigynia thin and scale-like, barely distended over the
          achene, lanceolate to narrowly ovate, to about 2.5 mm broad . . . . .
          . . . . .C. scoparia
      45 Perigynia ovate to suborbicular, firm and obviously distended
        over the achene.
        47 Leaves to 6.5 mm broad, their sheaths relatively loose;
          sterile culms frequent, with leaves clustered near the top;
          perigynia to 4 mm long, membranaceous, their beaks soon
          spreading or slightly recurving; scales tinged with yellowish
          brown; (Man. to SE James Bay) . . . . .C. normalis
        47 Leaves at most 4 mm broad, their sheaths tight; sterile leafy
          culms wanting or few; perigynia to 4 mm long, their beaks
          appressed; pistillate scales light reddish-brown to deep
          brown.
          48 Head to 4 cm long, its spikes rounded or somewhat
            tapering at base, the membranaceous perigynia to 5 mm
            long; (B.C. to P.E.I.) . . . . .C. tincta
          48 Head to 6 cm long, its spikes more clavate and gradually
            tapering to base, the subcoriaceous perigynia to 3.5 mm
            long; (Ont.) . . . . .C. festucacea
  41 Perigynia over 2 mm broad at maturity.
    49 Spikes of the head to 2.5 cm long, long-cylindric, acute; perigynia to
      1 cm long and 2.5 mm broad, narrowly lanceolate, thin and
      scale-like, finely nerved on both faces, about twice as long as the
      brownish scales; head to 8 cm long; leaves light green, firm, to 7
      mm broad, subcordate at the junction with the loose sheath; (SE
      Man. and s Ont.) . . . . .C. muskingumensis
    49 Spikes rarely over 1.5 cm long; perigynia, if more than 7 mm long,
      broadly ovate; leaf-sheaths tight (except in C. cumulata).
      50 Pistillate scales acuminate to awn-tipped.

51 Leaves scabrous-margined, to 5.5 mm broad; perigynia obovate to obovate-subrotund, broadly winged, abruptly contracted to the beak, obscurely nerved on both faces or nerveless ventrally; (s ?Ont.) . . . .[C. alata]

51 Leaves smooth except at tip, at most about 2.5 mm broad; perigynia distinctly finely many-nerved on both faces.

  52 Perigynia lance-ovate, tapering gradually to the beak; spikes rhomboid-elliptic, tapering gradually to summit and base, the terminal one tapering gradually to the short staminate base; (E Que. to Nfld. and N.S.) . . . .
. . . . .C. hormathodes

  52 Perigynia broadly ovate to subrotund, abruptly contracted to the beak; spikes more rounded, the terminal one rounded to the prolonged staminate base . . . . .C. straminea

50 Pistillate scales obtusish to merely acute or cuspidate.

  53 Perigynia elliptic to rhombic, obovate, or subrotund, broadest at or above the middle, to 4.5 mm long.

    54 Leaves to 6 mm broad, their sheaths loose; perigynia rhombic-ovate to subrotund, to about 3.5 mm broad, finely many-nerved dorsally, nerveless ventrally, the short broad beak tapering gradually into the body; (s Man. to N.S.) . . . . .C. cumulata

    54 Leaves to 3.5 mm broad, their sheaths tight; perigynia to 2.5 mm broad, finely nerved on both faces.

      55 Perigynium-body obovate to suborbicular, the beak rather abruptly narrowed above and markedly differentiated; heads erect or arching; leaves to 3.5 mm broad, relatively soft; culms smooth except at tip; (N.B. and N.S.) . . . . .C. albolutescens

      55 Perigynium-body elliptic to rhombic, tapering gradually to the short broad beak; heads stiffly erect; leaves to 3 mm broad, firmer and harsher; culms stiffer and stouter, more scabrous toward summit; (s Ont.) . . .C. longii

  53 Perigynium-body lanceolate to ovate or ovate-rotund, broadest below the middle.

    56 Perigynia lanceolate to narrowly ovate, to 7 mm long and about 2.5 mm broad, thin and scale-like, barely distended over the achene, many-nerved dorsally, several-nerved ventrally; leaves to 3 mm broad; (transcontinental) . . . . .
. . . . .C. scoparia

    56 Perigynia ovate to ovate-rotund.

      57 Leaves mostly 2 or 3 mm broad, their very long sheaths green-striate ventrally to mouth, not white-hyaline; perigynia to 5 mm long and 2.8 mm broad, faintly nerved dorsally, nerveless or nearly so ventrally, thin and scale-like, barely distended over the achene; (s ?Ont.) . . . .[C. suberecta]

      57 Leaves to over 3.5 mm broad, their sheaths strongly white-hyaline ventrally; perigynia more strongly nerved.

        58 Perigynia to about 7.5 mm long and 4.5 mm broad, membranaceous and thin, translucent, ovate to ovate-rotund, strongly nerved on both faces; (s Man. to sw Que.) . . . . .C. bicknellii

        58 Perigynia at most 5.5 mm long and 3.5 mm broad,

firm and opaque, well distended over the achene, strongly nerved.

59 Scales acuminate, reaching well up the perigynium-beak; perigynia broadly ovate to suborbicular, to 3.5 mm broad, subcoriaceous, becoming pale brown, their wings rarely nerved; spikes conical to slightly rounded at summit, more or less clavate at base; (B.C. to sw Que.) . . . .*C. brevior*

59 Scales mostly blunter and often reaching only to the base of the beak; spikes broadly rounded at summit and base, the lateral ones scarcely clavate at base.

60 Perigynia broadly ovate to subrotund, to 3.5 mm broad, stramineous, membranaceous, the translucent wings often with 1 or 2 nerves; leaves to about 4.5 mm broad; (s B.C. to sw Que.) . . . .*C. merritt-fernaldii*

60 Perigynia ovate, to 3 mm broad, pale green, somewhat firmer, their translucent narrow wings rarely nerved; leaves to about 3.5 mm broad; (?Sask. to s ?Ont.) . . . .*C. molesta*

Section *PALUDOSAE* (*see* p. 346)

1   Perigynia densely short-pubescent, prominently many-nerved, to over 1 cm long, longer than the acute scales, the beak with pubescent teeth to 2.5 mm long; leaves glabrous except for the scabrous margins, at most 8 mm broad; culms to over 1 m tall; (s Ont. and s Que.) . . . .*C. trichocarpa*

1   Perigynia glabrous.

2   Leaves to 12 mm broad, pilose beneath, their sheaths pilose; perigynia strongly nerved, lanceolate to lance-ovate, to 12 mm long, about equalled by their awn-tipped scales, their smooth, mostly recurving teeth to 3 mm long; culms to over 1 m tall; (B.C. to Que.) . . . .*C. atherodes*

2   Leaves and sheaths glabrous.

3   Teeth of perigynia usually 1 or 2 mm long; perigynia strongly nerved, ovoid, about 7 mm long; pistillate scales obtuse to acuminate, usually much shorter than the perigynia; leaves mostly not over 6 mm broad; culms to 8 dm tall; (Sask.–Man.) . . . .*C. laeviconica*

3   Teeth of perigynia at most 1 mm long.

4   Perigynia to 4 mm long, compressed-trigonous, dull and papillate, finely nerved, in slender spikes 6 or 7 mm thick and to over 5 cm long; pistillate scales purplish, rough-awned, longer than the perigynia; leaves glaucous, to 8 mm broad; culms to over 1 m tall; (s ?Ont.) . . . .[*C. acutiformis*]

4   Perigynia to 8 mm long, subterete or obscurely angled, smooth and more or less lustrous, in oblong to thick-cylindric spikes.

5   Spikes at most about 2 cm long and about 6 mm thick; perigynia 4 or 5 mm long, impressed-nerved, not much longer than their deep purple-red, lance-acuminate scales; leaves rigid, 2 or 3 mm broad; culm to 6 dm tall; (introd. near Montreal, Que.) . . . .*C. nutans*

5   Spikes to 1 dm long and 1.5 cm thick; perigynia to 8 mm long; leaves glaucous, to 1.5 cm broad; culm to over 1 m tall.

6   Mature perigynia distinctly many-nerved, mostly much longer than their blunt to awned scales; lower sheaths purple, bladeless, soon fibrillose on the inner side; leaves often prominently septate-nodulose; (Alta. to Nfld. and N.S.) . . . .*C. lacustris*

6   Mature perigynia obscurely impressed-nerved, about equalled by the

scale-awns; lower sheaths whitish or brownish, blade-bearing, usually
not fibrillose; leaves scarcely nodulose; (s Ont.) ....*C. hyalinolepis*

Section *PANICEAE* (see p. 348)

1   Perigynia with a long slender bidentate beak, nerveless or faintly nerved at maturity, to 5
    mm long; pistillate spikes loosely 3–20-flowered, all peduncled, their acute scales shorter
    than the perigynia; leaves to 5 mm broad; (transcontinental) ....*C. vaginata*
1   Perigynia beakless or with a very short oblique tip; pistillate spikes subsessile or more or
    less peduncled, generally more densely flowered.
    2   Leaves white-glaucous, at most 3.5 mm broad, soon folded or involute; perigynia
        finely nerved; (transcontinental) ....*C. livida*
    2   Leaves green or only slightly glaucous, flat (or revolute in age).
        3   Culms smooth throughout; peduncles smooth; leaves mostly basal, to 6 mm
            broad; perigynia obscurely striate, to 5 mm long; (introd. in s Ont., N.B., N.S.,
            and Nfld.) ....*C. panicea*
        3   Culms scabrous above.
            4   Lower (purple) sheaths largely bladeless; leaves thin, green, to 4 mm broad;
                pistillate spikes loosely alternate-flowered; perigynia lightly nerved, about
                4 mm long; (s Man. to Ont.) ....*C. woodii*
            4   Lower sheaths mostly blade-bearing; pistillate spikes denser.
                5   Leaves greyish and stiffish, to 7 mm broad; lower pistillate spikes to 1 cm
                    thick; perigynia strongly nerved, abruptly contracted to the short beak;
                    (Sask. to s Ont.) ....*C. meadii*
                5   Leaves green and softer, at most 5 mm broad; pistillate spikes at most
                    about 5 mm thick; perigynia finely many-nerved, strongly tapering to tip;
                    (s Man. and s Ont.) ....*C. tetanica*

Section *PANICULATAE* (see p. 344)

1   Inner band of leaf-sheath pale or only slightly copper-tinged; heads straight, usually
    compact, mostly not over 5 cm long, dark brown; perigynia dark brown, lustrous,
    soon wide-spreading, not wholly covered by the scales; leaves to 3 mm broad;
    (transcontinental) ....*C. diandra*
1   Inner band of leaf-sheath strongly yellowish brown or copper-tinged at summit; heads
    commonly somewhat flexuous and interrupted, to about 1 dm long; perigynia nearly
    concealed by the scales.
    2   Leaves to 3 mm broad, their ligules very short, their sheaths convex at summit;
        perigynia brownish at maturity, their scales tinged with reddish brown; (trans-
        continental) ....*C. prairea*
    2   Leaves to 6 mm broad, their ligules longer than broad, their sheaths concave at
        summit; perigynia brownish black at maturity, their scales tinged with chestnut-brown;
        (B.C.) ....*C. cusickii*

Section *PENDULINAE* (see pp. 345, 347)

A single species; (introd. in s Ont., sw Que., and N.S.) ....*C. flacca*

Section *PHYLLOSTACHYAE* (see p. 344)

1   Pistillate scales nearly all foliaceous and green throughout, mostly hiding the faintly
    nerved perigynia, the lowest to 6 mm broad; staminate scales about 3; leaves to 6 mm
    broad; (B.C. to N.B.) ....*C. backii*
1   Pistillate scales with hyaline margins, only the lowest 1 or 2 prolonged and foliaceous,
    these 1 or 2 mm broad, narrower than the essentially nerveless perigynia; leaves to 4 mm
    broad.

2   Perigynium-body oblong, tapering into a stout triangular serrate beak; staminate scales rarely more than 12, tapering to summit; (s Ont. and s Que.) . . . .*C. willdenowii*
2   Perigynium-body subglobose, abruptly contracted to a rather slender serrulate beak; staminate scales to about 20, truncate; (s Ont.) . . . .*C. jamesii*

Section *POLYTRICHOIDEA* (*see* p. 341)

A single species; (transcontinental) . . . .*C. leptalea*

· [Section *PRAECOCES*] (*see* p. 345)

A single species; (introd. in s ?Ont.) . . . .[*C. caryophyllea*]

Section *PSEUDO-CYPEREAE* (*see* p. 346)

1   Perigynia coriaceous, not inflated, somewhat compressed and 2-edged, at least the lower ones soon reflexed; leaves firm, strongly septate-nodose, to 1.5 cm broad, their ligules prolonged; culms in clumps.
    2   Perigynia to 5 mm long, their nearly straight teeth at most 1 mm long; pistillate spikes not much over 1 cm thick; (Alta. to Nfld. and N.S.) . . . .*C. pseudo-cyperus*
    2   Perigynia to 7 mm long, their curved-divergent teeth to 2 mm long; pistillate spikes to over 1.5 cm thick; (B.C. to N.S.) . . . .*C. comosa*
1   Perigynia membranous, inflated, subterete, not reflexed; leaves less conspicuously nodose, at most 1 cm broad, their ligules about as broad as long.
    3   Culms loosely caespitose, smooth, soft and easily compressed; pistillate spikes slenderly cylindric, at most 1.5 cm thick; perigynia to 7 mm long; staminate scales acute or cuspidate, smooth; (s Ont.) . . . .*C. schweinitzii*
    3   Culms in clumps, firm, sharply angled; pistillate spikes thick-cylindric to subglobose, to 2 cm thick; staminate scales rough-awned.
        4   Perigynia 15–20-nerved, ovoid, 6 or 7 mm long; leaves firm, to 1.5 cm broad; culm sharply angled, strongly scabrous above, to 1 m tall; (transcontinental) . . . .*C. hystricina*
        4   Perigynia 8–10-nerved; leaves weaker; (Ont. to N.S.).
            5   Pistillate spikes to over 7 cm long and mostly over 1.5 cm thick; perigynia to 9 mm long; leaves to 7 mm broad; culms rather obtusely angled, smooth . . . . .*C. lurida*
            5   Pistillate spikes at most about 4 cm long and 13 mm thick; perigynia about 6 mm long; leaves to about 4 mm broad; culms more sharply angled, more or less scabrous above . . . .*C. baileyi*

Section *RUPESTRES* (*see* p. 342)

A single species; (transcontinental) . . . .*C. rupestris*

Section *SCIRPINEAE* (*see* p. 341)

A single species; (transcontinental) . . . .*C. scirpoidea*

Section *SHORTIANAE* (*see* p. 347)

A single species; (s Ont.) . . . .*C. shortiana*

Section *SQUARROSAE* (*see* pp. 340, 346)

1   Culms obtusely angled, about equalled in length by the ribbon-like bracts; leaves deep green, to 1 cm broad; terminal spike usually staminate throughout (sometimes pistillate at summit); pistillate spikes cylindric, to 4 cm long and 12 mm thick, their rough-awned scales much surpassing the wide-spreading perigynia, these with 2 strong ribs and about 9 fine sharp nerves; (s Ont.) . . . .*C. frankii*

1   Culms acutely angled, much longer than the bracts; terminal spike pistillate except at base, to over 2 cm thick, the scales much shorter than the perigynia, these with 2 strong ribs, otherwise essentially nerveless; (Ont. and s Que.).

    2   Spikes solitary or at most rarely 4, ellipsoid to subglobose, the perigynia horizontally spreading to reflexed; pistillate scales sharp-pointed or awned; leaves warm green, to 6 mm broad . . . .*C. squarrosa*

    2   Spikes 1–6, subcylindric, the perigynia more or less ascending; pistillate scales blunt; leaves grey-green, to 1 cm broad . . . .*C. typhina*

Section STELLULATAE (see pp. 340, 343)

1   Spike solitary, either unisexual, or pistillate above and with a staminate base; leaves filiform-involute, wiry; perigynia about 3 mm long, distinctly few-nerved dorsally, obscurely nerved ventrally; (Ont. to Labrador, Nfld., and N.S.) . . . .*C. exilis*

1   Spikes normally 2 or more; leaves not wiry.

    2   Perigynia obscurely nerved or nerveless ventrally, commonly distinctly nerved dorsally, about 3 mm long.

        3   Perigynium-beak only minutely notched (the teeth usually less than 0.25 mm long); scales tawny or brown, about 1/2 as long as the perigynium-body; leaves to 3 mm broad; (transcontinental) . . . .*C. interior*

        3   Perigynium-beak sharply bidentate.

            4   Culms at most 1.5 mm thick at base; scales about 1/2 as long as the perigynia, these slender-beaked, faintly nerved dorsally; leaves less than 3 mm broad; (transcontinental) . . . .*C. muricata*

            4   Culms to 3.5 mm thick at base; scales about 2/3 as long as the perigynia, these tapering gradually to a broad beak, strongly nerved dorsally; leaves to 5 mm broad; (Ont. to Nfld. and N.S.) . . . .*C. wiegandii*

    2   Perigynia lightly to strongly nerved on both faces.

        5   Perigynia oblong-obovoid, to 4 mm long, broadest near the middle, their reddish-brown-tipped beaks only obscurely serrulate; scales yellowish brown with a green 3-nerved centre, shorter than the perigynia; leaves weak, flat, to 2 mm broad; (Alaska–B.C.) . . . .*C. laeviculmis*

        5   Perigynia lance-ovate to ovate-rotund, broadest near the base, their beaks more strongly serrulate.

            6   Perigynium-beak only minutely notched (the teeth commonly less than 0.25 mm long); perigynia to 5 mm long; scales whitish; leaves at most about 1 mm broad, soon convolute; (Maritime Provinces) . . . .*C. howei*

            6   Perigynium-beak sharply bidentate.

                7   Midrib of pistillate scales green, prominent nearly or quite to the sharp tip.

                    8   Perigynia broadly ovate to suborbicular, to 3 mm long and 2 mm broad; (Maritime Provinces and St-Pierre and Miquelon) . . . .*C. atlantica*

                    8   Perigynia lance-ovate to ovate (about twice as long as broad), to 4 mm long; (transcontinental) . . . .*C. muricata*

                7   Midrib of scales not reaching the flat hyaline tip.

                    9   Perigynia about 4 mm long and 1.5 mm broad, their ovate, obtuse, chestnut-brown, hyaline-margined scales about 2/3 as long as the perigynium-body; leaves flat or somewhat channelled, rather stiff, to 2.75 mm broad; (Alaska–B.C.) . . . .*C. phyllomanica*

                  9   Perigynia at most 3.3 mm long.

                      10 Culms to 3.5 mm thick at base; scales whitish to pale brown, obtuse to subacute, about 1/2 as long as the perigynia, these to 2.5 mm broad; leaves to 4 mm broad; (Maritime Provinces and St-Pierre and Miquelon) . . . .*C. atlantica*

                      10 Culms at most 1.5 mm thick at base; pistillate scales dark brown, the middle and upper ones accuminate and nearly as long as the perigynia, these about 1.5 mm broad; leaves to about 2.5 mm

broad, channelled or involute; (var. *sterilis*; Sask. to Nfld. and N.S.) ......
....*C. muricata*

Section *SYLVATICAE* (see pp. 345, 348)

1   Leaves and culms pilose, the leaves to 7.5 mm broad; pistillate spikes oblong-cylindric, to
    8 mm thick, at most about 2.5 mm long; perigynia glabrous, about 5 mm long, subsessile,
    finely several-nerved, the beak about half as long as the body; pistillate scales brown;
    (sw Man. to Nfld. and N.S.)..*C. castanea*
1   Leaves and culms glabrous; pistillate spikes linear-cylindric, rarely to 5 mm thick;
    perigynia noticeably nerved only toward base, their scales whitish to light brown.
    2   Pistillate spikes few-flowered, to 3 cm long; perigynia densely short-hispid, to 8 mm
        long, short-stipitate, obscurely nerved at base, the beak to nearly or quite as long as
        the body; achene sessile; (s Sask. and s Man.) ....*C. assiniboinensis*
    2   Pistillate spikes many-flowered, to over 6 cm long; perigynia glabrous, their beaks
        less than 1/2 the length of the body; (Ont. to Nfld. and N.S.).
        3   Pistillate scales awned or cuspidate; perigynia at most about 5 mm long, strongly
            stipitate, definitely 3-angled, prominently nerved at base; achene sessile, longer
            than the style; leaves of the sterile shoots to 1 cm broad, the culm-leaves mostly
            not over 5 mm broad ....*C. arctata*
        3   Pistillate scales rarely awned; perigynia to 7 mm long, subsessile, obscurely
            3-angled, faintly nerved at base; achene stipitate, shorter than the style; basal
            leaves less than 5 mm broad ....*C. debilis*

Section *TRIQUETRAE* (see p. 345)

A single species; (s Ont. to N.B. and N.S.) ....*C. hirtifolia*

Section *VESICARIAE* (see pp. 342, 346)

1   Stigmas normally 2 and the achenes lenticular; perigynia essentially nerveless, the very
    short beak emarginate or minutely toothed.
    2   Staminate spikes usually 2; lowermost pistillate spike long-peduncled and drooping;
        mature perigynia dark brown and shining, distinctly beaked, to 5 mm long; leaves to
        4 mm broad, revolute-margined; rhizomes elongate; (Alaska and B.C. to nw Sask.) ....
        ....*C. physocarpa*
    2   Staminate spike usually solitary; pistillate spikes short-peduncled, rarely drooping;
        perigynia dull, short-beaked; rhizomes shorter; (Dist. Mackenzie; Alta. to Nfld. and
        N.S.) ....*C. saxatilis*
1   Stigmas 3; achenes trigonous.
    3   Perigynia horizontally spreading or reflexed at maturity, to 1 cm long, much inflated,
        prominently nerved, the beak half as long as the body or longer, with slender stiff
        teeth; pistillate spikes thick-cylindric, peduncled; lower bracts many times longer than
        the inflorescence; leaves to 1 cm broad, septate-nodulose, dark green, their sheaths
        rather loose; (transcontinental) ....*C. retrorsa*
    3   Perigynia ascending to spreading-ascending (or squarrose at anthesis in *C.
        rhynchophysa*); leaf-sheaths mostly closer.
        4   Achene strongly indented on one side; perigynia thin and very papery, strongly
            inflated and nerved, to 1 cm long, mostly more than 5 mm thick, the sharply
            toothed beak about 1/3 as long as the body; pistillate spikes thick-cylindric; lowest
            bract up to several times longer than the inflorescence; leaves to 5 mm broad,
            green, not conspicuously nodulose; (Ont. to N.S.) ....*C. tuckermanii*
        4   Achene symmetrical; perigynia papery to firm, to 8 mm long and 5 mm thick;
            lowest bracts shorter than to usually not more than twice the length of the
            inflorescence.
            5   Culms thickish and spongy at base, mostly smooth and obtusely angled
                (except at summit); leaves strongly nodulose when dry, pale green or

glaucous, to 1.5 cm broad; perigynia strongly several-nerved, the beak about 1/3 as long as the body; staminate spikes 2 or more.

  6 Perigynia strongly spreading or squarrose at anthesis, much inflated, membranaceous, the body broadly obovoid or suborbicular; culms sharply triangular and scabrous at summit; ligules acute; (Alaska–Yukon–Dist. Mackenzie) . . . .*C. rhynchophysa*

  6 Perigynia ascending at anthesis, less inflated, firmer, the body broadly ovoid to oval-ovoid; culms bluntly triangular and smooth or only slightly scabrous at summit; ligules rounded at apex; (transcontinental) . . . . . .*C. rostrata*

5 Culms slender, more acutely angled and often harsh above; leaves not strongly nodulose; pistillate spikes 1–3, sessile or short-peduncled.

  7 Plants in dense tussocks, lacking prolonged stolons; culms sharply angled, usually harsh above, to about 1 m tall; perigynia rather soft, strongly nerved, with a distinct smooth beak.

    8 Perigynia to 1 cm long, tapering gradually to the beak, the teeth to 1.5 mm long; leaves to 7 mm broad; ligules about as broad as long; (B.C. to Sask.) . . . .*C. exsiccata*

    8 Perigynia commonly not over 8 mm long, more abruptly contracted to the beak, the teeth at most 1 mm long; ligules longer than broad; (transcontinental) . . . .*C. vesicaria*

  7 Plants strongly stoloniferous.

    9 Culms obtusely angled and smooth, rarely over 4 dm tall; perigynia obscurely nerved, to 4 mm long, the beak very short; pistillate spikes to 3 cm long; (transcontinental in arctic and subarctic regions).

      10 Leaves to 3 mm broad, involute; perigynia brownish to purplish-black, rather firm; lowest bract strongly divergent or pendulous . . . .*C. rotundata*

      10 Leaves to 5 mm broad, flat, their margins revolute; perigynia purplish black, papery; lowest bract ascending . . . .*C. membranacea*

    9 Culms sharply angled, more or less scabrous above, to about 1 m tall; perigynia firm, strongly nerved; pistillate spikes thick-cylindric to subglobose.

      11 Pistillate spikes 3–15-flowered, to 2 cm long; perigynia at most 7 mm long, broadly ovoid, gradually contracted to the short beak; staminate spike 1; leaves filiform-involute and wiry; culms filiform, stiff; (s Dist. Mackenzie and Alta. to Labrador, Nfld., and N.S.) . . . .*C. oligosperma*

      11 Pistillate spikes 20–40-flowered, to 5 cm long; perigynia to 1 cm long, narrowly ovoid, the serrulate beak to 3 mm long; staminate spikes 1–3; leaves to 5 mm broad, flat or channelled; culms slender; (N.S.) . . . .*C. bullata*

Section *VIRESCENTES* (see p. 345)

1 Terminal spike staminate throughout; perigynia glabrous.

  2 Perigynia beakless, nearly or quite nerveless, about equalling their oblong-ovate, acute or cuspidate scales; leaves to 5 mm broad; (Ont. to Nfld. and N.S.) . . . . .*C. pallescens*

  2 Perigynia abruptly short-beaked, finely many-ribbed, surpassing their suborbicular scales; leaves to about 3 mm broad; (B.C. to Man.) . . . .*C. torreyi*

1 Terminal spike with a clavate staminate base below a long pistillate summit; perigynia beakless, prominently nerved, mostly surpassing their scales; leaves to about 4 mm broad.

  3 Perigynia glabrous; pistillate spikes to 7 mm thick, thick-cylindric to subglobose, sessile or nearly so; (s Ont. and s Que.) . . . .*C. hirsutella*

  3 Perigynia pubescent; pistillate spikes rarely over 5 mm thick.

    4 Perigynia slightly pilose, strongly ribbed on the back; pistillate spikes linear-cylindric, to 4 cm long; anthers to 2.5 mm long; (s Ont., s Que., and N.S.) . . .*C. virescens*

Cyperaceae

4 Perigynia copiously pilose, more finely nerved; pistillate spikes thick-cylindric to subglobose, less than 2 cm long; anthers rarely over 1.5 mm long; (s Ont. to N.B. and N.S.) ....*C. swanii*

Section *VULPINAE* (see p. 343)

1 Perigynia ovate, nerved dorsally, essentially nerveless ventrally, at most only slightly surpassing their scales.
  2 Inner band of leaf-sheath dotted, not cross-puckered; leaves mostly not over 6 mm broad; head usually less than 5.5 cm long; perigynia to 4 mm long, distinctly stipitate; scales becoming brownish; (Sask. to Ont. and s Que.) ....*C. alopecoidea*
  2 Inner band of leaf-sheath closely cross-puckered; leaves to 1 cm broad; head to 7.5 cm long, usually interrupted below; perigynia to 5 mm long, subsessile; scales whitish; (sw ?Man. and s ?Ont.) ....[*C. conjuncta*]
1 Perigynia lance-subulate to narrowly ovate above the spongy-thickened truncate base, distinctly nerved on both faces, mostly considerably surpassing the scales.
  3 Inner band of leaf-sheath commonly cross-puckered, friable, the unthickened summit prolonged; heads paniculate-compound, to about 1 dm long; perigynia to 5 mm long, brown-nerved; leaves to 8 mm broad; (transcontinental) ....*C. stipata*
  3 Inner band of leaf-sheath usually without cross-puckering, firm, its concave to obliquely subtruncate summit cartilaginous-thickened; heads less paniculate, to 6 cm long; perigynia to 7 mm long, with pale nerves; leaves rarely over 6 mm broad; (s Ont.) .....
....*C. laevivaginata*

NOTE

For ease of reference, numbers have been given (in brackets) following the names of *Carex* species corresponding, whenever possible, to those used by Mackenzie (1931; 1935). In some cases, the lack of treatment by Mackenzie results in the absence of a number. In other cases, the species is dealt with by Mackenzie under another name or as a variety, this being indicated following the number. Illustrations of most species are also given by Mackenzie (1940).

C. ablata Bailey (366)
/T/W/ (Grh) Bogs and meadows (chiefly alpine) from s B.C. (Vancouver Is. (type from Mt. Mark); Garibaldi; Manning Provincial Park; Hope-Princeton; Chilliwack, Skagit, and Tulameen valleys) to N Calif. and Utah. [*C. luzulina* var. *ab.* (Bailey) Hermann; merged with *C. luzulina* by Hitchcock et al. 1969].

C. acuta L. (451)
/aST/E/GEA/ (Grh) Moist rocks, gravels, and turf near the coast from N Ont. (Albany, sw James Bay at ca. 52°10′N; Dutilly, Lepage, and Duman 1954; reports from farther west refer chiefly or wholly to other members of the Section *Acutae*) to Que. (N to Great Whale R., SE Hudson Bay at ca. 55°15′N, and the Côte-Nord), s Labrador (N to the Dead Is. at 52°48′N; GH; an isolated station at ca. 57°N according to Hultén's map), Nfld., N.B., P.E.I., and N.S., s to Mass. and R.I.; southernmost Greenland; Iceland; Europe; w Asia. [Var. *nigra* L. (*C. nigra* (L.) Reichard); *C. goodenowii (goodenoughii)* Gay; *C. vulgaris* Fries and its var. *strictiformis* Bailey (*C. nigra* var. *str.* (Bailey) Fern.)]. MAPS (*C. nigra*): Hultén 1958: map 107, p. 127; Meusel, Jaeger and Weinert 1965:70.

[*C. acutiformis* Ehrh.] (498)
[Eurasian; the report from s Ont. by Stroud (1941; Wellington Co.) requires confirmation. It is known definitely in N. America only in saline marshes of E Mass.]

C. adelostoma Krecz.
/S/(X)/EA/ (Grh) Wet or marshy places: Alaska (Alaska Range dist.; Ernest Lepage, Nat. can. (Que.) 36:70. 1959); Dist. Mackenzie (Great Bear L.; CAN); s Dist. Keewatin (Yathkyed L.; CAN) and N Man. (Churchill; CAN); N Ungava (Chimo, s Ungava Bay; CAN) and N Labrador (between

57°30′ and 58°13′N; *see* E Canada map by Raymond 1950*a*: map 5 (*C. adel.* and *C. morrisseyi*), p. 442); Iceland; Fennoscandia; cent. Asia (L. Baikal). [*C. buxbaumii* var. *alpicola* Hartm.; *C. morrisseyi* Porsild]. MAPS: Hultén 1968*b*:257, and 1958: map 18, p. 37; Raymond 1951: map 17, p. 11; Tolmachev 1952: map 22 (very inaccurate for N. America).

C. adusta Boott (192)
/T/X/ (Hs) Dry acid soils from B.C. (N to Hazelton, 55°15′N; reports from Dist. Mackenzie require confirmation) to Alta. (N to near Whitecourt), Sask. (N to Lac la Ronge), Man. (N to Wekusko L., NE of The Pas), Ont. (N to Sachigo L. at ca. 54°N, 93°W), Que. (N to the L. St. John R. at 50°36′N and the Gaspé Pen.), Nfld., N.B., and N.S. (not known from P.E.I.), s to Minn., Mich., N.Y., and Maine. [*C. albolutescens* var. *glomerata* Bailey; *C. pinguis* Bailey]. MAP (NE area): Marcel Raymond, Ann. ACFAS 19:89. 1953.

C. aenea Fern. (193)
/ST/X/ (Hs) Dry ground and open woods, the aggregate species from cent. Alaska–Yukon to L. Athabasca (Alta. and Sask.), Man. (N to Reindeer L. at 57°53′N), Ont. (N to the Severn R. at ca. 55°45′N), Que. (N to the Kaniapiskau R. at ca. 57°30′N), Labrador (N to the Hamilton R. basin), Nfld., N.B. (type from Kent Co.), P.E.I., and N.S., s to Mont., Idaho, S.Dak., Minn., Pa., and New Eng. MAP and synonymy: *see* below.
1 Perigynia elliptic-lanceolate, rather strongly nerved ventrally; [Ont.: reported from Isle St. Ignace, L. Superior, by F.K. Butters and E.C. Abbe, Rhodora 55 (652):130. 1953, and from w James Bay at 51°34′N by Dutilly, Lepage, and Duman 1954] . . . .
. . . . .f. *flumini-regalis* Butt. & Abbe
1 Perigynia ovate.
  2 Leaves at most 2.5 mm broad; perigynia rather strongly nerved ventrally; [reported from w James Bay, Ont., N to ca. 52°N, by Dutilly, Lepage, and Duman 1954] . . . .
  . . . . .f. *extrapolata* Butt. & Abbe
  2 Leaves to 4 mm broad; perigynia less strongly nerved ventrally; [*C. albolutescens* var. *sparsiflora* Olney, not *C. sparsiflora* Steud.; transcontinental; type from Kent Co., N.B.; MAP: Hultén 1968*b*:237] . . . . .f. *aenea*

[*C. alata* T. & G.] (185)
[Reports of this species of the E U.S.A. (N to Mich. and N.Y.) from s Man. and N.B. by John Macoun (1888; *C. straminea* var. *al.* Bailey) were later referred by him to *C. cumulata* (John Macoun 1890; *C. straminea* var. *cum.*, the relevant collection from Bass River, Kent Co., N.B., in CAN). The report from Lambton Co., s Ont., by Dodge (1915) requires confirmation, a search in Herb. MICH, where Dodge's main s Ont. collections are housed, failing to reveal the voucher-specimen]

C. albolutescens Schw. (184: *C. straminea sensu* Mackenzie 1931, not Willd.)
/T/EE/ (Hs) Swamps, bogs, and wet woods from s Mich. to N.B. (Kent Co.; NBM) and sw N.S. (Annapolis, Yarmouth, Shelburne, and Queens counties; not known from P.E.I.), s to Tex. and Fla. [*C. straminea* in part of Canadian reports in part, not Willd.].

C. albonigra Mack. (430)
/ST/W/ (Grh) Dry tundra and mountain slopes from Alaska (N to 69°22′N; CAN), s Yukon, and Dist. Mackenzie (Mackenzie R. Delta; Great Bear L.; Brintnell L.) through B.C. and sw Alta. (N to Jasper National Park) to Calif., Ariz., and Colo. MAPS: Hultén 1968*b*:259; Raymond 1951: map 12, p. 7; Raup 1947: pl. 18; D.F. Murray, Brittonia 21(1): fig. 4, p. 65. 1969.

[*C. alma* Bailey] (54)
[The report of this species of the w U.S.A. (Calif. to Nev. and Ariz.) from Yale, B.C., by John Macoun (1888, as *C. leiorhyncha* Mey.; taken up by Henry 1915) requires confirmation.]

C. alopecoidea Tuckerm. (79)
/T/EE/ (Hs) Calcareous meadows and moist thickets from Sask. (N to McKague, 52°37′N) to s Man. (N to about 50 mi N of Winnipeg), s Ont. (N to the Ottawa dist.), s Que. (N to St-Maurice Co. at L. St. Peter), and N New Eng., s to Iowa, Ind., and N.J.

C. amblyorhyncha Krecz. (88: *C. marina* Dewey; *C. heleonastes sensu* Mackenzie 1931, in part, not L. f.)
/aSs/X/GEA/ (Hsr) Wet boggy ground from the coasts of Alaska–Yukon–Dist. Mackenzie to Victoria Is., N Baffin Is., and N Que. (Diana Bay, ca. 61°N), s to s Alaska, Great Bear L., NE Man. (Churchill), and James Bay; w Greenland between ca. 67° and 73°N, E Greenland between ca. 70°30′ and 75°N; N Eurasia. MAPS: Hultén 1968*b*:239, and 1962: map 59, p. 69; Dutilly, Lepage, and Duman 1958: fig. 7. p. 63; Porsild 1957: map 81, p. 171; T.W. Böcher, Acta Arct. 5: fig. 12, p. 27. 1952 (*see* this paper for a discussion of the *C. amblyorhyncha-heleonastes* complex); Tolmachev 1952: maps 21 and 26.

According to G. Halliday and A.O. Chater (Feddes Repert. 80(2–3):105. 1969), the name *C. marina* Dewey (in the past misapplied to *C. glareosa* var. *amphigena* Fern.) has priority over the name *C. amblyorhyncha* Krecz.

C. amphibola Steud. (312)
/T/EE/ (Hs) Dry woodlands and calcareous slopes from Minn. to s Ont. (Essex, Kent, Lambton, and Elgin counties), sw Que. (Kingsmere, St. Helen's Is., Mt. Johnson, and Missisquoi Co.), and w N.B. (Woodstock, Carleton Co.), s to E Tex. and Ga.

Most Canadian material appears to belong to the completely intergrading var. *turgida* Fern. (perigynia relatively strongly inflated, round-angled; staminate spike often partly hidden among the upper pistillate ones, these to 8 mm thick rather than at most 6 mm; *C. grisea* of Canadian reports, not Wahl.).

C. amplifolia Boott (398)
/T/W/ (Hsr) Wet soil from s B.C. (Vernon, Nelson, Okanagan L., and Kootenay L.; CAN; reported from Vancouver by Eastham 1947) to Calif. and Idaho.

C. annectens Bickn. (62)
/T/EE/ (Hs) Dry or moist, usually sterile and often sandy soil from Iowa to Wisc., s Ont. (Lambton Co.; Gaiser and Moore 1966), sw Que. (St. Helen's Is., near Montreal, and Soulanges, St-Jean, and Mississquoi counties), and cent. Maine, s to Kans., Mo., and Va. [Incl. the completely intergrading var. *xanthocarpa* (Bickn.) Wieg. (*C. xan.* Bickn., not Degl.; *C. brachyglossa* Mack.)].

C. anthoxanthea Presl (6)
/sT/W/ (Grh) Grassy banks (chiefly coastal) from the Aleutian Is., s-cent. Alaska, s Yukon (Whitehorse; CAN), and s Dist. Mackenzie (Taltson R. s of Great Slave L.) to sw B.C. (type from Nootka Sound, Vancouver Is.). [*C. leiocarpa* Mey.]. MAP: Hultén 1968*b*:225.

C. aperta Boott (453)
/T/W/ (Grh) Swampy meadows and low grounds from B.C. (N to Stuart L. at ca. 54°30′N; CAN) and sw Alta. (Waterton Lakes; Breitung 1957*b*) to N Oreg.–Idaho–Mont. [*C. turgidula* Bailey].

C. aquatilis Wahl. (457 and 458 (*C. substricta*))
/AST/X/GEA/ (Grh (Hsr)) Wet ground and shallow water, the aggregate species from the coasts of Alaska–Yukon–Dist. Mackenzie throughout the Canadian Arctic Archipelago to northernmost Ellesmere Is., northernmost Ungava–Labrador, Nfld., N.B., P.E.I., and N.S., s to Calif., N.Mex., Nebr., Mo., Ind., Ohio, N.Y., and N.J.; w E Greenland N of the Arctic Circle; Eurasia. MAPS and synonymy: *see* below.
1 Leaves light green, usually shorter than the obtusely triangular, slightly scabrous culms; sheath of lowest bract usually black-auricled; staminate spike usually solitary; [var. *?epigeios* Laest.; *C. stans* Drej.; *C. concolor* R. Br., not *sensu* Mackenzie 1935; transcontinental, N to northernmost Ellesmere Is.; Greenland; MAPS: Porsild 1957: map 86 (*C. stans*), p. 171; Hultén 1962: map 65, p. 75, and 1968*b*:251] . . . . .var. *stans* (Drej.) Boott
1 Leaves glaucous, usually somewhat overtopping the inflorescence; sheath of lowest bract not black-auricled; staminate spikes 2 or more.
   2 Culms generally acutely triangular and scabrous at summit; leaves to about 8 mm broad; [var. *substricta* Kük. (*C. sub.* (Kük.) Mack.); *C. variabilis* vars. *elatior* Bailey and *altior* Rydb.; *C. ?rousseauii* Raymond; *C. suksdorfii* Kük.; transcontinental, N in

Ungava to the Larch R. at ca. 57°35′N (*see* NE Canada map by Dutilly, Lepage, and Duman 1953: fig. 10, p. 42); MAP: on the above-noted 1962 map by Hultén] . . . .
. . . . .var. *altior* (Rydb.) Mack.

2   Culms obtusely triangular and smooth except sometimes at summit; leaves to about 5 mm broad; [*C. variabilis* Bailey; transcontinental but largely replaced northwards by var. *stans*; MAPS: Hultén 1962: map 65, p. 75, and 1968*b*:250; Raup 1947: pl. 19 (aggregate species)] . . . . .var. *aquatilis*

C. arcta Boott   (98)

/sT/X/   (Hs)   Wet woods, alluvial thickets, and shores from s Alaska–Yukon (CAN) and N Alta. (N to Wood Buffalo National Park at 59°30′N) to Sask. (N to Amisk L., near Flin Flon), Man. (N to the Seal R. near Churchill), Ont. (N to the Severn R. at ca. 55°30′N; CAN; type material from "Lake Superior, Rainy Lake, Lake of the Woods"), Que. (N to Ungava Bay, L. St. John, and the Gaspé Pen.), Labrador (N to Goose Bay, 53°18′N), and N.B. (not known from Nfld., P.E.I., or N.S.), s to Calif., Idaho, Minn., Mich., and New Eng. [*C. canescens* vars. *oregana* Bailey and *polystachya* Boott]. MAPS: Hultén 1968*b*:245; Raup 1930: map 34 (now requiring considerable expansion), p. 203.

C. arctata Boott   (343)

/sT/EE/   (Hs)   Dry woods and thickets from Ont. (N to Lac Seul, near Sioux Lookout at 50°19′N; CAN) to Que. (N to the Rupert R. at ca. 51°30′N, L. Mistassini, and the Côte-Nord), Nfld., N.B., P.E.I., and N.S., s to Minn., Ohio, and Conn. [Incl. var. *faxonii* Bailey]. MAPS: Meusel, Jaeger, and Weinert 1965:76; Raymond 1950*a*: map 9, p. 443, and 1951: map 27 (E Canada), p. 151.

C. arenicola Schmidt   (18: *C. pansa*)

/t/W/eA/   (Grh)   Coastal sands and rocky bluffs from w B.C. (reported from Graham Is., Queen Charlotte Is., by Calder and Taylor 1968, and from Hope Is., off the N tip of Vancouver Is., by Boivin 1967*a*) to N Calif.

The N. American plant has been somewhat arbitrarily distinguished from the Asiatic one as ssp. *pansa* (Bailey) Koyama & Calder (*C. pansa* Bailey; pistillate scales usually dark chestnut-brown and with conspicuous broad hyaline margins over 2 mm wide rather than light brown, the yellowish to white hyaline margins usually less than 2 mm wide; inflorescence usually relatively congested).

C. argyrantha Tuckerm.   (194: *C. foenea sensu* Mackenzie 1935, but not as to Willdenow's type, which is *C. siccata* Dew. of the present treatment; *see* H.K. Svenson, Rhodora 40:(477) 325–31. 1938)

/sT/EE/   (Hs)   Dry woods and thickets from Man. (N to Flin Flon; CAN; reports from farther west are probably based upon *C. aenea* or other closely related members of the critical Section *Ovales*) to Ont. (N to near Sachigo L. at ca. 53°50′N; CAN), Que. (N to St-Pacome, Kamouraska Co.; CAN), Labrador (Goose Bay, 53°18′N; DAO), N.B., P.E.I., and N.S., s to Minn., Ohio, and N.C. [*C. foenea* of auth., not Willd.].

C. assiniboinensis Boott   (337)

/T/(WW)/   (Hs)   Moist open woods and shores from SE Sask. (Katepwa and Moose Jaw; Breitung 1957*a*) and s Man. (N to The Narrows of L. Manitoba; type from Assiniboine Rapids of the Assiniboine R.) to S.Dak., N Iowa, and Minn. MAPS: Meusel, Jaeger, and Weinert 1965:76; J.P. Bernard, Nat. can (Que.) 86: fig. 1, p. 13. 1959.

Forma *ambulans* Bernard (type from Otterburne, s Man.), characterized by sterile decumbent culms that develop secondary plants at their rooting tips, is known from Sask. and Man.

C. atherodes Spreng.   (502)

/sT/X/EA/   (Grh)   Calcerous meadows, marshes, and shores from s Alaska, s-cent. Yukon, and Great Slave L. to L. Athabasca (Alta. and Sask.), Man. (N to Gillam, about 165 mi s of Churchill), Ont. (N to the Attawapiskat R. at ca. 53°N; Dutilly, Lepage, and Duman 1954), and Que. (Boivin 1967*a*), s to Oreg., Utah, Mo., Ind., N.Y., and Maine; Eurasia. [*C. aristata* R. Br., not Honck.; *C. trichocarpa* var. *ar.* (R. Br.) Bailey; *C. mirata* Dew. and its var. *minor* Dew.]. MAPS: Hultén 1968*b*:280, and 1962: map 73, p. 83; Porsild 1966: map 20, p. 69; Raup 1930: map 30, p. 203.

C. athrostachya Olney   (195)
/T/WW/   (Grh)   Wet meadows and thickets from SE Alaska (Skagway and Anchorage; CAN) through B.C., Alta., and Sask. (Boivin 1967a) to Calif., Colo., and N.Dak. [*C. tenuirostris* Olney]. MAP: Hultén 1968*b*:232.

C. atlantica Bailey   (109)
/T/EE/   (Hs)   Swampy ground and bogs (ranges of Canadian taxa outlined below), s to Tex. and Fla.
1   Pistillate scales flat or very obscurely keeled, obtuse (the midrib not raised and not extending to the tip); leaves to 4 mm broad; culms to 3.5 mm thick at the obtusely triangular base; [*C. echinata* var. *conferta* Bailey; *C. sterilis* of Canadian auth. in part, not Willd.; N.B., P.E.I., N.S., and St-Pierre and Miquelon (Rouleau 1956)] . . . . .var. *atlantica*
1   Pistillate scales sharply keeled, acutish (the prominent midrib extending to the tip); leaves to 2.5 mm broad; culms to 2.5 mm thick at the sharply triangular base; [*C. incomperta* Bickn.; *C. stellulata (echinata)* var. *conferta* Chapm.; *C. sterilis* Willd. in part, not as to type; *C. sterilis (echinata; stellulata)* var. *excelsior* Bailey; reported from Wellington Co., s Ont., by Stroud 1941, and from between the Rupert R. and L. Mistassini, Que., by Arthème Dutilly and Ernest Lepage, Ann. ACFAS 11:93. 1945] . . . .
. . . . .var. *incomperta* (Bickn.) Herm.

C. atrata L.   (433, 425 (*C. atrosquama*), and 428 (*C. heteroneura*))
/aST/(X)/GEA/   (Grh (Hsr))   Damp tundra, alpine meadows, and rocky slopes (ranges of Canadian taxa outlined below), s to Calif., Ariz., Colo., Mich., and Vt.; Greenland; Iceland; Eurasia. MAPS and synonymy: *see* below.
1   Perigynia strongly compressed; [*C. heteroneura* Boott; cent. Alaska (McKinley Park; *see* Hultén 1942: map 286a, p. 409), B.C., and mts. of sw Alta.; Ont. (w James Bay at ca. 54°N); Que. (near Chimo, s Ungava Bay; Richmond Gulf, E Hudson Bay; Fort George region, E James Bay, ca. 54°N); w Greenland N to ca. 63°N, E Greenland N to near the Arctic Circle; MAPS: Hultén 1958: map 100, p. 119, and 1968*b*:259; Raymond 1951: map 12, p. 7; Meusel, Jaeger, and Weinert 1965: 71; Tolmachev 1952: maps 22 and 27] . . . .
. . . . .var. *atrata*
1   Perigynia scarcely compressed; [var. *nigra* of B.C. reports, not Olney; *C. atrosquama* Mack., the type from along the Smoky R., cent. Alta.; cent. Alaska–cent. Yukon–Great Bear L.–B.C.–sw Alta.; MAPS: Hultén 1968*b*:260; Porsild 1966: map 22, p. 69; D. F. Murray, Brittonia 21(1): fig. 4, p. 65. 1969; Raymond 1950: map 12, p. 7] . . . .
. . . . .var. *atrosquama* (Mack.) Cronq.

C. atratiformis Britt.   (434)
/ST/X/   (Grh)   Brooksides, ravines, and damp slopes (chiefly calcareous), (ranges of Canadian taxa outlined below, the western ssp. *ray.* not yet known from the U.S.A.), the eastern ssp. *atrat.* s to N Mich. and the mts. of N Maine and N New Eng. MAPS and synonymy: *see* below.
1   Spikes essentially one-toned in colour, their scales and weakly inflated perigynia mostly dark-brown to dark purplish-red, the midveins of the scales scarcely paler; tip of perigynium-beak reddish purple; reduced basal leaves bright reddish-purple; [*C. ovata* Rudge; *C. atrata* var. *ov.* (Rudge) Boott; Ont. (N shore of L. Superior near Thunder Bay and Schreiber; CAN) to Que. (N to s Ungava Bay, the Côte-Nord, Anticosti Is., and Gaspé Pen.), Labrador (N to ca. 56°30′N), Nfld. (type locality), N N.B. (CAN), and N.S.; MAP: J. A. Calder, Rhodora 54(646): fig. 1, p. 248. 1952] . . . . .ssp. *atratiformis*
1   Spikes two-toned, the moderately inflated perigynia greenish brown to castaneous, the scales light to dark purplish-red, with midveins and hyaline margins of lighter hue; tip of perigynium-beak brownish; reduced basal leaves light brownish-purple to purplish black; [*C. raymondii* Calder, the type from Gillam, Man.; cent. Alaska–Yukon and s Dist. Mackenzie to s Alta. (not yet known from B.C.), Sask. (s to ca. 52°30′N), and Man. (known only from Gillam, about 165 mi s of Churchill); MAPS: on the above-noted map by Calder; Hultén 1968*b*:260; J. G. Packer, Nat. can. (Que.) 98(2): fig. 3, p. 134. 1971] . . . .
. . . . .ssp. *raymondii* (Calder) Scoggan

C. atrofusca Schkuhr   (364)
/AST/X/GEA/   (Grh)   Wet tundra and calcareous soils from the coasts of Alaska–Yukon–Dist. Mackenzie to Banks Is., northernmost Ellesmere Is., Baffin Is., and northernmost Ungava–Labrador, s to s-cent. Alaska, Great Bear L., SE Dist. Keewatin, NE Man. (Churchill), and w and E James Bay; w and E Greenland between ca. 70° and 77°N; Eurasia. [*C. ustulata* Wahl.; incl. *C. karaginensis* (korag.) Meinsh. and the larger-dimensioned extreme, var. *nortoniana* Boivin]. MAPS: Hultén 1968*b*:272, and 1962: map 39, p. 47; Porsild 1966: map 21, p. 69, and 1957: map 94, p. 172; Meusel, Jaeger, and Weinert 1965:76; Tolmachev 1952: map 32 (incomplete).

   Forma *decolorata* (Porsild) Boivin (var. *dec.* Porsild, the type from Great Bear L.; scales of spike largely cinnamon-brown rather than blackish) is known only from the type locality.

C. aurea Nutt. (272)
/ST/X/   (Grh)   Wet meadows and slopes (usually calcareous) from N-cent. Alaska, s-cent. Yukon, and Dist. Mackenzie (N to 68°42′N) to Dist. Keewatin (Boivin 1967*a*), NE Man. (Churchill), northernmost Ont., Que. (N to E James Bay at ca. 54°N, L. Mistassini, and the Côte-Nord), Labrador (N to Turnavik, 55°16′N; CAN), Nfld., N.B., P.E.I., and N.S., s to Calif., N.Mex., Nebr., and Pa. [*C. mutica* R. Br.]. MAPS: Hultén 1968*b*:254; Raup 1947: pl. 18 (dots should be added for Labrador).

C. backii Boott (201 and 202 (*C. sax.*))
/T/X/   (Hs)   Dry rocky or sandy soil (ranges of Canadian taxa outlined below), s to Oreg., Utah, Colo., Nebr., Minn., Mich., and N N.J.   MAP and synonymy: *see* below.
1   Perigynium at least 4.5 mm long, empty at summit, its beak 2 or 3 mm long; [*C. durifolia* Bailey; s B.C. (Boston Bar; Manning Park) to Alta. (N to Ma-Me-O Beach, sw of Edmonton), Sask. (N to Waskesiu Lake, 53°55′N; type from Carlton House, about 40 mi sw of Prince Albert), Man. (N to Gillam, about 165 mi s of Churchill), Ont. (N to Sandy L. at ca. 53°N, 93°W), Que. (N to Rimouski Co. and the Gaspé Pen.), and N.B. (Albert Co; not known from P.E.I. or N.S.; reported from Nfld. by Reeks 1873, but not listed by Rouleau 1956); MAP: Raymond 1951: map 14 (incomplete northwards), p. 9] . . . . var. *backii*
1   Perigynium about 4 mm long, tightly filled by the achene, its beak not over 1 mm long; [*C. saximontana* Mack.; B.C. (N to Quesnel), Alta. (Waterton Lakes; Coleman), Sask. (N to Saskatoon), and Man. (N to Porcupine Mt.)] . . . . var. *saximontana* (Mack.) Boivin

C. baileyi Britt.   (527)
/T/EE/   (Hs)   Swampy woods and meadows from s Ont. (Ottawa dist.; Gillett 1958) to sw Que. (near Georgeville, Stanstead Co.; GH; reported from Missisquoi Co. by Marcel Raymond, Nat. can. (Que.) 70:264. 1943) and N.S. (Roland 1947; *C. lurida* var. *gr.*), s to Tenn. and Va. [*C. lurida* var. *gracilis* (Boott) Bailey].

[*C. barbarae* Dewey]   (459)
[Most of the citations of this species of the w U.S.A. (Oreg. to Calif.) from B.C. and Alaska by John Macoun (1888; 1890) and Henry (1915) probably refer to *C. sitchensis,* relevant collection in CAN from Vancouver Is. and Burnaby L., B.C.]

C. bebbii (Bailey) Fern.   (161)
/sT/X/   (Hs)   Swampy ground from SE Alaska (an isolated station also near Fairbanks, ca. 65°N) and s Dist. Mackenzie (N to Fort Simpson, ca. 62°N) to Alta. (N to Wood Buffalo National Park at 58°42′N), Sask. (N to Methy Portage on the Clearwater R. at 56°38′N), Man. (N to the Hayes R. about 120 mi sw of York Factory), Ont. (N to the Fawn R. at ca. 54°30′N), Que. (N to E James Bay at ca. 52°N, L. Mistassini, and the Côte-Nord), Nfld., N.B., P.E.I., and N.S., s to Oreg., Colo., Nebr., and N.J. MAP: Hultén 1968*b*:234.

[*C. bella* Bailey]   (429)
[A collection in CAN from McLeod L., B.C., ca. 55°N, has been placed here by Raup. However, it was originally named *C. vulgaris* var. *alpina* Boott *(C. bigelowii* Torr.) and, although lacking underground parts, appears otherwise better placed with *C. bigelowii. C. bella* is a species of the w U.S.A. (Utah, Colo., Ariz., and N.Mex.), perhaps best merged with *C. atrata*).]

C. bicknellii Britt.   (174)
/T/EE/   (Hs)   Dry slopes, thickets, and fields from s Man. (near Turtle Mt.; Winnipeg; Otterburne; reports from Sask. require confirmation) to s Ont. (N to Constance Bay, about 30 mi w of Ottawa; A.J. Breitung, Nat. can. (Que.) 84:82. 1957) and s Que. (Iberville, Rouville, Verchères, and Berthier counties; MT), s to N.Mex., Ark., Ill., and Del.

C. bicolor All.   (269)
/aST/X/GEA/   (Grh (Hsr))   Damp peaty or sandy (often calcareous) meadows and shores from N Alaska, cent. Yukon, and the coast of Dist. Mackenzie to s Baffin Is. and N Que. (coasts of James Bay–Hudson Bay N to ca. 61°N), s to SE Alaska, Great Bear L., N Sask. (Hasbala L. at 50°55′N), s Dist. Keewatin, N Man. (Churchill), and the Belcher Is. in James Bay, with isolated stations in the mts. of sw Alta. and NW Nfld. (St. Barbe and Pistolet Bay; CAN; GH; reports from E Que. and N.B. by various authors apparently referable to either *C. aurea* or *C. garberi*); w Greenland N to 71°23′N, E Greenland between ca. 69° and 75°N; Iceland; Eurasia. MAPS: Hultén 1968*b*:253, and 1962: map 31, p. 39; Porsild 1957: map 83, p. 171 (a dot should be added for NW Nfld.); Tolmachev 1952: map 34 (incomplete for Canada).

C. bigelowii Torr.   (438: *C. concolor* sensu Mackenzie 1935, not R. Br.)
/AST/(X)/GEA/   (Grh)   Damp tundra and shores, alpine meadows, and exposed places from Great Bear L. to Boothia Pen., Ellesmere Is. at ca. 80°N, northernmost Ungava–Labrador, and Nfld., s to Great Slave L., NE Sask. (G.W. Argus, Can. Field-Nat. 80(3):132. 1966), NE Man. (s to Churchill; not known from Ont.), James Bay (Belcher Is.), Que. (s to SE James Bay, the Côte-Nord, and Gaspé Pen.; not known from the Maritime Provinces), and the mts. of N.Y. and New Eng.; w and E Greenland N to ca. 78°N; Iceland; Eurasia. [*C. rigida* vars. *big.* (Torr.) Boott and *glacialis* Fries; *C. concolor* of American auth., not R. Br.; *C. ?drejeriana* Lange; *C. ?dubitata* Dew.; *C. ?fyllae* Holm; *C. ?hartzii* Gand.; *C. ?warmingii* Holm]. MAPS: Hultén 1968*b*:248, and 1962: map 43, p. 51; Porsild 1957: map 84, p. 171; Raymond 1951: map 8, p. 6.

The western limits of this species are obscure (as indicated in Hultén's 1962 map) through confusion with other members of the Section *Acutae,* particularly *C. lugens.* Forma *anguillata* (Drej.) Fern. (*C. anguillata* Drej.; perigynia relatively narrow and in very slender-peduncled spikes, the lowermost peduncles often arising from near the base of the culm) occurs throughout the range.

C. bolanderi Olney   (123)
/T/W/   (Grh)   Moist woods from B.C. (N to Hudson Hope, ca. 56°N; CAN) to Calif. and N.Mex. [*C. deweyana* var. *bol.* (Olney) Boott; scarcely separable from *C. deweyana; see* Hitchcock et al. 1969]. MAP: Meusel, Jaeger, and Weinert 1965:69.

C. bonanzensis Britt.   (94)
/Ss/W/eA/   (Hsr)   Damp tundra and shores of Alaska (N to Fairbanks; CAN), the Yukon (type from Bonanza Creek), and w Dist. Mackenzie (N to Aklavik and Inuvik); E Asia. [Incl. *C. praeceptorium* Mack.]. MAP: Hultén 1968*b*:242.

C. brevior (Dew.) Mack.   (167)
/T/X/   (Hs)   Dry open soil from s B.C. (Spences Bridge, Rogers Pass, and Trail; CAN) to s Alta. (near Redcliffe and Magrath; CAN), s Sask. (N to Saskatoon; Breitung 1957*a*), s Man. (N to Fort Ellice, about 70 mi NW of Brandon), s Ont. (N to the Ottawa dist.), and sw Que. (N to Rouville Co.), s to Oreg., N.Mex., Ark., Tenn., and Del. [*C. straminea* (*festucacea*) var. *brevior* Dewey].

C. bromoides Schkuhr   (120)
/T/EE/   (Hs)   Wet woods, swamps, and bogs from Wisc. to Ont. (N to the Ottawa dist.), Que. (N to Fond d'Ormes, Rimouski Co.; MT), N.B., and N.S. (not known from P.E.I.), s to La. and Fla. MAP: Meusel, Jaeger, and Weinert 1965:69.

C. brunnescens (Pers.) Poir.   (93)
/aST/X/GEA/   (Hs)   Boggy thickets and woods (ranges of Canadian taxa outlined below), s to

Oreg., Utah, Colo., Minn., Ohio, and N.C.; Greenland; Iceland; Eurasia. MAPS and synonymy: *see* below.

1 Head straight and stiffish, less than 4 cm long; spikes subglobose to ellipsoid, the lower ones not more than 1 cm apart; leaves and culms firm; [incl. ssp. *alaskana* and ssp. *pacifica* Kalela; *C. curta* var. *brun.* Pers.; *C. canescens* var. *alpicola* Wahl.; transcontinental: cent. Alaska–s Yukon–NW Dist. Mackenzie–B.C. to s-cent. Dist. Keewatin–Man., Ont. (N to the mouth of the Severn R., Hudson Bay, ca. 56°N), Que. (N to Ungava Bay at ca. 59°30′N), Labrador (N to Okkak, 57°40′N), Nfld., N.B., and N.S. (not known from P.E.I.); w and E Greenland N to ca. 67°N; MAPS: Hultén 1962: map 34, p. 41, and 1968b (combine the maps for ssp. *alask.*, p. 242, and ssp. *pac.*, p. 243)] . . . . .var. *brunnescens*

1 Heads flexuous, to 7 cm long, the lowest of the subglobose spikes to 2.5 cm apart; leaves and culms relatively soft; [*C. canescens* vars. *vulgaris* Bailey and *vitilis* (Fries) Carey (*C. vit.* Fries); transcontinental, the common form northwards: B.C., Alta. (N to 58°54′N), Sask. (N to L. Athabasca), Man. (N to Nueltin L. at 59°48′N), Ont. (N to Big Trout L. at ca. 53°45′N, 90°W), Que. (N to Fort George, E James Bay, ca. 53°45′N, and the Côte-Nord), Nfld., N.B., P.E.I., and N.S.] . . . . .var. *sphaerostachya* (Tuck.) Kük.

C. bullata Schkuhr   (521)
/T/E/   (Grh)   Acid meadows and bogs from w N.S. (E to Annapolis and Halifax counties; *see* N.S. map 128 by Roland 1947:239) to Tenn. and Ga. [Incl. var. *greenei* (Boeckl.) Fern.].

C. buxbaumii Wahl.   (437)
/aST/X/GEA/   (Grh (Hsr))   Wet shores, swamps, and bogs from s-cent. Alaska, s Yukon, and Great Bear L. to L. Athabasca (Alta. and Sask.), Man. (N to Norway House; John Macoun 1888), Ont. (N to w James Bay at ca. 53°N), Que. (N to the Clearwater R. at ca. 56°15′N, L. Mistassini, and the Côte-Nord), Nfld., N.B., and N.S. (not known from P.E.I.), s to Calif., Colo., Ark., Ky., and N.C.; SE Greenland at 61°10′N; Europe; w Asia. [Incl. var. *anticostensis* Raymond; *C. polygama* Schk., not Gmel.]. MAPS: Hultén 1968b:257, and 1958: map 254, p. 273; Meusel, Jaeger, and Weinert 1965:71.

Forma *dilutior* Kük. (scales of spikes whitish or pale brown rather than brown to purplish-black) is reported from Nfld. by Rouleau (1956). Forma *macrostachya* (Hartm.) Kük. (spikes of the head to 5 cm long rather than about 2 cm; *C. hartmanii* Caj.; MAP: Hultén 1958: map 114, p. 133) is reported from Anticosti Is., E Que., by Raymond (1950a). Forma *pedunculata* Raymond (terminal spike long-peduncled and entirely staminate rather than sessile and pistillate above) is known only from the type locality, Bellerive, Labelle Co., Que. Boivin (1967a) refers this species to *C. canescens* L., the current concept of which he refers to *C. curta* Good., as the earliest available legitimate name.

[*C. campylocarpa* Holm]   (444)
[The report of this obscure species of Wash. and Oreg. from s B.C. by Eastham (1947; Mt. Brent, near Penticton) requires confirmation. It is merged with *C. scopulorum* by Hitchcock et al. 1969.]

C. canescens L.   (96)
/aST/X/GEA/   (Hs)   Swamps, bogs, and wet meadows and woods, the aggregate species from N-cent. Alaska–Yukon and the Mackenzie R. Delta to L. Athabasca (Alta. and Sask.), s Dist. Keewatin, Ont.–Que.–Labrador, Nfld., N.B., P.E.I., and N.S., s to Calif., Ariz., Ind., Ohio, and Va.; w Greenland N to 69°25′N, E Greenland N to 63°32′N; Iceland; Eurasia. MAPS and synonymy: *see* below.

1 Scales somewhat enveloping the perigynia toward base and largely concealing them; spikes crowded; [*C. arctaeformis* Mack.; s ?Alaska to Vancouver Is. (type from Elgin, B.C.)] . . . . .ssp. *arctaeformis* (Mack.) Calder & Taylor

1 Scales not at all enveloping the perigynia; lower spikes usually remote . . . . .ssp. *canescens*

2 Spikes at most 7 mm long; perigynia not over 2 mm long, often quite smooth-beaked; head to 5 cm long; [var. *robustina* Macoun; *C. lapponica* Lang; transcontinental; MAP (*C. lapp.*): Hultén 1968b:241] . . . . .var. *subloliacea* Laestad.

2 Spikes to 12 mm long; perigynia to 3 mm long, often serrulate near summit.

3   Head to about 1.5 dm long, the two lower spikes up to 4 cm apart; [*C. disjuncta*
(Fern.) Bickn.; Ont. to Nfld. and N.S.] . . . . .var. *disjuncta* Fern.
3   Head to about 7 cm long, the lowest spikes at most 2.5 cm apart; [var. *brunnea*
Macoun; *C. ?curta* Good. (see *C. buxbaumii*); *C. brizoides* of American auth., not
L.; *C. richardii* Thuill.; transcontinental; MAPS: Hultén 1968*b*:241, and 1962: map
82, p. 91; Meusel, Jaeger and Weinert 1965:68; Raup 1947: pl. 17] . . . . .var. *canescens*

C. capillaris L.   (347)
/AST/X/GEA/   (Hs)   Damp mossy calcareous woods, thickets, meadows, and shores, the
aggregate species from the coasts of Alaska–Yukon–Dist. Mackenzie to Victoria Is., s Ellesmere
Is., Baffin Is., northernmost Ungava–Labrador, Nfld., N.B., and N.S. (not known from P.E.I.), s to
Oreg., Nev., Utah, N.Mex., s Sask., s Man., Minn., Mich., N.Y., and Vt.; w Greenland N to 78°45′N,
E Greenland N to 77°40′N; Iceland; Eurasia. MAPS and synonymy: see below.
1   Terminal spike typically staminate throughout; (transcontinental).
  2   Culms to about 2 dm tall, in dense tussocks; lateral (pistillate) spikes to 1 cm long,
      with at most 8 flowers; perigynia 2 or 3 mm long; [var. *minima* Beck; transcontinental;
      MAPS: Hultén 1968*b*:274, and 1962: map 47, p. 55; Porsild 1957: map 97, p. 173;
      Raup 1947: pl. 18] . . . . .ssp. *capillaris*
  2   Culms to over 5 dm tall, in looser tufts; lateral spikes to over 1.5 cm long, with up to
      about 20 flowers; perigynia to 4 mm long; [var. *elongata* Olney; var. *major* "Drej.";
      C. chlorostachys Stev.; *C. saskatchawana* Boeck.; MAP: Hultén 1962: map 47, p. 55] . . . . . .
      . . . . .ssp. *chlorostachys* (Stev.) Löve, Löve, & Raymond
1   Terminal spike typically gynaecandrous.
  3   Perigynia to about 3 mm long, smooth-edged; lateral spikes long-peduncled and
      nodding; [*C. krausei* Boeck., the type from Alaska, the only known locality;
      C. capillaris var. *k.* (Boeck.) Krantz; MAP: Löve and Löve 1956*b*: fig. 15 (circles),
      p. 175; Hultén 1968*b*:274, and 1962: map 48, p. 55 (both as *C. krausei* and probably
      including the following taxon)] . . . . .ssp. *krausei* (Boeck.) Böcher
  3   Perigynia less than 2 mm long, their edges often bearing spinules; lateral spikes
      erect; [incl. ssp. *robustior* (Drej.) Böcher (*C. boecheriana* Löve, Löve & Raymond, of
      Greenland); Alaska; N Hudson Bay; N Que.; Greenland] . . . .
      . . . . .ssp. *porsildiana* (Polunin) Böcher

C. capitata L.   (3)
/aST/X/GEA/   (Hsr)   Open tundra and slopes (ranges of Canadian taxa outlined below), s to
Calif., N Mexico, Utah, and ?Colo., and the mts. of N.H.; s S. America; w and E Greenland N to ca.
73°N; Iceland; Eurasia. MAPS and synonymy: see below.
1   Perigynia to 3 mm long, gradually tapering into a beak at most 0.5 mm long;
    [transcontinental: Alaska–Yukon–Dist. Mackenzie–s Dist. Keewatin–B.C.–Alta. to Sask.
    (s to ca. 52°N), Man. (s to ca. 55°N), northernmost Ont., Que. (N to ca. 58°N), Labrador (N
    to 59°15′N), and Nfld.; MAPS: Hultén 1958: map 20, p. 38, and 1968*b*:222; Meusel,
    Jaeger, and Weinert 1965:64; Raymond 1951: map 11 (NE area), p. 7] . . . . .var. *capitata*
1   Perigynia barely over 2 mm long, abruptly contracted into a beak to 0.75 mm long; [*C.
    arctogena* Sm.; at present known definitely from Man. (N to Baralzon L. at 60°N), s Baffin
    Is., Que. (N to s Ungava Bay; also known from Charlevoix Co. and mts. of the Gaspé
    Pen.), Labrador (N to the Komaktorvik R. at 59°15′N), and NW Nfld.; MAPS: Hultén 1958:
    map 19, p. 39; Porsild 1957: map 72 (*C. arct.*; indicating no stations w of Man.), p. 169;
    Meusel, Jaeger, and Weinert 1965:64]. Hultén (1958) states that, "As *C. arctogena* was
    recently distinguished as a separate taxon, its area is not yet clear. It might well occur in
    Siberia and Alaska although it has not yet been reported from these parts." Actually,
    Siberia and Alaska were included in the general range outlined in the paper in which the
    original description was published by H. Smith (Acta Phytogeogr. Suec. 13:191–200.
    1940), who outlined the w N. American distribution as from Alaska and the Yukon to Calif.
    and S. America, an area left "open" on Hultén's map. *See* Marcel Raymond (Contrib.
    Inst. Bot. Univ. Montréal 64:37–41. 1949) . . . . .var. *arctogena* (Sm.) Hult.

C. careyana Torr.   (285)
/T/EE/   (Hs)   Rich hardwoods from Mich. to s Ont. (Essex, York, and Wellington counties; GH; MT; TRT) and sw Que. (reported from Sweetsburg, Missisquoi Co., by Marcel Raymond, Ann. ACFAS 7:106. 1941, and from the Ottawa Valley by Raymond 1950*b*) to Mo., Ill., and Va.

Boivin (1967*a*) does not list this species for Canada and some or all of the above material may refer to other members of the Section *Laxiflorae*. Mackenzie (1935) and Fernald *in* Gray (1950) include Ont. in the range but Gleason (1958) does not. The hybrid, *C. digitalis* × *C. laxiculmis* (*see* below) may be involved.

[C. caryophyllea Latourr.]   (210)
[This Eurasian species is reported as introd. in s Ont. by Soper (1949) but it is not listed by Boivin (1967*a*) and no voucher-specimen has been seen.]

C. castanea Wahl.   (336)
/st/EE/   (Grh)   Wet meadows and swampy ground (chiefly calcareous) from SE Man. (Sandilands Forest Reserve; DAO) to Ont. (N to ca. 52°30′N; *see* James Bay map by Dutilly, Lepage, and Duman 1954: fig. 5, p. 49), Que. (N to E James Bay at 55°15′N, L. Mistassini, and Anticosti Is.), Nfld., N.B., and N.S. (not known from P.E.I.), s to Minn., Mich., and Conn. [*C. flexilis* Rudge]. MAP: Meusel, Jaeger, and Weinert 1965:76.

C. cephalophora Muhl.   (33)
/T/EE/   (Grh)   Dry or moist woods and open places from s Mich. to s Ont. (Essex, Lambton, Kent, Middlesex, Norfolk, Waterloo, and Wentworth counties), sw Que. (N to Montreal), and cent. Maine, s to Tex. and Fla.

C. chordorrhiza L. f.   (24)
/aST/X/EA/   (Hsr)   Wet sphagnum bogs and marshy ground from the coasts of Alaska–Yukon–Dist. Mackenzie to s Victoria Is., s Baffin Is., Que. (N to Hudson Strait, L. Mistassini, the Côte-Nord, and Anticosti Is.), Nfld. (Rouleau 1956), N.B., and P.E.I. (Boivin 1967*a*; not known from N.S.), s to B.C., Alta. (Nestow, 54°14′N; CAN), Sask., Man., N Iowa, Ind., and Vt.; Iceland; Eurasia. [Incl. var. *sphagnophila* Laestad.; *C. fulvicoma* Dew.]. MAPS: Hultén 1968*b*:229, and 1962: map 85, p. 95; Porsild 1957: map 79, p. 170; Meusel, Jaeger, and Weinert 1965:66.

C. circinata Meyer   (7)
/sT/W/   (Grh)   Aleutian Is. (type from Unalaska) and the coast of s Alaska (*see* Hultén 1942: map 234, p. 404) through coastal B.C. to the Olympic Pen., Wash., and reported from the Slims R. s of Kluane L. in sw Yukon by Porsild (1966). MAP: Hultén 1968*b*:226.

C. communis Bailey   (220)
/T/EE/   (Hs)   Dry woodlands (usually on rocky ledges) from SE Man. (Otterburne; Löve and Bernard 1959) to Ont. (N to the N shore of L. Superior), Que. (N to Rimouski Co. and the Gaspé Pen.), N.B., P.E.I., and N.S., s to Ark., Ky., S.C., and Ga. [*C. pilulifera* and its var. *longibracteata* of N.B. reports by M.L. Fernald, Proc. Am. Acad. Arts 37: 504. 1902, not L. nor Lange, respectively; *C. varia sensu* John Macoun 1888, at least in part (relevant collections from Truro and Pirate's Cove, N.S., and Shannonville, s Ont., in CAN), not Muhl. (nor Lumnitzer nor Host, whose authorship of the name predated that by Muhlenberg); the report of *C. communis* from B.C. by Henry 1915, probably refers to some other member of the Section *Montanae*].

C. comosa Boott   (495)
/T/X/   (Hs)   Swamps and shallow water from s B.C. (Oliver, Vernon, and Kootenay; CAN) to Calif. and Idaho and from Nebr. and Minn. to Ont. (N to the NW shore of L. Superior near Thunder Bay; CAN), Que. (N to the Gaspé Pen.), N.B., and N.S. (not known from P.E.I.), s to Calif., Idaho, Minn., La. and Fla. [*C. pseudo-cyperus* vars. *com.* Boott and *americana* Hochst.]. MAPS: Hultén 1958: map 149, p. 169; Meusel, Jaeger, and Weinert 1965:79.

C. concinna R. Br.   (244)
/ST/X/   (Hsr)   Cool calcareous woods and slopes from N-cent. Alaska and s Yukon to the

Cyperaceae

Mackenzie R. Delta, Great Bear L., Great Slave L., L. Athabasca (Alta. and Sask.), Man. (N to Churchill), northernmost Ont., Que. (N to the Larch R. at ca. 56°40′N, L. Mistassini, and the Côte-Nord), s Labrador (near Labrador City; CAN), Nfld., and N.B. (Restigouche R.; not known from P.E.I. or N.S.), s to B.C., Oreg., Colo., Wyo., S.Dak., Mich., and s Ont. The type is from NW Canada. MAPS: Hultén 1968b:267; Raup 1947: pl. 18.

C. concinnoides Mack.   (245)
/T/W/   (Hsr)   Dry woodlands from B.C. (N to Cariboo, ca. 53°N; CAN) and sw Alta. (N to Banff; CAN) to N Calif. and Idaho. [Perhaps best merged with *C. richardsonii*].

[C. conjuncta Boott]   (80)
[This species of the E U.S.A. (N to S.Dak. and N.Y.) is reported from s Man. by Löve and Bernard (1959) and from s Ont. by Soper (1949) but is so closely related to *C. alopecoidea* (found in the same areas) as to require further confirmation.]

C. conoidea Schkuhr   (311 and 310 (*C. katahdinensis*))
/T/EE/   (Hs (Grh))   Moist grassy places and gravelly or rocky shores from Minn. to Ont. (collections in CAN from Windsor, Essex Co., and Belleville, Hastings Co.; of f. *kat.* in RIM and CAN from the Missinaibi R. at ca. 50°N), Que. (N to Cap Rouge, near Quebec City; CAN), Nfld. (Bishop Falls, Exploits R.; GH; CAN), N.B. (Kent, Carleton, and Charlotte counties; CAN), and N.S. (not known from P.E.I.), s to Iowa, Ohio, and N.C. [*C. granularis sensu* Fowler 1885, as to the Bass River, N.B., plant, not Muhl.; *C. ?oligocarpa sensu* Fowler 1885, not Schkuhr]. MAP (E Canada): Raymond 1951: map 41, p. 19.

Forma *katahdinensis* (Fern.) Boivin (*C. katahdinensis* Fern.; upper pistillate spikes crowded about the sessile or short-peduncled staminate spike, this to about 1 cm long, the typical form with pistillate spikes well separated, the usually long-peduncled staminate spike to 2.5 cm long) is known from Ont. (Missinaibi R. at ca. 50°N; CAN), Que. (Montreal; L. St. John), Nfld. (Grand Falls and Rushy Pond, Exploits R.; GH; CAN), and Mt. Katahdin, Maine.

C. convoluta Mack.   (28)
/T/EE/   (Hs)   Dry woods and thickets from s Man. (Portage la Prairie and Winnipeg; CAN, detd. Mackenzie) to Ont. (N to the Ottawa dist.), sw Que. (N to the Montreal dist.), N.B. (Woodstock; GH), and N.S. (not known from P.E.I.), s to Kans., Ark., Ala., and S.C. [*C. rosea* var. *pusilla* Peck].

C. crawei Dewey   (306)
/T/X/   (Grh (Hsr))   Calcareous meadows, gravels, ledges, and shores from SE B.C. (Kootenay; Wapta; Golden; Big Bend; Yoho; Canal Flats) to s Alta. (N to Jasper National Park), Sask. (Oxbow), Man. (N to Bowsman, SE of Porcupine Mt.),Ont. (N to w James Bay at ca. 53°N; *see* James Bay map by Dutilly, Lepage, and Duman 1954: fig. 6, p. 51), Que. (N to the Harricanaw and Nottoway rivers at ca. 51°N and the Côte-Nord), w Nfld., N.B. (Ingleside, near St. John), and N.S. (not known from P.E.I.), s to Wash., Utah, Wyo., Kans., Mo., and N Ala. [*C. heterostachya* Torr.]. MAP: Marcel Raymond, Ann. ACFAS 19: fig. 1, p. 92. 1953.

C. crawfordii Fern.   (159)
/ST/X/   (Hs)   Damp to dry open ground from cent. Alaska (Circle; also known from White Pass) to Great Slave L., Alta. (N to L. Athabasca), Sask. (N to Meadow Lake, 54°08′N), Man. (N to Wekusko L., about 75 mi NE of The Pas), Ont. (N to the Fawn R. at ca. 54°30′N, 89°W), Que. (N to Fort George, E James Bay, 53°50′N), Labrador (N to Goose Bay, 53°18′N), Nfld., N.B., P.E.I., and N.S., s to Wash., Idaho, Mich., and Tenn. [Incl. var. *vigens* Fern.; *C. scoparia* var. *minor* Boott]. MAP: Hultén 1968b:235.

C. crinita Lam.   (475 and 474 (*C. gynandra*))
/sT/EE/   (Hs)   Damp to swampy woodlands and thickets (ranges of Canadian taxa outlined below), s to La. and Ga. MAP and synonymy: *see* below.
1   Leaf-sheaths smooth; perigynia nearly circular in cross-section, broadest above the middle, abruptly beaked; lower pistillate scales truncate or retuse at summit; terminal spikes usually staminate throughout; [incl. vars. *minor* Boott and *simulans* Fern.; s Man.

(Winnipeg; the report by Hooker 1839, "Canada to Norway House", requires confirmation) to Ont. (N to L. Nipigon), Que. (N to Rupert House, SE James Bay, 51°29′N, L. St. John, the Côte-Nord at Tadoussac, and the Gaspé Pen.; not known from Anticosti Is.), Nfld., N.B., P.E.I., and N.S.; MAP (E Canada): Raymond 1951: map 31, p. 16] . . . . .var. *crinita*

1 Leaf-sheaths minutely rough-hispid; perigynia strongly flattened, broadest at the middle, tapering to a minute beak; lower pistillate scales acute or acuminate; staminate spikes often pistillate at tip; [*C. gynandra* Schwein.; Ont. (N to Kapuskasing, 49°24′N), Que. (N to the Gaspé Pen.), Nfld., N.B., P.E.I., and N.S.] . . . . .var. *gynandra* (Schw.) Schw. and Torr.

C. cristatella Britt. (188)
/T/(X)/ (Hs) Swampy meadows and thickets from Alta. (Fort Saskatchewan; CAN, detd. Hermann), ?Sask. (a collection in CAN from between Prince Albert and Waskesiu has been placed here by Fraser; not listed by Breitung 1957a), S Man. (Sandilands Forest Reserve and Victoria Beach, L. Winnipeg; DAO), Ont. (N to near Thunder Bay; CAN; reported from the Nipigon R. N of L. Superior by John Macoun 1888), and Que. (N to L. St. Peter near Sorel; *see* Que. map by Robert Joyal, Nat. can. (Que.) 97(5): map D, fig. 1, p. 562. 1970; a collection in MT from Rimouski, Rimouski Co., may also belong here; not known from the Atlantic Provinces), S to Nebr., Mo., Ky., and Md. [*C. tribuloides* var. *cristata* (Schw.) Bailey (*C. cristata* Schw., not Clairv.)].

C. cumulata (Bailey) Mack. (181)
/T/EE/ (Hs) Dry or moist acid soils from S Man. (N to Fort Ellice, about 70 mi NW of Brandon; reports from Sask. require confirmation) to Ont. (N to the Sudbury dist.; TRT), Que. (N to the Montreal dist.), N.B., P.E.I., and N.S., S to Mich., Ohio, Pa., and N.J. [*C. straminea* var. *cum.* Bailey, the type from Kent Co., N.B.].

Forma *soluta* Fern. (head to 1 dm long, the spikes to 2 cm apart rather than crowded in a shorter head) is known from N.S. (Queens Co., the type from a beach near the mouth of the Broad R.).

C. cusickii Mack. (69)
/T/W/ (Hs) Wet meadows from B.C. (N to Lake House, ca. 58°N; CAN) to Calif. and Utah. [*C. teretiuscula (diandra)* var. *ampla* Bailey; *C. arizonensis* of B.C. reports, not Clarke].

C. davisii Schw. & Torr. (329)
/T/EE/ (Grh) Rich calcareous woods and shores from Minn. to SW Que. (Ottawa Valley and the Montreal dist.; Raymond 1950b) and W New Eng., S to Tex., Okla., Mo., Tenn., and Md.

C. debilis Michx. (340 and 341 (C. flexuosa))
/T/EE/ (Hs) Open woods, thickets, and meadows from Ont. (N to Batchawana Bay on the SE shore of L. Superior; CAN; the report from Norway House, Man., by Hooker 1838, is probably based upon *C. vaginata*) to Que. (N to E James Bay at 52°39′N and the Gaspé Pen.), Nfld., N.B., P.E.I., and N.S., S to Tex. and Fla. MAP: Meusel, Jaeger, and Weinert 1965:76.

The Canadian plant is referable to var. *rudgei* Bailey (*C. flexuosa* Muhl.; *C. tenuis* Rudge, not Gmel.; scales of spikes stramineous to rust-coloured rather than whitish, the perigynia at most 7 mm long rather than to 10 mm).

C. deflexa Hornem. (227)
/aST/X/G/ (Grh (Hsr)) Open woodlands and turfy slopes from cent. Alaska, S Yukon, and Great Bear L. to L. Athabasca (Alta. and Sask.), S Dist. Keewatin, Que. (N to the Korok R. at 58°50′N), Labrador (N to Okkak, 57°40′N), Nfld., N.B., P.E.I., and N.S., S to cent. B.C.-Alta.-Sask.-Man., Minn., Mich., N.Y., and Mass.; W Greenland N to 65°25′N, E Greenland N to 64°23′N, the type from Greenland. MAPS: Hultén 1968b:262; Meusel, Jaeger, and Weinert 1965:72; Raup 1947: pl. 18.

C. demissa Hornem. (Included in C. flava, 358, by Mackenzie 1935)
/T/EE/E/ (Hs) Boggy or peaty acid soils of Ont. (near Thunder Bay; MT), E Que. (Côte-Nord, Anticosti Is., Gaspé Pen., and Magdalen Is.), Nfld., S N.B., N.S., and Maine; Iceland; Europe.

[*C. tumidicarpa* And.; *C. flava* var. *graminis sensu* Robinson and Schrenk 1896, not Bailey; *C. oederi* of Canadian reports in part, not Retz.]. MAPS: E. W. Davies, Watsonia 3(1): fig. 1 (probably erroneously including most of Labrador and southernmost Greenland), p. 82. 1953; Raymond 1950*a*: map 16 (E Canada), p. 444.

C. deweyana Schw.   (121)
/sT/X/   (Hs)   Woods and thickets from sw Dist. Mackenzie (Liard Valley; CAN) and B.C.–Alta. to Sask. (N to Methy Portage, ca. 55°45′N), Man. (N to Knee L. at ca. 55°N), Ont. (N to Sachigo L. at ca. 54°N, 92°W), Que. (N to Rupert House, E James Bay, 51°29′N, and the Côte-Nord), Nfld., N.B., P.E.I., and N.S., s to Idaho, Colo., S.Dak., and Pa. [*C. remota* of most Canadian reports, not L.; *C. bolanderi* Olney and *C. leptopoda* Mack. are perhaps best merged here]. MAPS: Hultén 1968*b*:247; details of the E Canadian distribution are shown by Raymond 1950*a*: map 10, p. 443, and 1951: map 28, p. 15.

Plants with the spikes crowded in a head rarely over 3 cm long may be distinguished as var. *collectanea* Fern. (type from the Gaspé Pen., E Que.).

C. diandra Schrank   (67)
/ST/X/EA/   (Hs)   Swamps, peaty bogs, and wet meadows from N Alaska–Yukon–Dist. Mackenzie to L. Athabasca (Alta. and Sask.), Man. (N to Churchill), Ont. (N to Fort Severn, Hudson Bay, ca. 56°N), Que. (N to the Kaniapiscau R. at 56°31′N, L. Mistassini, and the Côte-Nord), s Labrador (Forteau, 51°28′N; GH), Nfld., N.B., P.E.I., and N.S., s to Calif., Colo., Nebr., Mo., and N.J.; Iceland; Eurasia. [*C. teretiuscula* Good.]. MAPS: Hultén 1968*b*:231, and 1962: map 120, p. 129; Raup 1947: pl. 17.

C. digitalis Willd.   (289)
/T/EE/   (Hs)   Dryish hardwoods from Wisc. to s Ont. (N to Northumberland and Hastings counties; CAN), sw Que. (St-Jean, Rouville, Huntingdon, and Deux-Montagnes counties; Herb. M. Raymond), and Maine, s to Mo., Miss., and Fla.

C. disperma Dewey   (82)
/ST/X/EA/   (Hsr (Grh))   Damp or boggy woods from N Alaska–Yukon–Dist. Mackenzie to L. Athabasca (Alta. and Sask.), s Dist. Keewatin, Que. (N to Chimo, s Ungava Bay), Labrador (N to near Nain, 56°33′N), Nfld., N.B., P.E.I., and N.S., s to Calif., N.Mex., S.Dak., Minn., and Pa.; Eurasia. [*C. tenella* Schkuhr; *C. ten.* f. *brachycarpa* Kük.]. MAPS: Hultén 1968*b*:243, and 1962: map 72, p. 81.

C. disticha Huds.
This Eurasian species is known in N. America only from s Ont. (in a peat bog near Belleville, Hastings Co., where taken by John Macoun in 1866 (GH, detd. Boott) and again in 1877 (CAN)) and sw Que. (Ile-Charron, Chambly Co., near Montreal; the Eurasian *C. nutans* also introd. at the same locality; *see* Frère Marie-Victorin, Contrib. Inst. Bot. Univ. Montréal 15:262–66. 1929). [*C. intermedia* Good., not Miégev. nor Suter].

M.L. Fernald (Rhodora 44(525): 282–84. 1942) believes that the peat bog habitat of the s Ont. plant argues in favour of its being native there. However, the fact that it is found elsewhere in N. America only in the seaport region of Montreal, where undoubtedly introd. in ballast, greatly weakens this assumption; (*see* discussion under *Luzula campestris*).

C. douglasii Boott   (15)
/T/WW/   (Grh)   Dry prairies and foothills from B.C. (N to Kamloops) to s Alta. (N to Calgary), s Sask. (Crane L.; Swift Current; Moose Jaw; Qu'Appelle Valley), and s Man. (Ste. Rose; Chatfield; Souris), s to Calif., N.Mex., Nebr., and Iowa.

C. eburnea Boott   (259)
/ST/X/   (Grh)   Calcareous ledges, gravels, and sands from cent. Alaska, s Yukon, and NW Dist. Mackenzie to L. Athabasca (Alta. and Sask.), Man. (N to the Nelson R. at ca. 57°N), northernmost Ont., Que. (N to s James Bay, L. Mistassini, and the Côte-Nord), Nfld., N.B., and N.S. (not known from P.E.I.), s to B.C., Nebr., Tex., Ala., and Va. MAPS: Hultén 1968*b*:268; Meusel, Jaeger, and

Weinert 1965:75; details of the ᴇ Canadian distribution are given by Raymond 1950a: map 11, p. 443, and 1951: map 29, p. 15.

C. eleusinoides Turcz.
/ST/W/eA/ (Grh) Rocky tundra and slopes of the Aleutian Is., Alaska, the Yukon, ɴ B.C. (Dease L. at ca. 58°30′N; CAN), sw Alta. (Mt. Edith Cavell, Jasper National Park; CAN), and ɴ Mont.; ᴇ Asia. [*C. enanderi* Hult.; *C. eurystachya* Hermann; *C. kokrinensis* Porsild; according to Polunin 1959, this species should probably be included in *C. caespitosa* L.]. ᴍᴀᴘs: combine the maps by Hultén 1968b:249 and 261 (*C. enanderi*); Porsild 1966: map 23, p. 69; Raymond 1951: map 12 (*C. enan.*), p. 7.

C. engelmannii Bailey (4)
/T/W/ (Grh) Open sunny slopes and summits from s B.C. (Cascade Mts. between Hope and Princeton; V) to Utah and Colo. [*C. breweri* var. *paddoensis* (Suksd.) Cronq.].

C. epapillosa Mack. (427)
/T/W/ (Hs) Alpine meadows from s B.C. (Old Glory Mt., near Rossland; CAN, detd. D.F. Murray) and sw Alta. ("w. of Edmonton"; CAN, detd. Murray) to Calif., Utah, and Wyo. [*C. heteroneura* var. *epap.* (Mack.) Hermann; *C. atrata* var. *erecta* Boott.]. ᴍᴀᴘ: D.F. Murray, Brittonia 21(1): fig. 7, p. 71. 1969.

C. exilis Dewey (100)
/sT/EE/ (Hs) Peaty bogs and wet meadows from Ont. (ɴ to Batchawana Bay, ᴇ end of L. Superior) to Que. (ɴ to the ᴇ Hudson Bay watershed at ca. 56°10′N, L. Mistassini, and the Côte-Nord), Labrador (ɴ to Makkovik, 55°10′N), Nfld., N.B., and N.S. (not known from P.E.I.), s to Minn., Mich., N.Y., and Del.

C. exsiccata Bailey (516)
/T/WW/ (Grh) Marshes or swamps from Dist. ?Mackenzie (Boivin 1967a) and B.C. (ɴ to Vanderhoof, ca. 54°N; not known from Alta.) to Sask. (ɴ to Lac la Ronge at ca. 55°N; DAO), s to cent. Calif. and Mont. [Incl. vars. *globosa* and *pungens* Bailey; *C. vesicaria* var. *major* Boott; according to Hultén 1942, the inclusion of sᴇ Alaska in the range by Mackenzie 1935, is based upon *C. rostrata*].

C. festucacea Schkuhr (165)
/T/EE (Hs) Moist open ground and woods from Iowa, Ill., and Mich. to Ont. (ɴ to Thunder Bay and Ottawa; CAN, detd. Fernald and Malte; not listed by Gillett 1958) and Mass., s to Okla., La., and Ga. [*C. straminea* var. *fest.* (Schk.) Gay, reports of which from N.B. by Fowler 1885, and N.B. and N.S. by John Macoun 1888, refer to *C. hormathodes* (relevant collections in CAN and NBM)].

Boivin (1967a) includes *C. bebbii* and other closely related taxa in his listing of *C. festucacea*, thus assigning it a transcontinental range.

C. filifolia Nutt. (205)
/sT/WW/ (Hs) Dry prairies and slopes from s-cent. Alaska, s Yukon, and Great Bear L. to B.C.–Alta., Sask. (ɴ to Langham and Saskatoon), and sw Man. (ɴ to St. Lazare, about 75 mi ɴw of Brandon), s to Oreg., Nev., N.Mex., Tex., Nebr., and Minn. [*C. elynoides* Holm (*C. fil.* var. *miser* Bailey); *Kobresia globularis* Dew.; *Uncinia breviseta* Torr.]. ᴍᴀᴘ: Hultén 1968b:223.

C. flacca Schreb. (400)
Eurasian; introd. in moist meadows of s Ont. (Puslinch, Wellington Co.; London, Middlesex Co., where taken by Burgess in 1884; CAN; TRT), sw Que. (Napierville; Montreal, where taken by Burgess in 1882; GH), and N.S. (Hants and Kings counties; CAN; GH). [*C. glauca* Scop.; *C. diversicolor* of Canadian reports in part, not Crantz].

C. flaccosperma Dewey (316 and 315 (*C. glaucodea*))
/t/EE/ (Hs) Calcareous woods, meadows, and swamps from Mo., Ill., Ind., Ohio, and s Ont.

(Leamington, Essex Co., GH, detd. Fernald; Amherstburg, Essex Co., and Clearville, Kent Co., TRT) to N.Y. and Mass., s to Tex. and Fla.

The Canadian plant is referable to var. *glaucodea* (Tuckerm.) Kük. (*C. gl.* Tuckerm.; staminate spike subsessile rather than usually elevated above the upper pistillate one, leaves to about 1 cm broad rather than 1.5 cm).

C. flava L.   (358, 357 (*C. laxior*), and 356 (*C. cryptolepis*))
/ST/X/EA/   (Hs)   Damp meadows, swampy ground, and shores, the aggregate species from SE Alaska and B.C. to Alta. (N to Jasper; not known from Sask.), Man. (N to Grand Rapids, near the NW end of L. Winnipeg), Ont. (N to L. Nipigon and w James Bay at ca. 53°N), Que. (N to E James Bay at ca. 54°N, L. Mistassini, and the Côte-Nord), Nfld., N.B., P.E.I., and N.S., s to Wash., Mont., Minn., Ind., and N.J.; Iceland; Eurasia. MAPS and synonymy: see below.
1   Pistillate scales pale; perigynia greenish, commonly not over 4 mm long, their beaks
    at most 1.5 mm long; leaves to 4 mm broad ....var. *fertilis* Peck
  2   Terminal spike entirely staminate; [var. *graminis* Bailey; range of f. *fertilis*] ....
      ....f. *graminis* (Bailey) Scoggan
  2   Terminal spike pistillate at least above; [*C. cryptolepis* Mack.; Ont. (N to Hearst),
      Que. (N to the Rupert R. and Anticosti Is.), Nfld., N.B., and N.S.] ....f. *fertilis*
1   Pistillate scales brown; perigynia yellowish, to 6 mm long, their beaks at least 1.5 mm
    long; leaves to 6 mm broad.
  3   Perigynia subulate, at most 1.2 mm thick; [Ont. (Bruce Co.; Krotkov 1940), Que. (L.
      Mistassini, Anticosti Is., and the Gaspé Pen.; type from banks of the Bonaventure
      R.), Nfld., and N.S. (Victoria Co., Cape Breton Is.)] ....var. *gaspensis* Fern.
  3   Perigynia lance-ovoid, to 2 mm thick; [incl. var. *rectirostra* Gaud. and *C. lepidocarpa*
      var. *laxior* Kük. (*C. laxior* (Kük.) Mack.); transcontinental; MAPS (aggregate species):
      Hultén 1958: map 43, p. 63; Meusel, Jaeger, and Weinert 1965:78; Raymond 1951:
      map 16, p. 10; E. W. Davies, Watsonia 3(1): fig. 1 (very inaccurate for N. America),p.
      82. 1953] ....var. *flava*

C. folliculata L.   (490)
/T/EE/   (Hs)   Peaty thickets and swampy woods from Wisc. to Ont. (N to the Ottawa dist.), Que. (N to near Quebec City; MT), St-Pierre and Miquelon, Nfld., N.B., P.E.I., and N.S., s to La. and Fla.

C. formosa Dewey   (327)
/T/EE/   (Hs)   Calcareous woods and meadows from Minn. to s Ont. (Essex, Lambton, and Hastings counties; CAN; TRT) and sw Que. (Montreal dist. and Rougemont, Rouville Co.; CAN; MT), s to Iowa, Wisc., Mich., N.Y., and Conn.

C. frankii Kunth   (507)
/t/EE/   (Hs)   Moist calcareous woods, meadows, and swampy ground from Kans., Mo., Ill., and Mich. to s Ont. (Pelee Is. (CAN) and Middle Is., the southernmost Canadian island of the Erie Archipelago, Essex Co.; Core 1948; a collection in OAC from Sturgeon L., Wellington Co., may also belong here), Ohio, N.Y., and Pa., s to Tex. and Ga.

C. garberi Fern.   (271: *C. hassei sensu* Mackenzie 1935, not Bailey)
/ST/X/   (Grh)   Calcareous sands, gravels, and ledges, the aggregate species from n-cent. Alaska–Yukon and the Mackenzie R. Delta to Great Bear L., Great Slave L., N Sask., N Man. (Churchill), Ont. (N to Fort Severn, Hudson Bay, ca. 56°N), Que. (N to L. Mistassini and the Gaspé Pen.), and N.B. (Restigouche and Victoria counties; not known from Nfld., P.E.I., or N.S.), s to s B.C.–Alta.–Sask.–Man., Ind., Mich., Ohio, N.Y., and Maine. MAP and synonymy: see below.
1   Culms stiffly erect, to 4 dm tall; leaves firm, to 5 mm broad; spikes crowded; perigynia to
    2.5 mm long; [*C. hassei sensu* Mackenzie 1935, not Bailey; *C. aurea* vars. *androgyna*
    and *celsa* at least in part; s Ont. (Bruce Pen. and islands of L. Huron; reports from
    elsewhere in Canada refer to var. *bifaria* or to *C. aurea* or *C. bicolor*); MAP: Fernald 1935:
    map 13, p. 255] ....var. *garberi*
1   Culms weaker, to 6 dm tall; leaves softer, to about 2.5 mm broad; spikes less crowded;
    perigynia to 3 mm long; [*C. bicolor sensu* John Macoun 1890, and M. L. Fernald, Rhodora

9(105):159. 1907, not All.; range of the species, the type from along the Ste-Anne-des-Monts R., Gaspé Pen., E Que.; MAP: on the above-noted map by Fernald (incomplete northwards)] . . . . .var. *bifaria* Fern.

C. geyeri Boott   (258)
/T/WW/   (Grh)   Dry open woods and slopes from SE B.C. (Moyie L., s of Cranbrook; Flathead; North Kootenay Pass) and sw Alta. (Waterton Lakes; CAN) to N Calif., Utah, and Colo.

C. glacialis Mack.   (256)
/AST/X/GEA/   (Hs)   Dry calcareous ledges, sands, and gravels, the aggregate species from N-cent. Alaska–Yukon and the Mackenzie R. Delta to Victoria Is., Southampton Is., Ellesmere Is. at ca. 79°N, Baffin Is., and northernmost Labrador, s to SE Alaska, L. Athabasca, NE Man. (Churchill), northernmost Ont. (Hudson Bay at ca. 56°50′N), Que. (s to E James Bay at ca. 53°N, L. Mistassini, and Knob Lake, 54°48′N), and Nfld.; w Greenland N to 73°25′N, E Greenland N to 74°58′N; Iceland; N Eurasia. MAPS and synonymy: see below.
1   Pistillate spikes subsessile to definitely peduncled, lax and open, the scales persistent after the perigynia fall; lowest bract subtruncate, usually with a short minutely hirsute blade; lower sheaths usually purplish at base; [*C. pedata* Wahl., not L.; transcontinental in arctic and subarctic regions; MAPS: Hultén 1962: map 23, p. 31, and 1968*b*:267; Tolmachev 1952: map 33; Porsild 1957: map 91, p. 172; Böcher 1954: fig. 33 (map 3; incomplete southwards), p. 135; Raup 1930: map 12 (incomplete), p. 202; Fernald 1925: map 69 (very incomplete northwards), p. 341] . . . . .var. *glacialis*
1   Pistillate spikes sessile or barely short-peduncled, densely few-flowered, the scales falling before the subpersistent perigynia; lowest bract obliquely sheathing, only rarely with a minute smoothish blade; lower sheaths drab or pale brown, sometimes slightly purplish at base; [*C. terrae-novae* Fern., the type from St. John Is., St. John Bay, Nfld.; also known from cent. Que. (Richmond Gulf, SE Hudson Bay, ca. 56°15′N, and adjacent islands; Clearwater R. at ca. 57°N; Mollie T. Lake and Knob Lake, both ca. 54°45′N, 67°W] . . . . .var. *terrae-novae* (Fern.) Boivin

C. glareosa Wahl.   (91, 90 (*C. pribylovensis*), and 89 (*C. marina sensu* Mackenzie 1931, not Dewey))
/AST/X/GEA/   (Hsr)   Brackish sandy or clayey shores (ranges of Canadian taxa outlined below; not known in the U.S.A.). MAPS and synonymy: see below.
1   Perigynia usually over 3 mm long and more than 2.6 times as long as broad, pale green and distinctly ribbed; [*C. bipartita* var. *glar.* (Wahl.) Polunin; Aleutian Is.–Alaska; Que. (northernmost Ungava; E James Bay; St. Lawrence R. estuary from Temiscouata Co. to the Côte-Nord and Gaspé Pen.); MAPS (all including var. *amphigena*): Hultén 1968*b*:239, and 1962: map 183, p. 195; G. Halliday and A. O. Chater, Feddes Repert. 80(2–3): fig. 1, facing p. 106. 1969] . . . . .ssp. *glareosa* var. *glareosa*
1   Perigynia usually less than 3 mm long, less than 2.6 times as long as broad, often brown and weakly nerved.
  2   Leaves 1.5–2.3 mm broad; pistillate scales 2.75–3.5 mm long; [*C. pribylovensis* Macoun; Aleutian Is.; Pribilof and ?Commander Is.; MAP: Hultén 1968*b*:240; G. Halliday and A. O. Chater, loc. cit., fig. 1, p. 94] . . . .
        . . . . .ssp. *pribylovensis* (Macoun) Halliday & Chater
  2   Leaves 1.0–1.9 mm broad; pistillate scales 2.25–3.0 mm long; [*C. bipartita* var. *amph.* (Fern.) Polunin; *C. amph.* (Fern.) Mack.; *C. marina* of auth., not Dewey (see note under *C. amblyorhyncha*); Alaska–Yukon–Dist. Mackenzie–w Victoria Is.; N Man. (Churchill); Ellesmere Is. at ca. 80°N; Devon Is., Baffin Is., and E Dist. Keewatin to Que. (s to s James Bay and the St. Lawrence R. estuary from St-Roch-des-Aulnets, l'Islet Co., to the Côte-Nord and Gaspé Pen.; type from Escuminac, Bonaventure Co., Gaspé Pen.), NE N.B. (Dalhousie), and N.S.; MAPS: Porsild 1957: map 82, p. 171; Schofield 1959: map 6, p. 111; Potter 1934: map 9 (very incomplete), p. 76] . . . .
        . . . . .ssp. *glareosa* var. *amphigena* Fern.

C. gmelinii H. & A.   (422)
/aST/W/eA/   (Grh)   Coasts of the Aleutian Is. and Alaska (N to Kotzebue Sound, s to the southernmost Alaska Panhandle); reported from B.C. by Mackenzie (1935) and Henry (1915; Cascades), and a collection in V from near Prince Rupert has been placed here; E Asia. MAPS: Hultén 1968*b*:258, and 1942: map 283, p. 408.

C. gracillima Schw.   (325)
/T/EE/   (Grh)   Woods, thickets, and meadows from Man. (N to Silver Falls, 50°31′N; reported N to Norway House by Hooker 1839) to Ont. (N to Thunder Bay and New Liskeard), Que. (N to the Gaspé Pen.), Nfld., N.B., P.E.I., and N.S., s to Mo., Tenn., and N.C.
　　　Var. *macerrima* Fern. (perigynia brown and acutely angled, less than 3 mm long, rather than greenish and obtusely angled, to 3.5 mm long; lateral spikes rarely over 3 cm long rather than up to 6 cm long; leaves at most about 5 mm broad rather than up to 9 mm), merely the reduced extreme and of little taxonomic significance, is known from the type locality, Bay of Islands, w Nfld.

C. granularis Muhl.   (305 and 303 (*C. haleana*))
/T/EE/   (Hs)   Calcareous woods, meadows, and shores (ranges of Canadian taxa outlined below), s to Kans., La., and Fla. MAP and synonymy: *see* below.
1　Perigynia broadly ovoid to subglobose, to 4 mm long and 2.5 mm thick; [s Ont. N to near Ottawa; MAP (aggregate species): Raymond 1951: map 40 (referring chiefly to var. *haleana*), p. 19] ....*var. granularis*
1　Perigynia at most 3 mm long and 1.5 mm thick; [*C. haleana* Olney; *C. shriveri* Britt.; E Sask. (collection in DAO from Spy Hill; not listed by Breitung 1957*a*), Man. (N to Swan River, N of Duck Mt.), Ont. (N to w James Bay at ca. 52°N), Que. (N to the Gaspé Pen.), N.B., and N.S.; intergrading with var. *gran.* through the nondescript var. *recta* Dew. and of doubtful taxonomic value; MAP: (*see* above)] ....*var. haleana* (Olney) Porter

C. gravida Bailey   (49)
/T/WW/   (Grh)   Prairies, swamps, and shores from Wyo. to SE Sask. (Roche Percée; DAO), s Man. (Otterburne, about 30 mi s of Winnipeg; WIN), and s Ont. (Walkerville, Essex Co.; GH, detd. Mackenzie), s to N.Mex., Tex., Okla., and Ark.

C. grayii Carey   (528)
/T/EE/   (Hs)   Calcareous meadows and alluvial woods from Iowa to s Ont. (N to the Ottawa dist.; Gillett 1958) and sw Que. (N to Buckingham and the Montreal dist.; *see* Que. map by Robert Joyal, Nat. can. (Que.) 97(5): map F, fig. 1, p. 562. 1970), s to Ark., Miss., and Ga. MAPS: M. L. Fernald, Rhodora 39(439): map 19, p. 343. 1937, and 44: 323. 1942.

C. gymnoclada Holm   (442)
/T/W/   (Grh)   Wet meadows, rocks, and shores from SE B.C. (Kootenay L. at Kaslo and Nelson; CAN, detd. Hermann and Eastham) and sw Alta. (Livingstone Falls, 41 mi N of Coleman; CAN, detd. Hermann) to Calif. and Colo.

C. gynocrates Wormsk.   (99)
/aST/X/GA/   (Grh)   Peaty soils and sphagnous bogs from the Aleutian Is., N Alaska, cent. Yukon, the Mackenzie R. Delta, and Great Bear L. to L. Athabasca (Alta. and Sask.), Man. (N to Churchill; an isolated station on Melville Pen., NE Dist. Keewatin), Ont. (N to Goose Creek, Hudson Bay, ca. 56°N), southernmost Baffin Is., N Ungava–Labrador, Nfld., N.B., and N.S. (not known from P.E.I.), s to B.C., Oreg., Colo., Minn., Mich., Pa., and N Maine; w Greenland (type locality) N to ca. 71°N; E Asia (the closely related *C. dioica* L. in Iceland and Eurasia). [*C. dioica* var. *gyn.* (Wormsk.) Ostenf.; *C. dioica* of early Alaska–Canada reports, not L.; *C. alascana* Boeck.]. MAPS: Hultén 1968*b*:222, and 1962: map 161, p. 171; Porsild 1957: map 73. p. 170; Raup 1947: pl. 17.

C. hallii Olney   (411)
/T/WW/   (Grh)   Moist prairies, foothills, and alpine meadows from Mont., ?Alta. (Mackenzie 1935; not listed by Moss 1959), ?Sask. (merged by Breitung 1957*a*, with *C. parrayana,* from which

it is scarcely separable), and s Man. (Brandon; Macgregor; Treesbank; Otterburne) to Colo. and S.Dak. [*C. parrayana* var. *h.* (Olney) Kük.]. MAP: D. F. Murray, Brittonia 21(1): fig. 7, p. 71. 1969.

[C. hassei Bailey]   (271: see *C. garberi*)
[Reports of this species from our area are chiefly based upon *C. garberi* Fern., considered by M. L. Fernald (Rhodora 37(439): 253–55. 1935) to be sufficiently distinct to warrant specific status. If this separation eventually proves to be untenable, the name *C. garberi* must be placed in synonymy under the prior name, *C. hassei.*

C. haydeniana Olney   (127: *C. nubicola*)
/T/W/   (Hs)   Mountain slopes from s B.C. (Vancouver Is.; Marble Range, NW of Clinton; Yoho; Mt. Paget, N of Hector; Windermere; Flathead) and sw Alta. (Crowsnest Pass; Waterton Lakes; Elbow Valley; Banff dist.; Laggan) to Calif., Nev., Utah, and Colo. [*C. festiva (macloviana)* var. *hay.* (Olney) Boott; *C. nubicola* Mack.].

C. haydenii Dewey   (463)
/T/EE/   (Grh)   Thickets, meadows, and swampy places from Ont. (N to Renison, s of James Bay at ca. 51°N; Hustich 1955; reports from B.C. require confirmation) to Que. (N to the Koksoak R. at ca. 57°50′N, L. Mistassini, and Anticosti Is.), N.B., and St-Pierre and Miquelon (not known from Nfld., P.E.I., or N.S.), s to Nebr., Mo., Ohio, and N.J. [*C. aperta* var. *minor* Olney; *C. stricta* vars. *decora* Bailey and *hay.* (Dew.) Kük.].

C. heleonastes Ehrh.   (88; see *C. amblyorhyncha*)
/ST/(X)/EA/   (Hsr)   Wet open places and shores (often calcareous) from cent. Alaska–Yukon–Dist. Mackenzie to NE Man. (Churchill), northernmost Ont., and Que. (James Bay–Hudson Bay watershed N to ca. 57°N), s to s-cent. B.C.–Alta.–Sask.–Man. and s James Bay; Iceland; Eurasia. MAPS and synonymy: see below.
1   Perigynia scabrous-beaked, not clasped by the margins of their scarcely hyaline-margined scales; culms stiffish; [incl. *C. amblyorhyncha* Krecz.; *C. carltonia* Dew.; *C. cryptantha* Holm; transcontinental; MAPS: Hultén 1968b:238, and 1962: map 59 (incl. the area of *C. ambly.*), p. 69; T. W. Böcher, Acta Arct. 5: fig. 12, p. 27. 1952; Löve and Löve 1956b: fig. 5, p. 134; Meusel, Jaeger, and Weinert 1965: 68] . . . . .ssp. *heleonastes*
1   Perigynia smooth-beaked, clasped by the margins of their distinctly hyaline-margined scales; culms relatively weak; [*C. neurochlaena* Holm, the type from Rink Rapid, Yukon Territory; Alaska (reported from Umiat by Hultén 1950, and from NE Alaska by Wiggins and Thomas 1962), s-cent. Yukon (see Hultén 1942: map 254, p. 406 (*C. neur.*), and Dist. Mackenzie (Mackenzie R. Delta, Eskimo L. basin, and Great Bear L.; Porsild 1943) . . . . . .
. . . . .ssp. *neurochlaena* (Holm) Böcher

C. hendersonii Bailey   (293)
/T/W/   (Hs)   Damp woods from sw B.C. (Vancouver Is., Vancouver, Agassiz, and Chilliwack R.; CAN; also reported from Yale, lower Fraser Valley, by Henry 1915) to N Calif. and Idaho.

C. hindsii Clarke   (450)
/sT/W/   (Grh)   Wet meadows along the coast from the Aleutian Is. and s Alaska (see Hultén 1942: map 274, p. 408) through B.C. to NW Calif. [*C. caespitosa sensu* Bongard 1833, not L.; *C. decidua* and *C. vulgaris* of w N. America reports in large part, not Boott nor Fries, respectively; *C. vulg. (lenticularis)* var. *limnophila* Holm; *C. interrupta* of Alaskan reports in part, not Boeck.]. MAPS: Hultén 1958: map 108, p. 127 (merged with *C. kelloggii* by Hultén 1968b); Meusel, Jaeger, and Weinert 1965:70.

C. hirsutella Mack.   (377)
/T/EE/   (Grh)   Fields, meadows, open woods, and clearings from Iowa and Mich. to s Ont. (Essex, Elgin, Lincoln, and Welland counties), sw Que. (Rouville, Iberville, and Missisquoi counties), and SE Maine, s to Tex. and Ala. [*C. hirsuta* Willd.; *C. triceps (complanata)* var. *hirsuta* (Willd.) Bailey].

C. hirta L. (386)
Eurasian; introd. in Que. (waste land at a factory in Limoilou, near Quebec City; MT), P.E.I. (damp field at Charlottetown; CAN; GH), and N.S. (sandy railway bank at Annapolis Royal, Annapolis Co.; CAN; GH).

C. hirtifolia Mack. (249)
/T/EE/ (Grh) Dry woods, thickets, and meadows (often calcareous) from Ont. (N to the Ottawa dist.), Que. (N to St-David, Lévis Co.), N.B. (Petitcodiac, Westmorland Co., and Woodstock, Carleton Co.; CAN; GH), and N.S. (Hants and Colchester counties; ACAD; GH; not known from P.E.I.) to E Kans., Mo., Ky., and Md. [C. pubescens Muhl., not Poir.]. MAPS: the distribution in Ont. and Que. is shown in maps by Rouleau 1945: fig. 9, p. 171, and Raymond 1950b: fig. 35, p. 85.

C. hitchcockiana Dewey (309)
/T/EE/ (Hs) Calcareous or rich woods from Wisc. to Ont. (N to the Ottawa dist.), sw Que. (N to the Montreal dist. and Hull; not known from the Atlantic Provinces), and Vt., s to Ark., Tenn., and Va.

C. holostoma Drejer (414)
/aST/X/GEA/ (Grh) Moist tundra and margins of ponds from the Seward Pen., Alaska (not known from the Yukon) and the coasts of Dist. Mackenzie and Dist. Keewatin to s-cent. Baffin Is., s to s-cent. Alaska, Great Bear L., s Dist. Keewatin, Southampton Is., and E Hudson Bay, Que., at ca. 58°N; w Greenland (type locality) between ca. 67° and 73°N; Iceland; N Scandinavia; widely scattered stations in N Asia. [C. alpina ssp. hol. (Drej.) Mela & Caj.]. MAPS: Hultén 1968b:255, and 1958: map 206 (indicating a station at Churchill, Man., this perhaps referable to C. media), p. 225; Porsild 1957: map 89, p. 172; Raymond 1950b: fig. 11, p. 18; Tolmachev 1952: map 35.

C. hoodii Boott (32)
/T/WW/ (Hs) Mountain meadows and slopes from B.C. (N to McLeod Lake, ca. 54°20′N), Alta. (N to Banff and Lake Louise), and sw Sask. (Cypress Hills; Breitung 1957a) to Calif., Colo., and S.Dak. [C. muricata var. confixa Bailey].

C. hookerana Dewey (39)
/T/WW/ (Hs) Plains, prairies, and dry banks from ?B.C. (Henry 1915) to Alta. (N to Fort Saskatchewan; CAN, detd. Hermann), Sask. (N to the type locality near Carleton, about 40 mi sw of Prince Albert), and sw Man. (Brandon and Macgregor; CAN; reported as a roadside introduction at Schreiber, N shore of L. Superior, Ont., by Fernald 1935), s to s Alta.–Sask.–Man. and N.Dak. [C. muricata var. gracilis Boott.].

C. hormathodes Fern. (179)
/T/E/ (Hs) Brackish to fresh marshes and sands (chiefly near the coast) from E Que. (l'Islet Co. to Anticosti Is., the Gaspé Pen., and Magdalen Is.) to w Nfld., N.B., P.E.I., and N.S., s to Va. (inland in Ind.; the report from B.C. by Henry 1915, probably refers to some other member of the Section Ovales). [C. straminea var. festucacea sensu Fowler 1885, and John Macoun 1888 (at least as to the N.B. and N.S. reports), not (Schk.) Gay, relevant collections in CAN and NBM].

Forma invisa (Boott) Fern. (var. inv. Boott; perigynia relatively small, in spikes less than 1 cm long rather than to 1.5 cm) is known from Nfld. and N.S.

C. hostiana DC. (351: C. fulvescens)
/T/E/E/ (Hs) Calcareous marshes and shores of E Canada; Europe. Ranges, MAPS, and synonymy: see below.
1   Perigynia about 3 mm long; pistillate spikes at most 1.5 cm long; staminate spike at most
    2 cm long; [C. hornschuchiana Hoppe; reported from St-Pierre and Miquelon by Fernald
    in Gray 1950, on the basis of a collection in GH; MAPS (all as C. hostiana except that by
    Hultén): Hultén 1958: map 136, p. 155; Meusel, Jaeger, and Weinert 1965: p. 77;
    Heslop-Harrison 1953: facing p. 111; Raymond 1951: map 15, p. 10; Marie-Victorin
    1929a: fig. 18, p. 48, and 1938: fig. 14, p. 506] . . . . .var. hostiana

1   Perigynia to 5 mm long; pistillate spikes to 2 cm long; staminate spike to 3 cm long;
     [C. horn. var. laur. F. & W., the type from Table Mt., w Nfld.; C. fulvescens Mack.; C. fulva
     Good.; E Que. (Anticosti Is.), St-Pierre and Miquelon, and w Nfld. (several localities in
     addition to the type locality); MAPS: all of the above-noted maps refer to this taxon, Hultén's
     map also indicating the station of var. hostiana by a solid dot] . . . .var. laurentiana Fern. & Wieg.

C. houghtonii Torr.   (383)
/T/(X)/   (Grh)   Dry acid sands, gravels, and clearings from Alta. (Athabasca Plains and Snuff
Mt., about 110 mi w of Whitecourt; CAN) to Sask. (N to Methy Portage, 56°38'N; John Macoun
1888), Man. (N to Horseshoe L. at 57°43'N; G. W. Scotter, Blue Jay 23(2):98. 1965), Ont. (N to
Sandy L. at ca. 53°N, 93°W), Que. (N to the St. John R. at 50°44'N, the Côte-Nord, and the Gaspé
Pen.), Nfld., N.B., and N.S. (not known from P.E.I.), s to Minn., Wisc., Mich., N.Y., and N New Eng.
[C. houghtoniana Torr.].

C. howei Mack.   (106)
/T/EE/   (Hs)   Sphagnous or mossy swamps and thickets from s Mich. and Ohio to N.B.
(Lepreau, Charlotte Co.; DAO), P.E.I. (McNeill's Mills, Prince Co.; NSPM), and N.S. (St. Paul Is.
and Digby, Yarmouth, Shelburne, and Inverness counties), s to La. and Fla. MAP: Fernald 1921:
map 1 (incomplete northwards), pl. 130, facing p. 120.

C. hyalinolepis Steud.   (500)
/t/EE/   (Grh)   Calcareous or brackish swamps and shores from Nebr. to Mich., Ohio, s Ont.
(Amherstburg, Essex Co.; TRT; Herb. M. Raymond), s Pa., and s N.J., s to Tex. and Fla.
[C. lacustris var. laxiflora Dew., not C. lax. Lam.; C. riparia var. impressa Wright; C. imp. (Wright)
Mack.]. MAP: Meusel, Jaeger, and Weinert 1965:79.

C. hystricina Muhl.   (493)
/T/X/   (Hsr (Grh))   Swamps, wet meadows, and shores from s B.C. (Lillooet and vicinity; CAN)
to s Alta. (De Winton and Medicine Hat; CAN), Sask. (N to near Saskatoon), s Man. (N to St. Lazare,
about 75 mi NW of Brandon), Ont. (N to L. Nipigon), Que. (N to Chicoutimi and the Gaspé Pen.),
Nfld., N.B., P.E.I., and N.S., s to Calif., N.Mex., Tex., Tenn., and Va. [C. hystericina, the original but
presumably derivatively incorrect spelling].

C. illota Bailey   (136)
/T/WW/   (Grh)   Gravelly shores and slopes from s B.C. (collections in CAN from Vancouver Is.,
Mayne Is., Chilliwack R., and mts. near Griffin L., Kamloops dist.; in DAO from Mt. Revelstoke,
Revelstoke) to Calif., Utah, and Colo. [C. dieckii Boeck.; C. bonplandii var. minor Boott; incl. C.
limnophila Hermann].

C. inops Bailey   (221)
/T/W/   (Grh)   Dry soil (chiefly in the Cascade Mts.) from s B.C. (Vancouver Is. and adjacent
islands, the Chilliwack R., and Yale, lower Fraser R.; CAN) to Calif. [C. pensylvanica var.
vespertina Bailey (C. vesp. (Bailey) Howell); C. pen. sensu John Macoun 1888, as to the B.C.
plant, not Lam., relevant collections in CAN].

C. interior Bailey   (105)
/sT/X/   (Hs)   Damp or wet, often calcareous, soils from s Yukon–Dist. Mackenzie and B.C.–Alta.
to Sask. (N to Windrum L. at ca. 56°N), Man. (N to Grand Rapids, near the NW end of L. Winnipeg),
Ont. (N to Goose Creek, Hudson Bay, ca. 56°N), Que. (N to E James Bay at ca. 54°N, L. Mistassini,
and the Côte-Nord), s Labrador (Forteau, 51°28'N), Nfld., N.B., P.E.I., and N.S., s to Calif., N
Mexico, Kans., Pa., and Del. [C. scirpoides Schkuhr in part]. MAP: Hultén 1968b:245.
        Forma keweenawensis (Hermann) Fern. (perigynia distinctly nerved rather than nerveless) is
known in Canada from Que., s Labrador, and Nfld.

[C. interrupta Boeckl.]   (454)
[This species of Wash.–Oreg.) is reported from sw B.C. by Henry (1915; Duncan, Vancouver Is.)

and a collection in V from about 45 mi N of Prince George has been placed here, but confirmation is required.]

C. intumescens Rudge   (529)
/T/EE/   (Hs)   Meadows, swamps, and alluvial woods (ranges of Canadian taxa outlined below), s to Tex. and Fla.
1   Perigynia conic-ovoid, to 8 mm thick at base; achenes broadest near the middle; [s Ont., sw Que., and N.S.; Fernald *in* Gray 1950] ....var. *intumescens*
1   Perigynia lanceolate to lance-ovoid; achenes broadest near the rounded summit ....
....var. *fernaldii* Bailey
   2   Perigynia to 8 mm thick; [Que., Nfld., and N.S.] ....f. *ventricosa* Fern.
   2   Perigynia at most about 5 mm thick; [Man. (N to Norway House, off the NE end of L. Winnipeg; Hooker 1839) to Ont. (N to the Nipigon R. N of L. Superior), Que. (N to the Rupert R. SE of James Bay at ca. 51°30'N, Anticosti Is., and the Gaspé Pen.), Nfld., N.B., P.E.I., and N.S.] ....f. *fernaldii*

C. jacobi-peteri Hult.
/a/W/   (Hs)   Known only from the type locality, Tin City, Seward Pen., coast of w Alaska. MAPS: Hultén 1968*b*:221, and 1942: map 225, p. 403.

C. jamesii Schw.   (200)
/t/EE/   (Hs)   Rich, mostly calcareous, woods from Iowa to Mich., s Ont. (Essex, Middlesex, Oxford, Welland, and Wellington counties; CAN; OAC; TRT; Herb. M. Raymond) and N.Y., s to Kans., Mo., Tenn., and Va. [*C. steudelii* Kunth].

C. kelloggii Boott   (449)
/sT/WW/   (Grh)   Wet meadows and swamps from s Alaska (*see* Hultén 1942: map 273, p. 407) through B.C. and the mts. of sw Alta. (Waterton Lakes; Castle Mts., near Banff) to Calif., Utah, and Colo. [*C. aleutica* Akiyama; merged with *C. lenticularis* by Hitchcock et al. 1969; *C. decidua sensu* John Macoun 1890, not Boott, relevant collections in CAN; *C. vulgaris* of B.C. reports in part, not Fries; *C. vulg.* var. *lipocarpa* Holm]. MAPS: Hultén 1968*b*:250, and 1958: map 108, p. 127: Meusel, Jaeger, and Weinert 1965: 70.

C. lachenalii Schkuhr   (87: *C. bipartita sensu* Mackenzie 1931, perhaps not All.)
/aST/X/GEA/   (Hsr)   Moist or wet (often calcareous) soils from the coasts of Alaska–Yukon–Dist. Mackenzie–Dist. Keewatin to cent. Baffin Is. and northernmost Ungava–Labrador, s to Great Bear L., SE Dist. Keewatin, NE Man. (Churchill; not known from Sask. or Ont.), Que. (s to E James Bay, Knob Lake at 54°48'N, and the Shickshock Mts. of the Gaspé Pen.), s Labrador, and N Nfld., and in the mts. of the West through B.C. and sw Alta. to Mont., Utah, and Colo.; w Greenland N to ca. 72°30'N, E Greenland, N to 74°37'N; Iceland; Eurasia. [*C. lagopina* Wahl.; *C. (Kobresia) bipartita* of American auth., perhaps not All.]. MAPS: Hultén 1968*b*:238, and 1962: map 44, p. 51; Porsild 1957: map 80, p. 170; Raup 1947: pl. 17 (*C. bip.*); G. E. Du Rietz, Acta Phytogeogr. Suec. 13: fig. 1, p. 216. 1940; Raymond 1951: map 6 (E Canada; *C. lag.*), p. 5; Young 1971: fig. 13, p. 88.

C. lacustris Willd.   (499)
/T/(X)/   (Grh)   Swamps and shallow water from ?Alta. (Moss 1959) to Sask. (N to Montreal L. at ca. 54°N; Breitung 1957*a*), Man. (N to Riding Mt.; CAN), Ont. (N to the N shore of L. Superior near Thunder Bay and Cochrane, ca. 49°N; the report of *C. riparia* as far N as Moose Factory, near James Bay, by John Macoun 1888, probably refers to *C. atherodes*), Nfld. (near Stephenville; CAN), N.B. (Bass River, Kent Co.; CAN), P.E.I. (Prince Co.; D. S. Erskine 1960), and N.S., s to Idaho, S.Dak., Mo., and Va. [*C. riparia* var. *lac.* (Willd.) Kük.; *C. riparia* of Canadian reports, not Curtis]. MAPS: Meusel, Jaeger, and Weinert 1965:79; Raymond 1951: map 33 (E Canada), p. 17.

C. laeviconica Dewey   (501)
/T/WW/   (Grh)   Calcareous marshes and wet prairies from Sask. (N to Waskesiu Lake, ca. 54°N; Breitung 1957*a*) and s Man. (Melita; Brandon; Otterburne) to Mont., Kans., Mo., and Ill. [*C. trichocarpa* var. *deweyi* Bailey].

C. laeviculmis Meinsh. (101)
/sT/W/eA/ (Hs) Moist meadows and slopes from s Alaska (*see* Hultén 1942: map 267, p. 407) through B.C. to Calif., Idaho, and Mont.; E Asia (the type material from Kamchatka and Sitka Is., Alaska). [*C. bolanderi (deweyana)* var. *sparsiflora* Olney; *C. elongata sensu* Hooker 1839, and Bongard 1833, not L.]. MAP: Hultén 1968b: 246.

C. laevivaginata (Kük) Mack. (77)
/t/EE/ (Hs) Boggy or swampy meadows and woods from Minn. to s Ont. (Kettle Point, Lambton Co.; OAC; reported from Hamilton, Wentworth Co., and "Delaware" (presumably Delamere, s of Sudbury), by Boivin 1967a) and Maine, s to Mo., Ala., and Fla. [*C. stipata* var. *laev.* Kük.].

C. langeana Fern. (*see* Mackenzie 1935: 469)
/aST/E/G/ (Grh) Reported from "Peaty limestone-barrens, Ingornachoix Bay, Nfld.; Greenl." by Fernald *in* Gray (1950; *see,* also, Fernald 1933:217–19), who notes that it may actually be a hybrid between *C. gynocrates* Wormsk. and *C. maritima* Gunn. (collections in CAN from N Alaska (Meade R. at ca. 70°42′N), Dist. Mackenzie (Great Bear L.), s Ellesmere Is. (Harbour Fjord), E Dist. Keewatin (Chesterfield Inlet), NE Man. (Churchill), Hudson Bay (Belcher Is. at ca. 56°30′N), and w and E Greenland have been placed here by Porsild, as also, tentatively, a collection from Moose Factory, s James Bay, Ont. (collections in DAO and MTJB from Payne Bay, N Que., ca. 60°N, have been referred here by Rousseau and it is reported from Makkovik, Labrador, ca. 55°N, by Hustich and Pettersson 1943). [*C. duriuscula* Lange, basionym (not Meyer), the type from Greenland].

C. lanuginosa Michx. (384)
/T/X/ (Grh) Swamps, marshes, and meadows from B.C. (N to Prince Rupert; CAN; introd. in gravel along a railway at Wasilla, cent. Alaska) to Alta. (N to Beaverlodge, 55°10′N), Sask. (N to Prince Albert), Man. (N to the Nelson R. about 150 mi s of Churchill), Ont. (N to the Shamattawa R. at ca. 55°N), Que. (N to E James Bay at ca. 52°N, L. Mistassini (type locality), and Anticosti Is.), Nfld., N.B., and N.S. (not known from P.E.I.), s to s Calif., Tex., and Va. [Incl. var. *oriens* Raymond; *C. lasiocarpa* ssp. *lan.* (Michx.) Clausen & Wahl; *C. filiformis (lasio.)* var. *latifolia* Boeck.; *C. pellita* Muhl.]. MAP: Hultén 1962: map 67, p. 77.

C. lasiocarpa Ehrh. (385)
/ST/X/EA/ (Grh) Bogs, peaty meadows, shores, and shallow water from s-cent. Alaska and w Dist. Mackenzie (near Great Bear L. and Great Slave L.; CAN) to B.C., L. Athabasca (Alta. and Sask.), Man. (N to Grand Rapids, near the NW end of L. Winnipeg), Ont. (N to Big Trout L. at ca. 53°45′N, 90°W), Que. (N to E James Bay at ca. 54°N and the Côte-Nord), Nfld., N.B., and N.S. (not known from P.E.I.), s to Wash., Idaho, Iowa, Ohio, and N.J.; Eurasia. [*C. filiformis* of American auth., not L.]. MAPS: Hultén 1968b:279, and 1962: map 67, p. 77.
   The American plant may be known as var. *americana* Fern. (type from Argyle, Yarmouth Co., N.S.; *C. lanuginosa* var. *amer.* (Fern.) Boivin), differing from the typical Eurasian form in the almost total absence of a sheath at the base of the foliaceous bract subtending the lowest spike, the more generally awn-tipped pistillate scales, and the shorter and relatively broader perigynia.

C. laxa Wahl.
/sT/W/EA/ (Grh) Known in N. America only from s-cent. Alaska (*see* Hultén 1942: map 307, p. 410) and NW Dist. Mackenzie (Mackenzie R. Delta; CAN, detd. Porsild); scattered stations in Eurasia. MAP: Hultén 1968b:270.

C. laxiculmis Schw. (290)
/T/EE/ (Hs) Dry or moist woods (chiefly calcareous) from Wisc. to s Ont. (N to Wellington, Peel, York, and Durham counties), sw Que. (Frelighsburg, Missisquoi Co.; Lionel Cinq-Mars, Nat. can. (Que.) 96: 159. 1969; a collection in MT from Chambly, near Montreal, requires confirmation) and s Maine, s to Mo., Tenn., N.C., and Long Is. [*C. retrocurva* Dewey].

C. laxiflora Lam. (296, 295 (*C. ormostachya*), 297 (*C. striatula*), 300 (*C. albursina*), 301 (*C. blanda*), and 302 (*C. gracilescens*))
/T/EE/ (Hs) Woods, thickets, meadows, and clearings, the aggregate species from s Man.

(Brandon; Whiteshell Forest Reserve, E of Winnipeg) to Ont. (N to Schreiber and Peninsula, N shore of L. Superior), Que. (N to L. St. John, the Côte-Nord, and Gaspé Pen.), N.B. (Charlotte Co.), and N.S. (not known from P.E.I.), s to Tex. and Fla.

1 Staminate spike usually long-peduncled, its base usually overtopping the uppermost pistillate spike; perigynia strongly outward-curving; upper bract rarely overtopping the inflorescence.

  2 Perigynia about 3 mm long, their short beaks very abruptly bent; sterile shoots forming conspicuous culms, their leaves to about 8 mm broad; fertile culms purple-tinged at base, their leaves to about 5 mm broad; [*C. gracilescens* Steud.; s Ont. and sw Que.] . . . . .var. *gracillima* (Boott) Robins. & Fern.

  2 Perigynia to 5.5 mm long, their relatively long beaks less bent; sterile shoots forming mere tufts of leaves, these to 1.5 cm broad; fertile culms green or brown-tinged at base, their leaves to 7 mm broad; [var. *michauxii* Bailey; *C. striatula* Michx.; s Ont. and N.S.] . . . . .var. *angustifolia* Dewey

1 Staminate spike usually subsessile or short-peduncled, partly hidden by the rather crowded upper pistillate spikes.

  3 Pistillate scales flabellate-obovate, truncate or emarginate and usually awnless; perigynia to 4 mm long, with short ascending or curving beaks; basal leaves to 4 cm broad; culm-leaves to 2 cm broad; upper bract broad and foliaceous, greatly overtopping the inflorescence; [*C. albursina* Sheldon; Ont. and Que.; *see* Que. map of *C. albursina* by Doyon and Lavoie 1966: fig. 1, p. 815] . . . . .var. *latifolia* Boott

  3 Pistillate scales mostly cuspidate or awned; culms not wing-angled.

    4 Perigynium-beak relatively long and slender; perigynia to 4.5 mm long, half as thick; pistillate scales pale; staminate scales usually with the midrib excurrent; basal leaves to 2.5 cm broad; culm-leaves to 1 cm broad; upper bract often equalling or surpassing the inflorescence; sterile shoots forming mere tufts of leaves; [var. *patulifolia* (Dewey) Carey; *C. anceps* Muhl.; Ont. to N.S. and ?N.B.] . . . . .var. *laxiflora*

    4 Perigynium-beak very short; sterile shoots forming conspicuous culms.

      5 Perigynium nearly straight, to 3.5 mm long, the beak suberect; basal leaves to 8 mm broad, often purple-tinged at base; culm-leaves to 6 mm broad; upper bract equalling or surpassing the inflorescence; [*C. ormostachya* Wieg.; Ont. to N.B. and N.S.] . . . . .var. *ormostachya* (Wieg.) Gl.

      5 Perigynium strongly outward-curving, to 4.5 mm long, its beak abruptly bent; basal leaves to 12 mm broad, brown at base; culm-leaves to 9 mm broad; upper bract usually much overtopping the inflorescence; [*C. blanda* Dewey; s Man. to Que.] . . . . .var. *blanda* (Dewey) Boott

C. leavenworthii Dewey   (34)
/t/EE/   (Grh)   Dry woods and open ground from Kans. to s Mich., s Ont. (Pelee Is., Essex Co., where taken by John Macoun in 1882; CAN; GH), Pa., and s N.J., s to Tex. and Fla. [*C. cephalophora* var. *angustifolia* Boott].

C. lenticularis Michx.   (447)
/sT/X/   (Hs)   Meadows, swampy ground, and shores, the aggregate species from B.C. (N to McLeod Lake, ca. 54°20′N) to Alta. (N to L. Athabasca), Sask. (N to Methy Portage, 56°38′N), Man. (N to the Cochrane R. at 58°13′N), Ont. (N to Sandy L., ca. 53°N, 90°W), Que. (N to s Ungava Bay, L. Mistassini, and the Côte-Nord), Labrador (N to Hopedale, 55°27′N), Nfld., N.B., and N.S. (not known from P.E.I.), s to Calif., Nev., Colo., Minn., Mich., and Mass.

1 Pistillate scales short-oblong to suborbicular, to 2 mm long; perigynia rounded at both ends; [Nfld.: type from the east Humber R.; also known from the Exploits R.] . . . . .var. *eucycla* Fern.

1 Pistillate scales oblong or elliptic, to 3.5 mm long; perigynia acutish at both ends.

  2 Terminal spike normally entirely staminate; [transcontinental] . . . . .var. *lenticularis*

  2 Terminal spike staminate only at base.

    3 Perigynia elliptic or oval, at most 3 mm long, with very short stipes and tips; [Ont. to Labrador and Nfld.] . . . . .var. *blakei* Dewey

3 Perigynia lance-ovate to subrhombic, to 3.5 mm long, tapering to long stipes and empty tips; [Labrador N to Hopedale, Nfld., and E Que.] . . . . .var. *albimontana* Dewey

C. lepidocarpa Tausch   (359)
/T/EE/E/   (Hs)   Calcareous bogs, swampy ground, and gravels of Ont. (Gogama, near Sudbury; CAN), Que. (N to L. Mistassini and the Côte-Nord), w Nfld., N.B. (Belledune and St. Andrews; CAN), and N.S. (Inverness Co., Cape Breton Is.; ACAD); Europe. [Incl. var. *nelmesiana* Raymond; *C. flava* var. *elatior* Schlecht. in part]. MAPS: Hultén 1958: map 157, p. 176; Meusel, Jaeger, and Weinert 1965:78; E. W. Davies, Watsonia 3(1): fig. 1 (Labrador stations should be deleted), p. 82. 1953.

C. leporina L.   (147)
Eurasian; introd. in wet meadows, pastures, and exposed slopes in s ?B.C. (*see* note under *C. tracyi*), Nfld. (Bay of Islands and St. John's; GH; CAN), P.E.I. (Charlottetown; Summerside, Prince Co.), and N.S. (Brier Is. and Digby, Yarmouth, and Shelburne counties). A collection in V from Okanagan, B.C., and the report from Batchawana Bay, L. Superior, Ont., by Hosie (1938) are probably based upon some other member of the Section *Ovales*. [*C. ovalis* Good.].

C. leptalea Wahl.   (198)
/ST/X/   (Hsr)   Mossy or wet woods and openings from N-cent. Alaska and s Yukon to Great Bear L., L. Athabasca (Alta. and Sask.), Man. (N to Churchill), northernmost Ont., Que. (N to s Ungava Bay, L. Mistassini, and the Côte-Nord), Labrador (N to Nain, ca. 56°30′N), Nfld., N.B., P.E.I., and N.S., s to Calif., Tex., and Fla. [Incl. ssp. *pacifica* Calder & Taylor; *C. polytrichoides* Muhl.; *C. microstachya* Michx., not Ehrh.]. MAPS: Hultén 1968b:225; Raup 1947: pl. 17.

C. leptonervia Fern.   (298)
/T/EE/   (Hs)   Damp woods, thickets, and clearings from N Minn. to Ont. (N to the N shore of L. Superior and the Missinaibi R. at ca. 50°N), Que. (N to Dyke L. at 54°25′N and the Côte-Nord), s Labrador (near Forteau, 51°28′N), Nfld., N.B., P.E.I., and N.S., s to Wisc., Mich., Ohio, Tenn., and N.C. [*C. laxiflora* vars. *lept.* Fern. and *varians* Bailey]. MAPS (E Canada): Raymond 1950a: map 12, p. 443, and Ann. ACFAS 17:159. 1951.

C. leptopoda Mack.   (122)
/T/W/   (Grh)   Woods and thickets from s B.C. (Vancouver Is.; Rossland; Nelson; Kootenay L.) to Calif. and Ariz. [*C. deweyana* var. *lept.* (Mack.) Boivin; scarcely separable from *C. deweyana; see* Hitchcock et al. 1969].

C. limosa L.   (407)
/ST/X/EA/   (Grh (Hsr))   Peaty bogs and pond-margins from N-cent. Alaska, s Yukon, and NW Dist. Mackenzie to s Dist. Keewatin, northernmost Ont., Que. (N to s Ungava Bay), Labrador (N to the mouth of the Fraser R. at ca. 57°N), Nfld., N.B., P.E.I., and N.S., s to Calif., Utah, Sask., Iowa, and Del.; Iceland; Eurasia. MAP: Hultén 1968b:269.

   [*C. pluriflora* Hult., known from the Aleutian Is. and s Alaska (*see* Hultén 1942: map 304, p. 410; type from Hinchinbrook Is., Alaska) to Wash. (collections in Herb. V from Queen Charlotte Is. and Vancouver Is. and adjacent islands, B.C., have been placed here), appears to be an obscure species combining the characters of *C. limosa* and *C. rariflora* and may be of this hybrid origin. (*C. rariflora* var. *pl.* (Hult.) Boivin). (See *C. stygia,* this perhaps a hybrid between *C. rariflora* and *C. paupercula*). MAP: Hultén 1968b:269.]

C. livida (Wahl.) Willd.   (273)
/ST/X/EA/   (Grh (Hsr))   Calcareous meadows and bogs, the aggregate species from the w Aleutian Is. and s-cent. Alaska to s Yukon, B.C. (an isolated station in NW Dist. Mackenzie), Alta. (Laggan and Nordegg; CAN), Sask. (N to Prince Albert; CAN), Man. (N to Churchill), Ont. (N to w James Bay at ca. 52°10′N), Que. (N to E James Bay at ca. 54°N, L. Mistassini, and the Côte-Nord), Labrador (N to Makkovik, 55°10′N), Nfld., N.B., P.E.I., and N.S., s to N Calif., Idaho, Minn., and N.J.; Iceland; Eurasia. MAPS and synonymy: *see* below.
1   Terminal spike pistillate above, staminate below; culms to about 1.5 dm tall; [known only

from E Que. (Mingan Is.; Anticosti Is.) and N Nfld. (type from the Strait of Belle Isle)] . . . .
. . . . .var. *rufinaeformis* Fern.
1 Terminal spike staminate throughout.
  2 Staminate and pistillate spikes to 2.5 cm long; perigynia about 4 mm long; culms to
    4.5 dm tall; [var. *grayana* (Dew.) Fern. (*C. grayana* Dew.); transcontinental in
    approximately the southern half of the area of var. *livida*; MAP: Hultén 1958: map 196
    (var. *gray.*), p. 215] . . . . .var. *radicaulis* Paine
  2 Staminate and pistillate spikes at most 1.5 cm long; perigynia about 3 mm long; culms
    to 3 dm tall; [*C. limosa* var. *liv.* Wahl.; transcontinental; MAPS: on the above-noted map
    by Hultén; Hultén 1968*b*:271] . . . . .var. *livida*

C. loliacea L. (85)
/ST/(X)/EA/ (Hsr) Swampy ground and mossy streambanks from cent. Alaska–Yukon–Dist.
Mackenzie and B.C.–Alta. to Sask. (N to L. Athabasca), s Dist Keewatin, and Ont. (N to the Fawn R.
at ca. 54°N, 89°W), s to cent. B.C.–Alta.–Sask., N Man. (Nueltin L. at 59°43′N; CAN), and s-cent.
Ont. (Kapuskasing; Baldwin 1958); N Eurasia. MAPS: Hultén 1968*b*:244, and 1962: map 69, p. 79.

C. longii Mack. (182)
/T/EE/ (Hs) Wet or damp, sandy, clayey, or peaty soils from Ind. to s Mich., s Ont. (an island in
Georgian Bay, L. Huron, near Pointe au Baril, where taken by Carter and Hamman in 1934, the
presumed basis of the inclusion of s Ont. in the range by Fernald *in* Gray 1950; GH), N.Y., and
Maine, s to Mexico, Tex., and Fla.; Guatemala; Bermuda. [*C. albolutescens* of American auth. in
part, not Schw.].

C. lugens Holm (462)
/aS/WW/eA/ (Grh) Turfy tundra-barrens from N Alaska (type from Kusilof) to the Mackenzie R.
Delta, Banks Is., Victoria Is., and sw-cent. Dist. Keewatin (Yathkyed L. at ca. 62°30′N, 98°W), s to s
Alaska, s-cent. Yukon, and N-cent. Dist Mackenzie; E Asia. [*C. ?consimilis* and *C. cyclocarpa*
Holm; *C. caespitosa* var. *filifolia* Boott; *C. nudata* var. *angustifolia* Bailey; *C. yukonensis* Britt.].
MAPS: Hultén 1968*b*:249; Porsild 1957: map 85, p. 171, and 1955: fig. 14, p. 51; Raymond 1951:
map 8, p. 6, and map 19, p. 12. Concerning *C. consimilis, see* Porsild (1943:23).

C. lupuliformis Sartwell (532)
/T/EE/ (Hsr) Swampy woodlands and meadows (chiefly calcareous) from Minn. to s Ont.
(collection in CAN, detd. Herriott, verified by Mackenzie, from Cambridge, Waterloo Co.; a
purported hybrid with *C. retrorsa* is known from Seymour, Northumberland Co., and Belleville,
Hastings Co.), sw Que. (N to Oka and Rigaud; MT), and Vt., s to E Tex., Ky., La., and Va.

C. lupulina Muhl. (531)
/T/EE/ (Hsr) Wet woods and swampy ground (often calcareous) from Minn. to Ont. (N to
Carleton and Stormont counties), Que. (N to near Quebec City), N.B. (Hampton and Norton; NBM;
ACAD), and N.S. (not known from P.E.I.), s to Tex. and Fla. MAP (E Canada): Raymond 1951: map
38, p. 18.
    Var. *pedunculata* Gray (pistillate spikes mostly peduncled rather than subsessile and crowded,
the lower peduncles to 12 cm long) is known from s Ont. and sw Que.

C. lurida Wahl. (526)
/T/EE/ (Hs) Swamps and wet woods from Minn. to Ont. (N to Renfrew and Carleton counties), s
Que. (N to Charlevoix Co.; MT), N.B., and N.S. (reports from P.E.I. require confirmation), s to E
Mexico, Tex., and Fla. [*C. tentaculata* Muhl.]. MAP (E Canada): Raymond 1951: map 39, p. 19.

[*C. luzulina* Olney] (367)
[The inclusion of B.C. in the range of this species of Oreg. and Calif. by Rydberg (1922) is based
upon *C. ablata,* according to Mackenzie (1935). Hitchcock et al. (1969), however, include B.C. as
part of the area.]

C. lyngbyei Hornem.   (477)
/aST/D/GEeA/   (Grh)   Brackish soil and coastal bluffs (inland in Wash.): Aleutian Is. and N-cent. Alaska (*see* Hultén 1942: map 279, p. 408) through coastal B.C. to NW Calif.; E Que. (Côte-Nord, Anticosti Is., and Gaspé Pen.) and S Labrador (Gready Is., ca. 53°45′N; GH); southernmost Greenland (?introd.); Iceland and the Faeroes; coastal E Asia. [*C. behringensis* Gand., not Clarke; *C. cryptocarpa* Mey. and its var. *pumila* Bailey; *C. cryptochlaena* Holm, a possible hybrid between *C. lyng.* and *C. ramenskii; C. macounii* Benn., not Dew.; *C. romanzowiana* Cham.; *C. salina* var. *robusta* Bailey; *C. scouleri* Torr.]. MAPS: Hultén 1968*b*:253, and 1958: map 273, p. 293; Löve and Löve 1956*b*: fig. 27, p. 234; Raymond 1951: map 20, p. 12.

C. mackenziei Krecz.   (92: *C. norvegica* Willd., not Retz.)
/aST/(X; coastal)/GEeA/   (Hsr)   Saline or brackish marshes and shores: coast of Alaska N to Kotzebue Sound; W Hudson Bay (SE Dist. Keewatin and Churchill, Man.) to James Bay in Ont. and Que.; E Que. (St. Lawrence R. estuary from l'Islet Co. to the Côte-Nord, Anticosti Is., Gaspé Pen., and Magdalen Is.), Labrador (N to Tikkoatokok Bay, ca. 57°N), Nfld., N.B., P.E.I., N.S., and Maine; southernmost Greenland; Iceland; Scandanavia; NE Asia. [*C. norvegica* Willd., not Retz.]. MAPS: Hultén 1968*b*:240, and 1958: map 274, p. 292; Fernald 1929: map 34 (*C. norv.*), p. 1502; Potter 1932: map 1 (E Canada; incomplete), p. 71.

C. macloviana d'Urv.   (129, 124 (*C. festivella*), 125 (*C. microptera*), and 144 (*G. pachystachya*))
/aST/(X)/GEeA/   (Grh)   Dry to moist open ground, slopes, and alpine meadows: Aleutian Is. and cent. Alaska to S Yukon, the Mackenzie R. Delta, and Great Bear L. through B.C.–Alta. and SW Sask. (Cypress Hills; CAN; DAO) to Calif., Mexico, Utah, and Colo.; northernmost Ungava–Labrador to S Labrador (isolated in the Shickshock Mts. of the Gaspé Pen., E Que.); S. America; Hawaii; W and E Greenland N to ca. 71°N; Iceland; Scandinavia; Kamchatka. MAPS and synonymy: *see* below.
1   Perigynia and scales copper-brown to purplish black at maturity; rhizomes brownish;
      [incl. *C. pachystachya* Cham. and its var. *gracilis* (Olney) Mack. (*C. gracilior* Mack. in
      part; *C. ?incondita* Herm.; *C. multimoda* Bailey; *C. olympica* Mack.; *C. soperi* Raup);
      C. pyrophila Gand.; range of the species; MAPS: Hultén 1968*b*:233 (ssp. *pachy.*), and
      1958: map 185 (*C. macl.* and *C. pach.*), p. 205; G. E. Du Rietz, Acta Phytogeogr. Suec.
      13: fig. 5, p. 220. 1940; Raymond 1951: map 13, p. 8] . . . . .var. *macloviana*
1   Perigynia and scales stramineous to brown at maturity; rhizomes blackish; [*C. festivella*
      and *C. microptera* Mack.; S Yukon (near Mackintosh), B.C., SW Alta. (Waterton Lakes),
      and SW Sask. (Cypress Hills; var. *macl.* also present); MAP: Hultén 1968*b*:236 (*C. mic.*)] . . . .
      . . . . .var. *microptera* (Mack.) Boivin

C. macrocephala Willd.   (81)
/sT/W/eA/   (Grh)   Coastal sands from S Alaska (*see* Hultén 1942: map 245, p. 405) through coastal B.C. (Queen Charlotte Is.; Vancouver Is. and adjacent islands and mainland; CAN; V) to Oreg.; coastal E Asia. [Incl. var. *bracteata* Holm and ssp. *anthericoides* (Presl) Hult. (*C. anth.* Presl); *C. brongniartii* var. *densa sensu* John Macoun 1888, not *C. densa* Bailey, Macoun erroneously listing *C. anthericoides* in synonymy]. MAP: Hultén 1968*b*:231.

C. macrochaeta Mey.   (409)
/sT/W/?eA/   (Grh)   Wet open places from the Aleutian Is. (type from Unalaska) and S Alaska–Yukon (*see* Hultén 1942: map 290, p. 409) through W B.C. (CAN; V) to Oreg.; E ?Asia (*see* Hultén 1942:358). [Incl. vars *emarginata* and *macrochlaena* Holm; *C. excurrens* Cham.; *C. kuehleweinii* Gand.]. MAP: Hultén 1968*b*:262.

C. maritima Gunn.   (14 (*C. incurva*) and 13 (*C. incurviformis*))
/AST/X/GEA/   (Grh)   Sandy, gravelly, or turfy places (chiefly along the coast), the aggregate species from the coasts of Alaska–Yukon–Dist. Mackenzie–Dist. Keewatin to Ellesmere Is. at ca. 83°N, northernmost Ungava–Labrador, and Nfld., S to S Alaska–Yukon–Dist. Mackenzie, NE Man. (Churchill and York Factory), James Bay, and E Que. (St. Augustin Is., Saguenay Co. of the Côte-Nord); circumgreenlandic (type from Greenland); Iceland; Spitsbergen; Eurasia. MAPS and synonymy: *see* below.

1 Plant very dwarf, the culms at most about 2 cm tall; leaves bristle-like, to 3.5 cm long; fruiting heads to 7 mm thick; perigynia about 3 mm long; [f. ?*inflata* (Simmons) Polunin; *C. incurva* var. *set.* Chr.; s-cent. Ellesmere Is.; James Bay, Ont.; Ingornachoix Bay, Nfld.; w Greenland between the Arctic Circle and ca. 71°N; MAP: Fernald 1933: map 4, p. 56] . . . . .var. *setina* (Chr.) Fern.
1 Plant taller, the culms to 2.5 dm tall; leaves to 1.5 dm long; fruiting heads to nearly 1.5 cm thick; perigynia to 5 mm long.
  2 Pistillate scales lance-ovate, narrowly hyaline-margined, acute to acuminate; perigynia oblong-oblanceolate, finely but conspicuously impressed-nerved on both faces, scarcely inflated; [*C. incurviformis* Mack., the type from Banff, Alta.; s Yukon (Porsild 1951*a*) and mts. of B.C. and sw Alta.; MAP: Meusel, Jaeger, and Weinert 1965:65 (*C. incurv.*)] . . . . .var. *incurviformis* (Mack.) Boivin
  2 Pistillate scales broadly ovate, silvery-hyaline-margined, obtuse or acutish; perigynia ovate-elliptic or ovate, nerveless ventrally, obscurely nerved dorsally, somewhat inflated; [incl. ssp. *yukonensis* Porsild; *C. incurva* Lightf.; not *C. maritima* Muell., which is *C. paleacea* Wahl.; transcontinental in chiefly arctic and subarctic regions; MAPS: Hultén 1968*b*:228, and 1962: map 41, p. 49; Porsild 1957: map 78, p. 170; Meusel, Jaeger, and Weinert 1965:65] . . . . .var. *maritima*

C. meadii Dewey   (276)
/T/EE/   (Grh)   Calcareous meadows, prairies, and depressions from s Sask. (File Hills and Qu'Appelle Valley; Breitung 1957*a*), s Man. (N to St. Lazare, about 75 mi NW of Brandon; CAN, detd. Mackenzie), s Ont. (Sarnia, Lambton Co., where taken by John Macoun in 1901; CAN), and N.J. to Tex. and Ga. [*C. tetanica* var. *m.* (Dew.) Bailey].

C. media R. Br.   (413: *C. vahlii sensu* Mackenzie 1935, in part, not Schkuhr)
/ST/X/EA/   (Grh (Hsr))   Mossy woods, thickets, and shores (often calcareous) from N Alaska–Yukon–w Dist. Mackenzie and Great Bear L. to Sask. (N to L. Athabasca), Man. (N to Nueltin L. at 59°43′N), northernmost Ont., Que. (N to s Ungava Bay and the Côte-Nord), Labrador (N to ca. 55°30′N), and N N.B. (Restigouche R.; not known from Nfld., P.E.I., or N.S.), s to s B.C., Idaho, Mont., N Minn., and N Mich.; Eurasia. [*C. angarae* Steud.; *C. alpina (vahlii)* var. *inferalpina sensu* Fernald 1933, not Wahl.; *C. vahlii sensu* Mackenzie 1935, in part, not Schk., the type of whose plant is *C. norvegica* Retz.; incl. *C. stevenii* Holm, intermediate between *C. media* and *C. norvegica*]. MAPS: Hultén 1968*b*:255, and 1962: map 29, p. 37; Meusel, Jaeger, and Weinert 1965:71 (*C. angarae*); Raup 1947: pl. 18; Fernald 1933: map 21 (*C. alp.* var. *inf.*), p. 222.

C. membranacea Hook.   (517)
/AST/X/eA/   (Grh)   Dryish turfy tundra from the coasts of Alaska–Yukon–Dist. Mackenzie–Dist. Keewatin to Banks Is., Devon Is., N Ellesmere Is. at 81°25′N, Baffin Is., and northernmost Ungava–Labrador, s to N B.C. (s to near Summit Pass at 58°31′N; CAN), Great Slave L., N Man. (Churchill), and NE James Bay, Que. (type from "Duke of York's Bay, north of Hudson Bay"); E Asia. [*C. compacta* R. Br.; *C. membranopacta* Bailey; *C. physochlaena* Holm]. MAPS: Hultén 1968*b*:278; Porsild 1957: map 101, p. 173, and 1951*b*: fig. 4, p. 142; Raup 1947: pl. 19; Raymond 1951: map 2 (E Canada), p. 4.

C. merritt-fernaldii Mack.   (173)
/T/X/   (Hs)   Dry meadows, thickets, and gravelly banks from s B.C. (Fernald *in* Gray 1950; not known from Alta. or Sask.) to s Man. (Melita, the Souris R. s of Brandon, and the Whiteshell Forest Reserve w of Winnipeg; CAN), Ont. (N to Kenora and Longlac; CAN, detd. Hermann and Lepage, respectively), Que. (N to L. St. Peter; Marcel Raymond, Rhodora 51(601):10. 1949), and New Eng., s to N Calif., Idaho, Kans., Minn., Mich., and N.Y.

C. mertensii Prescott   (435)
/ST/W/eA/   (Grh)   Turfy or rocky slopes from s-cent. Alaska (type from Sitka; *see* Hultén 1942: map 289, p. 409) and s Yukon through B.C. (CAN; V) to N Calif., Idaho, and Mont.; E Asia. MAP: Hultén 1968*b*:261.

C. michauxiana Boeckl.   (489)
/sT/EE/eA/   (Hs)   Bogs and wet meadows from Ont. (N to the N shore of L. Superior) to Que. (N to Seal L. at ca. 56°30′N, L. Mistassini, and the Côte-Nord), s Labrador (Goose Bay, 53°18′N), Nfld., N.B., and N.S., s to N Mich. and w Mass.; E Asia. [*C. abacta* Bailey; *C. ?subulata (C. collinsii* Nutt. of N.J.) *sensu* Hooker 1839, as to the Canadian plant, not Michx.].

C. microglochin Wahl.   (486)
/aST/X/GEA/   (Grh)   Peaty, often calcareous, soils from N Alaska, s-cent. Yukon, and Great Bear L. to the coast of Dist. Mackenzie (Coronation Gulf), s Baffin Is., and northernmost Ungava–Labrador, s to s B.C.–Alta. (isolated stations in Colo.), cent. Man. (s to Gillam, about 165 mi s of Churchill; not known from Sask.), James Bay, E Que. (Anticosti Is.), and NW Nfld.; s S. America; w Greenland N to ca. 71°N, E Greenland N to ca. 77°30′N; Iceland; Eurasia. [*C. lyonii* Boott]. MAPS: Hultén 1968*b*:227, and 1958: map 214, p. 233; Porsild 1957: map 76, p. 170; Raup 1947: pl. 19.

C. misandra R. Br.   (363)
/AST/X/GEA/   (Hs)   Dryish turfy and rocky tundra from the coasts of Alaska–Yukon–Dist. Mackenzie–Dist. Keewatin to Melville Is. (type locality), northernmost Ellesmere Is., Baffin Is., and northernmost Ungava–Labrador, s to the Alaska Panhandle, Great Bear L., SE Dist. Keewatin, NE Man. (Churchill), NE James Bay, and s-cent. Labrador; isolated stations in the mts. of sw Alta. (Jasper dist.), Wyo., Utah, and Colo.; w and E Greenland N of the Arctic Circle; N Eurasia. [Incl. var. *elatior* Lange and *C. stenocarpa* Turcz.; *C. fuliginosa* var. *mis.* (R. Br.) Lang]. MAPS: Hultén 1968*b*:272, and 1962: map 14, p. 21; Meusel, Jaeger, and Weinert 1965:75; Porsild 1957: map 95, p. 172; Raymond 1951: map 4 (Ungava–Labrador), p. 4; Young 1971: fig. 12, p. 87.
      Forma *flavida* Fern. (scales of spikes pale rather than castaneous to purplish black; type from NW Greenland) is known from Ellesmere Is. and Baffin Is. (Polunin 1940) and Greenland.

[C. miserabilis Mack.]   (446)
[The report of this obscure species of Wash., Idaho, and Oreg. from s B.C. by Eastham (1947; Mt. Brent, near Penticton) requires confirmation. It is merged with *C. scopulorum* by Hitchcock et al. (1969).]

C. molesta Mack.   (166)
/T/EE/   (Hs)   Dry open ground and borders of woods from ?Sask. (included in the range by Fernald *in* Gray 1950: not listed by Breitung 1957*a*) to Man. (collection in GH, detd. Fernald, from the "Winipeg Valley", where taken by Bourgeau in 1859; a collection in CAN from Whitewater L., near Boissevain, appears to belong here), s Ont. (St. Thomas, Elgin Co.; TRT), Vt., and Mass., s to Colo., Kans., Ark., Tenn., and Del.
      This species is admitted to our flora on rather tenuous grounds and some or all of the above reports may prove referable to other members of the Section *Ovales,* particularly *C. brevior.*

C. montanensis Bailey   (416)
/ST/W/eA/   (Grh)   Meadows and along mountain streams from N Alaska and cent. Yukon (*see* Hultén 1942: map 292, p. 409; also reported from Dist. Mackenzie by Hultén) through the mts. of B.C. (type from Kootenay Pass) and sw Alta. (Lake Louise and Laggan; CAN) to Idaho and Mont.; E Asia. [*C. venustula* Holm; not listed by Hultén 1968*b*, being scarcely separable from other members of the Section *Atratae* and merged with *C. spectabilis* by Hitchcock et al. 1969].

C. muhlenbergii Schkuhr   (44)
/T/EE/   (Grh)   Dry woods, clearings, and fields from Minn. to s Ont. (N to Bruce, Simcoe, and Peterborough counties), s Que. (N to L. St. Peter in St-Maurice Co.; Marcel Raymond, Rhodora 51(601): 10. 1949), and s Maine, s to Tex. and Fla.
      Var. *enervis* Boott (perigynia essentially nerveless rather than strongly nerved on both faces) is known from s Ont. (Pt. Abino, Welland Co.; GH).

C. muricata L. (112, 110 (*C. sterilis*), 117 (*C. angustior*), 118 (*C. laricina*), and 119 (*C. cephalantha*))
/sT/X/EA/ (Hs) Mossy or peaty soils and shores, the aggregate species from ?Alaska, s Yukon, Dist. Mackenzie and SE Dist. Keewatin (Boivin 1967a; island in ?James Bay) to Ont. (N to the Shamattawa R. at 54°13′N; *see* E Canada map by Lepage 1966: map 9 (*C. cephalantha*), p. 218), Que. (N to Knob Lake, 54°48′N, and the Côte-Nord), Nfld., N.B., P.E.I., and N.S., s to Calif., Nev., Colo., Ill., Tenn., and N.C.; (*C. echinata* is reported from Greenland by Fernald *in* Gray 1950, but neither Hultén's maps nor Joergensen, Soerensen and Westergaard 1958, indicate its occurrence there); Iceland; Eurasia. MAPS and synonymy: *see* below.
1 Perigynia obscurely nerved or nerveless ventrally; leaves to 2 mm broad.
    2 Midrib of the obtusish pistillate scales not reaching the flat hyaline tip; [*C. echinata* Murr.; *C. leersia* Willd.; *C. stellulata* Good.; Ont. to Labrador and Nfld.; MAPS: (*C. echinata*): Hultén 1968b:246, and 1958: map 122, p. 141] . . . . .var. *muricata*
    2 Midrib of the short-cuspidate pistillate scales prominent to tip; [*C. angustior* Mack.; often treated as a variety of *C. echinata, C. leersia, C. stellulata,* or *C. sterilis;* essentially transcontinental] . . . . .var. *angustata* Carey
1 Perigynia lightly to strongly nerved on both faces.
    3 Midrib of pistillate scales not reaching the flat hyaline tip; achene distinctly longer than broad; [*C. sterilis* Willd.; *C. scirpoides* Schkuhr in part; Sask. to Nfld. and N.S.] . . . .
    . . . . .var. *sterilis* (Carey) Gl.
    3 Midrib of pistillate scales green, prominent nearly or quite to the sharp tip; achene about as broad as long.
        4 Leaves strongly channelled, to about 1.5 mm broad; perigynia about 3 mm long, lightly nerved, slightly surpassing the scales; [*C. laricina* Mack.; Ont.] . . . .
        . . . . .var. *laricina* (Mack.) Gl.
        4 Leaves flat or slightly channelled, to 2.5 mm broad; perigynia to 4 mm long, strongly many-nerved, commonly slightly shorter than the scales; [*C. cephalantha* (Bailey) Bickn.; often treated as a variety of *C. echinata, C. leersia, C. stellulata,* or *C. sterilis;* transcontinental] . . . . .var. *cephalantha* Bailey

C. muskingumensis Schw. (189)
/T/EE/ (Grh (Hsr)) Moist or swampy ground from SE Man. (near Winnipeg; CAN; the report from as far N as Norway House, off the NE end of L. Winnipeg, by Hooker 1839, requires confirmation) to s Ont. (Amherstburg, Essex Co.; Walpole Is., Lambton Co.), s to Okla., Kans., Mich., Ohio, and Ky. [*C. arida* Schw. & Torr.].

C. nardina Fries (1 and 2 (*C. hepburnii*))
/AST/X/GEeA/ (Hs) Calcareous sands and gravels and dry, grassy or rocky places, the aggregate species from the N coast of Alaska to Victoria Is., Melville Pen., northernmost Ellesmere Is., Baffin Is., and northernmost Ungava–Labrador, s in the West through B.C. and sw Alta. (N to the Jasper dist.) to Wash., Mont., Wyo., Utah, and Colo., the s limits farther east being from Great Bear L. to SE Dist. Keewatin (not known from Sask., Man., or Ont.), Que. (s to SE James Bay, L. Mistassini, Knob Lake, 54°48′N, and the Shickshock Mts. of the Gaspé Pen.), and N Labrador (s to Ramah, 58°54′N); circumgreenlandic; Iceland; N Scandinavia; tip of NE Siberia. MAPS and synonymy: *see* below.
1 Achenes lenticular; pistillate scales obtuse, with a lighter-coloured midvein; spikes broadly ovate to suborbicular; leaf-sheaths tapering to the blade; [incl. var. *atriceps* Kük.; *C. elynaeformis* Porsild; transcontinental; MAPS (the one by Porsild for var. *atriceps,* the others for the aggregate species): Porsild 1957: map 71, p. 169; Hultén 1968b:220, and 1958: map 168, p. 187; Meusel, Jaeger, and Weinert 1965:64: Böcher 1954: fig. 33 (map 2), p. 135; Raymond 1951: map 7 (NE area), p. 5] . . . . .var. *nardina*
1 Achenes usually triangular in section; pistillate scales acutish to obtusish, with a conspicuous lighter-coloured centre; spikes ovate or obovate; sheaths abruptly contracted to the blade; [*C. hepburnii* Boott, the type taken by Drummond in the Rocky Mountains, perhaps of Alta.; s-cent. Alaska and w-cent. Dist. Mackenzie to the mts. of B.C. and sw Alta.; MAP: Raup 1947: pl. 17] . . . . .var. *hepburnii* (Boott) Kük.

Carex

[C. nebrascensis Dewey]   (452)
[Reports of this species of the w U.S.A. (N to Wash. and S.Dak.) from B.C. by Henry (1915; var. *praevia* Bailey) and from B.C. and Alta. by John Macoun (1888; *C. jamesii* T. & G., not Schw., the Alta. report taken up by Moss 1959) require confirmation.]

C. nesophila Holm   (418)
/ST/W/eA/   (Grh)   Meadows and damp tundra from the Aleutian Is., N-cent. Alaska (N to Kotzebue Sound; type from St. Paul Is.; *see* Hultén 1942: map 294, p. 409), and N Yukon–w Dist. Mackenzie (Richardson Mts.) to s Alaska–Yukon–w Dist. Mackenzie; E Asia (Commander Is.). [*C. microchaeta* Holm]. MAPS: Raup 1947: pl 18; combine the maps by Hultén 1968*b*:264 (*C. mic.*) and p. 265; Porsild 1966: map 25 (*C. mic.*), p. 70.

C. nigricans Meyer   (9)
/sT/W/eA/   (Grh)   Calcareous open slopes from the Aleutian Is. (type from Unalaska; *see* Hultén 1942: map 236, p. 404) and s Alaska through B.C. and sw Alta. (N to Jasper) to Calif., Utah, and Colo.; E Asia (Commander Is.). [*C. pyrenaica* var. *majuscula* Kurtz]. MAP: Hultén 1968*b*:227.

C. nigromarginata Schw.   (216, 211 (*C. peckii*), 213 (*C. artitecta*), and 214 (*C. albicans*))
/ST/X/   (Grh (Hsr))   Dry woods (often calcareous), thickets, rocky slopes, and clearings, the aggregate species from w-cent. Yukon (*see* Hultén 1942: map 298 (*C. peckii*), p. 410) and B.C. to Alta. (N to Waterways, ca. 56°40′N), Sask. (N to McKague, 52°37′N), Man. (N to Cross L., NE of L. Winnipeg), Ont. (N to L. Nipigon and Moose Factory, 51°15′N), Que. (N to the Gaspé Pen.), N.B., P.E.I., and N.S., s to s B.C.–Alta., S.Dak., Okla., Tenn., and Fla. MAP and synonymy: *see* below.
1   Pistillate scales rounded or blunt to sometimes short-pointed, shorter than the perigynium-body; perigynia to 4 mm long, their beaks usually about 1/3 the length of the body; culms to over 6 dm tall, conspicuously surpassing the leaves, these to 3 mm broad; [*C. emmonsii* var. *elliptica* Boott; *C. albicans* of auth. in part, not Willd.; *C. clivicola* Fern. & Weath.; *C. peckii* Howe; Alaska–B.C. to N.S.; MAP (*C. peckii*): Hultén 1968*b*:266] . . . .
. . . . .var. *elliptica* (Boott) Gl.
1   Pistillate scales acute to acuminate, about as long as the perigynia and largely concealing them; leaves less than 2 mm broad; culms rarely over 4.5 dm tall.
    2   Culms conspicuously surpassing the leaves; staminate scales typically blunt; [*C. artitecta* Mack.; *C. varia* of auth. in part, not Muhl.; s Ont. and sw Que.] . . . .
    . . . . .var. *muhlenbergii* (Gray) Gl.
    2   Culms typically much shorter than the leaves; staminate scales typically acute to acuminate or short-cuspidate; [*C. emmonsii* Dew.; *C. albicans* of auth in part, not Willd.; *C. davisii* Dew., not Schwein. & Torr.; Ont. to N.B. and N.S.] . . . .var. *minor* (Boott) Gl.

C. normalis Mack.   (164)
/T/EE/   (Hs)   Open woods and meadows from Man. (N to Norway House, off the NE end of L. Winnipeg; WIN) to Ont. (N to the Ottawa dist.; Gillett 1958), Que. (N to the Rupert R. SE of James Bay at ca. 51°30′N; Dutilly, Lepage, and Duman 1958), and Maine, s to Okla., Mo., Tenn., and N.C. [*C. mirabilis* Dew., not Host; *C. cristata (straminea)* var. *mir.* (Dew.) Boott].
    Forma *perlonga* Fern. (all but the terminal spikes remote rather than more or less crowded) is reported from Lambton Co., s Ont., by Gaiser and Moore (1966).

C. norvegica Retz.
/aST/EE/GEeA/   (Grh)   Dry turfy tundra and slopes on acidic rocks from cent. Dist. Keewatin, Baffin Is. (N to the Arctic Circle), and northernmost Ungava–Labrador to NE Man. (Churchill), northernmost Ont., Que. (s to E James Bay, Knob Lake at 54°48′N, and the Shickshock Mts. of the Gaspé Pen.), N Nfld., and Labrador (s to ca. 53°N); w and E Greenland N to ca. 74°N; Iceland; Europe; NE Siberia. [Not *C. norvegica* Willd., which is *C. mackenziei* Krecz.; *C. vahlii* Schk., not *sensu* Mackenzie 1935, No. 413; *C. alpina* Sw., not Schrank; *C. halleri* of auth., not Gunn. as to type; incl. ssp. *inserrulata* Kalela]. MAPS: Hultén 1958: map 74, p. 93; Porsild 1957: map 88, p. 171; Meusel, Jaeger, and Weinert 1965:71; D. F. Murray, Brittonia 21(1): fig. 7, p. 71. 1969; Fernald 1933: map 21 (*C. alpina*; now requiring considerable expansion), p. 222.

C. novae-angliae Schw. (212)
/T/EE/ (Grh (Hsr)) Woodlands and damp slopes from Ont. (N to the Ottawa dist.; Gillett 1958) to Que. (N to the Gaspé Pen. and Magdalen Is.), Nfld. (Holyrood and Rencontre West; CAN, detd. Raymond), N.B., P.E.I., and N.S., s to Wisc., Pa., and Conn.; (reports from B.C. by John Macoun 1888, and Carter and Newcombe 1921, are based upon *C. rossii,* relevant collections in CAN). [*C. varia* var. *nov.* (Schw.) Kük.].

C. nutans Host
Eurasian; known in N. America only in low damp ground of Ile Charron in the seaport dist. of Montreal, Que., where undoubtedly introd. in ballast. This locality (where the Eurasian *C. disticha* Huds. has also been taken in its only N. American station other than in s Ont.) is indicated in a map by Frère Marie-Victorin (Contrib. Inst. Bot. Univ. Montréal 15: fig. 5, p. 263. 1929).

C. obnupta Bailey (481)
/T/W/ (Grh) Swampy meadows from coastal B.C. (N to Queen Charlotte Is. and Prince Rupert; reports from Alaska refer to *C. sitchensis* according to Hultén 1942) to Calif. [*C. magnifica* Dew.; *C. sitchensis sensu* Hooker 1939, at least in part, not Prescott].

C. obtusata Lilj. (208)
/aST/WW/EA/ (Grh) Dry plains, bluffs, and rocky slopes from the N coast of Alaska (mouth of the Sadlerochit R.; CAN) to cent. Yukon, the Mackenzie R. Delta, and Great Bear L., s through B.C.–Alta.–Sask.–Man. to Mont., Utah, N.Mex., S.Dak., and Minn.; (the report from the Gaspé Pen., E Que., by M. L. Fernald, Rhodora 9(105):159. 1907, is based upon *C. rupestris* (relevant collections in GH and CAN), as also, probably, reports from Nfld. by Reeks 1873, and John Macoun 1888); Eurasia. [*C. backana* Dew.]. MAPS: Hultén 1968*b*:224, and 1962: map 137, p. 147; Raup 1930: map 21, p. 202; Meusel 1943: fig. 9c; Meusel, Jaeger, and Weinert 1965:64.

C. oligocarpa Schkuhr (308)
/T/EE/ (Hs) Calcareous woods and thickets from Kans. to s Mich., s Ont. (Essex, Elgin, Welland, and Prince Edward counties; CAN; TRT; reported N to the Ottawa dist. by John Macoun 1888, but not listed by Gillett 1958), sw Que. (N to the Montreal dist.; MT), and Vt., s to Tex. and NW Fla.

C. oligosperma Michx. (525)
/sT/(X)/ (Grh) Peat bogs, acid swamps, and shallow water from Great Bear L. and Great Slave L. to L. Athabasca (Alta. and Sask.), Man. (N to Tod L. at 56°35′N), Ont. (N to the Winisk R. at 55°12′N), Que. (N to the Wiachouan R. at 56°10′N, L. Mistassini, and the Côte-Nord), Labrador (N to Nain, ca. 56°30′N), Nfld., N.B., and N.S. (not known from P.E.I.), s to s Sask., Minn., Ohio, and Pa. [*C. depreauxii* Steud.].

C. pairaei Schultz (47)
Eurasian; introd. in fields and along roadsides of N. America and known in Canada from N.B. (St. John, St. John Co.; GH; reported from Kent Co. by Mackenzie 1931).

C. paleacea Wahl. (476)
/ST/EE/E/ (Grh) Saline or brackish marshes and shores of Hudson Bay–James Bay and the Atlantic Seaboard: Man. (York Factory; CAN), Ont. (w James Bay), Que. (E James Bay–Hudson Bay N to ca. 56°10′N; St. Lawrence R. estuary from l'Islet Co. to the Côte-Nord, Anticosti Is., Gaspé Pen., and Magdalen Is.), Labrador (N to Anatolak Bay at ca. 56°30′N), Nfld., N.B., P.E.I., and N.S. to Mass.; NW Scandinavia. [Incl. var. *transatlantica* Fern.; *C. maritima* Muell., not Gunn.]. MAPS: Hultén 1958: map 265, p. 285; Potter 1932: map 5 (*C. maritima*; incomplete northwards), p. 73; Raymond 1951: map 24 (E Canada), p. 15.

Reports from Greenland are now discredited, perhaps referring to *C. salina,* as also, perhaps, the report from Wakeham Bay, N Ungava, by Duman (1941). It is reported from Cumberland House, E-cent. Sask., by Hooker (1839), this indicated by dots on the above-noted maps by Hultén and Potter. It might possibly have persisted there in alkaline habitats as a relic from the former transgression of Hudson Bay. Forma *erectiuscula* Fern. (spikes at most about 2.5 cm long rather than to 7 or 8 cm, the lower peduncles rarely over 2 cm long rather than to about 8 cm) is known from Que. (Richmond Gulf, Hudson Bay; Dutilly and Lepage 1951*b*) and Nfld. (Rouleau 1956).

C. pallescens L.   (371)
/T/EE/EA/   (Hs)   Moist woods, thickets, and meadows from Ont. (N to Michipicoten, NE shore of L. Superior; CAN), Que. (N to Anticosti Is. and the Gaspé Pen.), Nfld., N.B., P.E.I., and N.S., s to Mich., Ohio, and N.J.; Iceland; Europe; w Asia. [Incl. var. *neogaea* Fern.]. MAPS: Hultén 1958: map 51, p. 71; Meusel, Jaeger, and Weinert 1965:72; Raymond 1951: map 32 (E Canada), p. 16.

C. panicea L.   (274)
Eurasian; introd. and locally abundant in meadows and grasslands of s Ont. (Sarnia, Lambton Co.; MT; TRT), N.B. (Bass River, Kent Co.; Fowler 1879 and 1885), St-Pierre and Miquelon (Boivin 1967*a*; also reported from Nfld. by Rouleau 1956), N.S., and Greenland. [Incl. var. *microcarpa* Sonder]. MAPS: Hultén 1958: map 102, p. 121; Meusel, Jaeger, and Weinert 1965:75.

C. parallela (Laest.) Sommerf.
/aST/-/GEA/   (Grh)   Bogs and grassy tundra of E Greenland (between ca. 70° and 75°N), Iceland, Spitsbergen, ?Scotland, N Scandinavia, and NW Asia. [*C. dioica* var. *par.* Laest.]. MAPS: Hultén 1958: map 63, p. 83.

C. parryana Dewey   (412)
/sT/WW/   (Grh)   Dry to damp prairies and slopes from SE Alaska (near Anchorage; CAN, detd. Porsild) and s Yukon (several localities; CAN) to Great Slave L., Sask. (N to near Flin Flon), and Man. (N to Gypsumville; the type material was collected by Richardson and labelled "Hudson's Bay", thus possibly originating from Manitoba but more likely from farther west in the broad domain of the Hudson's Bay Company), s through B.C.–Alta. to Idaho, Utah, and Colo. [*C. arctica* Dewey]. MAPS: Hultén 1968*b*:256; Porsild 1966: map 26, p. 70; D. F. Murray, Brittonia 21(1): fig. 7, p. 71. 1969.

C. pauciflora Lightf.   (487)
/ST/X/EA/   (Grh)   Acid peat and sphagnum-bogs from cent. Alaska and sw Yukon to B.C., Alta. (N to Fedorah, N of Edmonton; CAN; not known from Sask.), Man. (N to Reindeer L. at 57°37′N), Ont. (N to Hawley L. at 54°34′N), Que. (N to the Korok R., E Ungava Bay, at 58°35′N), Labrador (N to Anatolik Bay at 56°33′N), Nfld., N.B., and N.S. (not known from P.E.I.), s to Wash., Minn., N Ind., and Conn.; Eurasia. MAPS: Hultén 1968*b*:228, and 1962: map 77, p. 87; Raymond 1950*a*: map 2 (E Canada), p. 441.

C. paupercula Michx.   (407)
/aST/X/GEA/   (Grh)   Acid swamps and sphagnum-bogs, the aggregate species from N-cent. Alaska–Yukon and the Mackenzie R. Delta to Great Bear L., Great Slave L., L. Athabasca (Alta. and Sask.), Man. (N to Nueltin L. at 59°48′N), northernmost Ont., Que. (N to the Korok R., E Ungava Bay, at 58°35′N; type from near L. Mistassini), Labrador (N to near Nain at ca. 56°30′N), Nfld., N.B., P.E.I., and N.S., s to Wash., Utah, Colo., Minn., and New Eng.; w Greenland N to ca. 64°N; Iceland; Eurasia. MAPS and synonymy: *see* below.

   C. stygia Fries (Alaska through coastal B.C. to Wash.) is more or less intermediate between *C. paupercula* (var. ?*irrigua*) and *C. rariflora* and may be of hybrid origin through this parentage. (Compare *C. pluriflora,* this perhaps a hybrid between *C. limosa* and *C. rariflora*).
1   Culms smooth except rarely just below the head.
   2   Pistillate spikes less than 1 cm long; staminate spike at most 11 mm long; culms
      usually less than 2.5 dm tall; [incl. var. *brevisquama* Fern.; *C.* ?*magellanica* Lam.;
      transcontinental; MAP: G. E. Du Rietz, Acta Phytogeogr. Suec. 13: fig. 3 (incomplete
      for N. America), p. 218. 1940] . . . . .var. *paupercula*
   2   Pistillate spikes to about 2 cm long; staminate spikes to 18 mm long; culms to about 8
      dm tall; [*C. irrigua* Wahl.; *C. magellanica* Lam. var. *irr.* (Wahl.) BSP.; transcontinental;
      MAP: Hultén 1968*b*:270] . . . . .var. *irrigua* (Wahl.) Fern.
1   Culms scabrous for a distance of about 1 dm below the head, to about 8 dm tall; [Ont. to
   N.B. and N.S. (type from North Sydney)] . . . . .var. *pallens* Fern.

Cyperaceae

C. pedunculata Muhl.   (243)
/T/X/   (Hsr)   Rich woods and slopes (usually calcareous) from s Man. (N to Duck Mt.; CAN; the report N to Norway House and w to Cumberland House, E Sask., by Hooker 1839, is probably based upon *C. richardsonii*) to Ont. (N to Moose Factory, s James Bay, 51°15′N), Que. (N to the Harricanaw R. at ca. 50°N and Anticosti Is.), Nfld. (Table Mt.; GH; CAN), N.B., and N.S. (not known from P.E.I.), s to S.Dak., Iowa, Ohio, and N Ga. MAP (E Canada): Raymond 1950a: map 13, p. 444.

C. pensylvanica Lam.   (223, 222 (*C. heliophila*), and 224 (*C. lucorum*))
/T/X/   (Grh)   Dry open woods, plains, and prairies, the aggregate species from B.C. (N to Hudson Hope, ca. 56°N; CAN) to Alta. (N to Grande Prairie, ca. 55°N), Sask. (N to Prince Albert), Man. (N to Riding Mt.), Ont. (N to Renfrew and Carleton counties), Que. (N to L. St. Peter in St-Maurice Co.), N.B. (St. Stephen, Charlotte Co.; CAN; not known from P.E.I.), and N.S. (Kings, Queens, Cumberland, and Lunenburg counties), s to N Calif., N.Mex., Kans., Tenn., and S.C.
1   Leaves fairly soft and only slightly scabrous; perigynia obtusely angled;
    2   Perigynia at most 3 mm long, the short beak not over 1/4 the length of the body; [Man. to N.B. and ?N.S.] ....var. *pensylvanica*
    2   Perigynia to 4 mm long, the beak at least 2/3 as long as the body; [*C. lucorum* Willd.; Ont. to N.S.] ....var. *distans* Peck
1   Leaves relatively stiff and firm, strongly scabrous; perigynia subterete, about 3.5 mm long, the beak at least 1/3 as long as the body; [incl. vars. *pinicola* Jones and *vespertina* Bailey; *C. heliophila* Mack.; B.C. to s Ont.; type from Sask.] ....var. *digyna* Boeck.

C. petasata Dewey   (154)
/sT/WW/   (Hs)   Meadows and open woods from s Yukon (St. Elias Mts.; CAN, detd. Porsild; collections in CAN from near Mackintosh and at Miles' Canyon have also been placed here tentatively by Porsild) and B.C. to Alta. (N to the Banff dist.; CAN) and sw Sask. (Cypress Hills; CAN), s to Oreg., Nev., Utah, and Colo. [*C. liddonii* Boott; *C. rufovariegata* Boeck.; the transcontinental range assigned by Boivin 1967a, is based upon a more liberal interpretation of the species]. MAP: Hultén 1968b:236.

C. petricosa Dewey   (361, 362 (*C. franklinii*), and 370 (*C. misandroides*))
/aST/WW/   (Grh)   Calcareous barrens, crevices, slopes, and alluvia, the ranges of taxa outlined below (the species confined to Alaska–Canada), together with MAPS and synonymy.
1   Stigmas normally 2 (rarely 3); achenes normally lenticular (rarely trigonous); [*C. misandroides* Fern., the type from Table Mt., Nfld.; *C. franklinii* var. *mis.* (Fern.) Raymond; dry calcareous barrens, cliffs, and talus-slopes of Que. (L. Mistassini; Lac Pleureuse, N Gaspé Co.; Larch and Koksoak rivers between 57°–58°N and 69°–70°W), and sw Nfld.; reported from the Brooks Range near Wiseman, Alaska, at ca. 68°N, by Marcel Raymond, Can. Field-Nat. 66(4):102. 1952; MAPS: Raymond 1951: map 10 (*C. frank.* var. *mis.*), p. 7; Fernald 1924: map 6, p. 565, and 1925: map 16 (*C. mis.*; incomplete), p. 253] ....var. *misandroides* (Fern.) Boivin
1   Stigmas normally 3; achenes normally trigonous.
    2   Perigynia to 2.5 mm broad, abruptly minutely beaked; spikes relatively numerous, the terminal one strongly pistillate at base; culms to about 7 dm tall, 2 or 3 times longer than the leaves; [*C. franklinii* Boott, the type from the Rocky Mountains, presumably of Alta.; Alaska (2 stations near the Arctic Circle), the Yukon (N to ca. 63°N), and mts. of sw Alta. (Banff; Jasper; Nordegg); MAP: Porsild 1966: map 24, p. 69; the Alta. stations are indicated in the above-noted maps by Fernald] ....var. *franklinii* (Boott) Boivin
    2   Perigynia to 1.5 mm broad, tapering to the apex; spikes 2–4(5), the terminal one with only a few perigynia at base; culms to about 4 dm tall, less than twice as long as the leaves.
        3   Pistillate spikes loosely flowered, the more or less 2-ranked perigynia to 7 mm long; [*C. distichiflora* Boivin, the type from near the Canol Road at Sekwi R., w Dist. Mackenzie] ....var. *distichiflora* Boivin
        3   Pistillate spikes densely flowered, the perigynia to 6 mm long; [incl. vars. *edwardsii* and *nicholsonis* Boivin; Alaska (N to the N coast at ca. 70°N), the Yukon (N to ca. 65°N), w Dist. Mackenzie (N to ca. 70°N), NW Victoria Is., and

mts. of ?B.C. and sw Alta.; (type from the Canadian Rocky Mts., perhaps of Alta.); MAPS: Porsild 1957: map 96, p. 172, and 1955: fig. 10, p. 44; Hultén 1968b:273] . . . .
. . . . .var. *petricosa*

C. phaeocephala Piper   (149)
/sT/W/   (Grh)   Alpine meadows and slopes from SE Alaska (Juneau; CAN, detd. Beschel), s Yukon (CAN), and sw Dist. Mackenzie (Brintnell L., ca. 62°N; CAN) through B.C. and sw Alta. (N to the Banff dist.) to Calif., Nev., Utah, and Colo. [*C. leporina* var. *americana* Olney; *C. petasata* var. *pleiostachya* Kük.]. MAPS: Hultén 1968b:234; Porsild 1966: map 27, p. 70; Raup 1947: pl. 17.

C. phyllomanica Boott   (114)
/sT/W/   (Grh)   Sphagnum bogs from s Alaska (*see* Hultén 1942: map 266, p. 407) through w B.C. to N Calif. MAP: Hultén 1968b:247.

C. physocarpa Presl   (511)
/aST/X/EA/   (Grh)   Wet tundra and margins of pools from the Aleutian Is. and coasts of Alaska–Yukon–Dist. Mackenzie–Dist. Keewatin to Banks Is., Southampton Is., and northernmost Ungava–Labrador, s through B.C. (type from Vancouver Is.) and sw Alta. to Wash., Utah, and Colo., the s limits farther eastwards being NW Sask. (L. Athabasca), s Dist. Keewatin, NE ?Man. (forms intermediate between this species and *C. saxatilis* var. *miliaris* occur at Churchill), and Hudson Bay (Ont. and Que.); Eurasia. [*C. ambusta* Boott; *C. ochroleuca* Cham.; *C. pulla* var. *vesicarioides* Kurtz; *C. saxatilis* vars. *major* Olney and *laxa* (Trautv.) Ohwi; *C. compacta* of auth. in part, not Dewey]. MAPS: Hultén 1962: map 16 (*C. sax.* var. *laxa*), p. 23; Porsild 1957: map 98, p. 173; Raup 1947: pl. 19.

C. piperi Mack.   (152)
/T/W/   (Hs)   Damp meadows from s B.C. (Vancouver Is.; Vancouver; Rogers Pass; Nelson; Kicking Horse L.) to Oreg. [*C. pratensis* var. *furva* Bailey, basionym, the type from Goldstream, Vancouver Is., B.C.; *C. furva* (Bailey) Piper, not Webb; scarcely distinct from *C. praticola,* with which it is merged by Hitchcock et al. (1969)].

C. plantaginea Lam.   (284)
/T/EE/   (Hs)   Rich hardwoods from Ont. (N to the Ottawa dist.; reports from Sask. and Man. require clarification) to Que. (N to l'Ange-Gardien, about 15 mi NE of Quebec City; *see* Que. map by Doyon and Lavoie 1966: fig. 6, p. 816), w N.B., and N.S. (Colchester Co.; ACAD; not known from P.E.I.), s to Minn., Wisc., Tenn., and Ga.

C. platylepis Mack.   (153)
/T/W/   (Hs)   Dry conifer woods from s B.C. (McGillivray Creek, Cascade Range; CAN) and sw Alta. (N to Jasper) to Idaho and Wyo. [*C. piperi sensu* Rydberg 1922, not Mack.; merged with *C. (pachystachya) macloviana* by Hitchcock et al. (1969)].

C. platyphylla Carey   (286)
/T/EE/   (Hs)   Rich hardwoods and rocky slopes from Ont. (N to the Ottawa dist.; Gillett 1958) and sw Que. (N to near Quebec City; *see* Que. map by Doyon and Lavoie 1966: fig. 3, p. 815) to Tenn. and N.C.

C. podocarpa R. Br.   (419)
/ST/W/   (Grh)   Alpine meadows and grassy slopes from N-cent. Alaska–Yukon (*see* Hultén 1942: map 295, p. 409) and NW Dist. Mackenzie (CAN) through B.C. and sw Alta. (South Kootenay Pass and Waterton Lakes; CAN) to Oreg., Utah (Friedrich Hermann, Rhodora 39: 492. 1937), and Wyo. [*C. tolmiei* Boott, not Bailey; *C. ?paysonis* Clokey; *C. microchaeta* of auth., not Holm, which is apparently *C. nesophila; C. ?behringensis* Clarke; the type of *C. podocarpa* is from "Lat. 64°–69° northwestern Canada"]. MAPS: Hultén 1968b:263, and 1962: map 39, p. 47; Raup 1947: pl. 18 (incomplete southwards).
   Forma *pallidior* Lepage (perigynia yellowish rather than light green, blotched with purple, their

scales cinnamon-colour rather than purplish black) is known from the type locality, Eagle Summit, Alaska.

C. praegracilis Boott   (19)
/sT/WW/   (Grh)   Moist plains, prairies, and foothills from cent. Yukon (*see* Hultén 1942: map 242, p. 405) and B.C. to s-cent. Alta.–Sask., s Man. (N to Riding Mt.), and ?Ont. (Boivin 1967a), s to Calif., Mexico, Tex., Okla., and Iowa. [*C. marcida* Boott, not Gmel.; *C. mar.* var. *alterna* Bailey (*C. alt.* (Bailey) Clarke); *C. sartwellii* var. *occidentalis* Bailey; *C. ?gayana sensu* Henry 1915, not Desv.]. MAPS: Hultén 1968b:230; Porsild 1966: map 28, p. 70.

C. prairea Dewey   (68)
/T/X/   (Hs)   Wet calcareous meadows, bogs, and thickets from s B.C. (near Lillooet; CAN) to Alta. (N to Banff; CAN), Sask. (N to Prince Albert; CAN), Man. (N to Riverton; CAN), Ont. (N to Attawapiskat, w James Bay, ca. 53°N), Que. (N to Old Factory, SE James Bay, 52°37′N, and Ste-Luce, Rimouski Co.), NW N.B., and N.S. (Kings Co.; not known from P.E.I.), s to Nebr., Ohio, and N.J. [*C. teretiuscula* vars. *prairea* (Dew.) Britt. and *ramosa* Boott; *C. paradoxa sensu* Hooker 1839, not Willd.].

C. prasina Wahl.   (326)
/T/EE/   (Hs)   Rich moist woods and streambanks from Mich. to Ont. (N to the Ottawa dist.), Que. (N to Lévis Co.; *see* Que. map by Raymond 1950: fig. 34, p. 85), and Maine, s to Tenn. and S.C.

C. praticola Rydb.   (151)
/aST/X/G/   (Hs)   Open woods, meadows, prairies, and clearings from cent. Alaska–Yukon (*see* Hultén 1942: map 248, p. 405), sw Dist. Mackenzie, and B.C.–Alta. to Sask. (N to the Clearwater R. at ca. 57°N), Man. (N to Oxford L. at ca. 55°N), Ont. (N to w James Bay at ca. 53°N), Que. (N to s Ungava Bay and the Côte-Nord), Nfld., and N.S. (not known from N.B. or P.E.I.), s to Calif., Colo., N.Dak., N Mich., and N Maine; w Greenland N to ca. 64°N, E Greenland N to 66°44′N. [*C. pratensis* Drejer, basionym, the type from Godthaab, w Greenland; *C. adusta* var. *minor* Boott]. MAP: Hultén 1968b:237.

Var. *subcoriacea* Hermann (perigynia subcoriaceous and brown rather than membranaceous and greenish white, relatively short-beaked, the head more compact and erect than in the typical form) is known from the type locality, about 35 mi E of Jasper, Alta., and from Wyo.

C. preslii Steud.   (140)
/sT/W/   (Hs)   Alpine meadows and slopes from SE Alaska (near Anchorage; CAN, detd. Porsild) through B.C. and sw Alta. (N to Lake Louise; CAN) to Calif. and Idaho. [*C. leporina sensu* K. B. Presl, Rel. Haenk., vol. 1:203. Calve, Prague. 1828, not L., basionym, the type of Presl's plant from Nootka Sound, Vancouver Is., B.C.; merged with *C. (pachystachya) macloviana* by Hitchcock et al. 1969]. MAP: Hultén 1968b:233.

C. projecta Mack.   (187)
/sT/EE/   (Hs)   Wet meadows and swampy ground from Man. (N to The Pas) to Ont. (N to Sachigo L. at ca. 54°N, 92°W), Que. (N to E James Bay at ca. 52°30′N and the Côte-Nord), s Labrador (N to the Hamilton R. basin), Nfld., N.B., P.E.I., and N.S., s to Minn., Ill., and W.Va.; (the report from s B.C. by Eastham 1947, requires confirmation). [*C. lagopodioides* var. *moniliformis* Boott, basionym, the type from Kouchibouguac, N.B.; *C. tribuloides (cristata)* var. *reducta* Bailey].

C. pseudo-cyperus L.   (496)
/T/(X)/EA/   (Hs (Hel))   Swamps and shallow water from Alta. (two localities at ca. 54°N) to cent. Sask. (N to Amisk L., near Flin Flon), Man. (N to The Pas), Ont. (N to the N shore of L. Superior), Que. (N to the Gaspé Pen.), Nfld., N.B., P.E.I., and N.S., s to N.Dak., Ind., and Pa.; Eurasia; N Africa. MAPS: Hultén 1958: map 149, p. 169; Meusel, Jaeger, and Weinert 1965:79.

Forma *multispicula* Lepage (lower spikes subtended by 2 or more small spikelets 1 or 2 cm long, each of these subtended by a setaceous bract to 3 cm long) is known from the type locality, Rimouski, Rimouski Co., E Que., and also from Ste-Luce, near Rimouski.

C. pyrenaica Wahl. (8)
/ST/W/EeA/ (Hs) Calcareous alpine slopes (ranges of Canadian taxa outlined below), s to Oreg., Utah, and Colo.; S. Europe; E Asia. MAPS and synonymy: *see* below.
1 Stigmas 3; perigynia to 4 mm long; [s Alaska–s Yukon–w Dist. Mackenzie and mts. of
   B.C. and sw Alta.; MAPS (aggregate species): Hultén 1962: map 146, p. 155; Raup 1947:
   pl. 17] . . . . .ssp. *pyrenaica*
1 Stigma 2; perigynia to 3 mm long; [*C. micropoda* Mey., the type from Unalaska, Alaska;
   C. nivalis Cham., not Boott; Aleutian Is.–s Alaska; MAPS: Hultén 1968b:226, and 1942:
   map 235, p. 404] . . . . .ssp. *micropoda* (Mey.) Hult.

C. ramenskii Komarov
/ASs/W/eA/ (Grh) Brackish or saline marshes and shores of Alaska, islands in the Bering Sea, and E Asia.
1 Midrib of pistillate scales not excurrent; [range of the species; MAPS: Hultén 1968b:252,
   1962: map 184, p. 195, and 1942: map 278, p. 408] . . . . .var. *ramenskii*
1 Midrib of pistillate scales excurrent as an awn to 3 mm long; [type from St. Paul Is.,
   Alaska] . . . . .var. *caudata* Hult.

C. rariflora (Wahl.) Sm. (404)
/aST/X/GEA/ (Grh) Peaty bogs, barrens, and pond-margins, the aggregate species from the coasts of Alaska–Yukon–Dist. Mackenzie–Dist. Keewatin to cent. Victoria Is., Baffin Is. (N to the Arctic Circle), northernmost Ungava–Labrador, and Nfld., s to s Alaska, Great Bear L., s Dist. Keewatin, N Man. (s to Landing L., about 180 mi NE of The Pas), w and E James Bay, E Que. (Côte-Nord, Anticosti Is., Gaspé Pen., and Magdalen Is.), and Mt. Katahdin, Maine; w Greenland N to ca. 73°N, E Greenland N to ca. 74°30′N; Iceland; Eurasia. MAPS and synonymy: *see* below.
1 Terminal spike androgynous, bearing perigynia at base; [type from Atkinson Point, coast
   of Dist. Mackenzie] . . . . .var. *androgyna*
1 Terminal spike staminate throughout . . . . .var. *rariflora*
   2 Scales of spikes pale brown; [Churchill, Man.; Polunin 1940] . . . . .f. *pallidior* Blytt
   2 Scales purple-brown to blackish.
      3 Spikes on stout, permanently erect or suberect peduncles less than 1 cm long, the
        staminate spike often absent or rudimentary; [reported from the Canadian Arctic
        Archipelago by Polunin 1940, the type presumably from Greenland] . . . .
        . . . . .f. *erecta* Polunin
      3 At least the lower pistillate spikes on slender arching or recurving peduncles, the
        staminate spike long-peduncled and normal; [*C. limosa* var. *rar.* Wahl.; trans-
        continental; MAPS: Hultén 1962: map 8, p. 15, and 1968b:268; Porsild 1957: map
        92, p. 172]. Concerning possible hybridization with *C. limosa* and *C. paupercula,*
        see the discussion under those species . . . . .f. *rariflora*

C. raynoldsii Dewey (420)
/T/W/ (Grh) Mountain slopes and meadows from s B.C. (N to the Marble Mts. NW of Clinton; CAN) to sw Alta. (Waterton Lakes; CAN) and sw Sask. (Cypress Hills; Breitung 1957a), s to Calif., Utah, and Colo. [*C. lyallii* Boott; incl. *C. aboriginum* Jones]. MAP: D. F. Murray, Brittonia 21(1): fig. 1 (not indicating the sw Sask. station), p. 58. 1969.

C. retroflexa Muhl. (25)
/t/EE/ (Hs) Dry woods and thickets from s Ont. (Waterloo, Wellington, and Prince Edward counties; TRT; John Macoun 1888; Stroud 1941; Montgomery 1945) and Vt., s to Tex. and Fla. [*C. rosea* var. *ret.* (Muhl.) Torr.].

C. retrorsa Schw. (523)
/sT/X/ (Hs) Moist meadows, swamps, and alluvial woods from cent. Dist. Mackenzie (N to Fort Simpson, ca. 62°N; W. J. Cody, Can. Field-Nat. 75(2): 60. 1961) and B.C.–Alta. to Sask. (N to Waskesiu L. at ca. 54°N), Man. (N to Hill L., N of L. Winnipeg), Ont. (N to Big Trout L. at ca. 53°45′N, 90°W), Que. (N to L. Mistassini and Anticosti Is.), Nfld., N.B., P.E.I., and N.S., s to Oreg., Utah, Colo., S.Dak., and N.J. [Incl. var. *robinsonii* Fern.].

Forma *multispicula* Lepage (with clusters of up to 6 reduced spikes subtending the basal spike) is known from the type locality along the Missinaibi R., Ont., at 50°06′N.

C. rhynchophysa Meyer   (519: *C. laevirostris*)
/Ss/W/EA/   (Grh)   Marshes and swamps of Alaska (N to Fairbanks; CAN), the Yukon (N to Dawson; CAN), and s Dist. Mackenzie (s shore of Great Slave L. at the mouth of the Slave R.; CAN); Eurasia. [*C. laevirostris* (Blytt) Fries]. MAP: Hultén 1968*b*:277.

C. richardsonii R. Br.   (246)
/sT/(X)/   (Hsr)   Calcareous rocks, sands, barrens, plains, and prairies from s Dist. Mackenzie (Hay River; CAN; reported from Fort Smith, ca. 60°N, by Raup 1947) and B.C.–Alta. to Sask. (N to Cumberland House, ca. 54°N), Man. (N to Wekusko L., about 80 mi NE of The Pas), Ont. (N to w James Bay at ca. 52°N), and Que. (N to the Harricanaw R. at ca. 50°30′N, this station located by a dot on the Ont.–Que. map by Dutilly and Lepage 1951*b*: fig. 9, p. 271), s to s B.C.–Alta., S.Dak., Ohio, and N.Y. MAP (w Canada): Raup 1930: map 29, p. 203.
     According to Boott (*in* Hooker 1832), the type material was collected in the region between Norway House, Man., and Cumberland House, E-cent. Sask. Concerning a postglacial migration eastward as an explanation of the widely scattered stations in the East, *see* Stebbins (1935).

C. rosea Schkuhr   (27 and 29 (*C. radiata*))
/T/EE/   (Hs (Grh))   Swampy ground and low woodlands (ranges of Canadian taxa outlined below), s to La. and Ga.
1   Leaves pale green, to 2 mm broad; perigynia pale green, those of the lowest spikes to
    about 12 in number; stigmas flexuous or recurved; [s Man. (N to Seven Sisters Falls,
    about 65 mi NE of Winnipeg), Ont. (N to near Ottawa), Que. (N to the Gaspé Pen.), N.B.,
    P.E.I., and N.S.; reported from Nfld. by John Macoun 1888, but not listed by Rouleau
    1956] ....var. *rosea*
1   Leaves deep green, at most 1.2 mm broad; perigynia deep green, those of the lowest
    spikes not more than about 6; stigmas mostly tightly coiled; [*C. radiata* (Wahl.) Dew.;
    range of var. *rosea* but not yet known from N.B. or P.E.I.] ....var. *radiata* Wahl.

C. rossii Boott   (229, 228 (*C. brevipes*), and 232 (*C. brevicaulis*))
/ST/WW/   (Grh)   Dry prairies, sand-hills, open woods, and clearings from s Alaska, cent. Yukon, and s Dist. Mackenzie through B.C.–Alta. to Sask. (N to Carswell L. at 58°35′N), Man. (N to Gillam, about 165 mi s of Churchill), and w Ont. (N shore of L. Superior sw of Thunder Bay), s to Calif., Colo., S.Dak., and Mich. [*C. deflexa* var. *rossii* (Boott) Bailey; incl. *C. brevicaulis* Mack. and *C. brevipes* Boott; *C. novae-angliae sensu* John Macoun 1888, as to B.C. reports, not Schw., relevant collections in CAN; *C. globosa* of B.C. reports at least in part, not Boott]. MAPS: Hultén 1968*b*:265; Raup 1930: map 20 (w Canada), p. 202.

C. rostrata Stokes   (520)
/aST/X/GEA/   (Grh (Hel))   Swamps, wet shores, and shallow water, the aggregate species from N-cent. Alaska–Yukon–Dist. Mackenzie to Great Bear L., Great Slave L., L. Athabasca (Alta. and Sask.), s Dist. Keewatin, northernmost Ont., Que. (N to s Ungava Bay and the Côte-Nord), Labrador (N to Anatolak, 56°33′N), Nfld., N.B., P.E.I., and N.S., s to s Calif., N.Mex., Nebr., Ohio, Tenn., and Del.; southernmost Greenland; Iceland; Eurasia. MAPS and synonymy: *see* below.
1   Pistillate scales acuminate or short-awned, the lower ones often surpassing the
    perigynia.
    2   Culms barely 1 mm thick, at most 5 dm tall; leaves to 5 mm broad; pistillate spikes to
        2.5 cm long; [Que. (reported from the Gaspé Pen. by Fernald *in* Gray 1950, who also
        notes a report from the Rupert R. near James Bay) and N N.B. (type from South
        Tobique Lakes)] ....var. *ambigens* Fern.
    2   Culms coarse, to over 1 m tall; leaves to over 1 cm broad; pistillate spikes to 1.5 dm
        long; [*C. inflata* var. *utric.* (Boott) Bailey (*C. utriculata* Boott); range of the species but
        somewhat more southern and gradually merging with it northwards] ....
        ....var. *utriculata* (Boott) Bailey
1   Pistillate scales blunt to acute, shorter than the perigynia.

3 Perigynia barely inflated, attenuate to the minutely bidentate beak; pistillate spike
solitary, to about 2.5 cm long; leaves to 4 mm broad, channelled or involute; [type from
Anticosti Is., E Que.] . . . . .var. *anticostensis* Fern.
3 Perigynia strongly inflated, mostly abruptly narrowed to the prominently bidentate
beak; pistillate spikes 2–4, to 7 cm long; leaves to 8 mm broad, channelled or flat . . . .
. . . . .var. *rostrata*
    4 Perigynia more or less cup-shaped, their scales spreading; [mouth of the Beaver
    R., near James Bay, Que., at 53°25′N; Dutilly, Lepage, and Duman 1958] . . . .
    . . . . .f. *plumosa* Norman
    4 Perigynia ovoid, their scales ascending; [*C. ampullacea* Good.; *C. inflata* of auth.,
    not Huds.; range of the species, the common form northwards; MAPS (aggregate
    species): Hultén 1962: map 95, p. 105, and 1968b:276; Meusel, Jaeger, and
    Weinert 1965:79; Raymond 1950b: fig. 12 (based on a 1941 map by Tichomirov),
    p. 23] . . . . .f. *rostrata*

C. rotundata Wahl. (518)
/aST/WW/EA/ (Grh) Damp turfy or hummocky barrens and pond-margins from the coasts of
Alaska, the Yukon (Herschel Is.), and NW Dist. Mackenzie to Great Bear L. and E Dist. Keewatin, s
to s-cent. Alaska and N Man. (s to Lamprey, about 40 mi s of Churchill; CAN); N Eurasia.
[*C. saxatilis* var. *rot.* (Wahl.) Gelting; *C. melozitnensis* Porsild]. MAPS: Hultén 1968b:278; Marcel
Raymond, Nat. can. (Que.) 84(8–9): fig. 1, p. 176. 1957 (the N Que. area probably refers to
C. saxatilis var. *miliaris*).

C. rufina Drejer (268)
/aS/EE/GE/ (Hsr) Rocky barrens and sandy or gravelly shores of s Dist. Keewatin (Tha-anne
R. at ca. 61°N, 97°W; CAN; Porsild 1943) and N Man. (Nueltin L. at 59°48′N; CAN); w Greenland
(type from Godthaab) N to 69°25′N, E Greenland N to 71°14′N; Iceland; N Scandinavia. MAP: Hultén
1958: map 162, p. 181.

C. rupestris Bellardi (255 and 254 (*C. drummondiana*))
/AST/X/GEA/ (Grh) Arctic and alpine barrens and rocky or gravelly places, the aggregate
species from the coasts of Alaska–Yukon–Dist. Mackenzie to Banks Is., Victoria Is., northernmost
Ellesmere Is., Baffin Is., northernmost Ungava–Labrador, and Nfld., s to s Yukon, N Man.
(Churchill), and Que. (s to E James Bay, the Côte-Nord, Anticosti Is., and Gaspé Pen.); isolated in
the mts. of B.C. (Boivin 1967a) and sw Alta., as also in Wyo., Utah, and Colo.; w and E Greenland
between ca. 65°and 79°N; Iceland; Eurasia. MAPS and synonymy: see below.
1 Scales not enveloping the perigynia; spike slender, flexuous, the lower perigynia often
separate; leaves averaging about 1.5 mm broad; culms slender; [*C. attenuata* R. Br.;
transcontinental; MAPS (Alta. stations may refer to var. *drummondiana*): Porsild 1957:
map 75, p. 170; Hultén 1968b:224, and 1962: map 20, p. 27; Meusel, Jaeger, and
Weinert 1965:64; Böcher 1954: fig. 33 (map 2), p. 135; Raymond 1950b: fig. 14, p. 25,
and 1950a: map 8 (E Canada area), p. 442] . . . . .var. *rupestris*
1 Scales enveloping and largely concealing the perigynia; spike relatively stout and
straight, the perigynia not separated; leaves 2 or 3 mm broad; culms relatively stout;
[*C. drummondiana* Dewey, the type material from Fort Norman, Dist. Mackenzie, and
the Rocky Mts.; also reported from Waterton Lakes, Alta., by Breitung 1957b] . . . .
. . . . .var. *drummondiana* (Dew.) Bailey

C. sabulosa Turcz. (423: *C. leiophylla*)
/sT/W/EA/ (Grh) Known in N. America only from southernmost Yukon (Carcross dist. and the
St. Elias Mts., where abundant on dry sand flats and dunes and in blow-out areas; CAN; see Hultén
1942: map 284 (*C. leiophylla*), p. 408); NE Europe; Asia. MAPS (ssp. *leio.*): Hultén 1968b:256;
Porsild 1966: map 29, p. 70.
    The Yukon plant is referable to ssp. *leiophylla* (Mack.) Porsild (*C. leiophylla* Mack., the type
from Carcross, Yukon Territory; perigynia nearly or quite nerveless rather than strongly nerved, the
teeth of the bidentate beak about 1 mm long rather than 1.5 mm, the achene less distinctly beaked
than in the typical form).

C. salina Wahl. (478 (*C. lanceata*) and 480 (*C. recta*))
/aST/EE/GE/ (Grh) Saline or brackish shores and marshes from NE Man. (York Factory, ca. 57°N) to Ont. (shores of w James Bay), Que. (E James Bay–Hudson Bay N to ca. 57°N; s Ungava Bay; St. Lawrence R. estuary from l'Islet Co. to the Côte-Nord, Anticosti Is., Gaspé Pen., and Magdalen Is.), s Labrador (N to the Hamilton R. basin), Nfld., N.B., P.E.I., and N.S. to Mass.; southernmost Greenland; Iceland; Scandinavia; Novaya Zemlya. MAPS and synonymy: see below.
1   Spikes drooping, the lowest peduncle to 6 cm long; pistillate spikes to over 2 cm long; [E
    Que. (Anticosti Is. and the Mingan Is. of the Côte-Nord), s Labrador (opposite Blanc-
    Sablon, ca. 51°20′N), and Nfld.] . . . . .var. *pseudofilipendula* Kük.
1   Spikes erect, the lowest peduncle not over about 2 cm long.
    2   Culms to about 9 dm tall, subacutely angled; leaves to 9 mm broad; pistillate spikes to
        about 8 cm long; [*C. recta* Boott; reported by Fernald *in* Gray 1950, from Hudson Bay,
        Labrador, Nfld., and N.S.; MAP: Hultén 1958: map 264 (*C. recta*), p. 283] . . . .
        . . . . .var. *kattegatensis* (Fries) Almq.
    2   Culms mostly not over 3 dm tall, obtusely angled; leaves to 4 mm broad; pistillate
        spikes to about 3 cm long; [*C. cuspidata* Wahl.; *C. lanceata* Dew.; range of the
        species; MAPS: Hultén 1958: map 263, p. 283; Uno Saxén, Acta Bot. Fenn. 22: map 1,
        p. 8. 1938]. The report from Alaska by Fernald *in* Gray (1950) is perhaps based upon
        the very similar *C. ramenskii* Komarov. Hultén's map has a dot at the location of York
        Factory, Hudson Bay, Man., which should be verified. It may refer to *C. subspathacea*,
        known from somewhat farther north at Churchill . . . . .var. *salina*

C. sartwellii Dewey (20)
/sT/(X)/ (Grh) Calcareous bogs, marshes, and wet places from sw Dist. Mackenzie (near the w end of Great Slave L. at ca. 61°N; J. W. Thieret, Can. Field-Nat. 75(3): 114. 1961) and B.C.–Alta. to Sask. (N to the Clearwater R. at ca. 57°N), Man. (N to near Dauphin L.), Ont. (N to Moosonee, s James Bay, ca. 51°20′N), and Charlton Is., s James Bay, s to Mont., Colo., Mo., Ind., and N.Y. [*C. disticha* Huds. in part; *C. disticha* var. *sart.* Dewey]. MAP: Raup 1930: map 32 (now requiring considerable expansion), p. 203.

C. saxatilis L. (512, 510 (*C. miliaris*), and 513 (*C. mainensis*))
/AST/X/GEA/ (Grh) Peaty or gravelly damp soils and pond-margins (ranges of Canadian taxa outlined below), s to Wash., Nev., Colo., and Maine; w and E Greenland N to ca. 78°N; Iceland; Eurasia. MAPS and synonymy: see below.
1   Pistillate spikes purplish black; perigynia 3 or 4 mm long, empty at tip; leaves to 5 mm
    broad, flat or revolute-margined; [*C. pulla* Good.]. This taxon is generally regarded as
    being confined to Eurasia. However, Hultén's map indicates its possible presence in the
    eastern part of the Canadian Arctic Archipelago and in northernmost Ungava; it reaches
    79°N in E and w Greenland; MAPS: Hultén 1962: map 16, p. 22; Tolmachev 1952: maps 23
    and 30 (requiring considerable revision) . . . . .var. *saxatilis*
1   Pistillate spikes straw-colour to purplish; perigynia tightly filled by the achene.
    2   Leaves to 2.5 mm broad, involute in drying; perigynia rarely over 3.5 mm long; [var.
        obtusa Bailey; *C. miliaris* Michx., the type from the L. Mistassini region, Que.; NE
        Sask. (G. W. Argus, Can. Field-Nat. 80(3): 133. 1966), N Man. (known from between
        York Factory, Hudson Bay, and Baralzon L., 60°N), N Ont. (N to the mouth of the Black
        Duck R., Hudson Bay, ca. 56°50′N), N Ungava (N to Port Burwell, 60°28′N), s Baffin
        Is., N Labrador, and Nfld., s to Que. (E James Bay; L. Mistassini; L. St. John; the
        Côte-Nord; Anticosti Is.; Gaspé Pen.), Nfld., N.B., and N.S.; w Greenland N to ca.
        67°N, E Greenland N to ca. 61°N; MAPS: Hultén 1962: map 11 (*C. mil.*), p. 18; Porsild
        1957: map 99, p. 173] . . . . .var. *miliaris* (Michx.) Bailey
    2   Leaves to 4 mm broad, flat or revolute-margined; perigynia to 5 mm long . . . .
        . . . . .var.*rhomalea* Fern.
        3   Lower spikes on peduncles to over 2 dm long from the axils of normal foliage
            leaves; [known from the type locality on an island in James Bay at 54°10′N and
            from an adjacent island] . . . . .f. *longepedunculata* Lepage
        3   Lower spikes sessile or short-peduncled in the axils of bracts; [*C. miliaris* var.
            major Bailey, not *C. saxatilis* var. *major* Olney, which is *C. physocarpa* Presl;

Great Bear L. and N Sask. to Baffin Is., Labrador, and Nfld.; MAP: Porsild 1957: map 100, p. 173] . . . . .f. *rhomalea*

C. scabrata Schw.   (397)
/T/EE/   (Hsr)   Wet woods, meadows, and shores from Ont. (N to Batchawana Bay, E end of L. Superior; Hosie 1938) to Que. (N to the Gaspé Pen.), N.B., P.E.I. (Mackenzie 1935), and N.S., s to Mo., Ohio, Tenn., and S.C.

C. schweinitzii Dewey   (492)
/T/EE/   (Hsr)   Calcareous swamps, meadows, and low woods from Mich. to s Ont. (N to Wellington, Durham, and Leeds counties) and sw Vt., s to Mo., Tenn., and N.C.; (reports from Nfld. by Reeks 1873, and John Macoun 1888, probably refer to *C. pseudo-cyperus*).

C. scirpoidea Michx.   (237, 238 (*C. scirpiformis*), 239 (*C. stenochlaena*), and 240 (*C. pseudo-scirpoidea*))
/aST/X/GEeA/   (Grh (Hsr))   Rocky barrens, bogs, and shores, the aggregate species from the coasts of Alaska–Yukon–Dist. Mackenzie to Banks Is., Devon Is., Baffin Is., northernmost Ungava–Labrador, and Nfld., s to Calif., Colo., s Man., N Mich., s Ont. (Bruce Pen.), E Que. (s to the Gaspé Pen.; type locality between Hudson Bay and L. Mistassini), N.S. (Cape Breton Is.; not known from N.B. or P.E.I.), N.Y., and N New Eng.; w Greenland N to ca. 76°N, E Greenland N to 74°22′N; a single station in Norway; E Asia. MAPS and synonymy: see below.
1   Leaves convolute, at most 1.5 mm broad; [Bruce Pen. and Great Cloche Is., L. Huron, Ont.] . . . .
. . . . .var. *convoluta* Kük.
1   Leaves flat, 2–4 mm broad.
   2   Perigynia plano-convex in section, broadly lanceolate, to 4 mm long; achenes
      distinctly stipitate; [*C. stenochlaena* (Holm) Mack.; s Alaska–Yukon, B.C. (type from
      Chilliwack), and Alta.] . . . . .var. *stenochlaena* Holm
   2   Perigynia compressed-triangular in section, ovoid or oblong-ovoid, to 3 mm long;
      achenes very short-stipitate; [transcontinental].
      3   Scales with broad white hyaline margins; [*C. scirpiformis* Mack., the type from
         Banff, Alta.] . . . . .var. *scirpiformis* (Mack.) O'Neill & Duman
      3   Scales with very narrow hyaline margins; [var. *basigyna* Lange; *C. athabascensis*
         Hermann; *C. michauxii* Schw.; *C. pseudoscirpoidea* Rydb.; *C. wormskioldiana*
         Hornem.; MAPS (aggregate species): Porsild 1957: map 74, p. 170; Raup 1947:
         pl. 18; Hultén 1958: map 170, p. 189, and 1968b:223] . . . . .var. *scirpoidea*

C. scoparia Schkuhr   (160)
/T/X/   (Hs)   Woods, thickets, and open places from s B.C. (Vancouver Is. and adjacent islands; Trail) to Alta. (Edmonton; CAN), Sask. (N to Saskatoon; Breitung 1957a), Man. (N to Sasaginnigak L., about 125 mi NE of Winnipeg), Ont. (N to Moosonee, s James Bay, ca. 51°15′N), Que. (N to L. Mistassini and the Gaspé Pen.; reported from the Côte-Nord and Anticosti Is.), Nfld., N.B., P.E.I., and N.S., s to Oreg., N.Mex., Ark., and Fla.
1   Pistillate scales reddish-brown to blackish; perigynia 4 or 5 mm long; spikes crowded in a
   head to 2.5 cm long; [P.E.I. and N.S.] . . . . .var. *tesselata* Fern. & Wieg.
1   Pistillate scales stramineous or pale brown; perigynia to 7 mm long . . . . .var. *scoparia*
   2   Spikes top-shaped, broadly truncate at summit, in loose heads to 5 cm long; [Nfld.
      (type from Rushy Pond) and N.S.] . . . . .f. *subturbinata* (Fern. & Wieg.) Fern.
   2   Spikes acute or subacute.
      3   Tips of the crowded spikes acuminate or caudate; [Ont. and N.S. (type from Sand
         Beach, Yarmouth Co.)] . . . . .f. *peracuta* Fern.
      3   Tips of the spikes acute or subacute.
         4   At least the lower spikes remote in a loose flexuous head to 6 cm long; [E Que.,
            Nfld., P.E.I., and N.S.] . . . . .f. *moniliformis* (Tuckerm.) Kük.
         4   Spikes rather crowded, the head erect or slightly arching.
            5   Head to 3 cm long and 2.5 cm thick, some of the spikes divergent; [Nfld.,
               N.B., P.E.I., and N.S.] . . . . .f. *condensa* (Fern.) Kük.

Cyperaceae

> 5 Head to 5 cm long and 1.5 cm thick, the spikes all ascending; [trans-
> continental] ....f. *scoparia*

C. scopulorum Holm   (440)
/ST/W/   (Grh)   Alpine slopes and summits of the Yukon (isolated stations near the Alaska boundary at 66°25′N and in the Dawson Range at ca. 62°N, 138°W; CAN; Porsild 1951a), SE B.C. (Windermere, Whitehorse, and Yoho; Eastham 1947), and sw Alta. (Banff dist.; CAN) to Calif., Nev., Utah, and Colo. [*C. campylocarpa* Holm; *C. miserabilis* Mack.]. MAP: Hultén 1968b:248.

[*C. serotina* Mérat]
[This European species, closely related to *C. demissa* and *C. viridula* of the Section *Extensae,* is tentatively reported from the Magdalen Is. of E Que. by Raymond (1950a) and collections in GH from Nfld. and N.S. have been placed here. (*C. tumidicarpa* Anderss.).]

C. shortiana Dewey   (399)
/t/EE/   (Hs)   Rich woods and meadows (chiefly calcareous) from Iowa to s Ont. (Amherstburg, Essex Co., where taken by John Macoun in 1901; CAN), s to Okla., Mo., Tenn., and Va. MAPS: Meusel, Jaeger, and Weinert 1965:73; Meusel 1943: fig. 32c.

C. siccata Dewey   (22)
/sT/(X)/   (Grh)   Dry open soil and sandy prairies from s Yukon (Porsild 1951a) and s Dist. Mackenzie (Fort Simpson, ca. 63°N) to L. Athabasca (Alta. and Sask.), Man. (N to the Cochrane R. at 58°13′N), Ont. (N to Kakabeka Falls near Thunder Bay and Matheson, ca. 48°40′N), and sw Que. (Pontiac Bay, Pontiac Co.; Marcel Raymond, Nat. can. (Que.) 70:267. 1943), s to Wash., Ariz., N.Mex., Nebr., Ohio, and N.J. [*C. foenea* Willd. as to type, not *sensu* Mackenzie 1935, and many American auth., which is *C. argyrantha* Tuckerm.; *C. straminea* var. *foenea* (Willd.) Torr.]. MAP (*C. foenea*): Hultén 1968b:235.

C. silicea Olney   (183)
/T/E/   (Hs)   Coastal sands and rocks from E Que. (St. Lawrence R. estuary from l'Islet and Charlevoix counties to the Gaspé Pen. and Magdalen Is.) to sw Nfld. (St. George's; GH; CAN), N.B., P.E.I., N.S., and Md. [*C. straminea* vars. *sil.* Olney) Bailey and *moniliformis* Tuckerm. in small part; *C. foenea* var. *sabulonum* Gray]. MAP (E Canada): Raymond 1951: map 26, p. 15.

C. simulata Mack. (17)
/T/W/   (Grh)   Swampy ground from Wash. to s Alta. (Waterton Lakes; Elbow R.; Kananaskis; Nordegg) and Sask. (Cypress Hills, Pike Lake, and Yorkton; Breitung 1957a), s to Calif., N.Mex., and Colo. [*C. gayana* of auth., not Desv.].

C. sitchensis Prescott   (456)
/sT/W/   (Grh)   Swampy places, mostly near the coast, from s Alaska (*see* Hultén 1942: map 276, p. 408; type from Sitka) through B.C. to Calif. and Idaho. [*C. dives* Holm; *C. howellii* Bailey; *C. panda* Clarke; *C. barbarae sensu* John Macoun 1888 and 1890, and Henry 1915, at least in larger part, relevant collections from s B.C. in CAN]. MAP: Hultén 1968b:251.

C. sparganioides Muhl.   (53 and 51 (*C. cephaloidea*))
/T/EE/   (Grh)   Rich woods, thickets, and swampy ground (ranges of Canadian taxa outlined below), s to Kans., Mo., Tenn., and Va.
1   Head loose, to 1.5 dm long, the lower spikes well separated; scales acuminate, much narrower but not much shorter than the perigynia; perigynia narrowly wing-margined to base, essentially nerveless; ligules as broad as long; [s Ont. (N to near Ottawa) and sw Que. (N to near Montreal)] ....var. *sparganioides*
1   Head compact, rarely over 4 cm long; scales obtuse to acute, much shorter and narrower than the perigynia; perigynia narrowly winged only above the middle, obscurely nerved at base dorsally; ligules longer than broad; [*C. cephaloidea* Dew.; s Ont. (N to Wellington, Frontenac, and Leeds counties), Que. (N to Deschambault, Portneuf Co.; *see* Que. map

for *C. ceph.* by Doyon and Lavoie 1966: fig. 2, p. 815), and N.B. (Woodstock and vicinity, Carleton Co.)] . . . . .var. *cephaloidea* (Dew.) Carey

C. spectabilis Dewey   (417)
/ST/W/eA/   (Grh)   Meadows and alpine slopes from the Aleutian Is. and Alaska (N to the Seward Pen.; *see* Hultén 1942: map 293, p. 409) through B.C. and sw Alta. (N to the Jasper dist.) to Calif. and Mont.; E Asia (at least Kamchatka, U.S.S.R.). [Incl. vars. *elegantula, gelida,* and *superba* Holm; *C. invisa* Bailey; *C. nigella* Boott; *C. tolmiei* Boott and its vars. *longiuscula* Hult., *invisa* (Bailey) Kük., and *nigella* (Boott) Bailey]. MAP: Hultén 1968*b*:264.

C. spicata Huds.   (43)
Eurasian; introd. in dry fields and pastures and along roadsides of s Ont. (campus of Queen's University, Kingston; A. B. Klugh, Ont. Nat. Sci. Bull. 6:67. 1910) and N.S. (Hants, Pictou, Shelburne, and Victoria counties).

C. sprengelii Dewey   (350)
/T/X/   (Grh)   Open woods, meadows, and rocky places (mostly calcareous) from B.C. (N to Prince George; Eastham 1947) to Alta. (N to McMurray, 56°44′N; W. J. Cody, Can. Field-Nat. 70(3):108. 1956), Sask. (N to Saskatoon), Man. (N to Riding Mt.), Ont. (N to the Kaministiquia R. NW of Thunder Bay), Que. (reported N to St-David, Lévis Co., by Dominique Doyon and Richard Cayouette, Nat. can. (Que.) 97:448. 1970; *see* Que. maps by Robert Joyal (Nat. can. (Que.) 97(5): map E, fig. 1, p. 562. 1970, and Rouleau 1945: fig. 10, p. 52), and N.B. (Andover, Victoria Co., and Woodstock, Carleton Co.; CAN; not known from P.E.I. or N.S.), s to Mont., Colo., Nebr., and N.J. [*C. longirostris* Torr., not Krock., and its var. *microcystis* Boeck.; *C. deweyana* var. *long.* (Torr.) Boeck.].

C. squarrosa L.   (508)
/T/EE/   (Hs)   Calcareous moist woods and meadows from E Nebr. and Minn. to s Ont. (Essex and Wellington counties; TRT; Herb. M. Raymond) and sw Que. (N to Montebello, Papineau Co., and the Montreal dist.), s to Ark., Tenn., and N.C.

C. stenophylla Wahl.   (16: *C. eleocharis*)
/aST/WW/EA/   (Grh)   Dry plains and prairies from s Alaska–Yukon–Dist. Mackenzie and B.C.–Alta. to Sask. (N to Scott and Saskatoon) and Man. (N to Ethelbert, E of Duck Mt.), s to Oreg., Ariz., N.Mex., Kans., and Iowa; Eurasia. MAPS: Hultén 1962: map 138, p. 147; Porsild 1966: map 30, p. 70; Meusel, Jaeger, and Weinert 1965:66.

The N. American plant may be separated as var. *enervis* (Mey.) Kük. (ssp. *eleocharis* (Bailey) Hult.; *C. eleo.* Bailey, the type from "Saskatchewan Plains"; perigynia nerveless or nearly so rather than distinctly nerved, the spikes less crowded in the head than in the typical form).

C. stipata Muhl.   (75)
/sT/X/   (Hs)   Moist or swampy ground from s Alaska (*see* Hultén 1942: map 243, p. 405) and B.C.–Alta. to Sask. (N to N of Prince Albert), Man. (N to Swan Lake, N of Duck Mt.), Ont. (N to L. Nipigon and s James Bay), Que. (N to E James Bay at 52°16′N and the Côte-Nord), s Labrador (L. Melville, Hamilton R. basin; CAN), Nfld., N.B., P.E.I., and N.S., s to Calif., Tex., and Fla. [*C. ?crus-corvi sensu* Lindsay 1878, not Shuttlw.; *C. ?vulpina sensu* Cochran 1829, not L.]. MAPS: Hultén 1968*b*:230; Raymond 1950*a*: map 14 (E Canada), p. 444.

[C. straminea Willd.]   (184 and 177 (*C. feta*))
[The range of this species is confined to the U.S.A. according to Fernald *in* Gray (1950) but is extended to sw N.S. by Mackenzie (1931), who also reports *C. feta* Bailey (*C. stram.* var. *mixta* Bailey, basionym) from Vancouver Is., B.C. Boivin (1967*a*) includes *C. brevior, C. feta,* and *C. hormanthodes* (*C. stram.* var. *invisa* Boott) in his concept of the species. Further studies are required to clarify the complex.]

C. stricta Lam.   (466, 464, (*C. emoryi*), and 465 (*C. strictior*))
/T/EE/   (Grh (Hsr))   Marshy ground from E Man. (N to Sasaginnigak L., about 125 mi NE of

Winnipeg; reported N to York Factory) to Ont. (N to w James Bay at ca. 52°N), Que. (N to E James Bay at 52°23′N and Anticosti Is.), ?Nfld. (John Macoun 1888; not listed by Rouleau 1956), N.B., and N.S. (not known from P.E.I.), s to Colo., Tex., and Fla.

1   Ligules rounded at apex, as broad as long; pistillate spikes to 1(1.5) dm long, commonly overlapping; perigynia slightly granulose at tip; lower leaf-sheaths not fibrillose; plant freely stoloniferous; [*C. virginiana* var. *el.* Boeck.; *C. emoryi* Dew.; Man. (N to Sasaginnigak L., about 125 mi NE of Winnipeg)] ....var. *elongata* (Boeck.) Gl.

1   Ligules acutish, longer than broad; lower pistillate spikes usually remote; perigynia granular-papillate above the middle.

  2   Plant forming relatively small tussocks, freely stoloniferous; basal leaf-sheaths only sparingly fibrillose; [*C. strictior* Dew.; reported from Ont., Que., and N.S. by Fernald *in* Gray 1950] ....var. *strictior* (Dew.) Carey

  2   Plant forming dense and broad stools, horizontal stolons wanting or scarce; some or all of the basal leaf-sheaths strongly fibrillose ventrally ....var. *stricta*

    3   Pistillate spikes only 1 or 2 cm long and usually lacking a staminate apex; [*C. rousseauii* Raymond, this reported from near Cadillac, Que., by Marcel Raymond, Nat. can. (Que.) 82:31. 1955] ....f. *brevior* House

    3   Pistillate spikes to over 1 dm long, often staminate-tipped; [range of the species] ....
....f. *stricta*

C. stylosa Meyer   (415)

/aST/(X)/GeA/   (Grh)   Peaty, turfy, or gravelly places: Aleutian Is. (type from Unalaska) and N Alaska to s Yukon and Great Bear L., s along the coast through B.C. to NW Wash.; Que. (E James Bay–Hudson Bay N to ca. 56°15′N; Korok R. s of Ungava Bay at 58°50′N; E Saguenay Co. of the Côte-Nord; reports from the Gaspé Pen. require confirmation), coast of Labrador (N to Okkak, 57°40′N), and w Nfld.; w Greenland N to 62°13′N, E Greenland N to 63°05′N; NE Siberia and Kamchatka (reports from Norway require confirmation). [*C. beringiana* Cham.; *C. nigritella* Drej.; *C. parryana sensu* Hooker 1840, as to the plant of Sitka, Alaska, not Dewey]. MAPS: Hultén 1968b:258, and Bot. Not. (1943): fig. 2, p. 431. 1943; Böcher 1954: fig. 13 (bottom), p. 56; Raymond 1950b: fig. 2, p. 11, and Ann. ACFAS 15: 102. 1949.

[C. suberecta (Olney) Britt.]   (178)

[Reports of this species of the E U.S.A. (N to Minn. and Mo.) from s Ont. by Fernald *in* Gray (1950), Soper (1949), and Dodge (1915) are apparently based upon collections in MICH from Peach Is. in the Detroit R., Lambton Co. (O. A. Farwell in 1901; C. F. Wheeler in 1893). It is felt that further studies of this member of the very critical Section *Ovales* are required before admitting it definitely to our flora.]

[C. subfusca Boott]   (139)

[The report of this species of the w U.S.A. (Oreg. to Calif. and Ariz.) from B.C. by Eastham (1947) rests upon a 1941 collection by Eastham from Vancouver (duplicate in CAN, the identification either made or verified by Hermann). The concluding statement under *C. suberecta* also applies here.]

C. subspathacea Wormsk.   (479)

/aST/X/GEA/   (Grh)   Saline shores and salt marshes from the coasts of Alaska–Yukon–Dist. Mackenzie to Banks Is., Devon Is., Baffin Is., and northernmost Ungava–Labrador, s along the coasts of Hudson Bay to s James Bay and in the St Lawrence R. estuary from Ile-aux-Coudres, Charlevoix Co., to the Côte-Nord and Gaspé Pen.; w Greenland N to 70°20′N, E Greenland N to 76°49′N (type from Greenland); Iceland; Spitsbergen; coasts of Eurasia. [*C. salina* var. *sub.* (Wormsk.) Tuckerm.]. MAPS: Hultén 1968b:252, and 1962: map 184, p. 195; Porsild 1957: map 87, p. 171; Schofield 1958: map 1, p. 111.

The puzzling reports of this and another halophytic species, *C. paleacea (see* note under that species), from inland at Cumberland House, E-cent. Sask., by Hooker (1839) require clarification.

C. supina Wahl.   (209)

/aST/X/GEA/   (Grh)   Dry acidic rocks and sands from N Alaska, s-cent. Yukon, and the

Mackenzie R. Delta to Southampton Is., N Baffin Is., and N Ungava–Labrador (between ca. 57°and 60°N), s to s Alaska–Yukon–Dist. Mackenzie, N Sask. (L. Athabasca and Warren L. at ca. 59°45′N), and Man. (the main range s to Churchill; isolated stations in s Man. at Pine Ridge, Bowsman, and near Miami, the Pine Ridge collection verified by Porsild); isolated in N Minn. (F. K. Butters and E. C. Abbe, Rhodora 55(652):131. 1953); w Greenland N to 72°48′N, E Greenland N to ca. 77°N; Eurasia. [*C. obesa* var. *minor* Boott in part]. MAPS: Hultén 1968*b*:266, and 1962: map 178, p. 189; Porsild 1957: map 90, p. 172; Meusel, Jaeger, and Weinert 1965:73; Meusel 1943: fig. 63d; Böcher 1954: fig. 51 (bottom), p. 189.

The N. American plant may be distinguished as ssp. *spaniocarpa* (Steud.) Hult. (*C. span.* Steud., the type from Greenland; perigynia more prolonged and evenly tapering than in the typical form).

C. swanii (Fern.) Mack.   (373)
/T/EE/   (Hs)   Woods, thickets, and clearings from Wisc. to s Ont. (Lambton and Essex counties; CAN; TRT), sw Que. (Brome and Missisquoi counties; CAN; MT), sw N.B. (Grand Manan Is.; GH), and N.S. (Annapolis and Yarmouth counties; CAN; not known from P.E.I.), s to Ark., Tenn., and N.C. [*C. virescens* var. *swanii* Fern.]. MAP (E Canada): Raymond 1951: map 42, p. 19.

C. sychnocephala Carey   (197)
/sT/X/   (Hs)   Meadows, open woods, and clearings from s Yukon (Porsild 1951a) and Great Bear L. through B.C. and Alta. (N to L. Athabasca) to Sask. (N to McKague, ca. 52°45′N), Man. (N to Wekusko L., about 90 mi NE of The Pas), Ont. (N to the N shore of L. Superior and Kapuskasing, 49°24′N), and sw Que. (N to the Montreal dist.; *see* Que. maps by Rouleau 1945: fig. 8, p. 170, and Raymond 1951: map 35, p. 18), s to Wash., Mont., S.Dak., and N.Y. MAP: Hultén 1968*b*:232.

C. tenera Dewey   (162)
/sT/X/   (Hs)   Meadows, woods, and clearings from B.C. (N to Prince George; CAN) to Alta. (N to the Peace R. at 59°07′N; CAN), Sask. (N to McKague, ca. 52°45′N), Man. (N to Cross Lake, NE of L. Winnipeg), Ont. (N to Sandy L. at ca. 53°N, 93°W), Que. (N to L. St. John and the Gaspé Pen.), N.B., and N.S. (not known from P.E.I.), s to Mont., Wyo., Okla., Mo., Ohio, and N.C. [*C. straminea* vars. *ten.* (Dewey) Boott, *echinodes* Fern., and *moniliformis* Tuckerm. in part; *C. festucacea* var. *ten.* (Dew.) Carey].

C. tenuiflora Wahl.   (84)
/aST/X/EA/   (Hsr)   Mossy woods, bogs, and pond-margins from N Alaska, s Yukon, the Mackenzie R. Delta, and Great Bear L. to L. Athabasca (Alta. and Sask.), s Dist. Keewatin (Baralzon L. on the N Man. boundary), northernmost Ont., Que. (N to s Ungava Bay, the Côte-Nord, and Gaspé Pen.), Labrador (N to near Nain at ca. 56°30′N), Nfld., and N.B. (not known from P.E.I. or N.S.), s to B.C., Minn., N.Y., and Maine; Eurasia. MAPS: Hultén 1968*b*:244, and 1962: map 74, p. 83; Raup 1930: map 35 (incomplete northwards), p. 203.

C. tetanica Schkuhr   (275)
/T/EE/   (Grh)   Calcareous bogs, meadows, and wet woods from s Man. (N to Brandon and Virden; reports from Alta. and Sask. require confirmation) to s Ont. (N to Wellington and Hastings countries, with an isolated station on Manitoulin Is., N L. Huron), s to S Dak., Iowa, and Va. MAP: M. L. Fernald, Rhodora 43(514): map 4, p. 503. 1941.

C. tincta Fern.   (163)
/T/X/   (Hs)   Meadows and woodlands from s B.C. (Hector; CAN, detd. Mackenzie) to s Alta. (Moose Mt., near Calgary; CAN, detd. Mackenzie), Mich., Que. (N to the Gaspé Pen.; CAN), N.B., and P.E.I. (Charlottetown; CAN; not known from N.S.), s to N.Y. and w New Eng. [*C. mirabilis* var. *tincta* Fern.; perhaps a hybrid between *C. bebbii* and some other member of the Section *Ovales*].

C. tonsa (Fern.) Bickn.   (236)
/sT/(X)/   (Grh)   Dry sands and rocks, sterile fields, and open woods from L. Athabasca (Alta. and Sask.; reports from sw Dist. Mackenzie are referred to *C. rossii* by W. J. Cody, Can. Field-Nat. 77(2): 114. 1963) to Man. (N to Sasaginnigak L., about 125 mi NE of Winnipeg), Ont. (N to the N

shore of L. Superior), Que. (N to E James Bay at 52°23′N and the Gaspé Pen.), s Labrador (Goose Bay, Hamilton R. basin; not known from Nfld.), N.B., P.E.I., and N.S., s to Minn., Ind., and Va. [*C. umbellata* var. *tonsa* Fern.].

C. torreyi Tuckerm. (372)
/sT/WW/ (Hs) Thickets and moist meadows from NE B.C. (Boivin 1967*a*) to Alta. (N to Wood Buffalo National Park at 58°57′N), Sask. (N to Prince Albert), Man. (N to Riding Mt.), and Minn., s to Mont., Colo., and S.Dak. [*C. abbreviata* Prescott].

C. torta Boott (472)
/T/EE/ (Grh) Shallow water and margins of streams from Minn. to s Ont. (Peel and Elgin counties; CAN; TRT), Que. (N to the Gaspé Pen.; CAN; GH), N.B., and N.S. (not known from P.E.I.), s to Ark., Tenn., and Ga.

C. tracyi Mack. (148)
/T/W/ (Hs) Wet meadows and swampy grounds from B.C. (N to Quesnel, ca. 53°N; a collection in CAN by Dawson in 1887 at Telegraph Creek, on the Stikine R. at ca. 58°N, may also belong here, according to Porsild) to Calif. and Nev.
Hitchcock et al. (1969) merge this species with the Eurasian *C. leporina* L. If this concept is valid, the above B.C. material would probably have to be considered as introduced.

C. tribuloides Wahl. (186)
/T/EE/ (Hs) Low woodlands and moist grounds from Nebr. to Mich., Ont. (N to near Sioux Lookout at ca. 50°N; CAN), Que. (N to Anticosti Is. and the Gaspé Pen.), N.B., and N.S. (reports from P.E.I. require confirmation), s to Okla., La., and Fla. [*C. lagopodioides* Schkuhr].
Var. *sangamonensis* Clokey (perigynia 3 or 4 mm long rather than to 5 mm, the leaves narrower than in the typical form) is reported from Montreal, Que., by Ernest Rouleau (Ann. ACFAS 9: 103. 1943; St. Helen's Is.)

C. trichocarpa Muhl. (504)
/T/EE/ (Grh) Calcareous marshes and wet grounds from Minn. to s Ont. (N to Wellington, Ontario, and Hastings counties) and sw Que. (N to Oka and the Montreal dist.), s to N Iowa, Ohio, and Del. [Incl. var. *turbinata* Dewey].

C. trisperma Dewey (83)
/ST/X/ (Hsr (Grh)) Mossy woods, bogs, swampy grounds, and clearings from cent. Dist. Mackenzie (Porsild and Cody 1968) and B.C.–Alta. to Sask. (N to Candle L. at ca. 53°45′N), Man. (N to Wekusko L., about 90 mi NE of The Pas), Ont. (N to Sandy L. at ca. 53°N, 93°W), Que. (N to Chimo, s Ungava Bay; CAN, detd. Porsild), Labrador (N to Helen Falls, 58°08′N; DAO), Nfld., N.B., P.E.I., and N.S., s to Minn., Ill., and Va.
Var. *billingsii* Knight (the reduced extreme with perigynia at most 3.3 mm long and 1 or 2 in each spike rather than to 4 mm long and up to 5 in each spike, the leaves relatively narrow) is known from Ont., E Que. (Magdelen Is.), Nfld., N.B., P.E.I., and N.S.

C. tuckermanii Boott (522)
/T/EE/ (Grh) Rich or calcareous meadows and moist woods from Minn. to Ont. (N to the Pigeon R. sw of Thunder Bay near the Minnesota boundary; TRT), Que. (N to the Gaspé Pen. near Métis; RIM, detd. Lepage), N.B. (Kent and Westmorland counties), and N.S. (Hants and Cumberland counties; not known from P.E.I.), s to NE Iowa, Ohio, and N N.J.

C. typhina Michx. (509)
/T/EE/ (Hs) Calcareous meadows and moist woods from Wisc. to SE Ont. (N to the Ottawa dist.; Gillett 1958) and sw Que. (N to Buckingham and the Montreal dist.), s to La., Ala., and Ga. [*C. typhinoides* Schw.].

C. umbellata Schkuhr (234 and 235 (*C. rugosperma*))
/sT/X/ (Grh (Hsr)) Sandy grounds and dry sterile fields and open woods from B.C. (Victoria,

Vancouver Is.; CAN, detd. Mackenzie, as *C. abdita*) to Alta. (Banff; CAN), Sask. (N to near Hasbala L. at 59°55′N; CAN, detd. Calder), Man. (N to Tod L., about 300 mi sw of Churchill; Ritchie 1956), Ont. (N to the N shore of L. Superior and sw James Bay at 51°29′N), Que. (N to the Harricanaw R. SE of James Bay, the Côte-Nord, and Gaspé Pen.), s Labrador (Goose Bay, Hamilton R. basin; CAN), Nfld., N.B., P.E.I., and N.S., s to Mo., Tenn., and Va. [Incl. var. *brevirostris* Boott; *C. abdita* Bickn.; *C. rugosperma* Mack.].

Forma *vicina* (Dew.) Wieg. (some of the culms elongate and stiffly ascending rather than all crowded among the leaf-bases) is reported from New Liskeard, Ont., by Baldwin (1958).

C. unilateralis Mack.   (196)
/T/W/   (Hs)   Wet meadows and thickets from B.C. (Mackenzie 1935; reported from Glacier, in Rogers Pass, by Hitchcock et al. 1969, as, also, s ?Sask.) to Calif.; [Eastham (1947) believes that most records of *C. athrostachya* from the B.C. coast probably belong here].

C. ursina Dewey   (86)
/ASs/X/GEA/   (Hs)   Arctic and subarctic seacoasts of Alaska–Yukon–Dist. Mackenzie to Prince Patrick Is., Ellesmere Is. N to 81°25′N), Baffin Is., and northernmost Ungava–Labrador (Hudson Strait), s in Hudson Bay to NE Man. (Churchill) and w-cent. Que. (Cape Jones; not known from Ont.); w Greenland between the Arctic Circle and ca. 71°N, E Greenland between ca. 69° and 74°30′N; Spitsbergen; Novaya Zemlya; arctic Asia. [*C. glareosa* vars. *urs.* (Dew.) Bailey and *caespitosa* Boeck.]. MAPS: Hultén 1968*b*:221, and 1962: map 181, p. 193; Porsild 1957: map 77, p. 170; Raymond 1951: map 3 (E Canada), p. 4.

C. vaginata Tausch   (279: *C. saltuensis*)
/aST/X/GEA/   (Grh)   Mossy woods, calcareous swamps, and bogs from the coasts of Alaska–Yukon–Dist. Mackenzie to Victoria Is., s Baffin Is., northernmost Que. (Hudson Strait and Ungava Bay), Labrador (N to ca. 55°N), Nfld., and N.B. (St. Leonard and Belledune; not known from P.E.I. or N.S.), s to s B.C.–Alta., N Minn., N Mich., N.Y., and Vt.; an isolated station in E Greenland at 74°10′N; Iceland; Eurasia. [*C. saltuensis* Bailey; *C. phaeostachya sensu* Hooker 1839, in part, not Sm.]. MAPS: Hultén 1968*b*:271, and 1962: map 79, p. 89; Porsild 1957: map 93, p. 172; Raup 1947: pl. 18.

C. vesicaria L.   (515 and 514 (*C. raeana*))
/sT/X/EA/   (Grh)   Meadows, swampy grounds, and shores, the aggregate species from s B.C. (Vancouver Is. and adjacent mainland; Manning Provincial Park) to s Alta. (Moraine L., near Banff; CAN), Sask. (N to Methy Portage, on the Clearwater R. at ca. 56°30′N), Man. (N to Reindeer L. at ca. 57°45′N), Ont. (N to the James Bay watershed at 51°32′N; reported N to the Severn R. by John Macoun 1888), Que. (N to Knob Lake, 54°48′N, and the Côte-Nord), Labrador (N to Cartwright, 53°42′N), Nfld., N.B., P.E.I., and N.S., s to Calif., N.Mex., Mo., Ill., Ohio, and Del.; Eurasia. MAP (also showing the areas of the following varieties): Hultén 1962: map 60, p. 69.
1   Perigynia only slightly inflated, not over 2 mm thick; scales ovate; [*C. raeana* Boott; Sask. (type from Methy Portage, ca. 57°N); Que., Nfld., and N.B.] . . . . .var. *raeana* (Boott) Fern.
1   Perigynia strongly inflated, to 3.5 mm thick; scales relatively narrow.
　　2   Pistillate scales about equalling to surpassing the perigynia, purple; [E Que., Nfld. (type from the Highlands of St. John), N.B., and N.S. (Cape Breton Is.)] . . . .
　　 . . . . .var. *laurentiana* Fern.
　　2   Pistillate scales shorter than the perigynia, usually pale.
　　　　3   Perigynia not over half as thick as long, gradually tapering to the beak; spikes to 7.5 cm long; leaves to 7 mm broad; [*C. saxatilis* var. *alpigena* Fries; *C. vaseyi* Dew.; transcontinental] . . . . .var. *vesicaria*
　　　　3   Perigynia 1/2-2/3 as thick as long; leaves commonly less than 5 mm broad.
　　　　　　4   Pistillate spikes less than 1 cm thick; perigynia rarely over 5 mm long; [Ont. to Labrador and Nfld.] . . . . .var. *jejuna* Fern.
　　　　　　4   Pistillate spikes at least 1 cm thick; perigynia to 8 mm long.
　　　　　　　　5   Pistillate spikes 2 or 3, to 7.5 cm long; [*C. monile* Tuckerm.; ?B.C. (Henry 1915); Ont. to s Labrador, Nfld., and N.S.] . . . . .var. *monile* (Tuckerm.) Fern.

5 Pistillate spikes 1 or 2, not over 2.5 cm long; [*C. friesii* Blytt; Ont. to Nfld. and N.S.] ....var. *distenta* Fries

[C. vestita Willd.] (388)
[The report of this species of sandy woods and clearings of the E U.S.A. (N.Y. and Maine to Va.) from s Ont. by Stroud (1941; Wellington Co.) requires confirmation.]

C. virescens Muhl. (374)
/T/EE/ (Hs) Dry woods, thickets, and clearings from Ind. to s Ont. (Essex, Kent, Lincoln, and Welland counties), sw Que. (Freligsburg, Missisquoi Co.; MT), ?N.B. (Boivin 1967a; not known from P.E.I.), and N.S., s to Mo., Tenn., and Ga. [*C. costata* Schw., not Presl; *C. costellata* Britt.].

C. viridula Michx. (354 and 355 (*C. chlorophila*))
/aST/X/GeA/ (Hs) Damp shores and muddy places (often calcareous) from s Alaska–Yukon and Great Bear L. to Great Slave L., L. Athabasca (Alta. and Sask.), Man. (N to the N end of L. Winnipeg), Ont. (N to w James Bay at ca. 53°N), Que. (N to Knob Lake, 54°48′N, and the Côte-Nord), Nfld., N.B., P.E.I., and N.S., s to Calif., N.Mex., N.Dak., Pa., and N.J.; s Greenland; E Asia. MAPS and synonymy: *see* below.
1 Heads borne on reduced culms completely hidden among the bases of the leaves; [type from Lake Wickenden, Anticosti Is., Que.; also reported from Eastmain, E James Bay, Que., ca. 52°15′N, by Dutilly, Lepage, and Duman 1958; according to Marcel Raymond, Nat. can. (Que.) 77: 63. 1950, distinguishable from dwarf individuals of *C. demissa* by its narrower leaves and its terminal spike bearing some perigynia at the summit] ....
....f. *clandestina* Raymond
1 Heads borne on elevated clums.
2 Perigynia less than 2 mm long; plant dwarf; [f. *pygmaea* Lepage; *C. oederi* var. *rouss.* Vict., the type from Berthier-en-Bas, Montmagny Co., Que.; also several other localities in the province] ....f. *rousseauiana* (Vict.) Raymond
2 Perigynia mostly 2 or 3 mm long; [*C. flava* vars. *recterostrata* Bailey and *vir.* (Michx.) Bailey; *C. oederi* vars. *vir.* (Michx.) Kük. and *pumila* of auth., not Retz.; *C. chlorophila* Mack.; *C. urbanii* Boeck.; transcontinental, the type from NE of Montreal, Que.; MAPS: Hultén 1968b:276 (*C. oed.* ssp. *vir.*), and 1962: map 162, p. 171; E. W. Davies, Watsonia 3(1): fig. 2 (a very inaccurate outline), p. 83. 1953] ....f. *viridula*

C. vulpinoidea Michx. (64)
/T/X/ (Grh) Swamps and wet ground from s B.C. (Vancouver Is.; Yarrow; Creston; Similkameen R.) to Alta. (N to Lesser Slave L.), Sask. (N to Ile-à-la-Crosse, 55°27′N; Breitung 1957a), Man. (N to Playgreen L., N of L. Winnipeg), Ont. (N to Mattice, 49°37′N), Que. (N to the Côte-Nord), Nfld., N.B., P.E.I., and N.S., s to Oreg., Ariz., Tex., and Fla. [*C. multiflora* Muhl.; *C. setacea* Dew.].
Forma *segregata* (Farw.) Raymond (the spikes well separated in a moniliform head rather than relatively crowded) is reported from Que. by Marcel Raymond (Contrib. Inst. Bot. Univ. Montréal 48:80. 1943; Longueuil and Iberville).

C. wiegandii Mack. (111)
/T/EE/ (Hs) Boggy or peaty soils from Ont. (Fernald *in* Gray 1950) to Que. (N to the Romaine R. N of Anticosti Is. at 52°26′N; Lepage 1964), Nfld. (type from Curling), N.B., and N.S. (not known from P.E.I.), s to N.Y. and Mass.

C. willdenowii Schkuhr (199)
/T/EE/ (Hs) Rocky woods from Minn. to Ont. (Niagara, Lincoln Co.; TRT; reports from the Lake of the Woods region by John Macoun 1888, probably refer to *C. umbellata*) and sw Que. (Contrecoeur, Verchères Co.; CAN; MT), s to Tex. and Fla.

C. williamsii Britt. (348)
/aST/X/eA/ (Hs) Dryish rocky barrens from N Alaska, the Yukon (type from Dawson), and the coast of NW Dist. Mackenzie to E-cent. Dist. Keewatin, northernmost Que. (Sugluk, Hudson Strait),

and Labrador (N to Cutthroat Harbour, ca. 57°30′N), s to s-cent. Alaska–Yukon, N Man. (s to Churchill), w and E James Bay, and N-cent. Que. (s to the George R. SE of Ungava Bay at 56°50′N); E Asia. [*C. capillaris* var. *will.* (Britt.) Boivin]. MAPS: Hultén 1968*b*:275, and 1962: map 48, p. 55; Porsild 1957 (1964 revision), map 335, p. 202; Tolmachev 1952, maps 21 and 25 (very inaccurate).

C. woodii Dewey   (277)

/T/EE/   (Grh)   Dryish calcareous woodlands from s Man. (N to Riding Mt.) to Ont. (N to the Ottawa dist.; Gillett 1958), N.Y., and Conn., s to Mo., Ill., Ohio, and W.Va. [*C. tetanica* var. *woodii* (Dew.) Wood].

C. xerantica Bailey   (191)

/T/WW/   (Hs)   Dry prairies and hills from s B.C. (Penticton; Eastham 1947) to Alta. (N to near Dunvegan, ca. 55°15′N; Raup 1934), Sask. (N to McKague, 52°37′N; type from the File Hills), and s Man. (N to Birtle, about 60 mi NW of Brandon), s to N.Mex., S.Dak., and Minn.

Carex Hybrids

The following hybrids between various species of *Carex* have been reported from Alaska–Canada. However, it will be evident, from the highly technical and often controversial characters of many of the putative parents, that the existence of many of these hybrids should be regarded as merely suggested rather than firmly established. This is particularly so in those cases where two or more nothomorphs are proposed (these being hybrids of the same parentage but with different amounts of genetic infiltration from each parent) and in those cases where more than two parents are postulated. The names of the parents are given in the customary alphabetical sequence (modified, where necessary, to conform to the nomenclature used above), followed by the scientific binomial name (in brackets), if published. Many of these hybrids have been named or reported by Ernest Lepage (Nat. can. (Que.) 83: 105–56. 1956; 84: 37–62 and 89–103. 1957; 89: 75–79 and 113–19. 1962; 91: 165–74. 1964).

C. acuta (see *C. nigra,* under which name most hybrids have been reported).

C. aquatilis × *C. aquatilis* var. *stans*: Labrador (Hopedale).

C. aquatilis × *C. bigelowii* (× *C. limula* Fries): Labrador (Davis Inlet) and Que. (Tabletop Mt., Gaspé Pen.; Larch and Koksoak Rivers, Ungava).

C. aquatilis × *C. crinita* (× *C. crinitoides* Lepage, the type from Grand Falls, Nfld.): also known from St-Aubert, l'Islet Co., Que.

C. aquatilis × *C. nigra* (× *C. aquanigra* Boivin, the type from l'Ile d'Orléans, near Quebec City, Que.).

C. aquatilis × *C. paleacea* (× *C. neofilipendula* Lepage, the type from the Main R., Bonne Bay, Nfld.): also known from Que. (E James Bay; Harrington Harbour of the Côte-Nord).

C. aquatilis × *C. salina* (× *C. subsalina* Lepage, the type from Cape Jones, NE James Bay): Que. (E James Bay; Chimo) and Nfld. Hybrids between this hybrid and *C. aquatilis* and *C. bigelowii* are also reported from N Que. by Lepage.

C. aquatilis × *C. salina* var. *kattegatensis* (× *C. grantii* Benn.): Ont. (w James Bay), Que. (E James Bay; Gaspé Pen.), and Labrador (Gready Is.).

C. aquatilis × *C. salina* var. *pseudofilipendula*: Nfld. (M. L. Fernald, Rhodora 28(333): 166. 1926).

C. aquatilis × *C. salina* var *subspathacea* (× *C. flavicans* Nyl.; × *C. halophila* nm. *flav.* (Nyl.) Boivin): Ont. (Lake R., w James Bay) and Que. (island 12 mi s of Kipsawis R., E James Bay).

C. aquatilis × *C. saxatilis* var. *miliaris* (× *C. neomiliaris* Lepage, the type from the Opinaca R., E James Bay).

C. aquatilis × *C. stricta* (× *C. abitibiana* Lepage, the type from the Nottaway R., Que.).

C. aquatilis var. *stans* × *C. bigelowii* (× *C. nearctica* Raymond, the type from Southampton Is.): also known from northernmost Que. (Wakeham Bay) and Greenland.

C. aquatilis var. *stans* × *C. salina* (× *C. stansalina* Lepage, the type from Southampton Is.): N.W.T., Man. (Churchill), Ont. (NW James Bay), and Que. (E Hudson Bay).

C. aquatilis var. *stans* × *C. salina* var. *kattegatensis*: type from Lake R., James Bay (as *C. recta* × *C. stans*).

C. aquatilis var *stans* × *C. salina* var. *subspathacea* (× *C. substans* Lepage, the type from Dundas Harbour, Devon Is.): Baffin Is.; Devon Is.; Air Force Is.; Wakeham Bay, Ungava; Greenland.

C. arctata × *C. castanea* (× *C. knieskernii* Dew.): Ont. (Kakabeka Falls, about 20 mi w of Thunder Bay; John Macoun 1888); Que. (Gaspé Pen.); N.B. (Tobique R.).

C. atratiformis × *C. saxatilis* var. *miliaris* (× *C. patuensis* Lepage, the type from Patu L., Que.).

C. atratiformis × *C. norvegica* Retz. (× *C. quirponensis* Fern., the type from Quirpon Is., Nfld.): also known from N Que. (Manitou Gorge on the Kaniapiskau R.).

C. bigelowii × *C. (nigra) acuta* (× *C. groenlandica* Lange, the type from Greenland).

C. bigelowii × *C. lenticularis* (× *C. neobigelowii* Lepage, the type from the Romaine R., Que.).

C. bigelowii × *C. lyngbyei* (× *C. haematolepis* Drej.): Greenland.

C. bigelowii × *C. paleacea* (× *C. exsalina* Lepage, the type from the mouth of the Piagochiwi R., Que.): also known from E James Bay, Que.

C. bigelowii × *C. salina* (× *C. neorigida* Lepage): Labrador (Gready Is. and Cross Is.).

C. bigelowii × *C. salina* var. *subspathacea* (× *C. subreducta* Lepage, the type from the mouth of the Actamacow R., James Bay, Ont.): also known from E James Bay, Que.

C. bigelowii × *C. saxatilis* var. *rhomalea* (× *C. quebecensis* Lepage, the type from the Romaine R., Que.).

C. buxbaumii × *C. paleacea* (× *C. neopaleacea* Lepage, the type from Paint Hills Bay, James Bay, Que.).

C. canescens (curta) × *C. heleonastes* (× *C. calderi* Boivin, the type from Burns Lake, B.C.).

C. canescens × *C. mackenziei* (× *C. pseudohelvola* Kihlm.): Nfld., Que. (Côte-Nord and Magdalen Is.; the report from S Labrador by Fernald *in* Gray 1950: 313, is probably based upon a collection from Bonne Espérance, labeled Labrador but actually on the Côte-Nord), and N.B. (Bathurst and Grand Manan Is.).

C. canescens × *C. tenuiflora*: reported from Wells Gray Provincial Park, SE B.C., by L. Hämet-Ahti (Ann. Bot. Fenn. 2(2): 148. 1965).

C. cumulata × *C. scoparia*: type from Middleton, Annapolis Co., N.S. (M. L. Fernald, Rhodora 23(274):235. 1921; *C. albolutescens* var. *cumulata* × *C. scoparia*).

C. digitalis × *C. laxiculmis* (× *C. copulata* (Bailey) Mack.; *C. platyphylla* var. *pedunculata* Kük.): Ont. (Essex, Norfolk, and Elgin counties: M. L. Fernald, Rhodora 8(92):183. 1906, as *C. laxiculmis* var *copulata;* London, Middlesex Co.: John Macoun 1890, as *C. digitalis* var *copulata*).

C. exilis × *C. muricata* var. *cephalantha*: S Ont. (near Peterborough).

C. exilis × *C. muricata* var. *sterilis*: E Que. (× *C. minganinsularum* Raymond, the type from the Mingan Is. of the Côte-Nord; Mackenzie 1931).

C. flava × *C. hostiana* var. *laurentiana* (× *C. xanthina* Fern., the type from the mouth of the Main R., Bonne Bay, Nfld.): also known from Anticosti Is., E Que.

C. flava × *C. lepidocarpa* (× *C. pieperiana* Junge; incl. × *C. senayi* Raymond): Nfld. (Brig Bay and White Bay) and E Que. (Anticosti Is.).

C. flava × *C. viridula* (× *C. subviridula* (Kük.) Fern.): Ont. (w James Bay), E Que. (Anticosti Is.), and Nfld.

C. glareosa var. *amphigena* × *C. paleacea* (× *C. paleacoides* Lepage, the type from St-Simon, Rimouski Co., Que.).

C. hindsii × *C. rostrata* (× *C. oneillii* Lepage, the type from w Alaska).

C. hostiana var. *laurentiana* × *C. lepidocarpa* (× *C. pseudo-fulva* Fern., the type from Table Mt., Port au Port Bay, Nfld.; × *C. leutzii* Kneucker nm. *pseudofulva* (Fern.) Boivin): also known from Anticosti Is., E Que.

C. intumescens var. *fernaldii* × *C. retrorsa* (× *C. josephi-schmittii* Raymond, the type from Salmon River, Anticosti Is., E Que.).

C. kelloggii × *C. spectabilis:* B.C. (Mt. Revelstoke National Park; R. J. Moore and J. A. Calder, Can. J. Bot. 42(10): 1389. 1964).

C. limosa × *C. paleacea* (× *C. sublimosa* Lepage, the type from Old Factory, E James Bay): Que. to N.B. and N.S.

C. limosa × *C. paupercula* (× *C. connectens* Holmb.; *C. paupercula* var. *brevisquama* Fern., the type from Ile-aux-Coudres, Charlevoix Co., Que.): Que. (E James Bay; Charlevoix Co.; Magdalen Is.).

C. limosa × *C. rariflora* (× *C. firmior* (Norm.) Holmb.): E Que. (Anticosti Is.) and Nfld.

C. lupuliformis × *C. retrorsa* (× *C. macounii* Dewey, not *C. mac.* Benn.; *C. retrorsa* var. *mac.* Dew.; *C. ?canadensis* Dew. in part; *C. lurida* var. *divergens* Bailey): s Ont. (Seymour, Northumberland Co., and Belleville, Hastings Co.).

C. lurida × *C. retrorsa* (× *C. hartii* Dewey; *C. retrorsa* var. *hartii* (Dewey) Gray): s Ont. and E Que. (Anticosti Is.).

C. lyngbyei ssp. *cryptocarpa* × *C. ramenskii* (× *C. cryptochlaena* Holm, the type from Alaska).

C. maritima × *C. parallela*: E Greenland (Böcher, Holmen, and Jacobsen 1966).

C. nigra (*acuta*) × *C. aquatilis* (× *C. rollandii* Lepage): Que. (Ile Brion, Magdalen Is.).

C. nigra × *C. paleacea* (× *C. subnigra* Lepage, the type from Richmond Co., N.S.): also known from Que. (La Malbaie, Charlevoix Co.).

C. nigra × *C. salina* (× *C. spiculosa* Fries): N.S. (Canso, Guysborough Co.).

C. nigra × *C. salina* var. *kattegatensis* (× *C. super-goodenoughii* (Kük.) Lepage): E Que. (Charlevoix, Kamouraska, and Temiscouata counties), N.B. (Albert Co.), and N.S.

C. paleacea × *C. salina* (× *C. gardneri* Lepage, the type from Gready Is., Labrador): Ont. (w James Bay), Que. (E James Bay and E Hudson Bay), and Labrador.

C. paleacea × *C. salina* var. *kattegatensis* (× *C. saxenii* Raymond, the type from Anticosti Is.): Ont. (w James Bay) and Que. (E James Bay; Anticosti Is.).

C. paleacea × *C. salina* var *kattegatensis* (× *C. saxenii* nm. *ferruginea* Lepage, the type from Fort George, E James Bay): Que. (E James Bay; Anticosti Is.), Labrador (Goose Bay, Hamilton R. basin), and N.S. (Inverness Co., Cape Breton Is.).

C. paleacea × *C. salina* var. *subspathacea* (× *C. dumanii* Lepage, the type from Old Factory, SE James Bay): Ont. (sw James Bay), Que. (E James Bay), and Labrador (near Nain).

C. paleacea × *C. stylosa* var. *nigritella* (× *C. ungavensis* Lepage, the type from Fort George, E James Bay).

C. pensylvanica var. *distans* × *C. umbellata*: sw Que. (Contrecoeur, Verchéres Co.).

C. physocarpa × *C. rostrata* (× *C. physocarpioides* Lepage, the type from Alaska; also reported from Wells Gray Provincial Park, SE B.C., by L. Hämet-Ahti, Ann. Bot. Fenn. 2(2): 149. 1965).

C. projecta × *C. scoparia:* P.E.I. (Charlottetown).

C. ramenskii × *C. subspathacea* (× *C. kenaica* Lepage, the type from Alaska).

C. rariflora × *C. subspathacea* (× *C. soerensenii* Lepage; known only from the type locality, Greenland.

C. rostrata × *C. rotundifolia* (× *C. paludivagans* Drury, the type from Alaska).

C. rostrata × *C. saxatilis* var. *miliaris* (*C. inflata* (*rostrata*) var. *anticostensis* Fern.; *C. vesicaria* var. *grahamii sensu* Marie-Victorin, not Boott): Que. (Anticosti Is. and Gaspé Pen.) and St-Pierre and Miquelon.

C. rostrata × *C. saxatilis* var. *miliaris* (× *C. anticostensis* nm. *minor* Lepage, the type from Tabletop Mt., Gaspé Pen., E Que.).

C. rostrata × *C. saxatilis* var. *?rhomalea* (× *C. anticostensis* nm. *?inflatior* Lepage; × *C. mainensis* Porter in part): reported from the Roggan R., E James Bay, Que., by Lepage (1956), giving *C. saxatilis* var. *major* (presumably *sensu* Bailey, not Olney) as one of the parents. Olney's plant, as treated above, is *C. physocarpa*, a western species not found in Que.

C. rostrata var. *utriculata* × *C. miliaris* var *?rhomalea* (× *C. anticostensis* nm. *?longidens* Lepage, the type from Chimo, Ungava): reported by Lepage (1956) as partly derived from *C. miliaris* var. *major* (*see* above).

C. rostrata × *C. vesicaria* (× *C. pannewitziana* Figert): Que. (L. Chibougamou; CAN).

C. salina × *C. salina* var. *kattegatensis* (× *C. mendica* Lepage, the type from Moose Factory, s James Bay, Ont.): also known from Que. (E James Bay, and Kamouraska Co.) and Labrador.

C. salina × *C. salina* var. *subspathacea* (× *C. persalina* Lepage, the type from Old Factory, s James Bay, Que.): also known from w James Bay, Ont.

C. salina var. *kattegatensis* × *C. saxatilis* var. *miliaris* (× *C. nubens* Lepage, the type from Eastmain, E James Bay, Que.).

C. saxatilis var. *miliaris* × *C. vesicaria* (× *C. mainensis* Porter in part): Ont. (w James Bay), Que. (E James Bay; L. St. John; Anticosti Is.; Gaspé Pen.), Labrador, Nfld., St-Pierre and Miquelon, and N.B. (Kennebeckasis).

C. saxatilis var. *rhomalea* × *C. vesicaria* (× *C. mainensis* Porter in part): E Que. (Côte-Nord; St. John 1922).

C. tenuiflora × *C. trisperma* (× *C. trichina* Fern.): s Labrador and Nfld. (Fernald 1933).

CLADIUM P. Br. [489] Twig-rush

C. mariscoides (Muhl.) Torr.
/T/EE/ (Hel) Fresh or brackish swamps, marshes, and shores, the main area from Ont. (N to Batchawana Bay at the E end of L. Superior and the Ottawa dist.) to Que. (N to the Gaspé Pen.; *see* Que. map by Marcel Raymond, Nat. can. (Que.) 98: fig. 1, p. 736. 1971), Nfld., N.B. (St. Andrews, St. Stephen, and Grand Manan Is., Charlotte Co.; CAN; not known from P.E.I.), and N.S., s to Minn., Ill., Ala., and Fla.; a remarkable isolated station in s-cent. Sask. at Dahlton, 53°27'N, north of Little Quill L., where taken by Breitung in 1936 (DAO); *see* F. J. Hermann, Rhodora 39(462): 232. 1937; ?introd. by water fowl. [*Schoenus* Muhl.; *Mariscus* Ktze.]. MAPS: Hultén 1958: map 256, p. 275; Meusel, Jaeger, and Weinert 1965:62; Meusel 1943: fig. 32a.

Forma *congestum* Fern. (the inflorescence very compact, its rays very short or obsolete) is known from N.S. (type from Tiddville, Digby Co.; near Conquerall, Lunenburg Co.).

CYPERUS L. [459] Galingale, Umbrella-Sedge. Souchet

1 Styles 2-cleft; achenes lenticular; spikelets lance-oblong, to 3.5 mm broad, in loose subcapitate clusters; leaves usually less than 3 mm broad; annuals with fibrous roots; Ont.–Que.).
  2 Scales rather loose, thin and dull; styles cleft nearly to base, long-exserted, persistent; stamens usually 2; rays of umbel to about 6 cm long . . . .*C. diandrus*
  2 Scales close, firm and more or less shining; styles cleft only to near middle, short-exserted, usually deciduous; stamens 3; rays of umbel to over 1 dm long . . . .
. . . .*C. rivularis*
1 Styles 3-cleft; achenes trigonous; spikelets narrowly linear to lance-oblong, less than 2 mm broad.
  3 Annuals with fibrous roots.
    4 Scales loose, their slender tips conspicuously recurved; spikelets in loose subcapitate clusters, their rachillas persistent; leaves 1 or 2 mm broad; plant less than 2 dm tall, with the odour of sweet clover (*Melilotus*) when bruised; (s B.C. to N.B.) . . . .*C. inflexus*
    4 Scales close, their tips not strongly recurved; spikelets on a more or less elongated axis; plants to about 9 dm tall.
      5 Scales commonly less than 1.5 mm long, reddish brown, green-ribbed but otherwise nerveless; rachillas persistent, bearing deciduous chaff-like wings; leaves to 1 cm broad; roots red; (s Ont.) . . . .*C. erythrorhizos*
      5 Scales to 3.5 mm long, distinctly nerved; winged rachillas disarticulating into short segments.
        6 Tip of the thin yellowish-brown to reddish scale barely reaching the base of the next scale above on same side of spikelet; leaves to 5 mm broad; (s Ont.) . . . .*C. engelmannii*

6 Tip of scale distinctly surpassing the base of the next scale above on the same side; leaves to over 1 cm broad.
   7 Scales firm, more or less shining, yellowish to brownish, to about 3.5 mm long; mature achenes grey or blackish; (s ?Ont.) . . . .[C. odoratus]
   7 Scales thin, dull, less than 2.5 mm long; mature achenes rust-coloured or golden-brown; (s Ont. and sw Que.) . . . . .C. ferruginescens
3 Perennial from corm-like rhizomes or tuber-bearing stolons.
   8 Spikelets with at most 3 flowers, densely crowded and radiating in all directions in a subglobose head; leaves to about 8 mm broad; culms from short tuberous rhizomes; (s ?Ont.) . . . .[C. ovularis]
   8 Spikelets with up to 25 or more flowers, less densely crowded; leaves to over 1 cm broad.
      9 Spikelets on a more or less elongated axis.
         10 Culms from slender elongate stolons ending in small tubers; scales 2 or 3 mm long; rachillas continuous; (Ont. to N.B. and N.S.) . . . . .C. esculentus
         10 Culms from hard corm-like rhizomes; scales to 6 mm long; rachillas jointed at base, the mature spikelets deciduous; (s Sask. to sw Que.) . . . . . . .C. strigosus
      9 Spikelets in subcapitate clusters; leaves commonly less than 5 mm broad.
         11 Culms from slender elongate stolons ending in small tubers; scales 2 or 3 mm long; spikelets often altered into bulblets or leafy tufts; (Que., N.B., and N.S.) . . . . .C. dentatus
         11 Culms from hard corm-like rhizomes.
            12 Culms scabrous; leaves scabrous-margined; scales about 4 mm long and nearly as broad; achenes light brown, about 3 mm long; (s Alta. to sw Que.) . . . . .C. schweinitzii
            12 Culms smooth; scales less than 4 mm long.
               13 Scales roundish, less than 3 mm long; achenes dark brown, at most 1.5 mm long; leaves smooth; (s Man. to sw Que.) . . . . .C. houghtonii
               13 Scales oblong, to 3.5 mm long; achenes at least 1.5 mm long; leaves scabrous-margined; (Ont., sw Que., and N.S.) . . . . .C. filiculmis

C. dentatus Torr.
/T/EE/ (Hsr) Damp sandy ground and shores from Que. (N to St-Vallier, Bellechasse Co.; MT) to N.B. (Kings, Queens, and Charlotte counties; not known from P.E.I.), and N.S., s to NW Ind., W.Va., and N.C. MAP: Fassett 1928: fig. 1, pl. 10.

C. diandrus Torr.
/T/EE/ (T) Wet sandy, muddy, or peaty soils from Ont. (N to the Ottawa dist.) and sw Que. (N to L. St. Peter; MT; reports from N.B. by Fowler 1885, and John Macoun 1888, refer at least in part to *C. dentatus*, a relevant collection in CAN from Riverside) to cent. Maine, s to N.Mex. and S.C.

C. engelmannii Steud.
/T/EE/ (T) Moist or wet soils from Nebr. to Minn., s Ont. (York, Hastings, Frontenac, and Stormont counties), and Mass., s to Mo., Ill., and Va.

C. erythrorhizos Muhl.
/t/X/ (T) Alluvial or damp sandy soil from Wash. to Wyo., Minn., Mich., s Ont. (Kent, Lambton, Norfolk, and Waterloo counties), N.Y., and Mass., s to s Calif. and Fla. [*C. halei* Torr.].

C. esculentus L. Yellow Nut-grass. Amande de terre
/T/X/EA/ (Gst) Damp sandy soils and cult. fields from Wash. to Ont. (N to the Ottawa dist.; the inclusion of s Man. in the range by Fernald *in* Gray 1950, requires confirmation, as does a report from s Alaska), Que. (N to L. St. Peter; MT), N.B. (Fredericton, York Co., and Norton, Kings Co.; not known from P.E.I.), and N.S. (Somerset and Starr's Point, Kings Co.), s to Mexico, Tex., and Fla.; tropical America; Eurasia. [*C. phymatodes* Muhl.; *C. repens* Ell.].

Cyperaceae

C. ferruginescens Boeckl.
/T/X/ (T) Damp soils from Oreg. to Nebr., Minn., s Ont. (Kent, York, Waterloo, and Wellington counties), sw Que. (Boivin 1967a), N.Y., and Conn., s to Calif., Ariz., N.Mex., Tex., Ala., and Va.

C. filiculmis Vahl
/T/EE/ (Gst) Dry rocky, gravelly, or sandy grounds from Nebr. to Minn., Ont. (N to the Ottawa dist.; Gillett 1958), sw Que. (N to L. St. Peter; Marcel Raymond, Rhodora 51(601): 10. 1949), and N.S. (Antigonish Co.; D. S. Erskine, Rhodora 53(635): 267. 1951; not known from N.B. or P.E.I.), s to Tex. and Fla. [Incl. the more or less reduced but completely intergrading var. *macilentus* Fern.].

C. houghtonii Torr.
/T/EE/ (Gst) Light, usually dry sandy soil from s Man. (Victoria Beach; Marchand; Brokenhead) to Ont. (N to Renfrew and Carleton counties) and sw Que. (N to Pontiac and Verchères counties), s to Iowa, Pa., and Mass.

C. inflexus Muhl.
/T/X/ (T) Damp sandy soil from s B.C. (Vancouver Is.; Kamloops; Penticton; Osoyoos) to s Alta. (Redcliffe; Medicine Hat), Sask. (N to Amisk L. at ca. 54°45′N), s Man. (*C. aristatus* reported from L. Winnipeg by Hooker 1839, the occurrence in SE Man. confirmed by a 1959 collection by Sheila Anderson), Ont. (N to the Ottawa dist.; Gillett 1958), Que. (N to L. St. Peter; MT), and w N.B. (Fredericton and St. Mary's; CAN; GH; not known from P.E.I. or N.S.), s to Mexico, Tex., and Fla. [*C. aristatus* of American auth., perhaps not Rottb.].

[C. odoratus L.]
[The reports of this species of the U.S.A. and tropical America from Essex Co., s Ont. by Dodge (1914; *C. ferax*) and Core (1948) require confirmation, possibly being based upon the very similar and more northern *C. ferruginescens*. (*C. ferax* Richard).]

[C. ovularis (Michx.) Torr.]
[The report of this species of the E U.S.A. (N to Kans. and N.Y.) from s Ont. by Soper (1949) requires confirmation, perhaps being based upon the more northern *C. filiculmis.*]

C. rivularis Kunth
/T/X/ (T) Sandy, muddy, or peaty places from Calif. to Nebr., Minn., Ont. (N to the Ottawa dist.; Gillett 1958), Que. (N to the Gaspé Pen. at Ste-Flavie and Métis; QSA), and cent. Maine, s to Mexico, Tex., and Ga.; S. America. [*C. diandrus* var. *castaneus* Torr.]. MAP: Fassett 1928: map 4 (NE area; the above Gaspé stations should be indicated), pl. 9.

C. schweinitzii Torr.
/T/X/ (Gst) Dry or moist sandy soil from Alta. (Moss 1959) to s Sask. (Cut Knife and Elbow; Breitung 1957a), s Man. (N to St. Lazare, about 75 mi NW of Brandon), Ont. (N to the mouth of the Rainy R., Lake of the Woods; CAN), and sw Que. (Fernald *in* Gray 1950), s to Wash., Idaho, N.Mex., Mexico, Tex., Ind., and N.Y.

C. strigosus L.
/T/X/ (Gst) Damp grounds and shores from Wash. to s Sask. (Watrous, 52°07′N; Herbert Groh, Can. Field-Nat. 58(1): 17. 1944), s Man. (Wawanesa, 20 mi SE of Brandon, and near the Ont. boundary in SE Man.), Ont. (N to the Ottawa dist.; Gillett 1958), sw Que. (N to L. St. Peter; MT), and New Eng., s to Calif., Tex., and Fla.; tropical America.

Cyperus fuscus L.
This Eurasian species (not keyed out above) is recently reported as new to Canada by J. M. Gillett (Can. Field-Nat. 85: 190. 1971). It was taken in 1970 by L. Putnam at the edge of a pond in the St. Johns Conservation Area, Pelham Township, Welland Co., Ont. For keys and descriptions, *see* Fernald *in* Gray (1950), Gleason (1958), and Clapham, Tutin, and Warburg (1962).

430

[DICHROMENA Michx.]   [473]

[D. colorata (L.) Hitchc.]
[The report of this species of the coastal ᴇ U.S.A. (Va. to Fla. and Tex.) from s Ont. by Soper (1949) is probably based upon a species of *Cyperus*, possibly *C. filiculmis.*]

DULICHIUM Pers.   [458]

D. arundinaceum (L.) Britt.   Three-way Sedge
/T/X/   (Hel)   Swamps and margins of ponds and streams from B.C. (ɴ to Queen Charlotte Is.; not known from Alta. or Sask.) to Man. (Boivin 1967a), Ont. (ɴ to Timmins, 48°28′N), Que. (ɴ to the Nottaway R. sᴇ of James Bay at 51°09′N and the Côte-Nord), Nfld., N.B., P.E.I., and N.S., s to Calif., Tex., and Fla. [Incl. the reduced extreme, var. *boreale* Lepage; *Cyperus arund.* L.; *D. canadense* Pers.; *Schoenus* (*Cyperus; D.; Scirpus*) *spathaceus* L.].

ELEOCHARIS (*Heleocharis*) R. Br.   [469]   Spike-Rush

(Ref.: Svenson 1929–39)
1   Spikelet not much thicker than the top of the culm, this to over 9 dm tall; achenes with fine reticulation in vertical rows; scales finely striate, firmly persistent; tufted perennials.
  2   Culms terete, septate by cross-partitions 1–5 cm apart, to about 1 m tall and 5 mm thick; spikelet to 4 cm long; achenes biconvex, their reticulation consisting of transversely linear-rectangular areoles; stigmas 2 or 3; tubercle capping the achene conical, confluent with the achene, about 1 mm long, its broad sessile base about half as long; perianth-bristles none or weak and mostly shorter than the achene; (s Ont.) . . . .*E. equisetoides*
  2   Culms angled, not septate; reticulation of achenes more broadly rectangular; bristles equalling or surpassing the achene; tubercle separated from body of achene by a narrow constriction.
    3   Culms sharply 4-angled, to about 1 m tall and 5 mm thick; spikelets cylindric, to 6 mm long; tubercle conical, flattened, about twice as long as broad and up to half as long as the biconvex achene; stigmas 2 or 3; scales leathery, not keeled; (s Ont.) . . . .*E. quadrangulata*
    3   Culms sharply 3-angled, to about 7 dm tall and 2 mm thick, accompanied, when submersed, by numerous very elongate capillary sterile floating culms; spikelets subulate-lanceolate, at most about 2.5 cm long; tubercle subulate, flattened, much longer than broad; achenes compressed-trigonous; stigmas 2; scales herbaceous, keeled; (Ont. to N.B. and N.S.) . . . .*E. robbinsii*
1   Spikelet distinctly thicker than the top of the culm; scales deciduous.
  4   Tubercle conical, confluent with the summit of the trigonous or sometimes plano-convex achene; bristles mostly equalling or surpassing the achene; stigmas 3.
    5   Spikelet fusiform, many-flowered, to 2 cm long, its drab or brown scales leathery; reticulation of achene chiefly of vertically elongate areoles; culms more or less flattened, wiry, in firm tussocks, lacking stolons or tubers, to about 6 dm tall and 2 mm thick, the sterile ones often longer, arching and rooting at the tips; (B.C. and ?Alta.; s Ont. and sw N.S.) . . . .*E. rostellata*
    5   Spikelet ovoid, with less than 10 flowers, at most 7 mm long, flattened, its scales thin; culms soft and capillary, rarely over 3 dm tall, from tuber-bearing stolons; (transcontinental).
      6   Spikelets about 3 mm long; scales green or pale brown, less than 3 mm long; achenes to 1.5 mm long; annual, the densely tufted culms less than 1 dm tall, mat-forming . . . .*E. parvula*
      6   Spikelets to about 7 mm long; scales tinged with purple or brown, to about 8 mm long; achenes to 2.5 mm long; perennial, the culms to 4 dm tall, from creeping rhizomes . . . .*E. quinqueflora*
  4   Tubercle distinct from the achene and usually separated from it by a narrow constriction.

7 Achene with prominent straight longitudinal ridges and many transverse
cross-ridges, whitish or pearly, trigonous, about 1 mm long; stigmas 3; tubercle
minute, narrowly conical with a flange-like base, about 1/4 as long as the achene;
bristles 3 or 4, very delicate, or none; spikelet less than 1 cm long; perennials, the
culms tufted along slender rhizomes and stolons.

  8 Culms flattened, at least 1 mm broad (if apparently less, the edges inrolled), to
about 6 dm tall; spikelets terete, their scales in several rows, the middle ones
about 3 mm long; empty sterile scales subcoriaceous and larger than the fertile
ones; (?Sask.) . . . .[E. wolfii]

  8 Culms capillary, usually angular and grooved, commonly less than 3 dm tall;
spikelets flattened, their scales in 2 or 3 rows, uniformly membranaceous, the
middle ones about 2 mm long; (transcontinental) . . . . .E. acicularis

7 Achene smooth or more or less roughened or reticulate (but lacking prominent
ridges).

  9 Tubercle broadly ovate, about as long and broad as the achene; achenes
trigonous, with deep coarse honeycomb reticulation, iridescent; stigmas 3;
scales leathery; bristles coarse, reaching to about the middle of the tubercle;
spikelet to about 13 mm long; culms obtusely angled or subterete, to 8 dm
long; annual, in dense nonstoloniferous tussocks; (N.S.) . . . . .E. tuberculosa

  9 Tubercle much smaller than the smoothish, more or less lustrous achene.

    10 Tufted caespitose annuals (except the tufted perennial E. olivacea, with
soft filiform stolons often developed); achenes lenticular or plano-convex
and stigmas 2 (or achenes sometimes trigonous and stigmas 3); culms
mostly less than 4 dm tall.

      11 Achene dark purple to lustrous-black, about 1 mm long, capped by a
very depressed whitish spongy tubercle; bristles about 7, coarse, about
equalling the achene; spikelet ovoid to subglobose, to 5 mm long;
scales firm, brown to purplish; culm to about 2 dm tall; (s Ont.) . . . .
. . . . .E. geniculata

      11 Achene whitish to greenish or dark brown at maturity.

        12 Leaf-sheaths (at base of culm) prolonged at summit into a loose
white scarious tip; achenes olivaceous to dark brown, lustrous,
about 1 mm long; tubercle green, short-conic above a flange-like
base, about 0.25 mm high; spikelet to 9 mm long; tufted perennial or
with culms (to 4 dm tall) scattered along delicate stolons; (Ont.
to N.S.) . . . . .E. olivacea

        12 Leaf-sheaths with a firm close orifice; tufted annuals.

          13 Tubercle conic-subulate, up to half as long as the pale olive
achene, its flange-like base about 1/4 as broad as the achene;
bristles normally present, some of them overtopping the
tubercle; spikelet to 8 mm long; basal leaf-sheaths with an
oblique orifice with a projecting elongate lobe; culms capillary, to
4 dm tall; (Ont. and Que.) . . . . .E. intermedia

          13 Tubercle broader; achenes brown; sheath-orifices subtruncate,
with a short lobe.

            14 Tubercle triangular-ovate, separated from the summit of the
achene by a strong constriction below its flange-like base;
bristles much surpassing the tubercle; spikelet to 1 cm long;
culms weak and unequal, to 2.5 dm long; (sw Que.) . . . .
. . . .[E. macounii]

            14 Tubercle fitting closely to the summit of the achene.

              15 Tubercle very depressed and cap-like, nearly covering
the summit of the brown shining achene, rarely sur-
passed by the bristles (or these wanting); spikelets to
2 cm long; culms to 4 dm tall; (Alta. and sw Man.) . . . .
. . . . .E. engelmannii

15 Tubercle longer, usually surpassed by the bristles; spikelets less than 1.5 cm long.
    16 Tubercle deltoid-conic to conic-subulate, usually much longer than broad, its base about half covering the summit of the pale-brown achene; culms to 5 dm tall; (Ont. to Nfld. and N.S.) . . . .*E. ovata*
    16 Tubercle broadly deltoid to cap-like, much broader than long and nearly covering the summit of the pale to deep-brown achene; culms to 7 dm tall; (B.C. to N.S.) . . . .*E. obtusa*
10 Perennials with strong reddish to purplish-black rhizomes or stolons; achenes greenish yellow to light brown, usually separated from the spongy tubercle by a narrow constriction.
  17 Achenes trigonous to subterete, minutely wrinkled to deeply reticulate; stigmas 3; tubercle depressed; bristles soon deciduous or none; scales more or less deciduous; (essentially transcontinental).
    18 Surface of achene shallowly to deeply honeycomb-reticulate; tubercle broader than high; culms to 9 dm tall . . . .*E. tenuis*
    18 Surface of achene minutely wrinkled or roughened but not reticulate.
      19 Culms delicately capillary, 4-angled, to 1 dm tall; spikelets less than 5 mm long, their scales obtuse; achenes pale yellow to orange; tubercle a depressed saucer with central apiculation; rhizomes about 1 mm thick . . . .*E. nitida*
      19 Culms flat, wiry, to 5 dm tall; spikelets to over 1 cm long, their scales acuminate; achenes golden to brown; tubercle short-conic; rhizomes about 3 mm thick . . . .*E. compressa*
  17 Achenes biconvex, smoothish; stigmas 2; scales more or less persistent.
    20 Base of spikelet with usually 2 or 3 firm empty basal scales; spikelets to over 2 cm long; elongate perianth-bristles often present; culms relatively stout; (transcontinental).
      21 Tubercle lanceolate to conic-ovoid, much longer than broad; culms subterete . . . .*E. palustris*
      21 Tubercle broadly ovate to depressed-deltoid, as broad as or broader than long; achenes at most 1.6 mm long . . . .*E. smallii*
    20 Base of spikelet with a single spathiform empty scale nearly encircling the culm; spikelets less than 2 cm long; bristles usually wanting or rudimentary; culms nearly filiform, commonly 6 or 7 dm tall.
      22 Tubercle ovoid, nearly as long as the achene; spikelet dark brown, to about 12 mm long, loosely few-flowered; (transcontinental in subarctic regions) . . . .*E. kamtschatica*
      22 Tubercle depressed-deltoid to lanceolate, much shorter than the achene; spikelet to about 17 mm long.
        23 Tubercle depressed-deltoid to low-conical, about 1 mm long and broad, covering at least half of the summit of the achene; (transcontinental) . . . .*E. uniglumis*
        23 Tubercle slenderly conical or lanceolate, rarely half as broad as the achene.
          24 Spikelet loosely flowered, lanceolate to ovoid; fertile scales lustrous, loosely ascending, the lower and median ones to 5 mm long; achenes to 1.4 mm broad; (saline shores of Hudson Bay–James Bay and the Atlantic Ocean) . . . .
          . . . .*E. halophila*
          24 Spikelet closely many-flowered, linear-lanceolate to slenderly ovoid; fertile scales opaque, the lower and

median ones at most 3 mm long; achenes at most 1 mm
broad; (Alta. to N.S.) . . . .*E. erythropoda*

E. acicularis (L.) R. & S.
/aST/X/GEA/ (Grh (Hel)) Low grounds, damp shores, and shallow water from N-cent.
Alaska–Yukon and the Mackenzie R. Delta to Great Bear L., cent. Dist. Keewatin, s Baffin Is.,
northernmost Que. (Hudson Strait), Labrador (N to Makkovik, ca. 55°N), Nfld., N.B., P.E.I., and
N.S., s to s Calif., N Mexico, and Fla.; w Greenland between ca. 65° and 71°N; Iceland; Eurasia.
[*Scirpus* L.]. MAPS: Hultén 1968*b*:212, and 1962: map 105, p. 115; Porsild 1957: map 66, p. 169;
H. K. Svenson, Rhodora 41(483): map 55 (very incomplete northwards), p. 93. 1939.

Dwarf, usually sterile plants with culms lacking the longitudinal furrows characteristic of those
of the typical form, may be distinguished as var. *submersa* (Nilss.) Svenson. Submersed plants
with much-elongated delicate culms and rhizomes may be separated as f. *longicaulis* (Desmaz.)
Hegi (f. *inundata* Svenson).

E. compressa Sullivant
/T/X/ (Grh) Wet grounds from B.C. (N to Wells Gray Provincial Park, ca. 52°N; L. Hämet-Ahti,
Ann. Bot. Fenn. 2(2): 149. 1965) to ?Alta. (Moss 1959), Sask. (N to Nipawin, 53°22′N; Breitung
1957*a*), Man. (N to Porcupine Mt.), Ont. (N to Timmins, 48°28′N), Que. (N to about 50 mi N of
Montreal; *see* Que. map by Rouleau 1945: fig. 7, p. 169), and N N.Y., s to Colo., Tex., and Ga.
[*E. elliptica* var. *comp.* (Sull.) Drap. & Hohl.; *E. acuminata* of auth., not *Scirpus acum.* Muhl.]. MAP:
Svenson 1939: map 31 (incomplete), p. 53.

Forma *atrata* Svenson (*E. elliptica* f. *atr.* (Svenson) Drap. & Hohl.; *E. tenuis* var. *atr.* (Svenson)
Boivin; scales of spikelet blackish rather than brown or purplish brown) is known from Ont. (N to
Moosonee, sw James Bay, 51°16′N; *see* E Canada map by Dutilly and Lepage 1945: fig. 6, p. 211)
and Que. (E James Bay at ca. 51°30′N and Anticosti Is.).

E. engelmannii Steud.
/T/X/ (T) Wet sand, peat, or mud from Wash. to sw Alta. (collection in CAN, detd. Porsild, from
Granum, 49°52′N), ?Sask. (Fernald *in* Gray 1950; a dot for sw Sask. in Svenson's below-noted
map, but no locality given by Svenson 1929 or 1939), and sw Man. (Killarney, NE of Turtle Mt.,
where taken by John Macoun in 1896; CAN, detd. Fernald), s to s Calif., Tex., and Ga. MAP:
Svenson 1939: map 45 (incomplete for Canada), p. 75.

The above Granum, Alta., collection appears referable to var. *monticola* (Fern.) Svenson
(*E. mont.* Fern.; spikelet lanceolate rather than cylindric). The Killarney, Man., collection represents
var. *mont.* f. *leviseta* (Fern.) Svenson (perianth-bristles rudimentary or wanting rather than well
developed).

E. equisetoides (Ell.) Torr.
/t/EE/ (Hel) Shallow water from Wisc. to s Ont. (Long Point, Norfolk Co.; OAC), N.Y., and
Mass., s to E Tex. and N Fla. [*Scirpus* Ell.]. MAP: Svenson 1939: map 5 (the s Ont. station should be
indicated), p. 5.

E. erythropoda Steud.
/sT/(X)/eA/ (Grh) Wet ground from Alta. (Gull Lake, 52°34′N; CAN) to Sask. (Eagle Creek and
Prince Albert; CAN), Man. (N to Muskeg Creek, L. Winnipeg; CAN), Ont. (N to w James Bay at ca.
53°N), Que. (N to E James Bay at ca. 52°N and the Côte-Nord; not known from Labrador or Nfld.),
N.B., P.E.I., and N.S., s to N.Mex., Okla., Tenn., and Va.; Hawaii; E Asia (Fernald *in* Gray 1950). [*E.
calva* Torr. (*E. palustris* var. *calva* (Torr.) Gray) in large part; *E. palustris* var. *glaucescens* of auth.,
not *Scirpus gl.* Willd.]. MAPS (*E. calva*): Hultén 1962: map 165, p. 175; Svenson 1939: map 65, p.
98.

E. geniculata (L.) R. & S.
/t/X/ (T) Damp sands and gravels from s Ont. (Rondeau Provincial Park on the shores of L.
Erie, Kent Co; T.M.C. Taylor, Rhodora 37(442): 366. 1935; GH) to Mich. and N.C., s to s Calif.,
Tex., and Fla. [*Scirpus* L.; *E. caribaea* (Rottb.) Blake and its var. *dispar* (Hill) Blake; *E. capitata* of
American auth., not (L.) R. Br.].

The species-concept followed here is that of Fernald *in* Gray (1950). Svenson (1937; *see* his map 18, p. 259) limited its range to tropical America and his illustrations (pl. 463, p. 262) indicate a totally different plant from that illustrated by Fernald.

E. atropurpurea (Retz.) Kunth (*Scirpus* Retz.) is reported as sporadic in w N. America (as at L. Osoyoos, s B.C.) by Hitchcock et al. (1969). Its main area is from Colo. to Nebr., Iowa, and Ga., s to Mexico, Tex., and Fla. It may be distinguished from *E. geniculata* as follows:

1   Spikelet ovoid to subglobose, 2 or 3 mm thick, its scales firm; achenes about 1 mm long . . . .
. . . .*E. geniculata*
1   Spikelet relatively narrow in outline, at most 2 mm thick, its scales membranaceous;
achenes about 0.5 mm long; [*Scirpus* Retz.] . . . .*E. atropurpurea* (Retz.) Kunth

E. halophila Fern. & Brack.
/sT/EE/   (Grh (Hel))   Saline or brackish shores of w Hudson Bay (Churchill, Man.; CAN), southernmost James Bay (Ont. and Que.), Que. (St. Lawrence R. estuary from l'Islet Co. to the Côte-Nord, Anticosti Is., and Gaspé Pen.), s Labrador (Hamilton R. basin; CAN), Nfld., N.B., P.E.I., and N.S., s along the coast to Va. (an inland station in Cayuga Co., N.Y.). [*E. uniglumis* var. *hal.* F. & B., the type from the mouth of the Bonaventure R., Gaspé Pen., E Que.; *E. palustris* var. *glaucescens sensu* St. John 1921, not (Willd.) Gray, relevant collections in CAN]. MAP: Schofield 1959: map 3 (the above Man. and Labrador stations should be indicated), p. 111.

E. intermedia (Muhl.) Schultes
/T/EE/   (T)   Wet calcareous soils from Minn. to Ont. (N to the N shore of L. Superior near Thunder Bay; TRT) and Que. (N to the Gaspé Pen.; CAN; GH), s to Iowa, Ill., Tenn., and Md. [*Scirpus* Muhl.; *E. ?macounii* Fern.; *E. reclinata* Kunth]. MAP: Svenson 1937: map 19 (*E. reclinata*), p. 263.

E. kamtschatica (Mey.) Komarov
/ST/D(coastal)/eA/   (Grh (Hel))   Brackish or saline shores: coasts of Alaska (N to Norton Sound; *see* Hultén 1942: map 214, p. 403) and NW ?B.C.; coasts of James Bay (Ont. and Que.; N to s Hudson Bay, Ont., at ca. 55°30′N); Que. (Chimo, s Ungava Bay; DAO, detd. Svenson; also known from the mouth of the Romaine R., Saguenay Co., Côte-Nord); coast of s Labrador (mouth of the Paradise R. at ca. 53°30′N; H. K. Svenson, Rhodora 49: 66. 1947); E Asia. [*Scirpus* Mey.; *E. savatieri* Clarke]. MAPS: Hultén 1968*b*:211, and 1962: map 111, p. 121; Schofield 1959: map 5, p. 111; Dutilly, Lepage, and Duman 1958: fig. 8 (E Canada), p. 82.

[E. macounii Fern.]
[Known only from borders of marshes at the type station near Wakefield, Gatineau Co., sw Que., where taken by John Macoun in 1893 (CAN; GH). It is merged with *E. intermedia* by Gleason (1958).]

E. nitida Fern.
/sT/X/   (Grh)   Damp peaty, sandy, or rocky places from southernmost w Alaska (*see* Hultén 1942: map 217 (*Scirpus nit.*), p. 403) through B.C. and sw Alta. to Baja Calif. and Mexico, eastwards very locally to NE Minn., Que. (N to the Swampy R. at 55°21′N and the Gaspé Pen.; type from Parker's Station, Pontiac Co.), Nfld., N.S. (not known from N.B. or P.E.I.), and N N.H. [*Scirpus* Hult.]. MAPS: Hultén 1968*b*:212; Marie-Victorin and Rolland-Germain 1942: fig. 18 (E Canada), p. 33.

E. obtusa (Willd.) Schultes
/T/X/   (T)   Muddy or wet places (ranges of Canadian taxa outlined below), s to s Calif., Tex., and Fla.; Hawaii. MAP and synonymy: *see* below.

1   Perianth-bristles lacking; [Ont. (L. Nipissing; Fassett 1957) and Que. (N to St-Vallier,
Bellechasse Co.; GH)] . . . . .var. *peasei* Svenson
1   Perianth-bristles well developed.
2   Culms to about 2 dm long; spikelet to about 5 mm long, its scales more or less
spreading; [s Que. (Eastern Townships; Raymond 1950*b*) and N.S. (Yarmouth and
Cumberland counties)] . . . . .var. *jejuna* Fern.

2 Culms to 7 dm long; spikelet to over 1 cm long, its scales appressed; [*Scirpus* Willd.; incl. the tall extreme, var. *gigantea* (Clarke) Fern.; sw B.C.; SE Man. (near Otterburne, about 30 mi s of Winnipeg; Löve and Bernard 1959), Ont. (N to the NW and E shores of L. Superior), Que. (N to the Gaspé Pen.), N.B., P.E.I., and N.S.; MAP: Svenson 1939: map 43 (dots should be added for the SE Man. and Gaspé Pen. stations), p. 75] . . . .
. . . . .var. *obtusa*

E. olivacea Torr.

/T/EE/ (Grh) Wet sands and peats from Minn. to s Ont. (N to Renfrew and Carleton counties), Que. (N to Montcalm Co.; MT), and N.S. (not known from N.B. or P.E.I.), s to Ohio, Pa., and Fla. [*E. flavescens* (*flaccida*) var. *oliv.* (Torr.) Gl.]. MAP: Svenson 1939: map 18, p. 10.

E. ovata (Roth) R. & S.

/T/(X)/EA/ (T) Wet open places: NW Wash.; Minn. to Ont. (N to the NW and E shores of L. Superior), Que. (N to L. St. John and Rimouski Co.), Nfld., N.B., P.E.I. (Prince Co.), and N.S., s to N Ind. and Conn.; Hawaii; Eurasia. [*Scirpus* Roth; *E. diandra* Wright]. MAP: Svenson 1939: map 44, p. 75.

Var. *heuseri* Uechtr. (culms depressed or recurving rather than ascending, the scales of their very dark spikelets more or less spreading rather than appressed) is known from Que. (N to 88 mi NW of Mont-Laurier in Labelle Co.) and N.S. (Cumberland and Digby counties).

E. palustris (L.) R. & S.

/aST/X/GEA/ (Grh (Hel)) Wet grounds and shallow water from cent. Alaska–Yukon and the Mackenzie Delta region (Coyen's L.; CAN) to L. Athabasca (Alta. and Sask.), Man. (N to Churchill), northernmost Ont., Que. (N to E James Bay, L. St. John, and the Côte-Nord), s Labrador (N to the Hamilton R. basin), Nfld., N.B., P.E.I., and N.S., s to Oreg., Idaho, N.Dak., N Mich., and N New Eng.; southernmost Greenland; Eurasia. [*Scirpus* L.; *E. ?multicaulis sensu* Hooker 1839, not Sm.; incl. *E. macrostachya* Britt.]. MAPS: Hultén 1968*b*:210, and 1962: map 165, p. 175; Svenson 1939: map 64 (incomplete northwards), p. 98.

The high-grown extreme may be known as var. *major* Sonder (var. *vigens* Bailey; culms to 2 m tall rather than about 4 dm; lower and median fertile scales to over 5 mm long rather than at most 4 mm; anthers to 3 mm long rather than at most 2 mm).

E. parvula (R. & S.) Link

/T/X/EA/ (T) Wet saline or brackish grounds and shores from sw B.C. (Vancouver Is.) to Idaho, Wyo., S.Dak., Minn., Mich., s Ont. (Lambton Co.; Dodge 1915), Que. (N to the Gaspé Pen.), Nfld., N.B., P.E.I., and N.S., s to Calif., Mexico, Tex., and Fla.; S. America; W.I.; Europe; w and E Asia; ?Africa. [*Scirpus* R. & S.; *S. nanus* Spreng.; *E. pygmaea* Torr.]. MAPS: Hultén 1962: map 134, p. 143; Svenson 1934: map 3, p. 387; Charles Gilly, Am. Midl. Nat. 26(1): fig. 1, p. 67. 1941.

Forma *spongiosa* Fassett (with septate spongy culms to 1 mm thick and resembling a sterile *Sagittaria*) is known from Que. (Temiscouata Co.) and N.B.

E. quadrangulata (Michx.) R. & S.

/t/EE/ (Hel) Pools, pond-margins, and creeks (often tidal) from Wisc. to s Ont. (Kent, Lambton, and Norfolk counties), N.Y., and Conn., s to Tex. and N Fla. [*Scirpus* Michx.; incl. the robust extreme, var. *crassior* Fern., suspected by Svenson 1939, to result from growth in the absence of competition; according to M. L. Fernald, Rhodora 27(315): 37–40. 1925, this species has been confused with *E. mutata* (L.) R. & S., the report of which from Lambton Co., s Ont., by J. M. Macoun 1897, is thus probably based upon *E. quadrangulata*]. MAP: Svenson 1939: map 10, p. 5.

E. quinqueflora (Hartm.) Schwarz

/aST/X/GEA/ (Grh) Damp calcareous shores, ledges, and wet places from s-cent. Yukon and Great Bear L. (an isolated station at Manly Hot Springs, cent. Alaska) to Great Slave L., Sask. (N to L. Athabasca), Man. (N to Churchill), James Bay (Ont. and Que.), Que. (N to the Kaniapiskau R. at 57°32′N; Dutilly, Lepage, and Duman 1953; RIM), s Labrador (Hamilton R. basin; DAO), Nfld., N.B., P.E.I., and N.S., s to s Calif., Ariz., N.Mex., S.Dak., Iowa, Ohio, Pa., and N.J.; w Greenland N to ca. 67°30′N; Iceland; Eurasia. [*Scirpus* Hartm.; *E. (Scirpus) pauciflora* (Lightf.) Link and its vars.

fernaldii Svenson and *suksdorfiana* (Beauv.) Svenson (*E. suks.* Beauv.)]. MAPS (as *E. pauciflora*, except Hultén's 1968*b* map): Hultén 1968*b*:211, and 1958: map 236, p. 255; Porsild 1966: map 31, p. 70; Meusel, Jaeger, and Weinert 1965:62; T. W. Böcher, J. Ecol. 39: fig. 7, p. 389. 1951; Svenson 1934: map 1, p. 384.

E. robbinsii Oakes
/T/EE/ (Hel) Shallow ponds and peaty pools from N Wisc. to Ont. (N to the NE shore of L. Superior and the Timagami Forest Reserve N of Sudbury), Que. (St-Adolphe, Argenteuil Co.; MT), N.B., and N.S. (not known from P.E.I.), s to Ind., N.Y., and N Fla. MAP: Svenson 1939: map 3 (the Que. station should be indicated), p. 5.

E. rostellata Torr.
/T/X/ (Hsr) Saline, brackish, or limy marshes from s B.C. (Alberni, Vancouver Is., and Ainsworth, Kootenay L.; CAN) to ?Alta. (Moss 1959), Ill., Ind., Ohio, s Ont. (Bruce, Oxford, and Waterloo counties; CAN), and N.S. (Yarmouth and Shelburne counties; not known from Que., N.B., or P.E.I.), s to Mexico, Tex., Ohio, and N.Y.; S. America. [*Scirpus* Torr.; incl. var. *occidentalis* Wats.]. MAP: Svenson 1934: map 2, p. 385.

E. smallii Britt.
/sT/EE/ (Grh (Hel)) Wet places and shallow water from Man. (N to Churchill; reports from farther west refer chiefly or wholly to *E. palustris*) to Ont. (N to w James Bay at ca. 55°N), Que. (N to SE Hudson Bay at 54°45′N and the Côte-Nord), s Labrador (N to the Hamilton R. basin), Nfld., N.B., P.E.I., and N.S., s to s Calif., Mexico, Tex., La., Ill., and Va. MAPS: Hultén 1962: map 165, p. 175; Svenson 1939: map 66 (incomplete northwards), p. 98.

E. tenuis (Willd.) Schultes
/T/X/ (Grh) Wet or damp sands, gravels, and peats (ranges of Canadian taxa outlined below), s to Mont., Iowa, Ohio, Tenn., and S.C. MAPS and synonymy: *see* below.
1   Achenes olive or drab, with coarse reticulation; culms 4–5-angled; [*Scirpus* Willd.; N.B., P.E.I., and N.S.; MAP: Svenson 1939: map 33, p. 53] . . . . .var. *tenuis*
1   Achenes bright yellow or golden, with shallow reticulation; culms 6–8-angled; [*E. capitata* var. *bor.* Svenson, the type from Sand Beach, Yarmouth Co., N.S.; *E. elliptica* Kunth; *E. compressa sensu* Breitung 1957*a*, not Sulliv.; s B.C. (near Vancouver), Alta. (Mayerthorpe, ca. 54°N; CAN), Sask. (N to Nipawin, 53°22′N), Man. (N to the Red Deer R. at ca. 53°N), Que. (N to the Côte-Nord), Nfld., N.B. (not known from P.E.I.), and N.S.; MAP (*E. ellip.*): Svenson 1939: map 29, p. 53] . . . . .var. *borealis* (Svenson) Gl.

E. tuberculosa (Michx.) R. & S.
/T/EE/ (T) Wet sandy or peaty places from SE N.Y. to sw N.S. (Harper L., Shelburne Co.; GH; CAN; not known from N.B. or P.E.I.), s to E Tex. and Fla. [*Scirpus* Michx.]. MAPS: Svenson 1937: map 16, p. 248; Fernald 1921: fig. 14, pl. 130, facing p. 120.
    Forma *pubnicoensis* (Fern.) Svenson (var. *pub.* Fern., the type from Great Pubnico L., Yarmouth Co., N.S.; achene-bristles smooth rather than barbed) is known only from the type locality and from Shelburne Co., N.S., f. *retrorsa* Svenson (achene-bristles retrorsely rather than divergently barbed) also being known from Shelburne Co.

E. uniglumis (Link) Schultes
/ST/X/GEA/ (Grh) Wet places from cent. Alaska, s Yukon, and Great Bear L. to Alta. (N to Round L. at 59°21′N; not known from Sask.), Man. (N to Churchill), northernmost Ont., Que. (N to E James Bay at ca. 52°N and the Côte-Nord), s Labrador (N to the Hamilton R. basin), Nfld., and N.S. (not known from N.B. or P.E.I.), s to Oreg., Wyo., and N.Dak.; sw Greenland N to 61°13′N; Iceland; Eurasia. [*Scirpus* Link; *E. palustris* var. *glaucescens sensu* St. John 1922, in part, not (Willd.) Gray, relevant collections in CAN]. MAP: Hultén 1968*b*:210.

[*E. wolfii* Gray]
[The report of this species of the E U.S.A. (N to Kans. and Ind.) from Crane Lake, sw Sask., by

Cyperaceae

Svenson (1939; probable basis of the inclusion of Sask. in the range by Fernald *in* Gray 1950) is considered by Breitung (1957a) to be probably based upon an immature specimen of *E. acicularis.*]

ERIOPHORUM L.   [466]   Cotton-grass.   Linaigrette

1   Spikelets solitary, erect, not subtended by a leafy involucre; culm-leaves reduced to mostly bladeless sheaths; (transcontinental).
  2   Plants stoloniferous; culms usually solitary; empty basal scales mostly not more than 7.
    3   Anthers at most 1 mm long; perianth-bristles bright white; fruiting heads subglobose, the fertile scales with a narrow hyaline margin . . . .*E. scheuchzeri*
    3   Anthers at least 1.5 mm long.
      4   Fruiting heads globose, the fertile scales with a narrow hyaline margin; perianth-bristles cinnamon-colour; culm to 4 mm thick . . . .*E. chamissonis*
      4   Fruiting heads obovoid, the fertile scales with a broad hyaline margin; culm less than 2 mm thick . . . .*E. russeolum*
  2   Plants without stolons, densely tufted; empty basal scales usually at least 10.
    5   Scales and spathes with broad whitish margins, the basal ones divergent or reflexed at maturity; perianth-bristles white; upper sheaths conspicuously inflated above . . . .*E. vaginatum*
    5   Scales and spathes without conspicuous whitish margins, appressed-ascending.
      6   Sheaths scattered, the upper ones scarcely inflated, bladeless; lower scale with a broad ribless margin; perianth-bristles brownish-tinged; anthers to 2 mm long . . . .*E. brachyantherum*
      6   Sheaths mostly below the middle of the culm, the upper ones inflated and often with a short blade; lowest scale ribbed nearly to margin; perianth-bristles bright white; anthers at most 1 mm long . . . .*E. callitrix*
1   Spikelets normally more than 1, mostly on spreading or drooping peduncles, the inflorescence subtended by 1 or more leafy bracts.
  7   Involucral bract solitary; leaves at most about 1.5 mm broad, channelled to base; perianth-bristles bright white.
    8   Sheath of upper leaf longer than the round-tipped blade; culm smooth; achenes at most about 2 mm long; (transcontinental) . . . .*E. gracile*
    8   Sheath of upper leaf shorter than the sharp-pointed blade; culms scabrous above; scales greenish to reddish-brown; achenes at least 2.5 mm long; (Ont. to se Labrador, Nfld., and N.S.) . . . .*E. tenellum*
  7   Involucral bracts 2 or more; leaves averaging broader, flat at least below the middle.
    9   Spikelets mostly crowded, their greenish or straw-coloured scales with several strong ribs; perianth-bristles orange-brown or coppery at least at base; stamen solitary; (Ont. to s Labrador, Nfld., and N.S.) . . . .*E. virginicum*
    9   Spikelets mostly loosely peduncled, their scales 1-ribbed; perianth-bristles white or dull yellow; stamens 3; (transcontinental).
      10   Upper leaf-sheaths not dark-girdled at summit; leaves flat except at tip; midrib of scales extending to tip; anthers less than 1.5 mm long . . . .*E. viridicarinatum*
      10   Upper leaf-sheaths dark-girdled at summit; leaves channelled above middle; midrib of scales not extending to tip; anthers at least 2.5 mm long . . . .*E. angustifolium*

E. angustifolium Honckeny
/AST/X/GEA/   (Grh)   Damp or wet tundra, bogs, and shores, the aggregate species from the coasts of Alaska–Yukon–Dist. Mackenzie to northernmost Ellesmere Is., northernmost Ungava–Labrador, Nfld., N.B., P.E.I., and N.S., s to Oreg., N.Mex., Ill., and New Eng.; circumgreenlandic (with gaps); Iceland; Eurasia. MAPS and synonymy: *see* below.
  In addition to the following taxa, var. *coloratum* Hult. (perianth-bristles pale purplish rather than white) is reported from s Alaska by Hultén (1968a; type from Thompson Pass; "It may be supposed that gene-exchange with the taxa with rust-coloured bristles gave rise to this variation.").
1   Peduncles scabrous; spathes and scales black or blackish; culms relatively low; [ssp.

scabriusculum Hult.; *E. triste* (Fries) Hadač & Löve; transcontinental in arctic and subarctic regions: coasts of Alaska, the Yukon, and Dist. Mackenzie to northernmost Ellesmere Is., s to SE Alaska, Great Bear L., E-cent. Dist. Keewatin, and N Baffin Is.; w and E Greenland N of ca. 70°N; MAPS (the last four as *E. triste*): Hultén 1968*b*:199; Porsild 1955: fig. 18, p. 89, and 1957: map 59, p. 168; *Atlas of Canada* 1957: map 4, sheet 38; A. Löve 1950: fig. 5 (very incomplete for N. America), p. 34]. A hybrid between ssp. *scabriusculum* and *E. vaginatum* ssp. *spissum* (× *E. churchillianum* Lepage) is reported from the type locality, Umiat, Alaska, by Lepage 1957; one with *E. scheuchzeri* is reported from Greenland by Böcher, Holmen, and Jacobsen (1966) . . . . .var. *triste* Fries
1   Peduncles smooth; spathes and scales dull brown to lead-colour.
    2   Leaves to about 8 mm broad; culms to about 9 dm tall . . . . .var. *majus* Schultz
        3   Leaf-sheaths green; [var. *elatius* Mert. & Koch; var. *giganteum* Hult.; ssp. *?subarcticum* (Vasill.) Hult.; range of var. *angustifolium*; MAP: Hultén 1968*b*:198 (ssp. *subarct.*). A hybrid with *E. chamissonis* (× *E. beringianum* Raymond) is reported from the type locality, Norton Sound, Alaska, by Marcel Raymond (Nat. can. (Que.) 84: 184. 1957] . . . . .f. *majus*
        3   Leaf-sheaths bright red; [type from Harrington Harbour, Saguenay Co., Que., the only known locality] . . . . .f. *rubrovaginatum* Raymond
    2   Leaves to about 4 mm broad; culms usually less than 6 dm tall; [var. *alpinum* Gaud.; *E. polystachyon* of auth., not L.; transcontinental; MAPS: Porsild 1957: map 58, p. 168; Hultén 1962: map 51, p. 48; Meusel, Jaeger, and Weinert 1965: 59]. A hybrid with *E. scheuchzeri* (× *E. rousseauianum* Raymond; type from Povognituk, NE Hudson Bay, Ungava, at ca. 60°N) is known from Alaska and Ungava . . . . .var. *angustifolium*

E. brachyantherum Trautv.
/aST/X/EA/ (Hs) Bogs and wet places (often calcareous) from N Alaska–Yukon–Dist. Mackenzie–Dist. Keewatin to cent. Baffin Is., Que. (N to the Larch R. at 57°35′N), and Nfld. (not known from the Maritime Provinces), s to SE B.C. (Field; Hitchcock et al. 1969), ?Mont., s Alta.–Sask.–Man., Ont. (s to the N shore of L. Superior near Nipigon; reported from Hastings Co. by Marcel Raymond, Nat. can. (Que.) 78: 293. 1951), and Que. (s to Abitibi Co. and Anticosti Is.); Eurasia. [*E. opacum* (Bjorn.) Fern.]. MAPS: Hultén 1968*b*:204, and 1962: map 28, p. 35; Porsild 1957: map 63, p. 168; Raymond 1950*b*: fig. 17, p. 32, and Nat. can. (Que.) 78: 290. 1951; Raup 1947: pl. 17 (this and the following map requiring considerable expansion); Gunnar Seidenfaden and Thorv. Soerensen, Medd. Gronl. 101(1): fig. 5, p. 19. 1933.

    Var. *pellucidum* Lepage (more or less intermediate between this taxon and *E. vaginatum* ssp. *spissum* and possibly of hybrid origin through this parentage) is known from the type locality, Rupert House, James Bay, Que.

E. callitrix Cham.
/aST/X/GeA/ (Hs) Turfy tundra and bogs (often calcareous) from N Alaska (type from St. Lawrence Is., Bering Strait), s Yukon (CAN), and NW Dist. Mackenzie to Banks Is., Victoria Is., E-cent. Dist. Keewatin, N Baffin Is., northernmost Ungava–Labrador, and Nfld. (not known from s Que., N.B., P.E.I., or N.S., early reports from these places being chiefly based upon *E. vaginatum* ssp. *spissum,* relevant collections in CAN, NBM, and GH), s to s Alaska, N B.C. (Summit Pass, 58°13′N; CAN), Great Bear L., NE Man. (Churchill), and s James Bay; E Greenland between ca. 70° and 77°N; NE Asia. MAPS: Porsild 1957: map 62, p. 168; Hultén 1968*b*:202; Marcel Raymond, Nat. can. (Que.) 78: 295. 1951; Gunnar Seidenfaden and Thorv. Soerensen, Medd. Gronl. 101(1): fig. 4 (very incomplete), p. 18. 1933.

    Var. *moravium* Raymond (scales of spikelets pale brown rather than lead-coloured) is known from Churchill, Man., and the type station, Okkak, Labrador, 57°35′N. Var. *pallidus* Hult. (perianth-bristles yellowish rather than white) is reported from the type locality along the Sheenjek R., s Alaska, by Hultén (1968*a*).

E. chamissonis Meyer
/ST/X/eA/ (Grh) Wet peat from Alaska (N to the Seward Pen.), s-cent. Yukon (near Mayo; CAN), and B.C. to L. Athabasca (Alta. and Sask.), Man. (known only from Churchill), Ont. (sw James Bay; Prescott and Carleton counties), Que. (known only from the SE James Bay watershed),

and N Nfld. (Quirpon Is.), s in the West to Oreg., Utah, and Colo., and isolated in N Minn.; NE Siberia and Kamchatka. MAP: Marcel Raymond, Sven. Bot. Tidskr. 48(1): fig. 8, p. 75. 1954.

Reports from E Que., Labrador, s Nfld., and the Maritime Provinces, as on the maps by Raymond (1950b: fig. 20, p. 34) and A. Löve (Bot. Not. 1948:105. 1948), refer chiefly or wholly to *E. russeolum* (*see* the map for this species by Raymond, loc. cit., fig. 7: 74). Forma *turneri* Raymond (perianth-bristles whitish rather than reddish or brownish; type from near Edmonton, Alta.) occurs throughout the range. A hybrid with *E. vaginatum* ssp. *spissum* (× *E. porsildii* Raymond; *E. opacum* var. *cinnamomeum* Porsild; characterized by pale cinnamon-brown perianth-bristles) is known from the type locality, the Mer Bleue peat bog near Ottawa, Ont.

E. gracile Koch
/ST/X/EA/ (Grh) Wet peat and shores from s-cent. Alaska (*see* Hultén 1942: map 205, p. 402; not known from the Yukon) to Great Slave L., L. Athabasca (Alta. and Sask.), Man. (N to Churchill), northernmost Ont., Que. (N to NE James Bay and the Côte-Nord), s Labrador (Hamilton R. basin), Nfld., P.E.I. (Bristol, Kings Co.; MT; reports from N.B. by Fowler 1885, and John Macoun 1888, are based upon *E. tenellum,* relevant collections in NBM), and N.S., s to N Calif., Colo., Nebr., Minn., and Del.; Eurasia. MAPS: Hultén 1968b:200, and 1962: map 87, p. 97.

Var. *caurianum* Fern. (scales of the spikelets yellowish or brownish rather than lead-colour to blackish) is reported by Dutilly, Lepage, and Duman (1958; *see* their James Bay map, fig. 9, p. 84) as the common or only phase in the James Bay region of Ont.–Que.

E. russeolum Fries
/aST/X/EA/ (Grh) Wet peat and pond-margins (MAPS, ranges, and synonymy of individual taxa given below, the aggregate species confined in N. America to Alaska–Canada); Eurasia.

See note under *E. chamissonis.* A purported hybrid with *E. vaginatum* ssp. *spissum* (× *E. pylaieanum* Raymond, the type from St-Pierre and Miquelon; *E. spissum* (*callitrix*) var. *erubescens* Fern.) is known from E Que. (Saguenay Co. of the Côte-Nord), s Labrador, St-Pierre and Miquelon, Nfld., and N.S. (Scatari Is., Cape Breton Is.; CAN). One with *E. scheuchzeri* (× *E. medium* And.; *E. russeolum* var. *aquatile* Norman, not *E. chamissonis* var. *aq.* (Norman) Fern.) is believed by Marcel Raymond (Sven. Bot. Tidskr. 48(1): 80. 1954) to be confined to the Old World, contrary to the interpretations of the taxon by Hultén (1942) and Fernald in Gray (1950).

1 Achenes hairy toward summit; [*E. mandshuricum* Meinsh.; *E. russeolum sensu* Hultén 1942, not Fries, Hultén's map 207a, p. 402, applying here; Aleutian Is. and s Alaska; Que. (Korok R., E of Ungava Bay at 58°35′N); Asia. MAPS: Marcel Raymond, Sven. Bot. Tidskr. 48(1): fig. 7, p. 74. 1954; Hultén 1968b:202, and 1958: map 205, p. 225] . . . .
. . . . .var. *majus* Sommier
1 Achenes glabrous.
  2 Perianth-bristles white; [f. *leucothrix* Blomgr.; *E. chamissonis* f. *alb.* (Nyl.) Fern.; Aleutian Is. and the N coast of Alaska to Great Bear L., N Baffin Is., E Dist. Keewatin, and Southampton Is., s to NE Man. (Churchill; York Factory), Que. (NE James Bay; Côte-Nord; Gaspé Pen.), Nfld., N.B., and N.S.; N Norway; E Asia; MAPS: on the above-noted map by Raymond and the 1958 map by Hultén; Hultén 1968b:203]
. . . . .var. *albidum* Nyl.
  2 Perianth-bristles cinnamon-colour; [Ont. (w James Bay–Hudson Bay N to ca. 55°15′N), E Que. (s Ungava Bay to the Côte-Nord, Anticosti Is., Gaspé Pen., and Magdalen Is.), Labrador (entire coast), Nfld., St-Pierre and Miquelon, N.B., P.E.I., and N.S.; Iceland; Eurasia; MAPS: on the above-noted map by Raymond and the 1958 map by Hultén] . . . . .var. *russeolum*

E. scheuchzeri Hoppe
/AST/X/GEA/ (Grh) Damp tundra and wet peat from the coasts of Alaska–Yukon–Dist. Mackenzie–Dist. Keewatin throughout the Canadian Arctic Archipelago to northernmost Ellesmere Is. and northernmost Ungava–Labrador, s to N B.C. (isolated stations in the mts. of s B.C.–Alta., Utah, and Colo.; not known from Sask.), Great Slave L., N Man. (s to Gillam, about 165 mi s of Churchill), N Ont. (s to Moose Factory, sw James Bay, 51°16′N), N Que. (s to Fort George, E James Bay, 53°50′N), s Labrador, and NW Nfld.; circumgreenlandic; Iceland; Spitsbergen; Eurasia. [*E. capitatum* Host; incl. the slender narrow-leaved extreme, var. *tenuifolia* Ohwi]. MAPS: Hultén

1968b:201, and 1962: map 19, p. 27; Porsild 1957: map 60, p. 168; Meusel, Jaeger, and Weinert 1965:58; Raup 1947: pl. 17 (requiring expansion).

E. tenellum Nutt.
/T/EE/ (Grh) Peaty soil from Ont. (N to Timmins, ca. 48°30'N; CAN; the report from L. Athabasca, Sask., by Raup 1936, is based upon *E. viridicarinatum,* a relevant collection in CAN; reports from Man. by Jackson et al. 1922, and Lowe 1943, probably refer to *E. gracile*) to Que. (N to the Nottaway R. SE of James Bay at 51°05'N, the Côte-Nord, and Gaspé Pen.), sw Labrador (Ashuanipi L. at ca. 53°N; CAN), Nfld., N.B., P.E.I., and N.S., s to Ill., Mich., and N.J. [*E. gracile* var. *paucinervium* Engelm.; *E. ?strictum* R. Br.; incl. the reduced extreme bearing only one spikelet, var. *monticola* Fern.].

Forma *gorhamii* Raymond (perianth-bristles bluish rather than white) is known from the type locality, Halifax, N.S.

E. vaginatum L.
/aST/X/GEA/ (Hs) Acid bogs and peaty meadows and tundra (ranges of Canadian taxa outlined below), s to s Alta.–Sask.–Man., Minn., Ind., and N.J.; sw Greenland N to ca. 62°30'N; Eurasia. MAPS and synonymy: *see* below.
1 Heads oblong; scales lead-colour; anthers to 3 mm long; [*E. caespitosum* Host; N coast
   of Alaska and the Mackenzie R. Delta to Banks Is. and Victoria Is., s to NW B.C. (s to
   Prince George and the Peace River Dist.), Great Slave L., and s-cent. Sask. (s to the
   Saskatchewan R. at ca. 53°15'N); MAPS: Hultén 1968b:204, and 1962: map 163, p. 173;
   Porsild 1957: map 64, p. 169; Meusel, Jaeger, and Weinert 1965:58] . . . . .ssp. *vaginatum*
1 Heads depressed-globose; scales with a dark centre; anthers at most 2 mm long;
   [*E. spissum* Fern., the type from Sandy Cove, Nfld.; Porsild's map denotes a range from
   Melville Pen. and northernmost Baffin Is. to Sask., Ont., Que., and Nfld., it also occurring
   in N.B., P.E.I., and N.S.; however, the characters of the two taxa intergrade and Hultén's
   map indicates its probable occurrence in Alaska–Yukon–Dist. Mackenzie–w. Dist.
   Keewatin and in Alta.; MAPS: Hultén 1968b: 205, and 1962: map 163, p. 173; Porsild
   1957: map 64, p. 169] . . . . .ssp. *spissum* (Fern.) Hult.

E. virginicum L.   Tawny Cotton-grass
/sT/EE/ (Hsr) Bogs and peaty meadows from Ont. (N to the N shore of L. Superior and Oba L., s of Hearst; reports from Man. by Jackson et al. 1922, Lowe 1943, and Gleason 1958, require confirmation) to Que. (N to s James Bay, L. St. John, and the Côte-Nord), s Labrador (Hamilton R. basin; CAN), Nfld., N.B., P.E.I., and N.S., s to Minn., Wisc., Tenn., and Ga.

Forma *album* (Gray) Wieg. (perianth-bristles white rather than tawny to copper-brown) occurs throughout the range.

E. viridicarinatum (Engelm.) Fern.
/ST/X/ (Grh) Peats, wet meadows, and swampy woods from Great Slave L. (an isolated area in s Alaska) to Alta. (N to L. Athabasca), Sask. (N to Hasbala L. at 59°55'N), Man. (N to York Factory, ca. 57°N), northernmost Ont., Que. (N to E James Bay, Knob Lake, 54°48'N, and the Côte-Nord), s Labrador (N to the Hamilton R. basin), Nfld., N.B., P.E.I., and N.S., s to s B.C., Idaho, Colo., Iowa, Ohio, and Long Is. [*E. polystachion* of Canadian reports in part, not L.]. MAPS: Hultén 1968b:200; Meusel, Jaeger, and Weinert 1965:59.

Forma *fellowsii* Fern. (the spikelets all clustered in a dense inflorescence) is known from Ont. (Wellington Co. and Kapuskasing).

FIMBRISTYLIS Vahl [471]

1 Achenes 3-angled, less than 1 mm long, smooth or minutely roughened; styles 3-cleft;
   spikelets at most 1 cm long, brown or castaneous; leaves 1 or 2 mm broad; culms
   slender, flat, erect or diffuse, mostly less than 3 dm tall; soft-based tufted annual; (Ont.
   and sw Que.) . . . .*F. autumnalis*
1 Achenes lenticular, commonly about 1.5 mm long, minutely striate and reticulate; styles

2-cleft; spikelets to over 1.5 cm long, dark brown; leaves to 3 mm broad; culms wiry, to about 1 m tall; perennials.

2   Culms densely tufted, their bases covered with firm dark sheaths; scales of spikelets firm, lustrous, glabrous . . . .[F. castanea]

2   Culms densely to loosely tufted from a knotted rhizome, their basal sheaths pale and thin; scales soft, at least the outer ones minutely pubescent; (s Ont.) . . . . .F. spadicea

F. autumnalis (L.) R. & S.
/T/EE/   (T)   Moist sandy or peaty soils from s Ont. (N to the Chalk River and Ottawa districts; see the s Ont. map by Soper 1962: fig. 4, p. 10) to Que. (N to Lac Desmarais, Pontiac Co., at ca. 47°15′N; see Que. map by Marie-Victorin and Rolland-Germain 1942: fig. 17, p. 32) and cent. Maine, s to Tex. and Fla.; Cuba; Cent. America.

Forma brachyactis (Fern.) Blake (inflorescence very compact, the rays suppressed) is the phase represented by the above-noted Lac Desmarais, Que., collection.

[F. castanea (Michx.) Vahl]
[The report of this species of the coastal E U.S.A. (Long Is. to Fla. and Tex.) from Walpole Is., Lambton Co., s Ont., by J. M. Macoun (1897; probable basis of the listing by Soper 1949) is based upon F. spadicea, the relevant collection in CAN, taken by Dodge in 1894.]

F. spadicea (L.) Vahl
/t/EE/   (Grh)   Dry or moist woods, sterile meadows, and prairies from Mo., Ill., s Mich., s Ont. (known only from Walpole Is., Lambton Co.; see the above species), and N.J., s to E Tex. and Fla. [F. drummondii of Canadian reports, not Boeck.].

[FUIRENA Rottb.]   [467]

[F. pumila Torr.]   Umbrella-grass
[The map for F. squarrosa given by M. L. Fernald (Rhodora 33(386): map 32, p. 57. 1931; F. pumila Torr. as to the northern stations; see Fernald in Gray 1950:280) indicates a station along the Detroit R. in SE Mich. opposite Lambton Co., s Ont. Future search in that region will probably add it to our flora.]

HEMICARPHA Nees   [453]

H. micrantha (Vahl) Pax
/T/X/   (T)   Damp sandy grounds, the aggregate species from Wash. (the report from Victoria, B.C., by Eastham 1947, is based upon Scirpus setaceus (see under S. cernuus), the relevant collection in CAN, revised by Boivin) to Minn., Mich., s Ont. (known only from Amherstburg, Essex Co., where taken by John Macoun in 1891; var. drummondii taken by Macoun at the same locality in 1892), sw Que. (Philipsburg, Missisquoi Co.; MT), and cent. N.H., s to Mexico, Tex., and Fla.; tropical America. MAPS and synonymy: see below.

1   Scales of spikelets oblong to narrowly obovate, their short-pointed tips spreading or recurving; achenes brown, slightly reticulate and closely papillate; [Scirpus Vahl; H. subsquarrosa Nees; incl. var. minor (Schrad.) Friedland; s Ont. and sw Que.; MAPS: Solomon Friedland, Am. J. Bot. 28: fig. 2, p. 858. 1941; McLaughlin 1932: fig. 2 (aggregate species; very incomplete), p. 339] . . . . .var. micrantha

1   Scales of spikelets obovate to rhombic, their blunt tips appressed; achenes pale, scarcely reticulate, more sparingly papillate; [H. drummondii Nees; s Ont.; MAP: on the above-noted map by Friedland] . . . . .var. drummondii (Nees) Friedland

KOBRESIA (Cobresia) Willd.   [522]

1   Inflorescence a seemingly simple linear spike to 3 cm long; terminal spikelet staminate; lateral spikelets each with 1 staminate and 1 pistillate flower; culms to about 4 dm tall; (transcontinental) . . . . .K. bellardii

1   Inflorescence obviously compound, ovoid or ellipsoid.

2   Head to about 4 cm long and 7 cm thick; spikelets unisexual and 1-flowered, those of the terminal spike all staminate, the lateral spikes staminate above, pistillate below; culms commonly less than 2.5 dm tall; (transcontinental) . . . .*K. simpliciuscula*

2   Head commonly less than 2 cm long but up to about 8 mm thick; spikelets each with 1 staminate and 1 pistillate flower (or the terminal spikelet staminate); culms commonly taller; (Alaska and Banks Is. to Boothia Pen., Dist. Keewatin) . . . .*K. sibirica*

K. bellardii (All.) Degl.

/AST/X/GEA/ (Hr) Dryish tundra and barren ridges (often calcareous) from the coasts of Alaska–Yukon–Dist. Mackenzie to Banks Is., Victoria Is., northernmost Ellesmere Is., Baffin Is., and northernmost Ungava–Labrador (reports from Nfld. require confirmation), s in the West through the mts. of B.C.–Alta. to Calif., Utah, and Colo., farther eastwards s to Great Bear L., N Man. (N to Churchill), Southampton Is., N Que. (s to s Ungava Bay), and Labrador (s to ca. 55°N); circumgreenlandic; Iceland; Eurasia. [*Carex* Vill.; *Elyna* Fritsch; *C.* (*Elyna; Scirpus*) *bellardii* All.; *K. filiformis* Torr.; *C.* (*K.*) *myosuroides* Vill.; *K. scirpina* Willd.; *C. affinis* R. Br., not Nyl.]. MAPS: Hultén 1968b:213, and 1962: map 40, p. 47; Porsild 1957: map 68, p. 169; Meusel, Jaeger, and Weinert 1965:63; Meusel 1943: fig. 8f.

K. sibirica Turcz.

/aS/WW/eA/ (Hr) Dryish peaty tundra from the coasts of Alaska–Yukon–Dist. Mackenzie–Dist. Keewatin (Boothia Pen.), Banks Is., and Victoria Is., s to cent. Alaska, Great Bear L., and E-cent. Dist. Mackenzie; E Asia. [*K. arctica* Porsild, not Ivanova; *K. hyperborea* Porsild and its vars. *alaskana* and *lepagei* Duman]. MAPS (all but the first as *K. hyp.*): Hultén 1968b:214; Porsild 1957: map 69, p. 169, and 1955: fig. 12, p. 44; *Atlas of Canada* 1957: map 17, sheet 38.

K. simpliciuscula (Wahl.) Mack.

/AST/X/GEA/ (Hr) Damp to dryish tundra, gravels, and slopes (often calcareous) from the coasts of Alaska and Dist. Mackenzie (N to ca. 68°N in the Yukon) to Victoria Is., Ellesmere Is. (N to ca. 80°N), Baffin Is., northernmost Ungava–Labrador, and Nfld., s in the West through B.C. (Summit Pass, 58°31'N; CAN; isolated in the mts. of SE B.C. and sw Alta.; not known from Sask.) to Oreg., Idaho, and Colo., farther eastwards s to NE Man. (Churchill), N Ont. (Cape Henrietta Maria, NW James Bay; Dutilly, Lepage, and Duman 1954), and Que. (N to Hudson Strait and Akpatok Is., Ungava Bay; also known from the Côte-Nord and Anticosti Is.; *see* E Canada map by Raymond 1950a: map 1, p. 441); nearly circumgreenlandic; Spitsbergen; Eurasia. [*Carex* Wahl.; *K. bipartita* Dalla Torre, not (All.) Britt.; *Cobresia* (*K., Elyna*) *caricina* Willd.; incl. var. *americana* Duman]. MAPS: Hultén 1968b:214, and 1958: map 213, p. 233; Porsild 1957: map 70, p. 169; Meusel, Jaeger, and Weinert 1965:63.

RHYNCHOSPORA Vahl [492] Beak-rush

(Ref.: Gale 1944)

1   Perianth-bristles upwardly barbed, usually 6, 3 of them about equalling or surpassing the tubercle; achenes light brown, smooth; margins of tubercle minutely serrulate; flower-clusters deep brown or castaneous, terminal and from 1 or 2 upper axils, the terminal one often compound; leaves setaceous or involute; plant with slender stolons; (Ont. to Nfld. and N.S.) . . . .*R. fusca*

1   Perianth-bristles normally downwardly barbed; plants caespitose, often tussock-forming.

2   Perianth-bristles 8 or more, often minutely hairy at base; flower-clusters milk-white to whitish brown, terminal and long-peduncled from 1 or 2 upper axils, the terminal one often compound; achenes slightly roughened, brownish green with faint transverse brown lines; leaves to 2.5 mm broad; (transcontinental) . . . .*R. alba*

2   Perianth-bristles usually 6 (rarely 12), not hairy below; flower-clusters brown.

3   Achenes smooth, with a conspicuous pale wire-like margin; flower-clusters 2 or more, compact, all but the lowest short-peduncled; leaves to 3.5 mm broad; (Ont. to N.S.) . . . .*R. capitellata*

3   Achenes slightly roughened and more or less horizontally dark-striate, inconspicuously margined; flower-clusters 1 or commonly 2, rather loose, the terminal

one with up to about 10 spikelets on ascending pedicels; leaves setaceous or narrowly linear, at most 0.5 mm broad; (Alta. to Nfld. and N.S.) . . . .*R. capillacea*

R. alba (L.) Vahl

/sT/X/EA/ (Hs) Bogs and wet peats and sands from s Alaska and B.C. to Sask. (N to Nipawin, 53°22′N; not yet known from Alta. or Man. but Hultén's 1958 map indicates stations very close to the s and w boundaries of Alta.), Ont. (N to the Shamattawa R. at 53°44′N), Que. (N to Fort George, E James Bay, 53°40′N, L. Mistassini, and the Côte-Nord), SE Labrador (Hamilton R. basin; CAN), Nfld., N.B., P.E.I., and N.S., s to Calif., Idaho, Minn., Ohio, and N.C.; W.I.; Eurasia. [*Schoenus* L.]. MAPS: Hultén 1968b:213, and 1958: map 249, p. 269; Meusel, Jaeger, and Weinert 1965:63; Gale 1944: map 9 (incomplete northwards), p. 120.

Forma *leviseta* Gale (perianth-bristles smooth rather than barbed) is known from N.B. (Miscou Is.; CAN), St-Pierre and Miquelon, and Nfld. (type from French Is., Bay of Islands).

R. capillacea Torr.

/T/EE/ (Hs) Calcareous swamps, bogs, and shores from ?Alta. (Boivin 1967a) to Sask. (N to Nipawin, 53°22′N; also known from Wallwort), s Man. (Birds Hill, near Winnipeg; DAO), s Ont. (N to Manitoulin Is., N L. Huron; N. C. Fassett, Rhodora 35(420): 388. 1933), Que. (N to the mouth of the Bonaventure R., Gaspé Pen.; GH), Nfld., N.B., and N.S. (not known from P.E.I.), s to S.Dak., Ohio, Tenn., and Va. MAP: Gale 1944: map 10, p. 132.

Forma *leviseta* (Hill) Fern. (perianth-bristles smooth rather than barbed) is known from Ont., Nfld., and N.S.

R. capitellata (Michx.) Vahl

/T/EE/ (Hsr) Bogs and wet sands from Wisc. to s Ont. (N to Georgian Bay, L. Huron, and Renfrew Co.), sw Que. (N to Pont-Garneau, Lévis Co.; MT), N.B. (Bass River, Kent Co.; NBM; not known from P.E.I.), and N.S., s to E Tex. and N Fla. [*Schoenus* Michx.; *R. glomerata* vars. *cap.* (Michx.) Kük. and *minor* Britt.]. MAP: Gale 1944: map 1, p. 120.

Forma *discutiens* (Clarke) Gale (perianth-bristles smooth rather than barbed) is known from N.S.

R. fusca (L.) Ait. f.

/T/EE/ (Hsr) Bogs and wet peats and sands from Minn. to Ont. (N to the N shore of L. Superior at Schreiber; MT), Que. (N to Matane Co., Gaspé Pen.), Nfld., N.B., and N.S. (not known from P.E.I.), s to s Mich. and Del. [*Schoenus* L.]. MAPS: Hultén 1958: map 37, p. 56 (both this and the next map indicating an isolated station on the Saskatchewan R. near the Sask.–Man. boundary); Meusel, Jaeger, and Weinert 1965:63; Gale 1944: map 24, p. 132.

SCIRPUS L. [468] Bulrush

1   Involucre none or merely the modified outer scale of the solitary terminal spike or spikelet.
   2   Spikelets crowded into a terminal compound spike to about 2 cm long; scales dark reddish-brown, acuminate, folded; achenes stipitate, long-beaked; perianth-bristles wanting or very short; leaves to about 2 mm broad; plant with extensively creeping rhizomes; (transcontinental, chiefly in saline or brackish marshes) . . . .*S. rufus*
   2   Spikelet solitary, to about 7 mm long; achenes sessile or nearly so, beakless or short-beaked; leaves commonly about 1 mm broad.
      3   Culms sharply 3-angled, the angles scabrous above.
         4   Perianth-bristles terete, minutely barbed, shorter than to somewhat surpassing the achene, these rounded to a scarcely beaked summit; spikelet to 6 mm long, the denuded rachilla flexuous; (Alta. and Sask.; Ont. to N.B.) . . . .*S. clintonii*
         4   Perianth-bristles flat, smooth, finally crinkled, to 3 cm long; achene apiculate; spikelet to 7 mm long, the denuded rachilla straight; (transcontinental) . . . .
. . . .*S. hudsonianus*
      3   Culms terete, smooth.
         5   Spikelet to 4 mm long, the scales awnless; perianth-bristles none; achene

about 1.5 mm long, beakless; plant stoloniferous; (B.C. to Sask.; E Que.) . . . .
. . . .*S. pumilus*
  5 Spikelet to 6 mm long, the lower scales awn-tipped; perianth-bristles flat,
barbless, to about twice the length of the achene, this about 2 mm long,
beaked; plants in dense tussocks, nonstoloniferous; (transcontinental) . . . .
. . . .*S. caespitosus*
1 Involucre consisting of 1 to many leaf-like bracts.
  6 Inflorescence apparently lateral, its single firm erect bract simulating a continuation of
the culm, this naked or leafy only toward base.
    7 Spikelets solitary or up to about 15 in a sessile capitate cluster.
      8 Culm relatively stout, sharply 3-angled.
        9 Achene trigonous (3-angled), with a beak about as long as the body;
stigmas 3; scales about twice as long as broad; leaves 2 or 3, obliquely
rounded at apex; upper leaf-sheaths open or readily splitting; rhizome soft
and weak; (Alaska–B.C.; Man. to N.B.) . . . .*S. subterminalis*
        9 Achene normally plano-convex, short-beaked; scales minutely ciliate;
upper leaf-sheaths closed; rhizome hard.
          10 Culm wing-angled, hollow, to about 3 m tall; spikelets up to 12 in
number; achenes at most 2.3 mm long and less than 2 mm broad;
stigmas 2; involucral bract rarely over 5 cm long, blunt-tipped; upper
leaf-blade at most about 1.5 cm long, blunt-tipped; (N.S.) . . . .*S. olneyi*
          10 Culm not wing-angled, to about 1.5 m tall; spikelets commonly less than 8;
achenes to 3.25 mm long and 2.5 mm broad; involucral bract to about
1.5 dm long; upper leaves to 6 dm long, sharp-pointed; (transcontinental) . . . .
. . . .*S. americanus*
      8 Culm relatively slender, obscurely and obtusely 3-angled to oval or terete in
cross-section.
        11 Plants perennial from rhizomes and stolons.
          12 Spikelet solitary (rarely 2); scales becoming pale brown, with an
excurrent green midrib; achenes trigonous-obovoid, lustrous, abruptly
short-beaked; styles 3-cleft to middle; leaves very numerous, capillary
and flaccid in the usual submersed state, few and rigid or wanting in
emersed colonies; rhizome slender and soft, the capillary stolons
terminated by a tuber; (s Alaska and B.C.; Ont. to Nfld. and N.S.) . . . .
. . . .*S. subterminalis*
          12 Spikelets solitary or up to 5; scales walnut-brown, smooth and shining;
achenes plano-convex, minutely reticulate, scarcely beaked; style
2-cleft; leaves stiff; rhizome tough, deep in the soil; (B.C. to s Man.) . . . .
. . . .*S. nevadensis*
        11 Plants annual, fibrous-rooted, tufted.
          13 Perianth-bristles stout and persistent, some of them surpassing the
achene; achenes round-obovate, unequally biconvex, olivaceous to
dull black, minutely wrinkled and more or less pitted; stigmas 2 or 3;
culms 3-angled; (s Ont.) . . . .*S. juncoides*
          13 Perianth-bristles none or delicate and slender.
            14 Stigmas 2; achenes plano-convex, shining black and mostly
smooth; culms subterete or somewhat angled, to about 4 dm tall;
(s Ont. and sw Que.) . . . .*S. smithii*
            14 Stigmas 3; achenes trigonous, punctate, whitish becoming
brownish; culms filiform, to about 2 dm tall; (coast of sw B.C.) . . . .
. . . .*S. cernuus*
    7 Spikelets several to many in a spike-like or subsimple to twice-compound
irregularly umbelliform inflorescence; culms terete or subterete; plants perennial
from slender or stout rhizomes; (transcontinental).
      15 Inflorescence a solitary terminal spike, the reddish to dark-brown spikelets
overlapping in 2 vertical rows; achene long-beaked, stipitate; perianth-bristles

wanting or very short and soon deciduous; culms rarely over 4 dm tall; (chiefly saline and brackish marshes) . . . .*S. rufus*
15 Inflorescence irregularly umbelliform; perianth-bristles 2–6, downwardly barbed . . . .
. . . .*S. lacustris*
6 Inflorescence obviously terminal, its involucre consisting of at least 2 flat spreading leafy bracts; perennials with 3-angled culms.
16 Culms sharply 3-angled, solitary or scattered from corm-like portions of the elongate rhizomes; spikelets to 4 cm long, in a subsimple or twice-compound umbelliform inflorescence; scales puberulent.
17 Achenes equally and acutely 3-angled, conspicuously beaked, equalled or surpassed by the 6 persistent stiff downwardly-barbed bristles; stigmas 3; leaves to 2 cm broad; orifice of leaf-sheaths convex, the veins of the sheath continued nearly to the summit and abruptly divergent; (Alta. to N.B.) . . .*S. fluviatilis*
17 Achenes lenticular, plano-convex, or obscurely 3-angled; perianth-bristles weak and deciduous; leaves at most 1 cm broad.
18 Orifice of leaf-sheaths truncate or concave, the ventral nerves gradually divergent at a narrow angle well below the summit; spikelets pale brown to chestnut-brown; (transcontinental) . . . .*S. maritimus*
18 Orifice of leaf-sheaths convex, the veins continued nearly to the summit and abruptly divergent; spikelets reddish brown; (N.S.) . . . .*S. robustus*
16 Culms obtusely angled (sometimes sharply angled at summit), from close leafy crowns; spikelets mostly not over 1 cm long, very numerous in a much-branched umbelliform-paniculate inflorescence; scales glabrous.
19 Perianth-bristles downwardly barbed (rarely wanting); spikelets in capitate clusters; plants solitary or few from thick scaly stolons.
20 Sheaths not red-tinged, the orifice V-shaped, the ligule thin and friable; bristles barbed only above the middle, slender, shorter than or barely surpassing the achene; (Alta. to Nfld. and N.S.) . . . .*S. atrovirens*
20 At least the lower leaf-sheaths red-tinged at base and often at orifice; bristles barbed nearly to base, equalling or conspicuously surpassing the achene.
21 Achenes lenticular or plano-convex; stigmas 2; bristles 4; scales blunt or barely pointed; leaf-blades to 1.5 cm broad, most of their sheaths red-tinged, barely nodulose when dry; (transcontinental) . . . .*S. microcarpus*
21 Achenes compressed-trigonous; stigmas 3; bristles 3 or 6; scales subulate-tipped; leaf-blades hard and thick, to 2.5 cm broad, their leathery sheaths prominently nodulose when dry, only the lower ones red-tinged; (s ?Ont. and s ?Que.) . . . .[*S. expansus*]
19 Perianth-bristles bent or curled, smooth or with a few scattered ascending hairs; lateral spikelets of each cluster often distinctly pedicelled; achenes compressed-trigonous; stigmas 3; plants nonstoloniferous, their leaf-blades mostly not over 1 cm broad.
22 Mature perianth-bristles about twice as long as the achene, not much surpassing the scales.
23 Achenes brown or purplish brown; scales pale brown or reddish brown, their slender green keels excurrent as acuminate or subulate tips; central spikelet in each cluster usually sessile, the lateral ones usually distinctly pedicelled; (Ont. and sw Que.) . . . .*S. lineatus*
23 Achenes whitish, soft; scales rusty-black above the pale base, obtuse or merely acute; spikelets mostly sessile or very short-pedicelled; (sw Que.) . . . .*S. peckii*
22 Mature perianth-bristles several times longer than the achene, greatly surpassing and often hiding the scales.
24 Spikelets all or nearly all sessile in clusters of up to about 15; plants in dense tussocks, with many curving rigid narrowly linear basal leaves; (Man. to Nfld. and N.S.) . . . .*S. cyperinus*
24 Lateral spikelets of each cluster usually pedicelled.

25 Involucels and spikelets greenish black; leaves bright green,
   relatively soft, to 5 mm broad; (transcontinental) . . . .*S. atrocinctus*
25 Involucels and spikelets pale brown to rust-brown; leaves pale
   green, rather firm, to about 1 cm broad.
   26 Spikelets whitish brown to dull brown; longer rays of inflores-
      cence smooth except at tip; (Ont. to N.B. and N.S.) . . . .*S. pedicellatus*
   26 Spikelets rust-brown; longer rays of inflorescence scabrous
      except at base; (s ?Ont.) . . . .[*S. rubricosus*]

S. americanus Pers.

/sT/X/EA/ (Hel) Fresh, brackish, or saline shores and marshes, the aggregate species from
E-cent. Alaska (Circle Hot Springs; *see* Hultén 1942: map 212, p. 402) and B.C. (Vancouver Is. and
adjacent islands, Vancouver, Cariboo, Penticton, and Okanagan; V) to s Alta. (Winnifred, near
Medicine Hat; CAN), Sask. (N to Prince Albert; CAN), Man. (N to Wabowden, about 135 mi NE of
The Pas), Ont. (N to near Fort Severn, Hudson Bay, ca. 56°N), Que. (N to SE James Bay, the
Côte-Nord, and Gaspé Pen.), Nfld., N.B., P.E.I., and N.S., s to Calif., Mexico, Tex., and Fla.; S.
America; Eurasia. MAPS and synonymy: *see* below.
1 Floral scales reddish- to brownish-purple, the membranaceous midvein terminating in a
  short erect mucro not surpassing the small rounded apical teeth; [var. *polyphyllus sensu*
  Beetle in major part; B.C.; MAP: Koyama 1963: fig. 2, p. 1114] . . . .
  . . . . .ssp. *monophyllus* (Presl) Koyama
1 Floral scales yellowish- to reddish-brown, the thinly coriaceous midvein terminating in an
  excurrent mucro surpassing the acute apical teeth . . . . .ssp. *americanus*
  2 Stigmas 2; achenes plano-convex, longer than the perianth-bristles; spikelets ovate,
    rust-brown [*Schoenoplectus* Volkart; *Scirpus pungens* Vahl; *S. triqueter sensu*
    Hooker 1839, not L.; transcontinental; MAPS: Hultén 1968*b*:207, and 1958: map 200,
    p. 219; Koyama 1963: fig. 2, p. 1114] . . . . .var. *americanus*
  2 Stigmas 3; achenes obscurely 3-angled, shorter than the perianth-bristles; spikelets
    relatively narrow and pale; [var. *polyphyllus* (Boeck.) Beetle in part; B.C. (*see* map by
    Koyama 1963: fig. 2A, p. 1114) to SE James Bay, Que.] . . . . .var. *longispicatus* Britt.

S. atrocinctus Fern.

/sT/X/ (Hsr) Meadows and swamps from B.C. (Aldergrove, near Vancouver, and Revelstoke;
Herb. V, detd. Calder) to L. Athabasca (Alta. and Sask.), Man. (N to Reindeer L. at 57°48′N), Ont.
(N to Sandy L. at ca. 53°N), Que. (N to E James Bay at 53°40′N, L. St. John, and the Côte-Nord), SE
Labrador (Hamilton R. basin), Nfld., N.B., P.E.I., and N.S., s to Wash., Iowa, Mich., and N.J.
[Perhaps best merged with *S. cyperinus*; incl. *S. longii* Fern.].
   Forma *brachypodus* (Fern.) Blake (*S. cyperinus* var. *br.* (Fern.) Gilly; the inflorescence
consisting of 1 or few dense subglobose heads on reduced rays) occurs throughout the range.

S. atrovirens Willd.

/T/X/ (Hsr) Meadows and swampy ground, the aggregate species from Wash. to Alta. (near
Edmonton; Moss 1959), Sask. (Glen Ewan, near Elbow; A. J. Breitung, Am. Midl. Nat. 61: 511.
1959), Man. (N to Grand Beach, L. Winnipeg), Ont. (N to Renison, s of James Bay at ca. 51°N),
Que. (N to the Gaspé Pen.), Nfld., N.B., P.E.I., and N.S., s to Oreg., Ariz., N.Mex., Tex., Ark., Tenn.,
and Ga.
1 Leaves mostly less than 1.5 cm broad, scarcely nodulose; spikelets usually 2 or 3 mm
  long; scales dark green, becoming blackish; plant rather deep green . . . .
  . . . . .var. *georgianus* (Harper) Fern.
  2 Spikelets with normal flowers; [cent. Ont. to Nfld. and N.S.] . . . . .f. *georgianus*
  2 At least some of the spikelets reduced to leafy tufts; [sw Que., the type from
    Longueuil, near Montreal] . . . . .f. *viviparus* Vict.
1 Lower leaves to about 2 cm broad, they and their sheaths strongly nodulose when dry;
  spikelets usually over 4 mm long; plant pale green.
  3 Scales pale brown, becoming dark brown in age, 2 or 3 mm long, conspicuously
    subulate-tipped; [*S. pallidus* (Britt.) Fern.; Alta. to sw James Bay, Ont.] . . . .
    . . . . .var. *pallidus* Britt.

3 Scales dark green, becoming blackish in age, 1 or 2 mm long, barely subulate-tipped;
[Ont. to Que. and N.S.] . . . . .var. *atrovirens*

S. caespitosus L.
/aST/X/GEA/   (Hr)   Acidic bogs and peats and turfy tundra, the aggregate species from the
Aleutian Is. and coasts of Alaska and Dist. Mackenzie to cent. Dist. Keewatin, s Baffin Is.,
northernmost Ungava–Labrador, Nfld., N.B., P.E.I., and N.S., s to Oreg., Utah, Minn., Tenn., and
N.C.; w Greenland N to ca. 73°N, E Greenland N to ca. 70°N; Iceland; Eurasia. MAPS and synonymy:
see below.
1 Spikelets purple, to 8 mm long, with mostly more than 4 florets; perianth-bristles usually
upwardly barbed; upper sheaths scarious-margined; [*Baeothryon* Dietr.; *Eleocharis* Link;
Tricophorum Hartm.; Eurasia, reports from Alaska–Canada referring chiefly to var.
callosus; MAPS: Hultén, 1968*b*:206, and 1962: map 35 (*Tri. caesp.*), p. 43; Meusel,
Jaeger, and Weinert 1965:59] . . . . .var. *caespitosus*
1 Spikelets stramineous to brown, to 6 mm long, mostly 2–4-flowered; perianth-bristles
essentially barbless; upper sheaths coriaceous, barely scarious-margined.
    2 Culms rather soft and flexuous; basal sheaths dark grey to black; [Nfld.; E Que.
    (Gaspé Pen.), and N.S. (Inverness Co., Cape Breton Is.); MAP: on the above-noted
    1962 map by Hultén] . . . . .var. *delicatulus* Fern.
    2 Culms wiry, straightish; basal leaf-sheaths stramineous or drab; [ssp. *austriacus*
    (Palla) Asch. & Graebn.; N. American range of the species; MAPS (ssp. *austriacus*):
    Porsild 1957: map 67, p. 169; *Atlas of Canada* 1957: map 3, sheet 38; the N.
    American portions of the above-noted maps by Hultén and Meusel, Jaeger, and
    Weinert refer to this taxon combined with var. *delicatulus*] . . . . .var. *callosus*

S. cernuus Vahl
/t/W/EA/   (T)   Wet or marshy places near the coast from sw B.C. (Vancouver Is. and adjacent
mainland; CAN; V) to Calif. and N Mexico; S. America; Eurasia. [*S. riparius sensu* John Macoun
1888, not (R. Br.) Spreng.; incl. var. *californicus* (Torr.) Beetle].
        The very similar *S. setaceus* L. (*Isolepis* R. Br.) of Eurasia is known as introd. on Vancouver Is.,
B.C., where taken at Cadboro Bay by Malte in 1921 (collection in CAN, the basis of the report of
Hemicarpha micrantha from that locality by Eastham 1947). It may be distinguished from
S. cernuus as follows:
1 Floral bract shorter than or only slightly surpassing the usually solitary greenish or
reddish-brown spikelet; achenes smooth, reddish brown, dull . . . .*S. cernuus*
1 Floral bract much surpassing the usual cluster of 2 or 3 spikelets; achenes longitudinally
ribbed, dark brown, more or less lustrous . . . .*S. setaceus* L.

S. clintonii Gray
/T/(X)/   (Hs)   Clayey or slaty ledges, gravels, woods, and turfy shores from Alta. (Fernald *in*
Gray 1950; collection in CAN from near Edmonton, detd. Hermann) to Sask. (Meadow Lake,
54°08′N; DAO), Minn., Mich., Ont. (Timmins, 48°28′N; CAN), Que. (N to the Harricanaw R. SE of
James Bay at ca. 49°45′N and the Gaspé Pen.), and N.B. (Madawaska and Charlotte counties; not
known from P.E.I. or N.S.), s to w N.Y. and cent. Maine.

S. cyperinus (L.) Kunth
/T/EE/   (Hsr)   Wet meadows and swamps (ranges of Canadian taxa outlined below), s to Okla.
and N.C.
1 Involucels (secondary bracts at base of individual spikelet-clusters) and scales reddish
brown; [*Eriophorum* L.; *Trichophorum* Pers.; Man. (Boivin 1967*a*), Ont. (reported from
Bruce Co. by Krotkov 1940, and from Wellington Co. by Stroud 1941; a collection in OAC
from Rondeau Park, Kent Co., has also been placed here); reports from B.C. and
elsewhere in Canada probably refer to *S. atrocinctus*] . . . . .var. *cyperinus*
1 Involucels, scales, and bristles drab or greyish . . . . .var. *pelius* Fern.
    2 Spikelets all aggregated into 1 or more dense heads; [range of f. *pelius*] . . . .
    . . . . .f. *condensatus* (Fern.) Blake
    2 Spikelets terminating elongate branches of the inflorescence; [Ont. (N to the N shore

of L. Superior), Que. (N to the Gaspé Pen. and Magdalen Is.; reported, as *Eriophorum cyperinum*, from Tadoussac, Saguenay Co., by John Macoun 1888), Nfld., N.B., P.E.I., and N.S.; merging with the typical form southwards] . . . . .f. *pelius*

[S. *expansus* Fern.]
[This species of the E U.S.A. (Mich. to Maine, s to Ga.) is reported from s Ont. by Soper (1949) and from Que. by Marie-Victorin (1935; Ottawa Valley) and Raymond (1950*b*; St-Jérôme, N of Montreal). However, it is felt that its occurrence in our area should be confirmed by further collections. (S. *microcarpus* var. *bissellii* (Fern.) House). MAPS: Hultén 1958: map 158, p. 177; Meusel, Jaeger, and Weinert 1965:60.]

S. fluviatilis (Torr.) Gray   River-Bulrush
/T/X/eA/   (Hel)   Borders of lakes and streams (often calcareous) from Wash. to Alta. (N to near Edmonton at 53°20′N; CAN), Sask. (N to Pike Lake sw of Saskatoon), Man. (N to Arborg, about 70 mi N of Winnipeg), Ont. (N to L. Nipigon), Que. (N to the Rupert R. SE of James Bay at ca. 51°20′N), and s N.B. (Westfield and Burton; GH, detd. Fassett; not known from P.E.I. or N.S.), s to Calif., N.Mex., Kans., and Va.; E Asia (Marcel Raymond, Nat. can. (Que.) 88: 246. 1961). MAP (NE area): Fassett 1928: map 2, pl. 10.

S. hudsonianus (Michx.) Fern.
/ST/X/EA/   (Grh (Hel))   Wet meadows and gravels, peats, and marly swamps from s Alaska–Yukon to Great Slave L., L. Athabasca (Alta. and Sask.), Man. (N to Churchill), northernmost Ont., Que. (N to s Ungava Bay and the Côte-Nord), Labrador (N to Nain, 56°33′N), Nfld., N.B., and N.S. (not known from P.E.I.), s through B.C.–Alta. to ?Mont., Minn., and New Eng.; Eurasia. [*Eriophorum huds.* Michx., the type from Que., probably the L. Mistassini region according to Dutilly and Lepage 1945; *E. (Leucocoma; Trichophorum) alpinum* L.]. MAPS: Hultén 1968*b*:205, and 1962: map 36 (*E. alp.*), p. 43.

S. juncoides Roxb.
/t/EE/A/   (T)   Acid bogs and swamps and peaty or muddy shores from Minn. to s Ont. (Lambton, Waterloo, Lincoln, and Grey counties), N.Y., and Maine, s to Ala. and Ga.; Asia.
   The N. American plant may be distinguished as var. *williamsii* (Fern.) Koyama (S. *debilis* (*smithii*) var. *will.* Fern.; S. *purshianus* Fern.; perianth-bristles often absent (always present in the typical form), the spikelets averaging shorter and the midvein of the spikelet-scales generally less conspicuous than in the typical form). *See* Marcel Raymond (Nat. can. (Que.). 88: 246. 1961) and Koyama (1962).

S. lacustris L.
/ST/X/EA/   (Grh (Hel))   Marshes, shores, and pond-margins (ranges of Canadian taxa outlined below), s to Calif., N Mexico, Tex., Mo., Tenn., and Ga.; tropical America; Eurasia. MAPS and synonymy: *see* below.
1   Achenes compressed-trigonous, with an obtuse dorsal angle; stigmas 3; culms firm, not
    glaucous . . . . .ssp. *lacustris*
    2   Spikelets in groups of 2 or 3, sessile, elliptic-ovate to ovate; [Eurasia only, reports
        from Alaska–Canada referring to the following taxa; MAP: Hultén 1962: map 171,
        p. 183] . . . . .var. *lacustris*
    2   Spikelets solitary and pedicelled, lanceolate or lance-oblong; [S. *heterochaetus*
        Chase; SE Man. (Jessica L. in the Whiteshell Forest Reserve E of Winnipeg; Koyama
        1962), Ont. (reported from Aultsville, Dundas Co., by Koyama, and from Lambton Co.
        by Dodge 1915; a collection in TRT from near L. Nipissing has also been placed here),
        and Que. (N to the Montreal dist.; MT]. A hybrid with ssp. *glaucus* (× S. *oblongus*
        Koyama) is reported by Koyama from w Ont. (Ingolf) and Que. (Richelieu and
        Ste-Anne-de-Bellevue) . . . . .var. *tenuiculmis* Chase
1   Achenes plano-convex; stigmas 2; [transcontinental].
    3   Culms soft and easily compressed; scales about equalling the achenes, essentially
        glabrous except for the slightly pubescent tip and excurrent greenish midrib;
        [S. *(Schoenoplectus) val.* Vahl and its var. *creber* Fern., and f. *dutillyanus* Lepage

and f. *megastachyus* Fern. of the latter taxon; B.C. to Nfld., N.B., P.E.I., and N.S., rarely penetrating N to lat. 55°N; MAPS (*S. validus*): Hultén 1962: map 152, p. 161, and 1968b:208] . . . . ssp. *validus* (Vahl) Koyama

3   Culms firm; scales surpassing the achenes, more or less viscid-puberulent and conspicuously red-spotted, the relatively inconspicuous midrib extended as a short point; [*S. acutus* Muhl.; *S. lacustris* vars. *occidentalis* Wats. and *tabernaemontani* (Gmel.) Doell; *S. occidentalis* (Wats.) Chase; *S. tabernaemontani* Gmel.; s Alaska–cent. Yukon–cent. Dist. Mackenzie–SE Dist. Keewatin and B.C.–Alta.–Sask.–Man. to Ont. (N to Sandy L., ca. 53°N, 93°W), Que. (N to Old Factory, E James Bay, 52°37′N, L. Mistassini, the Gaspé Pen., and Magdalen Is.; reported from Anticosti Is. but not known from the Côte-Nord), Nfld., N.B., P.E.I. and N.S.; MAP: combine the maps by Hultén 1962: map 171 (*S. acutus*), p. 183 and map 152 (*S. tabern.*), p. 161] . . . . ssp. *glaucus* (Sm.) Hartm.

S. lineatus Michx.

/T/X/eA/   (Hsr)   Damp meadows and thickets from Oreg. (an isolated area) and Iowa to s Ont. (N to the Ottawa dist.; Gillett 1958), sw Que. (Iberville and Missisquoi counties; MT) and Maine, s to Tex., Ala., and Va.; E Asia (ssp. *wichurai*).

S. maritimus L.

/ST/X/EA/   (Hel (Grh))   Saline marshes and shores and alkaline marshes inland, the aggregate species from B.C. (N to Williams Lake; CAN; an isolated station at Anchorage, s Alaska; reported from Dist. Mackenzie by Boivin 1967a) to northernmost Alta., Sask. (N to Prince Albert), Man. (N to The Pas), Ont. (known only from the sw James Bay watershed), Que. (SE James Bay watershed; St. Lawrence R. estuary from l'Islet Co. to the Côte-Nord, Anticosti Is., and Gaspé Pen.), N.B., P.E.I., and N.S., s to Baja Calif., Mexico, Tex., Okla., Mo., Minn., N.Y., and Va.; Eurasia. MAPS and synonymy: *see* below.

1   Spikelets lanceolate, usually less than 6 mm thick, in clusters on well-elongated umbel-rays; [introd. in Que.; Koyama 1962] . . . . var. *tuberosus* (Desf.) R. & S.

1   Spikelets ovoid, usually over 6 mm thick.
  2   Spikelets blunt-tipped, relatively pale brown; [*S. paludosus* Nels. and its var. *atlanticus* Fern.; *S. campestris* var. *pal.* (Nels.) Fern.; *S. pacificus* And.; Alaska–Dist. Mackenzie–B.C. to N.S.; MAPS: Hultén 1968b:208, and 1962: map 169 (*S. paludosus* and its var. *altanticus*), p. 181] . . . . var. *paludosus* (Nels.) Kük.
  2   Spikelets acute-tipped, commonly deeper brown . . . . var. *maritimus*
    3   Inflorescence congested; spikelets to about 2.5 cm long; achenes mostly lenticular; [incl. var. *fernaldii* (Bickn.) Beetle (*S. fern.* Bickn.; *S. campestris* var. *fern.* (Bickn.) Bartl.) and its f. *agonus* Fern.; E Que. to N.B., P.E.I., N.S., and ?Nfld.; MAP: Hultén 1962: map 169 (*S. maritimus* and its var. *fernaldii*), p. 181] . . . . . f. *maritimus*
    3   Inflorescence open; spikelets usually less than 1.5 cm long; achenes mostly compressed-trigonous; [introd. in the U.S.A. but not yet reported from Canada] . . . . f. *cymosa* (Rchb.) Koyama

S. microcarpus Presl

/sT/X/eA/   (Grh (Hel))   Damp meadows, marshes, and thickets from cent. Alaska and B.C. to Great Slave L., L. Athabasca (Alta. and Sask.), Man. (N to Reindeer L. at 57°16′N), Ont. (N to the Fawn R. at ca. 54°30′N, 88°30′W), Que. (N to E James Bay at ca. 53°40′N, the Côte-Nord, Anticosti Is., and Gaspé Pen.), SE Labrador (N to the Hamilton R. basin), Nfld., N.B., P.E.I., and N.S., s to Calif., N.Mex., Nebr., Mich., and W.Va.; E Asia (Kamchatka). [*S. lenticularis* Torr.; *S. rubrotinctus* Fern.; *S. sylvaticus* var. *digynus* Boeck.]. MAP: Hultén 1968b:209.

S. nevadensis Wats.

/T/WW/   (Hsr)   Wet ground (usually alkaline) from B.C. (N to Cariboo, 52°51′N) to s Alta. (Milk River and s of Manyberries), Sask. (N to Redberry Lake, 52°43′N; Breitung 1957a), and Man. (Boivin 1967a), s to Calif., Utah, and N.Dak.

S. olneyi Gray
/T/X/ (Hel) Brackish or saline marshes from Oreg. to Idaho, Mo., N Ohio, w N.S. (several localities in Yarmouth Co.; GH; CAN), and s N.H., s to Calif., Mexico, and Fla., W.I.; S. America. [*Schoenoplectus* Palla].

S. peckii Britt.
/T/E/ (Hsr) Meadows and bogs from sw Que. (Brome, Laprairie, Terrebonne, Labelle, and Argenteuil counties) to N.Y., Maine, and Conn.
    According to A. E. Schuyler (Rhodora 63(753): 237–43. 1961), this species is probably a hybrid between *S. atrovirens* and either *S. atrocinctus* or *S. pedicellatus*.

S. pedicellatus Fern.
/T/EE/ (Hsr) Shores and wet thickets from Minn. to Ont. (N to Kapuskasing, 49°24′N; CAN), Que. (N to L. St. John and the Gaspé Pen.; GH), N.B., and N.S. (not known from P.E.I.), s to Iowa, Ohio, and N.J.
    Collections from Ont. and Que. have been referred to the nondescript var. *pullus* Fern., this approaching *S. atrocinctus* in characters and perhaps a hybrid between the two species.

S. pumilus Vahl
/T/D/EA/ (Hsr) Calcareous ledges, gravels, bogs, and shores: sw Yukon–Dist. Mackenzie (CAN), N B.C. (Mi 397, Alaska Highway; CAN and V, detd. Porsild), mts. of sw Alta. (Jasper; Banff; Laggan; Morley), and w Sask. (Sutherland, near Saskatoon; A. A. Beetle, Am. Midl. Nat. 41(2): 483. 1949), and reported from the mts. of Colo.; E Que. (Knob Lake dist. at ca. 54°50′N; Mingan Is. of the Côte-Nord; Anticosti Is.); Eurasia. [*Baeothyron* Löve & Löve; *Trichophorum* Schinz & Thell.; *S. alpinus* Schleich., not L.]. MAPS: Hultén 1968b:206, and 1958: map 227 (*T. pum.*), p. 247; Porsild 1966: map 32 (*S. rollandii*), p. 70; Marcel Raymond, Contrib. Inst. Bot. Univ. Montréal 70: fig. 1 (neither this nor Hultén's 1958 map indicating the Yukon–Dist. Mackenzie stations), p. 87. 1957.
    The N. American plant may be separated with difficulty as ssp. *rollandii* (Fern.) Raymond (*S. rol.* Fern., the type from the Mingan Is., E Que.; lowest scale of the spikelet with the midrib included rather than prolonged as a distinct mucro; achenes averaging slightly shorter but broader than those of the typical form).

S. robustus Pursh
/T/D (coastal)/ (Grh (Hel)) Brackish or saline marshes near the coast: sw ?B.C. (collections in V from Vancouver Is. have been placed here, as well as from inland at Kamloops and Okanagan, all perhaps referable to *S. maritimus* var. *paludosus*; Koyama (1962) assigns it a western range from Calif. to Mexico and S. America, this followed by Hultén in his below-noted map); N.S. (Koyama 1962; Bridgetown, Annapolis Co.; CAN, detd. Boivin) to Fla. MAP: Hultén 1962: map 169, p. 181.

[S. rubricosus Fern.]
[Reports of this species of the E U.S.A. (N to Ill. and Mass.) from s Ont. by Stroud (1941) and Soper (1949) require confirmation. (*S. eriophorum* Michx., illegitimate name; a very obscure species, perhaps best merged with *S. cyperinus*).]

S. rufus (Huds.) Schrad.
/ST/(X)/EA/ (Grh) Saline or brackish marshes (rarely in fresh peat): s Alaska (Matanuska; Hultén 1950) and sw Dist. Mackenzie (Porsild and Cody 1968); Sask. (Pike Lake and Sutherland, near Saskatoon; Breitung 1957a) and Man. (Red Deer R., L. Winnipegosis; York Factory; Churchill); James Bay (Ont. and Que.); E Que. (St. Lawrence R. estuary from l'Islet Co. to the Côte-Nord, Anticosti Is., Gaspé Pen., and Magdalen Is.), Nfld., N.B., P.E.I., and N.S. (not known in the U.S.A.); Eurasia. [*Schoenus* Huds.; *Blysmus* Link; incl. var. *neogaeus* Fern., with somewhat narrower and longer achenes than the typical form]. MAPS: Hultén 1968b:209, and 1962: map 78, p. 87; Meusel, Jaeger, and Weinert 1965:61; Schofield 1959: map 7, p. 119.

S. smithii Gray
/T/EE/eA/ (T) Sandy, peaty, or muddy shores (often tidal) from Minn. to s Ont. (Norfolk,

Waterloo, York, and Dundas counties and the Muskoka Dist. N of L. Huron) and Que. (N to St-Vallier, Montmagny Co.; CAN), s to Ill., Ohio, and Va.; E Asia (var. *leiocarpus* (Kom.) Koyama).

Forma *levisetus* (Fassett) Fern. (perianth-bristles smooth and elongate rather than wanting or smooth but rudimentary as in the typical form) is known from Que. (Portneuf, Québec, Lévis, Bellechasse, and Montmagny counties). Forma *setosus* Fern. (perianth-bristles retrorsely-barbed and elongate) is also known from Que. (Pontiac and Lotbinière counties).

S. subterminalis Torr.

/T/X/ (Hel) Ponds, deadwaters, quaking bogs, and peaty shores (ranges of Canadian taxa outlined below), s to Oreg., Mont., Mo., Ill., Ind., Mich., and Ga. MAP and synonymy: *see* below.

1   Culms terete, mostly submersed; spikelet solitary (rarely 2); [SE Alaska (near Wrangell) and B.C. (Vancouver Is.; Goose Is.; Revelstoke; Columbia Valley); Ont. (N to Oba L., s of Hearst, ca. 49°N, 84°W), Que. (N to Rimouski Co.; RIM; MT), Nfld., N.B., P.E.I., and N.S.; MAP: Hultén 1968*b*:207] . . . . .var. *subterminalis*

1   Culms 3-angled, emerged; spikelets 2–15; [*S. torreyi* Olney; s Man. (Victoria Beach, L. Winnipeg, and Delta, at the s end of L. Manitoba; WIN), Ont. (N to Quetico Park, about 100 mi w of Thunder Bay), Que. (N to Portneuf Co.), and N.B. (Nerepis, Kings Co.; Fox Creek L., Westmorland Co.)] . . . . .var. *cylindricus* (Torr.) Koyama

SCLERIA Bergius [515] Nut-rush

1   Achene smooth and shining, about 2.5 mm long; hypogonium (disk beneath achene) crustaceous, minutely pebbled; flower-clusters solitary or 1 or 2 lateral ones on erect peduncles, the terminal one often with 3 clusters of spikelets; leaves to 9 mm broad, firm, scabrous, their sheaths and midribs beneath sometimes pilose; culms coarse, to about 1 m tall, from a large, knotted and forking, woody rhizome; (s Ont.) . . . .*S. triglomerata*

1   Achene rough; flower-clusters sessile; leaves at most 3 mm broad; culms capillary, rarely over 6 cm tall.

  2   Achene about 3 mm long, transversely warty-papillate; hypogonium bearing 6 rounded tubercles; flower-clusters usually solitary, subtended by a leafy bract to 5 cm long; leaves to 3 mm broad, glabrous or sparingly and minutely hairy; perennial from an elongated woody forking rhizome . . . .[*S. pauciflora*]

  2   Achene about 1.5 mm long, transversely warty or ridged; hypogonium obsolete or low and stalk-like; spikelets 2 or more in setaceous-bracted capitate clusters in a greatly interrupted spike to 8 cm long; leaves at most 2 mm broad, soft, their sheaths usually long-villous; tufted fibrous-rooted annual; (s Ont.) . . . .*S. verticillata*

[*S. pauciflora* Muhl.]
[The report of this species of the E U.S.A. (N to Kans. and N.J.) from the delta of the St. Clair R. in Lambton Co., s Ont., by Dodge (1915; presumably the basis of the listing by Soper 1949) is probably based upon one of the following species. Relevant collections were not found in Herb. MICH, where Dodge's main Lambton Co. collection is housed, nor is the species cited from Ont. by E. L. Core (Brittonia 2: 1–105. 1936) in his monograph of the genus.]

S. triglomerata Michx.

/t/EE/ (Grh) Open or lightly wooded sandy grounds from Kans. to s Minn., Ohio, s Ont. (Essex, Lambton, Middlesex, and York counties), N.Y., and Mass., s to Tex. and Fla. MAP (s Ont.): Soper 1962: fig. 5, p. 11.

S. verticillata Muhl.

/T/EE/ (T) Calcareous shores, wet rocks, and bogs from Minn. to s Ont. (Lambton, Bruce, Northumberland, Hastings, and Leeds counties) and NW Conn., s to Tex. and Fla. MAP (s Ont.): Soper 1962: fig. 5, p. 11.

ARACEAE (Arum Family)

(Ref.: Marie-Victorin 1931)

Scapose herbs of moist or wet places. Leaves all basal or sub-basal, usually simple (compound in *Arisaema*), entire or merely undulate, their petioles sheathing at base. Flowers small, unisexual or perfect, crowded over all or part of a more or less fleshy spadix (spike with a fleshy axis), this subtended by and often partly or wholly enclosed in a leaf-like or much modified flat or marginally inrolled spathe (leaf-like bract). Perianth none or of 4 or 6 segments. Stamens 2–6. Ovary superior but more or less embedded in the spadix. Fruit commonly fleshy and berry-like.

1   Leaves compound, long-petioled, with 3–15 elliptic- or ovate-lanceolate leaflets; summit of spathe open and curved over the spadix, the summit of the latter not flower-bearing; plants from corms; (SE Man. to N.S.) . . . . *Arisaema*
1   Leaves simple; spadix flower-bearing nearly or quite to summit.
  2   Leaves narrow and sword-like; spathe green and resembling a foliage-leaf, to about 8 dm long; spadix to 9 cm long and 2 cm thick, diverging laterally from the 3-angled scape, this from stout rhizomes; (B.C. to N.S.) . . . . *Acorus*
  2   Leaves broad; spathe clearly differentiated, fleshy or petaloid; spadix terminal.
    3   Spathe finally open and flattish, exposing the spadix; leaves long-petioled, they and the scapes from fleshy thick rhizomes.
      4   Leaves cordate-ovate to -rotund, to about 1 dm long and broad; spathe ovate, to 6 cm long, abruptly long-acuminate, the upper surface white; spadix to 2.5 cm long, on a short thick stipe, shorter than the spathe; rhizome elongate, rooting at the nodes; (transcontinental) . . . . *Calla*
      4   Leaves elliptic to lance-oblong, acute at both ends, normally to about 7 dm long and 2.5 dm broad (often much larger); spathe yellowish, oblong-lanceolate, acute, to about 1.5 dm long; spadix to about 12 cm long and 6 cm thick; rhizome short; (Alaska–B.C.) . . . . *Lysichitum*
    3   Spathe permanently enveloping or partly enclosing the spadix, green and often striped or mottled with brown or purplish-brown.
      5   Leaves short-petioled, ovate-cordate, to about 6 dm long; spathe ovoid-shell-shaped, its margins inrolled; spadix ovoid to globose; plant from a stout erect rhizome; (SE Man. to N.S.) . . . . *Symplocarpus*
      5   Leaves long-petioled, typically ovate-hastate to broadly triangular-hastate, to over 2 dm long; spathe green with pale or whitish margins, leathery, tightly convolute throughout, to 2 dm long, the spadix nearly as long; plant from a thick fibrous or subtuberous root; (s Ont. and sw Que.) . . . . *Peltandra*

ACORUS L.   [694]   Sweetflag, Flagroot

A. calamus L.   Sweetflag.   Belle-Angélique or Rédote

/sT/X/EA/   (Hel)   Swamps and shallow water from Great Slave L. and B.C. (Shuswap L., NE of Kamloops; Eastham 1947) to Alta. (N to Fort Saskatchewan), Sask. (N to Windrum L. at ca. 56°N), Man. (N to McBride L. at 56°52′N), Ont. (N to w James Bay at ca. 53°N), Que. (N to Matane Co., Gaspé Pen.; introd. on Anticosti Is. according to Marie-Victorin and Rolland-Germain 1969), N.B., P.E.I., and N.S., s to Wash., Idaho, Mont., Tex., and Fla.; Eurasia (partly introd.; incl. several closely related taxa). MAPS: Hultén 1968b:280, and 1962: map 212, p. 223.

ARISAEMA Mart.   [786]   Jack-in-the-pulpit

1   Leaf usually solitary, pedately divided into up to 15 very unequal elliptic to oblanceolate leaflets; spadix tapering to a long slender point exserted up to over 1 dm beyond the oblong convolute-pointed summit of the greenish spathe; (s Ont. and sw Que.) . . . .
. . . . *A. dracontium*
1   Leaves usually 2, palmately divided into 3 (rarely 5) subequal elliptic to rhombic-ovate leaflets; spadix blunt, cylindric or clavate, included; summit of the green or purplish spathe flat, ovate, usually reflexed over the spadix.

2   Spathe conspicuously whitish-ridged lengthwise (especially at the tubular base), the tube passing by a gradual curve into the narrowly ovate attenuate hood; fruiting head to about 3.5 cm long; (Que. and the Maritime Provinces) . . . .*A. stewardsonii*
2   Spathe at most only shallowly ridged, the summit of the tube more or less flanged.
   3   Lateral leaflets strongly oblique at base; tube of spathe to over 7 cm long, the summit-flange to 8 mm broad; hood broadly oblong-ovate, abruptly acuminate, to about 6 cm broad; fruiting head usually over 3 cm long; (SE Man. to N.S.) . . . .
   . . . .*A. atrorubens*
   3   Lateral leaflets with sides subequally narrowed to base; tube of spathe less than 5 cm long, the summit-flange at most 2 mm broad; hood narrowly oblong to narrowly ovate, attenuate, to about 3 cm broad; fruiting head to about 2 cm long; (E U.S.A. only) . . . .*[A. triphyllum]*

A. atrorubens (Ait.) Blume   Jack-in-the-pulpit.   Petit prêcheur
/T/EE/   (Gst)   Rich woods and thickets from SE Man. (Carman; Dufferin; Emerson; Winnipeg) to Ont. (N to the N shore of L. Superior), Que. (N to Chicoutimi, E of L. St. John, and the mouth of the Grand Cascapédia R., Gaspé Pen.), N.B., and N.S. (according to D. S. Erskine 1960, early reports of *A. triphyllum* from P.E.I. are based upon *A. stewardsonii*), s to E Kans., Mo., and S.C. [*Arum* Ait.; *A. triphyllum* of Canadian reports, not (L.) Schott]. MAP: Marie-Victorin 1931: fig. 2 (*A. tri.;* the Canadian area applies here), p. 18.
    The typical form has the spathe and spadix rather uniformly purple throughout. Other forms reported from Canada are f. *viride* (Engl.) Fern. (spathe and spadix green throughout) and f. *zebrinum* (Sims) Fern. and f. *pallescens* (Sims) Raymond, these last two with the spathe longitudinally whitish-striped within (but f. *zebrinum* with an otherwise purple to bronze spathe and a dark-purple spadix, f. *pallescens* with a pale spathe, the spadix reddish only at summit).

A. dracontium (L.) Schott   Green Dragon, Dragon-root
/T/EE/   (Gst)   Rich or alluvial woods and thickets from Wisc. and Mich. to s Ont. (N to Huron and Waterloo counties), sw Que. (islands of the St. Lawrence R. between Montreal and Sorel; *see* Que. map by Robert Joyal, Nat. can. (Que.) 97(5): map G, fig. 1, p. 562. 1970), Vt., and N.H., s to Tex. and Fla. [*Arum* L.]. MAP: Marie-Victorin 1931: fig. 8, p. 27.

A. stewardsonii Britt.   Jack-in-the-pulpit
/T/EE/   (Gst)   Wet or swampy woods and thickets from Minn. to Que. (N to Grosse-Ile, NE of Quebec City in Montmagny Co.; Marcel Raymond, Contrib. Inst. Bot. Univ. Montréal 64: 48. 1949), N.B., P.E.I., and N.S., s to Pa. and N.C. [*A. triphyllum* var. *st.* (Britt.) Stevens]. MAPS: Marie-Victorin 1931: fig. 5 (incomplete northwards), p. 23; Marcel Raymond, Contrib. Inst. Bot. Univ. Montréal 64: fig. 1, p. 45. 1949.

[*A. triphyllum* (L.) Schott]
[Following the treatment by Fernald *in* Gray (1950), reports of this species of the E U.S.A. (N to N.Y. and Mass.) from Canada are referable to *A. atrorubens* (or perhaps in small part to *A. stewardsonii*). *See* M. L. Fernald (Rhodora 42(499): 247–54. 1940), Marcel Raymond (Contrib. Inst. Bot. Univ. Montréal 64: 43–50. 1949), D. G. Huttleston (Bull. Torrey Bot. Club 76(6): 407–13. 1949), and Gleason (1958).]

CALLA L.   [710]   Water-Arum

C. palustris L.   Wild Calla
/ST/X/EA/   (Hel)   Wet bogs, pond-margins, and ditches from N-cent. Alaska and s-cent. Yukon (*see* Hultén 1942: map 324, p. 412) to the Mackenzie R. Delta, Great Slave L., Sask. (N to L. Athabasca), Man. (N to the Seal R. at ca. 59°N), Ont. (N to Big Trout L. at ca. 54°N), Que. (N to E James Bay at ca. 53°N, the Côte-Nord, and Gaspé Pen.; not known from Anticosti Is.), s Labrador (N to the Hamilton R. basin), Nfld., N.B., P.E.I., and N.S., s to Colo., Tex., and Fla.; Eurasia. MAPS: Hultén 1968b:281, and 1962: map 76, p. 85; Meusel, Jaeger, and Weinert 1965:80; Lepage 1966: map 11 (Ont. eastwards), p. 220.
    Forma *polyspathacea* Vict. & Rousseau (the flowering branches bearing 2 or 3 spathes rather

than a soiltary one) is known from the type locality, Lanoraie, Que., and probably occurs throughout the range.

LYSICHITUM Schott   [707]   Skunk-cabbage

L. americanum Hult. & St. John   Western Skunk-cabbage
/sT/W/   (Grh)   Swamps and wet woods from s Alaska (*see* Hultén 1942: map 323, p. 412) through B.C. (E to the Columbia Valley) to Calif., Mont., Idaho, and ?Wyo. (the closely related *L. camtschatcense* (L.) Schott in E Asia). [*Symplocarpus (L.) kamtschaticus sensu* Bongard 1833, Hooker 1838, and other American auth., not *Dracontium kamtschatcense* L., basionym]. MAPS: Hultén 1968b:281; Eric Hultén, and Harold St. John, Sven. Bot. Tidskr. 25(4): fig. 5, p. 461. 1931; Marie-Victorin 1931: fig. 13 (*L. camt.*), p. 35.

PELTANDRA Raf.   [747]   Arrow-Arum

P. virginica (L.) Schott & Endl.   Tuckahoe.   Peltandre
/T/EE/   (Hel)   Swamps and shallow water from s Ont. (Welland, Prince Edward, and Frontenac counties) and sw Que. (Chateauguay, Yamaska, and Richelieu counties) to N N.Y. and N.H., s to Tex. and Fla. [*Arum* L.; *P. undulata* Raf.].

SYMPLOCARPUS Salisb.   [708]   Skunk-cabbage.   Symplocarpe

S. foetidus (L.) Nutt.   Eastern Skunk-cabbage.   Tabac du Diable or Chou puant
/T/EE/eA/   (Grh)   Wet meadows and swampy woods from SE Man. (relevant collection in Herb. Man. Prov. Mus., Winnipeg, from Winnipegosis, E of Duck Mt., where taken by Mrs. E. J. McMillan in 1924, confirming the report from SE Man. by Lowe 1943) to Ont. (N to Michipicoten Is., L. Superior; John Macoun 1888; *see* s Ont. map by J. H. Soper and A. S. Rao, Bull. Fed. Ont. Nat. 79: 18. 1958), Que. (N to Pointe-au-Père, Rimouski Co.; MT), N.B., and N.S. (not known from P.E.I.), s to Iowa, Tenn., and Ga.; a variety in E Asia. [*Dracontium* L.; *Pothos* Michx.; *Spathyema* Raf.]. MAPS: Polunin 1960: fig. 51, p. 190; Marie-Victorin 1931: fig. 11, p. 33; Fernald 1918b: map 12, pl. 12, and 1929: map 3, p. 1488.

[ORONTIUM L.]   [709]   Golden Club

[*O. aquaticum* L.]
[This species of the E U.S.A. (N to cent. N.Y. and Mass.; not keyed out above) is assigned the range "Canada to Florida" by A. Michaux (1803) and Pursh (1814), undoubtedly because of the vague political boundaries of that day. Its spathe consists merely of a leaf-sheath investing the lower part of the slender scape and bearing a small bract-like blade at summit. It has simple, oblong, entire leaves. The MAP by Braun (1937: fig. 11, p. 197) indicates no Canadian stations.]

LEMNACEAE (Duckweed Family)

Small free-floating or submersed aquatic plants consisting of a flat or rounded body (thallus) bearing the flowers in small marginal or dorsal pouches. Flowers unisexual, lacking a perianth, the 1 or 2 staminate flowers consisting of a single anther on a short filament, the solitary pistillate flower consisting of a single pistil. Fruit a utricle.

1 Roots none; thallus-body ellipsoid to globose, at most 1.5 mm long; reproductive pouch solitary; (Ont. and sw Que.) ....*Wolffia*
1 Roots present; thallus-body usually larger; reproductive-pouches 2; (essentially transcontinental).
  2 Root solitary; thallus green, with not more than 5 nerves ....*Lemna*
  2 Roots 2 or more; thallus commonly with more than 5 nerves, to 8 mm long, dark green above, purple beneath ....*Spirodela*

LEMNA L.   [795]   Duckweed, Duck's-meat.   Lenticule

1 Fronds elliptic-obovate to round-ovate, sessile, at most 5 mm long, quickly becoming detached and floating on the surface ....*L. minor*
1 Fronds oblong or oblong-lanceolate, long-stalked, to 1 cm long, remaining attached in a zigzag chain, mostly entirely submersed ....*L. trisulca*

L. minor L.   Duckweed.   Lentille d'eau or Merde de grenouille
/ST/X/EA/   (HH)   Floating on quiet waters from N-cent. Alaska, s Yukon (CAN), and Great Slave L. to L. Athabasca (Alta. and Sask.), Man. (N to Churchill), Ont. (N to w James Bay at ca. 53°30′N), Que. (N to the Côte-Nord and Anticosti Is.; not known from the Gaspé Pen.), s Nfld., N.B., P.E.I., and N.S., s to s Calif., Mexico, Tex., and Fla.; Eurasia. MAPS: Hultén 1968b:282, and 1962: map 121, p. 130; Meusel, Jaeger, and Weinert 1965:81; E. H. Daubs, Ill. Biol. Monogr. 34: 71. 1965; Porsild 1966: map 33, p. 71.

L. trisulca L.   Star-Duckweed
/ST/X/EA/   (HH)   Ponds and stream-margins from cent. Alaska, s Yukon, and the Mackenzie R. Delta to Great Bear L., Great Slave L., N-cent. Sask., Man. (N to Churchill), northernmost Ont., Que. (N to E James Bay at 52°37′N and Ste-Blandine, Rimouski Co.; not known from Nfld.), N.B., P.E.I., and N.S., s to Calif., Mexico, Tex., and Fla.; Uruguay; Eurasia. MAPS: Hultén 1968b:282, and 1962: map 122, p. 131; Porsild 1966: map 34, p. 71; E. H. Daubs, loc. cit., p. 69.

SPIRODELA Schleid.   [794]

S. polyrhiza (L.) Schleid.   Water-flaxseed.   Lentille d'eau
/T/X/EA/   (HH)   Ponds and stream-margins from s B.C. (Vancouver Is.; Agassiz; Shuswap L.; Sicamous) to Alta. (near Edmonton; High Prairie), Sask. (N to Meadow Lake, 54°08′N), Man. (N to Cross L., NE of L. Winnipeg), Ont. (N to New Liskeard, 47°31′N), Que. (N to L. Timiskaming at ca. 47°30′N), P.E.I., and N.S. (not known from N.B.), s to tropical America; Eurasia. [*Lemna* L.; *Salvinia ?natans sensu* Hooker 1840, not L.]. MAPS: Hultén 1962: map 127, p. 137; E. H. Daubs, Ill. Biol. Monogr. 34: 63. 1965.

WOLFFIA Horkel   [796]   Water-meal

1 Thallus not dotted, bearing about 3 papillae, the upper surface rounded and mostly under water; (Ont. and sw Que.) ....*W. arhiza*
1 Thallus brown-dotted, with a single papilla, the upper surface chiefly above water.
  2 Upper surface of thallus flattish, the elevated apical end prolonged into a terminal papilla; (Ont.) ....*W. punctata*
  2 Upper surface of thallus low-conical, with a prominent papilla near the centre ....[*W. papulifera*]

W. arhiza Wimm.

/T/EE/ (HH) Floating on quiet waters from Minn. (also reported from N Oreg. by Hitchcock et al. 1969) to Ont. (N to the Ottawa dist.; Gillett 1958; *see* s Ont. map by Soper 1962: fig. 7 (*W. col.*), p. 14) and sw Que. (N to the Montreal dist.), s to Tex., Fla., and tropical America. [*W. (Bruniera) columbiana* Karst.]. MAPS: E. H. Daubs, III. Biol. Monogr. 34: 93. 1965; W. G. Dore, Can. Field-Nat. 71(1): fig. 3 (*W. col.*), p. 13. 1957.

[W. papulifera Thompson]

[This species of the E U.S.A. (N to Ohio and Va.; perhaps inseparable from *W. punctata*) is reported by W. G. Dore (loc. cit.) from as far north as the shore of L. Erie E of Cleveland, Ohio, and should be searched for along the Ont. shores of that lake. MAP: E. H. Daubs, loc. cit., p. 95.]

W. punctata Griseb.

/T/EE/ (HH) Floating on quiet waters from Minn. (also reported from Wash. and Oreg. by Hitchcock et al. 1969) to Ont. (N to the Ottawa dist.; Gillett 1958; *see* s Ont. map by Soper 1962: fig. 8, p. 15; according to W. G. Dore, loc. cit, the report from Verdun, Que., by Wynne-Edwards (hence, presumably, the report from the Montreal dist. by Raymond 1950*b*:114) is based upon *W. (columbiana) arhiza*), s to Tex. and Fla.; W.I. [*Bruniera* Nieuwl.; *W. braziliensis sensu* John Macoun 1888, not Wedd.]. MAP: E. H. Daubs, loc. cit., p. 97.

XYRIDACEAE (Yellow-eyed Grass Family)

XYRIS L.   [826]   Yellow-eyed Grass

Herbs of wet peaty or sandy soil. Leaves in a basal cluster, narrow and grass-like, sheathing the base of the naked scape, this terminated by a solitary, broadly lanceolate to subglobose head-like spike to about 2 cm long. Flowers perfect, yellow, small, hypognyous. Sepals 3, the 2 lateral ones keeled and boat-shaped, the third broad and deciduous with the corolla. Petals 3. Stamens 3, opposite the petals and adnate to them. Ovary superior. Fruit a many-seeded 3-valved capsule.

1   Lateral sepals with essentially entire keels, their tips projecting beyond the tips or sides of
    their subtending bracts; mature spikes to 1 cm long and 7 mm thick; leaves to 2.5 mm
    broad, soft; scapes filiform, wiry, readily disarticulating at base, rarely over 4 dm tall; (Ont.
    to Nfld. and N.S.) . . . . .X. montana
1   Lateral sepals with distinctly erose, toothed, or ciliolate keels, their tips completely
    covered by the subtending bracts; scapes to 8 or 9 dm tall, not readily disarticulating at
    base.
    2   Lateral sepals with a minutely ciliate keel and usually also a terminal tuft of hairs;
        spikes to 18 mm long; leaves to 4 mm broad; scapes bulbous, the hard subglobose
        bulbs to 1 cm thick, commonly clustered in tufts of up to 50 . . . .[X. torta]
    2   Lateral sepals with keel toothed or erose (but not ciliate) above the middle; spikes less
        than 1.5 cm long; leaves to over 6 mm broad; scapes not bulbous-based; (Ont., N.B.,
        and N.S.) . . . . .X. caroliniana

X. caroliniana Walt.
/T/EE/   (Hr(r))   Wet peaty or sandy soils from Mich. to Ont. (N to Timagami, ca. 47°N; CAN, detd. Soper; Baldwin 1958; not known from Que.), N.B. (Charlotte Co., where taken by Vroom in 1881 and distributed as *X. flexuosa* var. *pusilla,* revised by C. A. Weatherby; NBM), and N.S. (not known from P.E.I.), s to Tex. and Fla.; tropical America. [*X. canadensis* Schnizl.; *X. bulbosa sensu* Lindsay 1878, and probably other early Canadian reports, not Kunth, a relevant collection from N.S. in NBM].

X. montana Ries
/T/EE/   (Hr(r))   Wet peaty or sandy soils from N Mich. to Ont. (N to the Timagami Forest Reserve NE of Sudbury; CAN), Que. (N to Chicoutimi Co. at 48°26′N and Magdalen Is.), Nfld. N.B., and N.S. (not known from P.E.I.), s to Pa. and N.J. [*X. flexuosa* var. *pusilla* Gray]. MAP: *Atlas of Canada* 1957: map 14, sheet 38.
    Forma *albiflora* Boivin (petals white rather than yellow) is known from the type locality, Shelburne, Shelburne Co., N.S.

[*X. torta* Sm.]
[The map by Cain (1944: fig. 37, p. 262) for this species of the E U.S.A. (N to Minn. and N.H.) indicates an occurrence along the shores of L. St. Clair in Minn. and of L. Ontario in N.Y. It should be searched for in the Great Lakes region of Ont. *Woodwardia areolata* is also indicated in the above two localities on Cain's map and he notes that, "this is also a Coastal Plain species which, like *X. torta* and a number of other species, found its way to the coastal plain and the Lakes from the old uplands".]

ERIOCAULACEAE (Pipewort Family)

ERIOCAULON L.   [828]   Pipewort

Scapose aquatics of shallow water, tidal flats, and muddy shores. Leaves linear-attenuate, very thin and often pellucid, commonly less than 1 dm long, all in a dense basal rosette. Scape filiform, angled, naked, terminated by a solitary button-shaped head of very small unisexual hypogynous flowers. Perianth consisting of 4 scarcely differentiated segments, the stamens as many or half as many. Style 1. Stigmas 2. Ovary superior. Fruit a 2-seeded capsule.

1   Mature heads hemispheric, the involucre scarcely reflexed; perianth and chaff of receptacle glabrous or very sparingly pubescent; scape obscurely 4-angled, commonly less than 1 dm tall; leaves at most about 6 cm long and 2 mm broad, lacking conspicuous cross-veinlets; (Que.) . . . .*E. parkeri*
1   Mature heads subglobose, nearly concealing the reflexed involucre; perianth and chaff of receptacle fringed at summit with abundant white hairs; scape with up to 7 striations, commonly over 2 dm tall; leaves to about 2.5 dm long and 4 mm broad, with conspicuous cross-veinlets; (Ont. to sE Labrador, Nfld., and N.S.) . . . .*E. septangulare*

E. parkeri Robins.
/T/E/   (Hel (HH))   Tidal (rarely fresh) mud and estuaries of Que. (Ottawa R. near Hull; Témiscamie R. E of L. Mistassini at ca. 51°N (type locality of *E. rollandii*); St. Lawrence R. estuary from L. St. Peter to Portneuf and Bellechasse counties) and along the coast from Maine to Va. [*E. septangulare* var. *park.* (Rob.) Boivin & Cayouette; incl. the dwarf extreme, *E. rollandii* Rousseau]. MAPS: Raymond 1950*b*: fig. 38, p. 105; M. L. Fernald, Rhodora 42(502): map 17, p. 378. 1940; Fassett 1928: fig. 3, pl. 11.

E. septangulare With.   White-buttons, Duckgrass
/sT/EE/E/   (Hel (HH))   Shallow pools and streams and muddy or peaty shores from Ont. (N to near Graham, NW of Thunder Bay at 49°14′N; CAN) to Que. (N to Duncan L. near James Bay at 53°33′N, the Côte-Nord, and Gaspé Pen.), s Labrador (Hamilton R. basin), Nfld., N.B., P.E.I., and N.S., s to Ind., N Ohio, Va., and Del.; E Ireland and w Scotland. [*E. articulatum* (Huds.) Morong; *E. pellucidum* Michx.; *E. decangulare* and *E. sexangulare* of Canadian reports, not L.]. MAPS: Hultén 1958: map 188, p. 207 (also citing a 1953 total-area map by Heslop-Harrison); Fernald 1929: map 40, p. 1505.

COMMELINACEAE (Spiderwort Family)

Herbs with simple entire alternate sessile leaves with sheathing bases. Flowers perfect, regular or irregular, hypogynous. Sepals and petals each 3. Fertile stamens 3 or 6. Style 1. Ovary superior. Fruit a capsule.

1 Corolla irregular, the larger pair of long-clawed petals to 1.5 cm long, pale violet-blue, the third one white and much smaller; fertile stamens 3; capsule 3-locular but the upper locule aborted, each locule 2-seeded; inflorescence cymose, subtended by a folded cordate-ovate spathe; leaves fleshy, the ovate-lanceolate blades to about 1 dm long; subglabrous annual with ascending or depressed stems and branches; (introd. in Ont. and sw Que.) . . . .*Commelina*
1 Corolla regular, the 3 petals all alike, sessile, blue, purple, or roseate; fertile stamens 6; capsule 3-locular, each locule with 1 or 2 seeds; inflorescence umbellate, subtended by elongate bracts similar to the foliage-leaves, these with lance-linear blades to about 6 dm long; perennials; (s Man. and s Ont.) . . . .*Tradescantia*

COMMELINA L.   [896]   Dayflower

C. communis L.   Common Dayflower
Asiatic; a garden-escape to woodlots, roadsides, and waste places in the E U.S.A. and in Ont. (N to Ottawa) and Que. (Montreal).

TRADESCANTIA L.   [911]   Spiderwort

(Ref.: Anderson and Woodson 1935)
1 Sepals glabrous or with only nonglandular hairs, to about 1.5 cm long.
  2 Pedicels glabrous, commonly less than 2.5 cm long; sepals glabrous or merely bearded at tip; leaves glaucous, to about 4.5 dm long; (s Ont.) . . . .*T. ohiensis*
  2 Pedicels to 3.5 cm long, they and the sepals conspicuously pubescent; leaves dull green, commonly less than 3.5 dm long; (introd. in s Ont.) . . . .*T. virginiana*
1 Sepals glandular-pubescent.
  3 Sepals and pedicels densely soft-villous with both glandular and nonglandular hairs to 1.5 mm long; sepals to about 1.5 cm long; petals to 2 cm long; plant green . . . .[*T. bracteata*]
  3 Sepals and pedicels sparingly pubescent with glandular hairs only, these commonly about 0.5 mm long; sepals rarely over 1 cm long; petals commonly less than 1.5 cm long; plant glaucous; (sw Man.) . . . .*T. occidentalis*

[T. bracteata Small]
[The report of this species of the w U.S.A. (N to Mont. and Mich.) from Routledge, sw Man., by Lowe in his undated supplement to his 1943 listing of Manitoba plants appears referable to *T. occidentalis* (relevant collection in WIN; the Virden collection also cited was not located). The MAP by Anderson and Woodson (1935: map II, p. 86) indicates no Canadian stations but their citation from the Red River Valley of Minn. calls for a search for it in s Man.]

T. occidentalis (Britt.) Smyth
/T/WW/   (Hp)   Sandy plains and prairies (particularly in dune and sand-hill areas) from Mont. to sw Man. (Routledge; ?Virden (see *T. bracteata;* at Routledge an extensive colony of both bluish-purple-flowered and roseate-flowered individuals observed by the writer in 1953; CAN), and 15 mi NE of Melita), s to Ariz., Mexico, Tex., and La. [*T. bracteata* of Man. reports at least in part, not Small; *T. ?virginiana sensu* Bourgeau, *in* Palliser 1863, not L.].

T. ohiensis Raf.
/t/EE/   (Hp)   Woods, thickets, meadows, and railway-ballast, etc., from Nebr. and Minn. to s Ont. (Courtland, Norfolk Co.; CAN; reported as abundant in dryish places of the St. Clair R. delta in

Lambton Co. by Dodge 1915; also taken on Squirrel Is., Lambton Co., by Soper and Dale in 1948; reported from Windsor, Essex Co., by C. M. Rogers, Can. Field-Nat. 80(4): 198. 1966) and Mass., s to Tex. and Fla. [*T. barbata, T. canaliculata,* and *T. reflexa* Raf.].

T. virginiana L.
This native of the E U.S.A. (Wisc. to Conn., s to Mo. and Ga.) is reported by Fernald *in* Gray (1950) as a garden-escape NE to Maine. As such, it is reported from s Ont. by Dodge (1915; Lambton Co.) and F. H. Montgomery (Can. Field-Nat. 62(2): 95. 1948; Wellington Co.).

PONTEDERIACEAE (Pickerelweed Family)

Aquatic or marshland herbs with perfect, regular or moderately irregular, hypogynous flowers from a spathe. Perianth corolla-like, the 6 segments coloured, united below into a tube. Stamens 3 or 6. Style 1. Ovary superior. Fruit a capsule or utricle. Plants perennial, from rhizomes.

1   Leaves linear, obtuse, translucent, to about 1.5 dm long and 6 mm broad, sessile and alternate on the submersed stem; flowers solitary in sessile axillary spathes, expanding on the surface; perianth regular, the thread-like tube elongating to about 7 cm, the pale-yellow limb with linear outer and lance-linear inner segments; stamens 3, their basifixed anthers sagittate; fruit a many-seeded capsule; (Ont. and s Que.) . . . . . .*Heteranthera*
1   Leaves varying from linear-lanceolate to broadly cordate, mostly long-petioled and basal (the flowering-stem with a single leaf in addition to the spathe), the blades to about 2 dm long, they and most of the stem emerged; flowers numerous in a terminal spike-like panicle; perianth funnelform, blue or violet, the limb moderately 2-lipped (the 3 upper lobes united about half their length, the middle upper lobe broader than the lateral and marked with yellow, the 3 lower lobes nearly distinct), the tube about 6 mm long; stamens 6, laterally hinged, 3 of them exserted; fruit a 1-seeded bladdery utricle; (Ont. to N.S.; introd. on Vancouver Is.) . . . .*Pontederia*

HETERANTHERA R. & P.   [924]   Mud-plantain

H. dubia (Jacq.) MacM.   Water-stargrass
/T/X/   (HH (Hel))   Streams and quiet waters or clayey or calcareous shores from Wash. to Minn., Ont. (N to the Ottawa dist.; CAN), and Que. (N to the NE end of L. St. Peter; MT; TRT), s to Calif., Mexico, Tex., and Fla.; tropical America. [*Commelina* Jacq.; *Zosterella* Small; *H. (Leptanthus; Schollera) graminea* of American auth., not Vahl].

The stranded phase, f. *terrestris* (Farw.) Vict. (stem relatively short and opaque, the rather fleshy leaves only 3 or 4 cm long) comprises part of the Ont. and Que. material.

PONTEDERIA L.   [922]

P. cordata L.   Pickerelweed
/T/EE/   (Hel)   Shallow water and muddy shores from Ont. (N to L. Nipissing and the Timagami Forest Reserve NE of Sudbury; CAN) to Que. (N to Montmorency Co. below Quebec City), N.B., P.E.I., and N.S., s to Tex., Mo., and N Fla.; (introd. on Vancouver Is., B.C., probably as muskrat food).

Forma *albiflora* (Raf.) House (flowers white rather than violet-blue) is tentatively reported from N.S. by Boivin (1967a). Forma *angustifolia* (Pursh) Solms (leaf-blades linear-lanceolate to narrowly deltoid rather than broadly ovate, tapering or truncate rather than broadly cordate at base) occurs throughout the range. Forma *bernardii* Lepage (leaves subrhomboid and with divergent basal lobes) is known from the type locality, Nominingue, Labelle Co., Que. Forma *latifolia* (Raf.) House (leaves broadly ovate as in the typical form but broadly rounded rather than cordate at base) occurs throughout the range.

JUNCACEAE (Rush Family)

(Ref.: Marie-Victorin 1929; Buchenau 1890)
Herbs with narrow grass-like alternate entire leaves commonly confined to the lower part of the stem, sometimes reduced to mere sheaths. Flowers small, greenish or brownish, perfect, hypogynous, solitary or in clusters of 2 or more in terminal, compact to loosely branched, cymose or umbellate inflorescences. Perianth-segments 6, chaff-like, essentially alike. Stamens 3 or 6. Style 1. Ovary superior. Fruit a capsule.

1   Capsule partially or wholly 3-locular, many-seeded, the seeds very small; inflorescence
    cymose (sometimes capitate); plants glabrous . . . .*Juncus*
1   Capsule 1-locular, 3-seeded, the seeds relatively large; inflorescence umbellate
    (sometimes spike-like); leaves and young stem often hairy . . . .*Luzula*

JUNCUS L.   [936]   Rush, Bog-Rush.   Jonc

1   Inflorescence appearing lateral, the solitary involucral leaf (bract) roundish in cross-section and
    resembling a continuation of the culm; flowers prophyllate (inserted singly on the branches of
    the inflorescence, each subtended by a pair of small bractlets in addition to the larger bractlet at
    the base of the pedicel); leaves all basal and nonseptate, or reduced to sheaths; perennials.
    2   Inflorescence with commonly not more than 3 flowers (sometimes 4 or 5); stamens 6, the
        anthers longer than their filaments; seeds with conspicuous whitish tail-like appendages at
        both ends; culms densely tufted; (mts. of B.C. and Alta.).
        3   Blade of the uppermost basal leaf-sheath well developed; capsules acute . . . . . .*J. parryi*
        3   Blade of the uppermost basal leaf-sheath reduced to a mere rudiment; capsules
            retuse at the broad summit . . . .*J. drummondii*
    2   Inflorescence many-flowered (except in depauperate individuals); seeds without
        tail-like appendages; basal leaf-sheaths bladeless (or rarely, in *J. filiformis,* the upper
        sheath arising with a short blade); culms arising from a stout creeping rhizome.
        4   Anthers commonly much longer than their filaments; stamens 6; flowers pale
            brown to blackish; leaf-blades wanting; culms arising in regular rows along the
            stout creeping rhizome, commonly over 2 mm thick at base.
            5   Flowers to 6 mm long; culms to 1 m tall; lowest leaf of the inflorescence often
                as long or nearly as long as the culm, spiny-tipped; (sw B.C.) . . . .*J. lesueurii*
            5   Flowers to 5 mm long; culms to about 8 dm tall; lowest leaf of the inflorescence
                to about 2 dm long; (transcontinental) . . . .*J. balticus*
        4   Anthers shorter than to about equalling their filaments.
            6   Stamens 3 (rarely 6); cyme many-flowered, compound, with forking branches;
                culms finely many-striate or coarsely furrowed, forming dense tussocks along
                the almost hidden rhizome; (B.C.; Ont. to Nfld. and N.S.) . . . .*J. effusus*
            6   Stamens 6; cyme few-flowered, with mostly simple branches; culms in rows
                along the rhizome; (transcontinental).
                7   Anthers much shorter than their filaments; flowers greenish; seeds about
                    0.5 mm long; culms filiform, commonly about 1 mm thick at base; rhizomes
                    slender and cord-like . . . .*J. filiformis*
                7   Anthers about equalling their filaments; flowers dark brown; seeds about
                    1 mm long; culms and rhizomes relatively stout . . . .*J. arcticus*
1   Inflorescence terminal (if sometimes appearing lateral, the short erect involucral leaf or
    leaves (bracts) flat or channelled on the upper side); culms with 1 or more blade-bearing
    sheaths.
    8   Leaves nodulose-septate (divided by cross-partitions), terete or, if flattened, strongly
        equitant (with one edge toward the stem, as in *Iris*); individual flowers eprophyllate
        (each subtended only by the single bractlet at the base of the very short pedicel),
        inserted on the branches of the cyme in clusters of 2 or more or in heads (only
        occasionally solitary), the clusters subtended by an involucre of hyaline bracts;
        perennials.
        9   Leaves flattened and strongly equitant, to over 4 mm broad, their cross-septa

463

incomplete; heads commonly rather numerous; perianth usually about 3 mm long; culms from creeping rhizomes; (chiefly B.C. and Alta.).

  10 Capsule tapering to a slender beak, surpassing the perianth; seeds not tailed; stamens 6; anthers about twice as long as their filaments; auricles of leaf-sheaths usually obscure; (s B.C.) .... *J. oxymeris*

  10 Capsule abruptly contracted to a short beak or mere apiculation, shorter than or only slightly surpassing the perianth; anthers mostly much shorter than their filaments.

    11 Styles surpassing the petals by about 1 mm in length; stamens 6; seeds with distinct tail-like appendages; leaf-sheaths with usually very short auricles; (SE Alta.) .... *J. tracyi*

    11 Styles slightly, if at all, surpassing the petals; seeds not tailed.

      12 Stamens 3; leaves to 5 mm broad; culms to 6 dm tall; (B.C. to sw Sask.; James Bay region) .... *J. ensifolius*

      12 Stamens 6.

        13 Heads commonly not more than 12; perianth dark brown; leaves usually less than 5 mm broad; culms to about 5 dm tall; (B.C. to sw Sask.) .... *J. saximontanus*

        13 Heads commonly more than 25; perianth usually straw-coloured; leaves to 1 cm broad; culms to about 8 dm tall; (w U.S.A. only) ....[*J. xiphioides*]

 9 Leaves terete or somewhat compressed but not equitant, commonly less than 2 mm broad, the cross-septa complete.

  14 Flowers in dense spherical heads; perianth-segments lance-subulate or narrowly lanceolate; anthers much shorter than their filaments; seeds lacking tail-like appendages; culms from creeping or matted rhizomes.

    15 Stamens 3; involucral bracts usually much shorter than the inflorescence.

      16 Heads commonly solitary or 2 or 3 (sometimes 5 or 6), the inflorescence usually congested; perianth to 3.5 mm long, brownish or reddish-brown; (?Vancouver Is.) .... [*J. bolanderi*]

      16 Heads several in an open paniculate inflorescence; perianth to 4 mm long, straw-colour; (s Ont.) .... *J. brachycarpus*

    15 Stamens 6; involucral bracts commonly overtopping the inflorescence.

      17 Heads usually solitary; perianth-segments about 4 mm long, dark brown; capsule narrowly oval, obtuse or emarginate at the mucronate apex; leaves about 1 mm thick; rhizomes densely matted; (B.C. and Alta.) .... *J. mertensianus*

      17 Heads commonly several or many; capsule subulate or lance-subulate; rhizomes elongate and creeping, often tuber-bearing; (essentially transcontinental).

        18 Sepals and petals subequal, reddish brown, to 4 mm long; leaves 1 or 2 mm thick, their sheaths with yellowish membranous auricles; culms rarely over 6 dm tall .... *J. nodosus*

        18 Sepals distinctly surpassing the petals, greenish to dull brown, to 5 mm long; leaves to nearly 5 mm thick, their sheaths with hyaline-scarious auricles; culms to 1 m tall .... *J. torreyi*

  14 Flowers few to many (sometimes solitary or only 2 or 3) in hemispherical or narrower heads.

    19 Seeds with definite white tail-like appendages; stamens 3.

      20 Seeds ellipsoid, their tails about 1/10 as long as the body; sepals and blunt petals soft and scarious-margined; cyme open or diffuse, the heads 2–5-flowered; (Ont. to N.S.) .... *J. brachycephalus*

      20 Seeds spindle-shaped, with longer tails; sepals and acute petals firm.

        21 Perianth equalling or only slightly shorter than the abruptly short-beaked capsule; heads 2–many-flowered; seeds with tails 2/3 to quite as long as the body; (Ont. to Nfld. and N.S.) ...... *J. canadensis*

        21 Perianth 2 or 3 mm long, much shorter than the tapering capsule.

22 Cyme strict, 3 or more times longer than broad, its heads
3–7-flowered; seeds with tails about 1/2 as long as the body;
(essentially transcontinental) ....*J. brevicaudatus*
22 Cyme loose and open, 1/3 to nearly as broad as long, its heads
5–20-flowered; seeds with tails about 1/3 as long as the body;
(w N.S.) ....*J. subcaudatus*
19 Seeds blunt or merely apiculate, lacking definite white tails.
23 Stamens 3; capsule included or barely exserted from the perianth.
24 Culms creeping or floating, they and the leaves capillary and weak,
the leaves barely septate; heads 1–8, few-flowered, remote, often
viviparous or leafy-tufted; petals blunt; cyme simple or but slightly
forking; (SE Nfld. and N.S.) ....*J. bulbosus*
24 Culms slender, erect, tufted, they and the leaves firm, the leaves
quill-like and strongly septate; heads rather few, 5–many-flowered,
rarely leafy-tufted; petals and sepals acute; cyme open, few-
many-branched; (s B.C.; Ont. to N.S.) ....*J. acuminatus*
23 Stamens 6.
25 Anthers much shorter than to about equalling their filaments;
capsule more or less exserted from the perianth.
26 Leaves dimorphic, the early (basal) ones elongated and
capillary, floating or submersed; stem-leaves much surpassing
the stem, about 1 mm thick, only obscurely septate; heads 2–6;
perianth about 4 mm long, light brown; (?Vancouver Is.). ....
....[*J. supiniformis*]
26 Leaves all uniform, conspicuously septate, mostly shorter than
the stem.
27 Inflorescence ample, divaricately branched, commonly
many-headed; perianth to 3 mm long; plant to 6 dm tall, the
rhizomes often coralline-tuberiferous; (transcontinental) ....
....*J. articulatus*
27 Inflorescence meagre, usually few-headed, its few branches
erect or slightly spreading; plants commonly less than 3 dm
tall.
28 Perianth to 2.5 mm long, its inner segments obtuse,
shorter than the outer segments; (transcontinental) ....
....*J. alpinus*
28 Perianth to over 5 mm long, the segments all acute;
(Alaska–B.C.) ....*J. oreganus*
25 Anthers distinctly longer than their filaments.
29 Stems creeping or floating (forming reddish mats when fresh),
capillary, bearing scattered clusters of reduced leaves and
axillary or terminal, sessile or short-pedicelled flowers; (Ont. to
Nfld. and N.B.) ....*J. subtilis*
29 Stems erect or ascending from a horizontal rhizome; anthers
markedly longer than their filaments; flowers commonly numerous in a branching inflorescence.
30 Stems very slender, from whitish rhizomes about 1 mm thick;
some or all of the flowers often replaced by promptly
deciduous slender bulblets consisting of reduced firm
leaves; perianth 2.5 mm long, greenish or tinged with red, its
segments obtuse; capsule tapering to an acuminate beak,
surpassing the perianth; (Ont. to s Labrador, Nfld., and N.S.) .....
....*J. pelocarpus*
30 Stems and rhizomes stouter; flowers not replaced by
bulblets; perianth 3 or 4 mm long, the segments acuminate.
31 Capsule tapering to an acuminate beak; (eastern
species).

32 Lower elongate tubular sheaths bladeless; stem-leaf usually solitary, terete, equalling or overtopping the cyme, this with stiffly ascending branches; perianth and capsule subequal; (Ont. to Nfld. and N.S.) . . . . .J. militaris

32 Lower sheaths short, open, with a short blade; stem-leaves usually at least 2, flattened; cyme very decompound, its slender branches divaricately spreading; mature capsule exserted from the perianth; (introd. in St-Pierre and Miquelon) . . . . .J. acutiflorus

31 Capsule abruptly acute, mucronate, shorter than or barely equalling the perianth; (western species).

33 Perianth pale brown; heads several to many in a usually open inflorescence; (Vancouver Is.) . . . . .J. columbianus

33 Perianth dark brown; heads often 1 or 2 (at most about 10) in a commonly congested inflorescence . . . .[J. nevadensis]

8 Leaves not nodulose-septate, terete or flattened (with the flat side facing the stem).

34 Individual flowers prophyllate (inserted singly on the branches of the inflorescence and each subtended by a pair of small bractlets in addition to the bractlet at the base of the pedicel); leaves at most about 1.5 mm broad.

35 Annual with soft base and fibrous roots; inflorescence commonly at least 1/3 of the entire height of the plant, the remote flowers scattered along its branches; seeds lacking tail-like appendages; leaves flat, their sheaths not auricled; (transcontinental) . . . . .J. bufonius

35 Perennials with elongate rhizomes or densely tufted culms; inflorescence rarely as much as 1/4 the height of the plant; leaf-sheaths auricled or prolonged at summit.

36 Inflorescence simple (unbranched), bearing at most 4 flowers, many times overtopped by the upper leaves; leaves filiform, mostly borne near the top of the culm, their sheaths with deeply cleft auricles; capsules very firm, long-beaked, the few short-tailed seeds angled; stems hard, densely tufted from matted creeping rhizomes; (s Baffin Is. and N Labrador to Nfld. and N.S.) . . . . .J. trifidus

36 Inflorescence branching, many-flowered; auricles of leaf-sheaths entire; capsules rather thin-walled, rounded at summit, many-seeded.

37 Leaf-sheaths extending about half-way up the culm; petals and sepals obtuse, the latter with incurved tips; seeds lacking tail-like appendages; rhizome horizontal, becoming slender and elongate.

38 Stamens about equalling the perianth-segments, the anthers about 3 times as long as their filaments; style elongate; capsule at most only slightly exserted; primary bract commonly shorter than the inflorescence; plant green; (s B.C.; Man. to Nfld. and N.S.) . . . . .J. gerardii

38 Stamens about half as long as the perianth-segments, the anthers and filaments subequal; style short; capsule distinctly exserted; primary bract commonly longer than the inflorescence; plant glaucous; (introd. from Man. to Nfld. and N.S.) . . . . .J. compressus

37 Leaf-sheaths confined to the base or lower third of the plant; sepals acute, their tips not incurved; rhizome short and erect.

39 Sepals erect, closely appressed to the capsule; seeds with white tails.

40 Petals sharp-pointed, with very narrow scarious margins; inflorescence pale stramineous; capsule shorter than the

perianth; seeds with white tails 1/4 the length of the brown
body; (?Alta.) . . . .[*J. oronensis*]
40 Petals blunt or merely acutish, with broad scarious margins;
capsules exserted.
    41 Inflorescence green or greenish-stamineous, subtended by
a stiffly erect bract to 8 cm long; capsule greenish; seeds
pale brown, with white tails commonly half as long as the
body; (transcontinental) . . . .*J. vaseyi*
    41 Inflorescence brown to castaneous when mature, subtended
by 1 or more flexuous bracts to 2.5 dm long; capsule
brownish, reddish, or castaneous; seeds reddish, with
minute white tails; (Ont. to N.S.) . . . .*J. greenei*
39 Sepals spreading-ascending, their tips not closely appressed to the
capsule; seeds at most 0.6 mm long, merely short-pointed or with
minute white tails
42 Auricles of leaf-sheaths short and rounded; leaf-blades mostly
less than half the height of the culm.
    43 Flowers about 5 mm long, chiefly clustered at the tips of the
floral branches; anthers slightly shorter than their filaments;
capsule imperfectly 3-locular; 1 or more involucral leaves
surpassing the inflorescence; auricles of leaf-sheaths
cartilaginous, yellowish or amber-coloured; (transcontinen-
tal) . . . .*J. dudleyi*
    43 Flowers less than 4 mm long, borne along the upper side of
the floral branches; anthers longer than their filaments;
capsule definitely 3-locular; involucral leaves commonly
shorter than the inflorescence; auricles membranous, pale . . . .
. . . .[*J. secundus*]
42 Auricles of leaf-sheaths delicate, white and scarious, the
uninjured ones prolonged to 3 mm beyond the base of the blade;
blades mostly over half the length of the clum; involucral leaves
commonly surpassing the inflorescence; anthers much shorter
than their filaments.
    44 Capsule completely 3-locular, nearly equalling the perianth;
inflorescence compact, commonly not over 2 cm long; (?B.C.
to Sask.) . . . .*J. confusus*
    44 Capsule incompletely 3-locular (the septa extending only
half-way to the centre), distinctly shorter than the perianth.
        45 Perianth-segments with white scarious margins, the
pale-green panicle to about 1.5 dm long; leaves
commonly over half the length of the culm;
(transcontinental) . . . .*J. tenuis*
        45 Perianth-segments with brown scarious margins, the
brownish panicle commonly not over 3 cm long; leaves
usually less than 1/2 the length of the culm; (?B.C.) . . . .
. . . .[*J. occidentalis*]
34 Individual flowers eprophyllate (each flower subtended only by the single bractlet
at the base of the very short pedicel), inserted in clusters of 2 or more in heads on
the branches of the inflorescence (only occasionally solitary); perennials.
46 Seeds with white tail-like appendages; heads rarely more than 4.
    47 Culms arising singly from tips of elongate stolons; perianths and capsules
brown to nearly black.
        48 Leaves firm, channelled, to about 1 dm long and 1 or 2 mm broad;
involucral leaf to 8 cm long; capsule acute, typically castaneous to
purple-black; (transcontinental) . . . .*J. castaneus*
        48 Leaves soft and flat, to about 2 dm long and 4 mm broad; involucral leaf

commonly poorly developed; capsule obtuse, dark brown above middle; (mts. of B.C.–Alta.) . . . .*J. regelii*

47 Culms nonstoloniferous, forming large or small tussocks; perianths and capsules pale; leaves terete or deeply channelled, 1 or 2 mm broad; involucral leaf rudimentary or rarely over 3 cm long; (transcontinental).

  49 Capsules to 9 mm long, distinctly pointed, strongly exserted; seeds to 4 mm long (including the long white tails); heads 1–2(4); plant to over 4 dm tall . . . . *J. stygius*

  49 Capsules about 4 mm long; seeds at most 3 mm long; heads solitary; plants lower.

    50 Lower involucral bract not much surpassing the lowest flower; capsule rounded to subtruncate at summit; seeds with relatively long white tails . . . . *J. triglumis*

    50 Lower involucral bract more elongate; capsule shallowly notched at summit; seeds with much shorter tails . . . . *J. biglumis*

46 Seeds merely short-pointed or with minute white tails.

  51 Leaves subulate, deeply channelled, commonly sharply reflexed above the sheathing base, usually all basal, from a very short erect rhizome, they and the stem wiry and rigid; flowers in clusters of 2 or 3 terminating the branches of the rather lax inflorescence; perianth-segments dark chestnut-brown, to 7 mm long, slightly surpassing the capsule; (s Greenland) . . . .*J. squarrosus*

  51 Leaves flat, to over 3 mm broad, they and the stem relatively soft.

    52 Anthers 3, reddish; capsule at most 3 mm long, beakless; perianth less than 4 mm long; heads up to 30 or more; leaves to 3 mm broad; culms tufted from a short thick often knotted rhizome, bulbous at base; (N.S.) . . . .*J. marginatus*

    52 Anthers 6; culms from slender creeping rhizomes.

      53 Junction of the leaf-sheath and blade well marked and the auricles well developed; heads 2–10; perianth-segments smooth and usually shining, about 6 mm long, drab or pale brown; capsule abruptly long-beaked; leaves to 4 mm broad; (essentially transcontinental) . . . .*J. longistylis*

      53 Junction of leaf-blade and sheath inconspicuous, the auricles rudimentary or wanting; perianth-segments minutely roughened on the back, brown to dark brown with a green midrib; capsule rounded or merely mucronate at apex; (B.C.).

        54 Heads usually many; inner perianth-segments distinctly longer than the outer ones; leaves to 5 mm broad . . . . *J. orthophyllus*

        54 Heads commonly solitary (sometimes 2 or 3); inner perianth-segments about equalling the outer ones; leaves to about 3 mm broad; (Alaska–B.C.) . . . . *J. falcatus*

J. acuminatus Michx.
/T/X/ (Hs) Damp soils of sw B.C. (Vancouver Is. and adjacent islands and mainland E to Agassiz; collections in CAN from New Westminster illustrate the characteristic proliferation of the heads in autumn, particularly when submersed, many of the capsules being replaced by tufts of elongate leaves functioning as bulblets) and from Minn. to Ont. (N to the E end of L. Superior at ca. 47°N; CAN), Que. (Laprairie, near Montreal; MT), and N.S. (not known from N.B.; the report from P.E.I. by Hurst 1952, is referred by D. S. Erskine 1960, to *J. articulatus,* to which the report from Nfld. by Reeks 1871, may also refer), s to Calif., Mexico, Tex., and Fla. [Var. *legitimus* Engelm.; *J. nevadensis sensu* John Macoun 1888, not Wats., relevant collections in CAN].

J. acutiflorus Ehrh.
Eurasian; reported from Nfld. as early as 1827 by Jean de Laharpe (Paris Soc. Hist. Nat. Mém. 3:128), the probable basis of its listing from there by Cormack (1856). Collections have been seen, however, only from St-Pierre and Miquelon (GH) and it is listed only from there by Rouleau (1956),

where rediscovered in 1942 by M. LeHors (Rhodora 46(548): 311. 1944). Fernald *in* Gray (1950) reports it as native in s Nfld. and St-Pierre and Miquelon but it is more likely to have been introd. in ship-ballast; *(see* note under *Luzula campestris).* MAPS: Hultén 1958: map 138, p. 157; Meusel, Jaeger, and Weinert 1965:85.

J. alpinus Vill.
/aST/X/GEA/ (Grh (Hsr)) Wet meadows and sandy or gravelly shores (often calcareous), the aggregate species from cent. Alaska–Yukon to Great Bear L., Great Slave L., L. Athabasca (Alta. and Sask.), Man. (N to near the Dist. Keewatin boundary at ca. 60°N), northernmost Ont., Que. (N to the Koksoak R. s of Ungava Bay at 57°42′N and the Côte-Nord), Labrador (Hamilton R. basin), Nfld., N.B., P.E.I., and N.S., s to Wash., Utah, Colo., Nebr., Minn., Mo., Ohio, and Vt.; w Greenland N to ca. 67°N; Iceland; Eurasia. MAPS and synonymy: *see* below.
1 Flowers sessile or subequally short-pedicelled in hemispherical heads.
    2 Cyme less than 1.5 dm long, bearing at most about 15 dark-brown heads, its branches erect; [incl. the dwarf northern extreme, f. *nanus* Neum. & Ahlf.; *J. alpinoarticulatus* Chaix, the correct name through priority according to Clapham, Tutin, and Warburg 1962, but *J. alpinus* here retained because of common past usage; essentially the southern part of the range of var. *rariflorus,* from B.C. to Nfld. and N N.B. (the Que. map by Dutilly, Lepage, and Duman 1953: fig. 11, p. 53, however, indicates stations for the typical form as far north in Que. as s Ungava Bay); MAPS: Hultén 1962: map 88, p. 97, and 1968*b*:292 (aggregate species); Bertil Lindquist, Acta Phytogeogr. Suec. 13: fig. 3, p. 125. 1940]. A hybrid with *J. articulatus* (× *J. alpiniformis* Fern. (*J. lamprocarpus* Ehrh.) is reported from Nfld. by Fernald 1933, and collections in CAN from New Westminster, B.C., and St. Andrews, N.B., have been referred to it. A hybrid with *J. nodosus* (× *J. nodosiformis* Fern.) is known from s Ont. (Bruce Pen.), Que. (Gaspé Pen.), Nfld., N.B., P.E.I., and N.S. . . . . . .var. *alpinus*
    2 Cyme to about 3 dm long, bearing up to over 20 greenish or pale-brown heads, its branches spreading-ascending; [B.C. (Vancouver Is.; J. M. Macoun 1913), Sask. (Fernald *in* Grey 1950), s Ont. (Waterloo Co.; Montgomery 1945), and collections from Que. as far N as L. St. John and the Côte-Nord and from Nfld. have been placed here] . . . . .var. *fuscescens* Fern.
1 Some of the central flowers of the ellipsoid or elongate-hemispherical heads elevated well above the others on long pedicels; [incl. the dwarf extreme, f. *pygmaeus* Lindq. (f. *?uniceps* (Hartm.) Krok & Lag.); vars. *alpestris* (C. H. Hartm.) C. J. Hartm. and *insignis* Fries; *J. affinis* R. Br.; *J. nodulosus* Wahl.; *J. richardsonianus* Schultes; *J. uliginosus* var. *?minor* Hook.; *J. acutiflorus sensu* Hooker 1838, in part, not Ehrh.; transcontinental; MAP: on the above-noted map by Lindquist]. A hybrid with *J. brevicaudatus* (× *J. gracilescens* Herm.) is known from the type locality, Petit-Rocher, N.B., and is reported from N L. Huron, Ont., by S. C. Wadmond (Rhodora 49(591): 120. 1947) . . . . .var. *rariflorus* Hartm.

J. arcticus Willd.
/AST/X/GEA/ (Grh) Wet sandy or clayey soils (mostly calcareous) from (according to Hultén's maps) the coasts of Alaska and Dist. Mackenzie to Victoria Is., s Dist. Keewatin, and cent. Baffin Is. (isolated in N Ellesmere Is.), s to N B.C.–Alta. (not known from Sask.), N Man. (Churchill), Ont. (Cape Henrietta Maria, s Hudson Bay), and N Que. (s to NE James Bay); w Greenland N to ca. 73°N, E Greenland N to ca. 75°N; N Eurasia. MAPS: Hultén 1962: map 18 (ssp. *arcticus*), p. 25; Meusel, Jaeger, and Weinert 1965:83; Porsild 1957: map 103, p. 173.
    There is much confusion between this species and *J. balticus*, Porsild's map indicating a western limit in the Thelon Game Sanctuary, w-cent. Dist. Keewatin, near the Dist. Mackenzie boundary. Hultén reduces *J. balticus* to subspecific rank under *J. arcticus.* The northernmost station in N Ellesmere Is. was recently reported by G. R. Brassard and R. E. Beschel (Can. Field-Nat. 82(2): 111. 1968). A hybrid with *J. filiformis* is reported from s Greenland by Böcher, Holmen, and Jacobsen (1966).

J. articulatus L.
/T/X/EA/ (Grh) Wet ground from s Alaska (Hultén 1968*a*), B.C. (N to Quesnel, ca. 53°N), and Wash. to Minn., Ont. (N to the w James Bay watershed at ca. 53°N), Que. (N to Anticosti Is. and the

Gaspé Pen.), Nfld., N.B., P.E.I., and N.S., s to Calif., N.Mex., N Ill., N Ind., and N.C.; Iceland; Eurasia. MAPS and synonymy: *see* below.

1   Flowers greenish; capsule pale brown, dullish; [Que. (N to the Magdalen Is.), Nfld., N.B., P.E.I., and N.S.] . . . . .var. *obtusatus* Engelm.
1   Flowers brown, to 3 mm long; capsule chestnut-brown, shining . . . . .var. *articulatus*
    2   Plant floating or more or less buried in mud, rooting at the nodes; [Que. (reported from Argenteuil and Brome counties by Frère Marie-Victorin, Nat. can. (Que.) 71:258. 1944, and from the Gaspé Pen. by Marcel Raymond, Nat. can. (Que.) 77: 70. 1950) and N.S. (Kings Co.; ACAD; NSPM)] . . . . .f. *stolonifera* (Wohlleb.) Raymond
    2   Plant erect; [range of the species; MAPS: Hultén 1962: map 209, p. 221, and 1968b:292; Meusel, Jaeger, and Weinert 1965:85]. A hybrid with *J. brevicaudatus* (× *J. fulvescens* Fern.) is known from Nfld., P.E.I., and N.S. Hybrids with apparent genetic infiltration from *J. canadensis* (× *J. ?lemieuxii* Boivin) and *J. nodosus* are known from sw Que. and N.S. . . . . .f. *articulatus*

J. balticus Willd.
/aST/X/GEA/   (Grh)   Wet fresh to brackish ground, the aggregate species from the coasts of Alaska–Yukon–Dist. Mackenzie to N Victoria Is., Dist. Keewatin, northernmost Ont., Que. (N to Ungava Bay and the Côte-Nord), Labrador (N to the Hamilton R. basin), Nfld., N.B., P.E.I., and N.S., s to Baja Calif., Mexico, Tex., Mo., and Pa.; Iceland; Eurasia. MAPS and synonymy: *see* below.

1   Capsule completely 3-locular.
    2   Capsule lance-ovoid, paler than the perianth; basal sheaths brown or reddish; [var. *?vallicola* Rydb.; *J. ater* Rydb.; *J. arcticus* ssp. *ater* (Rydb.) Hult.; Alaska–B.C. to Man.; MAPS: Hultén 1968b:286, and 1962: map 18 (*J. arcticus* ssp. *ater*), p. 25] . . . .
        . . . . .var. *montanus* Engelm.
    2   Capsule ovoid, long-mucronate, dark chestnut to black, darker than the perianth; [Que.: type from Bradore, Saguenay Co.; also reported from the Larch R. at ca. 57°N by Dutilly and Lepage 1951b] . . . . .var. *melanogenus* Fern. & Wieg.
1   Capsule incompletely 3-locular.
    3   Anthers about equalling their filaments; [ssp. *sitchensis* (Engelm.) Hult. (*J. arcticus* ssp. *sit.* Engelm.); var. *haenkei* (Mey.) Buch., possibly not *J. haenkei* Meyer; *J. arcticus* ssp. *alaskanus* Hult.; N Alaska–N B.C. to N Victoria Is. and E Dist. Mackenzie; MAPS: Porsild 1957: map 102, p. 173; Hultén 1968b:286 (*J. arct.* ssp. *al.*)] . . . .
        . . . . .var. *alaskanus* (Hult.) Porsild
    3   Anthers commonly 3 or 4 times as long as their filaments.
        4   Capsule ovoid, at most only slightly surpassing the perianth; [*J. ?glaucus sensu* Richardson 1823, not Ehrh.] . . . . .var. *littoralis* Engelm.
            5   Inflorescence relatively compact; [incl. f. *gracilis* Buch.; *J. arcticus* ssp. *litt.* (Engelm.) Hult.; transcontinental; MAPS: Porsild 1966: map 35, p. 71; Hultén 1962: map 18 (*J. arcticus* ssp. *litt.*), p. 25] . . . . .f. *littoralis*
            5   Inflorescence diffuse, the flowers remote; [Man. to Que. and N.S.] . . . .
                . . . . .f. *dissitiflorus* Engelm.
        4   Capsule lance-ovoid, distinctly surpassing the perianth . . . .
            . . . . .var. *stenocarpus* Buch. & Fern.
            6   Capsule brown; [incl. f. *opulentus* Lepage; *J. arcticus* ssp. *littoralis* var. *stenocarpus* (Buch. & Fern.) Hult.; N Ont. to s Labrador and E Que.; MAPS: Hultén 1962: map 18 (*J. arct.* ssp. *litt.* var. *sten.*), p. 24; Dutilly, Lepage, and Duman 1954: fig. 8, p. 61; Schofield 1959: map 2, p. 111]. A dot should be added to all of these maps to indicate the occurrence at Goose Bay, Labrador . . . . . .
                . . . . .f. *stenocarpus*
            6   Capsule buff-yellow; [type from Fort George, E James Bay, Que., the only known locality] . . . . .f. *flavidior* Lepage

J. biglumis L.
/AST/X/GEA/   (Grh (Hs))   Wet tundra and mossy margins of ponds and streams from the coasts of Alaska–Yukon–Dist. Mackenzie throughout the Canadian Arctic Archipelago (N to northernmost Ellesmere Is.) to northernmost Ungava–Labrador, s to N B.C. (ca. 58°30′N; isolated

in the mts. of w Alta. and Colo.), Great Bear L., s Dist. Keewatin, N Man. (Churchill), N Ont. (Cape Henrietta Maria, s Hudson Bay), and N Que. (s along the Hudson Bay coast to ca. 56°15′N); circumgreenlandic; Iceland; Eurasia. MAPS: Hultén 1968b:294, and 1962: map 37, p. 45; Porsild 1957: map 105, p. 174; Young 1971: fig. 9, p. 86.

[J. bolanderi Engelm.]
[This species of Oreg.–Calif. is reported from Vancouver Is., B.C., by John Macoun (1890; partly referring here his 1888 reports of *J. xiphioides* var. *macranthus* from B.C.), and a 1921 collection from Victoria has been placed here by Malte. It is felt, however, that further collections are advisable before admitting it to our flora.]

J. brachycarpus Engelm.
/t/EE/ (Grh) Damp clayey or peaty soils from Okla. to Mich., s Ont. (Sandwich, Essex Co., where taken by J. M. Bigelow in 1868; ?extinct; GH), NE Ohio, and Mass., s to Tex. and S.C. MAP: M. L. Fernald, Rhodora 39(465): map 21, p. 343. 1937.

J. brachycephalus (Engelm.) Buch.
/T/EE/ (Hs) Calcareous marshes, meadows, and shores from Ont. (N to Longlac, N of L. Superior at ca. 49°45′N; TRT) to Que. (N to an island near Fort George, E James Bay, at ca. 53°50′N; Dutilly, Lepage, and Duman 1958), N.B. (Woodstock; CAN), and N.S. (Yarmouth Co. and Cape Breton Is.; not known from P.E.I.), s to Ill., Ohio, and N.J. [*J. canadensis* var. *br.* Engelm.].

Forma *hexandrus* Martin (stamens 6 rather than 3) is known from s Ont. (Welland Co.; R. F. Martin, Rhodora 40(479): 460. 1938).

J. brevicaudatus (Engelm.) Fern.
/ST/(X)/ (Hs) Muddy or wet places from L. Athabasca (Alta. and Sask.) to Man. (N to the Cochrane R. at ca. 58°N), Ont. (N to w James Bay at 51°44′N), Que. (N to the Clearwater R. at ca. 56°15′N), Labrador (N to Makkovik, 55°10′N; CAN; GH), Nfld., N.B., P.E.I., and N.S., s to Minn., Pa., and N.C.; introd. in sw B.C. (Ucluelet, Vancouver Is., "with cranberry plants from the East"; J. M. Macoun 1913). [*J. canadensis* vars. *brev.* and *coarctatus* Engelm.].

J. bufonius L.   Toad-Rush
/aST/X/GEA/ (T) Damp or desiccated open grounds, ditches, and waste places, the aggregate species from N-cent. Alaska and cent. Yukon to Great Bear L., L. Athabasca (Alta. and Sask.), Man. (N to Churchill), northernmost Ont., Que. (N to E James Bay at 53°50′N and the Côte-Nord), Labrador (N to Pack Harbour, 53°50′N), Nfld., N.B., P.E.I., and N.S., s to Calif. and Fla.; w Greenland N to 67°49′N, E Greenland N to ca. 71°N; Iceland; Eurasia. MAPS and synonymy (together with a distinguishing key to the closely related *J. capitatus* Weigel): see below.

[*J. kelloggii* Engelm. of the w U.S.A. (Wash. to Calif.), an annual like *J. bufonius* (unique among our species in this respect) but lower, the 1–2(3) flowers terminating the scapose stems (rather than stems leafy, often floriferous nearly to base), the stamens 3 (rather than 6), is tentatively reported from sw B.C. by Boivin (1966b; Oak Bay, Vancouver Is.).]

1   Flowers in 1 or few terminal heads; seeds narrowly obovoid, about twice as long as broad; [native of Europe, the Canary Is., the Azores, Ethiopia, the Cameroons, and N Africa, and introd. in S. America and Australia; an early report by Bachelot de la Pylaie 1823, from Nfld., where either introd. and not taken since or misidentified; MAPS: Hultén 1958: map 140, p. 159; H. Weimarck, Sven. Bot. Tidskr. 40(2): fig. 9, p. 162. 1946; Meusel, Jaeger, and Weinert 1965:83] . . . .[*J. capitatus* Weigel]
1   Flowers in a much-branched leafy panicle occupying the greater part of the plant; seeds commonly broadly ovoid . . . .*J. bufonius*
    2   Flowers often in 2's or 3's, their inner segments (petals) obtuse; [essentially the range of the species] . . . . .var. *halophilus* Buch. & Fern.
    2   Flowers mostly solitary on the ultimate branches; sepals and petals pointed; [incl. var. *fasciculiflorus* Boiss.; *J. ranarius* Perr. & Song.; transcontinental; MAPS (aggregate species): Hultén 1962: map 192, p. 203, and 1968b:295] . . . . var. *bufonius*

Juncaceae

J. bulbosus L.

/T/E/E/ (Hel) Peaty margins of pools and streams (often floating): SE Nfld. (Jean de Laharpe, Paris Soc. Hist. Nat. Mém. 3: 135. 1827; CAN; GH); St-Pierre and Miquelon; N.S. (Sable Is.; CAN; GH); Iceland; Europe; N Africa; Canary Is.; the Azores; natzd. in New Zealand. [*J. supinus* Moench; *J. fluitans* Lam., not Michx.]. MAPS: Hultén 1958: map 112, p. 131; Meusel, Jaeger, and Weinert 1965:85; Böcher 1938: fig. 133, p. 247; M. L. Fernald, Rhodora 36(423): map 1 (the dot for Maine should be deleted), p. 93. 1934.

J. canadensis Gay

/T/EE/ (Grh) Marshy places, the aggregate species from Minn. to Ont. (N to Batchawana Bay, E end of L. Superior; Hosie 1938), Que. (N to L. St. John and the Côte-Nord), Nfld., N.B., P.E.I., and N.S., s to La., Tenn., and Ga.

An apparent hybrid between this and some other unspecified species (× *J. lemieuxii* Boivin; see *J. articulatus;* resembling *J. canadensis* but the flowers with 6 (rather than 3) anthers and the heads hemispheric (rather than often subglobose) is reported from the type locality, Cascade Point, Vaudreuil Co., Que., by Boivin (1967a).

1 Perianth to 4.5 mm long; cyme with stiffly erect branches and branchlets, commonly not over 1.5 dm long; [intergrading with the typical form and of doubtful taxonomic validity; Que. (Laurentide region N of Montreal; Côte-Nord; Magdalen Is.), Nfld., and N.S. (Digby, Queens, Kings, Halifax, and Cape Breton counties; type from Yarmouth Co.)] . . . . .var. *sparsiflorus* Fern.
1 Perianth rarely as much as 3.5 mm long; cyme with spreading-ascending branches and branchlets, to 3 dm long . . . . .var. *canadensis*
   2 Heads few-flowered; [N.S.: Yarmouth, Halifax, and Hants counties and Cape Breton Is.] . . . . .f. *apertus* Fern.
   2 Heads many-flowered.
      3 Heads densely crowded in sessile or short-stalked clusters or masses; [N.S.: Louisbourg, Cape Breton Is.]. . . . . .f. *conglobatus* Fern.
      3 Heads on elongate branches and branchlets of a loose open cyme; [incl. var. *longicaudatus* Engelm.; range of the species] . . . . .f. *canadensis*

J. castaneus Sm.

/AST/X/GEA/ (Grh (Hs)) Tundra, alpine meadows, gravelly beaches, clay banks, etc., the aggregate species from the Aleutian Is. and coasts of Alaska–Yukon–Dist. Mackenzie to N Victoria Is., Devon Is., Baffin Is., and northernmost Ungava–Labrador, s in the West through the mts. of sw B.C. and sw Alta. to Mont., Colo., and N.Mex., farther eastwards s to Great Slave L., NE Sask. (Hasbala L. at 59°55′N), Man. (s to the Hayes R. about 100 mi sw of York Factory), N Ont. (s to w James Bay at ca. 53°N), Que. (s to L. Mistassini, the Gaspé Pen., and Anticosti Is.), and Labrador (s to the Hamilton R. basin); w and E Greenland N to ca. 77°30′N; Iceland; Eurasia. MAPS and synonymy: see below.

1 Heads several, many-flowered, forming a branching inflorescence; perianth and acutish capsules pale, the capsule about double the length of the perianth; [*J. leucochlamys* Z. & K.; reported from s Alaska by Hultén 1943; see his map 339, p. 558; evidently intergrading with *J. castaneus*; MAPS: Hultén 1968b:289, and 1962: map 22, p. 29] . . . . .ssp. *leucochlamys* (Zing. & Kretch.) Hult.
1 Heads solitary or 1 or 2 additional smaller ones on erect branches, 1–several-flowered; perianth and obtusish capsules typically dark chestnut-brown, the capsule not greatly surpassing the perianth . . . . .ssp. *castaneus*
   2 Capsules whitish; [Man.: taken in 1910 at Churchill by J. M. Macoun and in 1949 by the writer on a gravel-cobble beach near the head of tide on the Nelson R. and on a clay beach of the Hayes R. about 20 mi sw of York Factory; CAN] . . . . . . .var. *pallidus* Hook.
   2 Capsules brown to purple-black; [range of the species; MAPS: Hultén 1968b:289, and 1962: map 22, p. 29; Porsild 1957: map 104, p. 173; Raup 1947: pl. 19] . . . . .var. *castaneus*

J. columbianus Coville

/t/W/ (Grh) Wet places from sw B.C. (near Victoria and Ucleulet, Vancouver Is.; CAN, detd.

472

Malte) to Oreg. and Mont. [Scarcely separable from *J. nevadensis*, with which it is merged by Hitchcock et al. (1969)].

J. compressus Jacq.
?Eurasian (considered native in N. America by Fernald *in* Gray 1950): known in N. America in fresh to brackish soils of Man. (N to Duck Bay, L. Winnipegosis), Ont. (N to Carleton Co.), Que. (N to La Malbaie, Charlevoix Co.; Marie-Victorin 1929*b*), Nfld., N.B. (Oak Bay, Charlotte Co.; ACAD), P.E.I. (Charlottetown), and N.S. MAPS: Hultén 1958: map 152, p. 171; Meusel, Jaeger, and Weinert 1965:81.

J. confusus Coville
/T/WW/ (Hs) Moist places from ?B.C. (Rydberg 1922) to Alta. (Pincher Creek and Milk River Ridge; CAN) and Sask. (Cypress Hills and Saskatoon; Breitung 1957*a*), s to Calif., N.Mex., and Nebr. [*J. tenuis congestus* Engelm. in part; scarcely separable from *J. tenuis*].

J. drummondii Meyer
/ST/W/ (Hs) Damp meadows and slopes from the Aleutian Is. (type from Unalaska), cent. Alaska, s Yukon (Porsild 1951*a*), and sw Dist. Mackenzie (Brintnell L.; Raup 1947) through B.C. and w Alta. to Calif. and N.Mex. [*J. pauperculus* Schwarz; *J. subtriflorus* (Mey.) Cov.]. MAPS: Hultén 1968*b*:287; Porsild 1966: map 36, p. 71; Raup 1947: pl. 19.

J. dudleyi Wieg.
/sT/X/ (Hs) Moist grounds from s Yukon (Porsild 1951*a*) and sw Dist. Mackenzie (Porsild and Cody 1968) to B.C., Alta. (N to L. Athabasca), Sask. (N to Prince Albert), Man. (N to Knee L. at ca. 55°N), Ont. (N to the James Bay watershed at ca. 53°N), Que. (N to the Nottaway R. SE of James Bay, L. St. John, and the Gaspé Pen.), Labrador (N to the Hamilton R. basin), Nfld., N.B., P.E.I., and N.S., s to Calif., Ariz., Tex., Okla., Mo., Tenn., and Va. [*J. tenuis* var. *dud.* (Wieg.) Hermann].

J. effusus L.   Soft Rush.   Jonc épars
/T/(X)/EA/ (Grh) Peaty swamps, thickets, bogs, and pond-margins, the aggregate species from the Alaska Panhandle through coastal B.C. (Vancouver Is. and adjacent islands and mainland) to s Calif. and Ariz., farther eastwards (following a large gap) from Ont. (N to the N shore of L. Superior; Hultén's 1958 map indicates a station at York Factory, Man., following the report from there by John Macoun 1888, this requiring confirmation) to Que. (N to the Nottaway R. SE of James Bay at 50°29′N and the Côte-Nord), Nfld., N.B., P.E.I., and N.S., s to Baja Calif. and Fla.; S. America; Eurasia. MAPS and synonymy: *see* below.

1   Sepals firm to subrigid (not wrinkling in drying), appressed or somewhat ascending, to over 4 mm long.
  2   Summit of uppermost leaf-sheath broad, usually emarginate, the edges much overlapping near or quite to tip, the cluster of veins very strongly converging at the summit; sheaths rarely paler toward summit; branches of inflorescence erect or strongly ascending; [s Alaska–w B.C.; MAP: Hultén 1958: map 153, p. 173] . . . .
. . . . .var. *pacificus* Fern. & Wieg.
  2   Summit of uppermost sheath somewhat narrowed, merely rounded, the edges not overlapping within 1 cm of the top, the veins much less strongly converging; upper sheaths much paler towards summit; inflorescence more open.
    3   Culms finely many-striate, soft; sepals and subequal petals with appressed tips, equalled or slightly surpassed by the capsule; [Ont. to Nfld. and N.S.; MAP: Hultén 1958: map 153, p. 173] . . . . .var. *solutus* Fern. & Wieg.
    3   Culms coarsely furrowed, firm; sepals not appressed.
      4   Perianth at least 3 mm long; sepals surpassing the petals and capsule; [*J. pylaie* Laharpe; Ont. to Nfld. and N.S.; MAP: Hultén 1958: map 153, p. 173] . . . .
. . . . .var. *pylaei* (Laharpe) Fern. & Wieg.
      4   Perianth at most 3 mm long; sepals, petals, and capsule subequal; [E Que., P.E.I., and N.S.; MAP: Hultén 1958: map 153, p. 173] . . . . .var. *costulatus* Fern.
1   Sepals soft and pliable (inclined to wrinkle in drying), at most 3 mm long.
  5   Sepals at least 2.5 mm long, rarely spreading; perianth-segments very dark brown

(each with 2 broad dark-brown lateral bands); [Vancouver Is. and adjacent islands and mainland of sw B.C.]

    6   Uppermost sheath close, coriaceous, lustrous, usually castaneous throughout . . . . .var. *gracilis* Hook.

    6   Uppermost sheath looser, membranous, dull greenish-brown above; [var. *hesperius* Piper] . . . . .var. *brunneus* Engelm.

  5 Sepals usually not over 2.5 mm long, mostly somewhat spreading from the base; perianth-segments paler (but usually with 2 distinct brown lateral bands).

    7   Culms coarsely furrowed near the inflorescence, soft; involucral bract dilated at base; inflorescence compact; [*J. conglomeratus* L.; *J. leersii* Marsson; N.B., N.S., St-Pierre and Miquelon, and Nfld.; MAPS (*J. congl.*): Hultén 1958: map 145, p. 165; Meusel, Jaeger, and Weinert 1965:83] . . . . .var. *conglomeratus* (L.) Engelm.

    7   Culms finely many-striate; involucral bract scarcely dilated at base.

      8   Culms at most about 3 mm thick at summit of uppermost sheath, this not over 12 cm long; capsules rarely emarginate; [Ont. to Nfld. and N.S.; MAP: Hultén 1958: map 153, p. 173] . . . . .var. *decipiens* Buch.

      8   Culms to 5 mm thick at summit of uppermost sheath, this to over 1.5 dm long; capsules emarginate at apex.

        9   Culms firm; inflorescence compact; capsules rather deep brown; [B.C. (introd. at Ucluelet, Vancouver Is.); Ont. to Nfld. and N.S.; MAP: Hultén 1958: map 153, p. 173] . . . . .var. *compactus* Lej. & Court.

        9   Culms easily compressed; branches of inflorescence to over 5 cm long; capsules pale brown; [P.E.I. and Nfld.; MAPS: Hultén 1958: map 153, p. 173, and 1968*b*:285 (aggregate species)] . . . . .var. *effusus*

J. ensifolius Wikstr.

/sT/W/ (Grh) Wet ground and margins of ponds and streams from the Aleutian Is. (type from Unalaska) and s Alaska–Yukon through coastal B.C. (isolated in the Cypress Hills of SE Alta. (Moss 1959) and sw Sask. (Breitung 1957*a*)) to s Calif., Utah, and N.Mex.; isolated in the s James Bay region of Ont. and Que. (*see* Ernest Lepage, Ann. ACFAS 24:86. 1958). [*J. xiphioides* var. *?triandrus* Engelm.]. MAP: Hultén 1968*b*:287.

J. falcatus Meyer

/sT/W/eA/ (Grh) Coastal sands and dunes (ranges of Canadian taxa outlined below), s to s Calif.; Japan; Australia; Tasmania. MAP and synonymy: *see* below.

1   Style long and slender; filaments much shorter than their anthers; [incl. var. *paniculatus* Engelm. and *J. covillei* Piper; B.C. (Queen Charlotte Is., Terrace, ca. 54°N, and Creston; CAN; V)] . . . . .var. *falcatus*

1   Style short; filaments about equalling their anthers; [var. *alaskensis* Cov.; Aleutian Is.-s Alaska (*see* Hultén 1943: map 337, p. 558; type from Sitka, Alaska); the report of var. *alask.* from sw B.C. by J. M. Macoun (1913; Vancouver Is.) is based upon a collection in CAN requiring confirmation; MAP: Hultén 1968*b*:288] . . . . .var. *sitchensis* Buch.

J. filiformis L.

/aST/X/GEA/ (Grh) Swampy ground, bogs, and shores from N-cent. Alaska–Yukon to Great Slave L., s Dist. Keewatin, northernmost Ont., Que. (N to the Payne R., w Ungava Bay, at 59°17′N; DAO), Labrador (N to ca. 56°20′N), Nfld., N.B., P.E.I., and N.S., s to Oreg., Utah, Colo., Minn., and W.Va.; w Greenland N to ca. 68°N, E Greenland N to ca. 64°N; Iceland; Eurasia. MAPS: Hultén 1968*b*:284, and 1962: map 94, p. 103.

J. gerardii Loisel.

/aST/X/GEA/ (Grh) Saline ground and marshes: sw B.C. (Vancouver Is. and adjacent mainland) and NW Wash.; isolated stations in the interior U.S.A., sometimes as a railway weed, from Minn. to Ind. and N.Y.; Man. (Brandon; Bernard Boivin, Rhodora 54(644): 202. 1952) to Ont. (Ottawa; Cochrane, where a railway weed according to Boivin), Que. (St. Lawrence R. estuary from l'Islet Co. to the Gaspé Pen.; reported from Lac Beaumont, Bellechasse Co., by John Macoun 1888), Nfld., N.B., P.E.I., and N.S., s along the coast to Va. (reported from Fla.); southernmost w

Greenland; Eurasia. [*J. fucensis* St. John]. MAPS: Hultén 1958: map 154 (the indicated occurrences at York Factory, Man., and along s James Bay require confirmation), p. 173; Meusel, Jaeger, and Weinert 1965:82; Potter 1932: map 7 (now requiring expansion), p. 75.

J. greenei Oakes & Tuckerm.
/T/EE/   (Hs)   Dry to moist soils from Ont. (N to Batchawana Bay, E end of L. Superior; CAN) to Que. (Trois-Rivières, St-Maurice Co.; Marcel Raymond, Rhodora 51: 9. 1949), P.E.I. (Wellington, Prince Co.; reports from N.B. evidently refer chiefly or wholly to *J. vaseyi*), and N.S. (Shelburne, Halifax, and Antigonish counties), s to Minn., N Ohio, and N.J.

J. lesueurii Boland.
/t/W/   (Grh)   Saline marshes and coastal sands from sw B.C. (Vancouver Is. and Cox Is.; CAN; V; reports from Alaska probably refer to *J. balticus* (*arcticus*) var. *alaskanus* according to Hultén 1943) to N Calif. [*J. lescurii*, orthographic variant].

J. longistylis Torr.
/sT/X/   (Grh)   Damp meadows and prairies and sandy or gravelly shores from B.C. (N to Lac la Hache, about 90 mi N of Kamloops) to Alta. (N to Edmonton), Sask. (Cypress Hills; Maple Creek; Redpath), Man. (N to Grand Rapids, near the NW end of L. Winnipeg), Ont. (Windsor, Essex Co.; Albany and Moose rivers of the w James Bay watershed; for these and other E Canada stations, see Dutilly, Lepage, and Duman 1954: fig. 9, p. 63), Que. (Nottaway R. SE of James Bay at 51°10′N; Trois-Pistoles, N Temiscouata Co.; not known from the Maritime Provinces), and w Nfld., s to Calif., N.Mex., Nebr., and N Mich. [*J. castaneus sensu* Ernest Lepage, Nat. can. (Que.) 69: 271. 1942, not Sm., relevant Trois-Pistoles, Que., collections in CAN and RIM].

J. marginatus Rostk.
/T/EE/   (Hs)   Moist clayey or peaty places from Kans. to Mo., Mich., ?Ont. (reported from Victoria Co. by John Macoun 1888, but not listed by Gillett 1958; a collection in CAN from Rockcliffe, near Ottawa, has been placed here), ?Que. (Chelsea, N of Hull; CAN), and N.S. (Yarmouth and Shelburne counties; not known from N.B.; early reports from P.E.I. require confirmation), s to E Tex. and Fla. [Incl. var. *paucicapitatus* Engelm.].

J. mertensianus Bong.
/ST/W/   (Hs)   Wet meadows and slopes from the Aleutian Is., s-cent. Alaska (type from Sitka), and s Yukon (Porsild 1951a) through B.C. and w Alta. to s Calif. and N.Mex. MAP: Hultén 1968b:288.

J. militaris Bigel.
/T/EE/   (Grh (Hel))   Shallow water or sandy, gravelly, or peaty margins of lakes and ponds from Ont. (Georgian Bay, L. Huron, and the Timagami Forest Reserve NE of Sudbury; TRT; Fernald *in* Gray 1950; reports from Man. require confirmation; not known from Que.) to NE Nfld. (CAN; GH), N.B., P.E.I., and N.S., s to Mich., N.Y., and Md. MAPS: *see* below.
1   Stem naked, lacking both a submedian leaf and an upper sheath; [Nfld. and N.S. (Digby and Yarmouth counties; type from Upper Cornwall, Lunenburg Co.)] . . . . .f. *subnudus* Fern.
1   Stem bearing an erect leaf near the middle, this submedian leaf overtopping the cyme.
  2   Stem-leaves 2 (the upper sheath blade-bearing); [N.S.: known from the type locality, Havelock, and from Craignish, Inverness Co.] . . . . .f. *bifrons* Fern.
  2   Stem-leaf usually solitary, the one below the inflorescence represented by a coloured, bladeless or nearly bladeless, scarious sheath; [range of the species; MAPS: Meusel, Jaeger, and Weinert 1965:84; H. Weimarck, Sven. Bot. Tidskr. 40(2): fig. 5, p. 154. 1946; Fernald 1918b: map 6, pl. 17] . . . . .f. *militaris*

[J. nevadensis Wats.]
[Reports of this species of the w U.S.A. (Wash. to Calif. and N.Mex.) from B.C. by John Macoun (1888; this taken up by Henry 1915, and Rydberg 1922) are based upon *J. acuminatus,* relevant collections in CAN.]

J. nodosus L.
/ST/X/ (Gst) Swamps and gravelly banks from cent. Alaska (near hot springs; *see* Hultén 1943: map 342, p. 558) and N B.C. to Great Bear L., L. Athabasca (Alta. and Sask.), Ont. (N to the Shamattawa R. at ca. 55°N), Que. (N to the E James Bay watershed at 53°50′N, Anticosti Is., and the Gaspé Pen.), Nfld., N.B., P.E.I., and N.S., s to Calif., N.Mex., Tex., Mo., and Va. [Var. *genuinus* Engelm.; *J. paradoxus* Mey.; *J. echinatus sensu* Richardson 1823, not Muhl.; *J. polycephalus* var. *tenuifolius sensu* Hooker 1838, not Michx.]. MAP: Hultén 1968*b*:290.

[J. occidentalis (Coville) Wieg.]
[This species of the sw U.S.A. is reported as common around Victoria, B.C., by Henry (1915) and collections in CAN from Vancouver Is., Osoyoos, and Kootenay Flats have been placed here. It seems scarcely separable from *J. tenuis,* with which it is merged by Hitchcock et al. (1969).]

J. oreganus Wats.
/sT/W/ (Grh) Coasts of s Alaska (*see* Hultén 1943: map 343, p. 558) and B.C. (inland to Chilliwack L.; CAN) to Oreg. [*J. paucicapitatus* Buch.; merged with *J. supiniformis* by Hitchcock et al. 1969]. MAP: Hultén 1968*b*:290.

[J. oronensis Fern.]
[This species of Maine is accredited to Alta. by Fernald *in* Gray (1950; possibly on the basis of an 1872 Macoun collection in CAN from Lesser Slave L., placed here by Raup but later referred to *J. tenuis* by Boivin).]

J. orthophyllus Coville
/t/W/ (Grh) Moist meadows and slopes from sw B.C. (Nanaimo, Vancouver Is.; CAN, detd. Malte; reported from the Koksilah R., Vancouver Is., by J. M. Macoun 1913) to Calif. and Nev. [*J. longistylis* var. *latifolius* Engelm.; *J. lat.* (Engelm.) Buch.].

J. oxymeris Engelm.
/t/W/ (Grh) Salt marshes from sw B.C. (Alberni, Vancouver Is.; CAN, detd. Malte; also reported from New Westminster by Henry 1915) to Calif. [*J. xiphioides* var. *littoralis sensu* John Macoun 1888, not Engelm., according to John Macoun 1890, the relevant collection in CAN].

J. parryi Engelm.
/T/W/ (Hs) Rocky slopes and alpine meadows from s B.C. (N to Lillooet, Kamloops, Nelson, and Windermere; CAN) and sw Alta. (Waterton Lakes and L. Agnes) to Calif. and Colo.

J. pelocarpus Meyer
/sT/EE/ (Grh) Damp shores, wet sands, and pools from Ont. (N to the mouth of the Albany R., James Bay, 52°12′N; Dutilly, Lepage, and Duman 1954) to Que. (N to the Nottaway R. SE of James Bay, the Côte-Nord, and Anticosti Is.), s Labrador (N to the Hamilton R. basin), Nfld., N.B., P.E.I., and N.S., s to Minn., N Ind., and Del. [*J. conradii* Tuckerm.; incl. the reduced extreme, var. *sabulonensis* St. John].

J. regelii Buch.
/T/W/ (Grh) Alpine meadows from B.C. (Vancouver Is.; Queen Charlotte Is.; Bare Is.; Chilliwack L.; Rogers Pass; Field) and sw Alta. (Laggan, Lake Louise, and Silver City; CAN) to Calif., Utah, and ?Wyo.

J. saximontanus Nels.
/T/W/ (Grh) Moist places from s B.C. (N to Glacier in Rogers Pass; CAN) to Alta. (N to Fort Saskatchewan; CAN) and sw Sask. (Cypress Hills; Breitung 1957*a*), s to Calif. and N.Mex. [*J. ensifolius* var. *major* Hook.; *J. mertensianus sensu* Fraser and Russell 1944, not Bong.; *J. xiphioides* (*ensifolius*) var. *montanus* Engelm.].

[J. secundus Beauv.]
[This species of the E U.S.A. (N to Ind. and Maine) is reported from Ont. by Gleason (1958) and

Boivin (1967a) but is not listed by Soper (1949) nor accredited to Ont. by Fernald *in* Gray (1950). The reports from Alta. and N.B. by John Macoun (1888; *J. tenuis* var. *sec.* (Beauv.) Engelm.) also require clarification.]

J. squarrosus L.
/aST/–/GE/ (Hr) Moist open places and bogs: southernmost Greenland; Iceland; Europe. MAPS: Hultén 1958: map 80, p. 99; Löve and Löve 1956b: fig. 6, p. 136; Böcher 1938: fig. 130 (Greenland), p. 245, and fig. 132, p. 247.

J. stygius L.
/ST/X/EA/ (Hs) Wet moss, bogs, and bog-pools from s-cent. Alaska (*see* Hultén 1943: map 344, p. 558; not known from the Yukon) and sw Dist. Mackenzie (J. W. Thieret, Can. Field-Nat. 75(3): 114. 1961) to N Sask. (L. Athabasca; not known from Man.), SE Dist. Keewatin (Eskimo Point, Hudson Bay, ca. 61°N), Ont. (N shore of L. Superior and Big Trout L., ca. 54°N, 90°W), Que. (N to Fort George, E James Bay, ca. 53°45′N, Anticosti Is., and the Gaspé Pen.), Labrador (N to the Hamilton R. basin), Nfld., N.B. (Madawaska, Restigouche, Westmorland, and Kent counties; not known from P.E.I.), and N.S. (Cape Breton Is.), s to B.C., sw Alta., Minn., N.Y., and Maine; Eurasia. MAPS: Hultén 1968b:294, and 1958: map 230, p. 249; Meusel, Jaeger, and Weinert 1965:84.

The plant of N. America and E Asia may be doubtfully separated as var. *americanus* Buch., based on its more evenly tapering capsules, longer seeds, and longer styles than those of the typical form.

J. subcaudatus (Engelm.) Coville & Blake
/T/EE/ (Grh) Mossy woods, bogs, and wet places from N.Y. and w N.S. (Kings, Digby, Yarmouth, Shelburne, Queens, Lunenburg, and Halifax counties; *see* N.S. map by Roland 1947: map 148, p. 254) to SE Mo. and Ga. [Incl. var. *planisepalus* Fern.].

J. subtilis Meyer
/aST/EE/G/ (Hsr) Margins and shores of ponds and streams from Ont. (N to the Shamattawa R. s of Hudson Bay at 54°24′N) to Que. (N to Fort McKenzie, s of Ungava Bay at ca. 56°50′N, the Côte-Nord, Anticosti Is., Gaspé Pen., and Magdalen Is.), Nfld., and N.B. (St. John R. about 30 mi N of St. John; CAN; not known from P.E.I. or N.S.), s to s-cent. Ont. (NW shore of L. Superior; TRT), s Que. (s to William L., Megantic Co.), and N-cent. Maine; w Greenland N to ca. 70°N, E Greenland at 65°39′N. [*J. uliginosus* (*pelocarpus*) var. *sub.* (Mey.) Hook.; *J. fluitans sensu* Michaux 1803, not Lam.; *J. verticillatus* Pursh in part, not Pers.]. MAPS: M. L. Fernald, Rhodora 36(423): map 1 (incomplete northwards), p. 93, 1934; Böcher 1938: fig. 133 (based on Fernald's map), p. 247; Lepage 1966: map 12 (northernmost stations in Canada), p. 224.

[J. supiniformis Engelm.]
[The report of this species of the w U.S.A. (Wash. to Calif.) from Vancouver Is., B.C., by John Macoun (1888; this taken up by Henry 1915) requires confirmation.]

J. tenuis Willd.
/sT/X/ (Hs) Wet to dry open places from SE Alaska and B.C. to s Alta., Sask. (N to Montreal L. at ca. 54°N; Breitung 1957a), Man. (N to Norway House, off the NE end of L. Winnipeg), Ont. (N to Sandy L. at ca. 53°N, 93°W), Que. (N to SE James Bay, L. Mistassini, the Côte-Nord, and Anticosti Is.), Labrador (N to the Hamilton R. basin), Nfld., N.B., P.E.I., and N.S., s to s Calif., Mexico, Tex., and Fla.; introd. in S. America, Europe, N Africa, Australia, New Zealand, and Japan. [Incl. vars. *anthelatus* Wieg., *multicornis* Mey., and *williamsii* Fern.; *J. macer* S. F. Gray and its f. *discretiflorus* Herm.]. MAPS: Hultén 1968b:291, and 1958: map 201, p. 221; Meusel, Jaeger, and Weinert 1965:82.

J. torreyi Coville
/T/X/ (Gst) Wet meadows and muddy or sandy places from s B.C. (N to Kamloops) to s Alta. (Taber, Granum, Milk River, and Manyberries; CAN), Sask. (N to Saskatoon), Man. (N to Swan River, N of Duck Mt.), s Ont. (N to Cornwall, Stormont Co.; TRT), sw Que. (St. John Co.; Lionel

Juncaceae

Cinq-Mars, Ann. ACFAS 18: 80. 1952), and N.Y., s to s Calif., N Mexico, Tex., Ala., and Va. (locally introd. along roadsides and railways to New Eng. and N.J.). [*J. nodosus* var. *megacephalus* Torr.].

J. tracyi Rydb.
/T/W/ (Grh) Moist montane habitats from Mont. and SE Alta. (Cypress Hills; R. D. Bird, Blue Jay (June), 1968; validating the listing, without locality, by Moss 1959) to Nev., Ariz., and Colo.

J. trifidus L.
/aST/E/GEA/ (Grh) Dry barrens, sands, and granitic or schistose ledges from southernmost Baffin Is. and northernmost Ungava–Labrador to N Nfld., s to SE Hudson Bay (Richmond Gulf), E Que. (Bic Mt., Rimouski Co.; Shickshock Mts. of the Gaspé Pen.; Côte-Nord), N N.S. (Inverness and Victoria counties, Cape Breton Is.), and the mts. of N N.Y. and N New Eng.; s half of w and E Greenland; Iceland; Eurasia. MAPS: Hultén 1958: map 28, p. 47, and 1937b: map 4, p. 127; Porsild 1957: map 107, p. 174; Meusel, Jaeger, and Weinert 1965:82; Meusel 1943: fig. 12f; Fernald 1925: map 70, p. 341.

J. triglumis L.
/AST/X/GEA/ (Grh (Hs)) Damp or wet tundra and margins of ponds (chiefly in calcareous clays and sands) from the Aleutian Is., N Alaska, cent. Yukon, and the coast of Dist. Mackenzie–Dist. Keewatin to Banks Is., Victoria Is., Ellesmere Is. (N to ca. 80°N), and northernmost Ungava–Labrador, s through the mts. of SE B.C. and sw Alta. to the mts. of Utah and Colo., farther eastwards s to Great Slave L., NE Sask. (Hasbala L. at 59°55′N), NE Man. (s to York Factory, Hudson Bay, 57°N), s James Bay (Ont. and Que.), E Que. (Tabletop Mt., Gaspé Pen.; Côte-Nord), and NW Nfld.; circumgreenlandic; Iceland; Spitsbergen; Eurasia. [Incl. the intergrading and scarcely separable var. *albescens* Lange (*J. alb.* (Lange) Fern.; *see* M. L. Fernald, Rhodora 26(311): 201–03. 1924, and Hultén 1962:48, and 1943:432]. MAPS: Hultén 1968b:293, and 1962: map 42, p. 49; Porsild 1957: map 106 (*J. alb.*), p. 174.

J. vaseyi Engelm.
/sT/X/ (Hs) Damp thickets, meadows, and shores from B.C. (N to Hudson Hope, ca. 56°N) and sw Dist. Mackenzie (N to Fort Simpson, 62°51′N; CAN) to L. Athabasca (Alta. and Sask.), Man. (N to Reindeer L. at 57°48′N), Ont. (N to L. Nipigon and the Albany R., James Bay, at ca. 52°12′N), Que. (N to the E James Bay watershed at ca. 53°50′N, L. St. John, and the Côte-Nord), SE Labrador (Hamilton R. basin), N.B., and N.S. (not known from P.E.I.), e to Idaho, Colo., Ill., and Maine. [*J. greenei* var. *vas.* (Engelm.) Boivin; *J. ?setaceus sensu* Pursh 1814, as to the report from Canada].

[J. xiphiodes Meyer]
[The typical form of this species is confined to the w U.S.A. according to Jepson (1951). Hultén (1943) refers Alaskan reports of it and its vars. *macranthus* and *triandrus* Engelm. to *J. ensifolius*, with which Rydberg (1922) considers var. *triandrus* identical. John Macoun (1890) refers his 1888 citations of var. *littoralis* Engelm. from Vancouver Is., B.C., to *J. oxymeris* and at least part of his var. *macranthus* citations to *J. bolanderi* (the actual occurrence of this species in B.C. requiring confirmation). Var. *montanus* Engelm. is generally accepted as identical with *J. saximontanus.*]

[J. slwookoörum Young]
[This new species (not keyed out above) has been recently reported from arctic Alaska by S. B. Young (Rhodora 72(792): 486–88. 1970; type from Boxer Bay, St. Lawrence Is., Bering Sea, 63°22′N, 171°35′W).]

LUZULA DC. [937] Woodrush

1 Flowers solitary (or 2, rarely 3) at the tips of the prolonged, spreading and drooping rays of the inflorescence; rhizomes usually well developed.
  2 Inflorescence a simple umbel, the rays normally unforked; seeds with a long curved tail-like appendage (the caruncle); leaves bearing long white hairs.

      3   Basal leaves to 12 mm broad; flowers to 4.5 mm long, on peduncles to about 4 cm long; seeds to 2.5 mm long; (Alta. to Nfld. and N.S.) . . . .*L. acuminata*

      3   Basal leaves commonly less than 5 mm broad; flowers to 3.5 mm long, on peduncles to 3 cm long; seeds about 1.5 mm long; (Alaska–Yukon–Dist. Mackenzie–N B.C.) . . . .*L. rufescens*

  2 Inflorescence a loose decompound cyme, the rays forking; flowers 2 or 3 mm long; seeds unappendaged; leaves essentially glabrous or bearing long white hairs only at the summit of the sheath.

      4   Stem-leaves 1 or 2, short, at most 2.5 mm broad (radical leaves to 4 mm broad); floral bracts lacerate and abundantly ciliate; stems tufted, usually less than 3 dm tall; (transcontinental) . . . .*L. spadicea*

      4   Stem-leaves usually more than 2, to about 1 cm broad; stems not tufted, commonly taller.

         5   Flowers about 2 mm long (in fruit to 3 mm); floral bracts entire or more or less lacerate, the upper ones essentially eciliate; leaves commonly less than 1 cm broad; (transcontinental) . . . .*L. parviflora*

         5   Flowers to 3.5 mm long; floral bracts ciliate; leaves to 12 mm broad; (mts. of B.C.–Alta.) . . . .*L. glabrata*

1 Flowers subsessile in dense spike-like clusters; rhizomes usually lacking or poorly developed.

  6 Flowers white or whitish, rarely more than 6 in a cluster, the clusters forming a diffuse inflorescence to 2 dm long; capsules reddish brown, about equalling the sepals; leaves to about 8 mm broad, tapering to long slender points; (introd. in Ont., sw Que., and N.S.) . . . .*L. luzuloides*

  6 Flowers pale greenish-brown to deep brown or brownish purple; inflorescence usually relatively compact (commonly open in *L. arcuata*); leaves rarely over 4 mm broad.

      7   Leaves tapering to long slender points, often involute; floral bracts ciliate-fringed; seeds not appendaged; (transcontinental).

         8   Sepals bristle-tipped; bracts silvery, longer than the flowers (the inflorescence hence somewhat silvery or whitish); inflorescence a dense or interrupted, arching or nodding, spike-like panicle subtended at base by elongate, very slender-pointed bracts . . . .*L. spicata*

         8   Sepals merely acute, not bristle-tipped; bracts not longer than the flowers.

            9   Culms stout, stiffly ascending, to over 3 dm tall; inflorescence spike-like or sparingly branched, the stout, erect or curved branches bearing dense many-flowered heads to 1 cm thick; flowers to 2.3 mm long; capsule equalling to slightly surpassing the perianth; floral bracts conspicuously fringed; leaves 2 or 3 mm broad; (transcontinental) . . . .*L. confusa*

            9   Culms slender, often curved, rarely over 1.5 dm tall; inflorescence diffuse, nodding, the numerous slender branches bearing flowers in small clusters rarely more than 0.5 mm thick; flowers 2 mm long; capsule shorter than the perianth; floral bracts obsolete; leaves 1 or 2 mm broad; (Alaska to sw Dist. Mackenzie; N B.C. and Alta.) . . . .*L. arcuata*

      7   Leaves with blunt callous tips, flat, often broader; floral bracts entire or merely lacerate; seeds with a commonly well-developed tail-like appendage.

        10 Plants loosely tufted, the small decumbent crowns connected by short horizontal stolons, the usually solitary flowering stems to about 2 dm tall; inflorescence consisting of up to 6 subglobose spikes, the lateral spikes on horizontal to recurving rays; anthers at least twice as long as their filaments; seeds long-appendaged; (introd. in Nfld.) . . . .*L. campestris*

        10 Plants more or less densely tufted, the 2 or more erect flowering stems to about 6 dm tall; spikes cylindric to ovoid or subglobose; anthers at most only slightly longer than their filaments.

           11 Inflorescence a commonly solitary subcapitate spike (occasionally with 1 or 2 lateral peduncled glomerules), the subtending bract typically short and inconspicuous; basal sheaths light brown; leaves flat, with a few long white

hairs along the margins, otherwise glabrous; culms densely tufted, to
about 4 dm tall; (transcontinental in arctic regions) . . . .*L. nivalis*
11 Inflorescence commonly consisting of at least 2 spikes, the subtending
bract usually prolonged; leaves channelled, their margins more or less
ciliate; culms less densely tufted.
12 Perianth-segments at least 2.5 mm long, the inner and outer series
equal; seeds usually over 1.1 mm long, the tail-like caruncle relatively
long; (transcontinental) . . . .*L. multiflora*
12 Perianth-segments rarely over 2.5 mm long, the inner series markedly
shorter than the outer one; seeds usually not over 1 mm long.
13 Perianth-segments mucronate; seeds small, with a caruncle to 0.3
mm long; inflorescence often with many small oblong pale-brown
heads; (introd.) . . . .*L. pallescens*
13 Perianth-segments pointed, not mucronate; seeds with a very short
or obsolete caruncle; leaves often purplish toward base.
14 Spikes often clustered, up to 5 of nearly equal size; lower
involucral bract erect and much surpassing the uppermost
spikes; style long-persistent on the capsule . . . .[*L. sudetica*]
14 One large sessile spike and often a few small short-stalked
lateral ones; lower involucral bract often projecting
obliquely from the stem; style soon deciduous; (trans-
continental) . . . .*L. groenlandica*

L. acuminata Raf.
/sT/(X)/ (Hsr) Rocky woods and clearings from Alta. (Boivin 1967a) to Sask. (N to Meadow
Lake, 54°08′N; Breitung 1957a), Man. (N to Duck Mt.; CAN), Ont. (N to L. Nipigon), Que. (N to
Rupert House, s of James Bay at 51°29′N, Anticosti Is., and the Gaspé Pen.), s Nfld., N.B., P.E.I.,
and N.S., s to S.Dak., Ind., N Ala., and N Ga. [*Juncus (Juncoides; L.) pilosus* of Canadian reports,
not L.; *L. saltuensis* Fern.; *L. carolinae (pilosa)* var. *salt.* Fern.]. MAPS (not indicating stations in
Alta. or Sask.): Hultén 1962: map 176, p. 187; J. E. Ebinger, Rhodora 64(757): fig. 1, p. 78. 1962.

L. arcuata Wahl.
/aST/W/EA/ (Hs) Moist or wet tundra, rocky slopes, and streambanks from the Aleutian Is. and
coast of Alaska to s Yukon and sw Dist. Mackenzie, s through the mts. of B.C. and sw Alta. (N to
Jasper) to Wash. (Mt. Rainier); Eurasia. [*Juncoides* Ktze.; incl. var. *unalaschkensis* Buch.]. MAPS:
Hultén 1968b:299; Raup 1947: pl. 19.

L. campestris (L.) DC.
European; known in N. America only from SE Nfld. (Ferryland and Murray's Pond, Avalon Pen.;
GH) and SE Mass. [*Juncus* L.]. MAP: Hultén 1958: map 139, p. 158.
    Reports from elsewhere in Canada are chiefly referable to the *L. multiflora* complex or
to *L. sudetica*. Although Fernald *in* Gray (1950) regards *L. campestris* as native in Nfld. (but introd.
in Mass.), it is now generally conceded to be entirely introd. in N. America, as is the case with the
following Old World taxa occurring in the New World (where considered native by Fernald) only in
St-Pierre and Miquelon (*Juncus acutiflorus; Alchemilla alpina*), only in Nfld. (*Potentilla sterilis;
Pedicularis sylvatica; Scrophularia nodosa; Galium saxatile*), only in Nfld. and a few other
restricted localities (*Festuca ovina* var. *capillata; Nardus stricta; Sieglingia decumbens; Luzula
multiflora* var. *congesta; L. pallescens; Rumex graminifolius; Ranunculus flammula; Cardamine
flexuosa; Alchemilla (vestita) minor; Fragaria vesca* var. *vesca; Potentilla anglica; P. erecta;
Lathyrus pratensis; Linum catharticum; Mimulus moschatus; Rhinanthus stenophyllus; Cirsium
palustre*), or only in a few other restricted areas of E Canada (*Atriplex laciniata*).

L. confusa Lindeberg
/AST/X/GEA/ (Hs) Turfy tundra, alpine slopes, and ledges from the coasts of Alaska–
Yukon–Dist. Mackenzie–Dist. Keewatin throughout the Canadian Arctic Archipelago to N Ellesmere
Is. (ca. 81°N), Baffin Is., and northernmost Ungava–Labrador, s to N B.C. (Summit Pass at 58°13′N;
CAN; isolated in the mts. of sw Alta.), Great Slave L., NE Sask. (Patterson L.), N Man. (s to

Churchill), Que. (s to NE James Bay, the Côte-Nord, and Shickshock Mts. of the Gaspé Pen.; not known from Ont. or the Atlantic Provinces), and s Labrador, and in the mts. to Maine and N.H.; circumgreenlandic; Iceland; N Eurasia. [*L. arcuata* and *L. hyperborea* of Canadian reports in large part, not Wahl. nor R. Br., respectively]. MAPS: Hultén 1968*b*:301, and 1962: map 15, p. 23; Porsild 1957: map 111, p. 174; Raup 1947: pl. 19.

Polunin (1940) notes possible hybridization of this species with *L. nivalis* in collections from Devon Is. and Port Burwell, N Ungava.

L. glabrata (Hoppe) Desv.
/T/W/E/ (Hs) Cliffs and slopes from s B.C. (N to Queest Creek, Shuswap L., NE of Kamloops) and sw Alta. (Banff; Waterton Lakes) to Oreg. and Wyo.; ?Europe. [*Juncus* Hoppe; *Juncoides* Piper; *Juncodes* Sheld.; *L. spadicea* var. *gl.* (Hoppe) Mey.; *L. piperi* of auth. not (Cov.) Jones].

L. groenlandica Böcher
/aST/X/G/ (Hs) Moist or wet tundra and sandy acidic soils from Alaska (N to the Seward Pen.; not known from the Yukon) to Dist. Mackenzie (Mackenzie R. Delta; Bathurst Inlet; Great Bear L.; Fort Liard, ca. 60°15′N), cent. Dist. Keewatin (Baker L.), NE Man. (Churchill), cent. Ont. (w James Bay; Dutilly, Lepage, and Duman 1954), Que. (E James Bay N to Great Whale R., SE Hudson Bay; Sugluk, Hudson Strait; s Ungava Bay), and Labrador (mouth of the Fraser R. at ca. 56°35′N; Goose Bay); w Greenland between ca. 63°30′ and 69°30′N, the type from ca. 66°N.

This species is scarcely separable from the *L. multiflora* complex, for a comparison with which *see* T. W. Böcher (Medd. Grønl. 147(7): 11–23. 1950). Var. *fuscoatra* Böcher (the heads dark brown, the perianth-segments lacking the broad pale margins characteristic of those of the typical form) was described from material taken in Labrador and w Greenland and appears to be general throughout the range.

L. luzuloides (Lam.) Dandy & Wilmott
Eurasian; introd. in lawns and waste places of Ont. (Norfolk Co.; OAC; reported from the Ottawa dist. by Gillett 1958, and from Queenston Heights, Lincoln Co., by A. B. Klugh, Ont. Nat. Sci. Bull. 2: 45. 1906), sw Que. (Kingsmere, Lac-Sept-Iles, and Senneterre; C. Rousseau 1968; *see* his Que. map 12, p. 65), and N.S. (Pictou; GH; M. L. Fernald, Rhodora 50(596): 213. 1948). [*L. nemorosa* Mey., not Baumg.].

L. multiflora (Retz.) Lejeune
/aST/X/GEA/ (Hs) Fields, meadows, and open woods, the aggregate species from the Aleutian Is., N Alaska, cent. Yukon and the coast of Dist. Mackenzie to E-cent. Dist. Keewatin, northernmost Ungava–Labrador, Nfld., N.B., P.E.I., and N.S., s to s Calif., Ariz., Colo., Minn., Ind., Ky., and Va.; w and E Greenland N to ca. 75°N; Iceland; Eurasia. MAPS and synonymy: *see* below.

1 Heads cylindrical (rarely globose), to 3 cm long; perianth to 4.5 mm long, equalling or surpassing the capsule; inflorescence loose or somewhat congested; [*L. (Juncoides) comosa* Meyer, the type from Vancouver Is., B.C.; *L. campestris* var. *comosa* (Mey.) Fern. & Wieg.; *L. subsessilis* Buch.; Alaska–B.C.; MAPS: Hultén 1962: map 64, p. 73, and 1968*b*:303] . . . . .ssp. *comosa* (Meyer) Hult.
1 Heads short-cylindrical or globose, usually not over about 1 cm long.
  2 Perianth usually longer than the capsule, its segments with pale margins (the spike thus rather pale brown); inflorescence condensed, many or all of the spikes sessile or subsessile, without or with some elongate rays; seeds to about 2 mm long . . . .
     . . . . .ssp. *multiflora*
    3 Inflorescence commonly with some stiff rays to about 3 cm long; spikes to 8 mm long and 7 mm thick; perianth to 4 mm long; stems less than 5 dm tall; [E Que. to N.B., P.E.I. (type from St. Peter's Bay), and N.S.; MAP: Hultén 1962: map 64, p. 73] . . . . .
       . . . . .var. *acadiensis* Fern.
    3 Inflorescence commonly capitate or the lateral spikes rarely on rays up to 9 cm long; spikes to 11 mm long and 9 mm thick; perianth to 4.5 mm long; stems to about 9 dm tall; [*L. campestris* var. *congesta* (Thuill.) Meyer; introd. in St-Pierre and Miquelon and Nfld.] . . . . .var. *congesta* (Thuill.) Koch

2   Perianth usually slightly shorter than the capsule, less than 3.5 mm long; [transcontinental].

    4   Perianth and capsule intensely castaneous to almost black; inflorescence congested, many or all of the spikes sessile or subsessile; seeds usually less than 1.5 mm long; [*L. campestris (sudetica)* var. *frigida* Buch., described from material collected in Alta. and Labrador; *L. frigida* (Buch.) Sam. and its var. *contracta* Sam.; incl. vars. *fusconigra* Čelak., *kjellmanniana* Satake, and *kobayasii* (Satake) Sam.; MAPS; Hultén 1968*b*:301 and 302, and 1962: map 64, p. 73] ....
....ssp. *frigida* (Buch.) Krecz.

    4   Perianth and capsule pale (reddish brown to moderately castaneous); inflorescence commonly consisting of a large sessile or subsessile central spike and lateral spikes on ascending rays of varying lengths (up to 6 cm); seeds to about 2 mm long; [*Juncus* Retz.; *L. campestris* var. *mult.* (Retz.) Čelak.; *L. intermedia* (Thuill.) Spenner; see T. W. Böcher, Medd. Gronl. 147(7): 11–23. 1950, and M. L. Fernald and K. M. Wiegand, Rhodora 15(170): 38–43. 1913); MAP: Hultén 1962: map 64, p. 73] ....ssp. *multiflora*

L. nivalis (Laest.) Beurl.

/AS/X/GEA/ (Hs)   Moist to dryish tundra from the Aleutian Is. and coasts of Alaska–Yukon–Dist. Mackenzie–Dist. Keewatin throughout the Canadian Arctic Archipelago to northernmost Ellesmere Is. and Baffin Is., s to s Yukon, Great Bear L., s-cent. Dist. Keewatin, (s to s Ungava Bay), and northernmost Labrador; w and E Greenland N of the Arctic Circle; N Eurasia. [Incl. f. *nana* Schol. and var. *latifolia* (Kjell.) Sam.; *L. arctica* Blytt; *L. tundricola* Gorod.]. MAPS: Hultén 1968*b*:300, and 1962: map 4, p. 10; Porsild 1957: map 109, p. 174.

L. pallescens (Wahl.) Bess.

Eurasian; introd. in clearings, meadows, and open woods of Sask. (Boivin 1967*a*), s Ont. (Belleville, Hastings Co.; John Macoun 1888), Que. (N to the Gaspé Pen. at New Richmond), Nfld. (Glenwood, Gander R.; CAN; GH), N.B. (St. Andrews and Newcastle), and N.S. (Eastern Harbour, Cape Breton Is.; CAN). [*Juncus* Wahl.; *L. campestris* var. *pall.* Wahl.]. MAP: Hultén 1958: map 121, p. 141 (the dot for Iceland should be deleted according to A. Löve, Rhodora 61(721): 32. 1959).

L. parviflora (Ehrh.) Desv.

/aST/X/GEA/ (Hs)   Damp woods, thickets, and slopes, the aggregate species from the Aleutian Is., N Alaska, cent. Yukon, and the Mackenzie R. Delta to Great Bear L., s Dist. Keewatin, northernmost Ont., northernmost Ungava–Labrador, Nfld., N.B., and N.S. (not known from P.E.I.), s to s Calif., Ariz., N.Mex., cent. Sask., s Man., N Mich., N.Y., and New Eng.; w Greenland N to ca. 70°N, E Greenland N to ca. 61°N; Eurasia. MAPS and synonymy: see below.

1   Capsule and perianth castaneous or brownish; cyme to 12 cm long, the flowers solitary or in groups of 2's or 3's at the ends of the branches; stems commonly not over 6 dm tall; [*Juncus* Ehrh.; *Juncoides* Cov.; *L. spadicea* var. *parv.* (Ehrh.) Mey.; *L. divaricata* of Alaska–B.C. reports, not Wats.; range of the species; MAPS: Hultén 1968*b*:298, and 1962: map 50 (aggregate species), p. 57]. This and the following taxon have also been reported by early authors as var. *fastigiata* Buch., *L. arida* Steud., *L. labradorica* Raf., *L. maxima* (Reich.) DC., *L. pilosa* Willd., *L. sylvatica* Gaud., and *L. vernalis* DC., several of these, also, as *Juncus* or *Juncoides* ....var. *parviflora*

1   Capsule brown to black; perianth stramineous or pale brown; cyme to about 2 dm long; culms to about 1 m tall; [*Juncus mel.* Michx.; *L. piperi* (Cov.) Jones; *L. spadicea* vars. *mel.* (Michx.) Mey. and *?subcongesta* (Buch.) Wats.; essentially the range of the species] ....var. *melanocarpa* (Michx.) Buch.

L. rufescens Fisch. & Mey.

/Ss/W/A/ (Hs)   Damp tundra and grassy slopes in Alaska (N to ca. 70°N), the Yukon (N to ca. 65°N), NW Dist. Mackenzie (Porsild and Cody 1968), and N B.C. (Sikanni R. at 57°14'N and Beatton R. at 57°05'N; CAN); Asia. [*L. pilosa* var. *ruf.* (F. & M.) Boivin; *L. japonica* of Alaska–Yukon reports, not Buch.]. MAPS: Hultén 1968*b*:296, and 1962: map 176, p. 187.

L. spadicea (All.) DC.
/aST/X/GEA/ (Hs) Tundra, bogs, and mossy margins of ponds and streams from the Aleutian Is., N Alaska, s-cent. Yukon, and the coasts of Dist. Mackenzie–Dist. Keewatin to Baffin Is. (N to near the Arctic Circle), s in the West through the mts. of B.C. and sw Alta. (N to the Jasper dist.) to Wash., Nev., and Idaho, farther eastwards s to cent. Dist. Mackenzie, s Dist. Keewatin, NE Man. (s to Churchill), N Que. (s to E Hudson Bay at 55°N and s Ungava Bay), and northernmost Labrador; a single station in E Greenland at ca. 75°N; Spitsbergen; N Eurasia. [*Juncus* All.; incl. *L. wahlenbergii* Rupr.]. MAPS (*L. wahl.*): Hultén 1968*b*:297, and 1962: map 24, p. 31; Porsild 1957: map 108, p. 174; Raup 1947: pl. 19 (somewhat incomplete).

L. spicata (L.) DC.
/aST/X/GEA/ (Hs) Dry tundra, slopes, and rocky ledges from the E Aleutian Is. and s-cent. Alaska–Yukon to the w coast of Dist. Mackenzie and Great Bear L., s through the mts. of B.C. and sw Alta. to Calif., Colo., and N.Mex., farther eastwards a large gap except for an isolated station in N Man. (Nueltin L. at ca. 59°45′N; CAN), then from Baffin Is. (N to near the Arctic Circle) to Que. (s along the coasts to NE James Bay and s Ungava Bay, with isolated stations in the Knob Lake dist. at ca. 55°N, the Côte-Nord, and N coast and Shickshock Mts. of the Gaspé Pen.), Nfld. (CAN; GH), N.S. (Victoria Co., Cape Breton Is.; not known from N.B.; according to D. S. Erskine 1960, the report from P.E.I. by Hurst 1952, probably refers to *L. multiflora*), and the mts. of N N.Y. and N New Eng.; w Greenland N to ca. 72°30′N, E Greenland N to 74°25′N; Iceland; Eurasia. [*Juncus* L.; *Juncoides* Ktze.; incl. var. *tenella* Meyer]. MAPS: Hultén 1968*b*:303, and 1958: map 218, p. 237; Porsild 1957: map 110, p. 174; Meusel, Jaeger, and Weinert 1965:88; Raup 1947: pl. 19.

[L. sudetica (Willd.) DC.]
[Hultén (1962:245) writes of this taxon that, "It is a very obscure species connecting with the *L. multiflora* subsp. *frigida* series." He ascribes it in N. America only to the mts. of Colo. M. L. Fernald (Rhodora 47(561): 265. 1945), however, reports the typical form from Nfld., treating the more widespread *L. campestris* var. *frigida* Buch. as a variety of *L. sudetica* rather than a subspecies of *L. multiflora* (as adopted here, following more recent authors). (*L. campestris* var. *?alpina* Gaud.; incl. more or less intermediate phases named *L. hybrida* Lindb. f.).]

LILIACEAE (Lily Family)

(Ref.: T.M.C. Taylor 1966a; Marie-Victorin 1929b)

Chiefly herbs (2 species of *Smilax* with woody thorny stems; *Yucca* with a woody caudex) with simple entire leaves, these mostly parallel-veined (net-veined in *Smilax* and *Trillium*; leaves reduced to thin dry scales in *Asparagus*), alternate, whorled, or basal, usually sessile or subsessile and more or less sheathing. Flowers mostly perfect, usually 6-merous (4-merous in *Maianthemum*), the sets of 3 sepals and 3 petals essentially alike (except in *Calochortus* and *Trillium*), the petals commonly distinct nearly or quite to base but in several genera united up to 3/4 their length or more. Stamens usually 6 (4 in *Maianthemum*), hypogynous or adnate to the perianth. Pistil 1. Styles 1 or 3, or the stigma sometimes sessile or subsessile. Ovary superior. Fruit a usually 3-locular capsule or berry. (Including Agavaceae, Alliaceae, Calochortaceae, Convallariaceae, Dracaenaceae, Smilacaceae, and Trilliaceae).

1 Stems relatively leafy, their leaves usually not much smaller than the basal ones (if these present); perianth-segments essentially alike (except in *Trillium*).
  2 Leaves reduced to thin dry scales (the modified green filiform branchlets functioning as leaves); flowers small, greenish yellow, long-pedicelled, solitary or paired in the branch-axils, their segments distinct or united; style slender, 3-lobed; fruit a red berry about 8 mm thick; stem bushy-branched, from thick matted rhizomes; (introd.) ....
  ....*Asparagus*
  2 Leaves with normal green blades.
    3 Leaves (at least some of them) whorled; perianth-segments 6, distinct; ovary 3-locular.
      4 Leaves ovate to broadly rhombic, in 1 or 2 whorls; style 3-cleft almost to base, the branches stigmatic along the inner side; fruit a berry.
        5 Leaves in a single whorl of 3 at the top of the stem, net-veined, ovate to broadly rhombic; flower solitary; sepals green, narrower than the white to pink, red, or purple petals, these to over 5 cm long; stem from a short tuber-like rhizome ....*Trillium*
        5 Leaves in 2 whorls, parallel-veined, the whorl near the middle of the stem with at least 5 oblong-oblanceolate leaves to 12 cm long, the involucral whorl with usually 3 ovate leaflets to 6 cm long; flowers pale greenish-yellow, slender-pedicelled in a terminal 3–9-flowered umbel; perianth-segments essentially alike, recurved, less than 1 cm long; stem from a horizontal white tuber; (Ont. to N.S.) ....*Medeola*
      4 Leaves linear to broadly lanceolate, the lower ones often alternate (the upper ones sometimes merely opposite and not whorled in *Fritillaria*); perianth-segments all alike, yellow to red or dark purple, commonly blotched or spotted; fruit a rather large capsule, the numerous seeds closely packed in 2 rows in each locule; stems from scaly bulbs.
        6 Flowers campanulate or deeply bowl-shaped, to about 3.5 cm long; anthers fixed near the base, slightly or not at all versatile; style simple or deeply 3-cleft; nectary a shallow pit at the base of the perianth-segments; capsule 3-angled; bulbs subtended by numerous "rice-grain" bulblets; (Alaska–Yukon–B.C.–sw Alta.) ....*Fritillaria*
        6 Flowers funnelform or funnelform-campanulate, mostly larger; anthers fixed near the middle and versatile; style simple, only the stigma 3-lobed; nectary a linear groove; capsule more or less 3-angled ....*Lilium*
    3 Leaves all alternate; sepals and petals alike.
      7 Leaves of the flowering-stem usually 2 (occasionally 1 or 3); stems from slender creeping rhizomes; flowers small, white, in simple racemes, their segments distinct; style short and thick; fruit a berry.
        8 Perianth-segments and stamens each 4; stigma 2-lobed; berry pale red, 2-locular; leaves ovate-cordate ....*Maianthemum*
        8 Perianth-segments and stamens each 6; stigma 3-lobed; berry dark red,

        3-locular; leaves lanceolate to oval or oblong, narrowed to base;
        (*S. trifolia*) . . . .*Smilacina*

  7 Leaves 3 or more; perianth-segments and stamens each 6; ovary 3-locular.
    9 Flowers large, the yellow to red distinct perianth-segments up to 9 cm long,
       blotched or spotted with purple-brown; style elongate or very short
       (dimorphic), 3-lobed; fruit a more or less 3-angled capsule closely packed
       with flat seeds; leaves linear to narrowly lanceolate; stem from a scaly bulb . . . . . .
       . . . .*Lilium*
    9 Flowers smaller.
      10 Perianth tubular, gamopetalous, shortly 6-lobed, greenish white or
         yellowish, to about 2 cm long; style slender, deciduous by a joint;
         peduncles axillary, 1–15-flowered; fruit a dark-blue to black, several-
         seeded berry; leaves subsessile or somewhat clasping, lance-elliptic to
         broadly oval, to 1.5 dm long; stems from creeping knotted rhizomes;
         (sᴇ Sask. to N.S.) . . . .*Polygonatum*
      10 Perianth-segments distinct or united only at base.
        11 Leaves long-petioled, narrowly ovate to rotund, to over 1 dm long,
           net-veined, often bearing tendrils in their axils; flowers small, yellow
           or greenish, in axillary peduncled umbels; stigma sessile or nearly
           so; fruit a 1–6-seeded blue-black or black berry to about 1 cm thick;
           mostly annual herbs (2 species with stems and branches woody
           and prickly), usually climbing by axillary tendrils . . . .*Smilax*
        11 Leaves sessile or subsessile, parallel-veined; plants herbaceous
           and without prickles, not climbing.
          12 Leaves linear, at most 1.5 cm broad; flowers long-pedicelled in a
             terminal umbel, often replaced by bulblets; style filiform, the
             stigma entire or only slightly 3-lobed; fruit a capsule; scape
             arising from a scaly bulb . . . .*Allium*
          12 Leaves lanceolate to broadly oval, wider; stem from a short or
             creeping rhizome.
             13 Flowers axillary, solitary or in pairs, greenish white to
               roseate or purplish, at most 1 cm long; fruit a red many-
               seeded berry . . . .*Streptopus*
             13 Flowers or inflorescence terminal (capsules appearing
               lateral in *Uvularia* through prolongation of the stem and
               branches).
               14 Flowers 1–few, to over 3 cm long; style elongate.
                  15 Fruit terminal, a yellow or orange to bright-red or
                    scarlet 3-angled berry; flowers whitish or greenish-
                    yellow, at most 2.5 cm long; style entire or shortly
                    3-lobed; leaves oblique at the sessile base . . . . . .*Disporum*
                  15 Fruit axillary, a 3-lobed capsule; flowers stramineous
                    to orange-yellow, to over 4 cm long; style deeply
                    3-lobed; leaves symmetrical at the clasping or per-
                    foliate base; (s Man. to N.S.) . . . .*Uvularia*
               14 Flowers several to many, in terminal racemes or panicles,
                at most 13 mm long.
                 16 Fruit a 1–2-seeded berry; flowers white, racemose or
                    paniculate, the segments to about 5 mm long; style
                    single, thick and very short; plants usually less than
                    1 m tall . . . .*Smilacina*
                 16 Fruit a many-seeded capsule; flowers in a large
                    pyramidal panicle to about 5 dm long, their segments
                    at least 8 mm long; styles 3, distinct to base; plant
                    more or less pubescent, to about 2 m tall . . . .*Veratrum*
1 Stems often more or less scapose, the principal leaves at or near the base (or at least
  sheathing the base), the upper leaves markedly reduced or wanting.

17 Flowers usually either solitary at the top of the scape or stem or in terminal umbels or irregularly branched clusters (if more than 1 in *Erythronium* and *Lloydia,* the inflorescence more or less racemiform); style (or stigma) solitary.

  18 Perianth-segments unlike, distinct, the 3 outer ones narrow and sepaloid, greenish white or purple-tinged, the 3 inner ones broader and petaloid; flowers solitary or up to 9 in a terminal umbel-like cluster; stigma nearly sessile, its 3 lobes recurved; fruit a septicidal capsule (dehiscing through the partitions between each locule); leaves linear; stem from a corm; (B.C. and Alta.) ....*Calochortus*

  18 Perianth-segments essentially alike, all petaloid; style elongate, the stigma entire or more or less 3-lobed.

    19 Perianth-segments united for about half their length; flowers white to blue or violet-purple, few to several in a terminal loose umbel subtended by membranous, linear or lanceolate bracts; capsule to about 5 mm thick, loculicidal (dehiscing into each locule through the dorsal suture); leaves narrowly linear, to 6 dm long; scape from a fibrous-coated co⁻ ⸱er 5 dm tall; (s B.C.) ....*Brodiaea*

    19 Perianth-segments free nearly or quite to base.

      20 Fruit a globose or short-ovoid berry to over 8 mm thick; flowers solitary and terminal or up to 8 in a terminal umbel, white or greenish yellow, to over 2 cm long, the segments downy-pubescent; leaves 2 or 3 (4, 5), all basal, narrowly elliptic to oblong or oval; scapes from slender creeping rhizomes ....*Clintonia*

      20 Fruit a loculicidal capsule.

        21 Perianth-segments about 1 dm long, tawny-orange, united below into a short tube, the 3 inner ones wavy-margined; flowers in a short irregularly branched cluster; leaves linear, to 2 cm broad; scape to over 1 m tall; perennial from a very short fibrous-rooted rhizome; (garden-escape) ....*Hemerocallis*

        21 Perianth-segments much smaller; scapes lower, from a solid corm or tunicated (onion-like) bulb.

          22 Flowers few to many in a terminal umbel (this enclosed before flowering in a spathe that splits into 1–3 ovate, usually papery or scarious bracts), to about 1 cm long, sometimes replaced by bulblets; leaves mostly linear; bulbs fibrous-rooted.

            23 Seeds usually only 1 or 2 in each locule of the capsule; flowers white, yellow, pink, roseate, or purplish; plants with odour of onion or garlic ....*Allium*

            23 Seeds usually at least 6 in each locule; flowers greenish white or white; scape to 3 dm tall; plants lacking onion odour; (?Vancouver Is.) ....*[Nothoscordum]*

          22 Flowers commonly solitary (occasionally 2 or more in a racemiform inflorescence); bulbs often arising from a creeping rhizome.

            24 Leaves several, narrowly linear and grass-like, to about 1 dm long and 2 mm broad; flowers creamy-white, purple-veined, tinged with rose on the back, about 1 cm long; not nodding; capsule usually less than 1 cm long; scapes to about 1.5 dm tall; (Alaska–Yukon–Dist. Mackenzie–B.C.) ....*Lloydia*

            24 Leaves a single subopposite pair (often solitary on flowerless plants), to about 2 dm long and 8 cm broad, often mottled; flowers white, yellow, roseate, or bluish, commonly at least 3 cm long; capsules mostly 2 or 3 cm long; scapes often taller ....*Erythronium*

17 Flowers in spike-like to open racemes or in panicles.

  25 Perianth-segments united into a tube for at least 3/4 of their length; style solitary, the stigma entire or obscurely 3-lobed.

    26 Flowers blue, tubular or urceolate, 4 or 5 mm long, short-lobed, nodding on slender short pedicels; raceme at anthesis compact, its bracts small and

scarious; fruit a capsule with 2 black roughish seeds in each locule; leaves linear-oblanceolate, few, to 2.5 dm long; scape to 2.5 dm tall, from a scaly bulb; perennial; (introd.) . . . .*Muscari*

  26 Flowers white; raceme looser, its bracts larger; leaves to about 2 dm long.

      27 Leaves 2 or 3, oblong; corolla campanulate, to 9 mm long, smooth, its 6 short broad lobes recurved; fruit a many-seeded berry (rarely developed); plant strongly stoloniferous; (introd.) . . . .*Convallaria*

      27 Leaves numerous, narrowly lanceolate or oblanceolate; corolla tubular, to 1 cm long, granulate, its 6 lanceolate lobes erect, to 2.5 mm long; fruit a many-seeded capsule; scape to 1 m tall; perennial from a short thick rhizome; (s Ont.) . . . .*Aletris*

25 Perianth-segments free or united only near the base; fruit a capsule.

  28 Style single, the stigma entire or obscurely 3-lobed; capsule loculicidal (opening into each locule through the dorsal suture); inflorescence bracted.

      29 Plants to 2 or 3 m tall, from a short woody caudex; flowers to 6 or 7 cm long, whitish or greenish-white, numerous in a paniculate inflorescence to over 1 m long; capsule to 7 cm long; leaves evergreen, thick and rigid, to 8 dm long . . . .*Yucca*

      29 Plants lower, from a bulb; flowers smaller, in racemes; capsules smaller; leaves not evergreen.

         30 Perianth-segments obscurely nerved, white, with a broad green median stripe on the back, to 3.5 cm long; style and stigma 3-angled; anther-filaments flattened; leaves linear, channelled, with a whitish stripe along the midrib; (garden-escape) . . . .*Ornithogalum*

         30 Perianth-segments strongly nerved longitudinally, coloured (rarely nearly white in *Camassia*); style and filaments filiform.

            31 Flowers several, cream-white to pale blue, violet, or bluish purple, ascending, their segments with at least 3 prominent nerves; pedicels jointed at the base of the flowers, to 2 cm long; leaves linear or lance-linear, to 4 dm long and 1 cm broad, acuminate; scapes to about 6 dm tall . . . .*Camassia*

            31 Flowers usually about 3, deep blue, horizontal or nodding, their segments about 12 mm long, with only the midnerve prominent; leaves linear-oblanceolate, subacute to obtuse; scapes to about 1.5 dm tall; (garden-escape) . . . .*[Scilla]*

  28 Styles 3, distinct nearly or quite to base; flowers several to many.

      32 Inflorescence a bractless spike-like raceme, the flowers white (drying yellowish); plant unisexual, the pistillate ones to over 1 m tall, the staminate ones lower and less leafy; staminate flowers on spreading pedicels in racemes to 12 cm long, their segments 3 or 4 mm long; pistillate flowers on erect or ascending pedicels in racemes to 3 dm long, their segments to about 1.5 cm long; capsule loculicidal, to about 1.5 cm long; basal leaves spatulate to obovate, the upper leaves linear; stem from a thick short tuberous rhizome; (s Ont.) . . . .*Chamaelirium*

      32 Inflorescence bracted at the bases of the pedicels; plants bisexual, the flowers perfect (but sometimes with an admixture of unisexual flowers); leaves narrowly linear to narrowly lanceolate or oblanceolate.

         33 Leaves equitant (enfolding each other lengthwise in 2 ranks); perianth-segments white or greenish, less than 5 mm long, lacking basal glands; anthers oblong or ovate, 2-locular; capsules septicidal, to about 8 mm long; plants from short or creeping rhizomes; (trans-continental species) . . . .*Tofieldia*

         33 Leaves not equitant; perianth-segments at least 6 mm long (commonly over 1 cm).

            34 Leaves tough and rather rigid, scabrous, pale green, linear-filiform, the basal ones very numerous in a dense tuft, to about 8 dm long and 6 mm broad; flowers white, to 1 cm long, numerous in a dense

raceme becoming about 6 dm long in fruit; pedicels erect in fruit (the raceme characteristically clavate as the lower flowers mature); capsules loculicidal, to about 6 mm long; stems to over 1.5 m tall, from a thick short woody rootstock; (mts. of s B.C. and sw Alta.) . . . .*Xerophyllum*

  34 Leaves softer, the basal ones less densely tufted and relatively wider; capsules septicidal, often over 1 cm long.

      35 Stems roughish-downy above and in the inflorescence, to about 1.5 m tall, from a thick rhizome; inflorescence an open pyramidal panicle to 4.5 dm long, the ascending to spreading branches being mostly simple racemes; flowers creamy (becoming green or purplish), scurfy outside; perianth-segments to 13 mm long, narrowed to a distinct claw, bearing a pair of glands at base; capsules to 18 mm long; seeds broadly winged; leaves firm, the basal ones broadly linear, attenuate, often as much as 3 dm long and 1 or 2 cm broad; (?Ont.) . . . .*[Melanthium]*

      35 Stems glabrous, lower, from a bulbous-thickened fibrous-rooted base; inflorescence a simple raceme (or somewhat compound, the lower branches often sparingly forking but relatively short); flowers greenish- or yellowish-white to bronze-colour or brownish-purple, their glabrous segments only obscurely clawed.

         36 Flowers mostly nodding on filiform pedicels (but the capsules erect in fruit), to about 1.5 cm long; perianth-segments narrowly lanceolate and with recurved tips, lacking a gland at base; capsule to about 1.5 cm long, its seeds winged; leaves to 3 dm long and 2 cm broad; (B.C.–sw Alta.) . . . .*Stenanthium*

         36 Flowers erect or ascending, to about 1 cm long; perianth-segments relatively broader, each bearing an obovate or obcordate gland at base, their tips not recurved; seeds wingless . . . .*Zigadenus*

ALETRIS L.   [1143]

A. farinosa L.   Unicorn-root

/t/EE/   (Grh)   Dry to moist, peaty or sandy soil from Mich. to s Ont. (Essex, Lambton, Middlesex, and Norfolk counties; CAN; MT; TRT; *see* s Ont. map by Soper 1962: fig. 26, p. 41), N.Y., and Maine, s to Tex. and Fla.

ALLIUM L.   [1049]   Onion, Garlic, Leek.   Oignon or Ail

1 Flowers all or mostly replaced by bulblets; leaves linear.

  2 Outer bulb-coats strongly fibrous-netted into diamond-shaped areas, the bulbs ovoid or conic-ovoid; perianth (when present) white or pink, the segments equalling or surpassing the stamens; leaves 2–5(7) mm broad, confined to the lower third of the stem.

    3 Ovary (and capsule) crestless; anther-filaments not dilated; leaves flat or slightly keeled; (Ont. to w N.B.) . . . .*A. canadense*

    3 Ovary inconspicuously crested with 6 low knobs; anther-filaments dilated below and united into a cup; leaves channelled; (B.C.–Alta.) . . . .*A. geyeri*

  2 Outer bulb-coats membranous or becoming only slightly fibrous; stems leafy nearly to the middle; (introd.).

    4 Leaves linear, subterete, hollow at least below and easily crushed, acute, to 2 mm broad; spathe a single scarious short-beaked bract; perianth (when present) greenish or purplish; stamens exserted, their filaments dilated, the inner 3 each with a pair of lateral appendages surpassing the anther; stems to over 1 m tall . . . .*A. vineale*

4 Leaves flattened at least toward the apex, minutely scabrous on the margins and keel; spathes green and herbaceous; stamens included.

    5 Leaves flat, not twisted, acute, to 1.5 cm broad; spathe a single long-beaked deciduous bract; perianth (when present) whitish or greenish; inner anther-filaments 3-cleft at the dilated base, the pair of lateral segments prolonged into flexuous awns; stems to over 1 m tall . . . .[A. sativum]

    5 Leaves thick and channelled at base, flattened above, twisted, round-tipped, to 4 mm broad; spathe consisting of 2 finally separate unequal persistent bracts, the longer one to 2 dm long; perianth (when present) yellowish brown, drying purplish; anther-filaments slender, unappendaged; stems to about 6 dm tall . . . .
. . . .[A. oleraceum]

1 Flowers rarely replaced by bulblets; bracts of spathe usually 2 (sometimes 3, rarely 4).

  6 Scapes inflated below the middle, hollow, to over 1 m tall; leaves all basal, hollow, in 2 rows; flowers numerous in a large umbel subtended by 2 or 3 reflexed bracts; bulb usually solitary; (introd.).

    7 Perianth-segments greenish white, spreading, subequal, to 5 mm long; ovary with 3 distinct nectar-pits; leaves semicircular in cross-section; bulb large, not attached to a rhizome; annual . . . .[A. cepa]

    7 Perianth-segments yellowish white, connivent, unequal, the larger ones to 8 mm long; ovary lacking conspicuous nectar-pits; leaves circular in cross-section; bulb attached to a rhizome; perennial . . . .[A. fistulosum]

  6 Scapes not inflated.

    8 Flowers bright yellow, on pedicels 2 or 3 times longer than the flowers; bulbs clustered, with membranous coats; leaves flat, to about 3 cm broad, about equalling the scape, this to about 4.5 dm tall; (introd. on Vancouver Is.) . . . .[A. moly]

    8 Flowers white, pink, roseate, or purple.

      9 Leaves slender-petioled, their lance-elliptic blades to 2 dm long and 6 cm broad, usually shrivelled before anthesis; capsule deeply 3-lobed, each locule 1-seeded; perianth white or whitish; bulbs to 6 cm long, with fleshy coats, clustered on a rhizomatous base; scapes to about 4 dm tall; (s Man. to N.S.) . . . . .A. tricoccum

      9 Leaves not definitely petioled, linear, at most 1.5 cm broad, present at anthesis; capsule usually only slightly lobed, each locule with 2 or more seeds; perianth white, pink, roseate, or purple.

        10 Pedicels much shorter than the flowers, the umbel capitate; ovary (and capsule) crestless; leaves terete, hollow; (transcontinental) . . . .
. . . . .A. schoenoprasum

        10 Pedicels equalling to several times longer than the flowers.

          11 Outer bulb-coats coarsely fibrous-netted, the bulb usually solitary; flowers white to pink or roseate, their segments acute or short-acuminate; stamens included; bracts 2(3), broadly ovate, abruptly acuminate, persistent.

            12 Ovary (and capsule) crestless . . . .[A. drummondii]

            12 Ovary (and capsule) with 6 more or less prominent crests at summit.

              13 Perianth-segments to 1 cm long; pedicels to 2.5 cm long; bracts of spathe to about 1.5 cm long; scapes to about 6 dm tall; (s B.C. to Sask.) . . . . .A. geyeri

              13 Perianth-segments to 7 mm long; pedicels rarely as much as 1.5 cm long; bracts of spathe about 1 cm long; scapes to about 3 dm tall; (Alta. to w Ont.) . . . . .A. textile

          11 Outer bulb-coats membranous or becoming only slightly fibrous (but often somewhat reticulate).

            14 Outer bulb-coats distinctly reticulate, the bulb usually solitary; leaves shorter than the scape; umbel erect; (s B.C.).

              15 Reticulation of the reddish outer bulb-coats consisting of transverse V-shaped meshes often appearing in vertical rows; perianth-segments to 8 mm long, acute or short-acuminate,

white or pinkish, entire, their tips not recurved; pedicels to 1.5 cm long; bracts scarcely 1 cm long; capsules distinctly 6-crested; leaves 2–4, slender and flexuous, soon withering; scapes to about 4.5 dm tall . . . . *A. amplectens*

15 Reticulation of the outer bulb-coats oblong or quadrate; scapes to about 3 dm tall.

16 Perianth-segments to over 1.5 cm long, long-acuminate, deep rose-purple, about 1/3 longer than the stamens, their tips recurved; pedicels to 2 cm long; bracts long-acuminate, to 2.5 cm long; capsules shallowly 3-crested near summit; leaves 2 or more, to 3 mm broad, soon withering . . . . *A. acuminatum*

16 Perianth-segments to 1 cm long, acute or acuminate, light roseate, equalled by the stamens, their tips not recurved; pedicels less than 1.5 cm long; bracts obtusish, about 1 cm long; capsules shallowly 6-crested near summit; leaves 2, to 1.5 cm broad, usually persisting at anthesis . . . . [*A. douglasii*]

14 Outer bulb-coats lacking distinct reticulation or the reticulation fine and in vertical lines; bracts of spathe mostly 1 or 2 cm long.

17 Umbel bent to one side or more or less nodding or deflexed in anthesis; perianth-segments white to roseate, acute to obtusish or rounded at apex, to 6 mm long; stamens exserted; pedicels to 3 cm long, loosely arching or flexuous; capsule 6-crested at the summit; bulbs usually clustered and crowning a barely persistent rhizome; leaves soft, to 6 mm broad, shorter than the scape, this to about 7.5 dm tall, angled; (B.C. to Ont.) . . . . *A. cernuum*

17 Umbels erect; bulbs usually solitary (or 2 or more on a rhizome in *A. stellatum* and *A. validum*); pedicels to about 1.5 cm long, barely arching; perianth-segments acutish to short-acuminate.

18 Stamens included in the rose-purple perianth, this to about 1 cm long; capsules crested at summit; leaves 2(1), flattened, basal, surpassing the 2-edged scape, this usually less than 1 dm tall; (s B.C.) . . . . *A. crenulatum*

18 Stamens equalling or finally exserted from the nearly white to roseate perianth; leaves usually several, shorter than to about equalling the scape, this to over 7 dm tall.

19 Capsule not crested at summit; leaves flat or slightly keeled, to 1.5 cm broad; scape slightly compressed and 2-edged; bulbs solitary or clustered; (s ?B.C.) . . . . . [*A. validum*]

19 Capsule 6-crested at summit; leaves thick, hard and firm, rounded on the back, to about 3 mm broad; scape terete except at summit; bulbs 1 or 2 on a rhizome; (B.C. to w Ont.) . . . . *A. stellatum*

A. acuminatum Hook.
/t/W/ (Gb) Rocky hillsides and prairies from sw B.C. (N to Pemberton and Lytton; see B.C. map by T.M.C. Taylor 1966*a*:96) to Calif. and w Colo.

A. amplectens Torr.
/t/W/ (Gb) Wet, often rocky slopes from sw B.C. (several localities in s Vancouver Is.; CAN) to s Calif. [*A. attenuifolium* Kell.; *A. nevii* of B.C. reports in part, not Wats., relevant collections from Vancouver Is. in CAN].

A. canadense L. Canada Garlic
/T/EE/ (Gb) Low woods, thickets, and meadows from Minn. to Ont. (N to Carleton Co.; TRT; not listed by Gillett 1958), Que. (N to the Montreal dist.), and w N.B. (Madawaska Co.; GH; not known

from P.E.I. or N.S.), s to Tex. and Fla. [Possibly merely the asexual phase of *A. mutabile* Michx. of the E U.S.A.].

[*A. cepa* L.]   Garden Onion
[Eurasian; a collection in CAN from N B.C. (Liard Hot Springs, 59°25′N) has been placed here by Porsild, with the note that it must have wintered in an abandoned garden plot. It cannot be considered an established member of our adventive flora.]

A. cernuum Roth   Nodding Onion
/T/X/   (Gb)   Dry woods, rocky banks, and prairies from B.C. (N to ca. 56°N) to Alta. (N to Dunvegan, 55°54′N), s Sask. (Cypress Hills and Carnduff; Breitung 1957*a*), Man. (N to Playgreen L., N of L. Winnipeg), Ont. (reported from Lake of the Woods by John Macoun 1888, from Pelee Is., Essex Co., by Dodge 1914, and from Wellington Co., by Stroud 1941), and N.Y., s to Ariz., Mexico, Tex., Mo., Ala., and Ga. [*A. recurvatum* Rydb.].

A. crenulatum Wieg.
/T/W/   (Gb)   Alpine slopes from sw B.C. (Mt. Arrowsmith and Bald Mt., Vancouver Is. CAN; Eastham 1947; *see* B.C. map by T.M.C. Taylor 1966*a*:96) to Wash. and Oreg. [*A. vancouverense* Macoun; *A. watsonii* Howell].

[*A. douglasii* Hook.]
[Reports of this species of the w U.S.A. (Wash. and Mont. to Oreg.) from s B.C. by J. M. Macoun (1894; Vancouver Is. and Botanie, near Spences Bridge) are based upon *A. acuminatum* (the Botanie plant) and *A. amplectens* (the Vancouver Is. plant), relevant collections in CAN. (Incl. var. *nevii* (Wats.) Ownbey & Mingrone (*A. nevii* Wats.)].

[*A. drummondii* Regel]
[A collection in CAN from Alta. (Malte and Watson, 3 June, 1925; Calgary) was referred to *A. nuttallii* Wats. (now generally merged with *A. drum.*) by Raup but, later, to *A. geyeri* Wats. by Boivin].

[*A. fistulosum* L.]   Welsh Onion
[Asiatic; a collection in CAN from an abandoned garden at Livengood, Alaska, has been placed here by J. P. Anderson. It does not become established in N. America.]

A. geyeri Wats.
/T/WW/   (Gb)   Plains and valleys from s B.C. (s Vancouver Is. and adjacent mainland N to near Spences Bridge; CAN; V; *see* B.C. map by T.M.C. Taylor 1966*a*:96) to s Alta. (Waterton Lakes; Calgary; Big Valley, 51°56′N) and Sask. (Boivin 1967*a*), s to Ariz., N.Mex., and Tex.
  With the exception of the Calgary plant, most of our material appears to belong to var. *tenerum* Jones (*A. fibrosum* Rydb.; *A. rubrum* Ost.; *A. rydbergii* Macbr.; most or all of the flowers replaced by bulblets). MAP: Marion Ownbey and H. C. Aase, Res. Stud. State Coll. Wash. 23: map 18, p. 87. 1955 (the typical form in map 16, p. 81).

[*A. moly* L.]   Lily Leek
[European; a collection in Herb. V from sw B.C. (clay cliff, Dallas Road, Victoria, Vancouver Is.) has been referred here. It does not become established in N. America.]

[*A. oleraceum* L.]   Wild Garlic
[European; reported as often found in waste heaps of s Ont. by Montgomery (1956). It does not become established in N. America]

[*A. sativum* L.]   Garlic
[Eurasian; a garden-escape in s Man. (near Otterburne; Löve and Bernard 1959) and s Ont. (Grey, Lambton, and Wellington counties; OAC), but scarcely becoming established.]

A. schoenoprasum L.   Chives.   Ciboulette or Brûlotte
/ST/X/EA/   (Gb)   Calcareous or basic rocks, gravels, and shores, the aggregate species from N-cent. Alaska–Yukon and the Mackenzie R. Delta to Great Bear L., Great Slave L., L. Athabasca (Alta. and Sask.), Man. (N to York Factory, Hudson Bay, 57°N), northernmost Ont., Que. (N to SE Hudson Bay, the Côte-Nord, Anticosti Is., and Gaspé Pen.), SE Labrador (Cartwright, 53°42′N), Nfld., N.B., and N.S. (not known from P.E.I.), s in the West to Oreg., Mont., and Colo., farther eastwards s to s-cent. Sask.–Man., Minn., Mich., and N New Eng.; Eurasia. MAPS and synonymy: *see* below.

1   Leaves slender, nearly equalling to overtopping the scape; bulbs crowded from a rhizomatous base; [Eurasian; a garden-escape in Ont. (Waterloo, Wellington, and Brant counties and near Timiskaming) and Que. (near Montreal; Amos); most early reports from Alaska–Canada refer to var. *sibiricum;* MAP (aggregate species): Hultén 1962: map 90, p. 99] . . . . .var. *schoenoprasum*
1   Leaves coarser, rarely reaching the umbel; bulbs few or solitary.
  2   Umbel usually at least 3.5 cm thick; perianth to nearly 1.5 cm long; sepals lance-attenuate; [*A. sibiricum* L.; transcontinental; MAPS: Porsild 1966: map 39, p. 71; Hultén 1968*b*:307; the N. American area shown in Hultén's 1962 map applies here] . . . . .var. *sibiricum* (L.) Hartm.
  2   Umbel less than 3.5 cm thick; perianth at most 1 cm long; sepals lance-ovate to ovate . . . . .var. *laurentianum* Fern.
    3   Flowers white; [Anticosti Is., E Que., the type locality] . . . . .f. *albiflorum* Rousseau
    3   Flowers deep pink; [Ont. (N shore of L. Huron), Que. (Rimouski Co., Gaspé Pen., and Anticosti Is.), Nfld. (type locality), N.B., and N.S.] . . . . .f. *laurentianum*

A. stellatum Nutt.   Prairie Onion
/T/WW/   (Gb)   Rocky or sandy prairies and hills from Sask. (N to Duck Lake, about 50 mi NE of Saskatoon; CAN; reports from B.C. and Alta. may be based upon *A. cernuum*) to Man. (N to Reader L., N of The Pas) and w Ont. (Rat Portage, N of Lake of the Woods, where taken by Fowler in 1887; CAN; collections in TRT from a railway embankment at Malachi, near Kenora, and from Guelph, Wellington Co., probably represent introductions), s to Wyo., Tex., Mo., and Ill.

A. textile Nels. & Macbr.   Wild Onion
/T/WW/   (Gb)   Dry prairies and rocky slopes from s Alta. (N to Veteran, 52°00′N) to Sask. (Cypress Hills, Old Wives Creek, Wood Mountain, and Moose Jaw; CAN) and s Man. (N to Birtle, about 60 mi NW of Brandon; reported from the N shore of L. Superior, Ont., by D. Löve (Can. J. Bot. 37:570. 1959; ?introd.)), s to Wash., Nev., N.Mex., Nebr., and Iowa. [*A. geyeri* var. *tex.* (N. & M.) Boivin; *A. reticulatum* Nutt., not J. & C. Presl; *A. angulosum sensu* Richardson 1823, perhaps not L.]. MAP: Marion Ownbey and H. C. Aase, Res. Stud. State Coll. Wash. 23: map 22, p. 95. 1955.

A. tricoccum Ait.   Wild Leek.   Ail des bois or Ail sauvage
/T/EE/   (Gb)   Rich woods and moist ground from s Man. (Morden, about 60 mi sw of Winnipeg; DAO) to Ont. (reported N to the Batchawana R. at the E end of L. Superior by John Macoun 1888), Que. (N to St-Joachim and Grosse-Ile, NE of Quebec City; *see* Que. map by Doyon and Lavoie 1966: fig. 17, p. 819), N.B., and N.S. (not known from P.E.I.), s to Iowa and Ga.

The similar *A. victorialis* ssp. *platyphyllum* Hult. of Eurasia is reported from the Aleutian Is. by Hultén (1937*a*; Attu; sterile specimens only).

[A. validum Wats.]   Pacific Garlic
[This species of the w U.S.A. (Wash. and Idaho to Calif. and Nev.) is reported from s B.C. by Eastham (1947; "Plentiful in running water at head of McIntyre Creek, Okanagan Falls, at 4500 ft."; Herb. V). It is felt that further collections are advisable before accepting it as a member of our flora.]

A. vineale L.   Field-Garlic
European; a noxious weed in lawns, fields, and meadows of the E U.S.A. and, as such, reported from s Ont. by F. H. Montgomery (Can. Field-Nat. 62(2):92. 1948; Huron Co.), Gaiser and Moore (1966; Lambton Co.), and Rousseau (Nat. can. (Que.) 98: 701. 1971; St. Catherines, Lincoln Co.).

ASPARAGUS L.   [1113]   Asparagus.   Asperge

A. officinalis L.   Garden Asparagus
European; known as a garden-escape in all provinces except Sask. (in Alta., N to Fort Vermilion, 58°24′N; Herbert Groh, Can. Field-Nat. 63(4): 125. 1949).

BRODIAEA Sm.   [1053]

1   Capsule sessile; flowers violet-purple, rarely more than 10, to 4 cm long, on firm, very unequal pedicels; fertile stamens 3, the anthers mostly 3–6 mm long, basifixed at the sagittate base, the filaments filiform; leaves about 2 mm broad, thick or somewhat terete; (sw B.C.) . . . . .B. coronaria
1   Capsule stipitate; flowers less than 3 cm long, commonly more numerous; fertile stamens 3, the anthers 1 or 2 mm long, the filaments dilated and somewhat united at base; leaves to about 1 cm broad; (sw B.C.).
   2   Flowers white or lilac, broadly campanulate, to 1.5 cm long; anthers yellow or purple, about 1 mm long, basifixed; capsules short-stipitate . . . . .B. hyacinthina
   2   Flowers blue, narrowly campanulate (concealing the ovary), to 2.5 cm long; anthers pale blue, about 2 mm long, more or less versatile, their filaments produced above the point of attachment as lateral teeth; capsule-stipe to 5 mm long . . . . .B. douglasii

B. coronaria (Salisb.) Jeps.   Harvest Lily
/t/W/   (Gb)   Rolling plains and foothills from sw B.C. (Vancouver Is. and adjacent islands and mainland; Herb. V; *see* B.C. map by T.M.C. Taylor 1966a:96) to Baja Calif. [*Hookera* Salisb.; *B. grandiflora* Sm., not *Triteleia gr.* Lindl., which is *B. douglasii* Wats.].

B. douglasii Wats.
/t/W/   (Gb)   Dry plains and hillsides from s B.C. (Vancouver Is. and adjacent islands; valleys of the Dry Interior E of the Coast Range N to Spences Bridge and Armstrong; Herb. V; *see* B.C. map by T.M.C. Taylor 1966a:102 (*Trit. gr.*)) to Oreg., Utah, and Wyo. [*Hookera* Wats.; *Triteleia grandiflora* Lindl., not *B. gr.* Sm., which is *B. coronaria* (Salisb.) Jeps.].

B. hyacinthina (Lindl.) Baker   White Brodiaea
/t/W/   (Gb)   Low moist grassy places from s B.C. (Vancouver Is. and adjacent islands and mainland E to near Chilliwack; Herb. V; *see* B.C. map by T.M.C. Taylor 1966a:102 (*Trit. hy.*)) to Calif. [*Hesperoscordum* Lindl.; *Triteleia* Greene; *Hookera* Ktze.; *Hesp. (B.) lacteum* Lindl.].

CALOCHORTUS Pursh   [1078]   Mariposa Lily

1   Petals not ciliate-fringed, purple with a median longitudinal green stripe and a yellow base, obovate, acute to acuminate, sparingly hairy within toward base, to 4.5 cm long; sepals to about 4.5 cm long, usually surpassing the petals, purple-tinged within; anthers to 12 mm long; capsule to 4.5 cm long and 8 mm thick, erect, its angles very narrowly winged; flowers rarely more than 3; stems relatively leafy, to about 5 dm tall, the narrowly linear leaves to about 1 dm long, revolute-margined, becoming recurved or curled; (B.C.) . . . . . .
. . . . .C. macrocarpus
1   Petals fringed with long cilia, to about 2.5 cm long; sepals to about 2 cm long, greenish white or purple-tinged at base; anthers to 8 mm long; capsules usually thicker and rather more broadly winged; leaf usually solitary and basal, to about 2 dm long, flat.
   2   Capsules erect, to 18 mm long; sepals and petals to about 1.5 cm long; petals narrowly ovate, acute, white tinged with bluish purple, the inner surface glabrous except for the doubly fringed basal scale; sepals greenish white; anthers to 5 mm long; flowers up to 8 in number; scape to 2 dm tall, the leaves about equalling or slightly surpassing it, less than 1 cm broad; (s B.C.) . . . . .C. lyallii
   2   Capsules finally nodding on recurved peduncles; petals obovate, obtuse, sparsely to densely hairy on the inner surface, longer than the sepals.
      3   Petals pale yellow, to 2.5 cm long, rather sparsely hairy on the inner surface, with

a very small basal gland; sepals greenish white, to 2 cm long; anthers to 8 mm long; capsules to 3 cm long; flowers up to 9 in number; leaves to 18 mm broad, commonly shorter than the scape, this to 4 dm tall; (B.C. and Alta.) . . . .*C. apiculatus*

3   Petals white, tinged with purple, to 2 cm long, densely hairy on the inner surface, with a broad basal gland; sepals greenish white, purple-tinged at base, to about 1.5 cm long; anthers to 5 mm long; capsules to 2 cm long; flowers rarely more than 3; leaves less than 1 cm broad, commonly surpassing the scape, this to about 2 dm tall . . . .[*C. elegans*]

C. apiculatus Baker
/T/W/   (Gb)   Woods and hillsides from the Rocky Mountain Trench of SE B.C. (N to Windermere; *see* B.C. map by T.M.C. Taylor 1966a:97) and its foothills in sw Alta. (Crowsnest Pass; Waterton Lakes) to Wash., Idaho, and Mont. MAP: Marion Ownbey, Ann. Mo. Bot. Gard. 27(4): map 2, p. 405. 1940.

[*C. elegans* Pursh]
[The report of this species of the w U.S.A. (Wash. and Mont. to Calif. and ?Utah) from B.C. by Henry (1915) is based upon *C. apiculatus*, according to Eastham (1947), as are, also, reports of its var. *nanus* Wood from B.C. by John Macoun (1888; relevant collections in CAN).]

C. lyallii Baker
/t/W/   (Gb)   Dry sandy or rocky woods and slopes from s B.C. (Similkameen R., close to the U.S.A. boundary s of Penticton) to N-cent. Wash. along the E slope of the Cascade Range. MAP: Marion Ownbey, loc. cit., map 4, p. 437.

C. macrocarpus Dougl.   Sagebrush Mariposa Lily
/T/W/   (Gb)   Dry sandy soils from B.C. (E of the Coast and Cascade ranges: in the Dry Interior N to near Williams Lake, ca. 52°N, and in the Rocky Mountain Trench N to 57 mi N of Kimberley; *see* B.C. map by T.M.C. Taylor 1966a:97) to Calif., Nev., and Mont. MAP: Marion Ownbey, loc. cit., map 6, p. 481. 1940.

CAMASSIA Lindl.   [1087]   Camass

(Ref.: Gould 1942)
1   Capsule subglobose, rarely over 1 cm long and 8 mm thick, its locules each bearing up to 5 seeds; flowers regular; perianth-segments white, blue, or blue-violet, 3-5-veined, to 1.5 cm long and 5 mm broad, usually withering separately at the base of the capsule after anthesis; pedicels to 3 cm long; leaves to 2 cm broad; scapes to 7 dm tall; (s Ont.) . . . .*C. scilloides*
1   Capsule ovate to oblong, to over 2.5 cm long and 1 cm thick, its locules bearing up to 10 or 12 seeds; perianth-segments to 3 or 4 cm long and 1 cm broad, with up to 9 longitudinal veins.
   2   Perianth-segments connivent and twisted together over the capsule after a short period of anthesis and early deciduous from the fruiting raceme; flowers regular; perianth-segments cream-white to blue-violet (rarely pure white or shades of pink or light violet); pedicels to 5 cm long; leaves to 2.5 cm broad; scapes to 13 dm tall; (sw B.C.) . . . .*C. leichtlinii*
   2   Perianth-segments withering separately but persistent at the base of the capsule in the fruiting raceme; flowers regular or slightly zygomorphic (5 perianth-segments curving upward, the 6th downward; both forms often occurring in the same raceme); perianth-segments typically pale to deep blue-violet (the 3 inner ones often slightly broader than the 3 outer); pedicels generally less than 2 cm long (occasionally to 3 cm); leaves to 2 cm broad; scapes to 8 dm tall; (s B.C. and sw Alta.) . . . .*C. quamash*

C. leichtlinii (Baker) Wats.   Great Camass
/t/W/   (Gb)   Wet flats and grassy plains from sw B.C. (s Vancouver Is. and adjacent islands; CAN; V; *see* B.C. map by T.M.C. Taylor 1966a:97; the report from the s mainland at Chilliwack by

Henry 1915, is presumably based upon a 1901 collection in CAN by J. M. Macoun, revised in 1940 to *C. quamash* ssp. *maxima* by Gould) to Calif. [*Chlorogalum* Baker]. MAP: Gould 1942: fig. 1, p. 713.

All of the B.C. material seen is referable to ssp. *suksdorfii* (Greenm.) Gould (*C.* (*Quamasia*) *suks*. Greenm.; perianth blue-violet (with occasional albino forms) rather than cream colour as in the typical form).

C. quamash (Pursh) Greene   Common Camass
/T/W/   (Gb)   Wet meadows (ranges of Canadian taxa outlined below), s to Calif. and Wyo. MAPS and synonymy: *see* below.
1   Perianth-segments to 10 mm broad and with up to 9 longitudinal veins (those of the outer whorl occasionally 3-veined); [sw B.C. (Vancouver Is. and adjacent islands; Chilliwack); MAP: Gould 1942: fig. 1, p. 713] . . . . .var. *maxima* (Gould) Boivin
1   Perianth-segments to about 5 mm broad, with 3 or 5 longitudinal veins . . . . .var. *quamash*
    2   Flowers white; [Vancouver Is., B.C.: type from Alberni; Victoria] . . . . .f. *albiflora* Henry
    2   Flowers blue-violet; [*Phalangium* Pursh; *C. esculenta* Lindl., not (Nutt.) Robins.; s B.C. (*see* B.C. map by T.M.C. Taylor 1966*a*:97, showing an isolated station near Haines, Alaska, where undoubtedly introd.) and sw Alta. (Crowsnest Forest Reserve; between Waterton Lakes and Pincher Creek); MAPS: Gould 1942: fig. 1, p. 713, and fig. 7, p. 729] . . . . .f. *quamash*

C. scilloides (Raf.) Cory   Eastern Camass, Wild Hyacinth
/t/EE/   (Gb)   Fields, meadows, and moist open woods from E Kans. and Iowa to Wisc. and s Ont. (Essex Co.: White Is. in the Detroit R. opposite Amherstburg, where taken by John Macoun in 1882; CAN; reported from Middle Sister Is. of the Erie Archipelago by Core 1948), s to Tex. and Ga. [*Cyanotris* Raf.; *Lemotrys* (*C.; Quamasia*) *hyacinthina* Raf.; *C.* (*Q.*) *esculenta* (Nutt.) Robins., not Lindl.; *C. fraseri* Torr.].

CHAMAELIRIUM Willd.   [950]   Devil's-bit

C. luteum (L.) Gray   Blazing-star, Fairy-wand
/t/EE/   (Grh)   Meadows, thickets, and rich woods from Ill. to Mich., Ohio, s Ont. (Brant, Norfolk, and Welland counties; CAN; OAC), N.Y., and Mass., s to Ark. and Fla. [*Veratrum* L.; *Helonias* Ait.; *H. ?dioica* Pursh].

CLINTONIA Raf.   [1117]

1   Flowers greenish yellow, commonly nodding, to 2 cm long, up to 8 in a loose umbel; leaves oblong to elliptic or obovate, glossy green, ciliate (otherwise glabrous), to about 3 dm long; scape relatively stout, usually surpassing the leaves; (Man. to Labrador, Nfld., and N.S.) . . . .*C. borealis*
1   Flowers white, erect, to 2.5 cm long, commonly solitary (rarely 2); leaves oblanceolate to obovate, more or less villous, to about 1.5 dm long; scape slender, rarely surpassing the leaves; (Alaska–B.C.–Alta.) . . . .*C. uniflora*

C. borealis (Ait.) Raf.   Bluebead-Lily, Corn-Lily
/sT/EE/   (Grh)   Coniferous forests from Man. (N to Bissett, about 90 mi NE of Winnipeg) to Ont. (N to the Severn R., Hudson Bay; John Macoun 1888), Que. (N to E James Bay at 52°15′N, L. Mistassini, and the Côte-Nord), Labrador (N to the Mealy Mts. at 53°49′N; Gillett 1954), Nfld., N.B., P.E.I., and N.S., s to Minn., Tenn., and Ga. [*Dracaena* Ait.; *Smilacina* Ker-Gawl].

Forma *lateralis* Peck (1 or 2 additional lateral umbels present in addition to the usually solitary terminal one) is reported from Que. by Marie-Victorin (1929*b*; Labelle and Montmorency counties).

C. uniflora (Schult.) Kunth   Queen's-cup
/T/W/   (Grh)   Coniferous forests from the southernmost Alaska Panhandle (*see* Hultén 1943: map 364, p. 560) through B.C. and sw Alta. (Waterton Lakes; South Kootenay Pass on the s B.C. boundary) to Calif. and Idaho. [*Smilacina uniflora* Menzies, the correct basionym according to

Hultén (1943), the earlier *S. borealis* var. *uniflora* Schultes having been published without an accompanying description]. MAP: Hultén 1968b:309.

CONVALLARIA L. [1128]

C. majalis L. Lily-of-the-valley. Muguet
Eurasian; a garden-escape or persisting for many years from old plantings in Ont., Que., N.B., P.E.I., and N.S., and reported from Nfld. by Rouleau (1956; ?garden-escape). MAP: Hultén 1958: map 240, p. 259.

DISPORUM Salisb. [1120] Mandarin

1 Stigma entire; style pubescent near the base or glabrous throughout; berry scarlet, with usually less than 6 seeds; flowers less than 1.5 cm long; leaves and stem copiously pubescent; (B.C. and sw Alta.) . . . .*D. hookeri*
1 Stigma with 3 short recurved-spreading lobes.
  2 Style woolly-pubescent throughout; perianth gibbously truncate at the broad base, to 2 cm long; berry yellow to bright salmon-red, with usually more than 6 seeds; plant short-woolly to nearly glabrate; (s B.C.) . . . .*D. smithii*
  2 Style glabrous or only slightly pubescent toward the base; berry red or scarlet.
    3 Leaves permanently crisp-pubescent beneath; flowers yellowish green, to 2.5 cm long; fruit oblong, pointed, with commonly 2 (occasionally 4 or 6) seeds; (s Ont.) . . . .
      . . . .*D. lanuginosum*
    3 Leaves sparingly pubescent to glabrate; flowers whitish, to about 1.5 cm long; fruit broadly obovate, rounded at summit, with up to over 12 seeds; (B.C. to Ont.) . . . .
      . . . .*D. trachycarpum*

D. hookeri (Torr.) Britt. Fairy Bells
/T/W/ (Grh) Shaded woods, ravines, and riverbanks from B.C. (N to the Stikine R. at Glenora, ca. 57°N; CAN; *see* B.C. map by T.M.C. Taylor 1966a:97; a report from SE Alaska requires confirmation) and sw Alta. (Waterton Lakes; Breitung 1957b) to Calif., Mont., and Idaho. [*Prosartes* Torr.]. MAP: Quentin Jones, Contrib. Gray Herb. Harv. Univ. 173: map 5, p. 23. 1951.
    Our material is referable to var. *oreganum* (Wats.) Jones (*Prosartes* (*D.*) *oreg.* Wats.; ovary and style pubescent rather than usually glabrous, the stamens usually distinctly exserted rather than included, the leaves mostly long-acuminate at apex rather than obtuse).

D. lanuginosum (Michx.) Nicholson Yellow Mandarin
/t/EE/ (Grh) Rich woods from s Ont. (Middlesex, Elgin, Oxford, Norfolk, Waterloo, Wentworth, and Dundas counties) to w N.Y., s to Ala. and Ga. [*Streptopus* Michx.; *Uvularia* Pers.]. MAP: Quentin Jones, loc. cit., map 2, p. 23.

D. smithii (Hook.) Piper Fairy Lantern
/t/W/ (Grh) Coastal woods and streambanks from sw B.C. (Vancouver Is.; CAN; V; the Vancouver Is. stations in the B.C. map by T.M.C. Taylor 1966a:98, should be extended northwards to include Nootka Sound, the type locality of *Uvularia smithii* Hook.) to cent. Calif. [*Uvularia* Hook.; *D. menziesii* Don]. MAP: Quentin Jones, loc. cit., map 4, p. 23.

D. trachycarpum (Wats.) B. & H.
/sT/(X)/ (Grh) Rich woods and thickets from cent. B.C. (N to Hudson Hope, ca. 56°N; a report from SE Alaska requires confirmation) to Alta. (N to near the mouth of the Firebag R. s of L. Athabasca at 58°43′N; CAN; GH), Sask. (N to Prince Albert), Man. (N to The Pas), and Ont. (Hearst, 49°42′N; Albany R. sw of James Bay at 52°11′N), s to Oreg., Ariz., N.Mex., and Nebr. [*Prosartes* Wats.; *Uvularia lanuginosa sensu* Hooker 1838, and Lowe 1943 (*D. lan.*), not Michx.; *U. ?puberula sensu* Richardson 1823, not Michx.]. MAP: Quentin Jones, loc. cit., map 3 (incomplete eastwards), p. 23.

ERYTHRONIUM L.   [1076]   Dog's-tooth-Violet, Fawn-Lily

(Ref.: Applegate 1935)
1   Flowers yellow (often darker coloured at base).
   2   Leaves commonly mottled with lighter areas, to about 5 cm broad; flower usually
      solitary; anthers yellow or purplish-brown; stigma essentially entire (or with perma-
      nently erect short lobes); (Ont. to N.S.) ....*E. americanum*
   2   Leaves not mottled, often broader; flower solitary to several; anthers white, golden-
      yellow, or purplish red; stigma 3-lobed, the lobes finally recurved; (mts. of B.C. and
      Alta.) ....*E. grandiflorum*
1   Flowers mostly white or whitish (rose-pink in *E. revolutum*), commonly solitary (but often
   2 or 3 or more).
   3   Flowers rich rose-pink; anthers golden-yellow; stigma deeply lobed, the lobes strongly
      recurved; leaves mottled; (sw B.C.) ....*E. revolutum*
   3   Flowers white or creamy-white (sometimes suffused with green, pink, or blue).
      4   Leaves rarely mottled; anthers golden-yellow.
         5   Stigma rather deeply lobed, the lobes finally recurved; flower usually solitary,
            normally bluish white (sometimes pinkish) and often suffused with green; (s Ont.) ...
            ....*E. albidum*
         5   Stigma essentially entire (or with permanently erect short lobes); flowers 1 to
            several, commonly pure white except for the yellow base, sometimes streaked
            with pink; (s B.C.) ....*E. montanum*
      4   Leaves commonly mottled with lighter areas; flowers 1 to several, white or
      creamy-white with a light-yellow base; (s B.C.).
         6   Stigma essentially entire (or with permanently erect short lobes); style short
            and strongly clavate; anthers white; (?B.C.) ....[*E. howellii*]
         6   Stigma rather deeply cleft, the lobes recurved; style long and filiform; anthers
            golden-yellow; (B.C.) ....*E. oregonum*

E. albidum Nutt.   White Dog's-tooth-Violet
/T/EE/   (Gb)   Rich woods and thickets from Minn. to s Ont. (N to the N shore of L. Ontario in
Hastings Co.; John Macoun 1888), s to Tex., Mo., Ky., and Ga. [*E. propullans* of s Ont. reports, not
Gray; *see* J. M. Macoun 1903].

E. americanum Ker   Yellow Adder's-tongue, Trout-Lily.   Ail doux
/T/EE/   (Gb)   Rich woods and thickets from Ont. (N to Michipicoten Is. in N L. Superior) to Que.
(N to the Ste-Anne-des-Monts R., Gaspé Pen.), N.B., and N.S. (not known from P.E.I.), s to Okla.,
Ala., and Fla. [*E. lanceolatum* Pursh]. MAP: *Atlas of Canada* 1957: map 13, sheet 38 (the line
should be extended northwards to include the sw Gaspé Pen., E Que.).
    Forma *backii* (Farw.) Dole (the anthers and the base of the perianth-segments purplish brown
rather than yellow) is known from Que. (Mt. Johnson, about 20 mi sw of Montreal; Rimouski and
Matane counties).

E. grandiflorum Pursh   Yellow Avalanche-Lily
/T/W/   (Gb)   Alpine meadows and slopes (ranges of Canadian taxa outlined below), s to Calif.
and Colo.
1   Anthers red; [incl. vars. *albiflorum* Purdy (not Hook.), *minus* Hook., *parviflorum* Wats.,
   and *giganteum* (Lindl.) Hook. (*E. giganteum* Lindl.); *E. obtusatum* Goodding; B.C. (N to
   near Dease L. at ca. 58°30'N; CAN; the B.C. map by T.M.C. Taylor 1966a:98, indicates
   a N limit at ca. 54°N) and w Alta. (N to Yellowhead Pass, near Jasper; CAN)] ....
   ....var. *grandiflorum*
1   Anthers white or yellow.
   2   Anthers white; [s B.C. and sw Alta.].............................var. *pallidum* St. John
   2   Anthers golden-yellow; [ssp. *chrysandrum* Applegate; SE B.C. (Kootenay L.; Eastham
      1947) and sw Alta. (Waterton Lakes; CAN)] ....var. *chrysandrum* (Applegate) Scoggan

Liliaceae

[E. howellii Wats.]
[The report of this obscure species of the w U.S.A. (Oreg. to Calif.; not listed by Hitchcock et al. 1969) from Vancouver Is. by J. M. Macoun (1913; this taken up by Henry 1915) requires confirmation, no relevant collections having been located by Applegate (1935) during his monographing of the genus in w N. America.]

E. montanum Wats.   White Avalanche-Lily
/T/W/   (Gb)   Alpine meadows from sw B.C. (the B.C. map by T.M.C. Taylor 1966a, indicates a single station, this on Mt. Waddington, about 175 mi NW of Vancouver; J.M. Macoun 1913, also reports it from Grand Central L., Vancouver Is.) to Oreg.

E. oregonum Applegate   White Fawn-Lily
/t/W/   (Gb)   Grassy ledges and open moist woods from sw B.C. (Vancouver Is. and adjacent islands and mainland of the lower Fraser Valley; CAN; V; *see* B.C. map by T.M.C. Taylor 1966a:98) to s Oreg. [*E. giganteum* of auth., not Lindl.; *E. grandiflorum* var. *albiflorum* Hook., not Purdy].

E. revolutum Sm.   Pink Fawn-Lily
/t/W/   (Gb)   Meadows and open moist woods from sw B.C. (Vancouver Is. and adjacent islands and mainland; CAN; V; *see* B.C. map by T.M.C. Taylor 1966a:98) to Calif. [*E. grandiflorum* var. *smithii* Hook.].

FRITILLARIA L.   [1074]

1   Flowers yellow or orange, purple-tinged, campanulate, to 2.5 cm long, nodding, usually solitary (sometimes 2); style 1, the stigma shallowly 3-lobed; stamens nearly equalling the perianth-segments; capsule obtusely angled, scarcely winged; leaves linear to narrowly oblanceolate, to about 1 dm long, scattered or somewhat whorled; stems slender, to about 2 dm tall; (B.C. and s Alta.) . . . .*F. pudica*
1   Flowers deep brownish-purple, usually more or less tinged with greenish yellow, to 3.5 cm long, commonly 2 or more; styles 3, united only near the base; stamens much shorter than the perianth-segments; leaves lanceolate to oblong- or ovate-lanceolate, usually in 2 or 3 whorls; stems stout, to about 6 dm tall.
  2   Flowers campanulate, unmottled, crowded near the top of the stem; capsule obtusely angled, scarcely winged; leaves rarely over 7 cm long; (Alaska–Yukon–B.C.) . . . .
      . . . .*F. camtschatcensis*
  2   Flowers deeply bowl-shaped, distinctly mottled, remote; capsule acutely angled and broadly winged; leaves to about 1 dm long; (s B.C.) . . . .*F. lanceolata*

F. camschatcensis (L.) Ker-Gawl.   Rice-root Lily
/ST/W/eA/   (Gb)   Coastal meadows and thickets from the Aleutian Is., s-cent. Alaska, and s Yukon (*see* Hultén 1943: map 362, p. 560) through coastal B.C. (and locally inland up the Skeena and Bulkley valleys as far E as Aleza L., near Prince George) to Wash.; Japan and Kamchatka, Siberia. [*Lilium* L.]. MAPS: Hultén 1968b:308; D. E. Beetle, Madroño 7(5): fig. 4, p. 154. 1944.

F. lanceolata Pursh   Chocolate-Lily, Mission-bells, Rice-root Lily
/t/W/   (Gb)   Coastal open woods, thickets, meadows, and slopes from s B.C. (Vancouver Is. and adjacent mainland, and the Dry Interior valleys N to Kamloops and Sicamous; CAN; V; *see* B.C. map by T.M.C. Taylor 1966a:99) to s Calif. [*F. mutica* Lindl.]. MAP: D. E. Beetle, loc. cit., fig. 4, p. 154.

F. pudica (Pursh) Spreng.   Yellow-bell, Yellow Snowdrop
/T/W/   (Gb)   Dry well-drained flats and slopes from s B.C. (valleys of the Dry Interior N to Kamloops and Sicamous; CAN; V; *see* B.C. map by T.M.C. Taylor 1966a:99) and sw Alta. (Milk River Ridge, about 50 mi SE of Lethbridge) to Calif., Utah, and ?N.Mex. [*Lilium* Pursh; *Ochrocodon* Rydb.]. MAP: D. E. Beetle, loc. cit., fig. 4, p. 154.

HEMEROCALLIS L.   [1019]   Day-Lily.   Lis d'un jour

1   Perianth dull orange, scentless, to 1 dm broad, its undulate segments reticulately veined; leaves to 2 cm broad; stem to 1 m tall, with a few scale-like leaves; roots thick but lacking swollen tubers . . . .*H. fulva*
1   Perianth lemon-yellow, to 8 cm broad, scented, its flat segments lacking transverse veins; leaves to 1 cm broad; stem to 8 dm tall, naked or nearly so; roots with swollen tubers . . . .*H. lilioasphodelus*

H. fulva L.   Orange Day-Lily.   Lis jaune or Lis d'un jour
Eurasian; a garden-escape or persisting from old plantings in N. America, as in Ont. (N to New Liskeard, 47°31′N), Que. (N to the Gaspé Pen. near the mouth of the Bonaventure R.; RIM), N.B., P.E.I. (D. S. Erskine 1960), and N.S. (Roland 1947).

H. lilioasphodelus L.   Yellow Day-Lily
Asiatic; a garden-escape in N. America, as from Ont. to N.B. according to Boivin (1967a). [*H. flava* L.].

LILIUM L.   [1072]   Lily.   Lis

1   Flowers erect; perianth-segments loosely ascending, narrowed below into claws.
  2   Perianth-segments pubescent within at base, lacking large spots, their veins bearing pale to dark pustules; leaves all alternate, the uppermost ones sometimes subtending axillary bulblets; summit of stem and peduncle sparingly pubescent; (garden-escape) . . . .*L. bulbiferum*
  2   Perianth-segments glabrous, blotched near base with large purple-brown spots, their veins lacking pustules; lower leaves alternate, the upper in 1 or 2 whorls; stem glabrous; (B.C. to Que.) . . . .*L. philadelphicum*
1   Flowers nodding, the peduncles becoming erect in fruit; perianth-segments not clawed.
  3   Leaves all alternate, the upper ones subtending dark bulbs; flowers pinkish- to orange- or brick-red, freely blackish-spotted within; stem cobwebby; (garden-escape) . . . .*L. tigrinum*
  3   Leaves mostly in whorls, not subtending bulbs; stem glabrous or minutely pubescent toward the tip.
    4   Flowers roseate to dull purple or purplish-red with purplish-black spots, their segments reflexed to the pedicel; anthers red; bulbs yellow; (garden-escape) . . . .[*L. martagon*]
    4   Flowers yellow to orange or orange-red with purplish spots, their segments spreading-recurved; anthers yellow; bulbs white; (Ont. to N.B. and N.S.) . . . .*L. canadense*

L. bulbiferum L.   Orange Lily
European; a garden escape or persisting from old plantings in Que. (between Beaumont and St-Vallier, Bellechasse Co.; Berthier-en-Bas, Montmagny Co.) and N.B. (near Moncton; CAN). [*L. croceum* Chaix].

L. canadense L.   Canada Lily
/T/D/   (Gb)   Meadows, thickets, and moist woods: var. *parviflorum* from s-cent. B.C. to N Calif. and Idaho; var. *canadense* (incl. the range of var. *umbelliferum*) from s Ont. (N to York Co.) to Que. (N to the Matapedia Valley, sw Gaspé Pen.), N.B., and N.S. (not known from P.E.I.), s to Ohio, Ala., Ky., and Va. MAPS and synonymy: see below.
1   Flowers relatively small, the bright reddish-orange perianth-segments to about 6 cm long and 12 mm broad; anthers to about 7 mm long; leaves to about 1 dm long; stems to about 12 dm tall; [*L. parv.* (Hook.) Holz.; *L. columbianum* Hanson; B.C. N to ca. 54°N; see B.C. map by T.M.C. Taylor 1966a:99 (*L. col.*)] . . . . .var. *parviflorum* Hook.
1   Flowers larger, the perianth-segments to 8 or 9 cm long and about 2 cm broad; leaves to over 1.5 dm long and 2 cm broad; stems to over 2 m tall; (eastern varieties).

2 Flowers orange or orange-red, the perianth-segments strongly reflexed (their tips reaching or surpassing the base of the perianth-tube, the stamens and style long-exserted); [*L. michiganense* Farw.; incl. *L. superbum* L.; *L. carolinianum* (*L. michauxii* Poir.) *sensu* John Macoun 1888, at least in part, not Michx.; s Ont.: N to Huron, Wellington, and York counties] . . . . .var. *umbelliferum* (Farw.) Boivin
2 Flowers typically yellow or orange-yellow, the perianth-segments recurved from near the middle but scarcely reflexed (the stamens and style included or barely exserted) . . . . .var. *canadense*
   3 Perianth-segments numerous (about 10 or 12); [type from St-Proper, Champlain Co., Que.] . . . . .f. *duplex* Boivin
   3 Perianth-segments 6.
      4 Flowers red; [Huntingdon Co., s Que.] . . . . .f. *rubrum* Britt.
      4 Flowers yellow or orange-yellow; [f. *flavum* (Waugh) Vict.; s Ont. to N.B. and N.S.; MAP: Fernald 1921: map 10, pl. 130 (facing p. 120, legend, p. 300; partial area); concerning an early report from Man., *see* Scoggan 1957:211] . . . . .f. *canadense*

[*L. martagon* L.] Martagon or Turk's-cap Lily
[Eurasian; reported from Que. by Rousseau (Nat. can. (Que.) 98: 701. 1971; Sillery, near Quebec City), where probably taken in an abandoned garden.]

L. philadelphicum L. Wood-Lily, Wild Orange-red Lily
/sT/X/ (Gb) Dry open woods, thickets, clearings, and grassy meadows (ranges of Canadian taxa outlined below), s to N.Mex., Nebr., Ky., and N.C.
1 Leaves in up to 6 whorls (but often a few scattered ones toward the base of the stem) . . . . .var. *philadelphicum*
   2 Flowers yellow, devoid of the usual dark spots; [near Moosehorn, Man.] . . . . .f. *flaviflorum* Williams
   2 Flowers orange to deep orange-red; [Man. (native southwards; ?introd. along a railway from Gillam to as far N as Churchill), Ont. (N to L. Nipigon, Longlac, and Hearst, N of L. Superior), and Que. (N to L. Timiskaming; Baldwin 1958)] . . . . .f. *philadelphicum*
1 Leaves mostly scattered (sometimes whorled at the upper 1–3 nodes) . . . . .var. *andinum* (Nutt.) Ker
   3 Flowers lemon-yellow, devoid of the usual dark spots; [known from the type locality, Jenkins L., N Alta.] . . . . .f. *immaculata* Raup
   3 Flowers reddish to scarlet; [*L. andinum* Nutt.; *L. montanum* Nels.; *L. umbellatum* Pursh; sw Dist. Mackenzie–B.C. to N-cent. Alta., Sask. (N to Prince Albert; provincial floral emblem), Man. (N to Norway House, off the NE end of L. Winnipeg), Ont. (N to w James Bay at ca. 52°30′N), and w Que. (Fernald *in* Gray 1950)] . . . . .f. *andinum*

L. tigrinum L. Tiger-Lily. Martagon or Matagon
Asiatic; a garden-escape or persisting from old plantings in N. America, as in s Ont. (Lambton and Wellington counties), Que. (N to the Gaspé Pen.; Raymond 1950*b*), N.B., and N.S.

LLOYDIA Salisb. [1077]

L. serotina (L.) Rchb. Alpine Lily
/aST/W/EA/ (Gb) Tundra and alpine meadows and slopes from the Aleutian Is. and the coasts of Alaska–Yukon (*see* Hultén 1943: map 363, p. 560) and mts. of NW Dist. Mackenzie (Richardson Mts., w of the Mackenzie R. Delta) through B.C. to Oreg., Nev., and N.Mex.; Eurasia. [*Bulbocodium* L.; *Anthericum* L.]. MAPS: Hultén 1968*b*:308; Porsild 1966: map 40, p. 71; *Atlas of Canada* 1957: map 6, sheet 38; Meusel, Jaeger, and Weinert 1965:96.
   Var. *flava* (Calder & Taylor) Boivin (ssp. *flava* C. & T.; petal-venation yellow or green rather than purplish) is known only from the type locality, Moresby Is., Queen Charlotte Is., B.C.

MAIANTHEMUM Weber [1119] Wild Lily-of-the-valley

1   Stem-leaves sessile or nearly so and often cordate-clasping, with a shallow and narrow basal sinus; berry pale red, speckled, about 4 mm thick; plant commonly less than 2 dm tall; (transcontinental) .... *M. canadense*
1   Stem-leaves petioled (the petiole of the lowest one commonly about equalling the blade), with a deep and open basal sinus; berry red, about 6 mm thick; plant to about 4 dm tall; (Alaska–w B.C.) .... *M. dilatatum*

M. canadense Desf.   Wild Lily-of-the-valley.   Muguet
/ST/X/   (Grh (Hpr))   Moist woods, thickets, and recent clearings (ranges of Canadian taxa outlined below), s to B.C.–Alta.–Sask.–Man., Iowa, Tenn., and Ga. MAPS and synonymy: *see* below.
1   Plant distinctly pubescent; [sw Dist. Mackenzie and E B.C. (s to ca. 54°N according to the B.C. map for *M. canadense* by T.M.C. Taylor 1966*a*:100, this actually referable to var. *interius;* replaced in w B.C. by *M. dilatatum*), Alta. (N to L. Athabasca), Sask. (N to L. Athabasca), Man. (N to Reindeer L. at ca. 58°N), and Ont. (N to Big Trout L. at ca. 54°N, 90°W); MAPS: Hultén 1962: map 175, p. 187; Meusel, Jaeger, and Weinert 1965:98] ....
....var. *interius* Fern.
1   Plant glabrous ....var. *canadense*
    2   Stem-leaves 3 or 4; [?Que.; Marie-Victorin 1929*b*; probably throughout the range of the species] ....f. *trifolium* (Farw.) Vict.
    2   Stem-leaves 2 (sometimes 1); [*Unifolium* Greene; *Smilacina* Pursh; *S. bifolia* var. *can.* (Desf.) Gray; *S. (M.) bifolia* of most Canadian reports, not Desf.; SE Man. (Whiteshell Forest Reserve; reports N to the Nelson R. at Oxford House by John Macoun 1888, probably refer to var. *interius*) to Ont. (N to the Severn R.; John Macoun 1888), Que. (N to E James Bay and L. Mistassini), Labrador (N to Nain, 56°33′N), Nfld., N.B., P.E.I., and N.S.; MAPS: on the above-noted maps by Hultén and Meusel, Jaeger, and Weinert]. In the legend to Hultén's map, the words "var. *canadense*" and "var. *interior*" (*interius*) should be interchanged to correspond with the areas outlined by the respective "line" symbols. The aggregate species is a characteristic element of the boreal coniferous forest and probably occurs in each province N to the N limits of black and white spruce ....f. *canadense*

M. dilatatum (Wood) Nels. & Macbr.
/sT/W/eA/   (Grh)   Moist shaded places from the Aleutian Is. and s Alaska (*see* Hultén 1943: map 367, p. 560) through coastal B.C. (*see* B.C. map by T.M.C. Taylor 1966*a*:100) to Calif. and Idaho; E Asia. [*M. bifolium* var. *dil.* Wood; *Unifolium* Howell]. MAPS: Hultén 1968*b*:311, and 1962: map 175, p. 187; Meusel, Jaeger, and Weinert 1965:98.

MEDEOLA L. [1135]

M. virginiana L.   Indian Cucumber-root.   Concombre sauvage or Jarnotte
/T/EE/   (Grh)   Rich woods from Minn. to Ont. (N to the NW shore of L. Superior near Thunder Bay; OAC), Que. (N to Ste-Anne-de-la-Pocatière, Kamouraska Co.; QSA), N.B., P.E.I., and N.S., s to La. and Fla. [*M. "virginica"* of Macoun 1888].

[MELANTHIUM L.] [971] Bunch-flower

[An Ontario collection by Goldie of *M. (Leimanthium) hybridum* Walt. is reported by Hooker (1838; "Upper Canada"), this taken up tentatively by John Macoun (1888), who saw no relevant collections, under the synonhmy of *M. virginicum* L. Neither of these species of the E U.S.A. is currently accredited a range northwards beyond s N.Y.]

MUSCARI Mill.   [1095]   Grape-Hyacinth

M. botryoides (L.) Mill.
European; a garden-ornamental in N. America and inclined to persist for many years, as in s Ont. (N to Frontenac Co.), Nfld. (Rouleau 1956), and N.S. (Mahone Bay, near Halifax; CAN).

[NOTHOSCORDUM Kunth]   [1050]   False Garlic

[N. bivalve (L.) Britt.]
[The report of this species of the E U.S.A. (Nebr. to Ohio and Va., s to Mexico, Tex., and Fla.) from sw B.C. by Carter and Newcombe (1921; Foul Bay Road, Victoria, Vancouver Is.) requires confirmation. It may possibly have been introd. there but scarcely established. (Ornithogalum L.).]

ORNITHOGALUM L.   [1089]   Star-of-Bethlehem

1   Flowers about 3 cm long, on pedicels not over 1 cm long, finally nodding in an elongate 1-sided raceme; anther-filaments not narrowed, terminating in two flat teeth that reach to about the tip of the anther; leaves to about 3 dm long and 6 mm broad; scape to about 6 dm tall . . . .[O. nutans]
1   Flowers to 2 cm long, on spreading-ascending pedicels to 6 cm long, in a corymbose raceme; anther-filaments tapering to apex; leaves to about 6 dm long and 1 cm broad; scape usually less than 2.5 dm tall . . . . .O. umbellatum

[O. nutans L.]
[Eurasian; a garden-escape in the E U.S.A. and reported from low wet woods of s Ont. by Montgomery (1956; a 1950 collection by Soper from Stamford Township, Welland Co., in TRT).]

O. umbellatum L.   Nap-at-noon
European; a garden-escape tending to persist in E N. America, as in s Ont. (Lambton, York, Wellington, Bruce, and Wentworth counties), Que. (Kingsmere, Gatineau Co.; Sillery, near Quebec City), N.S. (Yarmouth), and Nfld. (near St. John's).

POLYGONATUM Mill.   [1123]   Solomon's-seal.   Sceau-de-Salomon

(Ref.: Ownbey 1944)
1   Leaves glabrous on both surfaces; flowers to 2 cm long, up to about 10 on a peduncle; anther-filaments inserted near the middle of the perianth-tube; (SE Sask. to sw Que.) . . . .
. . . . .P. biflorum
1   Leaves minutely pilose or hirsute along the nerves beneath; flowers 1–3(5) on a peduncle, rarely over 1.5 cm long; anther-filaments inserted high in the perianth-tube.
    2   Stem, peduncles, and pedicels hirsute; (introd. in Ont. and Que.) . . . . .P. multiflorum
    2   Stem, peduncles, and pedicels glabrous; (Ont. to N.B. and N.S.) . . . . .P. pubescens

P. biflorum (Walt.) Ell.
/T/(X)/   (Grh)   Moist woods and thickets from SE Sask. (Boivin 1967a) to s Man. (N to the Lee R. about 85 mi NE of Winnipeg), Ont. (N to Wellington and York counties), sw Que. (N to the Montreal dist.), N.Y., and Conn., s to N Mexico, N.Mex., Tex., and Fla. [Incl. var. melleum (Farw.) Ownbey; Convallaria Walt.; P. canaliculatum of auth., perhaps not (Muhl.) Pursh; P. commutatum (Schult.) Dietr.; P. giganteum Dietr.]. MAP: combine the maps by Ownbey 1944: map 2 (P. bifl.), p. 395, and map 3 (P. comm.), p. 404 (Que. stations not indicated on either).

P. multiflorum (L.) All.
Eurasian; a garden-escape in E N. America and reported from E Ont. and s Que. by Boivin (1967a). [Convallaria L.]

P. pubescens (Willd.) Pursh
/T/EE/   (Grh)   Rich rocky woods and thickets from Ont. (N to Batchawana Falls, near the E end

of L. Superior, and L. Timiskaming at ca. 47°30'N) to Que. (N to near Rimouski, Rimouski Co.; RIM), N.B., and N.S. (not known from P.E.I.), s to Iowa, Tenn., and Ga. [*P. biflorum* of reports from N.B. and N.S., not (Walt.) Ell., relevant collections in CAN; concerning the report from Man. by Ownbey, *see* Scoggan 1957]. MAP: Ownbey 1944: map 1, p. 385.

[SCILLA L.] [1086] Squill

1 Flowers lavender to violet-blue, divergent, up to about 15 in a 1-sided tip-nodding raceme, each flower subtended by a pair of bluish bracts, the lower linear-lanceolate bracts surpassing the pedicel, the upper ones gradually reduced; scape to 5 dm tall . . . .[*S. nonscripta*]
1 Flowers bright blue, nodding, commonly about 3, minutely bracted; scape 1 or 2 dm tall . . . .[*S. sibirica*]

[*S. nonscripta* (L.) Hoffmgg. & Link]   English Bluebell or Harebell
[European; an occasional garden-escape in N. America, reported from sw B.C. by Boivin (1967a; presumably Vancouver Is.), where probably not established. (*Hyacinthus* L.; *Endymion* Garcke).]

[*S. sibirica* L.]
[Asiatic; an occasional garden-escape in N. America, reported from Que. by Marcel Raymond (Ann. ACFAS 15: 101. 1949; Rigaud, Vaudreuil Co.) and from Nfld. by Rouleau (1956), but scarcely established in our area.]

SMILACINA Desf.   [1118]   False Solomon's-seal

(Ref.: Galway 1945)
1 Flowers short-pedicelled in a racemose panicle; berries red, with purple dots; leaves finely pubescent beneath, several on a slightly zigzag stem to about 1 m tall; rhizome thick and fleshy; (B.C. to N.S.) . . . .*S. racemosa*
1 Flowers relatively long-pedicelled in a simple raceme; rhizomes relatively slender; (transcontinental).
    2 Leaves not more than 4, glabrous, on a stem commonly less than 2 dm tall; raceme rather long-peduncled; berries dark red, 5 or 6 mm thick . . . .*S. trifolia*
    2 Leaves up to 12, minutely downy beneath (sometimes glabrate), on a stem to over 6 dm tall; racemes sessile or short-peduncled; berries black, or green with black stripes . . . .*S. stellata*

S. racemosa (L.) Desf.   False Spikenard
/T/X/   (Grh)   Rich woods, thickets, and clearings (ranges of Canadian taxa outlined below), s to Calif., Ariz., Colo., Mo., Tenn., and Ga. MAPS and synonymy: *see* below.
1 Leaves merely acute or short-acuminate, commonly sessile and more or less clasping; style about 1 mm long; [incl. the broad-leaved extreme, var. *jenkensii* Boivin; *S. (Vagnera) amplexicaulis* Nutt.; *S. brachypetala* Rydb.; B.C. (N to Hudson Hope, ca. 56°N; *see* the B.C. map for *S. racemosa* by T.M.C. Taylor 1966a:100, this actually applicable to var. *amplex.*), Alta. (N to Lesser Slave L.), and sw Sask. (Cypress Hills; Breitung 1957a); the tentative report of the typical form from Wrangell, SE Alaska, by Hultén (1943; *see* his map 365, p. 560) would presumably be referable here if confirmed; MAP: Galway 1945: fig. 1, p. 649] . . . . .var. *amplexicaulis* (Nutt.) Wats.
1 Leaves acuminate, the lower ones contracted at base into distinct petioles; style about 0.5 mm long; [*Convallaria* L.; *Vagnera* Morong; incl. f. *foliosa* Vict. (panicle leafy below), var. *cylindrata* Fern. (*S. ?ciliata* Desf.), and the narrow-leaved extreme, var. *lanceolata* Boivin; SE Man. (Riding Mt.; Shoal L., near the Ont. boundary), Ont. (N to Matheson, 48°32'N), Que. (N to the Côte-Nord), N.B., P.E.I., and N.S.; the report from Nfld. by Waghorne 1888, requires confirmation; MAPS: Hultén 1968b:309; Galway 1945: fig. 1 (incomplete northwards), p. 649] . . . . .var. *racemosa*

S. stellata (L.) Desf.   Star False Solomon's-seal
/ST/X/   (Grh)   Thickets, meadows, and sandy or gravelly shores from s Alaska and s-cent. Yukon (*see* Hultén 1943: map 366, p. 560) to Great Bear L., northernmost Alta., Sask. (N to Prince Albert), Man. (N to the Hayes R. about 75 mi sw of York Factory), northernmost Ont., Que. (N to E James Bay at ca. 51°45′N, the Côte-Nord, Anticosti Is., and Gaspé Pen.), SE Labrador, Nfld., N.B., P.E.I., and N.S., s to s Calif., N.Mex., Ohio and Va. [Incl. vars. *crassa* Vict. and *sylvatica* Vict. & Rousseau; *Convallaria* L.; *Vagnera* Morong; *Unifolium* (S.; V.) *liliaceum* Greene; *S.* (*V.*) *sessilifolia* Nutt.]. MAPS: Hultén 1968*b*:310; Raup 1947: pl. 20; Galway 1945: fig. 2 (incomplete northwards), p. 650.

S. trifolia (L.) Desf.
/ST/X/   (Grh)   Bogs, mossy woods, and peaty shores from southernmost Yukon (Watson L. at 60°05′N) to Dist. Mackenzie (N to Norman Wells, 65°17′N; CAN), s Dist. Keewatin, northernmost Ont., Que. (N to Mollie T. Lake at ca. 55°N, the Côte-Nord, Anticosti Is., and Gaspé Pen.), Labrador (N to Makkovik, 55°10′N), Nfld., N.B., P.E.I., and N.S., s to s Alta., Minn., N Ill., Pa., and N.J. [*Convallaria* L.; *Vagnera* Morong]. MAPS: Hultén 1968*b*:310; Galway 1945: fig. 2 (incomplete northwards), p. 650.

SMILAX L.   [1151]   Greenbrier, Catbrier

1   Stem woody, it and the branches prickly; leaves glabrous; plants climbing by tendrils.
   2   Prickles rigid, broad-based, pale or merely dark-tipped; pedicels at most 7 mm long; peduncles rarely recurving, at most 1.5 cm long and often shorter than the subtending petiole; berry blue; stems flexuous, from long slender rhizomes; (s Ont. and N.S.) . . . . *S. rotundifolia*
   2   Prickles weak, slender-based, mostly blackish; pedicels to about 12 mm long; peduncles divergent to drooping, to 6 or 7 cm long and mostly surpassing the subtending petiole; berry black; stems not flexuous, from short knotted rhizomes; (Ont.) . . . . *S. tamnoides*
1   Stem annual, herbaceous, without prickles.
   3   Plant self-erect or nearly so, without or with only weak upper tendrils; peduncles mostly in the axils of bladeless bracts below the leaves; anthers and filaments subequal; (s Ont.) . . . . *S. ecirrhata*
   3   Plant climbing by tendrils borne from the axils of most of the middle and upper leaves; peduncles mostly in the axils of broad-bladed leaves; anthers shorter than their filaments; (s Sask. to Que. and w N.B.) . . . . *S. herbacea*

S. ecirrhata (Engelm.) Wats.
/T/EE/   (Grh)   Open woods from S.Dak. and Minn. to s Ont. (Essex, Lambton, and Ontario counties; OAC; TRT), s to Mo. and Tenn.

S. herbacea L.   Carrion-flower, Jacob's-ladder.   Raisin de couleuvre
/T/X/   (Hpr)   Alluvial thickets, low woods, and meadows (ranges of Canadian taxa outlined below), s to Mont., Colo., Mo., Tenn., Ala., and Ga.
1   Leaves bright green and shining beneath, minutely pilose on the veins beneath; peduncles to 10 times as long as the subtending petioles; flowers usually not more than 35; perianth to 6 mm long; [*S.* (*Nemexia*) *pul.* Michx.; reported from s Ont. by Soper 1949, and from s-cent. Sask. and w Ont. by Boivin 1967*a*] . . . . var. *pulverulenta* (Michx.) Gray
1   Leaves glaucous or pale beneath; perianth at most 5 mm long; umbels with up to over 100 flowers.
   2   Leaves minutely pubescent beneath, their petioles to about 9 cm long; peduncles commonly not more than twice as long as the petioles; bladeless lower bracts spreading-ascending; berry black; [*S.* (*Nemexia*) *las.* Hook.; Sask. (N to Runciman, ca. 53°N), Man. (N to Dauphin), and s Ont. (Lambton and Kent counties)] . . . . var. *lasioneura* (Hook.) A. DC.
   2   Leaves glabrous beneath, the longest petioles less than 5 cm long; peduncles to 8 times as long as the subtending petioles; bladeless lower bracts appressed-ascending; berry blue; [*S. peduncularis* Muhl.; Sask. (Boivin 1967*a*); Ont. (N to Thunder Bay); Que. (N to near Quebec City), and w N.B. (St. John R. valley)] . . . . var. *herbacea*

S. rotundifolia L.   Common Greenbrier, Bullbrier, Horsebrier
/T/EE/   (Mc (vine))   Moist to dryish woods and thickets from Okla. to Ohio, s Ont. (Point Pelee and near Leamington, Essex Co., where perhaps now extinct; *see* s Ont. map by Soper and Heimburger 1961:57), N.Y., s N.H., s Maine, and sw N.S. (Digby, Yarmouth, Shelburne, and Queens counties; *see* N.S. map by Roland 1947:266), s to E Tex. and Fla. [Incl. var. *quadrangularia* (Muhl.) Wood (*S. quad.* Muhl.); *S. ?caduca* Willd.].

S. tamnoides L.   Bristly Greenbrier, China-root
/T/EE/   (Mc (vine))   Moist woods and thickets from S.Dak. to Minn., s Ont. (N to the Bruce Pen., L. Huron, and L. Nipissing, E of Sudbury; *see* s Ont. map by Soper and Heimburger 1961:56), and N.Y., s to E Tex., Ark., Ala., and Fla. [Incl. var. *hispida* (Muhl.) Fern. (*S. hisp.* Muhl.)].

STENANTHIUM Gray   [957]

S. occidentale Gray   Mountain Bells
/T/W/   (Gb)   Damp wooded slopes from B.C. (along the coast N to Bella Coola, ca. 52°20′N, and in the Rocky Mountain Trench N to Vermilion Pass, ca. 52°N; *see* B.C. map by T.M.C. Taylor 1966a:100) and sw Alta. (N to Lake Louise and Banff) to Calif. and Mont. [*Stenanthella* Rydb.].

STREPTOPUS Michx.   [1121]   Twisted-stalk

(Ref.: Fassett 1935)
1   Leaves clasping at base, their margins not ciliate; perianth-segments wide-spreading or recurving from near the middle; style slender, the stigma subentire to merely lobed; anthers lance-subulate, entire, much longer than their filaments; (transcontinental) . . . .
. . . .*S. amplexifolius*
1   Leaves sessile but scarcely or only slightly clasping; stem more or less pubescent at the nodes.
  2   Perianth rotate (shallowly bowl-shaped, the lobes widely spreading), deep wine-colour with yellowish-green tips, the segments at most 2.5 mm long; style wanting, the conical stigma sessile; anthers suborbicular, subsessile, their filaments very short; leaves eciliate, to 5 cm long; stem simple, to about 1.5 dm tall; (Alaska–B.C.–Alta.) . . . .*S. streptopoides*
  2   Perianth campanulate, the segments pink or roseate, about 1 cm long, erect, with only the tips recurved; style slender, the stigma 3-lobed (lobes about 1 mm long); anthers narrowly ovate, 2-horned, shorter than to about equalling their filaments; leaves finely ciliate, to about 9 cm long; stem simple or branched, to about 6 dm tall; (Alaska–B.C.; Man. to s Labrador, Nfld., and N.S.) . . . .*S. roseus*

S. amplexifolius (L.) DC.   Liverberry, Scootberry
/aST/X/GEA/   (Grh)   Moist woods and thickets from the Aleutian Is. and cent. Alaska to s-cent. Yukon, B.C., sw Alta., Sask. (N to Meadow Lake, 54°08′N), Man. (N to Porcupine Mt.), Ont. (N to Renison, s of James Bay at ca. 51°N; Hustich 1955), Que. (N to the George R. s of Ungava Bay at ca. 58°N; CAN), Labrador (N to Turnavik, 55°16′N), Nfld., N.B., P.E.I., and N.S., s to Calif., Ariz., N.Mex., Minn., Wisc., Mich., and N.C.; w Greenland N to ca 63°45′N; Eurasia. MAPS and synonymy (together with a distinguishing key to the probable hybrid, *S. oreopolus*): *see* below.
1   Pedicels to over 1.5 cm long, usually longer than the flower or mature fruit; branches bearing up to 5 leaves beyond the uppermost flower (rarely only 1); [mts. of s Europe; MAPS (aggregate species): Hultén 1968b:311; Böcher 1954: fig. 17, p. 67] . . .var. *amplexifolius*
1   Pedicels rarely over 1 cm long, usually shorter than the flower or mature fruit; branches rarely bearing more than 3 leaves beyond the uppermost flower (or sometimes terminated by a flower).
  2   Perianth whitish green; berries scarlet, to about 2 cm thick; stem usually forking, its upper part and the lower leaf-surfaces glaucous and glabrous.
    3   Leaf-margins rather regularly denticulate (with at least 10 minute teeth per cm of length); [Aleutian Is., s Alaska, s Yukon, B.C., and sw Alta.; Ont. (N shore of L.

Superior and the Bruce Pen., L. Huron); MAP: Fassett 1935: map 2 (dots), p. 98] . . . .
. . . . .var. *denticulatus* Fassett
3   Leaf-margins entire or with not over 6 scattered minute teeth per cm of length.
    4   Leaves minutely but copiously papillate beneath; [s B.C. and sw Alta.; MAP:
        Fassett 1935: map 1 (dots), p. 97] . . . . .var. *chalazatus* Fassett
    4   Leaves not papillate beneath; [incl. the unbranched f. *indivisus* Lepage; *S.*
        distortus Michx.; transcontinental; MAPS: Porsild 1966: map 41, p. 72; Fassett
        1935: map 1 (circles; incomplete northwards), p. 97] . . . . . .var. *americanus* Schultes
2   Perianth roseate to deep purple; berries deep red, at most 1 cm thick; stem simple or
    forked, hispid, it and the lower leaf-surfaces greener; leaves distinctly ciliate-
    denticulate, smooth or minutely hairy beneath; [var. *oreopolus* (Fern.) Fassett;
    probably a hybrid between var. *americanus* and *S. roseus* (presumably var.
    perspectus); w Ont. (Boivin 1967a), E Que., and N Nfld.; MAPS (none indicating the w
    Ont. station): Fassett 1935: map 2 (circles), p. 98; Marie-Victorin 1929b: fig. 3, p. 22].
    See D. Löve and H. Harries, Rhodora 65(764): 310–17. 1963 . . . . × *S. oreopolus* Fern.

S. roseus Michx.
/sT/X/   (Grh)   Moist woods and thickets (ranges of Canadian taxa outlined below), s to Oreg.
(var. *curvipes*), farther eastwards other varieties s to Minn., Tenn., and Ga. MAPS and synonymy:
see below.
1   Rhizome matted, the short internodes usually obscured by the copious roots, the current
    year's stem appearing close beside the remains of that of the previous year; leaf-margins
    mostly with more than 30 cilia per cm of length; perianth-segments essentially glabrous;
    sepals with usually at least 9 nerves.
    2   Pedicels completely glabrous; [SE U.S.A. only; MAP (aggregate species): Meusel,
        Jaeger, and Weinert 1965:99] . . . . .var. *roseus*
    2   Pedicels more or less ciliate with multicellular hairs; [incl. f. *giganteus* Fassett;
        Alaska–B.C. (Boivin 1967a); Man. (Boivin 1967a) to Ont. (N to ca. 49°N), Que. (N to
        ca. 56°N), SE Labrador N to ca. 51°20′N), Nfld., N.B., P.E.I., and N.S.; MAP: Fassett
        1935: map 3 (circles; Ont. eastwards), p. 100] . . . . .var. *perspectus* Fassett
1   Rhizome slender, with small tufts of roots at nodes several cm apart; leaf-margins mostly
    with not more than 30 cilia per cm of length; perianth-segments minutely papillate-
    pubescent within; sepals with 3–5 nerves.
    3   Stem usually simple; leaves with rarely as many as 20 marginal cilia per cm; [*S.*
        curvipes Vail, the type from Asulkan Pass, B.C.; s Alaska–B.C.; MAPS: Hultén 1968b:
        312; Fassett 1935: map 3 (crosses), p. 100; Meusel, Jaeger, and Weinert 1965:99;
        the B.C. map for *S. roseus* given by T.M.C. Taylor 1966a, is applicable here] . . . .
        . . . . .var. *curvipes* (Vail) Fassett
    3   Stem often branched; leaves with mostly 20–30 marginal cilia per cm; [*S. longipes*
        Fern.; s Man. and Ont.; MAP: Fassett 1935: map 3 (dots), p. 100] . . . .
        . . . . .var. *longipes* (Fern.) Fassett

S. streptopoides (Ledeb.) Frye & Rigg
/sT/W/eA/   (Grh)   Moist woods and slopes from s Alaska (*see* Hultén 1943: map 370, p. 561)
through B.C. (an isolated station in the Swan Hills of w Alta. about 110 mi NW of Edmonton at
54°52′N; CAN) to Wash. and Idaho; Siberia and Japan. [*Smilacina* Ledeb.; *Kruhsea* Kearney].
MAPS: Hultén 1968b:312; Meusel, Jaeger, and Weinert 1965:99.
    The N. American plant differs from the typical form of E Asia in its eciliate (rather than marginally
ciliate) leaves and may be distinguished as var. *brevipes* (Baker) Fassett (*S. brevipes* Baker).

TOFIELDIA Huds.   [942]   False Asphodel

1   Scape and pedicels heavily glutinous with dark glands, the scape to about 5 dm tall,
    naked or often with a usually solitary bract near the middle; fruiting raceme to 2 cm thick,
    its pedicels mostly in groups of 3 or 4; leaves free at base; (transcontinental) . . . . . .*T. glutinosa*
1   Plants glabrous; fruiting raceme less than 1 cm thick, its pedicels solitary; leaves
    2-ranked and enfolding one another at base; (transcontinental).

2 Inflorescence yellowish green; capsules straw-coloured; scape to 3 dm tall, pale green, bractless or with only 1 small bract near base; leaves to 1 dm long . . . .*T. pusilla*
2 Inflorescence and capsules dark purple; scape at most 1 dm tall, purple, often with 1 or more bracts well above the base; leaves commonly less than 6 cm long . . . . .*T. coccinea*

T. coccinea Richards.
/aST/X/GA/ (Hr) Calcareous clays and gravels from the Aleutian Is. and coasts of Alaska–Yukon–Dist. Mackenzie (type from near Bathurst Inlet, Coronation Gulf) to Banks Is., Devon Is., and northernmost Que. (Hudson Strait and Ungava Bay), s to N B.C. (an isolated station at Eutsuk L., ca. 53°N; also isolated in the mts. of w Alta. along the Athabasca R. at ca. 53°30′N) and s Dist. Mackenzie–Dist. Keewatin; w and E Greenland between the Arctic Circle and ca. 78°N; Asia. [*T. nutans* of auth., perhaps not Willd.]. MAPS: Hultén 1968*b*:304; Porsild 1957: map 113, p. 175.

T. glutinosa (Michx.) Pers.
/sT/X/ (Hr (Grh)) Calcareous ledges, marshy ground, and shores from s Alaska–Yukon–Dist. Mackenzie to N Alta. (L. Athabasca), N Sask. (Carswell L. at 58°35′N), Man. (N to the Nelson R. at ca. 57°N), northernmost Ont., Que. (N to the Kaniapiscaw R. system at ca. 56°N; *see* NE Canada map by Lepage 1966: map 13, p. 224), SE Labrador (N to ca. 53°N), Nfld., N.B., and N.S. (not known from P.E.I.), s to Calif., Idaho, Wyo., Minn., N Ill., N Ohio, and Ga. [*Narthecium* Michx.; *Triantha* Baker; *T. intermedia* Rydb.; *T. occidentalis* Wats.; incl. ssp. *brevistyla* and ssp. *montana* Hitchc.]. MAPS: Hultén 1968*b*:305; C. L. Hitchcock, Am. Midl. Nat. 31(2):490 (incomplete northwards). 1944.

T. pusilla (Michx.) Pers.
/aST/X/GEA/ (Hr) Calcareous ledges, gravels, and marshy places from the coasts of Alaska–Yukon–Dist. Mackenzie to Victoria Is., Melville Pen., N-cent. Baffin Is., northernmost Ungava–Labrador, and Nfld., s in the West to SE B.C. and Mont., farther eastwards s to Great Slave L., L. Athabasca, N Man. (s to York Factory, Hudson Bay, ca. 57°N; the report from McCreary, E of Riding Mt., by Lowe 1943, tentatively accepted by Hultén in his 1962 map, requires confirmation, being more likely referable to *T. glutinosa*), Ont. (s to s James Bay; isolated along the N shore of L. Superior and at L. Nipigon), and Que. (s to the s James Bay watershed at ca. 51°N, L. Mistassini, the Côte-Nord, Anticosti Is., and N Gaspé Pen.); w Greenland N to ca. 78°N, E Greenland N to ca. 75°N; Iceland; Eurasia. [*Narthecium pus.* Michx., the type from near L. Mistassini, Que.; *T. borealis* Wahl.; *T. minima* (Hill) Druce; *T. palustris* of auth., not Huds.]. MAPS: Hultén 1968*b*:304, and 1962: map 33, p. 41; Porsild 1957: map 112, p. 174; Meusel, Jaeger, and Weinert 1965:88; Raup 1947: pl. 20 (*T. pal.*).

TRILLIUM L. [1138] Trillium, Wakerobin. Trille

1 Leaves distinctly petioled.
  2 Flower sessile, the petals erect or ascending, normally brown-purple, distinctly slender-clawed, 2 or 3 cm long; sepals abruptly reflexed between the petioles; anthers purple, longer than their filaments; leaves elliptic to subrotund, rarely over 1 dm long, their petioles 1 or 2 cm long; (introd. in s Ont.) . . . .[*T. recurvatum*]
  2 Flowers long-peduncled; sepals and petals spreading, the latter basically white, scarcely clawed; leaves elliptic to ovate-lanceolate or ovate, their petioles to about 1 cm long.
    3 Leaves to 5 cm long, they and the petals (white, ageing rose-pink) obtuse or subacute; anthers yellow, somewhat longer than their filaments; berry less than 1 cm long; plant at most about 1.5 dm tall . . . .[*T. nivale*]
    3 Leaves to over 1.5 dm long, acuminate; petals acute or acuminate, white streaked with purple toward base; anthers purple, shorter than their filaments; berry about 1.5 cm long; plant to about 5 dm tall; (Ont. to N.S.) . . . .*T. undulatum*
1 Leaves typically sessile or nearly so (petioles at most 5 mm long), mostly rhombic-ovate to subrotund, to over 1.5 dm long; flowers typically long-peduncled; berry acutely 6-angled or -winged.

4   Flower reflexed and hidden beneath the leaves; petals at most 2.5 cm long, normally white or pink; stigmatic style-branches short and stout, tapering from base to the recurved tip; anthers pink, typically less than 5 mm long, slightly longer than to about equalling their filaments; leaves acuminate; (Sask. to Nfld. and N.S.) . . . .*T. cernuum*

4   Flower held above the leaves; petals to over 5 cm long; anthers usually over 5 mm long.

    5   Petals purple or brown-purple; anthers purple, much shorter than their filaments; stigmatic style-branches short and stout, tapering from base to the recurved tip; (SE Man. to N.S.) . . . .*T. erectum*

    5   Petals normally white (often soon ageing pink to deep roseate); anthers yellow, about equalling to longer than their filaments; stigmatic style-branches slender, of uniform thickness, erect or spreading.

        6   Petals ascending at base, spreading above; leaves acuminate, to about 3 dm long; (Ont. to N.S.) . . . .*T. grandiflorum*

        6   Petals spreading from the base; leaves rarely over 1.5 dm long, acute or short-acuminate; (s B.C. and sw ?Alta.) . . . .*T. ovatum*

T. cernuum L.   Nodding Trillium

/T/EE/   (Grh)   Damp or peaty woods and thickets (ranges of Canadian taxa outlined below), s to Iowa, Ohio, Ala., and Ga.

1   Petals to about 1 cm broad; anthers to 4.5 mm long; peduncle to 2.5(3.5) cm long; [incl. var. *terrae-novae* Boivin; *T. erectum* var. *declinatum sensu* John Macoun 1888, in large part, not Gray; SE Sask. (Runnymede; Breitung 1957a) to s Man. (N to Riding Mt.), Ont. (N to Kapuskasing and the Pagwa R. at ca. 50°N), Que. (N to L. St. John and the Gaspé Pen.), Nfld., N.B., P.E.I., and N.S.] . . . . .var. *cernuum*

1   Petals to over 1.5 cm broad; anthers to over 6 mm long; peduncles to 4 cm long; [Sask. (reported from Carlton, about 40 mi sw of Prince Albert, by A. J. Eames and K. M. Wiegand, Rhodora 25(299): 191. 1923, the cited collection ("Hooker dup.") being undoubtedly the basis of the report of *T. cernuum* from "the Saskatchewan to Mackenzie River" by Hooker 1838) to SE Man. (near Otterburne; Löve and Bernard 1959) and Ont. (N to the N shore of L. Superior near Thunder Bay)] . . . . .var. *macranthum* Wieg.

T. erectum L.   Purple Trillium, Stinking Benjamin

/T/EE/   (Grh)   Rich woods and thickets from SE Man. (reported as rare by Lowe 1943; a single station reported from Otterburne by Löve and Bernard 1959, where evidently eradicated by the 1950 floods) to Ont. (N to the Ottawa dist.), Que. (N to L. St. John and Rimouski and Matapédia counties), N.B., and N.S. (reported from P.E.I. by Hurst 1952, but evidently now extinct), s to Tenn. and Ga. [Incl. vars. *giganteum* and *horizontale* Louis-Marie]. The following forms occur in Canada (note the triple use of lead 4):

1   Flower subsessile, the peduncle at most 1.5 cm long; [f. *sessiloides* Boivin; known from the Ottawa dist., Ont., and the type locality, La Trappe, sw Que.] . . . . .f. *brevipedicellatum* Louis-Marie

1   Flower long-peduncled.

    2   Leaves and floral-parts each at least 4; [Ottawa, Ont., and the presumed type locality, Que.] . . . . .f. *polymerum* Vict.

    2   Leaves and floral-parts each 3.

        3   Petals white, yellowish, or greenish.

            4   Petals white; [var. *album* (Michx.) Pers.; Ont., Que., and N.S.] . . .f. *albiflorum* Hoffm.

            4   Petals yellowish; [type from La Trappe, sw Que.] . . . . .f. *luteum* Louis-Marie

            4   Petals greenish; [s Ont. and sw Que.] . . . . .f. *viridiflorum* (Hook.) Peattie

        3   Petals crimson to purplish-black.

            5   Anther-connectives yellow, almost transparent; [type from Rimouski, Rimouski Co., E Que.] . . . . .f. *pallidandrum* Vict. & Rousseau

            5   Anther-connectives violet, opaque.

                6   Petals red-streaked; [type from La Trappe, sw Que.] . . . . .f. *rubrostriatum* Louis-Marie

                6   Petals uniformly coloured.

7 Leaves at most about 4 cm long and broad; flowers small, the sepals and petals at most 2 cm long; [type from La Trappe, sw Que.] . . . .
. . . . .f. *parvum* Louis-Marie
7 Leaves to nearly 2 dm long.
   8 Flowers small, the petals (and sepals) less than 2 cm long; [type from La Trappe, sw Que.] . . . . .f. *parviflorum* Louis-Marie
   8 Flowers larger, the petals to over 5 cm long; [var. *atropurpureum* Michx.; f. *nigrescens* Louis-Marie; *T. obovatum* Pursh] . . . . .f. *erectum*

T. grandiflorum (Michx.) Salisb.   White Trillium

/T/EE/   (Grh)   Rich woods and thickets from Minn. and Mich. to s Ont. (N to the Goulais R. N of Sault Ste. Marie, the s end of L. Nipissing, and Chalk River, Renfrew Co.; *see* s Ont. map by J. H. Soper, Fed. Ont. Nat., Bull. 92:6. 1961; provincial floral emblem) and Que. (N to Grosse-Ile, Montmagny Co.; reported from N.B. by Hooker 1839, but not listed by Fowler 1885, or later authors; there is a collection in CAN from Truro, N.S., where taken by John Macoun in 1883, but the species is apparently now extinct in the Maritime Provinces), s to Ark. and Ga. [*T. rhomboideum* var. *gr.* Michx.] The following forms occur in Canada:

1 Leaves and floral-parts not each the usual 3 in number.
   2 Leaves, sepals, petals, and carpels each 2; [type from La Trappe, sw Que.] . . . .
. . . . .f. *dimerum* Louis-Marie
   2 Leaves and floral-parts each at least 4; [Ont. (Norfolk Co.; OAC; reported from the Ottawa dist. by Gillett 1958) and sw Que. (type from Mt. Royal, Montreal)] . . . .
. . . . .f. *polymerum* Vict.
1 Leaves and floral-parts each 3.
   3 Leaves variously lobed or divided, sessile or petioled; [type from La Trappe, sw Que.] . . . . . .
. . . . .f. *divisum* Louis-Marie
   3 Leaves entire.
      4 Leaves petioled; [Ont. (Norfolk, Waterloo, and Brant counties) and sw Que.] . . . .
. . . . .f. *lirioides* (Raf.) Vict.
      4 Leaves sessile or nearly so.
         5 Petals green or green-striped.
            6 Stamens gradually modified into carpels; [type from Caughnawaga, near Montreal, Que.] . . . . .f. *capellare* Dansereau
            6 Stamens normal.
               7 Petals green throughout; [Que.; Marie-Victorin 1929*b*] . . . . .f. *viride* Farw.
               7 Petals merely striped with green; [type from La Trappe, sw Que.] . . . .
. . . . .f. *striatum* Louis-Marie
         5 Petals white (often turning pink in age).
            8 Carpels and stamens modified into petals or leaves.
               9 These changed to petals; [type from La Trappe, sw Que.] . . . .
. . . . .f. *petalosum* Louis-Marie
               9 These (and the petals) changed to leaves; [type from La Trappe, sw Que.] . . . . .f. *regressum* Louis-Marie
            8 Carpels and stamens normal.
            10 Leaves, sepals, and petals relatively narrow (leaves narrowly elliptic; sepals linear-lanceolate; petals lanceolate); [type from La Trappe, sw Que.] . . . . .f. *elongatum* Louis-Marie
            10 Leaves broadly ovate or round-ovate to subrhombic-oval; sepals lanceolate; petals oblong or narrowly obovate to suborbicular . . . .
. . . . .f. *grandiflorum*

[*T. luteum* (Muhl.) Harbison]   Toadshade
[This species (not keyed out above but readily recognized by its strongly mottled, sessile leaves and its erect flower and greenish- to buttercup-yellow petals) is reported from s Ont. by R. E. Beschel (Blue Bill 17(2): 23. 1970; near Kingston, Frontenac Co., where growing along the Cataraqui R. in clay-rich soil on limestone bedrock inside the forest edge). If native, this is a remarkable outlier, the normally accepted range being from Mo. to Ky. and N.C., s to Ark. and Ala.

Liliaceae

However, Beschel suspects that the plants may have been introduced, at least one Ont. nursery offering the species commercially.]

[T. nivale Riddell]   Dwarf Trillium
[This species of the E U.S.A. (N to Minn. and Pa.) is represented in CAN by a collection, verified by S. J. Smith, bearing the data, "Collected by Lou Chapham and grown from root sent to Galt from Massey, Algoma, 1906. In bloom at Galt Mar. 20th, '08. W. Herriot." It was undoubtedly cultivated at the original Algoma locality, N of L. Huron, Ont.]

T. ovatum Pursh   Western Wakerobin, Wood Lily
/t/W/   (Grh)   Rich damp woods from s B.C. (N to ca. 50°20'N on Vancouver Is. and the s mainland; see B.C. map by T.M.C. Taylor 1966a:102) and sw ?Alta. (reported from Waterton Lakes by Moss 1959, but not listed by Breitung 1957b) to cent. Calif. and Colo. [*T. scouleri* Rydb.; *T. grandiflorum sensu* Hooker 1838, as to w Canada reports, not (Michx.) Salisb.; *T. obovatum sensu* Hooker 1838, not Pursh].

[T. recurvatum Beck]   Prairie Trillium
[Reports of this species of the E U.S.A. (N to Iowa and Ohio) from Nfld. by John Bell (Can. Nat. 4 (ser. 2): 260. 1869), A. C. Waghorne (*A Summary Account of the Wild Berries and other Edible Fruits of Newfoundland and Labrador.* 11 pp. Mercury, St. John's, Nfld. 1888), and J. W. Harshberger (*Phytogeographic Survey of North America,* p. 354. Engelmann, Leipzig. 1911.) are probably all based upon *T. cernuum.* A collection in OAC taken in 1901 by Wm. Herriot in Dickson's Woods, Cambridge, Waterloo Co., s Ont., is annotated, "from roots sent from Chicago."]

T. undulatum Willd.   Painted Trillium
/T/EE/   (Grh)   Rich moist woods and thickets from Ont. (N to North Bay on L. Nipissing and Algonquin Park, Renfrew Co.; CAN; the report from Renison, s of James Bay at ca. 51°N, by Hustich 1955, requires confirmation, as does the inclusion of E Man. (?extinct) in the range by Fernald *in* Gray 1950) to Que. (N to L. St. John and the Gaspé Pen.; reported from Lac-la-Loutre on the Côte-Nord by André Lafond, Ann. ACFAS 13: 92. 1947), N.B., P.E.I., and N.S., s to Wisc., Mich., Tenn., and Ga. [*T. erythrocarpum* Michx.; *T. pictum* Pursh]. MAP: Braun 1937: fig. 30 (incomplete northwards), p. 202.
   Forma *polymerum* Vict. (floral-parts each 4 or more) is reported from Ont. by Gillett (1958; Ottawa dist.) and from the type locality, St-Augustin, near Oka, sw Que., by Pére Louis-Marie (Inst. Agric. d'Oka Rev. 14: 152. 1940). Forma *striatum* Louis-Marie (sepals striped with white along the midvein) is also known from St-Augustin, the type locality.

UVULARIA L.   [966]   Bellwort

1   Leaves sessile but not perfoliate, glaucous beneath, their scarious margins minutely papillose-denticulate; capsule acute at both ends, with 3 winged angles, the walls smooth within (or at least not densely pebbled); perianth to 2.5 cm long; rhizome elongate; (s Man. to N.S.) . . . .*U. sessilifolia*
1   Leaves perfoliate (their bases connate around the stem), with smooth scarious margins; capsule truncate at summit, obtusely 3-lobed, the walls densely pebbled within; rhizome short.
   2   Perianth at most 3.5 cm long, its segments conspicuously granular-papillate within; stamens shorter than the style, the tip of the connective acuminate; leaves glaucous and essentially glabrous, usually 3 or 4 present below the fork of the stem; upper bladeless stem-sheath less than 6 cm long; (s Ont.) . . . .*U. perfoliata*
   2   Perianth to about 5 cm long, its segments glabrous; stamens surpassing the style, obtusely tipped; leaves usually at least moderately puberulent beneath, 2, 1, or none below the fork of the stem; upper bladeless stem-sheath to about 12 cm long; (Ont. and Que.) . . . .*U. grandiflora*

U. grandiflora Sm.
/T/EE/   (Grh)   Rich woods and thickets from N.Dak. to s Ont. (N to Renfrew and Carleton

counties; *see* s Ont. map by J. H. Soper, Rhodora 54(639): fig. 1, p. 61. 1952), Que. (N to St-Augustin, near Quebec City; MT; *see* Que. map by Doyon and Lavoie 1966: fig. 8, p. 816), and Maine, s to Okla., Ark., Ala., and Ga. [*Oakesia* Wats.]. MAP: R. L. Wilbur, Rhodora 65(762): fig. 5 (not validating the inclusion of SE Man. in the range assigned by Soper), p. 178. 1963.

Forma *latifolia* Louis-Marie (leaves suborbicular rather than broadly oval to oblong) is known from the type locality, La Trappe, sw Que.

U. perfoliata L.

/t/EE/ (Grh) Rich woods and thickets from s Ont. (Lincoln, Welland, and Wentworth counties; *see* s Ont. map by Soper, loc. cit., fig. 2, p. 65; the report from Victoria Beach, s Man., by Lowe 1943, requires confirmation) to Vt. and Mass., s to La. and NW Fla. [*Oakesia* Wats.]. MAP: R. L. Wilbur, Rhodora 65(762): fig. 6, p. 180. 1963.

U. sessilifolia L.

/T/EE/ (Grh) Rich woods and thickets from s Man. (known only from Roseisle, 15 mi s of Portage la Prairie, where taken by Marshall in 1942 and 1949, validating the report from Roseisle by Lowe 1943; DAO) to Ont. (N to the Ottawa dist.; *see* s Ont. map by Soper, loc. cit., fig. 3, p. 65), Que. (N to Matapédia, sw Gaspé Pen.; CAN), N.B., and N.S. (not known from P.E.I.), s to Ark., Ala., and Ga. [*Oakesia* Wats.; *Oakesiella* Small]. MAP: R. L. Wilbur, Rhodora 65(762): fig. 4 (the occurrence in s Man. should be indicated), p. 176. 1963.

VERATRUM L.   [960]   False Hellebore.   Vérâtre or Varaire

1   Panicle spike-like or with short upright branches; leaves pubescent only on the nerves of
    the lower surface; (ssp. *oxysepalum;* Aleutian Is. and Alaska) ....*V. album*
1   Panicle open, the often long branches spreading-ascending to drooping; leaves rather
    evenly pubescent over the lower surface; (transcontinental) ....*V. viride*

V. album L.

/ST/W/EA/ (Grh) Grassy depressions and dryish slopes of the Aleutian Is. and Alaska (N to the Seward Pen.; *see* Hultén 1943: map 358, p. 560); Eurasia. MAP: Hultén 1968*b*:306.

The Alaskan plant may be distinguished as ssp. *oxysepalum* (Turcz.) Hult. (*V. ox.* Turcz.; *V. parviflorum sensu* Bongard 1833, not Michx.; plant relatively low-grown, the panicle at most about 1.5 dm long, densely beset with very short-pedicelled flowers on peduncles 1 or 2 cm long).

V. viride Ait.   White Hellebore, Indian Poke.   Tabac du diable

/sT/D/ (Grh) Grassy meadows and swamps: s-cent. Alaska, cent. Yukon, and sw Dist. Mackenzie through B.C. and the foothills and mts. of sw Alta. (N to Jasper) to Oreg. and Mont.; Minn. to ?Ont. (Boivin 1967*a*; occurrence doubted), Que. (N to Ellen L. at ca. 52°36′N, near the Labrador boundary, the Côte-Nord, and Gaspé Pen.; *see* Que. map by Löve, Kucyniak, and Johnston 1958: fig. 4, p. 61), Labrador (N to Carol L., near the Que. boundary at ca. 53°N; CAN), and N.B. (not known from P.E.I.; reports from N.S. require confirmation), s to Tenn. and Ga. [*V. album sensu* Michaux 1803, not L.]. MAP: Hultén 1968*b*:306.

The western plant may be doubtfully separated as var. *eschscholtzii* (Gray) Breitung (*V. esch.* Gray (type from Sitka, Alaska) and its var. *incriminatum* Boivin; *V. lobelianum* var. *eschscholtzianum* R. & S.; *V. viride sensu* Hooker 1838, and John Macoun 1888 (as to western citations), not L.; branches of inflorescence relatively long and drooping, stamens relatively long, leaves relatively hairy). MAPS: Hultén 1968*b*:306; Porsild 1966: map 42 (*V. esch.*), p. 72.

XEROPHYLLUM Michx.   [948]

X. tenax (Pursh) Nutt.   Bear-Grass, Elk-Grass, Fire-Lily, Bear-Lily

/T/W/ (Hrr) Dry ridges and rocky slopes from the mts. of SE B.C. (s of 50°N; *see* B.C. map by T.M.C. Taylor 1966*a*:102) and sw Alta. (Waterton Lakes; Castle Mt., NW of Banff) to cent. Calif., Idaho, Mont., and ?Wyo. [*Helonias* Pursh].

YUCCA L.   [1103]   Beargrass, Spanish-bayonet

1   Leaves to about 1 dm broad, much overtopped by the peduncle; panicle freely branched,
    to 2 m long (the whole plant to about 3 m tall); stigmas slender; capsules to 4.5 cm long,
    the seeds 6 or 7 mm long; (persisting following cultivation in s Ont.) . . . .*Y. filamentosa*
1   Leaves less than 1.5 cm broad, nearly or quite reaching the bottom of the panicle, this
    rather simple and racemose (or with a few long branches near base), to about 1 m long;
    stigmas swollen; capsules to 7 cm long, the seeds commonly at least 1 cm long; (s Alta.) . . . . . .
    . . . .*Y. glauca*

Y. filamentosa L.   Silkgrass, Spoon-leaf Yucca
A Coastal Plain species of the E U.S.A., native from s N.J. to Ga. Dodge (1915) reports it as
"cultivated and inclined to persist in sandy ground for many years" in Lambton Co., s Ont.

Y. glauca Nutt.   Soapweed
/T/WW/   (Hr)   Dry plains and sandhills from Mont. to SE Alta. (Onefour, near the U.S.A.
boundary about 65 mi s of Medicine Hat; Moss 1959), N.Dak., and Iowa, s to Ariz., N.Mex., Tex.,
and w Mo.

ZIGADENUS Michx.   [958]

1   Perianth-segments to 1.5 cm long, not clawed, the basal glands to about half as long;
    ovary party inferior, adnate to the perianth at base; raceme simple or branched; leaves to
    1.5 cm broad; stems to over 1 m tall; (Alaska–B.C. to N.B.) . . . .*Z. elegans*
1   Perianth-segments usually less than 8 mm long, the basal glands relatively smaller;
    ovary wholly superior; leaves to about 6 mm broad; stems usually lower; (western
    species).
    2   Inflorescence commonly a simple open raceme to about 2 dm long (occasionally with
        one or more basal branches); perianth-segments obtuse, all distinctly and about
        equally clawed; upper stem-leaves lacking a sheath, their bases somewhat clasping;
        (B.C. to sw Sask.) . . . .*Z. venenosus*
    2   Inflorescence a rather dense panicle to about 1.5 dm long; perianth-segments acute
        or acuminate, only the petals distinctly clawed; upper stem-leaves with conspicuous
        papery sheaths; (?B.C.–Alta.) . . . .[*Z. paniculatus*]

Z. elegans Pursh   White Camass, Alkali-grass
/aST/X/   (Gb)   Damp meadows, prairies, open woods, shores, and bogs from the coasts of
Alaska–Yukon–NW Dist. Mackenzie to Great Bear L., Great Slave L., Sask. (N to Prince Albert),
Man. (N to The Pas), s Ont. (N to Cloche Is. in N L. Huron; OAC), Que. (St. Lawrence R. estuary
from l'Islet and Charlevoix counties to the Côte-Nord, Anticosti Is., and Gaspé Pen.), and N N.B.
(Belledune Point, Gloucester Co.; CAN; ACAD; GH; Fowler 1885; not known from P.E.I. or N.S.), s
to Oreg., Ariz., N Mexico, Tex., Minn., Mo., Ohio, and N.Y., and the mts. of Va. and N.C. [*Anticlea*
Rydb.; *Z. (A.) chloranthus* Rich.; *Z. (A.) glaucus* Nutt.; *see* Fernald 1935:256–58]. MAPS (w area,
Z. glaucus being considered a distinct eastern species): Hultén 1968*b*:305; Porsild 1951*b*: fig. 6,
p. 143; Raup 1947: pl. 20.

[*Z. paniculatus* (Nutt.) Wats.]
[This species of the w U.S.A. (Wash. and Mont. to Calif. and N.Mex.) is doubtfully distinct from
Z. venenosus, to which reports from B.C. by Eastham (1947; Anarchist Mt., near Osoyoos, and
Wilmer, Columbia Valley) and from Alta. by Henry (1915) probably refer. (*Helonias* Nutt.;
Toxicoscordion Rydb.).]

Z. venenosus Wats.   Death Camass
/T/WW/   (Gb)   Wet meadows and moist rocky ledges (ranges of Canadian taxa outlined below),
s to Baja Calif., Utah, Colo., and Nebr.
1   Upper stem-leaves sheathless, their bases merely somewhat clasping; perianth-

segments equally long-clawed; [*Toxicoscordion* Rydb.; s B.C. (*see* B.C. map by T.M.C. Taylor 1966*a*:103) and sw Alta. (Crowsnest Forest Reserve; CAN)] . . . . .var. *venenosus*

1 Upper stem-leaves with conspicuous papery sheaths; outer perianth-segments sessile or short-clawed, only the inner ones long-clawed; [*Z. gram.* and *Z. intermedius* Rydb.; *Toxicoscordion gram.* Rydb.; B.C. (N to ca. 54°N; *see* B.C. map for *Z. gramineus* by T.M.C. Taylor 1966*a*:103) to s Alta. and sw Sask.; the report of *Tox. gramineum* from sw Man. by Lowe 1943, requires confirmation] .var. *gramineus* (Rydb.) Walsh

HAEMODORACEAE (Bloodwort Family)

LACHNANTHES Ell. [1161] Redroot

Stem stout, to about 9 dm tall, tomentose above and into the terminal compound dense cyme, from rhizomes and long slender stolons with red juice. Leaves linear, to 4 dm long and 1 cm broad, mostly near the base of the stem, the upper ones much reduced. Flowers dingy yellow and loosely woolly, epigynous, about 1 cm long, the segments united at base, the 3 long-exserted stamens opposite the 3 larger inner divisions (petals). Anthers introrse. Ovary inferior. Fruit a subglobose capsule with a few reddish-brown seeds 2 or 3 mm thick; (N.S.).

L. tinctoria (Walt.) Ell. Redroot
/T/EE/ (Hsr) Sandy or peaty shores and swamps of sw N.S. (Beartrap L., a branch of Ponhook L., Queens Co., and other localities along Ponhook L.; CAN; GH) and from Mass. and Del. to Fla. and La.; W.I. [*Gyrotheca* Morong].

DIOSCOREACEAE (Yam Family)

DIOSCOREA L. [1252] Yam

Stem twining, to about 5 m long, from a large tuberous root. Leaves alternate (or the lower often opposite), cordate-ovate, abruptly acuminate, to about 1 dm long. Flowers unisexual, the minute staminate ones in short axillary spike-like racemes along the branches of loose axillary panicles, the small pistillate ones solitary along the axis of 5–10-flowered axillary spikes. Perianth white to greenish-yellow, 6-parted. Stamens 3 or 6. Styles 3. Ovary inferior. Fruit a 3-winged capsule to about 2.5 cm long, each of the thin valves semiobovate to semicircular. Seeds very flat, about 1 cm broad, with a pale-brown hyaline wing around a dark centre. (s Ont.).

D. villosa L. Wild Yam
/t/EE/ (Grh) Wet woods, thickets, and swamps from Minn. to s Ont. (N to s Huron, s Waterloo, Wentworth, and Welland counties; *see* s Ont. map by Soper 1956: fig. 3, p. 74) and s New Eng., s to Tex., Ark., Tenn., Ohio, Va., and w Fla. [*D. paniculata* Michx.].

Forma *glabrifolia* (Bartl.) Fern. (leaves glabrous rather than short-hairy on the nerves beneath) is also known from s Ont. (Essex, Norfolk, and Waterloo counties; TRT; Montgomery 1945).

AMARYLLIDACEAE (Amaryllis Family)

Perennials from bulbs, corms, or rhizomes. Leaves linear, commonly basal. Flowers perfect, regular or nearly so. Perianth-segments and stamens each 6, the anthers introrse. Style single. Ovary partly or wholly inferior. Fruit a 3-locular capsule. (Including Hypoxidaceae).

1   Perianth densely woolly outside and at the base inside, golden yellow, about 1 cm long and broad; inflorescence a terminal woolly compound corymb; ovary half-inferior; stem to about 7 dm tall, from rhizomes, leafy, the leaves subglabrous; (w N.S.) . . . .*Lophiola*
1   Perianth glabrous or glabrate; flowers terminating leafless scapes; ovary completely inferior; plants from a bulb or corm-like rhizome.
  2   Perianth with a slender tube terminated by a campanulate or cup-shaped crown (corona) above the insertion of the widely spreading segments; flowers solitary, large; (garden-escape) . . . .*[Narcissus]*
  2   Perianth-segments distinct nearly to base, lacking a corona; flowers at most about 2 cm long, solitary or in 2–many-flowered umbels.
    3   Perianth-segments yellow, widely spreading during anthesis, short-pilose but soon glabrate on the back; leaves to 1 cm broad, they and the long pedicels pilose; scape to 3.5 dm tall, filiform; (s Sask. to s Ont.) . . . .*Hypoxis*
    3   Perianth-segments white or tinged with green, glabrous, ascending; plants glabrous; (garden-escapes).
      4   Flower solitary, declined or nodding on a slender pedicel, the outer segments white, veiny, to about 2.5 cm long, the inner about half as long, tipped with green; leaves 2 or 3, to about 6 dm long and 6 mm broad; scape solid, usually less than 3 dm tall . . . .*[Galanthus]*
      4   Flowers up to 8 or more, on drooping pedicels, the perianth-segments equal and all tipped with green; leaves to about 4.5 dm long and 1.5 cm broad; scape hollow, to about 8 dm tall, reclining in fruit . . . .*Leucojum*

[GALANTHUS L.]   [1172]   Snowdrop

[*G. nivalis* L.]   Common Snowdrop
[Eurasian; a garden-escape in the U.S.A. and, presumably as such, reported from Lambton Co., s Ont., by Gaiser and Moore (1966) and from Nfld. by Rouleau (1956), but, at most, only slightly persisting outside of cultivation.]

HYPOXIS L.   [1230]   Stargrass

H. hirsuta (L.) Cov.   Common Stargrass
/T/EE/   (Gst (Gb))   Open woods and meadows from s Sask. (Buchanan, Yorkton, and near Moose Mountain) to s Man. (N to Moosehorn, about 90 mi N of Portage la Prairie), s Ont. (N to the shore of L. Ontario in Hastings Co.; John Macoun 1888), Ohio, and sw Maine, s to Tex. and Fla. [*H. erecta* L.].

LEUCOJUM L.   [1174]   Snowflake

L. aestivum L.   Summer-Snowflake
European; according to Fernald (1921), this species is thoroughly naturalized and considered a troublesome weed in an old field at Yarmouth, N.S.; (GH).

LOPHIOLA Ker   [1240]   Golden-crest

L. americana (Pursh) Wood
/T/E/   (Gb)   Bogs and peaty shores of w N.S. (Digby, Queens, and Lunenburg counties) and the Pine Barrens of N.J. (apparently extinct in N Del.), s to Fla. and Miss. [*Conostylis* Pursh; *L. aurea* Ker-Gawl; *L. septentrionalis* Fern.]. MAP (NE area): Fernald 1929: map 24, p. 1499.

[NARCISSUS L.] [1201] Narcissus

1   Flowers predominantly white; corolla-crown less than 1/4 as long as the perianth-
    segments, white or yellow with a red margin ....[*N. poeticus*]
1   Flowers wholly yellow; corolla-crown about equalling the perianth-segments, usually
    yellow throughout ....[*N. pseudo-narcissus*]

[N. poeticus L.]   Poets' Narcissus
[European; a garden-escape in s Ont. (Elgin and Lambton counties), ?Que. (Boivin 1967a), Nfld. (Rouleau 1956; ?escaped), and N.B. (near Dalhousie; CAN), but scarcely established.]

[N. pseudo-narcissus L.]   Daffodil
[European; a garden-escape in N. America, reported from s Ont. by Gaiser and Moore (1966; Lambton Co.) and from Nfld. by Rouleau (1956), but scarcely established.]

IRIDACEAE (Iris Family)

(Ref.: Marie-Victorin 1929*b*)
Perennials with linear, 2-ranked and equitant, mostly basal leaves. Flowers perfect, regular, epigynous, the perianth 6-parted. Stamens 3, opposite the outer perianth-segments. Style single, usually 3-cleft. Ovary inferior. Fruit a 3-locular capsule.

1   Aerial stem none, the white or lilac, often purple-striped flowers produced at the surface among or before the leaves by elongation of the perianth-tube; perianth-segments subequal, to over 3.5 cm long; anthers lemon-yellow, their filaments white; plant from a corm; (garden-escape) ....[*Crocus*]
1   Stem (or scape) present.
   2   Flowers large; sepals spreading or recurved, commonly longer than the erect or spreading petals; style-branches petal-like, opposite to and arching over the stamens; leaves relatively broad, sword-shaped; stems from stout creeping rhizomes ....*Iris*
   2   Flowers about 1 cm long, the perianth nearly rotate, its spreading segments subequal; style-branches filiform, alternate with the stamens; leaves less than 6 mm broad; stems 2-edged or 2-winged, from fibrous roots ....*Sisyrinchium*

[CROCUS L.] [1259] Crocus

[C. vernus Wulfen]   Common Crocus
[European; reported from Nfld. by Rouleau (1956), but scarcely established there or elsewhere in N. America.]

IRIS L. [1264] Iris. Fleur-de-lis

1   Stem short or almost wanting, rarely more than about 1 dm tall, bearing a single flower.
   2   Sepals (the petal-like reflexed "falls") densely long-bearded within toward base along the median line, commonly yellow or lilac (sometimes purple); perianth-tube green, filiform, about 5 cm long; leaves to about 12 cm long and 6 mm broad; (garden-escape) ....
      ....[*I. pumila*]
   2   Sepals not long-bearded medially, with a white and purple patch at the base of the orange-white, fringed or lacerate crest; perianth-tube dull yellow, 1 or 2 cm long; leaves to over 1.5 dm long and to 1.5 cm broad; (s Ont.) ....*I. lacustris*
1   Stem to over 5 dm tall, with often more than 1 flower; sepals lilac or pale blue to deep bluish-purple.
   3   Sepals densely long-bearded within toward base along the median line, various shades of violet or purple, not much longer than the petals, to over 3.5 cm broad; leaves to 3.5 cm broad; stems to about 2.5 m tall, usually 2-flowered at summit and commonly with two 1-flowered branches; (garden-escape) ....*I. germanica*
   3   Sepals not long-bearded medially (sometimes pubescent toward base).
      4   Leaves not over 1 cm broad.
         5   Capsule (and ovary) sharply 3-angled, almost winged, to 5 cm long; flower blue or blue-violet; perianth-tube about 3 mm long; leaves to 9 mm broad; rhizomes and stolons near the ground-surface, usually less than 5 mm thick; (N.S.) ....
            ....*I. prismatica*
         5   Capsule (and ovary) obtusely 3-angled.
            6   Leaves all basal or nearly so, to 1 cm broad; sepals pale blue, to about 6 cm long; perianth-tube to 12 mm long; stems to about 6 dm tall, from stout rootstocks; (sw Alta.) ....*I. missouriensis*
            6   Leaves distributed upward beyond the middle of the stem.
               7   Leaves to about 8 mm broad; spathes bearing up to 5 lilac to blue-violet flowers; perianth-tube none or very short; stems to 6 dm tall, from a short rootstock; (garden-escape) ....*I. sibirica*
               7   Leaves commonly not over 3 mm broad, stiff, paler beneath; spathes

1-flowered; perianth-tube about 1 cm long; stems to 3 dm tall, from slender
rootstocks; (?Vancouver Is.) ....[*I. thompsonii*]
4   Leaves mostly over 1 cm broad; rootstocks stout.
8   Perianth mostly bright yellow, its green tube not constricted above the ovary;
capsule (and ovary) obtusely 6-angled, to 8.5 cm long; stems to about 1 m tall,
from pink-fleshed rhizomes; (garden-escape) ....*I. pseudacorus*
8   Perianth lavender or blue to light violet or blue-violet, its tube usually
noticeably constricted above the ovary.
9   Petals bristle-tipped, involute or tubular, 1 or 2 cm long; flowers blue or
blue-violet; capsule thin-walled, beakless; seeds pear-shaped; leaves not
over 1.5 cm broad; (Alaska–Yukon; coasts of E Que., s Labrador, Nfld.,
and N.S.) ....*I. setosa*
9   Petals not bristle-tipped, oblanceolate to obovate, longer; capsule
firm-walled, usually beaked; seeds not pear-shaped; leaves to about 3 cm
broad.
10 Capsule (and ovary) 6-angled, to 5 cm long, indehiscent; lower or
axillary spathes sessile or very short-stalked; leaves flaccid; stem
flexuous, compressed, weak or reclining; (s Ont.) ....*I. brevicaulis*
10 Capsule (and ovary) 3-angled, dehiscent; lower axillary spathes, when
present, terminating long branches; leaves firm.
11 Fresh leaf-tufts purplish at base; spathes mostly scarious; capsule
symmetrical, its inner surface lustrous; seed-coat firm, finely and
regularly pitted; (Man. to s Labrador, Nfld., and N.S.) ....*I. versicolor*
11 Fresh leaf-tufts buff or pale brown at base; spathes firm; capsule
often asymmetrical, its inner surface dull; seed-coat brittle and
corky, deeply and irregularly pitted; (s Ont. and sw Que.) ......*I. virginica*

I. brevicaulis Raf.   Lamance Iris
/t/EE/   (Grh)   Wet ground and borders of rich woods from E Kans., Mo., Ill., Ind., and Ohio to s
Ont. (about 3 mi SE of Leamington at the base of Pelee Point, Essex Co.; J. K. Shields, and M. E.
Shields, Rhodora 56(664): 80. 1954), s to E Tex. and Ala.

I. germanica L.   German Iris
European; persisting in gardens or an occasional escape in N. America; reported from SE Man. by
Löve and Bernard (1959; edge of an orchard at Otterburne, 30 mi s of Winnipeg, where originating
"from rejected rootstocks thrown away some years ago") and from s Ont. by Dodge (1915;
Lambton Co., "inclined to escape and persist"). The aggregate species is probably a series of
hybrids of uncertain parentage.

I. lacustris Nutt.   Dwarf Lake Iris
/T/EE/   (Grh)   Beaches, sandy woods, and bogs along the shores of L. Huron, Ont. (Bruce Pen.
and Southampton, Bruce Co., and Manitoulin Is.) to the shores of s L. Superior in Wisc.

I. missouriensis Nutt.   Western Blue Flag
/T/WW/   (Grh)   Moist places, meadows, wet flats, and slopes from Wash. and Mont. to sw Alta.
(Whiskey Gap and Carway, both s of Lethbridge near the U.S.A. boundary; introd. in N B.C.) and
N.Dak., s to Calif., N Mexico, N.Mex., and S.Dak.
The Canadian stations are indicated in a map by B. de Vries (Can. Field-Nat. 80(3): fig. 1,
p. 159. 1966), who believes that the species is probably native in Alta. because, "Well established
populations occur at the margin of shallow depressions in open grassland at low elevations
apparently uninfluenced by man." He cites four stations in B.C. at which the plant was obviously
introduced (Prince George, ca. 54°N; Quick Station, ca. 55°N; Atlin, ca. 59°35′N; Bennett, ca.
59°50′N), the first two being on a railway line, the plant having probably been brought in to the last
two by early miners and settlers. He notes that, "Tremendous quantities of rhizomes were used in
the days of herbal remedies and quack medicines."

I. prismatica Pursh
/T/E/ (Grh) Saline to fresh marshes, meadows, and shores from NE N.S. (Louisbourg, Cape Breton Is., where taken in wet meadows by John Macoun in 1883 and considered native by Fernald *in* Gray 1950; CAN; reported as introd. in Ont. by Boivin 1967a) to sw Maine, Md., and Del.; var. *austrina* Fern. s in acid swamps and wet barrens to Ga., inland to the mts. of Tenn. [*I. virginica sensu* John Macoun 1888, as to the Cape Breton Is. plant, not L., the relevant collection in CAN; the report from Nfld. by John Macoun 1890, is probably based upon *I. setosa* var. *canadensis*].

I. pseudacorus L.   Yellow Iris of Europe
European; a garden-escape or persisting from old plantings in s B.C. (Duncan, Vancouver Is.; Lulu Is.; Chilliwack; Sardis), s Man. (roadside ditch at St-Francois-Xavier, 12 mi w of Winnipeg; CAN), Ont. (N to the Ottawa dist.), s Que. (Stanstead and St-Maurice counties; MT), Nfld. (Avalon Pen.; CAN; GH), N.B., P.E.I., and N.S.

A horticultural phase (var. *mandschurica* Hort.; flowers pale yellow rather than bright yellow) is reported from Ont. by W. J. Cody (Can. Field-Nat. 75(3): 140. 1961; Carleton and Russell counties).

[I. pumila L.]   Dwarf Iris
[Eurasian; an occasional garden-escape in N. America but not becoming established; a collection from s Ont. (Puslinch L., Wellington Co.; OAC) has been placed here. The species is probably of hybrid origin.]

I. setosa Pallas   Beachhead-Iris or -Flag
/ST/D/eA/ (Grh) Chiefly coastal (var. *interior* inland in cent. Alaska–Yukon) on turfy headlands, ledges, upper beaches, and dunes (Alaska–Canada ranges given below), s in the East along the coast of Maine; E Asia. MAPS and synonymy: *see* below.
1   Stem often branched; [western varieties].
  2   Pedicels mostly shorter than the floral bracts; [*I. arctica* Eastw.; Aleutian Is. and coastal Alaska (N to Kotzebue Sound); to be expected along the coast of N B.C.; MAPS: Edgar Anderson, Ann. Mo. Bot. Gard. 23(3): map 1, p. 460. 1936; Cain 1944: fig. 31, p. 237; Hultén 1943: map 371a, p. 561, and 1968b:313] . . . . .var. *setosa*
  2   Pedicels often surpassing the floral bracts; [incl. var. *platyrhyncha* Hult.; interior Alaska N to ca. 67°N, the type from Fort Gibbon; Snag, Yukon Territory; MAPS: Hultén 1968b:313, and 1943: map 371b, p. 561, and on the above-noted maps by Anderson and Cain] . . . . .var. *interior* Anderson
1   Stem usually unbranched; [eastern phase] . . . . .var. *canadensis* Foster
  3   Flowers white or whitish; [E Que. (Rimouski Co. and Magdalen Is.) and Nfld. (type from Flower Cove)] . . . . .f. *pallidiflora* Fern.
  3   Flowers deep blue-violet.
    4   Leaves with transverse white or yellowish-white bands; [E Que. (Magdalen Is.) and Nfld. (type from Bay St. George)] . . . . .f. *zonalis* Eames
    4   Leaves uniformly green; [*I. hookeri* Penny; *I. caurina, I. graminea, I. setosa, I. tenax,* and *I. tridentata* of all or most reports from coastal E Canada, not Hook., L., Pallas, Dougl., nor Pursh, respectively; E Que. (coast and estuary shores from Montmagny and Charlevoix counties eastwards), Labrador (N to Turnavik, 55°17'N; CAN; M. L. Fernald, Rhodora 49(584): 213. 1947, notes a report near 58°N, presumably in the vicinity of Hebron), Nfld., N.B., P.E.I., and N.S.; MAPS: on the above-noted maps by Anderson and Cain]. A purported hybrid with *I. versicolor* (× *I. sancti-cyrii* Rousseau; petals intermediate in length between those of the putative parents) is known from the type locality, Anticosti Is., E Que. . . . .
. . . . .f. *canadensis*

I. sibirica L.   Siberian Iris
Eurasian; a garden-escape or persisting from old plantings in N. America, as in B.C. (Atlin, ca. 59°35'N; Herb. V) and s Ont. (flats of the Thames R. about 15 mi below London, Middlesex Co.; OAC; *see* W. W. Judd, Bull. Fed. Ont. Nat. 84: 16. 1959).

[I. thompsonii Foster]
[A collection from sw B.C. (Oak Bay, near Victoria, Vancouver Is.; Herb. V) has been placed here but, if correctly identified, may have originated from a cultivated plant or a casual escape. The species is native in the w U.S.A.]

I. versicolor L.   Blue Flag.   Clajeux or Iris sauvage
/sT/EE/ (Grh) Meadows, marshes, and turfy shores from s Man. (N to Hecla Is., L. Winnipeg; reported by J. M. Macoun 1913, as introd. with cranberry plants from the East at Ucluelet, Vancouver Is., B.C.) to Ont. (N to the Fawn R. at ca. 54°N, 89°W), Que. (N to E James Bay at 53°54′N and the Côte-Nord), Labrador (N to ca. 53°40′N; R. H. Woodworth, Rhodora 29(340): 56. 1927), Nfld., N.B., P.E.I., and N.S., s to Minn. and Va. [*I. virginica sensu* Hooker 1839, in large part, not L.]. MAP: Edgar Anderson, Ann. Mo. Bot. Gard. 23(3): map 2 (incomplete northwards), p. 464. 1936.

A hybrid with *I. virginica* (presumably its var. *shrevei*; × *I. robusta* And.) is reported from s Ont. by Soper (1949). The typical form has bluish or purplish petals, sepals, and stigmas. Forma *albocaerulea* Rousseau (the petals and petaloid stigmas white but the petaloid sepals bluish or purplish as in the typical form) is known from the type locality, Anticosti Is., E Que. Forma *murrayana* Fern. (petals, sepals, and stigmas white; type from near the mouth of the Salmonier R., Nfld.) is known from Ont., Nfld., P.E.I., and N.S.

I. virginica L.   Southern Blue Flag
/T/EE/ (Grh) Marshes, shores, and shallow water from Minn. to s Ont. (N to the Bruce Pen. and Georgian Bay, L. Huron) and sw Que. (Fernald *in* Gray 1950), s to E Tex. and Fla. [Incl. *I. shrevei* Small]. MAP: Edgar Anderson, Ann. Mo. Bot. Gard. 23(3): map 3 (Que. stations not shown), p. 465. 1936.

SISYRINCHIUM L.   [1286]   Blue-eyed Grass.   Bermudienne

1 Flowers bright yellow; anther-filaments free to near the middle; spathe solitary at the tip of the broad-winged scape, this to 6 mm broad; (sw B.C.) . . . .*S. californicum*
1 Flowers whitish or lilac to various shades of blue or purple; scapes rarely over 3 mm broad.
  2 Anther-filaments free to below the middle; flowers red-tinged, usually 1 or 2 borne in the solitary terminal spathe; scape more or less flattened but scarcely winged; (s B.C.).
    3 Filament-tube with a slight gradual enlargement near the base; capsules to 11 mm long; perianth dark magneta-purple . . . .*S. douglasii*
    3 Filament-tube sharply inflated near the base; capsules to 8 mm long; perianth pinkish purple (atypically white) . . . .*S. inflatum*
  2 Anther-filaments united into a tube to near the top; flowers various shades of lilac, blue, or bluish purple (without a reddish cast); scape flattened and usually more or less winged.
    4 Perianth-segments about 2 cm long; spathe usually solitary and terminal, with up to 4 flowers; leaves to 3 mm broad; scapes to 1.5 mm broad; (sw B.C.) . . . .[*S. macounii*]
    4 Perianth-segments smaller.
      5 Spathes peduncled from the axils of leaf-like bracts; spathe-bracts subequal; (eastern species).
        6 Old leaf-bases fibrous and persistent as crowded erect bristles; spathe to 2 cm long; perianth blue-violet; capsules to 5 mm long, on erect or ascending pedicels; (w N.S.) . . . .[*S. arenicola*]
        6 Old leaf-bases soon deciduous or persistent merely as loose irregular shreds.
          7 Stems broadly winged, to 6 mm broad; peduncles (terminated by spathes) flat and winged; spathe to 2 cm long; perianth pale blue, becoming violet; capsules to 6 mm long, on arching or recurving pedicels; plant drying blackish; (s Ont. to N.S.) . . . .*S. angustifolium*
          7 Stem more slender, narrowly winged, at most 3 mm broad; peduncles

filiform; spathe to 1.5 cm long; perianth blue-violet; capsules less than 5 mm long, on ascending pedicels; plant pale green or glaucous, scarcely blackening in drying; (N.S.) . . . .[*S. atlanticum*]

5 Spathes sessile or very short-peduncled, mostly subtended by an erect leaf-like bract.

   8 Spathes usually 2 together (rarely 1 or 3), the inner bracts less than 2.5 cm long, commonly about half as long as the erect outer leaf-like bract; perianth white or pale violet; stems flattened and slightly winged, to 3 mm broad; (s ?Ont.) . . . . . .
. . . .[*S. albidum*]

   8 Spathe usually solitary.

      9 Perianth white or pale blue (rarely yellow); outer spathe-bract with margins free to the base, the inner bract gibbous at base, half as long as the outer one; stem slightly winged, to 3 mm broad; (s ?Man.) . . . .
. . . .[*S. campestre*]

      9 Perianth usually deeper blue to bluish-purple; outer spathe-bract with margins united near base.

         10 Outer spathe-bract to about 6 cm long, rarely more than a third longer than the inner one; leaves to 3.5 mm broad; stems to 2 mm broad, distinctly winged; (s B.C.) . . . .*S. idahoense*

         10 Outer spathe-bract sometimes 2 or 3 times longer than the inner one.

            11 Pedicels soon strongly spreading to recurving, much surpassing the inner bract; capsules 3 or 4 mm long; leaves 1 or 2 mm broad; stems to 1.5 mm broad, scarcely winged; (Man. to s Ont.) . . . . .
. . . .*S. mucronatum*

            11 Pedicels erect or ascending, only slightly surpassing the inner bract; capsules to 6 mm long; leaves to 3 mm broad; stems to 3 mm broad, distinctly winged; (transcontinental) . . . .*S. montanum*

[S. albidum Raf.]
["Prairies and dry, often sandy, open soil or thin woodland, Ga. and Ala. to La., N to N.S., s Ont., Ohio, s Mich., s Wisc. and Mo.; casual northeastw." (Fernald *in* Gray 1950). Also reported from s Ont. by Montgomery (1945; Waterloo Co.), Soper (1949), and Gillett (1958; Ottawa dist.). Collections in OAC from s Ont. (Lambton, Waterloo, and Lincoln counties) have been placed here but the only so-named collection in CAN (Sandwich, Essex Co., where taken by John Macoun in 1901; identification confirmed by Fernald) appears referable to *S. mucronatum* and may be the basis of Fernald's inclusion of s Ont. in the range.]

S. angustifolium Mill.
/T/EE/ (Hsr) Meadows, damp woods, thickets, and shores from E Kans. to Mo., Ill., Ind., Ohio, s Ont. (N to the Ottawa dist.; Gillett 1958), Que. (perhaps not much farther N than Montreal; the inclusion of SE Nfld. in the range by Fernald *in* Gray 1950, may be based upon *S. montanum*), N.B., ?P.E.I., and N.S., s to E Tex. and Fla. [*S. gramineum* Curtis and of Lam. in part; *S. graminoides* and *S. ?intermedium* Bickn.; *S. bermudiana sensu* L. H. Shinners, Rhodora 59(702): 159. 1957, not L.].

   S. montanum as interpreted below has long been referred by many authors to *S. angustifolium*, making it impossible to delimit the range in Canada other than as sketched above. *See* D. B. Ward (Taxon 17(3): 270–76. 1968) and Michel Guédès (Taxon 18(5): 542–45. 1969).

[S. arenicola Bickn.]
[This species of the E U.S.A. (N to Mich. and Mass.) is reported from N.S. by Fernald (1921) and Fernald *in* Gray (1950); on the basis of collections in GH from Halifax, Yarmouth, and Annapolis counties), this taken up by Roland (1947). Further studies are advisable, however, before accepting this critical species as a member of our flora. (*S. farwellii* Bickn.).]

[S. atlanticum Bickn.]
[Fernald *in* Gray (1950) includes w N.S. in the range of this species of the E U.S.A. (N to Mich. and Maine). However, it seems scarcely separable from *S. angustifolium* Mill. (*See* Ward, loc. cit.).]

S. californicum (Ker-Gawl.)   Dryander, Golden-eyed Grass
/t/W/   (Hsr)   Shores and wet places from sw B.C. (Vancouver Is. and adjacent islands; John Macoun 1888; Henry 1915; Estham 1947) to s Calif. [*Marica* Ker-Gawl.; *Hydastylus (S.) borealis* and *H. (S.) brachypus* Bickn.]

[S. campestre Bickn.]
[The report of this obscure species of the E U.S.A. (Wisc. to Tex. and La.) from s Man. by Shimek (1927; possible basis of the inclusion of Man. in the range by Fernald *in* Gray 1950) requires confirmation.]

S. douglasii Dietr.
/t/W/   (Hsr)   Moist places and mountain slopes from sw B.C. (Victoria district, Vancouver Is.; CAN) to Calif. [*Olsynium* Bickn.; *S. grandiflorum* Dougl., not Cav.]. MAP: Harold St. John, B.C. Prov. Mus. Nat. Hist., Rep. (1930): pl. 11. 1931 (the Vancouver Is. station should be indicated by a solid dot rather than the triangle denoting *S. inflatum*).

S. idahoense Bickn.
/sT/W/   (Hsr)   Coast of s Alaska (*see* Hultén 1943: map 372 (*S. lit.*), p. 561) through B.C. (E to Windermere, Columbia Valley) to Wash., Idaho, and Wyo. [*S. birameum* Piper; *S. littorale* Greene; *S. segetum* Bickn.]. MAP: Hultén 1968*b*:314 (*S. lit.*).

S. inflatum (Suksd.) St. John
/T/W/   (Hsr)   Grassy places from s B.C. (Adams Lake, about 30 mi NE of Kamloops; CAN; V; St. John, loc. cit.; *see* his MAP, pl. 11, on which the solid-dot symbols (*S. douglasii*) indicating the Adams Lake stations should be replaced by the triangle symbol specified for *S. inflatum*) to Calif., Idaho, and Utah. [*Olsynium* Suksd.].

[S. macounii Bickn.]
[The type of this obscure taxon is an 1893 Macoun collection in sw B.C. (Comox, Vancouver Is.). It is included in the *S. angustifolium* complex by Hitchcock et al. (1969).]

S. montanum Greene
/aST/X/   (Hsr)   Meadows and sandy open ground from s Yukon and w-cent. Dist. Mackenzie to L. Athabasca (Alta. and Sask.), Man. (N to the Churchill R. at ca. 57°30'N), northernmost Ont., Que. (N to E James Bay at ca. 52°N and the Côte-Nord), Labrador (N to the Hamilton R. basin), Nfld., N.B., P.E.I., and N.S., s to Baja Calif., Colo., Ind., Pa., and mts. of w Va.; w Greenland between ca. 62° and 65°25'N (?introd.); introd. in Europe. [Incl. *S. bellum* Wats., *S. halophilum* Greene, *S. groenlandicum* Böcher, and *S. septentrionale* Bickn.; *S. sarmentosum* Suksd. in part; *S. angustifolium sensu* Bickn. and other American auth. in large part, not Mill.]. MAPS: Hultén 1968*b*:314, and 1958: map 189, p. 209; T. W. Böcher, Medd. Gronl. 147 (3): fig. 5, p. 14. 1948.

    According to T. W. Böcher (Bot. Tidsskr. 61(4): 283. 1966), "*S. montanum* is clearly a very complex entity and seems to include forms at three ploidy levels. It is certainly not a good species, but rather an aggregate of closely related species." The extension of the range of *S. angustifolium* westwards to s Alaska–Baja Calif. by Hitchcock et al. (1969) arises through their merging with it of several taxa herein merged with *S. idahoense* (*S. birameum, S. littorale, S. segetum*) and *S. montanum* (*S. bellum, S. halophilum, S. sarmentosum,* and *S. septentrionale*).

S. mucronatum Michx.
/T/EE/   (Hsr)   Meadows, fields, and open woods from Sask. (N to Prince Albert; CAN, detd. Bicknell) to s Man. (Woodmore, about 50 mi s of Winnipeg; Löve and Bernard 1959, confirming earlier reports from the province), s Ont. (N to Manitoulin Is., N L. Huron; TRT; CAN), and NE Maine, s to Minn., Wisc., and N.C. [*S. anceps* Cav., a prior but illegitimate name].

NOTE

Since the preparation of the above treatment of *Sisyrinchium,* Theodore Mosquin (Madroño 20(5): 269–75. 1970) has proposed a new, much simplified classification of the North American taxa. He recognizes as more or less distinct species only *S. douglasii* Dietr., *S. arenicola, S. atlanticum,* and *S. capillare* Bickn., *S. bermudiana* L., *S. mucronatum* Greene, and possibly *S. bellum* Wats. He merges in the widespread *S. bermudiana* complex such obscure taxa as *S. angustifolium* Mill., *S. campestre, S. idahoense,* and *S. graminoides* Bickn., *S. littorale,* and *S. montanum* Greene, *S. groenlandicum* Böcher, *S. sarmentosum* Suksd., and, by inference, probably *S. albidum* Raf. These he considers to represent three different chromosome-races of the complex. His MAP (fig. 1, p. 274) showing the North American distribution of *S. bermudiana* corresponds closely to the range and maps cited above for *S. montanum,* the Alaskan area presumably being that cited above for *S. idahoense.*

ORCHIDACEAE (Orchis Family)

(Ref.: Correll 1950; Szczawinski 1959)

Perennial herbs, commonly from corms or coral-like roots. Leaves entire, variously disposed. Flowers solitary or in spikes or racemes, perfect, epigynous. Perianth usually of 6 divisions of similar texture (or the sepals 1 and a fused pair). Petals 3, one of them (the lip; really the posterior petal, but by a half-turn of the pedicel or ovary usually directed downward and apparently anterior) differing from the others in shape and sometimes spurred at base (*Habenaria* and *Orchis*). Fertile stamen 1 (in *Cypripedium*, 2), united with the style to form the column at the base of the lip. Ovary inferior. Fruit a many-seeded 1-locular capsule, the seeds minute.

1 Flowers large and showy, spurless, one of the petals (the lip) a much inflated nearly closed pouch up to 6 cm long; fertile anthers 2; leaves lanceolate to ovate, plaited lengthwise; plants more or less glandular-pubescent ....*Cypripedium*
1 Flowers mostly smaller; lip various, if slightly pouch-like, much smaller; fertile anther 1.
  2 Flower normally solitary, spurless; lip bearded or crested.
    3 Stem a leafless scape, the solitary, oval to round-ovate, long-petioled leaf basal and from a separate bud; lip about 2 cm long, shaped like a sugarscoop, whitish with yellow tip and marked with red-brown, crested with 3 rows of yellow hairs; plant to about 2 dm tall; (transcontinental) ....*Calypso*
    3 Stem leafy, the leaves sessile.
      4 Leaves usually at least 5 in a whorl at the summit of the stem, oblong-obovate to rhombic-elliptic, to about 1 dm long; lateral petals yellowish green; lip green, with a median crest, 3-lobed, the terminal lobe greenish yellow, the lateral lobes striped with purple; plant to about 3.5 dm tall; (s Ont.; extinct) ....*Isotria*
      4 Leaf solitary.
        5 Leaf plaited lengthwise, linear-attenuate, appressed-ascending, less than 5 mm broad, borne below the middle of the stem, developing after anthesis; flower subtended by a pair of small bracts; perianth magenta-pink, the pinkish-white lip shallowly notched, usually spotted and streaked with purple, with 3 or 5 yellow or white fringed crests; plant to about 3 dm tall; (Sask. to s Labrador, Nfld., and N.S.) ....*Arethusa*
        5 Leaf flat, narrowly lanceolate to elliptic, commonly at least 1 cm broad, borne near the middle of the stem, well developed at anthesis; flower subtended by a solitary leaf-like bract; perianth various shades of pink, the lip entire, veined with red and with a beard of 3 rows of yellow hairs; plant to about 4 dm tall; (Ont. to Nfld. and N.S.) ....*Pogonia*
  2 Flowers 2 or more, in spikes or racemes (those of *Triphora* in the axils of leaf-like upper bracts).
    6 Flowers with a distinct slender and elongate spur at the base of the lip (*Corallorhiza mertensiana* of B.C. may be sought here); plants glabrous.
      7 Perianth bicoloured, the lip white or white and purple-dotted, the sepals and lateral petals lilac to roseate; leaves (in our native species) 1 or 2, elliptic to obovate, sub-basal ....*Orchis*
      7 Perianth essentially uniformly coloured; leaves 1 to several, basal or cauline ....*Habenaria*
    6 Flowers spurless or the lip merely with a small basal pouch.
      8 Plants saprophytic, within green foliage (*Spiranthes* may often be sought here), the leaves reduced to white or coloured sheathing scales, the whole plant white, yellow, or purple, glabrous.
        9 Plant white, to 5 dm tall, from a creeping rootstock; flowers to about 1.5 cm long, nearly sessile in terminal racemes to 2 dm long, the perianth-segments not gibbous at base; (s B.C.) ....*Eburophyton*
        9 Plants yellow to purple, from branched and toothed, coral-like rhizomes; flowers subsessile or short-petioled, the perianth-segments somewhat gibbous at base ....*Corallorhiza*

8 Leaves or a leaf present (in basal-leaved species of *Spiranthes,* soon shrivelling), they and the stem with normal green colour.

  10 Leaves in a basal rosette or paired at the base of the stem.

    11 Leaves 2, ascending, sheathing the stem near base; plant from a large corm (solid bulb) .... *Liparis*

    11 Leaves several in a spreading basal rosette.

      12 Leaves thin, soon shrivelling, uniformly green; flowers sessile in a more or less spirally twisted elongate spike; lip not inflated or saccate; plant from commonly clustered roots .... *Spiranthes*

      12 Leaves more or less fleshy, often with white reticulations or veins; flowers subsessile in a loosely to tightly spiralling spike-like raceme; lip conspicuously saccate or pouch-like at base; plant from a somewhat fleshy creeping fibrous-rooted rhizome .... *Goodyera*

  10 Leaves cauline or, if basal or sub-basal, the leaf commonly solitary.

    13 Leaves a single subopposite pair sessile near the middle of the stem; roots fibrous; stems more or less glandular-pubescent above .... *Listera*

    13 Leaves scattered and alternate, or the leaf solitary.

      14 Leaf solitary; flowers racemose.

        15 Leaf basal, its elliptic blade to about 1.5 dm long, plaited lengthwise; perianth purplish toward base, brown toward summit, the white, violet-marked lip about 1.5 cm long, with 3 low parallel ridges near centre; plant to about 6 dm tall, from glutinous corm-like tubers connected by slender stolons; (Ont. and sw Que.) .... *Aplectrum*

        15 Leaf sub-basal to midway on the stem, smaller; plant from a single solid tuber or corm.

          16 Flowers very small, green or yellowish-green; lip not bearded; leaf relatively broad, located from near base to about midway on the stem .... *Malaxis*

          16 Flowers commonly at least 2.5 cm broad, pink-purple, apparently inverted (the lip uppermost, densely bearded along the inner side); leaf linear to narrowly oblong, many times as long as broad, sub-basal; (SE Man. to s Labrador, Nfld., and N.S.) .... *Calopogon*

      14 Leaves 2 or more, alternate, sessile-clasping; lip neither bearded nor crested.

        17 Flowers relatively large, slender-pedicelled in an open raceme.

          18 Flowers few, white to pale pink, from the axils of scarcely reduced upper leaves, the sepals and lateral petals lanceolate, to 2 cm long; lip 3-lobed, with 3 green median lines; leaves few, ovate, at most about 2 cm long; glabrous woodland plant to about 3 dm tall; (s Ont.) .... *Triphora*

          18 Flowers more numerous in a bracted raceme to about 3 dm long, green, suffused with purple, the strongly saccate pointed lip with 2 tubercles at base; leaves several, lanceolate to ovate, to about 1 dm long; plant to about 9 dm tall, more or less pubescent above .... *Epipactis*

        17 Flowers small, white or greenish to yellow or yellow-green, subsessile or short-pedicelled in a spike-like raceme.

          19 Stem scaly-bracted between the linear to narrowly oblanceolate foliage-leaves and the more or less spirally twisted raceme; plants commonly more or less pubescent above, from tuberous-thickened roots .... *Spiranthes*

          19 Stem naked between the lance-ovate foliage-leaves and the raceme; plant glabrous, from a solid bulb; (*M. paludosa*) .... *Malaxis*

APLECTRUM (Nutt.) Torr.   [1642]

A. hyemale (Muhl.) Torr.   Putty-root, Adam-and-Eve
/T/EE/   (Gst)   Rich woods from Ont. (N to Ottawa; CAN; John Macoun 1888) and sw Que. (N to Oka; Marie-Victorin 1935; reports from Sask. and Man. require confirmation) to Ark., Tenn., and Ga. [*Cymbidium* Muhl.; *A. spicatum* (Walt.) BSP.]

ARETHUSA L.   [1474]

A. bulbosa L.   Arethusa, Swamp-pink.   Aréthuse
/sT/EE/   (Gst)   Sphagnous bogs and peaty meadows from Sask. (known only from the s shore of L. Athabaska) to Man. (N to Schist L., near Flin Flon at ca. 55°N), Ont. (N to Kenora, Thunder Bay, and Nikip L. at ca. 53°N, 92°W), Que. (N to St-Fabien, Rimouski Co., and Anticosti Is.), Labrador (N to Goose Bay, 53°18'N), Nfld., N.B., P.E.I., and N.S., s to N Ind. and N.J. and in the mts. to S.C.; reported from La. MAP: Raymond 1950*b*: fig. 22 (somewhat incomplete), p. 56.

CALOPOGON R. Br.   [1534]

C. pulchellus (Salisb.) R. Br.   Swamp-pink
/sT/EE/   (Gst)   Sphagnous bogs, peaty meadows, and wet shores from SE Man. (bogs about 30 mi and 40 mi SE of Winnipeg) to Ont. (N to L. Nipigon; TRT), Que. (N to Ste-Luce, Rimouski Co.; RIM), s Labrador (Goose Bay, 53°18'N), Nfld., N.B., P.E.I., and N.S., s to Tex. and Fla. [*Limodorum* Salisb.; *Cymbidium* Willd.; *Cal. tuberosus sensu* BSP., perhaps not *L. tub.* L.]. MAP: M. L. Fernald, Rhodora 13: map 7 (incomplete northwards), pl. 89, p. 139. 1911.

   Forma *latifolius* St. John (the relatively broad leaves often paired and often much surpassing the scape, this usually less than 2 dm tall; the typical form with its usually solitary leaf much surpassed by the scape, this usually over 2.5 dm tall) is known from E Que. (Magdalen Is.) and N.S. (type from Sable Is.). Forma *albiflorus* (Britt.) Fern. (flowers white rather than lilac to rose-pink) is known from s Ont. (Port Franks, Lambton Co.; Gaiser and Moore 1966) and N.S. (Yarmouth, Guysborough, and Richmond counties).

CALYPSO Salisb.   [1559]

C. bulbosa (L.) Oakes   Calypso
/ST/X/EA/   (Gst)   Cool mossy woods and shingle beaches (chiefly calcareous) from cent. Alaska–Yukon (*see* Hultén 1943: map 399, p. 563) to Great Bear L., Great Slave L., L. Athabasca (Alta. and Sask.), Man. (N to Churchill), Ont. (N to sw James Bay at ca. 52°45'N), Que. (N to L. Mistassini, the Côte-Nord, Anticosti Is., and Gaspé Pen.), Nfld., and N.B. (not known from P.E.I.; reported from N.S. by Pursh 1814, but not listed by Roland 1947, and probably extinct), s to Calif., Ariz., Colo., Minn., Mich., and N New Eng.; Eurasia. [*Cypripedium* L.; *Cal. borealis* (Sw.) Salisb.; incl. f. *occidentalis* Holz.]. MAPS: Hultén 1968*b*:331, and 1962: map 71, p. 81.

CORALLORHIZA Chat.   [1548]   Coral-root

1   Lip 3-lobed, the terminal lobe the largest; spur nearly or quite completely fused with the ovary; perianth at most 1 cm long, often red-spotted, not striped; (transcontinental).
   2   Lateral sub-basal lobes of lip prominent; spur a prominent swelling near the summit of the ovary; capsule yellowish brown or brownish; rhizome brownish; scape to about 6 dm tall, the sheaths extending to well above its middle ....*C. maculata*
   2   Lateral lobes of lip small; spur inconspicuous; capsule greenish; rhizome whitish; scape at most 3 dm tall, with 3 or 4 close sheaths toward base ....*C. trifida*
1   Lip entire or merely retuse at apex.
   3   Spur of lip with a free tip 1–5 mm long; perianth pinkish yellow or whitish, tinged and striped with purple, to 1 cm long; sepals reflexed; (B.C.) ....*C. mertensiana*
   3   Spur wanting or nearly or quite completely fused with the ovary; sepals and petals connivent as a hood or the sepals somewhat spreading (but not reflexed).

4   Perianth to 2 cm long, its segments conspicuously purple-veined; lip tongue-
    shaped; capsule to 2 cm long; scape to 4 dm tall; (B.C. to E Que.) . . . .*C. striata*
4   Perianth at most about 7 mm long, not obviously striped; lip broadly oval to
    roundish; capsule at most 12 mm long.
     5   Perianth 3 or 4 mm long, somewhat hood-like; lip at most 3 mm long; scape to
       about 2 dm tall; (s Ont.) . . . .*C. odontorhiza*
     5   Perianth about 7 mm long, its segments more or less spreading; lip about
       5 mm long; scape to about 4 dm tall; (s ?Ont.) . . . .[*C. wisteriana*]

C. maculata Raf.   Spotted Coral-root
/T/X/   (Grh)   Dry to wet coniferous forests from B.C. (N to the Beatton R. at ca. 57°N; CAN; *see*
B.C. map by Szczawinski 1959:103) to Alta. (N to Lac la Biche, 54°46′N; W. J. Cody, Can.
Field-Nat. 70(3): 109. 1956), Sask. (N to McKague, ca. 52°45′N; Breitung 1947), Man. (N to Grand
Rapids, near the NW end of L. Winnipeg), Ont. (N to Sandy L., ca. 53°N, 93°W), Que. (N to L. St.
John, Anticosti Is., and the Gaspé Pen.), Nfld., N.B., P.E.I., and N.S., s to Calif., N.Mex., S.Dak.,
Ind., Tenn., and N.C. [*C. multiflora* Nutt.; incl. var. *occidentalis* (Lindl.) Cock.; *Neottia nidus-avis*
sensu Lindsay 1878, not L., the relevant collection in NSPM].
   Forma *flavida* (Peck) Farw. (perianth yellow as in the typical form but the lip unspotted rather
than spotted with red or purple) is known from s Ont. (Bruce Pen., L. Huron) and N.S. (Cape Breton
Is.). Forma *punicea* (Bartl.) Weath. & Adams (perianth reddish purple rather than yellow, the lip
darker-spotted, the scape and leaf-sheaths also reddish purple) is known from N.B., P.E.I., and
N.S.

C. mertensiana Bong.   Pacific Coral-root
/sT/W/   (Grh)   Coniferous forests from the N Alaska Panhandle (type from Sitka; *see* Hultén
1943: map 395, p. 563) through B.C. (*see* B.C. map by Szczawinski 1959:106) to Calif., Idaho, and
Wyo. [*C. maculata* ssp. *mert.* (Bong.) Calder & Taylor; *C. vancouveriana* Finet]. MAP (*C. mac.* ssp.
mert.): Hultén 1968*b*:329.

C. odontorhiza (Willd.) Nutt.   Late or Autumn Coral-root
/t/EE/   (Grh)   Dry to moist woods and ravines from Minn. to s Ont. (Kent, Middlesex, Norfolk,
Waterloo, Lincoln, and Wentworth counties) and sw Maine, s to Mo., Miss., Ala., and Ga.
[*Cymbidium* Willd.].

C. striata Lindl.   Striped Coral-root
/T/X/   (Grh)   Moist woods (often calcareous) from B.C. (N to the Peace River dist. at ca.
55°40′N; *see* B.C. map by Szczawinski 1959:108) to s Alta. (N to Castle Mt., NW of Banff), s Sask.
(N to Kelliher, 51°15′N; Breitung 1957*a*), Man. (N to Duck Mt.), Ont. (N to Kapuskasing; Baldwin
1958), Que. (N to L. St. John and Bonaventure Is., Gaspé Pen.; the report from N.B. by Correll
1950, requires confirmation), and sw N.Y., s to Calif., Mexico, Tex., N.Dak., Mich., s Ont., and sw
Que. [*C. macraei* Gray].
   Forma *fulva* Fern. (perianth, leaf-sheaths, and scape yellowish to orange-brown rather than
purplish) is known from E Que. (Bic, Rimouski Co.; Percé Mt. (type locality) and Bonaventure Is.,
Gaspé Pen.).

C. trifida Chat.   Early or Pale Coral-root
/aST/X/GEA/   (Grh)   Bogs, thickets, and woods from the E Aleutian Is., N Alaska, and cent.
Yukon (*see* Hultén 1943: map 396, p. 563) to NW Dist. Mackenzie, cent. Dist. Keewatin,
northernmost Ont., Que. (N to Hudson Strait and the Côte-Nord), Labrador (N to ca. 55°30′N), Nfld.,
N.B., P.E.I., and N.S., s to Oreg., Colo., S.Dak., Mo., Ohio, Tenn., and Ga.; w Greenland N to ca.
69°30′N; Iceland; Eurasia. [Incl. var. *verna* (Nutt.) Fern., the perianth-lip uniformly white or with
only a few red or purplish dots]. MAPS: Hultén 1968*b*:329, and 1962: map 102, p. 111; Meusel,
Jaeger, and Weinert 1965:112; Raup 1947: pl. 20.

[*C. wisteriana* Conrad]
[The report of this species of the E U.S.A. (N to S.Dak., Mo., Ind., and Pa.) from s Ont. by F.H.C.

Dempsey (Can. Field-Nat. 57(4/5): 93. 1943; Bala, E of Georgian Bay, L. Huron) requires confirmation.]

CYPRIPEDIUM L.   [1391]   Lady's-slipper, Moccasin-flower.   Sabot de la Vierge

1 Leaves 2(3), basal or sub-basal, the scape or scapose stem terminated by a solitary flower (rarely 2); lateral sepals united nearly or quite to the apex.
  2 Corolla-lip to 6 cm long, crimson-pink with rose veins (occasionally pure white), copiously puberulent externally, with a deep fissure extending down the front; sepals and lateral petals yellowish-green to greenish-brown, to 5 cm long; leaves opposite and basal, broadly elliptic or oblong-elliptic, to about 2 dm long and 1.5 dm broad; scapes to over 5 dm tall; (Alta. to Nfld. and N.S.) ....C. acaule
  2 Corolla-lip to 2.5 cm long, white or yellowish-green, commonly blotched with purple, glabrous externally, not fissured in front but with a rounded opening near the base of the upper face; sepals and lateral petals white or greenish-white, the dorsal sepal to 2.8 cm long, the others less than 2 cm long; leaves approximate a short distance above the base of the scapose stem, elliptic-lanceolate to elliptic, to about 1.5 dm long and 6 cm broad; (Alaska–?B.C. to w Dist. Mackenzie) ....C. guttatum
1 Leaves 3 or more (very rarely 2), well distributed along the stem.
  3 Lateral sepals separate and spreading, linear, the upper sepal lanceolate, the sepals and the lance-linear lateral petals greenish-brown, to 2.5 cm long; lip whitish but strongly netted with purplish veins, to about 2 cm long, irregularly triangular (tapering to a greenish or yellowish conical tip); flower solitary; leaves commonly 3 or 4, often folded, lanceolate to elliptic, finely ciliate, otherwise glabrous, to about 1 dm long; (Sask. to N.S.) ....C. arietinum
  3 Lateral sepals usually united nearly or quite their full length; lip less strongly veined, broadly rounded.
    4 Lip yellow, to 5 cm long; sepals and lateral petals greenish yellow to purplish brown, to 8 or 9 cm long; leaves ovate-lanceolate to oval, more or less glandular-pubescent; flowers usually 1 or 2; (transcontinental) ....C. calceolus
    4 Lip prevailingly white but with coloured venation.
      5 Sepals and lateral petals white, to 4.5 cm long; upper sepal elliptic-oval to suborbicular, the lower fused pair and the petals narrower; lip to 5 cm long, with broad pink or roseate bands in front between the narrow vertical white furrows; flowers often 1 or 2 (but up to 4); leaves ovate, to about 2 dm long and 1.5 dm broad, strongly ribbed and densely glandular-hirsute (the hairs sometimes causing serious eczema); (Man. to Nfld. and N.S.) ....C. reginae
      5 Sepals greenish to brown-purple; flowers usually solitary but sometimes 2 (or even 3 in C. montanum); leaves commonly less than 1.5 dm long and 7 cm broad.
        6 Sepals, lateral petals, and lip each rarely over 1.5 cm long, the lateral sepals commonly more or less free above the middle (or even nearly to base); petals white, oblong, rounded at apex; leaves to 5 cm broad; plant to about 2.5 dm tall, more or less densely villous-pubescent throughout; (Alaska–B.C. to SE James Bay; L. Superior, Ont.; E Que.) ....C. passerinum
        6 Sepals and petals each commonly over 2 cm long, lance-acuminate, the lateral sepals usually fused nearly or quite to the tip; lip 1.5 cm long or more; plants mostly taller.
          7 Sepals and petals brown-purple, to 6 cm long; lip to 3 cm long, white with purple veins; leaves to about 7 cm broad; plant usually rather copiously glandular-puberulent, usually over 3.5 dm tall; (SE Alaska–B.C.–sw Alta.) ....C. montanum
          7 Sepals and lateral petals greenish yellow, often purple-lined, commonly less than 3 cm long; lip to 2.5 cm long, wax-white, smooth and shining outside, purple-veined within at base; leaves to about 4 cm broad; plant sparingly glandular-puberulent throughout, rarely over 3 dm tall; (SE Sask. to s Ont.) ....C. candidum

C. acaule Ait.   Stemless Lady's-slipper
/sT/(X)/   (Grh)   Acid soils of bogs, mossy places, and dry to wet woodlands from L. Athabasca (Alta. and Sask.; also reported from Great Bear L., Dist. Mackenzie) to Man. (N to Wekusko L., about 75 mi NE of The Pas), Ont. (N to the Severn R. sw of Hudson Bay; John Macoun 1888), Que. (N to L. Mistassini, the Côte-Nord, and Shickshock Mts. of the Gaspé Pen.), Nfld., N.B., P.E.I. (provincial floral emblem), and N.S., s to Minn., N Ind., Ala., and Ga. [*Fissipes* Small; *C. humile* Pursh].

Forma *albiflorum* Rand & Redf. (the flower with a white lip and pale sepals and petals) is known from Ont., Que., N.B., and N.S.

C. arietinum R. Br.   Ram's-head Lady's-slipper
/T/EE/   (Grh)   Bogs and damp or mossy woods from Sask. (N to Prince Albert; Breitung 1957a) to Man. (N to the N end of L. Winnipegosis), Ont. (N to near Kenora and Ottawa), Que. (reported N to the Saguenay R. by John Macoun 1888), and N.S. (known only from around gypsum quarries in Hants Co.: reported from Newport by H. How, Jr., Proc. Trans. N.S. Inst. Sci. 4: 318. 1878, and from Wentworth by J. S. Erskine, Rhodora 56(669): 203. 1954; collections in ACAD and CAN from Wentworth and St. Croix; not known from N.B. or P.E.I.; a puzzling report from Nfld. by Reeks 1873), s to Minn., Mich., N.Y., and Mass. [*Criosanthes* House].

C. calceolus L.   Yellow Lady's-slipper
/ST/X/EA/   (Grh)   Bogs (chiefly calcareous), swamps, and wet woods, the aggregate species from the Yukon (N to ca. 66°N near the Alaska boundary) to Great Bear L., s-cent. Alta.–Sask., Man. (N to the Hayes R. about 100 mi sw of York Factory), Ont. (N to the Shamattawa R. at ca. 55°N), Que. (N to the Harricanaw R. SE of James Bay and the Côte-Nord), Nfld., N.B., P.E.I., and N.S., s to Oreg., Utah, Colo., Tex., Mo., Tenn., Ala., and Ga.; Eurasia. MAPS and synonymy: *see* below.

Three hybrids with *C. candidum* (× *C. andrewsii* Fuller) are reported from Canada by Bernard Boivin (Nat. can. (Que.) 87: 32. 1960): × *C. andrewsii* nm. *andrewsii* (*C. calceolus* var. *parviflorum* × *C. candidum*; Brandon, Man.); × *C. andrewsii* nm. *favillianum* (Curtis) Boivin (× *C. fav.* Curtis; *C. calc.* var. *pubescens* × *C. candidum*; Simcoe and Turkey Point, Norfolk Co., s Ont.); and × *C. andrewsii* nm. *landonii* (Garay) Boivin (× *C. landonii* Garay; × *C. andrewsii* nm. *fav.* × *C. calc.* var. *parv.*; Turkey Point, Norfolk Co., s Ont.). *See* H. H. Marshall, A.T.H. Grass, and G. A. Stevenson, Rhodora 68(773): 53–58. 1966.

1 Upper sepal merely acute, usually rounded at base, at most 4 cm long; petals straw-colour to purple, oblong-lanceolate, flat or undulate, to about 4 cm long; stem-leaves rarely more than 4, mostly less than 1 dm long; [*C. parviflorum* var. *plan.* Fern.; *C. calceolus* f. *rupestris* Vict. & Rousseau; E Que. and Nfld. (type from Savage Cove); MAP: Hultén 1958: map 261, p. 281] . . . . .var. *planipetalum* (Fern.) Vict. & Rousseau
1 Upper sepal long-acuminate, more or less narrowed at base, to 7 cm long; petals yellowish, often purple-striped, narrowly lanceolate, spirally twisted; stem-leaves commonly 5 or 6, to about 2 dm long.
   2 Lip to 5 cm long; upper sepal to 7 cm long, the lateral ones united below or distinct; petals to 9 cm long; largest leaves to over 1 dm broad; [*C. pubescens* Willd.; s Sask. (Breitung 1957a) to N.B. and N.S.; MAP: on the above-noted map by Hultén] . . . .
    . . . . .var. *pubescens* (Willd.) Correll
   2 Lip at most 4 cm long; upper sepal to 5 cm long, the lateral ones united nearly to apex; petals to about 5 cm long; leaves less than 1 dm broad; [*C. parviflorum* Salisb.; Yukon–B.C. to Nfld. and N.S.; MAPS: on the above-noted map by Hultén, and Hultén 1968b:316; Meusel, Jaeger, and Weinert 1965:104] . . . . .var. *parviflorum* (Salisb.) Fern.

C. candidum Muhl.   Small White Lady's-slipper
/T/EE/   (Grh)   Calcareous meadows, marly bogs, swamps, and moist prairies from SE Sask. (Boivin 1967a) to s Man. (N to about 30 mi NW of Winnipeg), s Ont. (Kent, Lambton, Norfolk, and Bruce counties), and N.Y., s to S.Dak., Mo., Ky., and N.J.; (a puzzling report from Nfld. by Reeks 1873).

C. guttatum Sw.

/ST/W/eEA/ (Grh)   Meadows, open woods, and slopes from N-cent. Alaska–Yukon (*see* Hultén 1943: map 373, p. 561) to NW Dist. Mackenzie (N to Norman Wells, 68°14′N), s to the Aleutian Is. and Great Slave L. (the inclusion of B.C. in the range by Correll 1950, requires confirmation); Eurasia. (The accrediting of *C. fasciculatum* Kell. to B.C. by Hitchcock et al. 1969, may refer here). MAPS: Hultén 1968*b*:315; Raup 1930: map 14, p. 202.

The eastern Asiatic ssp. *yatabeanum* (Makino) Hult. (*C. yat.* Mak.) is reported from Attu Is. of the Aleutian Is. by Hultén (1950), who notes that, "It differs from *C. guttatum* in the spotted yellowish green flowers, in the greenish white sepals and in the lateral petals, which do not taper to the apex but end in a rounded disc."

C. montanum Dougl.   Mountain Lady's-slipper

/sT/WW/ (Grh)   Cool rich humus on subalpine slopes (occasionally in wooded or swampy places at lower elevations) from the Alaska Panhandle (N to near or on the Yukon boundary; *see* Hultén 1943: map 374, p. 561) through B.C. (*see* B.C. map by Szczawinski 1959: 27) and sw Alta. (Waterton Lakes; CAN; Breitung 1957*b*) to Calif., Idaho, and Wyo. [*C. occidentale* Wats.]. MAP: Hultén 1968*b*:317.

C. passerinum Richards.   Sparrow's-egg Lady's-slipper

/ST/(X)/ (Grh)   Moist coniferous forests, deep ravines, streambanks, lake-margins, and (usually calcareous) gravel outwashes and talus-slopes from N Alaska and cent. Yukon (*see* Hultén 1943: map 375, p. 561) to Great Bear L., Great Slave L., Sask. (N to Hasbala L. at ca. 59°N), Man. (N to Churchill), northernmost Ont., s James Bay (Charlton Is.), and E Que. (Mingan Is. of the Côte-Nord), s to s B.C.–Alta., N Mont., SE Sask. (Cypress Hills), s Man. (Stony Mountain, about 10 mi N of Winnipeg), and cent. Ont. (s to Moose Factory, sw James Bay, and the Pic R., N shore of L. Superior, near Marathon, whence reported by John Macoun 1888, and rediscovered by Soper and Fraser in 1964; MTMG; TRT). MAPS: Hultén 1968*b*:317; Marie-Victorin, Contrib. Inst. Bot. Univ. Montréal 12: fig. 1., p. 170. 1928; Raup 1947: pl. 20.

The plant of the Mingan Is., E Que., has been separated as var. *minganense* Vict. (type from Grande-Île-à-la-Vache-Marine; upper petal essentially nerveless rather than strongly nerved).

C. reginae Walt.   Showy Lady's-slipper

/T/EE/ (Grh)   Chiefly calcareous swamps, bogs, moist woods, and shores from Man. (N to Dawson Bay, L. Winnipegosis; the report from Sask. by Correll 1950, requires confirmation) to Ont. (N to the N shore of L. Superior near Thunder Bay), Que. (N to the N Gaspé Pen.; *C. hirsutum* reported from Anticosti Is. by Schmitt 1904), Nfld., N.B., P.E.I., and N.S., s to N.Dak., Mo., and Ga. [*C. canadense* Michx.; *C. spectabile* Salisb.; *C. hirsutum* of Canadian reports, not Mill.]. MAP: Fernald 1933: map 2, p. 8.

Forma *albolabium* Fern. & Schub. (corolla-lip wholly white rather than strongly suffused with broad pink to roseate bands on the front) is known from s Ont. (Inverhuron Beach, Bruce Pen.; TRT).

EBUROPHYTON Heller   [1481]

E. austiniae (Gray) Heller   Phantom Orchid

/t/W/ (Grh)   Moist coniferous forests (lowland to subalpine) from sw B.C. (near Agassiz and Chilliwack; *see* B.C. map by Szczawinski 1959:81) to s Calif. [*Chloraea* Gray; *Cephalanthera* Heller; *Ceph. oregana* Rchb.].

EPIPACTIS Sw.   [1482]   Helleborine

1   Bracts small, only the lowest 1 or 2 about equalling or slightly surpassing their subtended flower; flowers less than 20; sepals dark brownish-purple outside; terminal lobe of corolla-lip deep reddish-violet; ovary pubescent; leaves distichous, commonly less than 8 cm long and 5 cm broad; (introd. at Montreal, Que.; extinct) . . . .[*E. atrorubens*]

1   At least several of the lowest bracts usually leaf-like and often much surpassing their subtended flower; sepals greenish (rarely purplish) outside; terminal lobe of lip

greenish-white to purplish or red; ovary essentially glabrous; leaves spirally arranged at least below, to over 1.5 dm long and about 1 dm broad.

2 Raceme with at most 15 flowers, its bracts mostly leaf-like and only gradually reduced upwardly; lip to 9 mm long, deeply 3-lobed (mid-lobe much smaller than the triangular-ovate to suborbicular lateral lobes); (s B.C.) . . . .*E. gigantea*

2 Raceme with up to 50 or more flowers, usually only its lower bracts much surpassing the flowers, the upper bracts greatly reduced; lip to 11 mm long, only slightly constricted at about the middle; (introd. in B.C., Ont., and Que.) . . . .*E. helleborine*

[E. atrorubens (Hoffm.) Schultes]
[Eurasian; apparently known in N. America only from limestone cliffs of Mt. Royal, Montreal, Que., where taken by Mousley in 1925 and now extinct. (*E. atropurpurea* Raf.; *E. (Amesia) rubiginosa* (Cr.) Koch; *see* Henry Mousley, Can. Field-Nat. 41(1): 1–6. 1927; description: Clapham, Tutin, and Warburg 1962:1021).]

E. gigantea Dougl.   Giant Helleborine, Stream Orchid
/T/WW/   (Grh)   Wet prairies, meadows, marshy places, and streambanks from s B.C. (N to Enderby, s of Shuswap L., and Fairmont Hot Springs, Columbia Valley, both ca. 51°15′N; *see* B.C. map by Szczawinski 1959:76) to Baja Calif., Mexico, and Tex. [*Serapias* Eat.].

E. helleborine (L.) Crantz   Helleborine
Eurasian; introd. in N. America and spreading rapidly in woods, ravines, and alluvia near settled areas, as in sw B.C. (s Vancouver Is.), Ont. (N to Chalk River and Ottawa; *see* s Ont. maps by J. H. Soper, Can. J. Bot. 42(8): fig. 7, p. 1096. 1964, J. H. Soper and L. A. Garay, Bull. Fed. Ont. Nat. 65: 6. 1954, and F. H. Montgomery 1957: fig. 5, p. 101, and Rhodora 50(597): 237. 1948), and Que. (N to Grosse-Ile, near Quebec City, and l'Islet, l'Islet Co.; *see* Que. maps by Dominique Doyon and Richard Cayouette, Nat. can. (Que.) 93: 172. 1966, and C. Rousseau 1968: map 13, p. 65). [*Serapias* L.; *E. (Amesia) latifolia* All.].

Forma *alba* (Webster) Boivin (*E. latifolia* f. *alba* Webster; flowers white rather than green suffused with madder-purple) is known from sw Que. (Montreal; Henry Mousley, Can. Field-Nat. 41(1): 30. 1927). Forma *monotropoides* (Mousley) Scoggan (*Amesia lat.* f. *mon.* Mousley; plant white, devoid of chlorophyll) and forma *variegata* (Webster) Boivin (*E. lat.* f. *var.* Webster; leaves with long streaks or patches of yellow along the veins) are also reported from Montreal by Mousley.

GOODYERA R. Br.   [1504]   Rattlesnake-plantain

1 Raceme densely flowered, cylindric, to about 1 dm long; beak of lip less than half as long as the body; anther blunt; leaf-blades whitish-reticulate, to about 6 cm long; (Ont. and Que.) . . . .*G. pubescens*

1 Raceme loosely flowered, spiral or 1-sided; beak of lip usually at least about 2/3 as long as the body.

2 Perianth to 1 cm long; beak of lip about 2/3 as long as the body, slightly deflexed and with upturned sides; anther long-acuminate; raceme to 12 cm long; leaf-blades green above, often with a broad white midrib or sometimes mottled but scarcely reticulate, to 1 dm long; (transcontinental) . . . .*G. oblongifolia*

2 Perianth 4 or 5 mm long; beak of lip about as long as or longer than the body.

3 Lip rather shallowly saccate, the pouch about half as deep as long, the beak with upturned sides, straight or slightly deflexed; anther abruptly accuminate; raceme to 1 dm long; leaf-blades whitish-reticulate, to about 5 cm long; (Ont. to Nfld. and N.S.) . . . .*G. tesselata*

3 Lip deeply saccate, the pouch about as deep as long, the beak recurved; anther blunt; raceme rarely over 6 cm long; leaf-blades to about 3 cm long; (transcontinental) . . . .*G. repens*

G. oblongifolia Raf.   Giant Rattlesnake-plantain
/sT/X/   (Hrr)   Dry coniferous or mixed woods from the Alaska Panhandle (*see* Hultén 1943: map 393, p. 563; *G. decip.*) and B.C. to sw Alta. (Crowsnest Forest Reserve; Waterton Lakes; Castle

Mt., near Banff), sw Sask. (Cypress Hills; Breitung 1957a; not known from Man.), Ont. (N to the N shore of L. Huron; CAN), Que. (N to L. St. John and the Gaspé Pen.), N N.B. (Madawaska and Restigouche counties; not known from P.E.I.), and N.S. (Inverness and Victoria counties, Cape Breton Is.), s to Calif., Mexico, N.Mex., Minn., Wisc., and N Maine. [*Spiranthes* (*Goodyera; Epipactis; Peramium*) *decipiens* Hook.; *G.* (*Peramium*) *menziesii* Lindl.]. MAPS: Hultén 1968b:328; Fernald 1925: map 20 (*E. decip.*), p. 255.

G. pubescens (Willd.) R. Br.   Downy Rattlesnake-plantain
/T/EE/   (Hrr)   Dry or moist coniferous or mixed woods from Ont. (N to the E end of L. Superior at Batchawana Bay; CAN) to Que. (N to Temiscouata Co.; MT) and Maine, s to Tenn., Ala., and Fla. [*Neottia* Willd.; *Epipactis* Eat.; *Peramium* MacM.].

G. repens (L.) R. Br.   Dwarf Rattlesnake-plantain
/ST/X/EA/   (Hrr)   Damp mossy coniferous or mixed woods, the aggregate species from N-cent. Alaska and s-cent. Yukon (*see* Hultén 1943: map 394, p. 563) to Great Bear L., L. Athabasca (Alta. and Sask.), Man. (N to Reindeer L. at 57°14′N), Ont. (N to Sachigo L. at ca. 54°N, 92°W), Que. (N to the Kaniapiscau R. at 56°44′N, L. Mistassini, and the Côte-Nord), Labrador (N to the Hamilton R. basin), Nfld., N.B., P.E.I., and N.S., s to s B.C.–Alta.–Sask. (isolated in the mts. of Ariz., Colo., and N.Mex.), S.Dak., Tenn., and N.C.; Eurasia. MAPS and synonymy: *see* below.
1   Leaves green, with dark veins; [*Satyrium* L.; *Epipactis* Crantz; *Neottia* Sw.; *Peramium* Salisb.; transcontinental but somewhat more southern than var. *ophioides,* ranging northwards to N B.C., L. Superior, Ont., and Anticosti Is., E Que.; MAP (aggregate species): Hultén 1962: map 154, p. 163] . . . . .var. *repens*
1   Leaves prominently whitish-reticulate; [*Epipactis rep.* var. *oph.* (Fern.) Eat.; *Peramium oph.* (Fern.) Rydb.; transcontinental; MAPS: Hultén 1968b:328; also the Alaska–Canada area in the above-noted 1962 map by Hultén] . . . . .var. *ophioides* Fern.

G. tesselata Lodd.   Checkered Rattlesnake-plantain
/T/EE/   (Hrr)   Dry to moist coniferous or mixed forests from Ont. (N to Longlac, N of L. Superior at ca. 49°45′N; reported from Renison, s of James Bay at ca. 51°N, by Hustich 1955) to Que. (N to the Côte-Nord, Anticosti Is., and Gaspé Pen.), Nfld. (GH), N.B., P.E.I., and N.S., s to Minn., Mich., Ohio, N.Y., and Conn. [*Epipactis* Eat.; *Peramium* Heller].
    The inclusion of E Man. in the range by Fernald *in* Gray (1950) requires confirmation. Early reports of *G. pubescens* from Nfld. (taken up by Gleason 1958) are probably largely based upon *G. tesselata*, as are its reports from Petitcodiac, N.B., by Fowler (1885; relevant collection in NBM) and from Truro, N.S., by Lindsay (1878; relevant collection in NSPM).

HABENARIA Willd.   [1422]   Rein-Orchis, Fringed Orchis

1   Corolla-lip fringed, simple or deeply 3-parted; stem leafy, the upper leaves reduced.
    2   Lip simple, entire or only slightly toothed at apex; spur to 2.5 cm long; raceme to 1.5 dm long.
        3   Flowers bright white or creamy; lip at most 11 mm long, its fringe-segments to 1.5 mm long; lower leaves linear to lance-linear, to 2 dm long and 2 cm broad; (Ont. to Nfld. and N.S.) . . . .*H. blephariglottis*
        3   Flowers orange-yellow; lip to about 1.5 cm long, its fringe-segments 3 mm long or more; lower leaves lanceolate, to 1.5 dm long and 4 cm broad; (s Ont.) . . . .*H. ciliaris*
    2   Lip deeply 3-parted.
        4   Flowers pale lilac to deep rose-purple; middle lobe of the narrowly fan-shaped lip as broad as or broader than long; (Ont. to Nfld. and N.S.) . . . .*H. psycodes*
        4   Flowers greenish white, yellowish green, or creamy; middle lobe of lip cuneate to a slender claw; lower leaves to 3.5 cm broad.
            5   Perianth to 1 cm long; lip to 2 cm long; spur at least 2 cm long; raceme to 2 dm long; plant to over 1 m tall; (s Ont. to N.B.) . . . .*H. leucophaea*
            5   Perianth about 5 mm long; lip not over 1.5 cm long; spur less than 2 cm long; (Ont. to Nfld. and N.S.) . . . .*H. lacera*
1   Corolla-lip not fringed, entire or merely toothed, not 3-parted.

6 Leaves all basal or essentially so; flowers mostly greenish or greenish white (rarely yellow), short-pedicelled.
  7 Leaf solitary, oblanceolate or spatulate to obovate; scape to 4 dm tall; (trans-continental) . . . .*H. obtusata*
  7 Leaves normally 2 or 3.
    8 Flowers sessile in spikes; spur directed downward; scape naked or with a few small bracts.
      9 Spikes at least 2 cm thick; spur to over 2 cm long; sepals with 3 or 5 nerves; leaves oblong-obovate to broadly elliptic or rotund, thick, flat on the ground, to about 1.5 dm long, less than twice as long as broad; scapes to 4 dm tall, bractless; (SE Man. to Nfld. and N.S.) . . . .*H. hookeri*
      9 Spikes at most about 1 cm thick; spur to 5 mm long.
        10 Scapes to about 9 dm tall, from fleshy-fibrous roots (including a pair of ovoid fleshy tuberoids up to 4 cm long and 1 cm thick), usually bearing a few small bracts; basal leaves usually 2 or 3, linear-lanceolate to oblanceolate (rarely obovate), obtuse to subacuminate, to 1.5 dm long and 3 cm broad, usually withering by flowering-time; sepals 1-nerved; (Alaska–B.C.–sw Alta.; cent. Ont. and E Que.) . . . .*H. unalascensis*
        10 Scapes rarely as much as 1 dm tall, from several fusiform root-like tubers, bractless or with a small bract near the middle; basal leaves usually 2 (sometimes 1), lanceolate to subrotund, obtuse to acute, to about 4 cm long and 2 cm broad, more persistent; sepals 3-nerved; (Alaska–B.C.) . . . .*H. chorisiana*
    8 Flowers pedicelled in loose racemes; spur directed more or less transversely; scape bracted nearly to summit; leaves broadly elliptic to rotund, to 2.5 dm long and broad.
      11 Perianth to 2 cm long, whitish; raceme to 3.5 dm long; spur to over 4 cm long; lip to over 2 cm long; upper sepal broadly ovate, the two lateral ones ascending; (Ont. to Nfld. and N.S.) . . . .*H. macrophylla*
      11 Perianth to 12 mm long, greenish white; lip to 1.5 cm long; upper sepal subrotund, the two lateral ones drooping or reflexed; (transcontinental) . . . .
. . . .*H. orbiculata*
6 Leaves cauline.
  12 Foliage-leaf solitary (rarely 2), the upper leaves reduced to bracts; flowers green, greenish white, or greenish yellow, in an open few-flowered spike to about 6 cm long; (Ont. to Nfld. and N.S.) . . . .*H. clavellata*
  12 Foliage-leaves 3 or more.
    13 Flowers milk-white to creamy-white or stramineous.
      14 Spur 2 or 3 mm long; lip to 5 mm long, broadly cuneate and 3-lobed at apex; flowers creamy-white or straw-colour; leaves oblong-obovate, to about 7 cm long and 3.5 cm broad; stem less than 4 dm tall; (SE Hudson Bay and N Nfld.) . . . .*H. albida*
      14 Spur to 8 mm long; lip to 8 mm long, dilated at base, its apex entire; flowers milk-white; leaves linear to lanceolate, to about 2 dm long and 4 cm broad; stem to about 1 m tall; (transcontinental) . . . .*H. dilatata*
    13 Flowers green to greenish-yellow.
      15 Lip entire, gradually narrowed to the blunt apex; only the lowest bracts equalling or surpassing the flowers.
        16 Spur saccate; spike relatively loose and few-flowered, the bracts erect; (Alaska–B.C.–Alta.) . . . .*H. saccata*
        16 Spur cylindric to clavate; spike densely flowered, the bracts spreading; (transcontinental) . . . .*H. hyperborea*
      15 Lip toothed near the broadened apex.
        17 Lip to 1 cm long, with 2 lateral teeth at apex and sometimes a smaller middle one; larger leaves usually about 5 in number, oblong to oblong-obovate, blunt; (transcontinental) . . . .*H. viridis*
        17 Lip about 5 mm long, with a prominent elongate median tubercle at the

hastate base, the margin near apex with a few small teeth; larger leaves lanceolate to lance-oblong; only the lower floral bracts equalling or surpassing the flowers; (Ont. to N.S.) . . . .*H. flava*

H. albida (L.) R. Br.
/aST/E/GEwA/ (Grt) Hilly meadows and turfy or peaty places of w-cent. Que. (Richmond Gulf region of sᴇ Hudson Bay between 55°30′ and 56°15′N; CAN; MT) and ɴw Nfld. (CAN; GH); w Greenland ɴ to 69°32′N, ᴇ Greenland ɴ to 66°08′N; Iceland; Europe; w Asia. [*Satyrium* L.; *Leucorchis* Mey.]. MAPS (the one by Fernald as *H. straminea*, the others as *Leuc. alb.*): Hultén 1958: map 98, p. 117; Meusel, Jaeger, and Weinert 1965:107; Böcher 1954: fig. 21 (top), p. 79; A. Löve 1950: fig. 7, p. 36; Fernald 1929: map 33, p. 1502.

The plant of ᴇ N. America, Greenland, Iceland, and the Faeroes may be rather arbitrarily separated as var. *straminea* (Fern.) Morris (*H. straminea* Fern., the type from ɴw Nfld.), differing from the typical European form in its somewhat thicker spikes, slightly longer bracts, longer and more evidently nerved stramineous (rather than whitish) sepals, and longer and more evidently nerved petals. However, Hultén (1958) points out that very similar plants occur in Scandinavia, these treated by him as *Leuc. alb.* var. *subalpina* (Neum.) Hyl.

H. blephariglottis (Willd.) Hook. White Fringed Orchis
/T/EE/ (Grt) Wet boggy or peaty soils from Ont. (ɴ to the Mer Bleue, Ottawa; John Macoun 1888) to Que. (ɴ to Nominingue, Labelle Co., and the Quebec City dist.; MT), Nfld. (CAN), N.B., P.E.I., and N.S., s to Miss. and Fla. [*Orchis* Willd.; *Blephariglottis* Rydb.; *Platanthera* Lindl.; *P. holopetala* Lindl.].

H. chorisiana Cham.
/sT/W/eA/ (Grt) Coastal swamps and sphagnous bogs in the Aleutian Is. (*see* Hultén 1943: map 379 (*Plat. chor.*), p. 561; type from Unalaska), s Alaska (Juneau; Szczawinski 1959), B.C. Queen Charlotte Is. and Vancouver Is.; *see* B.C. map by Szczawinski 1959:38), and Wash.; Kamchatka, Sakhalin, and ɴ Japan. [*Limnorchis* And.; *Platanthera* Rchb.; *Peristylus* Lindl.]. MAP (*Plat. ch.*): Hultén 1968b:324.

H. ciliaris (L.) R. Br. Yellow Fringed orchis, Orange-plume
/t/EE/ (Grt) Acid swamps and bogs and sandy woodlands and thickets from Mo. to s Wisc., s Mich., s Ont. (Leamington and Windsor, Essex Co.; CAN; John Macoun 1888), and N.Y., s to ᴇ Tex. and Fla. [*Orchis* L.; *Blephariglottis* Rydb.; *Platanthera* Lindl.].

H. clavellata (Michx.) Spreng. Green Woodland Orchis
/T/EE/ (Grt) Acid bogs and wet soil from Ont. (ɴ to the ᴇ shore of L. Superior at Batchawana Bay; CAN) to Que. (ɴ to the Côte-Nord, Anticosti Is., and Gaspé Pen.), Nfld., N.B., P.E.I., and N.S., s to Tex. and Fla. [*Orchis* Michx.; *Denslovia* and *Gymnadeniopsis* Rydb.; *O. (H.; Gymnadenia) tridentata* Muhl.].

The typical southern phase is reported "very locally to N.B." by Fernald *in* Gray (1950). The plant of the remaining part of our area may be distinguished by its relatively broader and shorter leaves as var. *ophioglossoides* Fern. (type from Yarmouth Co., N.S.).

H. behringiana (Rydb.) Ames is known from the w Aleutian Is. It will key out with *H. clavellata* in the above key to *Habenaria* (corolla-lip not fringed; stem bearing a solitary (rarely 2) foliage-leaf) and may be distinguished from it as follows:

1 Lip narrowly cuneate-oblong, truncate and sinuately 3-toothed at summit, to 7 mm long and 4 mm broad; flowers yellowish white or greenish . . . .*H. clavellata*
1 Lip elliptic-lanceolate to elliptic (broadest near or below the middle), obtuse at the entire summit, to 8 mm long and 3 mm broad; flowers reputedly purplish; [*Limnorchis* Rydb.; *Platanthera (Orchis) tipuloides* (L. f.) Lindl. var. *behr.* (Rydb.) Hult.; w Aleutian Is. (Attu Is.) and ᴇ Asia; MAP: Hultén 1968b:323 (*P. tip.* var. *behr.*)] . . . .*H. behringiana* (Rydb.) Ames.

H. dilatata (Pursh) Hook. Bog-candle, Scent-bottle
/sT/X/eA/ (Grt) Meadows, bogs, and wet woods, the aggregate species from the Aleutian Is. and s Alaska–Yukon (*see* Hultén 1943: map 381, p. 562; *Plat. dil.*) to sw Dist. Mackenzie, sw Alta.,

Sask. (N to Prince Albert; a collection in CAN from Windrum L. at ca. 56°N may also belong here), Man. (N to Porcupine Mt.; John Macoun 1888; a collection in DAO from Churchill has been placed here but may prove to be *H. hyperborea*), Ont. (N to w James Bay at ca. 53°N), Que. (N to the Korok R., Ungava Bay, at 58°35'N), Labrador (N to the Hamilton R. basin; reported from Nain, 56°33'N), Nfld., N.B., P.E.I., and N.S., s to Calif., N.Mex., S.Dak., and N.J.; (reports from Disko, w Greenland, may refer to *H. hyperborea*); E Asia (Commander Is., Bering Sea; Hultén 1943). MAPS and synonymy: *see* below.

1 Spur to about twice as long as the lip; [*H. leuc.* (Lindl.) Wats.; *Limnorchis leuc.* (Lindl.)
   Rydb. and its var. *robusta* Rydb.; s ?Alaska–B.C. (*see* B.C. map by Szczawinski 1959:44)] . . . .
   . . . . .var. *leucostachys* (Lindl.) Ames
1 Spur at most only slightly longer than the lip.
   2 Spur to 2/3 the length of the lip, often distinctly clavate; [*H. borealis* var. *albiflora*
      Cham., the type from Unalaska, Alaska; s Alaska–B.C. (*see* B.C. map by Szczawinski
      1959:42)] . . . . .var. *albiflora* (Cham.) Correll
   2 Spur equalling or slightly longer than the lip; [*Orchis dilatata* Pursh, the type from
      Labrador; *Limnorchis* Rydb.; *Platanthera dil.* Lindl. and its vars. *angustifolia* Hook.
      and *chlorantha* Hult.; *H. borealis* Cham.; *H. graminifolia* and *H. leptoceratitis* Rydb.;
      transcontinental; MAPS (aggregate species): Porsild 1966: map 44, p. 72; Hultén
      1968b:322]. A purported hybrid with *H. hyperborea* var. *huronensis* (× *H. media*
      (Rydb.) Niles; *P. dil.* var. *media* (Rydb.) Ames) is reported by Hultén 1960, from the
      Aleutian Is. and has been taken in s Ont. (Bruce Co.), Que. (Rupert House, s James
      Bay; Anticosti Is.; Gaspé Pen.; Saguenay and Megantic counties), Nfld., and N.S.
      (Cape Breton Is.) . . . . .var. *dilatata*

H. flava (L.) R. Br.   Pale Green Orchis
/T/EE/ (Grt) Swampy woods and wet ground from Ont. (N to the Ottawa dist.) to sw Que. (N to L. St. Peter in St-Maurice Co.), N.B. (Oromocto R., near Fredericton; GH), and N.S. (not known from P.E.I.), s to E Tex. and Fla. [*Orchis* L.; *Perularia* Farw.].
The typical southern phase occurs in N.S. according to Fernald *in* Gray (1950; Yarmouth and Queens counties). The commoner phase of our area may be distinguished as var. *herbiola* (R. Br.) Ames & Correll (*H. herbiola* R. Br.; *Orchis fuscescens* L., not Pursh; var. *virescens* of early Canadian reports, not *O. virescens* Muhl.; stem bearing up to 5 leaves rather than 2(3), the spike relatively compact).

H. hookeri Torr.   Hooker's Orchis
/T/EE/ (Grt) Rich moist or dryish woods from s Man. (N to L. Winnipegosis; Jackson et al. 1922) to Ont. (N to Oba L., about 100 mi NE of L. Superior; John Macoun 1888), Que. (N to Tadoussac, Saguenay Co., and the Gaspé Pen.), Nfld., N.B., P.E.I., and N.S., s to Iowa, W.Va., and New Eng. [*Lysias* Rydb.; *Platanthera* Lindl.].
The Nfld. plant may be distinguished as var. *abbreviata* Fern. (type from limestone-barrens of NW Nfld.; the small-dimensioned extreme with the upper sepal ovate and obtuse rather than lance-acuminate, the spur less than 1.5 cm long rather than to 2.5 cm).

H. hyperborea (L.) R. Br.   Northern Green Orchis
/aST/X/GEA/ (Grt) Bogs and wet woods from the Aleutian Is., N Alaska, and cent. Yukon (*see* Hultén 1943: map 382, p. 562; *Plat. hyp.*) to Great Bear L., Great Slave L., L. Athabasca (Alta. and Sask.), Man. (N to Churchill), Ont., Que. (N to s Ungava Bay and the Côte-Nord), Labrador (N to the Hamilton R. basin), Nfld., N.B., P.E.I., and N.S., s to Calif., N.Mex., Nebr., Ohio, and New Eng.; Greenland; Iceland; NE Asia. [*Orchis* L.; *Limnorchis* Rydb.; *Platanthera* Lindl.; *L. brachypetala* Rydb.; *H. viridiflora* Cham.; incl. the high-grown coarse extreme, var. *huronensis* (Nutt.) Farw. (*Platanthera* (*H.; Limnorchis*) *hur.* Lindl.; *P. convallariaefolia* (Fisch.) Lindl., at least in part; *P. koenigii sensu* Hooker 1839, perhaps not Lindl.)]. MAPS: Hultén 1968b:321; Raup 1947: pl. 20.

H. lacera (Michx.) Lodd.   Ragged Orchis
/T/EE/ (Grt) Dry to wet meadows, thickets, and woods from Ont. (N to the SE end of L. Superior at Batchawana Bay) to Que. (N to Magdalen Is.), St-Pierre and Miquelon, Nfld., N.B., P.E.I., and N.S., s to Tex. and Fla. [*Orchis* Michx.; *Blephariglottis* Farw.; *H. leucophaea sensu* John Macoun

1888, as to the Baddeck and North Sydney, N.S., citations, not (Nutt.) Gray, the relevant collections in CAN; incl. the reduced northern extreme, var. *terrae-novae* Fern. (× *H. ?andrewsii* White; *see* M. L. Fernald, Rhodora 28(326): 21. 1926)].

A hybrid with *H. psycodes* (× *H. andrewsii* White) is known from ?Nfld., Que., N.B., P.E.I., and N.S.

H. leucophaea (Nutt.) Gray Prairie White Fringed Orchis
/T/EE/ (Grt) Bogs and marshes from N.Dak. to Minn., Mich., s Ont. (Kent, Lambton, Middlesex, Wentworth, and Bruce counties), ?Que. (Boivin 1967a), and cent. N.Y., s to Kans., La., Ill., Ohio, and Maine. [*Orchis* Nutt.; *Blephariglottis* Farw.].

Concerning reports from N.S., see *H. lacera*. The report from N.B. by Boivin (1967a) may also be based upon that species.

H. macrophylla Goldie
/T/EE/ (Grt) Rich woods from Ont. (N to Cochrane, ca. 49°N; TRT) to Que. (N to L. Nominingue, Labelle Co., and Montmagny, Montmagny Co.; type from Montreal), Nfld. (Whitbourne and Salmonier; GH), N.B., and N.S. (not known from P.E.I.), s to Wisc., N.Y., and Conn. [*Lysias* House; *H. orbiculata* var. *mac.* (Goldie) Boivin; *see* M. L. Fernald, Rhodora 52(616): 61–65. 1950].

H. obtusata (Pursh) Richards. Blunt-leaf Orchis, One-leaf Rein-Orchis
/aST/X/EA/ (Grt) Mossy woods and tundra from N Alaska–Yukon (*see* Hultén 1943: map 383, p. 562; *Plat. ob.*) and the coast of Dist. Mackenzie (Bathurst Inlet, Coronation Gulf) to cent. Dist. Keewatin (s of Chesterfield Inlet), Que. (N to Cape Smith, NE Hudson Bay at ca. 60°45′N, and the Côte-Nord), Labrador (N to ca. 56°N), Nfld., N.B., P.E.I., and N.S., s through B.C.–Alta.–Sask. to Oreg., Utah, Colo., Minn., Mich., N.Y., and New Eng.; N Norway and Asia (ssp. *oligantha* (Turcz.) Hult.). [*Orchis obtusata* Pursh, the type from Fort Albany, sw James Bay, Ont.; *Lysiella* Rydb.; *Platanthera* Lindl.]. MAPS: Hultén 1968b:325, 1962: map 150, p. 159, and Fauna Flora (1943): fig. 3, p. 172. 1943 (all as *Plat. ob.*); Raup 1947: pl. 20.

The reduced extreme may be distinguished as var. *collectanea* Fern. (type from w Nfld.; scape less than 1.5 dm tall, nearly equalled by the leaf, the densely flowered raceme less than 5 cm long; largely replacing the typical form northwards).

H. orbiculata (Pursh) Torr. Round-leaved Orchis
/T/X/ (Grt) Dry to moist woods, the aggregate species from the southernmost Alaska Panhandle (*see* Hultén 1943: map 384, p. 562), sw Dist. Mackenzie (CAN; not known from the Yukon), and B.C.–Alta. to Sask. (N to Amisk L., near Flin Flon at ca. 55°N; Breitung 1957a), Man. (N to The Pas), Ont. (N to Sandy L. at ca. 53°N; 93°W), Que. (N to the Gaspé Pen.), Nfld., N.B., P.E.I., and N.S., s to Oreg., Mont., Minn., Ill., Tenn., and S.C. MAPS and synonymy: *see* below.

1 Leaves to 2 dm long and broad; scape to about 5 dm tall, with up to 5 bracts; raceme lax, to 2.5 dm long; pedicels filiform, usually at least 6 mm long; spur to about 1.5 cm long; [*Orchis* Pursh; *Lysias* Rydb.; *Platanthera* Lindl.; transcontinental; MAPS (aggregate species): Hultén 1968b:320 (*Plat. orb.*); Fernald 1933: map 9 (incomplete northwards), p. 87] . . . . .var. *orbiculata*
1 Leaves less than 1 dm long and 6 cm broad; scape commonly less than 2 dm tall, with usually 2 bracts; raceme dense, less than 6 cm long; pedicels stouter, less than 5 mm long; spur to about 1.5 cm long; [open heathy barrens of St-Pierre and Miquelon (type locality) and w Nfld.]. M. L. Fernald, Rhodora 52(616): 61–65. 1950, points out that both var. *lehorsii* and the western var. *menziesii* (Lindl.) Fern. (*Platanthera* (*H.; Lysias*) *men.* Lindl.; s ?Alaska–B.C.) differ from the typical form in their smaller stature, smaller leaves, denser racemes, and smaller floral dimensions. Var. *lehorsii* is the small-dimensioned densely flowered extreme, var. *menziesii* being transitional between it and the typical form. Taking into account the unsheltered habitat, var. *lehorsii* seems to be merely an ecological form of little taxonomic significance . . . . .var. *lehorsii* Fern.

H. psycodes (L.) Spreng. Small Purple Fringed Orchis, Soldier's-plume
/T/EE/ (Grt) Meadows, thickets, and moist or wet woods, the aggregate species from SE ?Man. (*see* Scoggan 1957) to Ont. (N to Renison, s of James Bay at ca. 51°N; Hustich 1955), Que. (N to L.

Albanel, near L. Mistassini, Anticosti Is., and the Gaspé Pen.; reported from the Mingan Is. of the Côte-Nord by Saint-Cyr 1887), Nfld., N.B., P.E.I., and N.S., s to Iowa, Tenn., and Ga.

1   Stem to about 9 dm tall; leaves to about 5 cm broad; raceme to about 1.5 dm long and 5 cm thick; sepals at most 8 mm long; lip less than 1.5 cm long; spur at most 2.5 cm long . . . . .var. *psycodes*

    2   Flowers lilac-pink to deep rose-purple; [*Orchis* L., *Platanthera* Lindl.; *Blephariglottis* Rydb.; incl. var. *fernaldii* Rousseau & Rouleau; Ont. to Nfld. and N.S.; concerning reports from Man., *see* Scoggan 1957:221] . . . . .f. *psycodes*

    2   Flowers white; (Que., P.E.I., and N.S.) . . . . .f. *albiflora* (Bigel.) Hoffm.

1   Stem to 1.5 m tall; leaves to 1 dm broad; raceme to 3 dm long and 6 cm thick; sepals to 11 mm long; lip to about 2 cm long; spur to 3.5 cm long . . . . .var. *grandiflora* (Bigel.) Gray

    3   Flowers lilac-pink to deep rose-purple; [*Orchis* (*Blephariglottis*) gr. Bigel.; *Orchis* (*H.; Platanthera*) *fimbriata* Ait.; Ont. to Nfld. and N.S.] . . . . .f. *grandiflora*

    3   Flowers white; [*H. fimbriata* f. *albiflora* Rand. & Redf.; s Ont. (Welland Co.), sw Que. (Montreal dist.), and N.B. (Victoria Co. and Grand Manan Is.)] . . . . .f. *leucophaeopsis* Boivin

H. saccata Greene   Slender Bog-orchis

/sT/WW/   (Grt)   Peaty places at low to rather high elevations from the E Aleutian Is. and s Alaska (N to near or on the Yukon boundary; *see* Hultén 1943: map 385, p. 562; *Plat. stricta*) through B.C. (*see* B.C. map by Szczawinski 1959:54) and sw Alta. (Waterton Lakes; Breitung 1957b) to Calif. and N.Mex. [*Platanthera* Hult.; *P.* (*H.; Limnorchis*) *gracilis* Lindl.; *P.* (*H.; L.*) *stricta* Lindl.]. MAP: Hultén 1968b:322 (*Plat. sac.*).

H. unalascensis (Spreng.) Wats.

/sT/D/   (Grt)   Dry or moist soil on grassy open slopes, along streambanks, and in open woodlands, at low to high elevations (ranges of Canadian taxa outlined below), s to Baja Calif., Mexico, Colo., and S.Dak.; isolated in the Great Lakes region of Ont. and on Anticosti Is., E Que. MAPS and synonymy: *see* below.

1   Spur at most only slightly longer than the lip; raceme rather loosely flowered, usually not over 1 cm thick; [*Spiranthes* Spreng.; *Piperia* and *Montolivaea* Rydb.; *Platanthera* Kurtz; Aleutian Is. (type locality), Alaska, B.C., and sw Alta.; L. Huron, Ont. (Bruce Pen. and Manitoulin Is.; also reported from L. Superior, Ont., by Fernald *in* Gray 1950); E Que. (Anticosti Is.); MAPS: Hultén 1968b:324 (*Plat. unal.*); Fernald 1925: map 19, p. 255] . . . . .var. *unalascensis*

1   Spur at least twice as long as the lip; raceme usually at least 1.5 cm thick.

    2   Raceme laxly or densely flowered, usually less than 2 cm thick; flowers greenish; plant typically slender and tall; [*H. elegans* var. *elata* Jeps.; *H. michaelii* Greene; s B.C.: N to ca. 51°N; *see* B.C. map by Szczawinski 1959:58] . . . . . .var. *elata* (Jeps.) Correll

    2   Raceme densely flowered, at least 2 cm thick; flowers white; plant typically short and stout; [*H. maritima* Greene, not Raf.; *H. greenei* Jeps.; Queen Charlotte Is., B.C.] . . . . .var. *maritima* (Greene) Correll

H. viridis (L.) R. Br.   Frog-Orchis

/aST/X/EA/   (Grt)   Rich (often calcareous) meadows, thickets, and woodlands, the aggregate species from the Aleutian Is. and Alaska (N to the Seward Pen.; not known from the Yukon) to w Dist. Mackenzie (N to Norman Wells, 65°17′N), northernmost Alta., N-cent. Sask., Man. (N to York Factory, Hudson Bay, ca. 57°N), northernmost Ont., Que. (N to L. Nominingue, Labelle Co., and the Gaspé Pen.; reported from Anticosti Is. by Verrill 1865), Nfld., N.B., P.E.I., and N.S., s to s B.C., ?Wash., Mont., N.Mex., Nebr., Iowa, Ohio, Tenn., and N.C.; Iceland; Eurasia. MAPS and synonymy: *see* below.

1   Lower and median bracts up to 6 times as long as the subtended flowers; upper bracts up to twice as long as the subtended flowers; [*Orchis viridis sensu* Pursh 1814, not *Satyrium viride* L., basionym; *Orchis* (*Coeloglossum; Habenaria; Peristylus*) *bracteata* Muhl.; *Orchis* (*H.*) *virescens* Muhl.; *O. fuscescens* Pursh, not L.; *Platanthera ?herbiola* Lindl.; transcontinental; MAPS: Hultén 1968b:319, and 1962: map 68, p. 77]. Fernald *in* Gray (1950) points out that this taxon is almost exactly intermediate between the following one and typical *H. viridis* of Eurasia . . . . .var. *bracteata* (Muhl.) Gray

1   Lower bracts at most about 3 times as long as the subtended flowers; middle and upper
     bracts shorter than to barely surpassing the subtended flowers; [essentially the range of
     var. *bracteata*] . . . . .var. *interjecta* Fern.

[ISOTRIA Raf.] [1467]

[I. verticillata (Willd.) Raf.]   Whorled Pogonia
[This species of moist woodlands of the E U.S.A. (Mo. to Mich., Ohio, N.Y., and Maine, s to E Tex.
and Fla.) is known in Canada only through collections from s Ont. (Komoka, near London,
Middlesex Co.), where first taken by Burgess in 1879 and now apparently extinct; CAN; TRT).
(*Arethusa* Willd.; *Pogonia* Nutt.).]

LIPARIS Richard   [1556]   Twayblade

1   Pedicels divergent, filiform, to 1 cm long; perianth-segments about 1 cm long; sepals
     greenish white, lance-oblong but usually tightly inrolled and appearing thread-like; lateral
     petals drooping and often crossed, greenish to pale purple, the flattish lip usually
     brown-purple; (s Ont.) . . . .*L. lilifolia*
1   Pedicels ascending, at most 5 mm long; perianth-segments all greenish yellow and at
     most 6 mm long; sepals narrowly oblanceolate to oblong; lateral petals thread-like, the
     broadly ovate lip with upturned margins; (Sask. to N.S.) . . . .*L. loeselii*

L. lilifolia (L.) Richard   Lilia-leaved Twayblade
/t/EE/   (Hr)   Rich woods, ravines, and clearings from s Minn. to s Ont. (an 1864 collection by
Day from Fort Erie, Welland Co., reported by Zenkert 1934; a count of 83 plants in 1960 at
Komoka, near London, Middlesex Co., reported by H. Andrews, Rhodora 63(750): 175. 1961;
CAN; OAC) and New Eng., s to Mo., Ala., and Ga. [*Ophrys* L.; *Leptorchis* Ktze.; *Malaxis* Sw.].

L. loeselii (L.) Richard   Bog-Twayblade, Yellow Twayblade
/T/X/EA/   (Hr)   Bogs, peaty meadows, damp thickets, and alluvial shores from Wash. and ?B.C.
(Hultén's 1958 map indicates two stations, the southern one, at ca. 50°N, representing the report
from Nicola by Eastham 1947, this discounted by Taylor 1959; the northern station, at ca. 54°N, is
not commented upon by Hultén; not known from Alta.) to Sask. (Dahlton and McKague; Breitung
1957a), Man. (N to The Pas), Ont. (N to near Cochrane at ca. 49°N), Que. (N to the Gaspé Pen.;
reported from Anticosti Is. by Schmitt 1904), N.B., P.E.I., and N.S., s to Nebr., Mo., Ala., and N.J.;
Eurasia. [*Ophrys* L.; *Leptorchis* MacM.]. MAPS: Hultén 1958: map 41, p. 61, and 1937b: map 4, p.
126; Meusel, Jaeger, and Weinert 1965:111; Meusel 1943: fig. 27e.

LISTERA R. Br.   [1494]   Twayblade

1   Lip cleft to middle or below into 2 linear segments; column (of united stamen and style) to
     0.5 mm long.
     2   Leaves round-ovate, truncate or subcordate at base, to 3 cm long; floral-axis and
          pedicels glabrous; column minute; lip purplish green, 4 or 5 mm long, not auricled but
          with a pair of laterally projecting horn-like teeth about 1.5 mm long near the base;
          (transcontinental) . . . .*L. cordata*
     2   Leaves ovate, rounded at base, to 2.5 cm long; axis and pedicels glandular; column
          about 0.5 mm long; lip dull red, to 1 cm long, with only a rudimentary basal horn but
          with a small rounded tooth in the sinus and a pair of short backwardly-directed basal
          auricles; (s Ont. and Que.; see *L. ovata,* at end) . . . .*L. australis*
1   Lip with a broad or dilated summit, cleft at most 1/3 of its length, watery-white or greenish
     white; column 2 or 3 mm long.
     3   Lip auricled at base, oblong, more or less ciliate, lacking lateral basal teeth; pedicels
          and ovaries glabrous.
          4   Lip cleft to 1/3 its length, the sinus narrow, the short basal auricles
               incurved-clasping; leaves oval or broadly ovate, to over 6 cm long; (Ont. to s
               Labrador, Nfld., and N.B.) . . . .*L. auriculata*

4   Lip merely shallowly notched, the sinus broad and open; basal auricles divergent and about 1.5 mm long and broad; leaves narrowly ovate, to about 4 cm long; (transcontinental) . . . .*L. borealis*

3   Lip not auricled at base, cuneate-obovate, with 2 broadly rounded lobes at the shallowly notched apex; pedicels glandular; leaves oval to roundish, to about 7 cm long.

5   Lip about 9 mm long, short-stalked, more or less ciliate, the lateral basal teeth obsolete; ovary glandular; bracts sometimes equalling or slightly surpassing the pedicels; (transcontinental) . . . .*L. convallarioides*

5   Lip about 5 mm long, sessile, 2-toothed at base, not ciliate; ovary glabrous; bracts averaging about 1/3 as long as the pedicels; (SE Alaska, B.C., and sw Alta.) . . . .*L. banksiana*

L. auriculata Wieg.   Auricled Twayblade

/T/EE/   (Grh)   Wet woods (often cedar swamps), thickets, and alluvial soils from Ont. (N to Big Trout L. at 53°49′N; CAN; the report from Churchill, Man., by Radforth, Natl. Res. Counc. Canada, Tech. Mem. 16: C4. 1950, is probably based upon *L. borealis*) to Que. (the type locality, as the first region cited; N to the Gaspé Pen. and Mingan Is. of the Côte-Nord), Labrador (N to Goose Bay, 53°18′N; DAO), Nfld., and s N.B. (Deer Is. and Grand Manan Is., Charlotte Co.; not known from P.E.I. or N.S.), s to Minn., Mich., N.Y., and N New Eng. [*Ophrys* House].

L. australis Lindl.   Southern Twayblade

/T/EE/   (Grh)   Damp woods, thickets, and sphagnous bogs from Ont. (Mer Bleue bog, Ottawa, where taken by Fletcher in 1893; CAN) to Que. (N to Grosse-Ile, near Quebec City; Marcel Raymond, Contrib. Inst. Bot. Univ. Montréal 64: 49. 1949; also known from Ste-Dorothée, near Montreal, and from Hatley, Compton Co.) and NE N.S. (near Inverness, Inverness Co.; CAN), s to La. and Fla. [*Ophrys* House].

L. banksiana Lindl.

/T/W/   (Grh)   Dense moist coniferous forests and wooded slopes up to rather high elevations from the Alaska Panhandle (*see* Hultén 1943: map 390, p. 562; *L. caur.*) through B.C. (type from Banks Is., E of Queen Charlotte Is.) and sw Alta. (Waterton Lakes; Breitung 1957b; *L. caur.*) to N Calif. and Mont. [*L. (Ophrys) caurina* Piper; *see* M. L. Fernald and B. G. Schubert, Rhodora 50(597): 231–33. 1948]. MAP (*L. caur.*): Hultén 1968b:326.

L. borealis Morong   Northern Twayblade

/ST/X/   (Grh)   Dryish humus of coniferous forests and thickets (chiefly or wholly in calcareous regions) from cent. Alaska–Yukon (*see* Hultén 1943: map 389, p. 562) to the Mackenzie R. Delta, Great Bear L., Alta. (N to L. Athabasca), Sask. (N to Waskesiu Lake, 53°55′N), Man. (N to Churchill), northernmost Ont., Que. (N to the Larch R. at 57°35′N, the Côte-Nord, Anticosti Is., and Gaspé Pen.; *see* NE Canada map by Dutilly, Lepage, and Duman 1953: fig. 12, p. 56), and N Nfld. (not known from the Maritime Provinces), s in the West to s B.C., N Wash., Idaho, Mont., Wyo., and Utah. [*Ophrys* Rydb.]. MAPS: Hultén 1968b:327; Raymond 1950b: fig. 24, p. 65; Raup 1947: pl. 20; Marie-Victorin 1938: fig. 58 (very incomplete), p. 545.

The type of *L. borealis* is from Fort Smith, on the Dist. Mackenzie–Alta. boundary at 60°N. Forma *trifolia* Lepage (a leaf present below the raceme in addition to the two subopposite midstem leaves normally comprising the complete foliage) is known from the type locality in the Paint Hills of the James Bay watershed, Que., at 52°58′N.

L. convallarioides (Sw.) Nutt.   Broad-lipped Twayblade

/T/X/eA/   (Grh)   Damp peaty or mossy woods, thickets, and swampy places from the E Aleutian Is. (Atka Is. and Unalaska Is.; *see* Hultén 1943: map 391, p. 562) and B.C. (N to ca. 55°N; *see* B.C. map by Szczawinski 1959:70) to sw Alta. (Banff and Waterton Lakes; not known from Sask.; the report from L. Winnipegosis, Man., by John Macoun 1888, is based upon *L. borealis* (relevant collection in CAN), as are, perhaps, other reports from that province), Ont. (N to the N shore of L. Superior; CAN), Que. (N to L. Mistassini, the Côte-Nord, Anticosti Is., and Gaspé Pen.), Nfld., N.B., P.E.I., and N.S., s to Calif., Ariz., Colo., Minn., Wisc., Mich., Tenn., and N.C.; E Asia (Bering Is.).

[*Epipactis* Sw.; *Ophrys* Wight; *L. eschscholziana* Cham.; *see* M. L. Fernald and B. G. Schubert, Rhodora 50(597): 231–33. 1948]. MAP: Hultén 1968b:326.

L. cordata (L.) R. Br.   Heartleaf Twayblade
/aST/X/GEA/   (Grh)   Damp mossy woods and openings from the Aleutian Is., s Alaska (*see* Hultén 1943: map 392, p. 563), and s Yukon–Dist. Mackenzie (CAN) to Sask. (N to McKague, 52°37′N; Breitung 1957a), Man. (N to Churchill), Ont. (N to Big Trout L. at ca. 54°N, 90°W), Que. (N to the James Bay watershed at ca. 56°15′N, the Côte-Nord, Anticosti Is., and Gaspé Pen.), Nfld., N.B., P.E.I., and N.S., s to Calif., N.Mex., Minn., Mich., and N.C.; w Greenland N to 69°32′N, E Greenland N to 63°22′N; Iceland; Eurasia. [*Ophrys* L.; incl. var. *nephrophylla* (Rydb.) Hult. (*L.* (*Ophrys*) *neph.* Rydb.), the phase with relatively broad and more apically-rounded leaves, to which Hultén refers his Alaskan material]. MAPS: Hultén 1968b:327, and 1958: map 233, p. 253; Meusel, Jaeger, and Weinert 1965:105.

Forma *disjuncta* Lepage (the two leaves well separated along the stem rather than subopposite, the upper leaf distinctly the smaller) is known from Que. (Anticosti Is.; type from Fort George, E James Bay, ca. 53°50′N).

L. ovata (L.) R. Br.
Eurasian (not keyed out above); the first collection of this species in N. America is apparently one taken by J. F. Alex et al. in 1968 in s Ont. (Red Bay, Bruce Co., L. Huron; "About 15 plants scattered through 20 m area recently cleared of brush, deadfall, etc., under cedar with ash, poplar and birch. Soil moderately drained organic sand.").

It has the leaf shape and lip structure of *L. australis* but the lip is yellow-green rather than dull red and it is a taller plant (to over 6 dm tall) with a raceme to 2.5 dm long rather than about 1 dm.

MALAXIS Sw.   [1552]   Adder's-mouth

1   Leaves 2 or more, ovate-lanceolate, alternate, to about 3 cm long, near the base of the scape-like stem and eventually subtending new tubers; lip erect, green-nerved, about 1.5 mm long, much shorter than the sepals; flowers resupinate (inverted on their pedicels, the lip uppermost), very small, short-pedicelled in a very slender spike-like raceme; (Alaska–Dist. Mackenzie–B.C. to Sask.; Ont.) . . . .*M. paludosa*
1   Leaf solitary (rarely 2), elliptic to oval, to over 6 cm long; lip 2 or 3 mm long.
    2   Lip drooping, abruptly long-pointed; pedicels 1 or 2 mm long, the slender raceme less than 1 cm thick; leaf inserted near base of scape; (transcontinental) . . . . . .*M. monophyllos*
    2   Lip finally ascending, cleft to below middle into 2 lanceolate lobes; pedicels to 8 mm long, much longer than the flowers, the oblong-cylindric raceme over 1 cm thick; leaf inserted near middle of scape; (Man. to s Labrador, Nfld., and N.S.) . . . .*M. unifolia*

M. monophyllos (L.) Sw.
/sT/X/EA/   (Gst)   Damp woods and bogs (chiefly calcareous) from the E Aleutian Is. and s Alaska (*see* Hultén 1943: map 398, p. 563; *Micro. mon.*) to B.C., Alta. (Moss 1959), Sask. (N to Waskesiu Lake, 53°55′N), Man. (N to L. Winnipegosis; John Macoun 1888), Ont. (N to near L. Timagami at ca. 47°N; TRT), Que. (N to the Côte-Nord, Anticosti Is., and Gaspé Pen.), SE Labrador (Blanc Sablon R. at 51°27′N; GH; Fernald *in* Gray 1950), Nfld., N.B. (Grand Manan, Charlotte Co.; GH; not known from P.E.I.), and N.S., s to Calif., Tex., Ind., and N.J., and in the mts. to Tenn.; Eurasia. [*Ophrys* L.; *Microstylis* Lindl.; *Mal.* (*Mic.*) *diphyllos* Cham.; incl. var. *brachypoda* (Gray) Morris & Eames (*Mic.* (*Mal.*) *brach.* Gray), the phase with slightly smaller perianths, capsules, and pedicels, to which the eastern plant is often referred]. MAPS: Hultén 1968b:330, and 1962: map 155, p. 165.

Forma *bifolia* Mousley (a second smaller leaf present below the usually solitary one) is known from the type locality, St-François-de-Sales, near Montreal, Que.

M. paludosa (L.) Sw.   Bog Adder's-mouth
/sT/(X)/EA/   (Gst)   Very local in wet sphagnous bogs of the Alaska Panhandle (*see* Hultén 1943: map 397, p. 563), sw Dist. Mackenzie (near Fort Liard, at 60°12′N; W. J. Cody, Can. Field-Nat. 77(2): 115. 1963), cent. B.C. (Queen Charlotte Is., Prince Rupert, and Aleza Lake, all

between ca. 53° and 54°15′N; *see* B.C. map by Szczawinski 1959: 96), Alta. (reported from Glenevis, 53°48′N, by Cody, validating the listing by Moss 1959), Sask. (Boivin 1967a; not listed by Breitung 1957a), N Minn., and cent. Ont. (N shore of L. Superior near Thunder Bay; CAN; MT; TRT; reported from Ogoki, about 70 mi NE of L. Nipigon, by W.K.W. Baldwin, Can. Field-Nat. 75(2): 74. 1961); Eurasia. [*Ophrys* L.; *Hammarbya* Ktze.]. MAPS: Hultén 1968b:330, and 1958: map 252, p. 271; Meusel, Jaeger, and Weinert 1965:111.

M. unifolia Michx.   Green Adder's-mouth
/T/EE/   (Gst)   Dry or moist woods, margins of ponds and bogs, and gravelly slopes from s Man. (*Mic. oph.* reported from the Winnipeg district by Bourgeau, *in* Palliser 1863; collection in CAN from L. Winnipegosis, where taken by John Macoun in 1881; taken in a cedar bog in the Sandilands Forest Reserve SE of Winnipeg by Gibson in 1965) to Ont. (N to the N shore of L. Superior near Thunder Bay; CAN), Que. (N to the Côte-Nord, Anticosti Is., and Gaspé Pen.), s Labrador (Goose Bay, 53°18′N; DAO), Nfld., N.B., P.E.I., and N.S., s to Tex. and Fla. [*Microstylis* BSP.; *Mal* (*Mic.*) *ophioglossoides* Muhl.].

Forma *bifolia* Mousley (a secondary leaf present above the usually solitary one) is reported from the type locality, the Laurentide region N of Montreal, Que., by Henry Mousley (Can. Field-Nat. 43(5): 99. 1929). The inclusion of Sask. in the range assigned to the species by Fernald *in* Gray (1950) may be based upon the report of *Microstylis ophioglossoides* from "Canada, to the Saskatchewan" by Hooker (1839). The Saskatchewan R. empties into Cedar L., N of L. Winnipegosis, Man., the plant evidently reaching its western limits at this locality.

ORCHIS L.   [1396]   Orchis, Orchid.   Orchide

1   Stem leafy, from fleshy-fibrous roots and fan-shaped clusters of more or less palmately lobed or divided, thickened, lanceolate tubers; lower leaves oblanceolate to broadly oblong-lanceolate; corolla-lip rhombic.
  2   Flowers usually rich deep purple, the lip usually less than 9 mm broad, the spur less than 8 mm long; leaves rarely over 12 cm long, the longest averaging about 2 cm broad; (introd. in Ont.) . . . .[*O. purpurella*]
  2   Flowers usually light magenta to violet-purple, the lip to 11 mm broad, the spur at least 1 cm long; leaves to nearly 1.5 dm long and 4 cm broad; (Aleutian Is.–s Alaska) . . . .
       . . . .*O. aristata*
1   Stem scapose, the 1 or 2 elliptic to ovate or obovate leaves sub-basal.
  3   Leaves 2 (rarely 3), rather fleshy, broadly elliptic to narrowly obovate; scape stout; flowers subtended by leafy oblong-lanceolate bracts to 5 cm long; sepals and lateral petals pink to pale purple, the lateral sepals ascending and connivent with the dorsal sepal and the petals; lip entire, white (rarely pink), tongue-shaped to broadly ovate, not notched at apex, to 2 cm long, about equalled by the spur; (s Ont., s Que., and w N.B.) . . . .*O. spectabilis*
  3   Leaf solitary, elliptic to roundish; scape slender; flowers subtended by linear-lanceolate bracts to 1.5 cm long; sepals and lateral petals white to pale purple; lip hastate (with a pair of projecting lobes at base), white with purple dots or stripes, notched at apex, less than 1 cm long, much longer than the spur; (transcontinental) . . . .
       . . . .*O. rotundifolia*

O. aristata Fisch.
/sT/W/eA/   (Grt)   Sheltered or exposed rocky tundra of the Aleutian Is. (type from Unalaska; *see* Hultén 1943: map 376, p. 561) and s Alaska; E Asia. [*Dactylorhiza* Soó; *O. latifolia* var. *beeringiana* Cham.]. MAP: Hultén 1968b: 318 (*Dact. ar.*).

Forma *perbracteata* Lepage (the floral organs transformed into up to 40 or 50 bracts) is known from the type locality, Kodiak Is., Alaska.

[*O. purpurella* T. & T. A. Steph.]   Spotted Orchis of Europe
[European; apparently known in N. America only from Timmins, Ont., about 150 mi N of Sudbury, where reported by H. Andrews (Rhodora 63(750): 176. 1961) as comprising, in 1960, a stand of

about 20 plants growing along a lake margin being filled in for a roadway. The source of the plant was evidently discarded packing material. (*Dactylorchis* Vermeul.).]

O. rotundifolia Banks   Small Round-leaved Orchis
/aST/X/G/ (Grt) Mossy calcareous swamps and woods from N-cent. Alaska and cent. Yukon (*see* Hultén 1943: map 377, p. 561) to the Mackenzie R. Delta, Great Bear L., Great Slave L., L. Athabasca (Alta. and Sask.), northernmost Man.–Ont., Que. (N to SE James Bay, L. Mistassini, the Côte-Nord, Anticosti Is., and Gaspé Pen.), w Nfld., and N.B. (Madawaska and Gloucester counties; not known from Labrador, P.E.I., or N.S.), s to s B.C., SE Alta. (Cypress Hills), Mont., Wyo., Minn., Wisc., and N New Eng.; w Greenland N to the Arctic Circle. [*Amerorchis* Hult.; *Habenaria* Rich.; *Platanthera* Lindl.]. MAPS: Hultén 1968b:318 (*Amer. rot.*); Porsild 1966: map 45, p. 72.

The type locality is noted by Pursh (1814) as "On Hudson's Bay." Forma *angustifolia* Rousseau (the leaf narrowly lanceolate rather than ovate-oblong) is known from the type locality, Ile-Couture, L. Mistassini, Que. Forma *beckettii* Boivin (the petals uniformly white and unspotted rather than the upper petals roseate, the white lip spotted with magenta or purple) is known from the type locality, Churchill, Man. Forma *lineata* (Mousley) Voss (the corolla-lip striped with purple rather than merely spotted) is known from SE Alta. (type from the Cypress Hills) and Ont. (N shore of L. Superior near Thunder Bay; Frontenac Co.).

O. spectabilis L.   Showy Orchis
T/EE/ (Grt) Rich (mostly calcareous) woods from Ont. (N to Carleton and Russell counties) to Que. (N to l'Ange-Gardien, a few mi NE of Quebec City in Montmorency Co.; *see* Que. map by Doyon and Lavoie 1966: fig. 7, p. 816) and w N.B. (York and Carleton counties; not known from P.E.I.; an early report from N.S., where evidently now extinct, by Cochran 1829), s to Kans., Ark., Tenn., and Ga. [*Galeorchis* Rydb.; incl. var. *lilacina* Ames].

POGONIA Juss.   [1464]

P. ophioglossoides (L.) Ker   Pogonia, Beard-flower
/T/EE/ (Grt) Open bogs and wet mossy shores, the aggregate species from Ont. (N to the Rainy R. near Lake of the Woods; TRT) to Que. (N to L. Nominingue, Labelle Co., and Magdalen Is.), Nfld., N.B., P.E.I., and N.S., s to Tex. and Fla.

Var. *brachypogon* Fern. (the beard of the corolla-lip almost wanting or consisting of very short knobs rather than of 3 rows of elongate yellowish or brownish hairs, the sepals and petals ascending rather than spreading) is known from gravelly beaches of ponds in w N.S. (type from Cedar L., Yarmouth Co.). Forma *albiflora* Rand & Redf. (flowers white rather than pale to deep pink) is reported from Que. and N.B. by Boivin (1967a).

SPIRANTHES Richard   [1490] Ladies'-tresses

1 Raceme loosely flowered, the flowers in a single spiral rank.
   2 Perianth creamy or yellowish, to about 1 cm long; raceme finely pubescent; leaves linear or lance-linear, often extending slightly up the stem; (sw Que.) . . . .*S. vernalis*
   2 Perianth white, 4 or 5 mm long; raceme subglabrous; leaves ovate or ovate-oblong, all basal; (Sask. to N.S.) . . . .*S. lacera*
1 Raceme closely flowered, the flowers usually in 2 or 3 spiral ranks.
   3 Lip pandurate (somewhat "fiddle"-shaped), strongly constricted below the roundish terminal lobe, to 11 mm long, its basal callosities minute and obscure; flowers white or creamy, to 12 mm long; leaves linear to oblong-lanceolate or oblanceolate; (transcontinental) . . . .*S. romanzoffiana*
   3 Lip not (or only slightly) constricted above the middle.
      4 Callosities at base of lip minute and obscure, the lip with nearly parallel sides, undulate-crenulate above the middle; flowers white, to about 7 mm long; leaves elliptic-lanceolate to oblong-elliptic, blunt to subacuminate; (Ont. to N.S.) . . . . . .*S. lucida*
      4 Callosities at base of lip prominent (usually about 0.5 mm long); flowers to over 1 cm long.
         5 Flowers typically white, usually fragrant; lip to 1 cm long, ovate-oblong (the

sides nearly parallel), the margins crisped or erose; floral bracts to 1.5 cm long; leaves linear to linear-lanceolate, acute to acuminate; plant downy-pubescent above; (Ont. to N.S.) ....*S. cernua*

5 Flowers greenish white or creamy; lip to 8.5 mm long, broadly ovate-lanceolate (tapering to the slightly dilated apex), the margins essentially entire; floral bracts to 2 cm long; leaves linear to elliptic-lanceolate or oblanceolate, obtuse to acute; plant essentially glabrous throughout ....*[S. porrifolia]*

S. cernua (L.) Richard   Nodding Ladies'-tresses, Screw-auger
/T/EE/   (Grt)   Moist meadows and open woods from Minn. to Ont. (N to near Sudbury; MT), Que. (N to the Montreal dist.), N.B. (near Sussex, Kings Co.; CAN), P.E.I., and N.S., s to E Tex. and Fla. [*Ophrys* L.; *Gyrostachys* Ktze.; *Ibidium* House; *Neottia* Willd.; incl. var. *ochroleuca* (Rydb.) Ames and var. *odorata sensu* Henry Mousley, Can. Field-Nat. 55(6): 79. 1941, perhaps not (Nutt.) Correll].

This species is very difficult to distinguish in dried material from *S. romanzoffiana* and reports of it from Canada elsewhere than as outlined above refer largely to that species, such reports including Man. (Lowe 1943), E Que. (Verrill 1865; Saint-Cyr 1887; John Macoun 1888), Labrador (Hustich and Pettersson 1943; Makkovik, ca. 55°N), and Nfld. (Hooker 1839; Reeks 1873). A purported hybrid with *S. romanzoffiana* (× *S. steigeri* Correll) is reported from N.S. by Correll (1950; Yarmouth Co.).

S. lacera Raf.   Northern Slender Ladies'-tresses
/T/EE/   (Grt)   Dry to moist clayey or peaty places, barren fields, and thickets from cent. Sask. (*S. gracilis* reported from Prince Albert, Waskesiu Lake, and Meadow Lake by Breitung 1957a) to Man. (N to Flin Flon; Lowe 1943; collection in CAN from the N end of L. Winnipegosis), Ont. (N to Sioux Lookout, about 175 mi NW of Thunder Bay; OAC), Que. (N to L. Nominingue, Labelle Co., and Magdalen Is.; *S. gracilis* reported N to Mingan, Saguenay Co. of the Côte-Nord, by John Macoun 1888), N.B., P.E.I., and N.S., s to s Man., Minn., Wisc., Ill., Ind., Tenn., and N.C. [*S. gracilis* of Canadian reports, not *Neottia* (*Gyrostachys*; *Ibidium*) *gracilis* Bigel.; the report of *S. gracilis* from the Mackenzie R. by Hooker 1839, undoubtedly refers to *S. romanzoffiana*].

S. lucida (Eat.) Ames
/T/EE/   (Grt)   Damp thickets, meadows, and shores (often calcareous) from Minn. to Ont. (N to the Ottawa dist.), Que. (N to the Montreal dist.; a collection in MT from near Quebec City may also belong here), N.B., P.E.I. (D. S. Erskine 1960), and N.S. (CAN), s to Mo., Tenn., and N.C. [*Neottia* Eat.; *S.* (*Gyrostachys*) *latifolia* Torr.; *S.* (*Ibidium*) *plantaginea* Raf.].

[*S. porrifolia* Lindl.]
[A collection in CAN from Sidney, Vancouver Is., B.C., has been placed here by John Macoun but later referred to *S. romanzoffiana* by Boivin. Correll (1950) assigns it a range from Wash. to Calif. and Utah. (*Ibidium* Rydb.; *S. romanzoffiana* var. *por.* (Lindl.) Ames & Correll).]

S. romanzoffiana Cham.   Hooded Ladies'-tresses
/ST/X/E/   (Grt)   Meadows, swampy places, thickets, and clearings from the Aleutian Is. and N-cent. Alaska–Yukon (*see* Hultén 1943: map 388, p. 562) to the Mackenzie R. Delta, Great Bear L., Great Slave L., L. Athabasca (Alta. and Sask.), Man. (N to Churchill), northernmost Ont., Que. (N to N James Bay, L. Mistassini, the Côte-Nord, Anticosti Is., and Gaspé Pen.), Labrador (N to the Hamilton R. basin), Nfld., N.B., P.E.I., and N.S., s to Calif., N.Mex., Nebr., Ohio, and New Eng.; a relic area in sw Ireland and w Scotland. [*Ibidium* (*S.*; *Gyrostachys*) *strictum* House]. MAPS: Hultén 1968b:325; 1958: map 190, p. 209; and 1937b: map 6, p. 126; Fernald 1929: map 39, p. 1505.

S. vernalis Engelm. & Gray   Spring Ladies'-tresses
/T/EE/   (Grt)   Dry to moist clayey or gravelly fields and meadows, the main area from E Kans., Tenn., Ohio, and Md. to E Tex. and Fla., with an isolated area in sw Que. (Ste-Dorothée, a suburb of Montreal; CAN; reported from Hatley, Compton Co., and from 20 mi N of Montreal by Henry Mousley, Can. Field-Nat. 55(6): 79. 1941, and 56: 1–2. 1942; reported from St-Chrysostome, Chateauguay Co., by Lionel Cinq-Mars, Ann. ACFAS 18: 82. 1952; a collection from St-Grégoire,

Iberville Co., has also been placed here; some of these reports may refer to *S. cernua* but the species is cited from Hatley by Correll 1950). [*Ibidium* House; *Gyrostachys* Ktze.; *S. neglecta* Ames].

TRIPHORA Nutt.   [1466]

T. trianthophora (Sw.) Rydb.   Nodding Pogonia
/t/EE/   (Grt)   Moist rich deciduous woods from Iowa to Ohio, s Ont. (Wheatley, straddling the Essex–Kent counties boundary; OAC; reported from SE Essex Co. by C. H. Zavitz and L. O. Gaiser, Rhodora 58(686): 31. 1956; living plants were seen by the writer in 1960 at Rondeau Provincial Park, Kent Co.), and Maine, s to E Tex. and Fla. [*Arethusa* Sw.; *Pogonia* BSP.; *P. pendula* (Muhl.) Lindl.]. MAP (s Ont.): Soper 1962: fig. 26, p. 41.

In spite of the great rarity of this plant in Canada, there is an early citation by Hooker (1839) of a collection by Goldie in Canada. *See* Zavitz and Gaiser, loc. cit.